C0-CED-368

QUATERNARY AND GLACIAL GEOLOGY

QUATERNARY AND GLACIAL GEOLOGY

JÜRGEN EHLERS
Geologisches Landesamt, Germany

Translated from *Allgemeine und historische Quartärgeologie*

English version by Philip L. Gibbard

JOHN WILEY & SONS
Chichester · New York · Brisbane · Toronto · Singapore

Published 1996 by John Wiley & Sons Ltd
Baffins Lane, Chichester
West Sussex PO19 1UD, England

National 01243 779777
International (+44) 1243 779777
e-mail (for orders and customer service enquiries): cs-books@wiley.co.uk
Visit our Home Page on http://www.wiley.co.uk
or http://www.wiley.com

Other Wiley Editorial Offices

John Wiley & Sons, Inc., 605 Third Avenue,
New York, NY 10158-0012, USA

Jacaranda Wiley Ltd, 33 Park Road, Milton,
Queensland 4064, Australia

John Wiley & Sons (Canada) Ltd, 22 Worcester Road,
Rexdale, Ontario M9W 1L1, Canada

John Wiley & Sons (Asia) Pte Ltd, 2 Clementi Loop #02-01,
Jin Xing Distripark, Singapore 0512

Library of Congress Cataloging-in-Publication Data

Ehlers, Jürgen.
[Allgemeine und historische Quartärgeologie. English]
Quaternary and glacial geology / Jürgen Ehlers ; translated from Allgemeine und historische Quartärgeologie ; English version by Philip L. Gibbard.
p. cm.
Includes bibliographical references (p. –) and index.
ISBN 0-471-95576-0 (cloth ; acid-free paper)
1. Geology, Stratigraphic—Quaternary. I. Title.
QE696.E2813 1996
551.7′9—dc20 95-52107
CIP

British Library Cataloguing in Publication Data

A catalogue record for this book is available from the British Library

ISBN 0-471-95576-0

Typeset in 10/12pt Times from author's disks by MHL Typesetting Limited, Coventry
Printed and bound in Great Britain by Bookcraft (Bath) Ltd
This book is printed on acid-free paper responsibly manufactured from sustainable forestation, for which at least two trees are planted for each one used for paper production.

Contents

Preface

I wish to thank James S. Aber (Emporia, USA), Peter J. Barnett (Sudbury, Canada), Helmut Brückner (Marburg, Germany), John A. Catt (Rothamsted, UK), Joakim Donner (Helsinki, Finland), Lynda Dredge (Ottawa, Canada), Aleksis Dreimanis (London, Canada), Matti Eronen (Helsinki, Finland), Leon Follmer (Champaign, USA), Mebus A. Geyh (Hannover, Germany), Philip L. Gibbard (Cambridge, UK), Walter Grottenthaler (München, Germany), Eberhard Grüger (Göttingen, Germany), Ardith Hansel (Champaign, USA), Helmut Heuberger (Salzburg, Austria), Christian Hoselmann (Hannover, Germany), Michael Houmark-Nielsen (København, Denmark), Hilton Johnson (Champaign, USA), Paul Karrow (Waterloo, Canada), Else Kolstrup (Sønderborg, Denmark), Eduard A. Koster (Utrecht, The Netherlands), Stefan Kozarski (Poznań, Poland), Grahame Larson (East Lansing, USA), Jan Mangerud (Bergen, Norway), Leszek Marks (Warszawa, Poland), Klaus-Dieter Meyer (Hannover, Germany), Douglas Peacock (Edinburgh, UK), Jan Piotrowski (Kiel, Germany), Mikko Punkari (Edinburgh, UK), Bertil Ringberg (Stockholm, Sweden), Veli-Pekka Salonen (Turku, Finland), James Scourse (Bangor, UK), Nick J. Shackleton (Cambridge, UK), Johan Ludvig Sollid (Oslo, Norway), Ralph Stea (Halifax, Canada), Dirk van Husen (Wien, Austria), Stefan Wansa (Halle, Germany), Herbert E. Wright, Jr. (Minneapolis, USA) for corrections and comments on the manuscript. Photographs were provided by Helmut Heuberger (Salzburg, Austria), Karl-Dieter Meier (Garbsen, Germany), Reginald Morrison (Halifax, Canada), Christian Schlüchter (Zürich, Switzerland), Ralph Stea (Halifax, Canada) and Hansjörg Streif (Hannover, Germany). Erhard Bibus (Tübingen, Germany) provided unpublished information about the South German Pleistocene stratigraphy and Erik Lagerlund (Lund, Sweden) supplied details about the history of the Scandinavian ice-sheet model. Without the help provided by these colleagues it would have been impossible to prepare the book in its present form. Numerous others have helped by stimulating discussions or by sending reprints. The manuscript was provisionally translated with the help of Theodor von Bassewitz (Köln, Germany) and Alex Chepstow-Lusty (Cambridge, UK) and brought into its final form by Philip L. Gibbard (Cambridge, UK), the discussions with whom have largely contributed to its quality. Thanks to them all.

Hamburg, August 1995 JÜRGEN EHLERS

1

Introduction

1.1 TRACES OF ICE AGES IN THE EARTH'S HISTORY

We tend to interpret natural phenomena that we do not understand by applying processes familiar to us. The concept of 'ice ages' was alien to the scientists of the early 19th century. However, they did know that from time to time during the earth's history large parts of the land surface had been inundated by the sea. It was therefore inevitable that the Quaternary deposits, especially the large erratic blocks, were interpreted as the result of a major flood. That this was the biblical deluge was only believed by a minority. One of this minority, Reverend William Buckland from Oxford University, in 1823 introduced the term 'Diluvium' into the literature.

One of the earliest attempts to find a natural explanation for the occurrence of erratics far from their source areas was the 'Rollstein' or mudflood theory, favoured especially by Leopold von Buch (1815), and also by Alexander von Humboldt (1845) and the Swedish physician and scientist Nils Gabriel Sefström (1836, 1838). They suggested that the erratic blocks had been spread by enormous water masses, the so-called 'Petridelaunic Flood'. A plausible cause for the sudden outburst of such water masses from the Alps and the Scandinavian mountains, however, could not be provided.

A later interpretation offered was the drift theory, proposed by the English geologist Charles Lyell (1840a, b), which gained importance towards the middle of the 19th century. According to this concept, icebergs drifting in a cold sea could have scratched and striated the ground and transported the erratics. Proponents of the drift theory included Charles Darwin and the physicist Hermann von Helmholtz. They had to accept that glaciers formerly covered much greater areas in order to explain the occurrence of the numerous icebergs, but they firmly refuted the idea of continental ice sheets. Geologist Bernhard Cotta (1848) wrote: 'It surpasses the limits of the thinkable, to assume that glaciers extended from the Norwegian mountains to the Elbe River and to Moscow, and even to the coasts of England, which, covered with morainic material, should have moved across this hardly sloping but uneven terrain.' In Britain the drift theory was only accepted for a brief lifespan, but it succeeded and remained dominant in North Germany and parts of northern Europe for many years (e.g. Cotta 1867).

The lecture by Louis Agassiz at the Naturforschende Gesellschaft in Neuchâtel on 24 July 1837 is generally regarded as marking the birth of glacial theory. In this lecture Agassiz explained the origin of the erratics by glacier transport. It is well known that others before him had come to similar conclusions (including Hutton 1795, Esmark 1824, Venetz 1830, Bernhardi

1832 and De Charpentier 1834); although none of them had had the opportunity from an influential position (the then 30-year-old Agassiz had been professor for five years and was president of the Naturforschende Gesellschaft) to address a major part of the scientific community in a single lecture. Nevertheless, there was great reluctance in accepting the new theory. Glaciers of such dimensions were beyond comprehension. For instance it is necessary to remember that Antarctica was then practically unknown, and that it was not until 1852 that a scientific expedition could provide evidence that the glaciers of Greenland formed part of one single ice sheet (Imbrie & Imbrie 1986).

In North America, as in Europe, people had started to discuss the possible origin of the post-Tertiary deposits. A Peter Dobson (1826) from Connecticut had observed that boulders dug out from the ground were worn on the underside. He concluded '... that they have been worn by being suspended and carried in ice, over rocks and earth, under water.' In 1839 Timothy Conrad, a palaeontologist, reported in a short note on Agassiz's Neuchâtel lecture and the glacial theory and added: 'In the same manner I would account for the polished surfaces of rocks in Western New York.' Two years later Edward Hitchcock, State Geologist of Massachusetts, presented the glacial theory before the newly formed Association of American Geologists (Hitchcock 1841). In 1846 Agassiz arrived in North America, where he soon received a professorship at Harvard. During this period glacial theory became more and more accepted. In 1849 Dana put forward the hypothesis that the Canadian Cordillera had been more extensively glaciated in the past, which would explain the morphological difference between the straight coastlines of Oregon and Washington and the fjord landscape of what was British and Russian North America (Canada and Alaska). In 1863 Dana's textbook that included the glacial theory appeared, and from then on the theory became well established in North America. An overview of the historical evolution of US Quaternary research has been presented by Flint (1965); the research history of the Cordilleran Ice Sheet was summarised by Jackson & Clague (1991), and Fulton (1989) has given a brief overview of the Quaternary research history of Canada.

In Britain, Agassiz presented the glacial theory at the annual meeting of the British Society for the Advancement of Science in Glasgow in September 1840. On a subsequent excursion, he managed to convince Buckland of his ideas. In the same year, Lyell also became a convert to the new theory.

In North Germany the glacial theory was accepted particularly late. De Charpentier (1842) had postulated the existence of a northwest European glaciation as far as England, the Netherlands, the Harz Mountains, Saxony, Poland, and 'almost to Moscow'. However, neither he, nor Bernhardi (1832) or Von Morlot (1844, 1847), could convince their fellow scientists. When eventually the Swedish geologist Otto Torell (on 3 November 1875) identified the striae on the Muschelkalk rocks at Rüdersdorf, a change of opinion was overdue. By this time glacial theory had been established in Britain and North America for over ten years (Dana 1863, Lyell 1863). Moreover, Torell (1865, 1872, 1873) had published his concept of a Scandinavian ice-sheet model and his ideas on the extension of ice-age glaciers into North Germany years before. However, it was only with Albrecht Penck's article on 'The erratic-bearing formation of North Germany' ('Die Geschiebeformation Norddeutschlands') (1879) that final doubts were eliminated.

The Quaternary is traditionally subdivided into two units: the Pleistocene (the ice-age period proper) and the Holocene (or Recent). The term Quaternary for the ice-age era was introduced by Von Morlot (1855). The term Pleistocene was first used by Forbes (1846); he defined it as the period during which the ice-age deposits had formed. A single period of ice-sheet expansion is called a 'glaciation'. The adjective 'glacial' is used whenever the temporal aspect of an event or deposit is intended; where genesis by a glacier is to be expressed, the term 'glacigenic' is used.

Stratigraphic subdivision of the Quaternary is largely based on climate. Cold periods are referred to as 'cold stages', 'glacials' or 'glaciations'. The intervening temperate periods are 'warm stages', 'interglacials' or 'interglaciations'. Minor climatic oscillations within these stages are called 'stadials' or 'stades' (cold) and 'interstadials' or 'interstades' (temperate).

The onset of the Quaternary does not represent a radical change of climate, but rather a gradual transition. As a consequence, the boundary between the Tertiary and Quaternary eras must be drawn more or less arbitrarily, and a number of criteria could potentially be applied.

Geologists generally tend to draw stratigraphical boundaries in marine deposits because they are more likely to represent continuous sedimentation and relatively consistent environments in comparison to terrestrial sediments. However, marine deposits from the period in question are relatively rarely exposed at the surface. According to a conclusion of the International Geological Congress 1948, the Tertiary/Quaternary boundary was defined as the base of the marine deposits of the Calabrian in southern Italy. In the Calabrian sediments fossils are found in the Mediterranean region that reflect the first distinct climatic cooling (amongst others the foraminifer *Hyalinea baltica*). This climatic change roughly coincides with a reversal of the earth's magnetic field (see chapter 10.2); it is situated at the upper boundary of what is called the Olduvai Event (Backman *et al.* 1983). Consequently, it is relatively easy to identify; its age is today estimated at 1.77 million years (Shackleton *et al.* 1990). The position of this boundary was reconfirmed at the INQUA Congress in Moscow 1982.

However, in contrast to the older parts of the earth's history, the significant changes within the Quaternary are not changes in faunal composition but changes in climate. For reasons of long-term climatic evolution the base of the Calabrian is not a very suitable global boundary. Its adoption excludes some of the major glaciations from the Quaternary. Therefore, in the Lower Rhine area, for instance, another Tertiary/Quaternary boundary is in use. In the sediments there the most significant climatic change is already recorded as far back as the Gauss/Matuyama magnetic reversal (about 2.44 million years ago). At this time the catchment area of the Rhine extended into the Alpine foreland, resulting in a dramatic change in heavy-mineral composition of the sediments (Boenigk 1982). At the same time the typical Pliocene pollen spectra changed to cooler Quaternary spectra (Zagwijn 1957). Similarly important changes are found in gravel composition and in the mollusc assemblages (Boenigk *et al.* 1974). Although all of these changes did not occur simultaneously, there can be no doubt that they happened in a very short period of time. Today these changes are also used to mark the Tertiary/Quaternary boundary in the British Isles and the northern Alpine area (e.g. West 1980a, Brunnacker 1986, Gibbard *et al.* 1991, Schreiner 1992, Jerz 1993). This boundary can be easily identified worldwide, both biostratigraphically and magnetostratigraphically (Azzaroli *et al.* 1988, Driever 1988). Moreover, the Chinese loess record began at about that time (Ding *et al.* 1992). Therefore this boundary is also used in this book.

It became clear at an early stage of research that the Pleistocene ice age was not unique in earth history. At a time when North German scientists still believed in the Drift Ice, traces of an older, Permo-Carboniferous glaciation had been identified in India (1856), Australia (1859) and South Africa (1868). In 1871 another major glacial period was identified, reaching even farther back into the earth's history, in the Late Precambrian; the so-called Vendian, some 600 million years ago. Today glaciation is also known to have occurred during the Ordovician and within the earlier Precambrian, some 2000–2800 million years ago (Hambrey & Harland 1981, Harland *et al.* 1990).

These great earlier glacial periods are exceptional within earth history. The spatial distribution of glacial sediments from these periods is rather well known nowadays, although their correlation and their exact position with regard to the contemporaneous pole are not in all cases well established. It is obvious, however, that the ancient ice ages also comprised numerous individual glaciations. In the youngest Precambrian tillite series of Scotland (Port Askeig Formation) alone more than 40 layers of consolidated till (tillite) have been identified, representing 17 ice advances (Schwarzbach 1974).

The oldest Precambrian glaciations have been identified in North America (Canadian Shield and Montana), South America (Brazil) and South Africa. From Europe, only the glacial deposits of the youngest Precambrian are known (from Scotland, Norway and Finland); respective deposits are found in Greenland, Asia, Africa and Australia. The glaciation at the end of the Ordovician has so far been identified in South America, South Africa and Morocco. Traces of the Permo-Carboniferous glaciation are widespread on the southern continents (the former Gondwanaland supercontinent), especially well exposed in South Africa. These glaciations from the stages earlier in the earth's history exhibit the same assemblage of landforms and sediments as the Quaternary glaciations of North America and Europe (see for instance Wright & Moseley 1975, Hambrey & Harland 1981, Deynoux *et al.* 1994).

The oldest comprehensive overview of the Quaternary ice age was provided by Geikie (1874, 1894) in his book *The Great Ice Age*. In Britain it was followed by Lewis (1894), W.B. Wright (1914, 1937) and Zeuner (1959). Some monographs have concentrated on special topics. Quaternary stratigraphy has been dealt with by Bowen (1978), Quaternary palaeoclimatology by Bradley (1985), Quaternary environments by Goudie (1977, 1993) and Williams *et al.* (1993). The British Quaternary has been summarised in recent textbooks by Lowe & Walker (1984) and Jones & Keen (1993). The latest discussion of the Quaternary in Scandinavia is provided by Donner (1995).

In North America, G.F. Wright's (1890) *The Ice Age in North America*, which was modelled after Geikie's book, provided a first synthesis. The most recent regional overviews of the North American Quaternary appeared in the 'Decade of North American Geology' series (Ruddiman & Wright 1987, Fulton 1989, Morrison 1991). In Germany, Woldstedt (1929) published the first edition of his book *Das Eiszeitalter*, a comprehensive textbook, providing, in the author's words, 'Fundamentals of a geology of the Diluvium'. Because of the immense expansion in knowledge, the second edition had to be issued in three volumes (Woldstedt 1958, 1961, 1965).

For the Alpine region, Penck & Brückner's three volumes on *Die Alpen im Eiszeitalter* (1901/1909) provided an early geological and geomorphological synthesis. No comprehensive summary of the Alpine glaciations has since been published. Hantke's *Eiszeitalter* (1978, 1980, 1983) comes closest, dealing with the Quaternary of Switzerland and neighbouring areas (east to the Inn Glacier). A slightly revised version has been published in a single volume (Hantke 1992). For Bavaria, Jerz (1993) gives an up-to-date summary of the recent state of research.

For North Germany Wahnschaffe's (1891) *Die Oberflächengestaltung des Norddeutschen Flachlandes* provided the first regional review. Until the fourth edition (Wahnschaffe & Schucht 1921) this remained the standard reference book for North German Quaternary Geology. It was replaced in 1955 by Woldstedt's book *Norddeutschland und angrenzende Gebiete im Eiszeitalter*, the third edition of which appeared 20 years ago (Woldstedt & Duphorn 1974). An overview of the Quaternary morphology in the region of the Nordic glaciations in North Germany and adjoining areas is given by Liedtke (1975, with a second, much improved edition in 1981).

International reviews are provided by the North American textbook of Flint (1947; third edition 1971), the English textbooks of Charlesworth (1957), West (1968; second edition 1977) and the Swedish textbook of Nilsson (1972; enlarged English language edition 1983). A new European overview (in Polish) has been provided by Mojski (1993). A well-illustrated book on *The Ice Age World* has recently been published by Andersen & Borns (1994); a good textbook on glacial morphology, apart from Sugden & John's (1976) superbly illustrated *Glaciers and Landscape*, is Hambrey's (1994) book on *Glacial Environments.*

1.2 CAUSES OF ICE AGES

We live in an ice age. Even if Central Europe and North America are free of major ice sheets today, the present 'warm stage' still belongs to the cold episodes of the earth's history. During overwhelming periods of geological time the polar regions have been ice-free, and in the middle latitudes climate was milder than today.

Based on the investigation of deep-sea cores and lake deposits the climatic cycles of the ice age are fairly well known. About 60 oxygen-isotope stages can be distinguished within the Pleistocene, i.e. about 30 warm and 30 cold stages (cf. chapter 10.1). With the aid of palaeomagnetic investigations it has been possible to date this sequence of cold and warm events with sufficient precision to determine the duration of the individual climatic cycles. During the last 600,000 years a cold stage/warm stage cycle had a period of about 100,000 years. Before this, a 40,000-year cycle was dominant.

The causes for the observed climatic oscillations lie in periodic changes of the earth's orbit around the sun. Croll (1864) had already assumed that the periodic variations in eccentricity of the earth's orbit around the sun caused by gravitational pull of the other planets must bear an influence on the climate on earth. Today it is known that things are slightly more complicated. In fact, radiation influx from the sun (insolation) is controlled by three factors that undergo cyclic changes. This concept, the fundamentals of which had been outlined by Milankovitch (1941), was substantiated only 25 years later by concrete scientific evidence (Hays *et al.* 1976). The following factors influence the intensity of solar radiation (Figs 1 and 2):

1. *Eccentricity.* The earth's orbit around the sun forms an ellipse, which constantly changes its form. The deviation from a circular orbit is called eccentricity. It varies between 0.5 and 6.0% within a period of about 100,000 years. Strong eccentricity enhances the contrast in temperature between summers and winters on both hemispheres, weak eccentricity smoothes the differences.
2. *Tilt of the earth's axis.* The earth's axis stands not at right angles to the orbit but is tilted at a certain angle. This tilt changes within a period of ca. 40,000 years between 22.1° and 24.5°. The smaller the angle, the lower is the radiation that the polar regions receive in summer. The effects are equally felt on both hemispheres.
3. *Precession.* The earth's axis rotates slowly around the pole. Simultaneously with the orbital ellipse this circular movement determines the date of the perihelion, i.e. the time when the earth comes closest to the sun. The perihelion shifts with a periodicity of ca. 20,000 years. Its precession enhances the annual temperature gradient on one hemisphere and smoothes it on the other.

The overall radiation budget of the earth is hardly influenced by the changes of the orbital

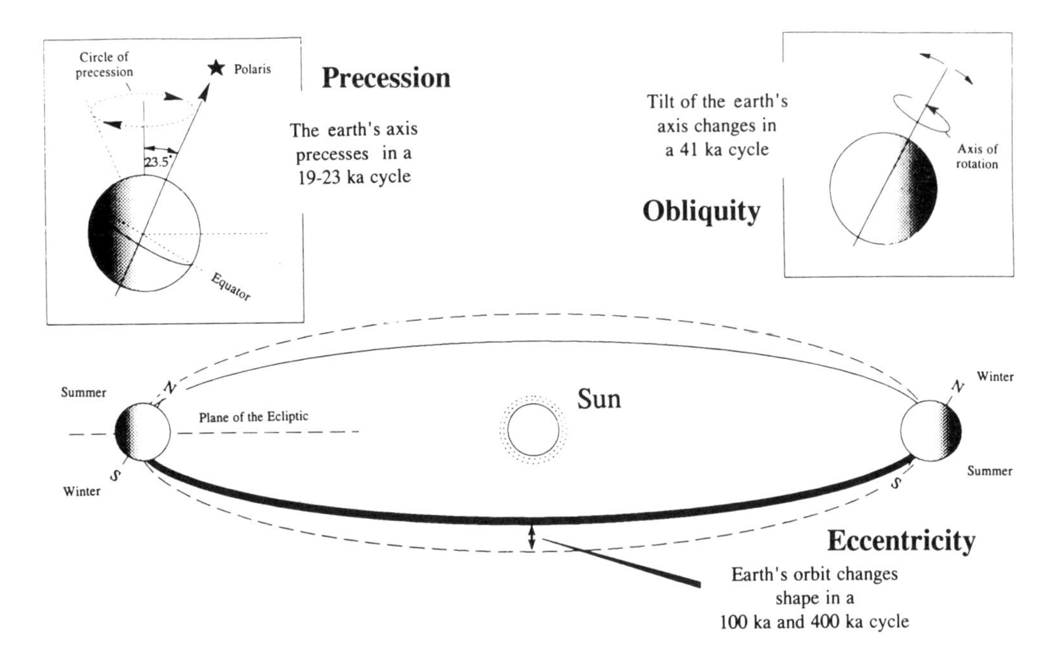

FIGURE 1 Orbital eccentricity, obliquity and precession, the three astronomical cycles involved in solar input and climate variation (from Patience & Kroon 1991)

parameters. Not even the hemispheric heat balance is affected: lower radiation in one season, in which the earth is farther away from the sun, is completely balanced by higher intake during the other season. The decisive point climatologically is that the seasonal distribution of radiation is altered. Lower summer radiation in high northern latitudes lessens the potential to melt the snow that accumulated during the preceding winter and thus increases the potential for the formation of glaciers. Between 45 and 65° almost no continents occur in the southern hemisphere, whereas in the northern hemisphere the main land masses of Eurasia and North America are situated in just this belt. Accordingly a decrease in summer radiation is felt much more strongly in the northern hemisphere.

The good temporal agreement between the climatic cycles of the Quaternary recorded in the deep-sea sediments and the Milankovitch cycles does show that there must be a relationship. However, the precise underlying mechanism is not well known. Broecker & Denton (1990) point out that global cooling and almost synchronous formation of mountain glaciers and ice sheets cannot be explained by the Milankovitch cycles alone. The earth's reaction to radiation changes is more complex; it includes changes in carbon dioxide content of the atmosphere and changes in atmospheric and oceanic circulation and in the formation of oceanic deep water (Dawson 1992).

As the changes of the orbital parameters were not restricted to the Quaternary, climatic cycles, though of lesser intensity, existed during all geological periods (e.g. the Cretaceous). The climatic cycles therefore are just the pacemakers, not the causes of the ice age (Hays *et al.* 1976).

Because ice ages have occurred repeatedly during the course of the earth's history, the question arises whether a common triggering mechanism can be found. These events seem to have been repeated at an interval of about 250–300 million years, so there might be a

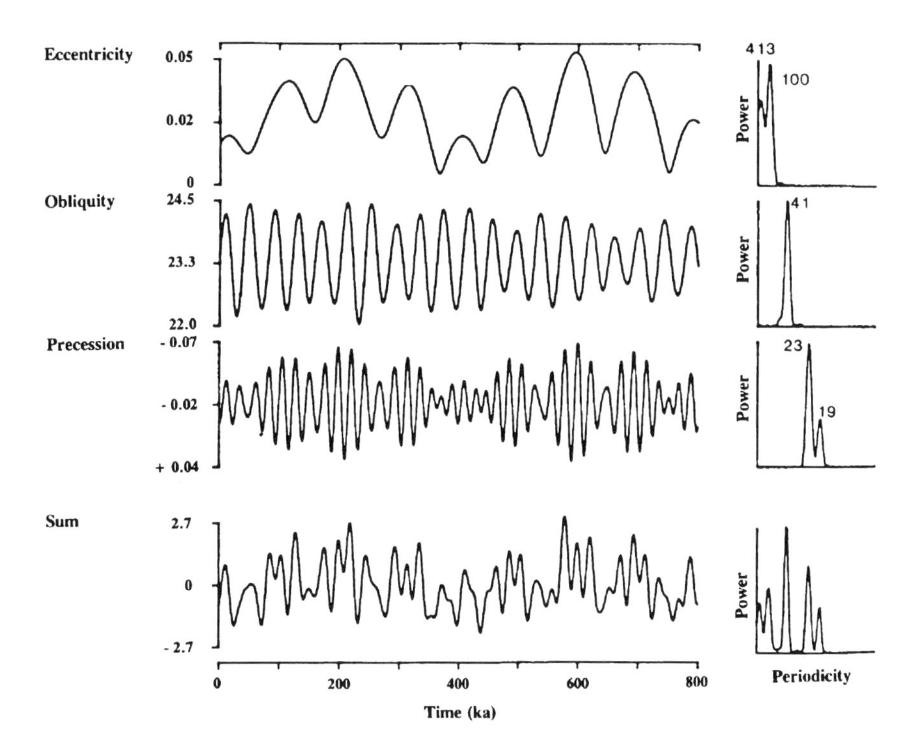

FIGURE 2 Graphical illustration of the calculated variation in eccentricity, obliquity and precession (0–800 ka). The upper three plots on the left are the result of calculations of solar insolation through the last 800 ka (Berger 1977). The lower plot is a composite (sum) curve by Imbrie *et al.* (1984). Scale for individual cycles in degrees, for the 'sum' curve in standard deviation units. Right: variance spectra calculated from these time series, with dominant periods (in ka) of conspicuous peaks indicated (after Imbrie *et al.* 1984, from Patience & Kroon 1991)

coincidence with the circulation of the galaxies. McCrea (1975) speculated that the solar system at this interval passes through dark clouds in one of the spiral arms of the galaxies, so that solar radiation input is reduced. Based on the results of more recent datings, Harland (1981), however, says that the earth's ice ages occurred at irregular intervals. The interval between the Ordovician and Permo-Carboniferous glaciations was only about 110 million years.

Most authors today assume that the onset of ice ages may have had terrestrial causes. One possible triggering factor considered for the generation of ice ages may have been the distribution of major land masses on the earth (Schwarzbach 1974). Major continents near the poles provide favourable conditions for the formation of ice sheets. In fact, our recent ice age began when Antarctica had drifted to the south pole, and during the Permo-Carboniferous glaciation the southern supercontinent Gondwana was in a polar position. The situation is less clear for the other ice ages.

Major changes in oceanic circulation resulting from the opening or closing of important straits have been considered as possible causes. Amongst these are the uplift of Panama in the early Pliocene (Keigwin 1978, 1982), the opening of the Bering Strait in the middle Pliocene (Einarsson *et al.* 1967), isolation and even temporal drying-up of the Mediterranean 5–6 million years ago (Seibold & Berger 1993), or the development of the Iceland–Færöer Ridge (Strauch

1983). In addition, uplift of the continental margins bordering the North Atlantic since Tertiary times might have favoured the formation of major ice sheets in the northern hemisphere (Eyles 1993).

Volcanic eruptions are known to have an influence on global climate. Rampino & Self (1982) have shown that the historical eruptions of the Indonesian volcanoes Tambora (1815), Krakatau (1883) and Agung (1963) resulted in the decrease of surface temperatures of a few tenths of a degree. A similar effect was felt after the recent eruption of Mount Pinatubo (1991) in the Philippines. Dust and volatiles injected into the stratosphere after major volcanic eruptions, like that of Toba volcano at the transition of Oxygen Isotope Stage 4, might initiate a 'volcanic winter' and lead to accelerated glaciation (Rampino & Self 1992, 1993). However, they may be able to influence climate but they seem not to be the primary cause of the Quaternary ice age.

Schwarzbach (1974) also assumed that major changes in the earth's relief (orogenetic phases) may have triggered the ice ages. Indeed Kuhle (1989) thinks that glaciation of the Tibetan Plateau had a decisive influence on global cooling during the Pleistocene. After extensive field investigations he came to the conclusion that during the Pleistocene cold stages the snow line in Tibet had been lowered by as much as 1200–1500 m, so that an ice sheet of 2.4 million km^2 could form on the plateau. After the plateau had been uplifted to a sufficient height during the Early Quaternary, it is assumed to have strongly influenced northern hemisphere atmospheric circulation, which in co-operation with the orbital cycles might have sufficed to trigger glaciations on a continental scale (Kuhle 1989). Gupta & Sharma (1992) think that, based on oxygen-isotope measurements on an ice core taken from Dunde Glacier in northern Tibet, they have found evidence to confirm Kuhle's results. However, the existence of extensive Pleistocene ice sheets on the Tibetan Plateau has been seriously questioned by other workers (Zheng 1989, Burbank & Cheng 1991, Derbyshire *et al.* 1991, Li Shijie & Shi Yafeng 1992). Further investigations are needed before this question can be solved.

There can be no doubt that even without a Tibetan ice sheet the formation of new mountain ranges and uplands several kilometres high have had an impact on global climate. Ruddiman & Raymo (1988), Ruddiman & Kutzbach (1990) and Raymo & Ruddiman (1992, 1993) pointed out that the uplift of Tibet and the mountainous areas of western North America must have had a strong influence on atmospheric circulation. However, modelling has shown that this influence alone is not sufficient to trigger ice ages. Nevertheless, uplift of major parts of the continents to such elevations also must have strongly enhanced weathering and erosion. Increased chemical weathering potentially reduces atmospheric CO_2, turning the 'greenhouse earth' into an 'ice-house earth'. The idea that this process might be a cause of the ice ages had been originally raised by Chamberlin (1899). Recently it has been made more sophisticated by Raymo *et al.* (1988), Saltzman & Maasch (1990) and Raymo (1991). The details of this process, however, are still poorly understood. In fact, reduced CO_2 levels have also been reconstructed for the Permo-Carboniferous ice age but not for that during the Ordovician which, with a duration of only about 0.5–1 million years, was also the shortest and may have had other causes (Brenchley *et al.* 1994).

From the above it seems that we may have come closer to answering the question of what triggers the ice ages. But the subject is far from settled, and further investigations must be awaited.

SECTION I

GENERAL QUATERNARY GEOLOGY

2

Glacier Dynamics

2.1 FORMATION OF GLACIERS

The area covered by glaciers worldwide is estimated at 15.8 million km^2, an equivalent of 10% of the total land surface. Antarctica has a share of 13.6 million km^2, Greenland 1.7 million km^2. The remaining 3.4% are distributed over the rest of the globe as numerous mountain glaciers and small ice caps (Table 1). Melting of the Antarctic Ice Sheet would result in a worldwide sea-level rise of 85 m; melting of the Greenland Ice Sheet alone would only raise the sea level about 7 m (Drewry 1991). During the last glacial maximum the total volume of glacier ice was approximately 2.5 times larger than the modern volume (Denton & Hughes 1981) and resulted in a lowering of sea level of about 130 m.

Glaciology, the science of glaciers, is a highly specialised discipline, dealing largely with the physics of glaciers, and, in a broader sense, also with ice and snow. A well-illustrated introduction has been published by Hambrey & Alean (1992) and a profound, easy to read overview is given by Sharp (1988). An introduction to the physics of glaciers has been provided by Paterson (1994); Lliboutry (1964/65) and Hutter (1983) give far more detail. Quaternary research to a large degree has to deal with deposits and landforms left behind by former glaciers. Glacial geology and geomorphology are comprehensively introduced by Sugden & John (1976), Drewry (1986) and Hambrey (1994). The following paragraphs can only provide a very brief overview.

A basic precondition for glaciers to form is that the winter snowfall exceeds melt during the subsequent summer, so that the snow of the following winters adds to the volume. This requires low summer temperatures, which determine the proportion of precipitation that is lost as meltwater. With increasing distance from the poles, low temperatures are restricted to higher altitudes, so that glaciers near the equator are limited to the highest mountains. The only three glaciated summits of Mexico are all higher than 5000 m. The glaciation limit in the subtropics is even higher. Low precipitation at the margins of the South American Atacama desert, for instance, leaves mountains of over 6000 m height unglaciated.

Apart from temperature, another precondition for glacier formation is sufficient snowfall. High evaporation and low precipitation can lead to the formation of polar cold deserts instead of ice sheets. As a rule, the west coasts in high latitudes provide extremely favourable conditions for glacier formation. The westerly winds supply ample precipitation. The decisive factor is the seasonal distribution of the precipitation. Summer precipitation often falls as rain

Table 1 Extent of recent glaciers (after World Glacier Monitoring Service 1989, from Hambrey 1994)

Region	Area (km^2)
South Polar Region	
Subantarctic islands	7,000
Antarctica	13,586,310
	13,593,310
North America	
Greenland	1,726,400
Canada	200,806
USA (including Alaska)	75,283
Mexico	11
	2,002,500
South America	25,908
Europe	
Scandinavia (including Jan Mayen)	3,174
Svalbard (Spitsbergen)	36,612
Iceland	11,260
Alps	2,909
Pyrenees/Mediterranean Mountains	12
	53,967
Asia	
Commonwealth of Independent States	77,223
Turkey/Iran/Afghanistan	4,000
Pakistan/India	40,000
Nepal/Bhutan	7,500
China	56,481
Indonesia	7
	185,211
Africa	10
Pacific Region	
New Zealand	860
Total	15,861,766

and contributes little to the glacial mass balance. Most of it does not freeze but drains rapidly together with the glacial meltwater on the glacier surface or through tunnels in and under the ice (Sugden & John 1976).

For the formation of large ice sheets high precipitation is required; for their maintenance comparatively little snow is sufficient, provided that summer temperatures remain low. The interior areas of large ice shields figure amongst the most arid parts of the world. In the northern parts of the Greenland Ice Sheet precipitation is below 150 mm per annum. However, all of this falls as snow and contributes to the formation of glacier ice.

Glaciers form preferentially in areas with a favourable relief. This is, for instance, the case in lee-side positions in high mountains. In Scandinavia most of the modern glaciers are found on northeast-facing slopes, which are both wind-protected and cold (Østrem *et al.* 1973). The interaction of several factors together determines glacier formation and distribution. This becomes obvious when considering the fact that within large glaciated areas there are ice-free oases like the dry valleys of Victoria Land (Antarctica). A major example from the Pleistocene glaciations is the Driftless Area in Wisconsin and Illinois (Willman *et al.* 1989).

In the upper reaches of a glacier more snow accumulates than can be melted. This part is called the **accumulation area**. In the lower part of the glacier, the **ablation area**, melting prevails. In modern Alpine glaciers the accumulation area is approximately twice as large as the ablation area (Hoinkes 1970) (Fig. 3). **Snow line** is a descriptive term used for the actual lower boundary of the snow cover on a glacier. The boundary between accumulation area and ablation area is referred to as the **equilibrium line**. The equilibrium line is defined as the line connecting all points on the glacier surface where the mass balance is zero at the end of a budget year. It mostly has a complicated course and cannot be reconstructed from maps. As the height of the equilibrium line changes from year to year, measurements are made over a longer time period to determine average values. Such an equilibrium line is equivalent to the 'local' or 'orographic snow line' of the older literature (Gross *et al.* 1978, Sharp 1988).

Freshly fallen snow that does not melt is gradually transformed into ice. At first it changes into old snow. The change occurs rather rapidly (mostly within a few days), for the temperature of the snow crystals is mostly near zero, so that the molecules are relatively mobile. Migration of the molecules tends to reduce the crystal surface and to minimise the free energy. In this way snow crystals with their relatively complicated shapes are gradually altered to spherical nodules. Additionally, large crystals tend to grow at the expense of smaller ones, further minimising the free energy. Simultaneously a general settling occurs, reducing the cavities between the particles. Intergranular bridges form at the contact points between the grains (Paterson 1994).

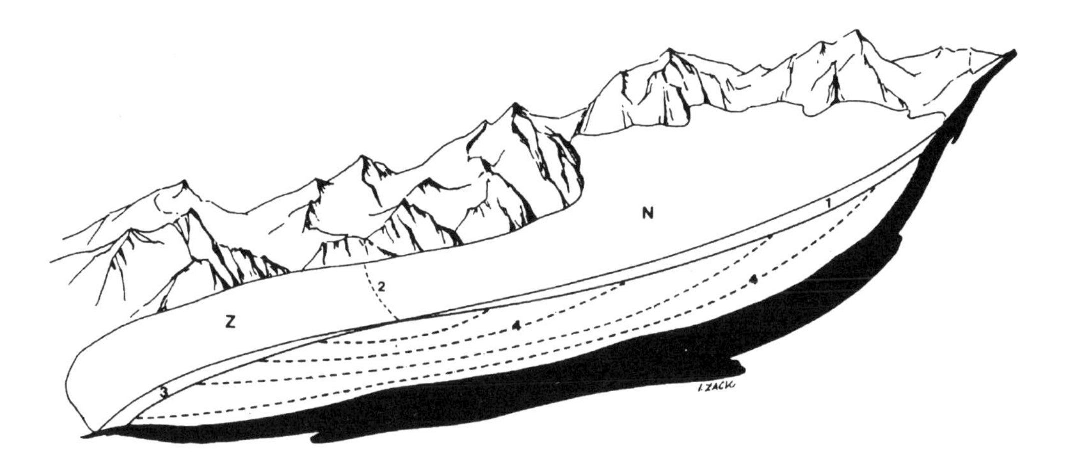

FIGURE 3 Longitudinal section through a valley glacier. In the upper part of the glacier, the accumulation zone (N), especially in the upper reaches (1) new ice is formed that flows downvalley, following the gradient. The ice formed in the uppermost parts of the glacier moves near the glacier sole (see flow lines 4), as the ice moves in a laminar style without mixing (in contrast to turbulent flow in rivers). Formation of new ice decreases continuously down to the equilibrium line (2). Downglacier the ablation zone (Z) is found. Here the ablation increases continuously down to the glacier terminus (3) (from Van Husen 1987)

Old snow of the preceding year has undergone this process; it is called **firn**. The lower limit of the firn cover of a glacier is called the firn line (Hoinkes 1970). Molecular diffusion leads to recrystallisation and further consolidation of the firn. When its density reaches a value of about 0.8 g/cm^3, the pores between the crystals are sealed. Trapped air stays behind in the form of gas bubbles. The firn has turned into ice (Paterson 1994).

2.2 GLACIER FLOW

Glacier flow largely depends on the temperatures within a glacier. Under pressure, water freezes at lower temperatures. Thus beneath a zone of 10–15 m depth many glaciers, especially in lower latitudes, are at the **pressure-melting** temperature. Additionally, each glacier receives a permanent heat supply from below. Most glaciers are actually much thicker than 15 m. They avoid complete melting of their lower parts by the fact that melting of 1 gram of ice requires 80 calories of heat. This can only be supplied from the ice, so that melting actually cools the remaining ice. In the polar regions, surface temperatures are so low that some glaciers are frozen throughout.

Based on a concept dating back to Ahlmann (1935) two types of glaciers are distinguished: temperate and polar glaciers. The basal part of a **temperate glacier** is continuously at the pressure-melting point, and on its surface in the ablation area – at least in non-polar areas – ice and snow melt perennially. On a **polar glacier** there is no surface melt, and its base is firmly frozen to the bed. Earlier it was thought that the glaciers of the Alps all belonged to the temperate type (e.g. Hess 1904). Today it is known that at least in the higher reaches of some Alpine glaciers the bottom temperatures remain continuously below zero. Haeberli (1975b) measured ice temperatures of −2 to −3°C on the Monte Rosa Plateau. A large glacier may be partly temperate, partly polar. The ice sheets of Greenland and Antarctica are frozen to the base only in their marginal areas, whereas the central parts are at pressure-melting (Sharp 1988). A similar zonation has also been assumed for some Pleistocene ice sheets. However, field evidence suggests that the central area of the Scandinavian Ice Sheet was cold-based and frozen to its bed (Sollid & Sørbel 1994). Also in the Atlantic Appalachian region of North America till-covered deeply weathered bedrock and woolsack tors suggest the presence of cold-based ice (Gauthier 1980, Grant 1989). The same seems to have been true for the small Late Wisconsinan ice caps of the Canadian Arctic Islands (Dyke 1993), but not for the Laurentide Ice Sheet.

When the ice reaches a minimum thickness of about 60 m it starts to flow. Traditionally, glacier flow has been regarded as composed of two components, **internal deformation** and **basal sliding** (Sharp 1988). A third component, **bed deformation**, has been added in the last decades (see below).

Where the glacier sole is at the pressure-melting point, basal sliding occurs in addition to the internal deformation. This is most effective where the glacier can slide on a water film (Weertman 1964). For the flow velocity, therefore, water-filled cavities at the glacier sole play a major role. With increasing hydrostatic pressure the water-filled cavities can expand, lowering the basal friction and thus increasing the flow velocity (Lliboutry 1968). In this way glaciers flow faster in summer than in winter, at least in their ablation areas (Sugden & John 1976). In cold-based glaciers no basal sliding occurs, and a major part of the deformation may occur in the substrate, far below the ice/ground interface.

Horizontal flow of a glacier (Fig. 4) is not completely uniform but is determined by the mass balance and the shape of the glacier bed. Nye (1952) has pointed out that principally two

FIGURE 4 Flow behaviour of a valley glacier. Over a zone of plastic flow (1) lies a zone of brittle ice (2) in which deformation results in crevasses. The thickness of this zone depends on the temperature /pressure conditions within the glacier. The zone is thickest in extremely cold regions. In recent Alpine glaciers it is up to several tens of metres thick. In addition to the frequent crevasses over irregularities and steps of the glacier bed (3), also short marginal crevasses are to be seen, pointing upglacier. They result from the contrast between higher flow velocities in the central part of the glacier than near the margin, when they are slowed by increased friction over the rough glacier bed (from Van Husen 1987)

different types of glacier flow can be distinguished: **compressive** and **extending flow**. Compressive flow reduces glacier movement. It occurs where strong ablation lowers the overburden of ice, or where the glacier sole dips up-ice. Compressive flow causes upward-directed flow paths within the glacier, whereas extending flow has the opposite effect.

In most cases the flow velocities of glaciers show relatively little variation (mostly a few metres to some tens of metres per year). However, considerable periodic changes can occur. In **surges**, a special type of rapid glacier advance, flow velocity of the ice can increase by ten to a hundred times the normal value. Within the rapidly advancing ice three zones can be distinguished: (1) a wave of thickening ice moving under compressive flow conditions, (2) a zone of high velocities with a strongly fractured ice surface behind the wave crest, and (3) a zone of ice extension and decreasing velocity (Sugden & John 1976). The surge wave does not reach the terminus of a glacier tongue in all cases. Where this occurs, however, it can result in a catastrophic glacier advance. For example, Brúarjökull in Iceland advanced over a distance of 8 km in its 1963/64 surge, reaching maximum velocities of up to 5 m/h (Thorarinsson 1969). At the end of such a surge the advanced glacier tongue often stagnates until sufficient ice has built up again to support further advance. According to investigations on North American glaciers, surges occur once in 10–100 years (Meier & Post 1969). However, not all glaciers do surge.

The surges are caused where meltwater supply at the glacier sole exceeds the drainage capacity. They are normally triggered in winter, when throughflow is low and capacity of the conduits at a minimum. In areas of higher ice pressure water can be dammed up. Water pressure finally starts to lift the upglacier part of the ice. Reduced basal friction causes rapid ice movement, so that this part of the glacier pushes against the ice dam. Finally pressure exceeds the shear strength of the ice dam, and the surge starts. Only lateral friction and bed roughness limit the flow velocity. The surge continues until ice supply in the accumulation area is exhausted. In a surging glacier the ice moves almost entirely by basal sliding. Thus basal velocity is nearly equal to the advance rate. Moreover, in contrast to normal glacier flow, the whole cross section moves at a nearly equal velocity (Kamb *et al.* 1985, Kamb 1987).

Surges are not restricted to small glaciers. During deglaciation in North America, for instance, surging was a major factor. The receding ice sheet was fronted by extensive glacial lakes, which made the margin dynamically very unstable.

Whereas mountain glaciers normally move in large tongues involving the entire glacier terminus, this is not necessarily the case with ice sheets. Along the Antarctic coastline three major types of ice margin can be distinguished: **shelf ice**, **ice walls** and **ice streams**. An ice shelf consists of large plates of ice, up to several hundred metres thick, which float on the sea but are connected to the ice sheet and fed by it. Their landward boundary against the ice sheet is the grounding line. They advance at a rate of several hundred metres per year. Ice walls form the seaward, near-vertical limit of a grounded ice sheet. They have a flow velocity of only 5–50 m/year. Ice streams are parts of the ice sheet that flow far more rapidly than their surroundings (Ehrmann 1994).

Major ice streams permanently have considerable flow velocities (Clarke 1987):

Rutford Ice Stream (Antarctica)	over 400 m/year
Ice Stream B (Antarctica)	827 m/year
Jakobshavns Glacier (West Greenland)	8360 m/year

The catchment area of such an ice stream can be a thousand times larger than that of a typical rapidly advancing mountain glacier. Hughes (1992) points out that about 90% of the drainage of the Greenland and Antarctic ice sheets into the sea occurs by means of such ice streams. Fast-flowing ice streams are partly held responsible for the rapid decay of the Pleistocene ice sheets. There is increasing evidence from ground surveys and mapping in North America that ice streams also controlled major parts of the Laurentide Ice Sheet. Examples have been documented in the Arctic, in Ontario, in Manitoba and in the prairies (Dredge, personal communication).

2.3 FORMATION OF ALPINE-TYPE ICE-STREAM NETWORKS

For the formation of Alpine glaciers the local topography is of utmost importance. Steep rock walls prevent glacier formation. On gently inclined slopes, however, firn accumulates in depressions. The process of erosion under the influence of freeze–thaw cycles around snow patches is called **nivation**. Nivation niches gradually develop into deeper depressions, in which **cirque glaciers** can form. With a continuously positive mass balance, cirque glaciers can finally develop into **valley glaciers**. Especially favourable conditions for glaciers are found where accumulation areas can form ice caps on old plain remnants situated above the snow line. Backward erosion of cirque or valley glaciers leads to the formation of sharp-crested ridges, so-called **arêtes**.

There is a major difference between the dendritic ice-stream pattern of the North American Cordilleran glaciation and the ice-stream network of the Alps. In the Alps the main valleys trend parallel to the mountain range, which hampers ice flow. The thick ice of the central Alps could drain through the narrow valleys only insufficiently. Congestion in some areas led to transfluence across Alpine cols. Thus the ice tongues in the Ammer, Loisach and Isar valleys at the northern Alpine margin grew much larger than could possibly be supported by the rather small catchment areas of these rivers (Van Husen 1985b, 1987). In contrast, the Iller and Lech valleys, where no transfluence occurred, developed only minor glaciers.

In the Alps, a marked asymmetry is observed between glaciation of the northern and southern slopes. On the south side of the Alps the snow line is about 200 m higher than in the north (Von Klebelsberg 1948/49). This contrast is further enhanced by the different altitudes of the two Alpine forelands (about 500 m in the north as opposed to about 100 m in the south). As a consequence, Quaternary glaciers have advanced far into the northern foreland of the Alps, whereas in Italy they terminated as soon as they had reached the foot of the mountains.

During the last glacial maximum the Alpine snow line was about 1200 m lower than today. As a consequence, the accumulation areas were much enlarged. The Alpine valleys filled with powerful ice streams, which eventually coalesced to form an ice-stream network, surmounted only by very few summits. At Bozen, for instance, the distance between the summits flanking the Etsch valley glacier was about 40 km (Von Klebelsberg 1948/49).

The maps of Jäckli (1970), Hantke (1983) and Van Husen (1987) indicate the maximum extent of the last glaciation in the Alps, the Würm Glaciation (Fig. 5). The figure shows that not all parts of the Alps were equally affected by glaciation. In the western and central Alps only the highest summits towered above the ice-stream network, the surface of which was at an elevation of about 2500 m a.s.l. and higher. In the east the glaciated area grades into the large ice tongues of the Enns and Mur valleys and an ice lobe in the Drau valley. Further east not only are the mountains lower, but also increased continentality results in less precipitation. Here there are only a few major and minor local glaciers and ice caps formed.

2.4 FORMATION OF CONTINENTAL ICE SHEETS

In a mountain area a glacier that developed from a cirque glacier into a valley glacier at first flows downvalley, following the natural gradient. Strong ice supply leads to the merging of individual glacier tongues into major ice streams. Adjoining ice streams grow into extensive piedmont glaciers at the foot of the mountains. Where ice supply permanently exceeds the valley's capacity for drainage, adjoining mountain ranges can be surmounted, and the glaciers of neighbouring valleys may unite to form a major ice shield. Further growth of the ice mass can lead to the formation of an ice sheet several hundred kilometres in diameter.

The major Pleistocene ice sheets were not singular ice masses but normally initiated from several independent glaciation centres, the dynamics of which might differ considerably. Already at the end of the last century Tyrrell (1898) came to the conclusion that the last glaciation of North America did not consist of one single ice shield but was initiated from two different centres, one in Keewatin and the other in Labrador. Later he added a third centre in Patricia (south of Hudson Bay). Flint (1943) was of the opinion that the centre in Patricia did not exist, and that also the other two centres had been rather short-lived and of only local importance. His concept of a single-domed, major Laurentian Ice Sheet was adopted by most scientists (e.g. Woldstedt 1965, Denton & Hughes 1981). Today, however, major doubts have arisen about this concept. Prest & Grant (1969) and Prest *et al.* (1972) after air photographic analysis of glacial landforms defined multi-domed Laurentide and Appalachian ice complexes.

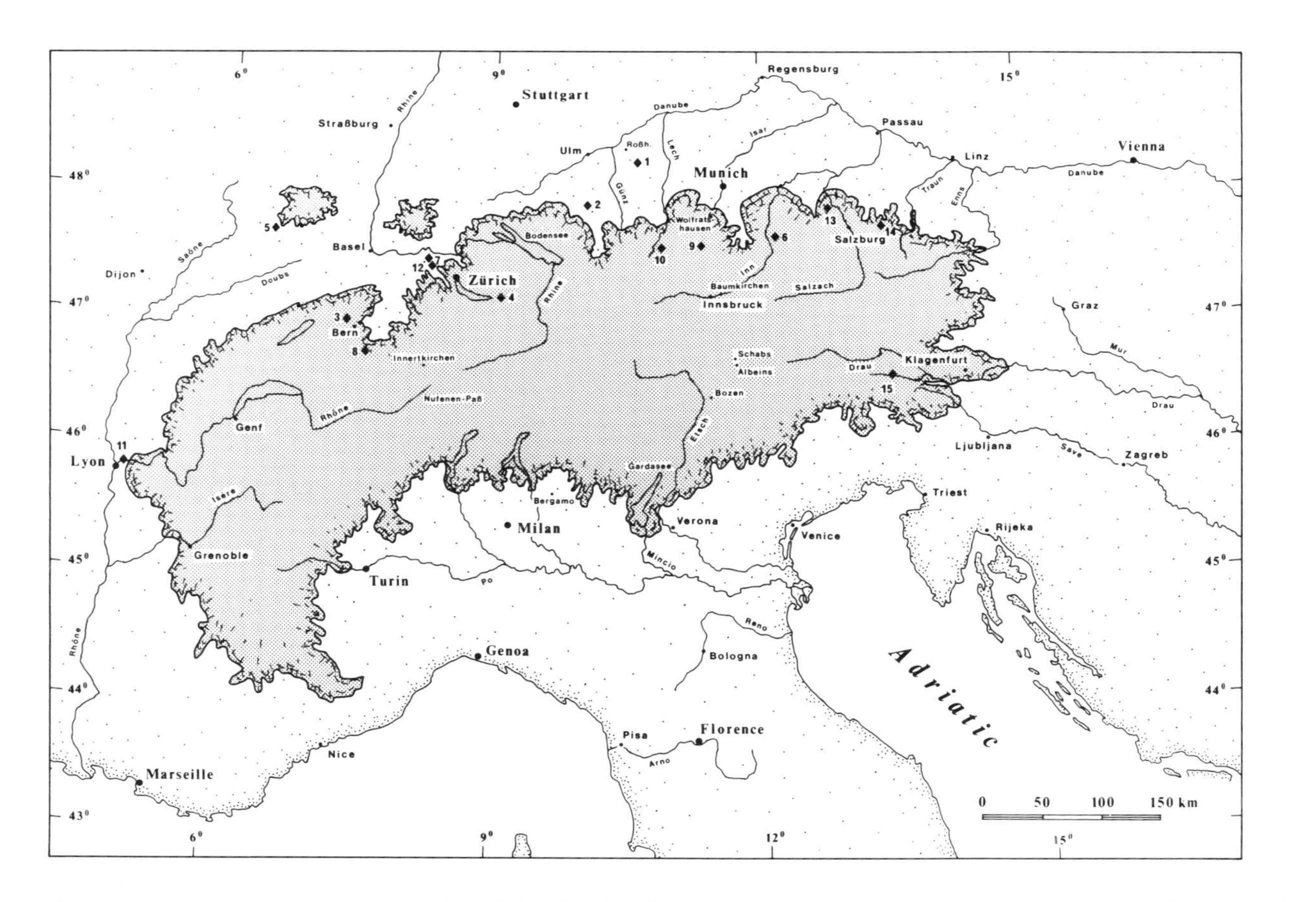

FIGURE 5 Extent of the Würm Glaciation in the Alps (after Hantke 1978 and Van Husen 1987), and important interglacial and interstadial sites. 1 = Uhlenberg; 2 = Unterpfauzenwald; 3 = Meikirch; 4 = Uznach; 5 = Grand Pile; 6 = Samerberg; 7 = Niederweningen; 8 = Thalgut; 9 = Großweil; 10 = Pfefferbichl; 11 = Les Echets; 12 = Sulzberg; 13 = Zeifen; 14 = Mondsee; 15 = Nieselach. Roßh. = Roßhaupten

Following investigations of indicator clasts, Shilts (1980) came to the conclusion that Flint's (1943) glaciation centre over the Hudson Bay could not have existed, but that instead two individual centres in Keewatin and Nouveau Quebec/Labrador had determined the ice-sheet development (Fig. 6). However, if and how long this was actually the case, is still a matter of debate (Dredge & Cowan 1989).

In addition to these large glaciation centres of the Laurentide Ice Sheet, the Appalachian region (Maritime Provinces of Canada and Newfoundland) harboured separate ice caps during at least the last glaciation (in North America: Wisconsinan). Chalmers (1895) was the first to recognise these Appalachian ice centres. Much of his work that is based on mapping striations is still valid today. The Appalachian ice sheets were collectively nearly as large as the Fennoscandian Ice Sheet. Major ice divides existed over northern Maine and New Brunswick (Chalmers 1895, Rappol 1989, Lowell *et al.* 1990a), the Gulf of St Lawrence (Rampton *et al.*

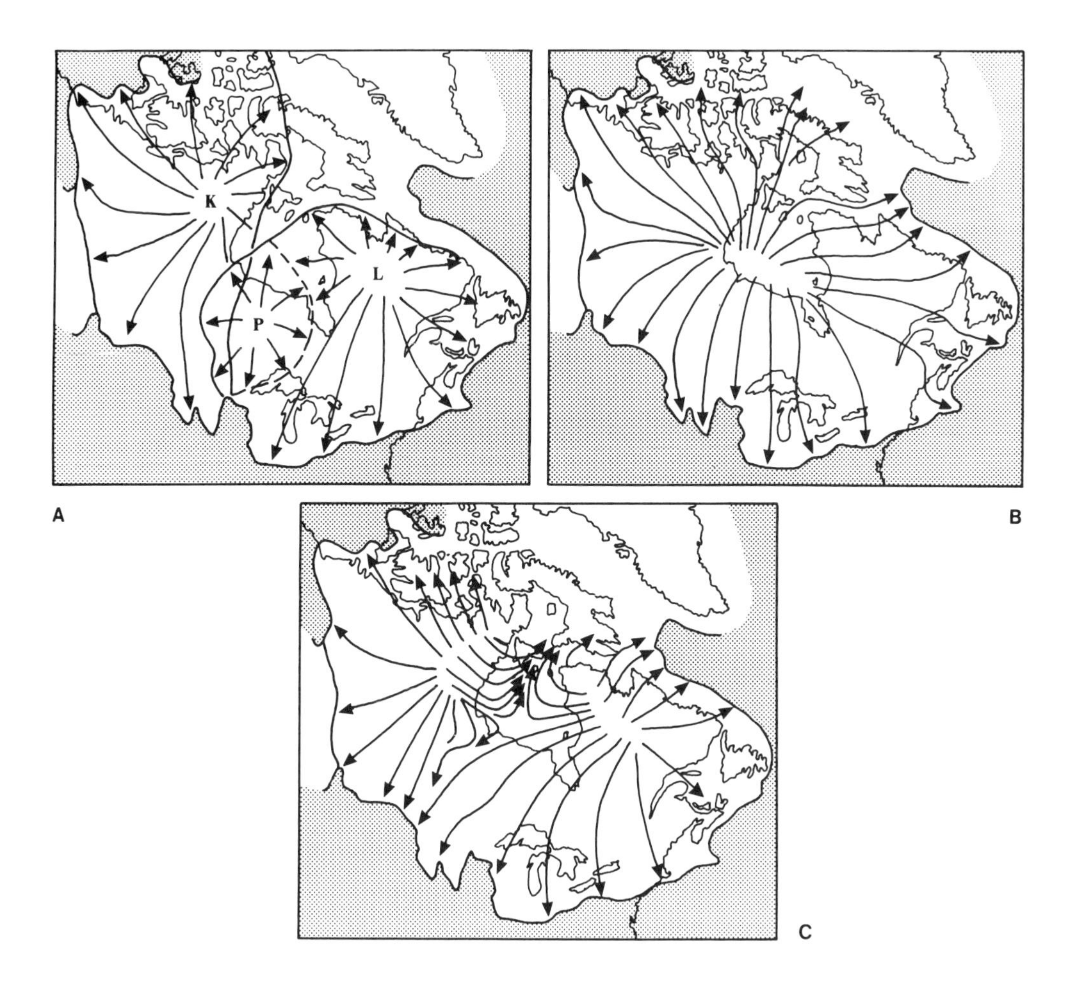

FIGURE 6 Glaciation centres in North America. K = Keewatin; P = Patricia; L = Laurentia (A = after Tyrrell 1913; B = after Flint 1943; C = after Shilts 1980). These maps leave out the Appalachian ice centres and the centre in the Foxe Basin (see text)

1984, Stea *et al.* 1992), Newfoundland (Brookes 1970) and Nova Scotia at various times during the Wisconsinan. Another, smaller North American ice centre seems to have been situated in the Foxe Basin region, northwest of Hudson Bay. It had a diameter of about 1000 km (Ives & Andrews 1963, Andrews 1989).

The Cordilleran Ice Sheet in North America formed an independent glaciation that had only weak contact with the Laurentide Ice Sheet. Both ice masses were separated over a major distance by an ice-free corridor, the exact extent of which is still unknown. Some authors think it had a length of a few hundred kilometres, whereas others assume a gap of more than one thousand kilometres – both at the northern and at the southern end of the contact zone (Prest 1984).

The north European glaciated area also spread from several glaciation centres. Both on the British Isles and in the east in the Ural–Timan area independent ice sheets developed which to a greater or lesser extent came into close contact with the Scandinavian Ice Sheet, and there was an ice sheet on the Barents Sea shelf. Apart from these, Lagerlund (1980) assumes an additional marginal ice dome in the southern Baltic Sea for the last glaciation (in Europe: Late Weichselian). This view, however, could not be substantiated by more recent investigations (Houmark-Nielsen 1987, Ringberg 1988).

Knowledge of the global extent of glaciation during the last cold stage has been summarised by Denton & Hughes (1981). However, a number of major uncertainties still remain, especially in areas like Siberia or the Tibetan Plateau. Even opinions on the form of the North European Ice Sheet have changed considerably during the last few decades. When Aseev (1968) attempted a reconstruction of the Late Weichselian ice sheet, he assumed a maximum height of 2500 m a.s.l. However, in southern Norway some summits have been proved to have remained above the Late Weichselian ice surface. Based on these maximum altitudes, cross sections through the ice sheet were constructed (Nesje & Sejrup 1988). They show that the ice was considerably thinner than previously assumed (Fig. 7). From the areas further east no ice thicknesses are known. However, in contrast to Aseev, it is well-known today that the ice reached much further north and east. It is regarded as certain now that the Barents Sea was

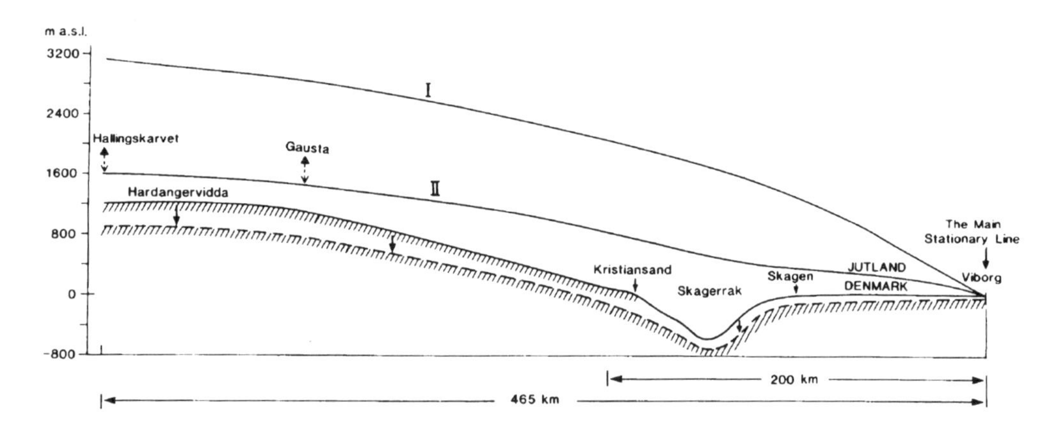

FIGURE 7 North–south profile through the Late Weichselian Scandinavian Ice Sheet from Hardangervidda (Norway) south to Jutland. I = profile after Vorren 1977; II = present interpretation. The tops of Hallingskarvet and Gausta stood above the ice as nunataks. The profile is based on an isostatic depression of about 300 m in the Hardangervidda region (from Nesje & Sejrup 1988)

covered by a marine ice sheet and that the Scandinavian Ice Sheet was in contact with the Ural–Timan glaciation (see chapters 7 and 15).

In contrast to the Alpine glaciers, the North European Ice Sheet was not confined to valleys. During the glacial maximum it was able to spread relatively freely in all directions, controlled largely by its mass balance. Precipitation was supplied by the westerly winds and snow accumulated mainly in the lee of the mountains. Because of the steep slope of the continental margin in the north and the several hundred metres deep Norwegian Channel in the south, no extensive glaciation could form west of the Scandinavian mountains. Thus an asymmetric ice shield was formed, the crest of which was displaced considerably towards the northwest of the centre.

Two hypotheses have been discussed with regard to ice sheet inception and growth.

1. *Highland origin and windward growth* (Enquist 1916). Traditionally it has been assumed that the large Pleistocene ice sheets developed from local mountain glaciers which gradually advanced into the foreland, coalesced to form a piedmont glacier and finally increased in volume until they formed a major ice sheet. Flint (1943, 1971) also proposed that in North America glaciers started from corries in highland areas, such as local ice caps in the Labrador plateaux and Laurentians in Quebec, and grew against the main wind direction (westward), supplied by moisture sources from the Gulf of Mexico and Atlantic.
2. *Instantaneous glacierisation* (Ives *et al.* 1975). The previous model may sound convincing for the Scandinavian and British ice sheets, but not for the Laurentide Ice Sheet. In Canada the mountains in question are situated at the outermost eastern margin of the ice sheet. Moreover, the raised margin of the Canadian Shield forms uplands rather than mountains, descending rather abruptly towards the sea but dipping only very gently towards the interior. No ice-movement indicators of a continental glaciation spread from this area. Ives *et al.* (1975) therefore envisioned that the Laurentide Ice Sheet formed rather spontaneously by a process they called 'instantaneous glacierisation'. They assumed that extensive and simultaneous accumulation of snow over large areas of the Canadian Shield within a few centuries would result in the formation of a large ice sheet, starting from the plateaux of Labrador–Ungava, Baffin Island, District of Keewatin and Ontario. In fact, during the Little Ice Age, when the snow line was lowered considerably, permanent snow cover resulted in the formation of thin ice fields covering major parts of the plateau of Baffin Island (Ives 1962) – an event which Lamb & Woodroffe (1970) referred to as 'abortive glaciation'. Instantaneous glacierisation thus is regarded as the most likely explanation for the formation of the Laurentide Ice Sheet (e.g. Dyke *et al.* 1989, Clark & Lea 1992). The same model, to a certain degree, might apply to the North European Ice Sheet.

The major ice sheets of the Pleistocene cold stages spread very rapidly. This is best documented for the Late Weichselian glaciation. As late as 24,000 BP the major Alpine valleys were still all ice-free. The same is true of southern Sweden at least as far north as Göteborg (Lundqvist 1983a). If the glacial maximum in Europe was reached at about the same time as in North America, i.e. about 20,000 BP, the ice sheet must have spread at a rate of no less than 75 m/year. This calculation only takes into account the progradation of the ice margin from Sweden to North Germany. If one considers that the erratic composition of the Late Weichselian tills indicates that ice from the glaciation centre advanced to the areas south of the Baltic Sea, and that melting occurred at the margin, velocities of 100–150 m/year must be assumed. Based on varve analyses, Eissmann (1975) has reconstructed ice-advance rates of 250 m/year for the first Elsterian and 140 m/year for the second Saalian ice advances in the Leipzig region, Germany. By radiometric dating of tree trunks overridden by the advancing ice, Goldthwait (1959) demonstrated that the Wisconsinan ice in Ohio advanced at rates between 17 and 119 m/year. Much higher rates, in the order of 700 m/year, have been postulated by Mickelson *et al.* (1981) and Clayton *et al.* (1985).

Whereas a rapid advance of the Alpine glaciers seems plausible because of the short distances and the special character of the glaciation (ice-stream network), the rapid advances of the North European and North American ice sheets are hard to comprehend using traditional explanations. Other possibilities must be taken into consideration.

By glacier investigations in a tunnel under an Icelandic glacier, Boulton & Jones (1979) showed that not only the glacier was moving but also the sediment at its base. Such a possibility had been speculated about earlier, e.g. by Gripp (1974), but it had never been substantiated. In 1986 Alley *et al.* (1986) published the spectacular results of their investigations of Ice Stream B in western Antarctica. At the base of this ice stream between ice and bedrock they had found a layer about 6 m thick of till – the '**deformable bed**'. This discovery and its consequences caused Boulton (1986) to speak of a 'paradigm shift in glaciology': the high flow rates of certain recent and ancient glaciers had found a simple explanation. Alley (1991) thinks that deformation by pervasive shearing can contribute considerably to ice movement and sediment transport (Fig. 8). Consideration of the glacier dynamics of the Scandinavian Ice Sheet (chapter 2.8) suggests that at least temporarily a deformable bed may have played a part in glacier movement.

The till material dragged along at the base of the ice is not part of the glacier, because it is ice-free. It gains its mobility from the high pore-water pressure at the glacier base, which in a relatively impermeable substrate (like till) can hardly be relieved. Especially under major ice shields there is a strong possibility that considerable water pressure may build up, as long as the hydraulic conductivity of subglacial sediments is not sufficient to drain the excess meltwater. Channel systems, according to Weertman (1972), can only remain stable as long as considerable amounts of meltwater are supplied either from the glacier surface or by basal melting.

In the case of the Antarctic Ice Stream B, apparently most till is transported by subglacial deformation and only a small percentage within the ice (Alley *et al.* 1987). In order to maintain such a movement, material must be eroded at the base of the deforming layer and be incorporated into the movement. Alley *et al.* (1987) regard this as an explanation for the well-known fact that tills of earlier glaciations are transformed by later ice advances in such a way that their fabric sometimes reflects the younger ice-movement direction. MacClintock & Dreimanis (1964) had first described this phenomenon from the St Lawrence valley, where reorientation reached down to a depth of 10 m. To a lesser degree the same phenomenon has been observed elsewhere, e.g. in the north German till areas (Ehlers 1990a). This phenomenon so far had been explained by freezing-on of older material at the glacier sole. This process did not result in complete mixing with the younger till, as is typical for glacial erosion, because the velocity vectors of the deforming bed form an exponential velocity-profile curve. This phenomenon has not been found everywhere. But subglacial deformation need not necessarily have extended down into older tills, and in many cases it may have been restricted to recently deposited till of the later glaciation (Alley *et al.* 1987).

Whereas traditionally ice thicknesses of over 2500 m had been assumed for the Scandinavian Ice Sheet (Aseev 1968), Boulton *et al.* (1985) under the assumption of subglacial deformation modelled maximum ice thicknesses of about 2000 m. Where Aseev had

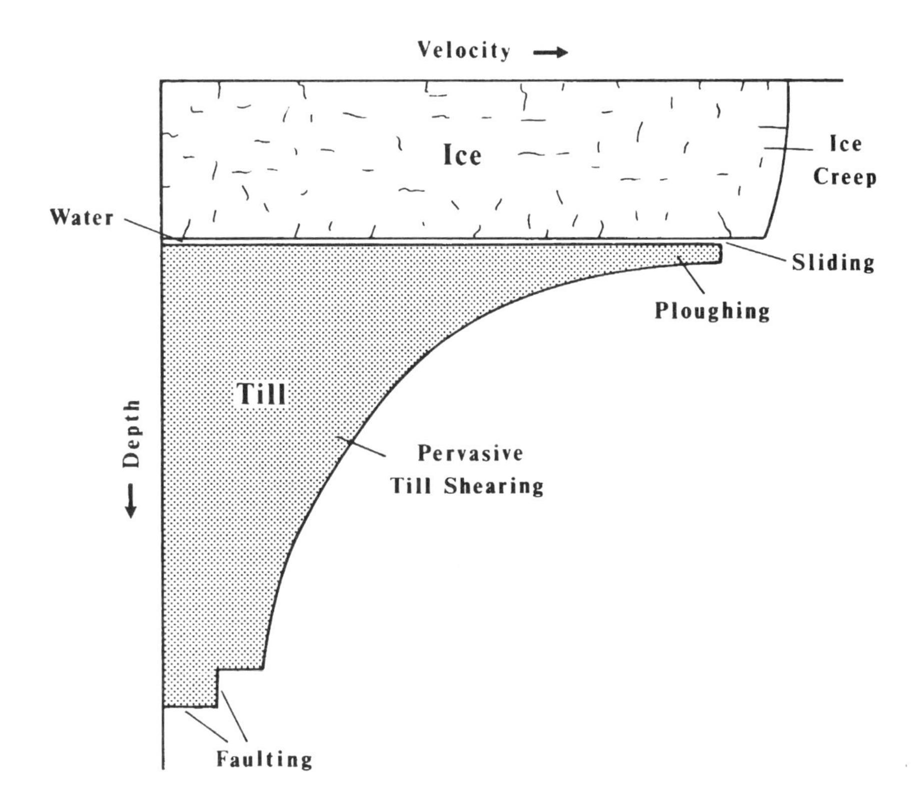

FIGURE 8 Possible mechanisms of ice motion on a bed of unconsolidated sediment, with deformation being the principal velocity component in wet-based glaciers (after Alley 1989b). It is assumed that deformation may result in a substantial flux of sediment to the glacier terminus (Alley 1991)

assumed 1000 m of ice around Copenhagen, Boulton *et al.* needed no more than 500 m to maintain their ice sheet. The resulting model of ice-movement directions and ice thicknesses coincides rather convincingly with the actual observations made in the field (cf. chapter 2.8).

2.5 FORMATION OF MARINE ICE SHEETS

Major parts of the recent ice sheets in Greenland, Antarctica and Spitsbergen drain directly into the sea. They are partially surrounded by ice shelves, i.e. large masses of ice floating on the sea. Other parts of the ice sheets that are grounded below sea level are referred to as **marine ice sheets**. A drop in sea level, as occurs during the build-up phase of continental ice sheets in a cold stage, will result in grounding of major parts of floating ice shelves and pack ice and thus result in the formation of additional marine ice sheets. This is assumed to have happened in the Barents Sea region, where an independent glacial centre developed during the Pleistocene glaciations. It also seems to have occurred in the Kara Sea region.

The only major recent marine ice sheet is found in West Antarctica. There have been speculations about its stability. Hughes (1973) suggested that the West Antarctic Ice Sheet might disintegrate. Mercer (1978) assumed that this might happen as a consequence of the greenhouse effect and continued global warming. However, the controlling mechanisms are not yet well enough understood to decide whether this is possible or not (Paterson 1994).

2.6 GLACIAL EROSION

Even under arctic conditions, subaerial erosion is a relatively slow process. By measuring sedimentation rates in a lake on Spitsbergen, Svendsen *et al.* (1989) reconstructed Holocene erosion to be in the order of 15 mm/1000 years. In contrast, Elverhøi *et al.* (1995) calculate denudation rates of 400 mm/1000 years for the Storfjorden drainage basin on Spitsbergen for the last 2.5 million years. The difference is the result of glacial erosion. As the areas in question were only ice-covered for limited periods within this time span, the actual erosion rate during glaciations was much higher. In areas of a rapidly sliding glacier on Spitsbergen, Elverhøi *et al.* (1983) found denudation rates of 1000 t/km^2 per annum.

Erosion by glaciers and ice sheets represents the combined effects of glacial abrasion and meltwater erosion at the base of the glacier. Both processes depend on the temperature at the glacier sole. Whereas temperate glaciers slide over their bed and produce ample meltwater that drains subglacially, cold glaciers move by internal deformation that may hardly affect the underlying substratum.

Erosion by the ice is mainly achieved by two different processes, plucking and abrasion. Abrasion is the result of scoring where rock fragments in the ice are moved across a rock

FIGURE 9 Roches moutonnées in the forefield of Nigardsbreen, Norway; ice movement from left to right (Photograph: Ehlers 1980)

surface. Where a glacier moves over an irregularly-shaped bed, the increased pressure in front of an obstacle will produce meltwater, which at pressure release on the lee side of the obstacle refreezes. Refreezing may result in freeze-on of rock fragments which is particularly the case in areas with jointed bedrock. Freeze–thaw cycles at the glacier bed may loosen further blocks. Thus on the upglacier side the obstacle will become striated and smoothed by the overriding glacier, whilst the lee side often attains a rather irregular shape. The features thus produced are called **roches moutonnées** (Fig. 9), a term that was originated by De Saussure (1779–96).

Generally, glacial erosion produces smooth, striated, sometimes even polished surfaces, whereas troughs, channels and occasionally deep hollows result mainly from meltwater erosion. Strong meltwater erosion can result in plastically-shaped surfaces, the so-called 'p-forms' of Dahl (1965) (Fig. 10). In mountainous areas, combined erosion by glaciers and meltwater result in the formation of deep U-shaped valleys (Fig. 11). Higher terrain or individual peaks may rise above the ice surface as nunataks. Under favourable conditions, the upper limit of glacial erosion, the 'schliffgrenze' can be identified in the field (e.g. Sollid & Reite 1983), or by measuring the degree of rock surface weathering (McCarroll & Nesje 1993). In contrast to river valleys, the floor of glacially-sculptured valleys is often rather irregular, forming thresholds and basins. Where such valleys are drowned by a subsequent transgression, they are referred to as **fjords**. These landforms of glacial erosion are normally the result of multiple glaciation.

Under an ice sheet, maximum glacial erosion is found neither in the central nor in the peripheral parts. Near the centre, flow velocities are small and the major northern hemisphere ice

FIGURE 10 Plastically moulded features on glaciated bedrock surface (p-forms) at Scarisdale at the southern shore of Loch na Keal, Isle of Mull, Scotland (Photograph: Ehlers 1992)

FIGURE 11 U-shaped tributary valley, Sognefjord in Norway (Photograph: Ehlers 1980)

sheets may have been cold-based. Thus in many cases older deposits and landforms have survived the last glaciation. This applies to the mountain plateaux of Scandinavia (e.g. Andersen 1981), as well as to the Laurentide Ice Sheet. The marginal zones, on the other hand, are areas of net deposition, although locally signs of strong erosion occur. It is the region intermediate between the two extremes, where under a warm-based ice cover maximum erosion takes place, including the formation of fjords and channels (Hughes 1981, Andersen & Nesje 1992).

2.7 GLACIAL TRANSPORT

Even before the complete acceptance of the glacial theory at the end of the 19th century, the North German erratics were regarded as witnesses of a former southward-directed long-distance transport from the Scandinavian mountains. Investigations in Scandinavia, however, have revealed that long-distance transport over hundreds of kilometres plays a comparatively minor role in glacigenic deposition. Usually a small-scale alternation of uptake and redeposition prevails. The composition of tills in Scandinavia normally reflects the composition of the local bedrock (e.g. Lindén 1975) (Fig. 12). Strobel & Faure (1987) found that the

FIGURE 12 (opposite) Changing composition of the clast fraction (5.6–60 mm) of a Weichselian till over changing bedrock lithologies in Sweden. The profile strikes approximately N–S, parallel to the ice-movement direction. North is at the left. The site is several kilometres west of Uppsala (Sweden). Length of profile: 48 km. The numbers indicate sampling sites (from Lindén 1975)

20 - 60 mm

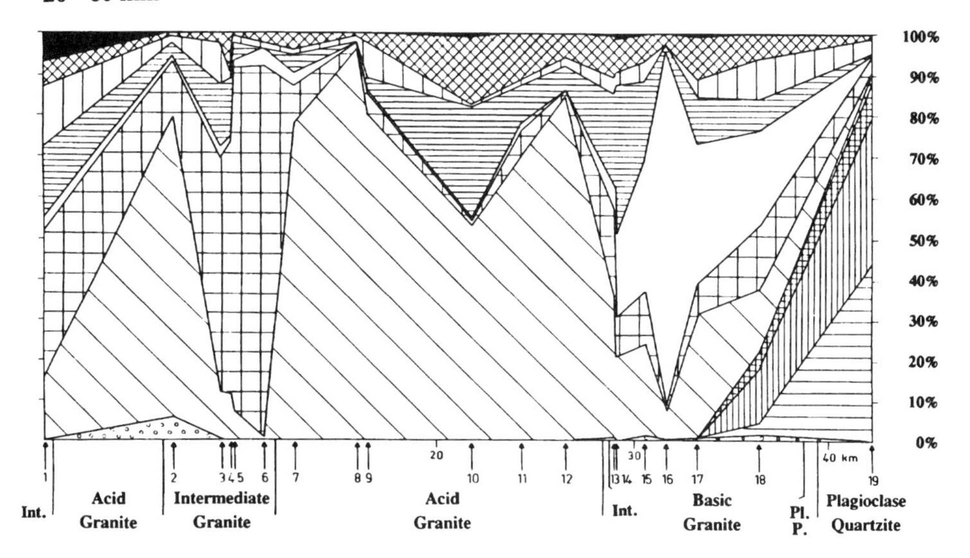

5.6 - 20.0 mm

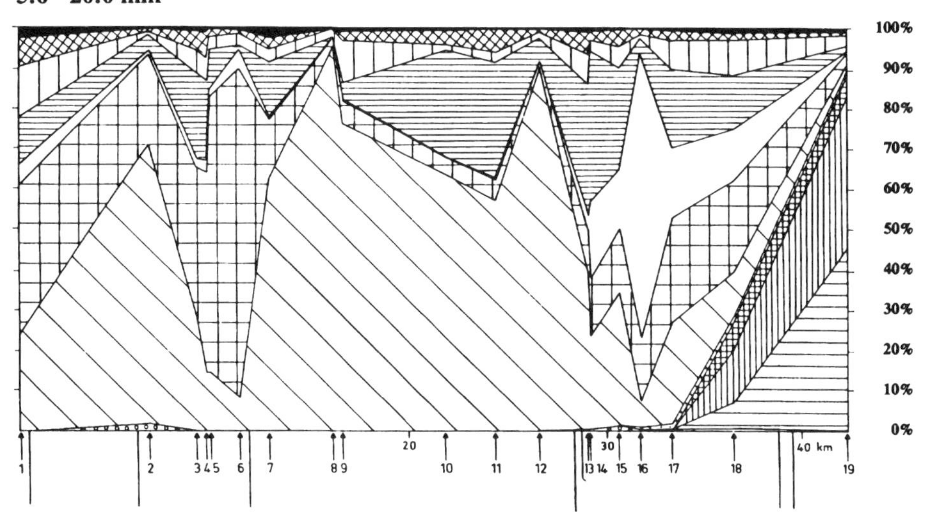

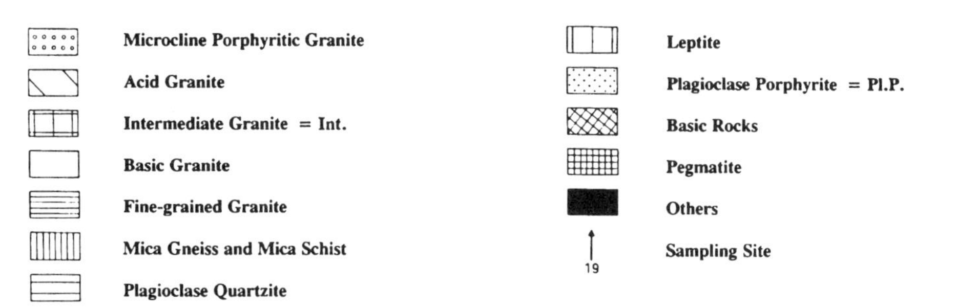

abundance of erratics from the Precambrian Shield in Wisconsinan till in Ontario at first decreases exponentially downglacier. After a distance of 50–100 km, however, it remains nearly constant at about 4% over a distance of more than 800 km all the way to the ice margins in Ohio and Indiana. However, recent investigations in Canada have revealed that in areas where ice streams occurred, vast quantities of glacial debris may be transported over long distances, with very little lateral dilution. This is a change from previous thinking and strongly affects drift prospecting (Dredge, personal communication).

In North Germany most tills also contain a very high proportion of local material, though deposited close to the former ice margin in a zone of net deposition. The striking erratic blocks and clasts of Scandinavian origin not only dominate in the clast fraction but form a minor proportion of the overall till composition. They are surrounded by a matrix of completely different composition. The fraction over 2 mm in diameter as a rule represents no more than 3% of the North German till. The remaining 97% are sand, silt and clay consisting largely of local material, of which Tertiary quartz sands and clays or in Westphalia also Cretaceous sands and clays form a major component. As the Miocene and Pliocene deposits contain only very few fine gravel and almost no coarser particles, it is no surprise that local clasts are apparently missing from the coarser fractions. That the local components nevertheless play a major role becomes obvious when glacial deposits from Saxony and Thuringia are compared with those of Lower Saxony or Westphalia. In the first case, the glaciers could erode hard rock of 'southern', local origin. In these cases the composition of the coarser fraction shows a very sudden change, and in many cases the 'southern' components become dominant (e.g. Ehlers 1983b).

Transport of rock debris largely occurred in the basal parts of the Pleistocene ice sheets. Debris on the ice surface ('obermoräne') only plays a significant role in mountain glaciations, as in the Alps or Scotland or the North American Cordillera. In the areas of the large ice sheets, debris only becomes enriched on the ice surface in the most marginal areas, where melting from the top occurs (Dreimanis 1990). During basal transport repeated alteration of deposition and reworking leads to a mixture of all rock types crossed by the ice on its way. This also applies to cases where the underlying substratum consists of older till. For this reason tills, like those in North Germany, normally form a mixture of very different rock types, part of which was redeposited during more than one glaciation.

Most englacially transported debris moves in the basal zone of the ice, immediately above the glacier sole. This debris-rich layer can reach a thickness of several metres (Drewry 1986). The debris concentration is normally about 25% by volume, but it can vary in individual layers from fractions of 1% to over 90% (Lawson 1979). The main proportion of glacial abrasion and comminution of the rock particles occurs in this zone (Dreimanis 1990). The basal zone is overlain by ice that is largely debris-free, containing only isolated particles or rock fragments. In areas of compressive flow, however, debris-rich ice can be lifted into this zone. As further breakdown and abrasion in the higher parts of the ice is minimal, clasts in this zone can easily survive transport over very long distances (Dreimanis 1976). This may explain how entire till layers have been transported unmixed with foreign material from Scandinavia into North Germany.

The red tills of the late ice advances in North Germany, in contrast to most other tills, consist of relatively pure far-travelled material and contain next to no local components (K.-D. Meyer 1983b). This is not only proved by indicator pebble counts but also by fine gravel analyses, which showed that local material (quartz) and medium-transport material (flint) are largely lacking (Ehlers 1990a, 1992). In order to explain the origin of this morainic material free from influences of the underlying local bedrock, either transport in the higher parts of the glacier or ice streaming must be invoked.

In spite of this mixing, the clast inventory of a till unit normally shows a specific composition which, within a certain range of variation, is characteristic of the till. This applies to fine gravel as well as to larger clasts. The main difference is only that in coarser fractions specific sedimentary and crystalline rock types can be differentiated according to their source areas, whereas fine gravel analysis is restricted to the determination of major rock groups. The specific indicator assemblage of the tills allows correlation between boreholes or exposures. Under favourable conditions, correlations over great distances are possible, in some cases up to more than 100 km. If this is attempted, possible changes in clast composition perpendicular to ice movement direction must be taken into consideration. The specific composition is always restricted to one lithofacies of till, and does not include the whole assemblage of morainic deposits of a particular age, like the Weichselian or Saalian. Therefore it requires additional information to be applied for chronostratigraphical interpretations. The glacigenic deposits of a certain age normally consist of a sequence of different till facies, the changing composition of which reflects changes in glacier dynamics and ice-movement direction. Therefore it is crucial always to take samples for provenence analyses from the same till facies.

Within the Alpine ice-stream network glacier movement was largely controlled by topography. As a rule, only marginal contact between the individual glacier tongues occurred. Nevertheless their catchment areas have changed occasionally through the course of the glaciations. During the glacial maxima under favourable conditions ice drained from one valley into a neighbouring valley. Such transfluences have affected clast composition (Doppler 1980, Dreesbach 1985). Indicator counts have been used to delineate the influence of the individual glaciers involved.

2.8 GLACIER DYNAMICS AS EXEMPLIFIED BY THE LATE WEICHSELIAN SCANDINAVIAN ICE SHEET

The dynamics of the large continental glaciations can only be reconstructed if it is possible to determine the sequence of events. Although in the central area of the Scandinavian Ice Sheet a broad picture of the last Weichselian ice-movement directions was obtained relatively early, comparable information from the marginal regions was not readily available. The regional reconstruction of the ice-movement directions in North Germany began with Richter's (1936b) map of long-axis orientation measurements from the Weichselian till in Pomerania. In the old morainic landscape the situation was less favourable, and Woldstedt (1938) could only base his reconstruction of ice-movement directions on morphological indications. In the meantime, however, the situation has changed. Ehlers & Stephan (1983) compiled evidence from numerous individual publications in order to achieve an overall picture of the ice-movement directions based on till-fabric measurements. More recently, Houmark-Nielsen (1987, 1989) presented a comprehensive synthesis of till stratigraphy and ice movement in Denmark, and from the Netherlands Rappol (1983, 1987) has published the results of numerous measurements. An updated version of the Ehlers & Stephan (1983) maps by Ehlers (1990a, b) includes the most recent results.

The typical development of a North European Ice Sheet can be best illustrated using the Late Weichselian as an example. During the initial phase of the Weichselian glaciation two ice advances occurred in Denmark: an advance of Norwegian ice, which reached the northern margins of the Danish islands, and an advance of the so-called Old Baltic ice, which reached the southern margin of the Danish islands (Sjørring 1983, Houmark-Nielsen 1987). The deposits of these events do not overlap. Sjørring (1983) and Ringberg (1988) assumed that the Norwegian ice advance was the oldest, whilst Houmark-Nielsen (1987) assumed that the Norwegian ice succeeded the Old Baltic ice advance. All authors agree that the Norwegian and Old Baltic ice sheets were no longer present in Denmark and Skåne, southernmost Sweden, when the later Main Ice Advance crossed the area.

The advance of Norwegian ice is depicted in Figure 13A. The crucial point is that such an advance of Norwegian ice far into Denmark was only possible early in the glaciation, when the ice divide was still far enough to the west.

During the subsequent Weichselian maximum the Baltic Sea depression did not control ice-movement directions, but the ice flowed radially from the centre to the margins. This is not only shown by till-fabric measurements in eastern Schleswig-Holstein (Stephan 1987a, Ehlers, unpublished data), but it is also based on evaluation of lithological data for Berlin and East Germany (Böse 1990). In addition, these results are confirmed by the investigations in Denmark (Houmark-Nielsen 1987) and Skåne (Ringberg 1988). Simulation of this situation is most closely modelled under the assumption that at that period no deformable bed existed. Figure 13B deviates in five respects from the model published by Boulton *et al.* (1985):

1. At present it is assumed that Scandinavian and British ice did meet in the North Sea during the Weichselian (Sejrup *et al.* 1994). This and the glaciation of the Barents Sea area are omitted here.
2. For both Denmark (Houmark-Nielsen 1987) and Schleswig-Holstein the ice-flow lines calculated by the model do not coincide with those established from the field evidence (fabric measurements and glaciotectonic investigations). The very uniform ice-movement directions recorded, particularly in Denmark, indicate that this difference cannot result from local effects. Therefore presumably the ice-sheet surface contours were more concentric in form and less parallel to the ice margins than was envisaged by Boulton *et al.* (1985).
3. In western Norway, the ice shield has been flattened in order to comply with the assumption that Hallingskarvet (1933 m a.s.l.) and the mountain summits in Møre remained ice free (Sollid & Reite 1983, Nesje & Sejrup 1988).
4. The position of the ice divide has been considerably adjusted to the west in order to match the Swedish field evidence (cf. J. Lundqvist 1983a, Persson 1986) and the situation in Norway (cf. Bergersen & Garnes 1983).
5. Ice-movement directions on the Kola Peninsula, as reconstructed by Evzerov & Koshechkin (1981), differ from the Boulton *et al.* model in that there always seems to have been divergence over the peninsula.

The situation during the later Pomeranian Phase, however, is again strongly controlled by the shape of the Baltic Sea depression (Stephan & Menke 1977, Stephan *et al.* 1983). During this phase a Baltic ice stream advanced from the south into the Danish islands (Houmark-Nielsen 1983, 1987). In contrast to the situation during the Brandenburg Phase, the ice divide was then situated further to the east. The map in Figure 13C takes into account that this situation was not reached during but after the glacial maximum. The map does match the ice-movement directions documented by striae and transport directions of erratics (cf. J. Lundqvist 1983a).

The rapid ice movement during the early and final stages of the Weichselian is best explained by deforming water-saturated sediments at the glacier sole. The resulting thin ice

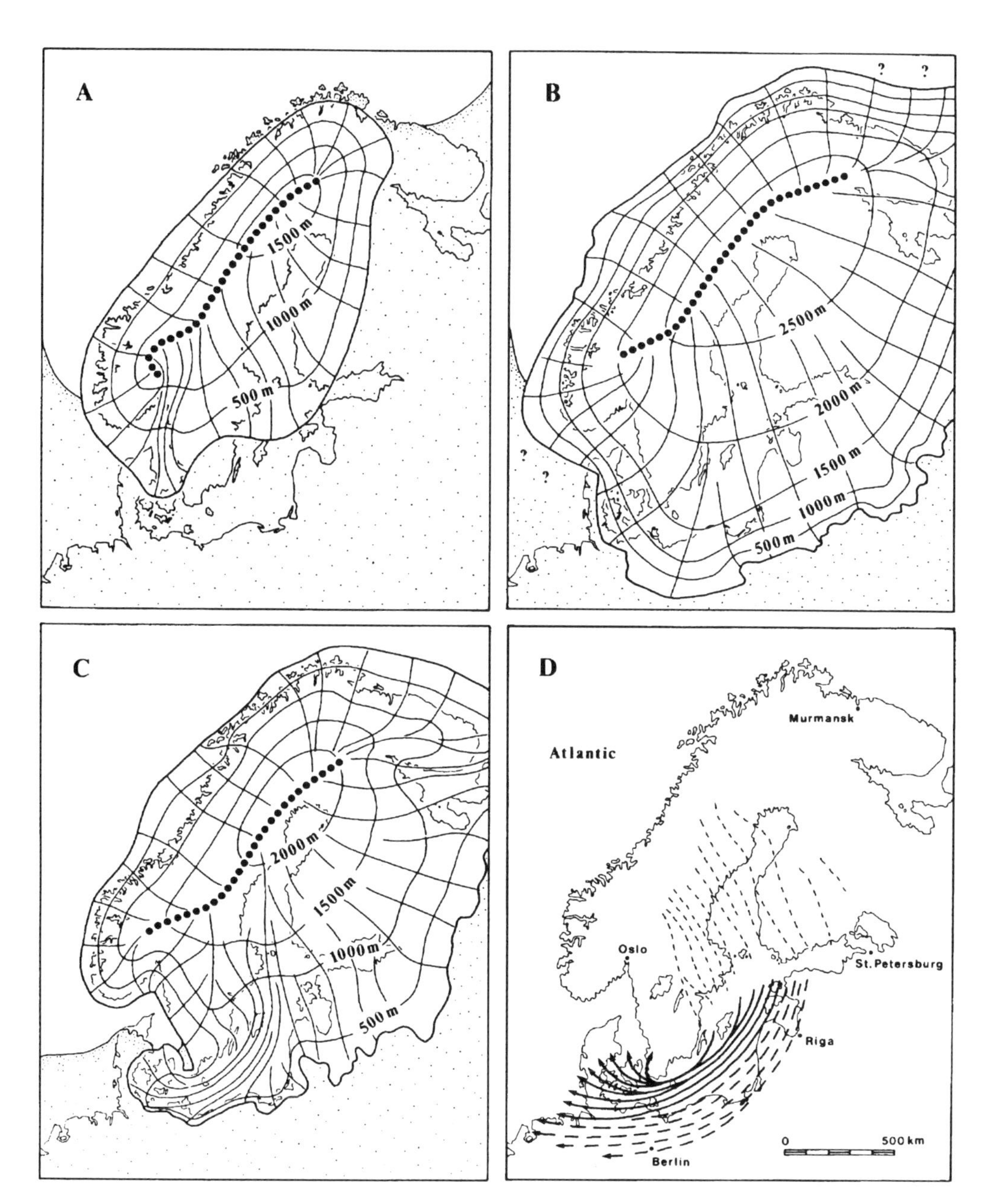

FIGURE 13 A–C: Ice thickness and glaciodynamics of the Late Weichselian Scandinavian ice sheet. A = early phase, B = Weichselian maximum (Brandenburg Phase), C = Pomeranian Phase (from Ehlers 1990b); D = ice-movement directions during the Late Saalian (schematic); dashed lines = East Baltic material, solid lines = Baltic material (after Stephan in Ehlers *et al.* 1984)

stream would have been able to bring about rapid build-up of the ice sheet as well as rapid deglaciation. The changes in glacier dynamics as outlined above were accompanied by shifting ice-movement directions, leading to the deposition of a sequence of differently composed tills. Ehlers (1990a, b) assumes that a similar mechanism operated during the Elsterian and Saalian glaciations. In this way the tills with Baltic and East Baltic indicator spectra far in the west, deposited towards the final phases of the glaciations, might be explained (Fig. 13D; cf. Ehlers *et al.* 1984). However, this interpretation has been questioned by Rappol & Stoltenberg (1985) and Kluiving *et al.* (1991).

A characteristic feature of the eastern parts of the Late Weichselian ice sheets is that during the retreat phase the formerly relatively straight ice margin was replaced by numerous adjoining ice lobes. This occurred first during the Luga Stadial (about 13,000 BP) and continued throughout the ice retreat into Finland (Punkari 1980, Dreimanis & Zelčs 1995).

2.9 METHODS OF RECONSTRUCTING GLACIER DYNAMICS

2.9.1 Landforms

The flow pattern of glaciers and ice sheets adjusts to changes in mass balance and temperature regime during the course of the glaciation. This results in shifting ice divides (J. Lundqvist 1983a) and alternating advances of different parts of an ice sheet, as known, for example, from Denmark (advances of Norwegian, Swedish and Baltic ice; cf. Houmark-Nielsen 1987). The key to the reconstruction of these changes in glacial dynamics is the evaluation of glacial deposits and landforms.

An overriding glacier sculpts the land surface and creates forms that strike in the direction of ice movement. This is best seen in the formerly glaciated areas of the far north, where the absence of vegetation cover allows unhampered investigation by air photograph or satellite imagery (e.g. Punkari 1985). Major landforms in hardrock areas cannot be sculptured in one glaciation. Their present appearance is the result of multiple glaciations during several cold stages, during which the ice followed largely the same flow lines. The first glaciation advancing across deeply weathered terrain possibly has a greater effect on the orientation of large landforms than the subsequent ones. All major glacigenic landforms in Scandinavia, for instance, result from multiple overriding by ice sheets, flowing from the high Scandinavian mountains towards lower ground (J. Lundqvist 1990). The formation of the south Swedish lake fan, apart from the overall tectonic setting, results from the repetition of a similar ice-flow pattern repeatedly during all glaciations.

In areas with net accumulation, streamlined landforms like drumlins and flutings indicate the direction of former ice flow (see chapter 3.4.1). End moraines are mostly aligned at right angles to glacier flow directions. They can also be used for the reconstruction of ice movement. Melting active glaciers tend to deposit **annual moraines**, i.e. small end-moraine ramparts formed during winter advance phases. Krüger (1987) mapped such landforms in the forefield of Myrdalsjökull on Iceland (Fig. 14). Figure 15 shows annual moraines in relation to other landforms of the glacier forefield. Amongst these are also streamlined features like flutes and drumlins. Drumlins are often found associated with so-called Rogen moraine, landforms of a slightly irregular shape striking mostly transverse to ice movement (Aario 1977b). Not all elongate features allow direct conclusions about ice-marginal positions. Transverse ridges are amongst the more problematic examples. They include the so-called 'De Geer moraines'

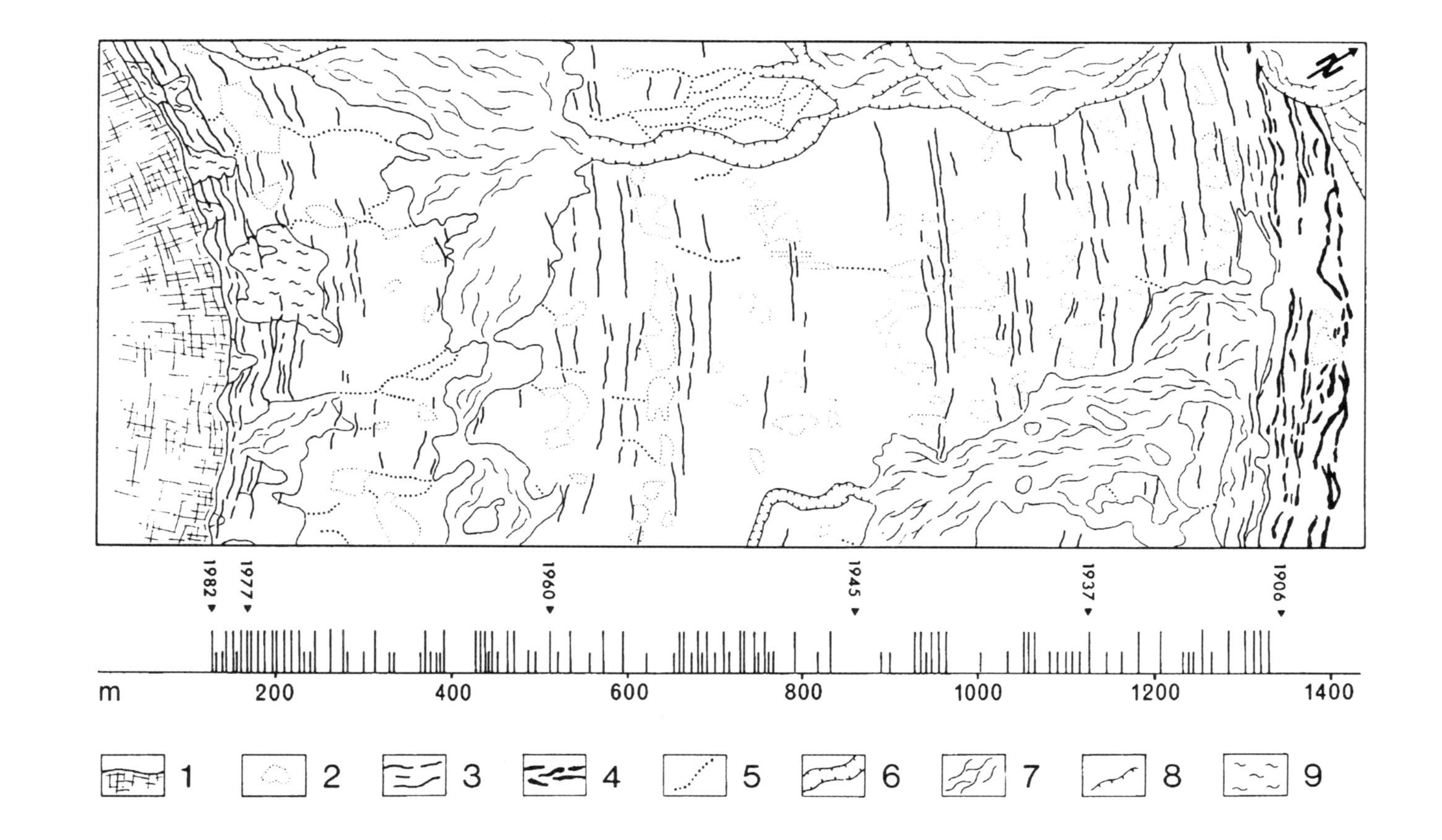

FIGURE 14 Synthetic map of the morainic landscape between the modern Myrdalsjökull ice margin and its 1890 end moraine. Below the map all clearly visible (long lines) and hardly detectable (short lines) annual moraines are recorded that could be identified in the field. 1 = ice margin, 2 = till with closed depression, 3 = till plain with annual moraines, 4 = end moraines, 5 = minor meltwater channel, 6 = major meltwater channel, 7 = sandur plain, 8 = erosive scarp, 9 = lake (from Krüger 1987)

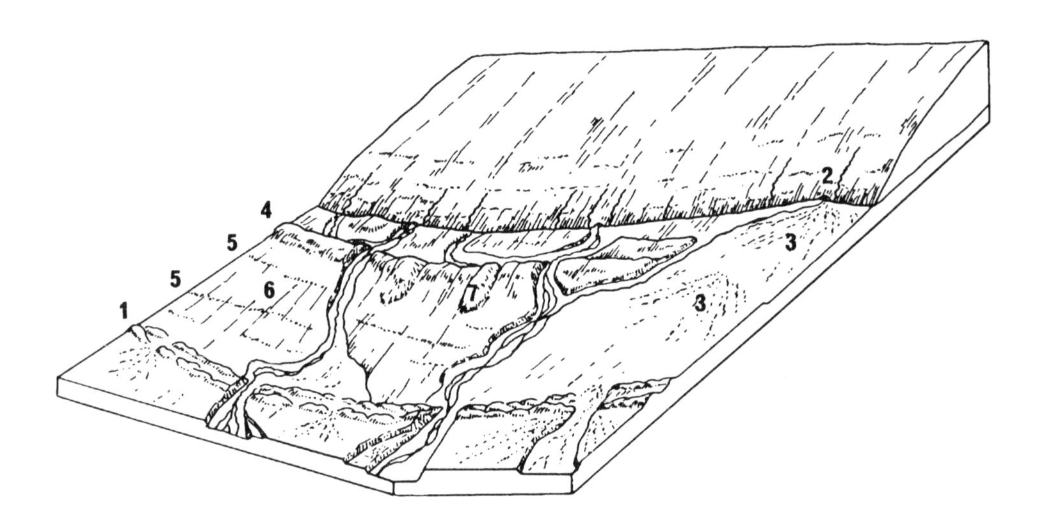

FIGURE 15 Glacier forefield of an actively retreating glacier. The till landscape is modified by meltwater draining from subglacial and supraglacial streams. 1 = thrust moraine, 2 = outlet of subglacial stream, 3 = sandur cone, 4 = old overridden end moraine, 5 = annual moraines, 6 = flutes, 7 = drumlin (from Krüger 1987)

(Hoppe 1959), which in contrast to De Geer's (1940) original interpretation are not end moraines but former crevasse fillings (J. Lundqvist 1990).

Glaciofluvial deposits can also indicate former ice-movement directions. Meltwater sediments deposited in tunnels under the ice, so-called eskers, tend to trend parallel to the ice-flow direction. The same applies to landforms of subglacial meltwater erosion (tunnel channels). Ice-marginal meltwater accumulations (especially the ice-marginal terraces of Scandinavia) have a steep proximal slope that trends roughly at right angles to the former ice-movement direction. Examples of the investigation of glacial morphology from aerial photography are given by Kujansuu (1990). The distribution of glacigenic landforms is not restricted to the present land surface. Underwater major and minor landforms can be identified with sidescan sonar. In Hudson Bay Josenhans & Zevenhuizen (1990) were thus able to reconstruct former ice-movement directions. The availability of satellite imagery in recent years has allowed the evaluation of landform assemblages of entire glaciation areas. Thus, Boulton & Clark (1990) attempted an evaluation of parts of the Laurentide Ice Sheet, in which overlapping landform assemblages of different age could be used to establish the sequence of events in that area (Fig. 16). Investigations of this type must be carefully checked with evidence on the ground.

In some regions of the large Pleistocene ice sheets, landforms from earlier phases of glaciation or even older glacial stages have been preserved. For example, Lagerbäck & Robertsson (1988) described landforms from north Sweden that were generated during the first Weichselian ice advance and were only slightly modified by subsequent ice advances. Stea (1994) describes two generations of streamlined landforms (drumlins) from Nova Scotia. Also the parallel-striking glacigenic landforms of the Stader Geest, Syker Geest and Oldenburgisch–Ostfriesische Geest, in North Germany and the Drentse Plateau in the Netherlands belong to

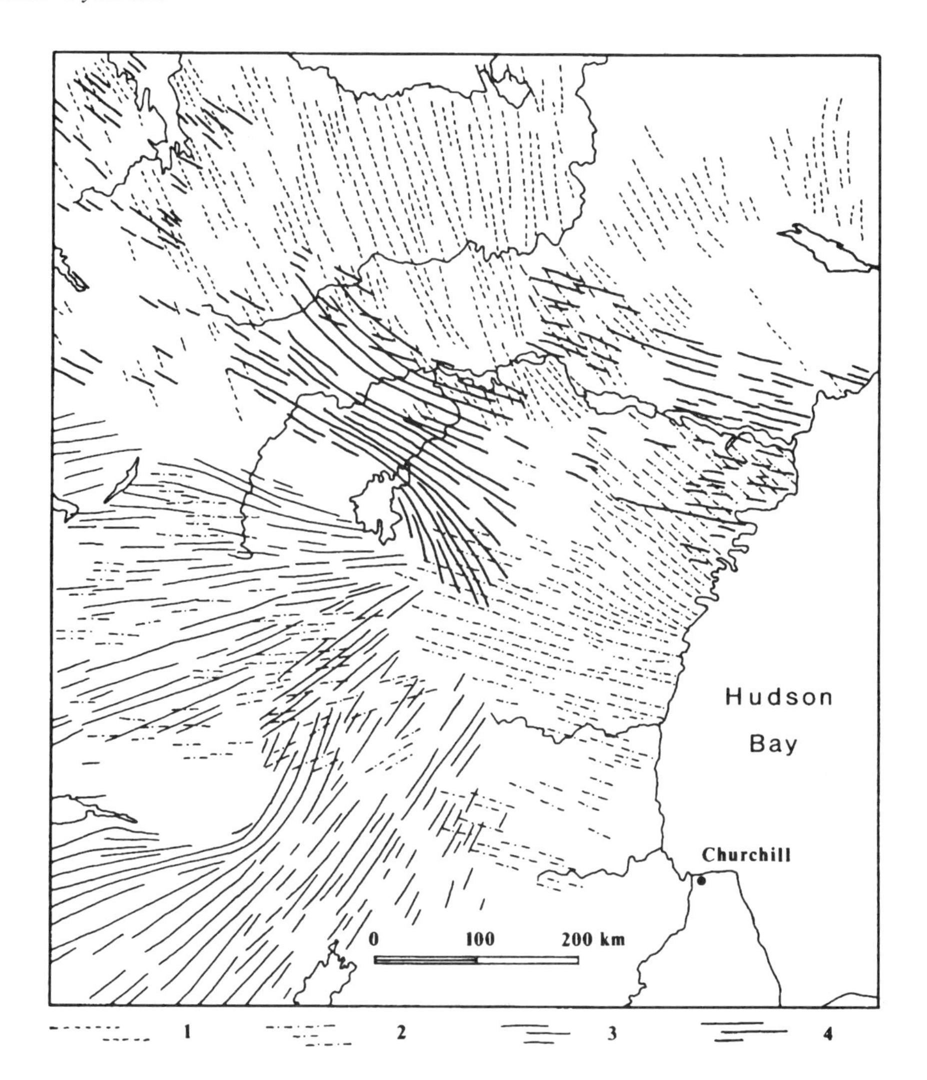

FIGURE 16 Reconstructed sequence of ice-movement directions based on satellite-image interpretation of landforms striking parallel to ice movement. Laurentide Ice Sheet, western marginal area of Hudson Bay. Sequence of landform groups: 1 = oldest, 2 = next oldest, 3 = third, 4 = youngest direction (after Boulton & Clark 1990)

this group. All of these cannot have been formed simultaneously. The strike direction of the Syker Geest landforms, for instance, deviates by about 70° from the landforms of the adjoining areas. Ehlers (1990a) and Höfle (1991) both assumed that the NNW–SSE-striking features are the older ones. Höfle thought they were conserved because in this area part of the ice sheet was trapped motionless in front of the central German uplands, whilst the active ice flowed around it towards the southwest.

2.9.2 Striae

Striae can only be identified on hard rock. Generally they form very small, parallel-striking grooves that have been scored by stones in the base of a glacier (Fig. 17). Individual lines may be more than a metre long, but can be much shorter. Short, blunt striae, deep at one end and tapering at the other, are referred to as nailhead-striae. They usually indicate ice flow away from the tapering end. On surfaces sloping upglacier, however, the opposite may be the case, as a result of gouging. Where the scorer is crushed in the process, a single stria is replaced by a set of parallel striae. Such 'forked striae' are good directional indicators of ice movement (Charlesworth 1957, Sharp 1988).

Even on flat surfaces rotation of the gouging stone results in striae that deviate markedly from the ice-flow direction. Demorest (1938) has also shown that strike of striations differs in stoss- and lee-side positions. The ice is always moving from areas of higher pressure to areas with lower pressure, i.e. more or less around obstacles. Therefore sites for measurements must be chosen carefully in order to avoid local effects. A large number of striae should be measured in order to obtain statistically significant results (Hambrey 1994).

Striae follow the same direction as elongated glacigenic landforms, they trend in the direction of ice flow. Whilst the strike of landforms can be determined from aerial photographs or satellite imagery, a survey of glacial striae requires fieldwork. The drawback of the laborious information gathering is more than made up for by the opportunity to reconstruct sequences of ice-movement directions. Whilst evaluation of landforms normally allows the reconstruction of no more than two generations of features (e.g. small drumlins or flutes superimposed on major older forms), frequently three or more generations of striae can be distinguished (Fig. 18). A sequence of different ice-movement directions is reflected in several systems of striae, the younger ones crossing and finally destroying the older ones.

The observation that striae can be used for the reconstruction of ice-movement directions was already known to Torell (1865). Ljungner (1949) refined the method of striae evaluation. An overview of the present technique and problems of striae interpretation is given by Kleman (1990). The quality of the striae to a large degree depends on bedrock lithology. Striae form on all types of bedrock. At the base of temperate glaciers, which always carry sufficient basal debris, they are formed in all instances where the ice moves across bedrock. Polar glaciers normally deform internally, without gliding over the underlying rock surface. Based on theoretical assumptions, however, Shreve (1984) concluded that even under cold glaciers some abrasion can occur. However, the power to striate a bedrock surface is certainly reduced under these conditions more than at the base of temperate glaciers.

The preservation potential of striae varies from place to place. For example, in Scandinavia the entire assemblage of striae of the last glaciation may only be preserved in those areas that have been recently isostatically uplifted above the sea. However, they will only be found where rock surfaces have been protected beneath a covering of till. Amongst these areas are the small islands of eastern Sweden and western Finland. Kleman (1990) points out that most striae measurements have been conducted on freshly exposed surfaces (mostly in road cuttings), where the striae had been exhumed from under a cover of 0.4–2 m of till. A till sheet of 0.5 m thickness very effectively preserves the rock surface from weathering. Striae found under a till sheet therefore tend to date from earlier events than those found on barren rock surfaces. Where tills of different age have been preserved, striae of older glaciations may also be found.

The distribution pattern of striae does not reflect true flow lines, because the striae did not

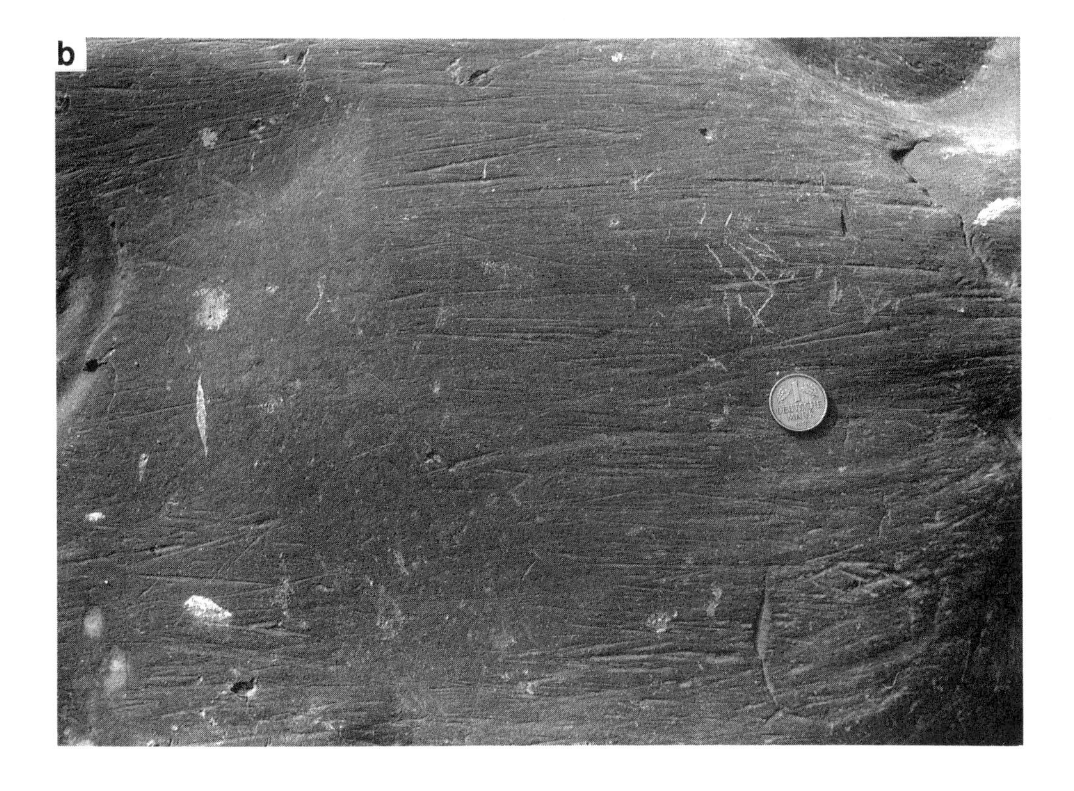

FIGURE 17 (a) Erratic boulder with striae and glacier polish on the Lake Erie shore; (b) striated boulder on the Baltic Sea shore at Heiligenhafen, North Germany (Photographs: Ehlers 1993 and 1979)

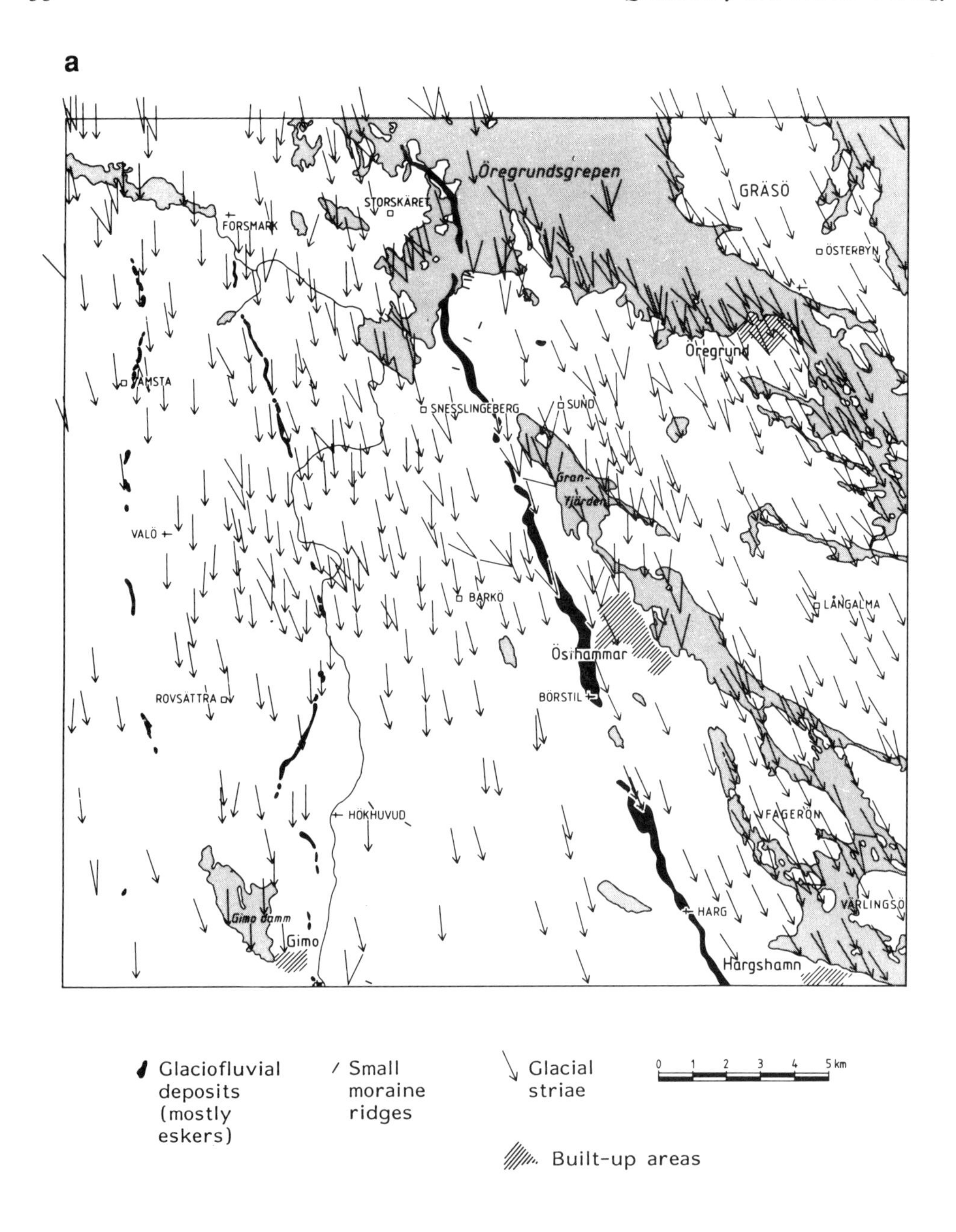

FIGURE 18 Reconstruction of glacier dynamics in hardrock areas: (a) all striae and eskers; (b) sequence of formation of crossing striae of different age; (c) sequence of ice-movement directions and estimated ice-marginal positions during retreat. Östhammar, north of Uppsala, Sweden (from Persson 1985)

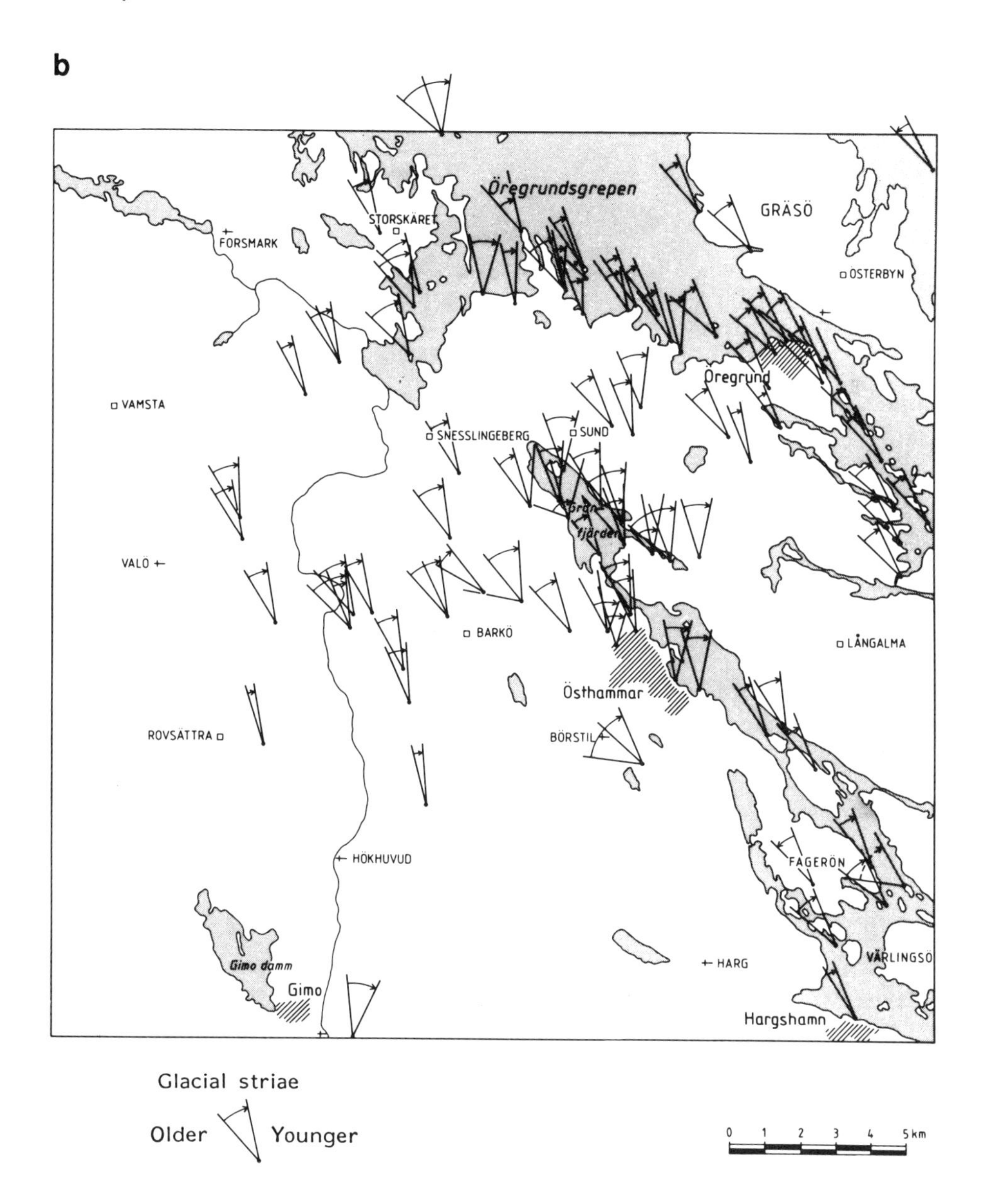

FIGURE 18 (*continued*)

c

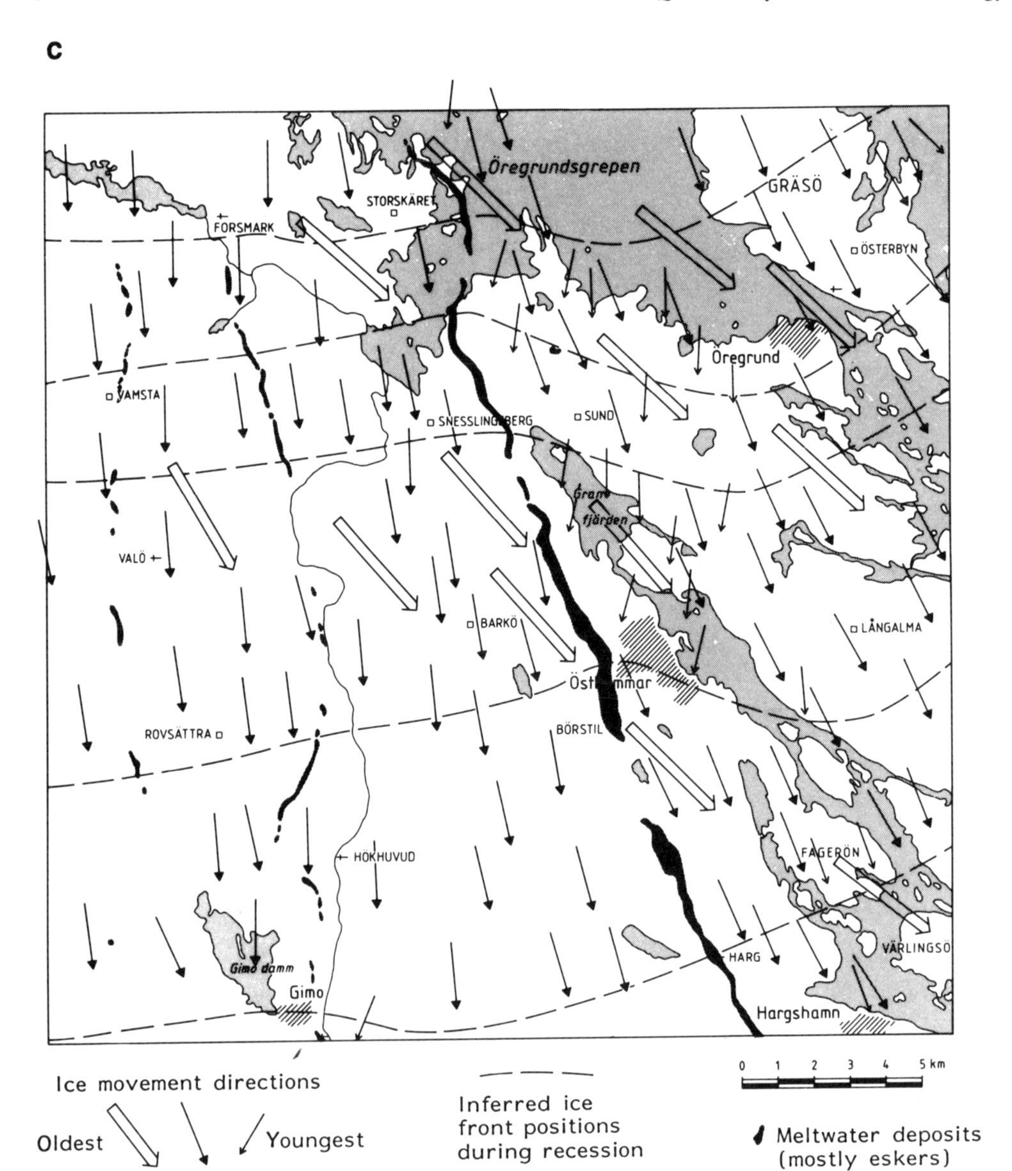

FIGURE 18 *(continued)*

form simultaneously. Even if considerations are restricted to the youngest set of each area, they represent a time-transgressive sequence of directional indicators. In particular near the glaciation centres where minor shifts of the ice divide or changes in configuration of the ice sheet may result in considerable changes of ice-movement direction, the evaluation of striae of different ages can result in faulty interpretations.

Striae are one of the keys for reconstructing the complicated course of glaciation in the Canadian Shield areas. For instance, Klassen (1983a, b) postulated an ice dispersal centre between Churchill River and the Gulf of St Lawrence based on the evaluation of striae. The striae predate the Late Wisconsinan northeastward, eastward and southeastward ice flow and may be of early Late Wisconsinan age. However, their exact age is unknown (Vincent 1989).

Apart from striated boulder pavements or major erratics in till (e.g. Seifert 1952), in areas of net glacial deposition, like North Germany, their occurrence therefore is largely limited to bedrock rises. The striae on the porphyry knolls of the Hohburger Schweiz area at Wurzen were interpreted by Von Morlot (1844) as evidence that Scandinavian ice reached the northern margin of the Erzgebirge (Ore Mountains) in Germany. After initial doubts, Naumann (1848) followed this interpretation. However, this was vehemently refuted by the eminent Alpine geologist Heim (1870, 1874). A few years later another set of striae gained decisive importance. Glacial striae on the Muschelkalk rocks of Rüdersdorf near Berlin allowed Torell (1875) to prove that northern Germany had in fact been glaciated. Only after the glacial theory became generally accepted were the Saxonian sites also accepted as truly striated (see Eissmann 1974).

In addition to glacial striae, small glacigenic chatter marks can be used for reconstructing ice-movement directions. First descriptions date back to Chamberlin (1888). Amongst such forms are the crescentic marks. Basically, three different types of crescentic marks occur: crescentic gouges, lunate fractures ('sichelwannen') and crescentic fractures (Fig. 19) (Ljungner 1930, Dreimanis 1953). More recently, Wintges & Heuberger (1982) and Wintges (1984) have dealt in detail with the interpetation of these small-scale features.

2.9.3 Long-axis orientation of clasts

That stones in till are not randomly distributed but adopt a certain orientation was already known to early investigators (e.g. Miller 1884). This knowledge, however, was later forgotten. Richter (1932, 1936a) demonstrated that elongate clasts are orientated in active ice in such a

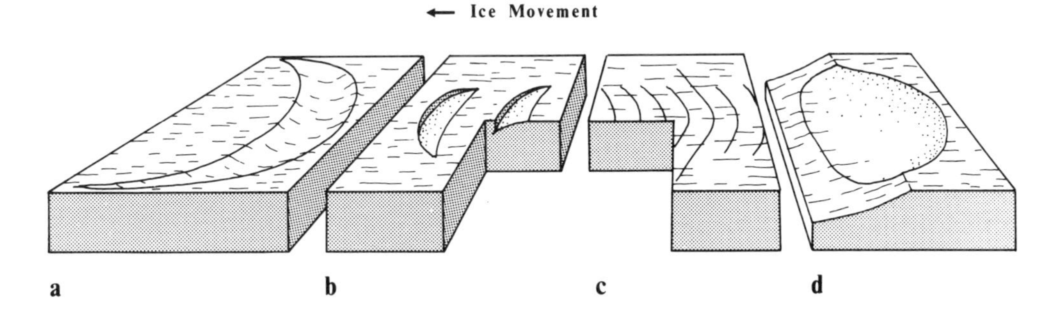

FIGURE 19 Marks produced by ice on rock surfaces: (a) 'Sichelwanne', (b) crescentic gouges, (c) crescentic cross fractures, (d) conchoidal fractures (after Ljungner 1930, from Charlesworth 1957)

way that their long axes strike parallel to the ice movement. This direction is largely retained when the glacial debris is finally deposited as till. The till fabric therefore allows reconstruction of the ice-movement direction. This was later confirmed by thorough investigations of Holmes (1941), who was unaware of Richter's results. The measurement of clast orientation in till is today a standard method of reconstructing ice movement. The sets measured by Krüger & Thomsen (1981) on a till plain in Iceland demonstrate the good reproducibility of the results and general agreement between landform orientation and ice-movement direction (Fig. 20).

About 50–100 measurements are needed to obtain a clear maximum. The result is not unequivocal in all cases: a secondary, so-called B-maximum at right angles to the ice-movement direction may occur. As a rule it is weaker than the main maximum, but in thrust zones and under compressive flow it can become the main maximum (Boulton 1971). Post-depositional deformation can also alter the original fabric. Where older till has been overridden by a younger ice advance, it can be partially or completely remobilised by the new ice movement (deformable bed *sensu* Boulton & Jones 1979). Then its fabric reorients in the direction of the new ice movement (MacClintock & Dreimanis 1964, Ramsden & Westgate 1971, Stea *et al.* 1986).

The orientation of clast long axes strongly depends on clast shape. Stones with a marked long axis are better orientated than oblong clasts (Holmes 1941, Krüger 1970). Ellipsoidal or columnar stones are especially well aligned. The grain size is also important. Large clasts tend to be better orientated than pebbles and sand grains. The clasts more or less 'float' within a fine-grained matrix, whereas the finer particles are often in contact with each other and thus hindered from perfect orientation. Nevertheless, in thin sections of till samples it is possible to reconstruct the ice-movement direction even from sand-grain orientation (Dapples & Rominger 1945, Seifert 1954, Ostry & Deane 1963). Measurement of fine grain sizes is particularly useful when orientated samples from drill cores or hand specimens need to be investigated (Dreimanis 1959). The orientation of fine grains can be determined on thin sections or X-ray radiographs (Piotrowski & Vahldiek 1991). The orientation of magnetic particles within till can be also used for reconstructing the ice movement (Fuller 1962, Boulton 1971, Gravenor & Wong 1987). Measuring smaller particles allows a stronger differentiation within the individual strata. On the other hand, more measurements are required to obtain a comprehensive picture of the fabric of the entire till unit.

Long-axis measurements are mainly applied where there are doubts about the former ice-movement direction. This is the case particularly in areas affected by the major Pleistocene ice sheets. In Alpine areas and upland regions ice movement was largely controlled by the topography. Fabric measurements can also be used to distinguish genetically different types of deposit (Boulton 1971, Dowdeswell & Sharp 1986). In lodgement till a strong fabric is generally found, with the long axes normally dipping slightly upglacier. In meltout till also a strong fabric is found. It may, however, be weakened by the meltout process and the dip direction of the long axes may not be clearly upglacier (e.g. Piotrowski 1992). In flow tills, if the clasts show a preferred orientation, it may be either parallel or transverse to the flow direction, depending on the position within a mud slump. Often it is random (Dreimanis 1989). Domack & Lawson (1985) showed that waterlain, ice-rafted diamicton mainly has a nearly random fabric, often with greater numbers of stones dipping at angles greater than 45°. Rose (1974) has used fabric measurements to distinguish slump deposits from lodgement till. However, in most cases fabrics alone do not allow firm conclusions on the origin of a deposit to be drawn.

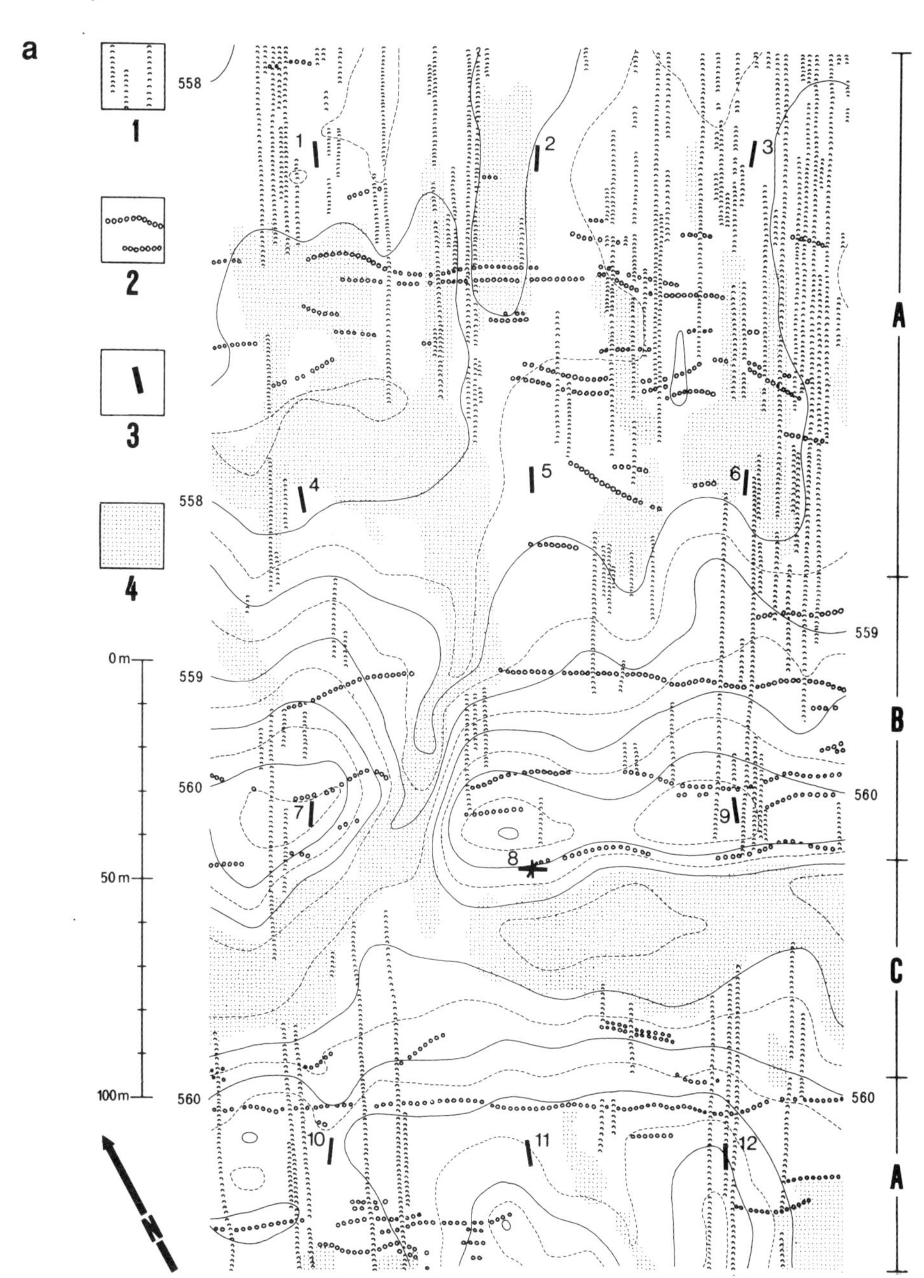

FIGURE 20 (a) Reconstruction of ice-movement direction from landforms and till fabric in the morainic landscape of the Myrdalsjökull forefield. Contour interval 0.25 m. Altitudes in metres above sea level. 1 = flutes, 2 = annual moraines, 3 = fabric maxima, 4 = depressions with meltwater or slopewash deposits; A = till plain, B = older, overridden end moraine, C = infilled meltwater lake. Ice movement from southwest (from Krüger & Thomsen 1981). (b) Long-axis orientation at the positions marked in Figure 20a, shown in a Wulff net. The arrow indicates orientation maxima. Ice movement from southwest (lower left) from (Krüger & Thomsen 1981)

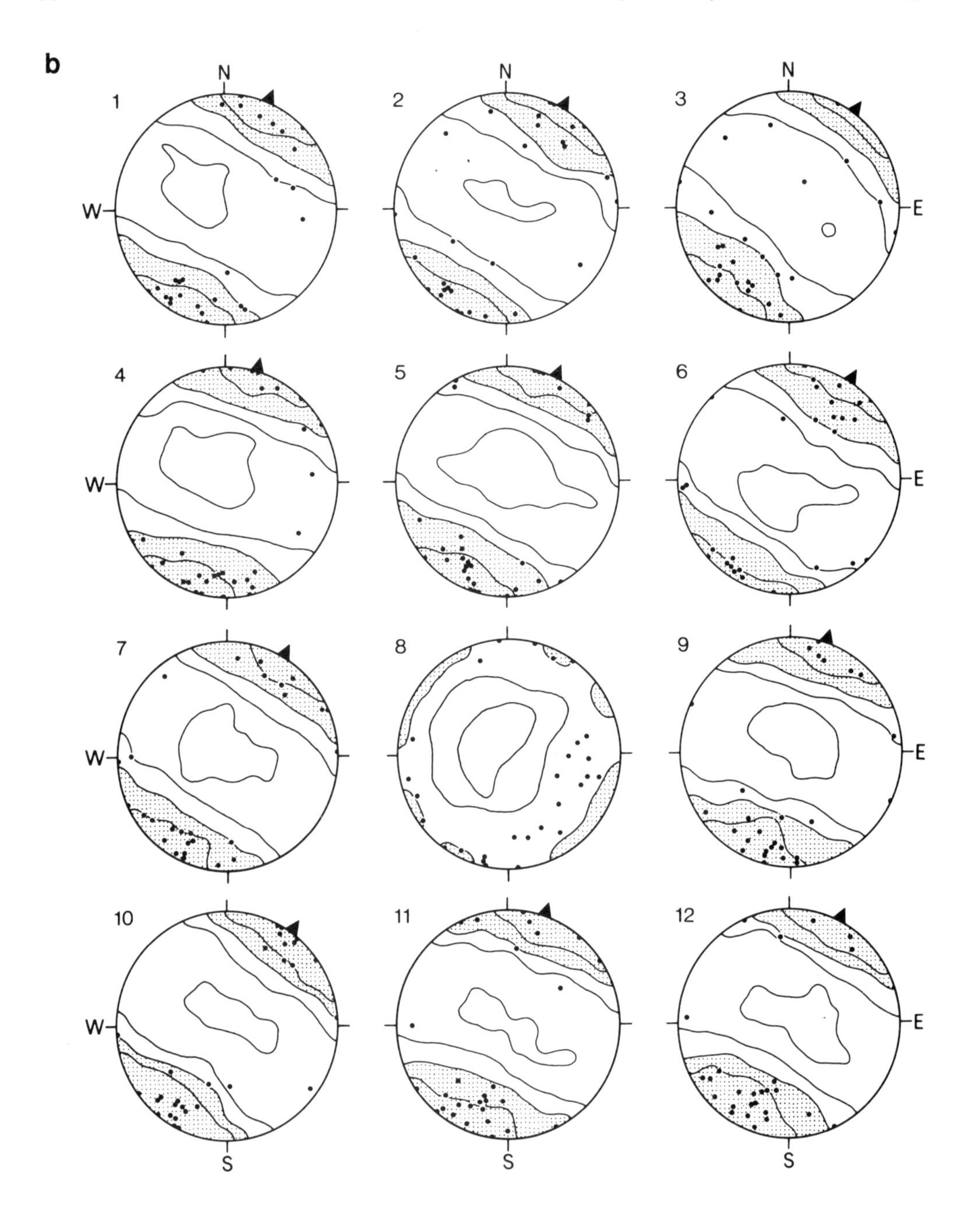

FIGURE 20 *(continued)*

Statistical analyses can be applied to quantify the strength of orientation maxima and to determine the precise orientation maximum. With the use of computer programs, eigenvalue/eigenvector analyses have proved to be very useful in this respect (e.g. Lawson 1979, Dowdeswell *et al.* 1985).

2.9.4 Joint systems

In many exposures in till, well-developed joint systems can be seen in the exposure face, especially if the surface is dry. Apart from secondary joints which form as a result of soil formation or spalling from the exposure wall, most of the joints represent tension planes that have formed under the influence of actively moving or stagnant and downwasting ice. The type of joint pattern depends on the mode of deposition of the till. It is only later that they are transformed into visible fissures (Stephan in Ehlers & Stephan 1983). Orientation measurements have shown that joints in till tend to be most strongly developed parallel to the former ice-movement direction (Fig. 21, C and F). However, as seen in Figure 21, the joint pattern is not as suitable for the reconstruction of ice movement as clast orientation. In many cases a second set of joints occurs, striking at right angles to the direction of ice advance (Fig. 21, example B). In addition there are also cases in which the joint directions have no obvious relationship to former ice movement (examples A, D and E). This may be for several reasons:

1. The clast orientation may not exactly represent the ice-movement direction.
2. The joint system tends to reflect the last tension field affecting the till sheet. If this was only active during the very last phase of ice movement, it possibly lacked the power to reorientate the clasts. In some cases a weak secondary maximum in this direction can be observed in the long-axis orientation diagram (perhaps in example E).
3. A joint system may be reorientated by a younger tension field. In this case often old and new elements are observed adjacent to each other. Diagrams A and D may serve as examples. In these diagrams the strike direction of part of the joints coincides with the clast orientation, whereas a second, stronger maximum shows no similarity to the clast fabric. The reorientation does not necessarily occur because of a deviating ice-movement direction of this or a later glacier. A younger joint system can also form during the downwasting of stagnant ice. The underlying topography is of decisive importance for such a joint system.
4. Compressive deformation results in a primary joint system that strikes neither parallel nor at right angles to the ice movement. Diagrams A and D show examples of this type. In both cases only one of two possible joint sets are developed.

Joint measurements may provide a valuable supplement to other fabric investigations. This is especially the case where not only ice movement but also the origin of the deposit are in question. The measurements require excellent exposures. Natural exposures such as coastal cliffs or steep sections along river banks are most suitable, because the joints are most visible in faces that undergo natural retreat and weathering.

2.9.5 Thrust directions

Where advancing glaciers override older strata or their own deposits, the latter may become glaciotectonically deformed. In many areas where neither a till cover nor characteristic landforms have been preserved, measurements of thrust strata can serve to reconstruct the

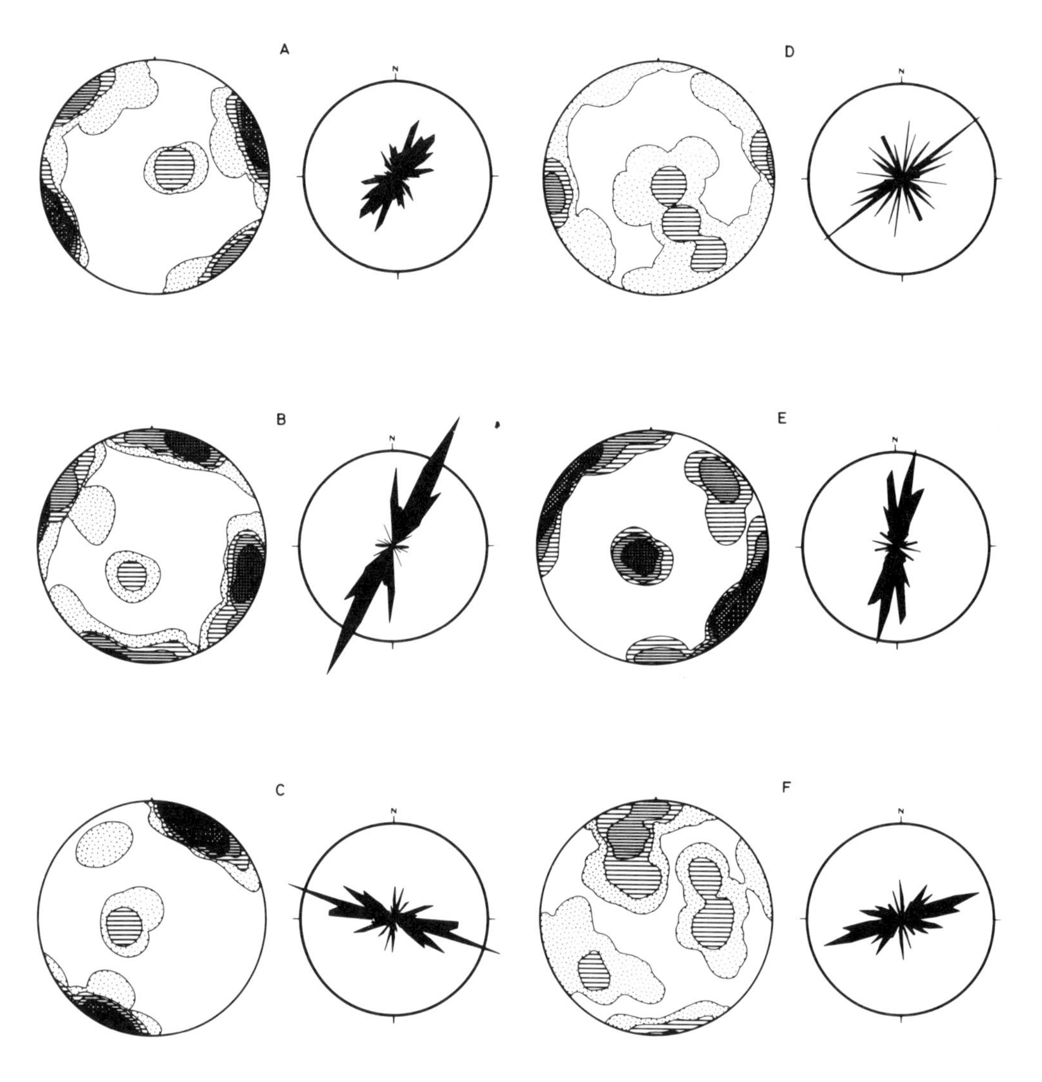

FIGURE 21 Examples of till-fabric measurements from Schleswig-Holstein, Germany. Circles show the strike and dip of joints given on the lower hemisphere of a Schmidt Net. The rose diagrams show the long-axis orientation of clasts at the same sites. (A) Brodtener Ufer, lower till, Late Weichselian Fehmarn Advance, 44/139 measurements; (B) Flintbek, Blumenthal Advance of the Late Weichselian, 57/84 measurements; (C) Vaale, Vaale Advance of the Saalian Glaciation, 50/100 measurements; (D) Pinneberg, Niendorf Advance of the Saalian Glaciation, 69/80 measurements; (E) Kuden, Kuden Advance of the Saalian Glaciation, 20/109 measurements; (F) Burg (Dithmarschen), Burg Advance of the Saalian Glaciation, 67/138 measurements (from Ehlers & Stephan 1983)

former ice-movement direction. This has been done successfully in various places in North America or, for example, in the marginal areas of the Saalian Glaciation in the Netherlands.

Under favourable conditions, a sequence of ice advances of different ages can be reconstructed from the deformation pattern conserved in the sediments. Berthelsen (1975) has introduced the term **kinetostratigraphy** for this kind of investigation. The respective

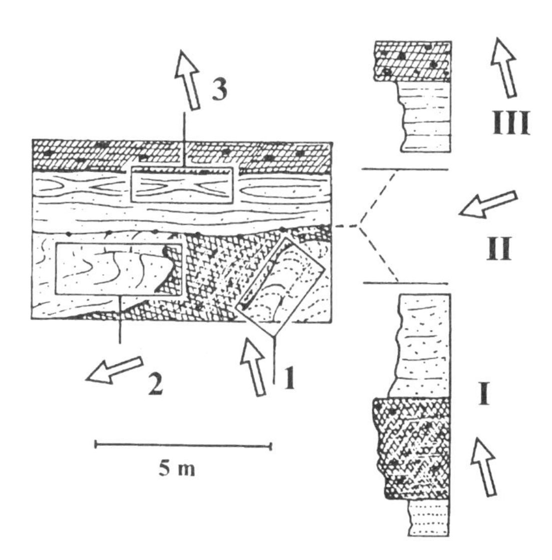

FIGURE 22 Sketch to illustrate the kinetostratigraphic evaluation of an exposure. The profile includes just two stratigraphic units. The existence of a missing third unit can be deduced from deformation of the overridden strata (from Berthelsen 1978)

movement pattern is not only imprinted on the sediments of the ice advance (internal sediment deformation) but also on part of the underlying deposits (external sediment deformation). Even in cases where the deposits of a certain ice advance have been completely removed afterwards, it may be possible to reconstruct its direction from the remaining external sediment deformation (Berthelsen 1978) (Fig. 22).

Like other investigation methods in geology, kinetostratigraphy is an empirical method. Before local events may be interpreted, it is important to clarify the regional context. The method requires exact lithostratigraphical investigations. It works best in cases where different ice sheets or lobes have advanced from different directions, as in major parts of Denmark. For example, Houmark-Nielsen & Berthelsen (1981) have applied this method on the cliff exposures of the little island of Samsø.

First the cliff, about 4 km long, was surveyed and mapped at a scale 1:500 (Fig. 23). The survey shows that four different tills are exposed: a lower till (L), two middle tills (Lm and Um), and an upper till (U). Next, lithological columns are drawn for representative cliff sections. These columns are later used to construct a kinetostratigraphic/lithostratigraphic profile (Fig. 24) in which all the important information about the exposure is summarised. It comprises the results of multiple investigations which all contribute to the stratigraphical interpretation of the section.

During the mapping, fabrics were measured. The profile shows that the cliff exposure is dominated by large-scale, recumbent folds, the axes of which strike NW–SE. The folds are tilted towards the southwest. Respective small-scale folds have also been observed. Apart from these obvious thrust features, which predate the upper layer of the middle till (Um), a second set of disturbances occurs: recumbent folds with SW–NE striking axes. These have incorporated the strata underlying the uppermost till (U) but do not reach to the foot of the

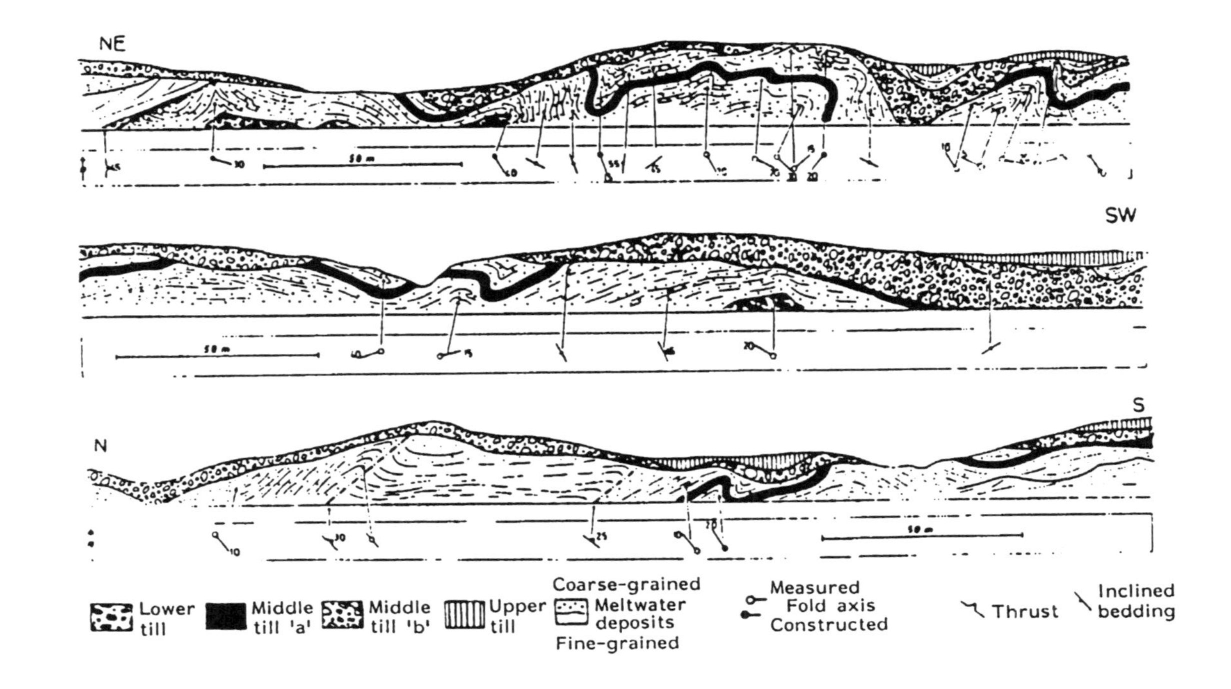

FIGURE 23 Two cliff sections from the north coast of Samsø, Denmark, mapped for kinetostratigraphic investigations (from Houmark-Nielsen & Berthelsen 1981)

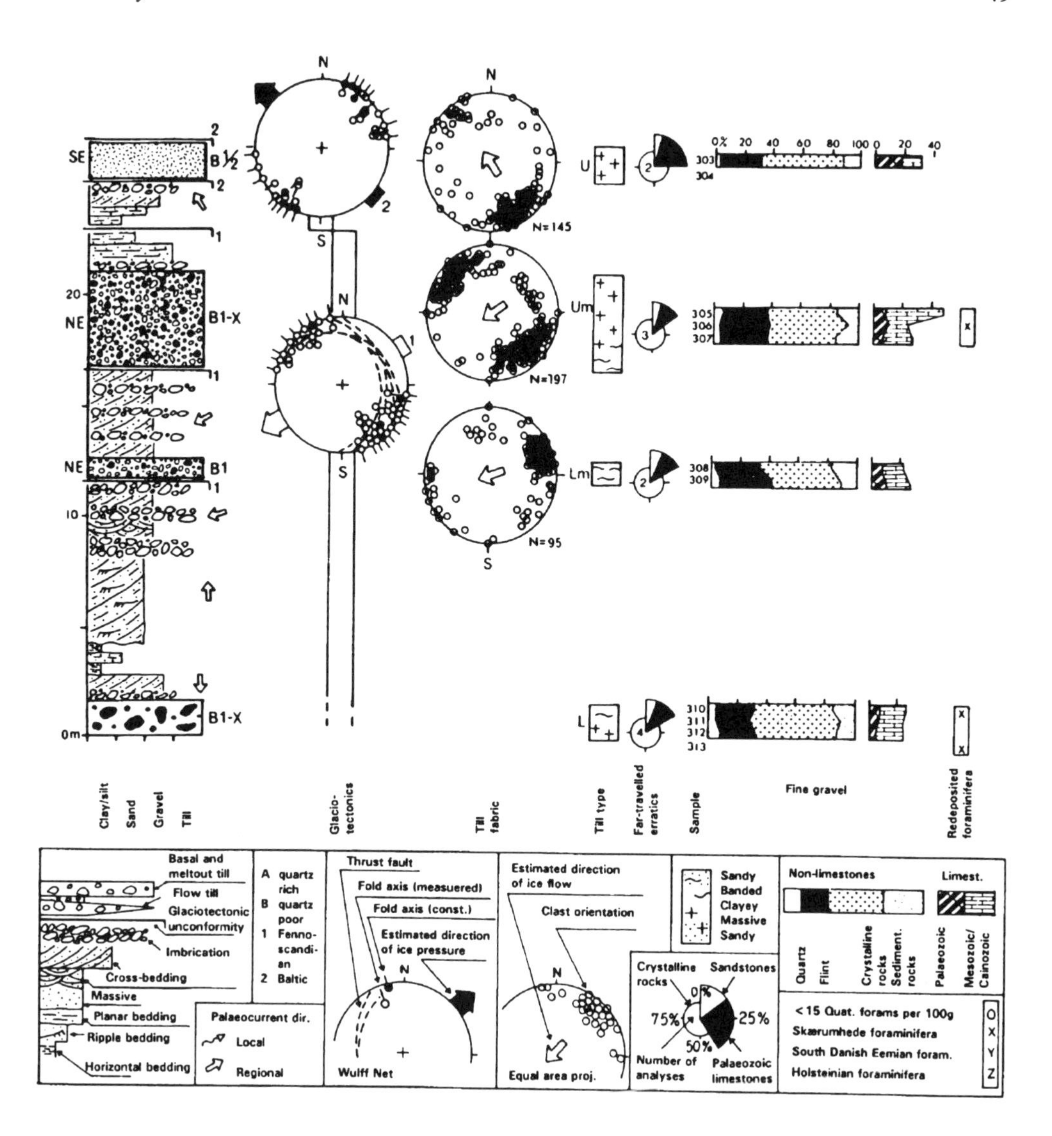

FIGURE 24 Stratigraphic diagram of the Samsø north coast. Kinetostratigraphic units, lithology, bedding directions, glaciotectonics, clast orientation, clast assemblage, fine-gravel composition, and foraminifera content are given (from Houmark-Nielsen 1987)

cliff. They are largely restricted to the strata above the middle till. Both sets of disturbances have been caused by two successive ice advances, an older one from the northwest and a younger from the southeast.

Three-dimensional clast-orientation measurements reveal the relationship between the tills and the glaciotectonic features. As the lowermost till was too strongly disturbed, no measurements could be taken in it. The two middle tills both yielded fabric maxima striking NE–SW, contrasting to the SE–NW orientation found in the uppermost till. Lithological considerations indicate that the lowermost till is the 'Norwegian Till'. It contains redeposited foraminifera from the Early Weichselian Skærumhede Series, which confirms its Weichselian age. The middle tills represent the typical Central Danish Till deposited by an ice advance from the northeast. The uppermost till, in spite of its relatively high flint content, has to be regarded as Young Baltic (East Jylland) till (Houmark-Nielsen 1987).

Apart from major glaciotectonic disturbances of the subsurface that are visible in the field, very minute dislocations also occur, some of which are only visible in thin sections under the microscope (e.g. Van der Meer 1993) or SEM images. Feeser (1988) evaluated the glacigenic deformation of the Lauenburg Clay in several exposures in North Germany, including microscopic features. Likewise, Bruns (1989) investigated the microtectonic structure of the Miocene Mica Clay in Hamburg in order to reconstruct its glaciotectonic deformation history.

Thrust measurements do not always yield unequivocal results. This may be due to the fact that in some cases the deforming ice margin did not form a straight line but was differentiated into lobes and tongues. Thus depending on the position of each measurement with regard to the ice lobe, very different directions may be measured, representing only local movement. Ice thrust is also not restricted to the ice front. Many glaciotectonic deformations are formed under the glacier, and their orientation results from the local stress directions, which are often unrelated to the ice margin. Stephan (1985), for example, has demonstrated that lateral thrust has played a major role. A well-documented example is the Heiligenhafen moraine on the North German Baltic Sea coast.

2.9.6 Moraine elevations

The surface profile of a glacier or ice sheet can be calculated from the elevations of its marginal moraines. Such studies have been made for Alpine glaciers by Flint (1937) in Washington and by Mathews (1967) in New Zealand, where the upper limits of erratics and identification of unglaciated mountain tops (nunataks) at varying distances from the former ice margin allowed the reconstruction of Pleistocene glacier profiles. Because the shape of a glacier tongue is convex, the true ice profile cannot be directly measured on the ice margins but must be calculated.

Whereas in valley glaciers the ice-marginal features show a relatively close resemblance to the ice profile, the situation is different with major ice sheets. For instance, Mathews (1974) has calculated the ice surface of the southwestern part of the Laurentide Ice Sheet, based on the outer limit altitudes of the ice sheet measured in Montana and on the foothills of the Rocky Mountains in Alberta. He found that the Laurentide Ice Sheet had had an extremely low profile, considerably lower than that of the Greenland or Antarctic ice sheets. Beget (1987) came to similar conclusions for the northwesternmost lobe of the Laurentide Ice Sheet in Canada.

Because the shape of the large Pleistocene ice sheets is important for estimations of ice volume and sea-level change, various methods of reconstruction have been attempted. On the base of directional indicators of ice flow, Denton & Hughes (1981) and Hughes (1987) assumed a single-domed Laurentide Ice Sheet centred over Hudson Bay with a surface morphology similar to the present Antarctic Ice Sheet. On the other hand, Dyke & Prest (1987) advocated a multi-domed ice sheet with several dispersal centres, a concept also favoured by the glaciological model of Boulton *et al.* (1985).

The actual ice sheet surface geometry largely depends on flow lines. In most cases flow lines are perpendicular to ice surface contours (Reeh 1982). Strongly divergent flow lines indicate steep surface slopes, nearly parallel flow lines mark gentle slopes. Clark (1992) calculated the ice surface morphology of several lobes of the Laurentide Ice Sheet, using moraine altitudes and indicators of ice-flow direction. His reconstructed ice surfaces are similar to those estimated by Mathews (1974). Clark (1992) thus concludes that the Laurentide Ice Sheet was thin and multi-domed as suggested by Dyke & Prest (1987) and Andrews (1987). In its southern, western and northwestern parts it advanced rapidly over subglacially deforming sediments (Beget 1987, Clark 1992).

SECTION II

QUATERNARY DEPOSITS AND LANDFORMS

3

Glacigenic Deposits and Landforms

3.1 DEPOSITS OF RECENT GLACIERS

As a rule, deposits of glaciers consist of a poorly sorted mixture of clay, silt, sand and gravel called **till** (Fig. 25). The Scottish word till originally meant a stony, loamy soil. As Scotland was one of the early centres of glacial research, the local term has been transferred into the scientific literature. Geikie (1863) already used it as both a descriptive and genetic term; in North America it was introduced by Chamberlin (1877). Today, till should only be used as a genetic term. It refers to the origin of the deposit, not to its composition. Where the sediment is to be characterised by its composition, or where the genesis of a deposit is not known, it is

FIGURE 25 Freshly deposited till in the forefield of an Icelandic glacier (Photograph: Ehlers 1990)

referred to by the descriptive term **diamicton** (Flint *et al.* 1960). Tills of older glaciations, such as those of Precambrian or Permo-Carboniferous age, have undergone diagenesis and are thus referred to as **tillites**. Glaciers leave behind a variety of different types of deposit. Classification of glacigenic deposits was the subject of investigations of the INQUA Commission on Genesis and Lithology of Quaternary Deposits. The final report of this commission was presented at the XII INQUA Congress in Ottawa (Dreimanis 1989).

3.1.1 Subglacial till

Till produced directly by glacial deposition and not modified by secondary processes is referred to as '**primary till**'. Because till deposition occurs under the glacier, the process cannot be observed directly, it can only be deduced from the textural properties of the deposits. Basically three types of subglacial till can be distinguished: lodgement till, subglacial meltout till and deformation till. During the formation of **lodgement till** glacial debris is deposited at the glacier sole from the actively moving ice. This occurs either grain by grain as originally envisioned by Flint (1957), or may also include deposition of sheets of debris-rich ice, which melts out subsequently (Dreimanis 1989). **Meltout till**, on the other hand, is the sediment of a stagnant glacier deposited when basal melting of the ice slowly lowers the entrained debris to the ground (Boulton 1970). Both types of till tend to occur together in most areas glaciated during the Quaternary. They generally differ very little in outer appearance. However, some types of meltout till exist that can be easily distinguished from other tills. Both the Kalix till (Beskow 1935) and the Sveg till (J. Lundqvist 1969) of Scandinavia clearly show the influence of water during the ice melting. The Kalix till is largely composed of sorted materials, with only a thin cover of diamicton and minor beds of diamicton within. It is frequently found in the northern coastal regions of the Gulf of Bothnia. J. Lundqvist (1969, 1983b) regards it as transitional between dead-ice deposits and ice-lake sediments. The Sveg till consists of diamicton interspersed with numerous fine lenses and laminae of sorted materials. It is found mainly where subglacial drainage of glacial lakes occurred. The characteristics of meltout till are discussed in detail by Haldorsen & Shaw (1982).

During the formation of lodgement till larger grains remain entrained in the moving ice longer than smaller ones. A small sand grain that comes into contact with the underlying, already deposited till will be deposited immediately. If a larger clast comes into contact with the till, it may remain in movement, because most of its surface is still enclosed in ice. Increased basal friction, however, leads to a slowing down of the movement and furrowing of the till surface. As the ice moves on, the grain may either topple (Ehlers 1990a) or try to rise out of the furrow. In both cases it will be left in an imbricate position, dipping slightly upglacier. This orientation is observed in many tills (Lindsay 1970, Hambrey 1994). The rate of lodgement till deposition depends on the basal temperature, the content of glacial debris and on the ice velocity, which also determines whether glacial debris is deposited as a till sheet or rather on the stoss sides and squeezed into low-pressure zones in the lee side of obstacles (Fig. 26).

At the till base small ribs or undulations can occasionally be observed. When carefully excavated, they are found to be elongate features that formed in the lee of major erratics and strike parallel to the ice movement (Westgate 1968, Ehlers & Stephan 1979, Clark & Hansel 1989). Those forms are good indicators that the associated till was formed by lodgement and not by gradual meltout from the glacier bed (Ehlers 1990a, Krzyszkowski 1994).

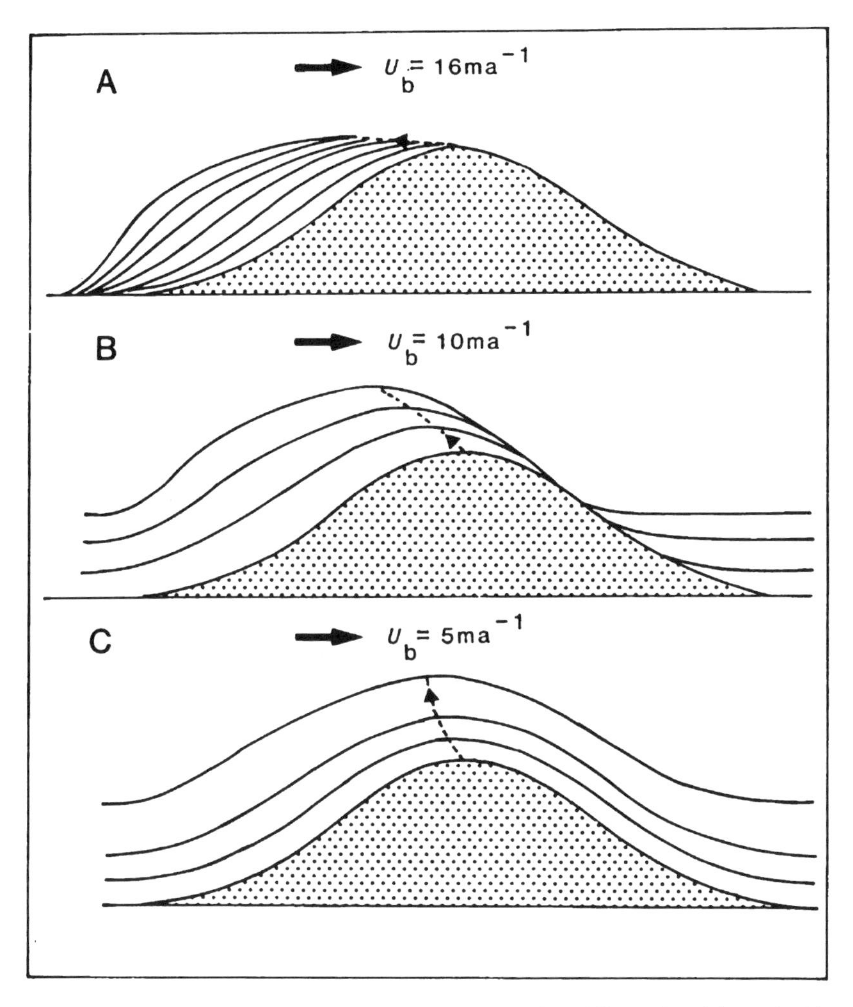

FIGURE 26 Accretion of lodgement till to a bed hummock for different basal ice velocities. At high velocities (U_b) till material is only deposited at the stoss side of the obstacle, at lower velocities also in the lee and on top of the obstacle (from Drewry 1986, after Boulton 1982)

Glaciers with the basal temperature at the pressure-melting point, release subglacial meltout till. The geothermal heat flux and the temperature gradient within the glacier result in deposition of meltout till predominantly in depressions of the glacier bed (Fig. 27). Under long-term extremely cold and arid conditions, such as in parts of Antarctica, till is also released by sublimation of debris-rich ice. This is referred to as sublimation till (Shaw 1989b).

Subglacial deformation of older sediments can result in the formation of another genetic variety of subglacial till. Mixing of materials transported under the sole of a glacier can lead to the formation of a diamicton, that may be referred to as **deformation till** (Elson 1961). A good example of this facies is the Sunnybrook diamicton described by Hicock & Dreimanis (1992). However, this term should only be applied where subglacial deformation has resulted in a true diamicton.

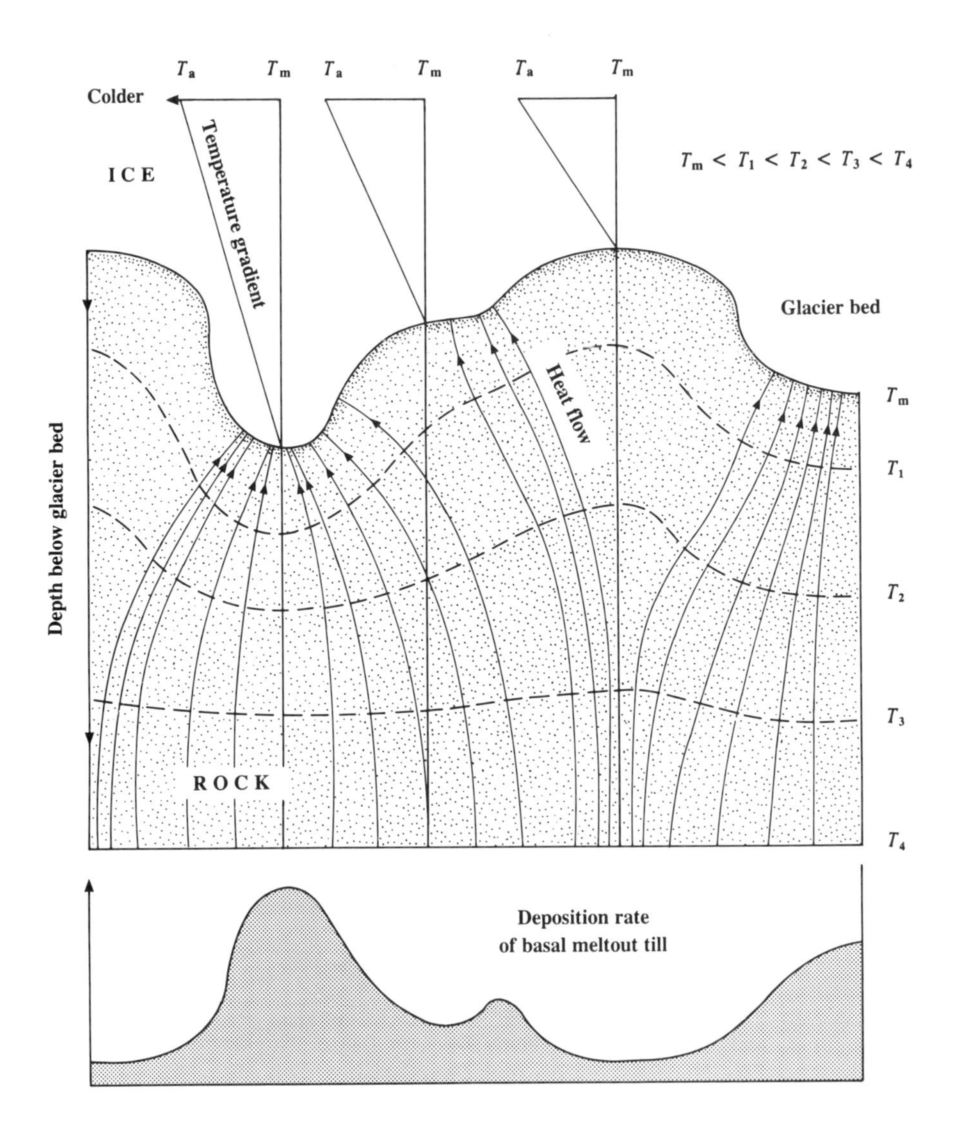

FIGURE 27 Accretion of basal meltout till depending on heat flow in the underlying subglacial rock. Temperature isolines are designated in increasing magnitude T_1–T_4. Note how with depth the isolines damp out the influence of the irregular rock surface. The temperature at the glacier surface is T_a. The ice–rock interface is at the pressure-melting point T_m. Heat from the earth flows approximately perpendicular to the isolines. Where there are bedrock peaks the isolines diverge, giving rise to a lower surface heat flux than in depressions, where the heat flow lines converge. This results in enhanced deposition of meltout till in depressions. Additionally, under the assumption of a level glacier surface the temperature gradient in the ice is steeper over the peaks than over the depressions. The steeper gradient is more effective in draining away geothermal heat, further reducing the heat available for the formation of meltout till (after Drewry 1986)

Meltout of glacial debris is not restricted to the glacier sole but also occurs on the ice surface. This leads to the formation of **supraglacial ablation till** (supraglacial meltout till). In areas of continental glaciations, subglacial tills are rarely covered by supraglacial ablation till. One reason is that in the large ice sheets debris transport was largely restricted to the basal parts, so that little glacial debris was available for the formation of supraglacial ablation till. On the other hand, the upper portions of till profiles as a rule have been either removed by later ice advances or altered by periglacial processes and subsequent soil formation. Therefore in most cases supraglacial ablation tills, if they ever existed, can no longer be identified. One area where ablation till has been found is the ice-decay landscape immediately south of the Baltic Sea. In these areas lodgement till from the ice-advance phase may in extreme cases be restricted to a few basal centimetres or even millimetres, whereas the bulk of the material consists of subglacial and supraglacial meltout till (Stephan & Ehlers 1983).

3.1.2 Other morainic deposits

Apart from the till types mentioned above, two other types of morainic deposit occur, namely flow till and subaquatic till. These sediments, in contrast to the tills *sensu stricto*, are referred to as **secondary tills**, because they have not been released directly from the glacier but have undergone reworking, which has altered their texture (Dreimanis 1989, Lawson 1989).

Flow till is glacial debris that has been redeposited in a solifluction-like process on the ice front (Boulton 1968). When Gripp (1929) originally described such sediments from Spitsbergen, he referred to them as mudflows ('schlammströme'), not as tills. Schlüchter (1977) called them 'schlammoräne' (mud till). Although for practical considerations it may be advisable to include these deposits in the group of 'tills', because in boreholes they cannot be distinguished from true tills, it must be taken into account that genetically they are the result of mass movement (sediment flows) and not really tills in the genetic sense (Lawson 1979).

Subaquatic till (or waterlain till) is glacial debris released from the ice front or from the glacier sole into standing or flowing water (Dreimanis 1989). The environment at an ice margin is often rather water-soaked, and where the glacier terminates in water a variety of different sedimentological processes take place simultaneously. A special case of waterlain sediment is ice-rafted diamicton, originating by rain-out from icebergs.

3.2 PLEISTOCENE GLACIGENIC DEPOSITS

The origin of tills in formerly glaciated areas has been a matter of debate since the emergence of glacial theory. In North Germany, in the late 19th century, tills were first regarded as supraglacial morainic deposits of the Pleistocene ice sheets. This concept was only refuted after Nansen had crossed the Greenland Ice Sheet. Nowhere on the inland ice did he and his expedition observe supraglacial debris. Only in the very marginal parts of the ice sheet is glacial debris transported to the ice surface (Mohn & Nansen 1893). Penck (1894) assumed that a till sheet several metres thick could be moved under the ice. After his investigations in Greenland, Von Drygalski (1897) concluded that the glacial debris was transported entirely within the ice, namely in its basal parts. This concept is still valid, although it has been discovered recently that to a certain degree subglacial movement of glacial debris is also possible (cf. chapter 2.4).

In Quaternary geology there is always the danger that too much effort is spent on stratigraphical correlation and too little on the origin of the deposits. Where the genesis of a sediment is unclear, however, pointless correlations of genetically unrelated deposits may occur. Even the application of standard laboratory tests (e.g. grain-size analyses, clay mineralogy) are no guarantee of meaningful stratigraphical correlations. With these aims in mind, Eyles *et al.* (1983) have developed a lithofacies scheme for clastic sediments (Table 2). In order to record the different lithofacies types, mapping of vertical profiles of all relevant sections of an exposure is required. Eyles *et al.* (1983) demonstrated the application of their method using the example of Scarborough Bluffs of Lake Ontario (Fig. 28); Piotrowski (1992, 1994b) has considered the different facies types of morainic deposits in North Germany. However, it must be kept in mind that the interpretation of a particular sequence may differ from author to author. Unbiased, detailed description of the deposits and their sedimentological characteristics should ideally be used for stratigraphic interpretations (Sharpe & Barnett 1985).

It is not always easy to define when precisely a deposit should be called till. In many cases older sediments beneath morainic deposits are not entirely undisturbed, but part of the sequence

TABLE 2 Four-part lithofacies code and symbols for the characterisation of diamicton (from Eyles *et al.* 1983)

Lithofacies code	Symbols
D = Diamicton Dm = matrix supported Dc = clast supported D - m = massive D - s = stratified D - g = graded Genetic interpretation: D - - (r) = redeposited D - - (c) = current reworked D - - (s) = sheared	two types of symbols, proportional to clast size stratified sheared jointed
	Gravel Sand Laminations (spacing proportional to thickness) - with silt and clay clasts - with dropstones - with loading structures
S = Sand Sr = rippled St = trough cross-bedded Sh = horizontally laminated Sm = massive Sg = graded Sd = soft-sediment deformed	
F = Fine-grained sediments Fl = laminated Fm = massive F - d = with dropstones	Contacts: Erosional Conformable Loaded Interbedded

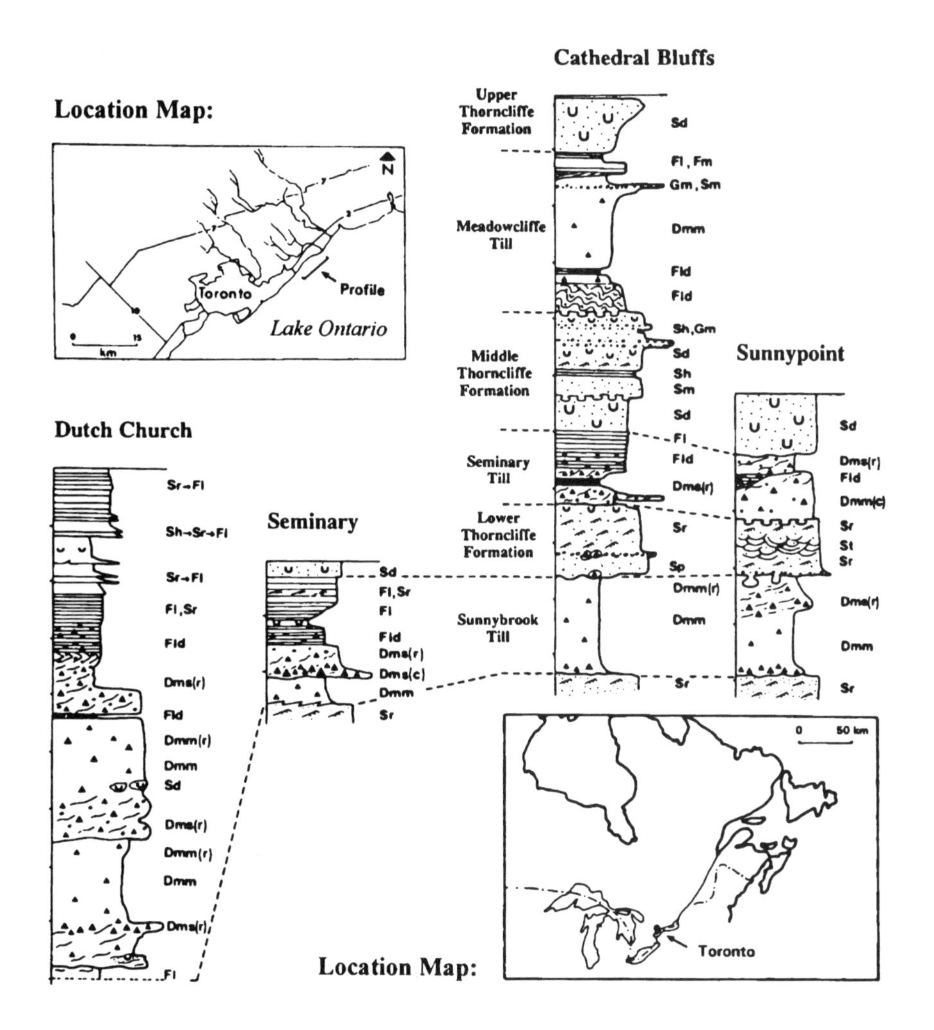

FIGURE 28 Characteristic profiles from the Scarborough Bluffs section, Ontario (Canada). Diamictons are differentiated according to the classification in Table 2. Lithofacies types of sandur deposits: Gm = gravel, massive or crudely bedded; Sp = sandur deposits with planar cross-bedding. The former stratigraphic subdivision (after Karrow 1967) did not include facies relationships (from Eyles *et al.* 1983)

has been incorporated by the ice movement (Fig. 29; Stephan 1981). At the base of many tills a layer of deformed sediments several decimetres thick is found, with a grain-size composition that resembles that of the underlying sequence, but with a structure that comes closer to till. This 'sole till' (Grube 1979) consists largely of locally eroded material that has not been completely reworked and incorporated into the moving glacier. It belongs to the deformed bed *sensu* Alley *et al.* (1987). Some authors suggest that anything moved by a glacier, even for a few centimetres, should be called till. However, as such deposits show neither marked deformation nor the textural composition of a true till ('diamicton'), they should be better referred to as 'subglacially deformed sediments' (Stephan & Ehlers 1983).

FIGURE 29 Differing degrees of subglacial deformation of meltwater sediments. (a) Tilted sands underlying an Older Saalian till are cut by a horizontal shear plane parallel to the till base (Photograph: Ehlers 1977). (b) Till of the Late Weichselian Fehmarn Advance overlies older ice-pushed sediments. At the till base a zone of intensive shearing is visible (Photograph: Stephen 1981) (from Stephan & Ehlers 1983)

Supraglacial flow tills are frequently found at the terminus of recent glaciers (e.g. Lawson 1979), but they are less often identified in the Pleistocene sedimentary sequence. Most flow tills that originally formed at the front of an advancing glacier were apparently removed immediately afterwards when being overridden by the ice sheet. On the other hand, flow tills from the decay phase of an ice sheet in many cases have been altered considerably by post-depositional processes and are no longer distinguishable from other weathered diamicton. Only thick flow tills deposited in protected positions (e.g. small depressions) have a high probability of preservation such as described by Wansa (1991) from the margins of Elsterian buried channels exposed in an opencast brown-coal mine at Gräfenhainichen (north of Bitterfeld, Germany).

For instance glacigenic deposition into water, from a floating ice shelf, has often been referred to as 'subaquatic till'. However, genetically these deposits are marine or, in the case of ice-dammed lakes, lacustrine sediments rather than tills. They are further discussed in chapter 7.3.2.

Ice-dammed lakes formed in front of all major ice sheets. In North America, immense areas in front of the Laurentide Ice Sheet were covered by lakes. Consequently, subaquatic diamictons are widely distributed. They are not only associated with the Great Lakes but are also found in large areas of central Canada, which were temporarily covered by glacial lakes Agassiz and Barlow-Ojibway. The deposits are characterised by distinctive facies sequences and textures, and most of the diamictons deposited in this environment are stratified (Dredge, personal communication).

The largest ice-dammed lake in Europe was the late Weichselian Baltic Ice Lake (see chapter 11.18.2). Another large ice-dammed lake existed in North Germany in front of the melting Elsterian ice sheet in incompletely filled subglacial channels. The basal part of the lake sediments frequently consists of a diamicton (subaquatic till), which grades into partly laminated glaciolacustrine silts and clays ('Lauenburg Clay') towards the top (cf. Ehlers & Linke 1989).

In contrast to the North European glaciation area, the Alpine Region is characterised by a pronounced lack of tills within their soft rock sequences. In the Swiss 'Mittelland' region, tills several tens of metres thick occur, but closer to the centres of the glaciations very little till is found. Schlüchter (1980b) emphasises that these are found largely in places where the advancing glacier encountered rock obstacles, so that because of the compressive flow no crevasses could form. This had the effect that the subglacial till was largely protected from meltwater erosion. In lee of the obstacles, however, extending flow allowed unhampered meltwater drainage and erosion.

Because of the glacial overburden, tills deposited under the ice often show a certain degree of **overconsolidation**. As a consequence these deposits react like a massive rock. They are very vulnerable to rewetting after desiccation, which results in the formation of gullies with steep slopes and in extreme cases the formation of earth pillars (e.g. at Bozen in Südtirol, at Pont-Haut south of Grenoble or at Euseigne in the Valais) (Van Husen 1981). However, this is only the case if the pore water could escape, and if the material was not dilated. Most deformation tills are therefore not overconsolidated. In many cases transmissivity of the tills was also too low to allow pore-water escape (Piotrowski, personal communication).

3.3 POST-DEPOSITIONAL ALTERATION

Many differences between tills that can be observed in the field are not primary but result from post-sedimentary processes. The older tills, especially of the Illinoian and pre-Illinoian glaciations of North America or the Elsterian and Saalian glaciations of Europe, which lay at the ground surface throughout the last glaciation, were under the influence of periglacial climate. The till sheets were dissected by ice-wedge polygons, and their upper portions were cryoturbated and on slopes removed by solifluction. In general, summer thaw determined the depth of these alterations, which include easily identified cryoturbations as well as isolated involutions and the formation of stone layers within apparently undisturbed till profiles. During the Weichselian the depth of the active layer in the Hamburg region does not seem to have exceeded 2 m (Ehlers 1978). Ice wedges, however, penetrated much deeper.

More difficult to identify than the periglacial alterations are the changes that occurred during the course of soil-forming processes (pedogenesis). Because the latter are largely also controlled by substrate-independent factors like exposure, slope gradient, vegetation, precipitation and ground water, a broad range of alterations can occur in all tills (Felix-Henningsen 1979, 1983). Decalcification, iron and manganese redistribution and clay translocation can cause changes in colour, texture and grain-size composition, which may give the impression that two different tills are involved where there is only one. Where colour and grain-size composition are regularly used for characterising tills, special caution is recommended where tills above the groundwater table are involved, because this is the region where the strongest pedogenetic processes take place.

Caution is also advised where carbonate content is used to characterise tills. Pedogenic decalcification from above leads to a secondary enrichment in carbonate further down-profile. For example, in a till in western Hamburg Cornelius (1984) found a secondary pedogenetic enrichment in carbonate by as much as 10%.

3.4 GLACIGENIC LANDFORMS

3.4.1 Drumlins

The landscape overridden by a glacier in many cases is altered into a relief with numerous larger and smaller more or less streamlined landforms, the long axes of which strike in the direction of ice movement. In many cases this process has generated characteristic landform assemblages, which include small-scale features such as flutes and large-scale features such as drumlins. **Flutes** are mostly less than 2 m high and can attain a width of up to 50 m and a length of several hundred metres. In many cases they are hard to detect on a vegetated ground surface, but aerial photographs may help (Flint 1971).

Drumlins consist predominantly of till. They are mostly several hundred metres long and less than half as wide, although forms up to 3 km long occur in extreme cases. Extremely long drumlins are referred to as 'drumlinoids' by some authors (e.g. Prest 1983). The term drumlin (after the Gaelic word 'druim' = hill) was introduced by Close (1867) into the scientific literature. The term originated in Ireland, where extensive drumlin swarms extending over several hundred kilometres attracted the attention of geologists at an early stage (Coxon & Browne 1991, McCabe 1991, 1993, McCabe *et al.* 1992). Several symposia have been held focusing particularly on the origin of drumlins. Menzies (1984) has published a comprehensive drumlin bibliography.

Drumlins include well-developed streamlined features with commonly steep stoss ends and gently sloping lee ends as well as symmetrical forms, which may be either subtle or steep. Some drumlins contain a rock core, others are completely built up of waterlain sediments. The rock-cored drumlins resemble roches moutonnées, i.e. streamlined features of hard rock resulting from glacial erosion.

As early as 1886, Brückner reported the occurrence of drumlins from the Alpine Salzach Glacier area, whilst in northern Europe Doss (1896) gave a detailed description of the drumlin fields of Latvia. Thorough investigations into the drumlins of the Alpine foreland were conducted by Ebers (1925, 1937). She concluded that they were formed by a two-phase process: sculpturing of the streamlined drumlin cores was followed by later covering by a till blanket. Eberl (1930) in contrast was of the opinion that the drumlins were mostly erosional landforms. Today it is well established that both erosional and depositional drumlins occur (e.g. De Jong *et al.* 1982).

Drumlins tend to occur in groups called 'drumlin swarms' or 'drumlin fields', often comprising more than a hundred individual forms. Famous drumlin swarms occur, for instance, in upper New York, in the Green Bay Lobe area of Wisconsin, in parts of New England (e.g. Boston Harbor), on Nova Scotia, and in Ireland. The Alpine glaciation area in Europe includes the large south German drumlin fields (Habbe 1988) (Fig. 30c).

Individual drumlinoid landforms sculptured by the glacier include the features on Fehmarn described by Stephan (1987a) (Fig. 30d). Drumlin formation seems to be restricted to a subglacial zone close to the ice margin. Habbe (1988) concluded that they are relatively late forms caused by a lowering of the permafrost table towards the end of the glacial maximum. Some authors assume that drumlins are not rigid landforms but that they migrate towards the ice margin (e.g. Boulton 1987, Piotrowski 1989). This does not refer to sand-cored drumlins. In many cases movement seems to have been rather limited, because the cores of many drumlins consist of undisturbed older sediments (e.g. Krüger & Thomsen 1984).

Because of their striking shape, drumlins have always attracted the attention of Quaternary scientists. They are one type of bedform created at the interface between moving ice and immobile bedrock. If mobile sediments at the glacier base are involved, two interfaces must be envisioned: an upper interface at the ice/mobile sediment contact, and a lower interface between mobile and immobile sediment (Menzies 1987). Numerous interpretations have been attempted. Boulton (1987) subdivides these into two major groups.

The first group of theories assumes that subglacial meltwater plays a decisive role in drumlin formation. Dardis *et al.* (1984) thus explain the occurrence of vast amounts of bedded sediments at the distal ends of many drumlins. In places, water-throughflow structures have been observed in drumlins that have resulted in large-scale reorganisation of the diamictons involved (McCabe & Dardis 1994). Shaw (1983) and Shaw & Kvill (1984) point out that many drumlin swarms resemble forms eroded by flowing water. Accordingly they assume that drumlins were formed by sheetflood-like outbursts of subglacial meltwater which eroded cavities into the basal ice. These cavities were then filled with glaciofluvial sand and gravel to form drumlins.

The second group of theories assumes a glacigenic origin of the drumlins. Aario (1977a) holds a spiral-shaped sediment-transport mechanism within the glacier responsible for drumlin formation. He regards the drumlins as accumulative features. On the contrary, Whittecar & Mickelson (1977, 1979) interpret drumlins as erosional features. In some cases underlying sediment has been intruded into the drumlin core under the overburden of the ice. Boyce & Eyles (1991) suggest drumlins were carved by deforming till streams under the ice. Stanford & Mickelson (1985) point out that drumlins often contain folds striking parallel to their long axis. They explain these folds as a result of lateral sediment transport under the influence of

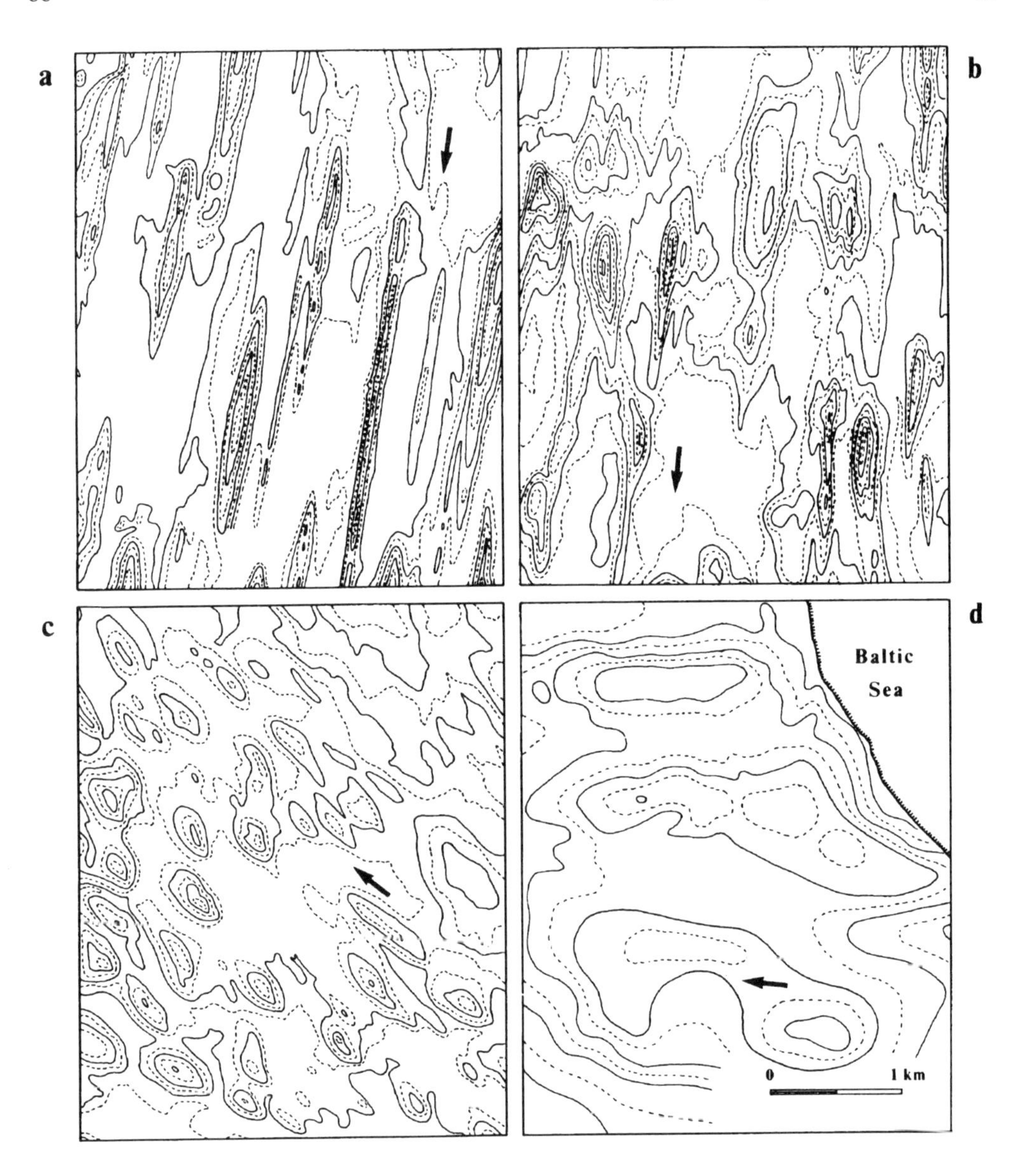

FIGURE 30 Types of drumlins: (a) extremely elongated drumlins in Wisconsin (Clyman Quadrangle); (b) elongated drumlins in Wisconsin (Helenville Quadrangle); (c) roundish drumlins on the Bodanrück peninsula (western Rhine Glacier area); (d) low drumlins on Fehmarn. Contour intervals: a, b = 15.0 m; c = 10.0 m; d = 2.5 m

differences in overburden pressure. Both hypotheses can explain the origin of streamlined features under the ice but not the triggering mechanism that initiates this process.

A comprehensive interpretation has been attempted by Smalley & Unwin (1968) and further developed by Smalley & Piotrowski (1987). In glacial debris with a high gravel content a much higher shear stress is required to initiate shear deformation than is required to maintain this process. If the tension between the glacier and its bed passes a critical value or if the glacial debris passes a critical gravel content, sediment is accumulated under the glacier to form stable obstacles, which are shaped by the moving ice into streamlined drumlin fields. Drumlin formation thus depends on ice thickness. Where the ice thickness is too great, shear stress is insufficient to deform the sediment. Where on the contrary the required shear stress is much greater than the shear strength of the sediment, no stable obstacles can form. Drumlin formation occurs only where the ratio of shear stress to shear strength is near 1 (Fig. 31). This corresponds with the observation that drumlins occur at a certain distance from the outermost ice margin. Undisturbed meltwater sediments in drumlin cores or at the distal ends are in most cases remnants left by erosion.

None of the above theories is able to explain all of the observed features. Subglacial meltwater action seems to be an important factor in drumlin formation. The subglacial deformation of unconsolidated sediments as envisaged by Boulton (1987) and Menzies (1987, 1989), however, should help to explain some of the other observed processes.

3.4.2 End moraines

All debris taken up by a glacier and not being deposited somewhere on its way is transported to the ice margin. Where the ice margin comes to a halt for a period of time, the melting-out glacial debris will accumulate as a rampart, a **depositional end moraine** (Fig. 32). The process has been described, for instance, by Goldthwait (1951). At a retreating ice margin this process can form a series of annual moraines. The debris accumulated at the ice margin is termed frontal or lateral end moraine (Fig. 33). The term end moraine dates originally from Agassiz (1840) and comes from the Alpine glaciation area. Early workers in the marginal areas of the north European glaciations had attempted to identify features similar to those known from the mountain glaciers. Depositional end moraines were the first such features mapped in North America (Gilbert 1871, Chamberlin 1877). Also Gottsche (1897b) mapped the end moraines of Schleswig-Holstein as accumulations of coarse-grained debris ('blockpackungen'). Today we know that many of the 'blockpackungen' mapped this way are the result of meltwater activity, i.e. sediments deposited immediately at the ice margin, where the meltwater stream left the glacier. They are referred to as 'heads of outwash' in North America.

Accumulative end moraines have a steep distal face and a more gently sloping ice-contact (proximal) face. In valleys they tend to have a convex downglacier form, and in lowlands they are also often lobate features. In most cases, major parts of the moraine ridges are destroyed by meltwater erosion. Well-developed sequences of recessive end moraines are found in front of many modern valley glaciers. Prominent Quaternary end moraines include, for instance, the morainic amphitheatres at the southern ends of the large north Italian lakes (Hambrey 1994).

In many cases former ice-marginal positions are characterised by a low till plain bordering against a higher sandur plain. This landform assemblage, which was first described from the Weichselian ice margin of the Brandenburg Phase (Franz & Weisse 1965), has been observed elsewhere as well. The core of the Harburger Berge hills south of Hamburg is such a 'substitute end moraine' (Ehlers 1978). Another well-developed example is the Hauerseter sandur described by Sørensen (1983) from southern Norway.

Undisturbed sediments do not always accumulate at the ice margin. Ice advance often causes deformation, resulting in the formation of **thrust moraines**. The oldest descriptions of Alpine thrust moraines date back to the 16th/17th centuries. Further cases were described by De

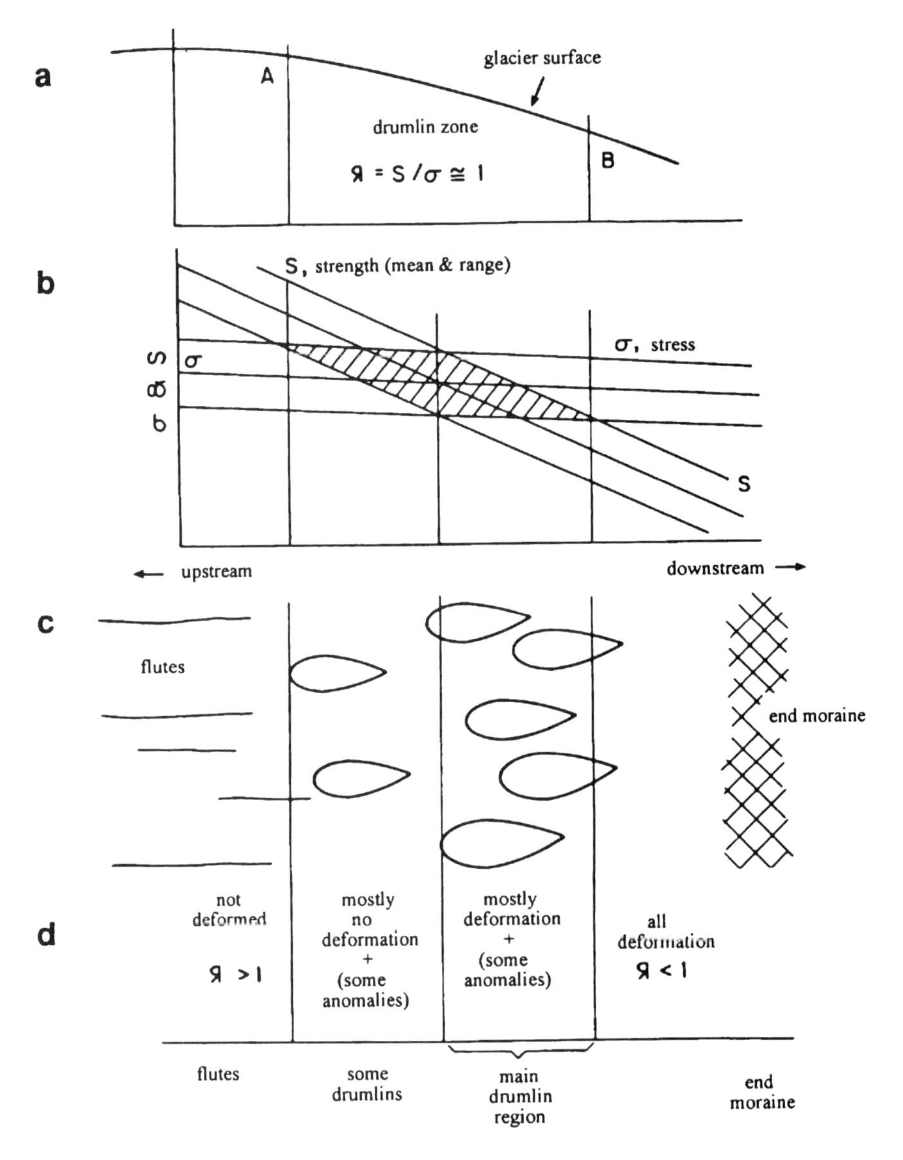

FIGURE 31 Conditions for the formation of drumlins (a) The zone of drumlin formation lies near the edge of an ice sheet, in an area of decreasing ice thickness. (b) In the central area of the ice sheet the shear strength (S) is larger than the shear stress (σ), but near the ice margin the two curves cross; this is the area of drumlin formation. (c) Within the area of drumlin formation the number of drumlins increases towards the ice margin. (d) Drumlin formation is favoured in areas of strong deformation; further towards the ice margin shear stress predominates so that no forms can be preserved (from Smalley & Piotrowski 1987)

Charpentier (1841). That glacial pushing also played a part in the Nordic glaciation areas was already postulated by Johnstrup (1874). He observed that the disturbances of the chalk rocks exposed on the Isle of Møn in the Baltic Sea (Denmark) did not continue at depth. He concluded that they had been moved by glacier thrust. Wahnschaffe (1882) also described ice-

FIGURE 32 Minor, unnamed glacier at the Stockhorn north slope near Zermatt, Switzerland; 1991 end moraine of the 1979–91 advance (Photograph: Schlüchter 1992)

pushed sequences from North Germany. Only after detailed investigations of recent glaciers and their foreland on Spitsbergen was the great importance of the thrust moraines realised for the sculpturing of the North German landscape (Gripp 1938). In Britain the works of Slater (1926, 1927a, 1931) drew attention to thrust sequences and Banham (1975) discussed the various types of glaciotectonics. Spectacular glaciotectonic features are exposed on the north Norfolk coast; a major thrust moraine is the Bride Moraine at the northern end of the Isle of Man. A recent overview of such features has been provided by Allen (1991).

In North America large-scale glaciotectonics was first described by Hopkins (1923) and Slater (1927b). The most spectacular features are found across Alberta, Saskatchewan and North Dakota (Kupsch 1962, Moran *et al.* 1980, Bluemle & Clayton 1984, Fenton 1987, Aber

FIGURE 33 Lateral moraines and end moraine of Findelen Glacier near Zermatt, representing all ice advances of the last 3000 years. All advances had approximately the same dimension as the 'Little Ice Age' advance (Photograph: Schlüchter 1992)

& Bluemle 1991). Others include the coastal moraines of Massachusetts (Oldale & O'Hara 1984).

A special type of glaciotectonic disturbance involves the dislocation of major rafts of bedrock. Minor examples include the dislocated chalk of Møn (Denmark), Rügen (Germany) or Skåne (Sweden). On the Canadian Prairies much larger glacial rafts have been found, extending over tens of square kilometres. Stalker (1973) was the first to realise that they were in fact underlain by glacial deposits and not part of the bedrock. Some of these **megablocks** (Stalker 1974) have been transported over distances of about 300 km (Prest & Nielsen 1987). The largest megablock so far described has been found at Esterhazy in Saskatchewan. There a block of Cretaceous shale over 1000 km^2 was found, disturbed at the base and locally underlain by till. Its maximum thickness is about 100 m (Christiansen 1971b). This raft seems to have only been moved over a short distance, but it is clearly not *in situ*. As proglacial formation of such a raft seems impossible, the most likely explanation would be that it had been frozen onto the base of the ice sheet (Aber *et al.* 1989). In comparison, the largest known single ice-pushed hill complex in Europe, the Elbląg Heights in Poland, cover less than 400 km^2 (Aber, personal communication).

According to their position with respect to the ice, three types of glacier-induced glaciotectonic disturbances can be distinguished (Houmark-Nielsen 1988):

1. *Proglacial structures.* Those include recumbent folds, reverse listric disturbances (shovel-shaped, curved tectonic movement planes), nappes, reverse upthrows and normal listric disturbances, which form in the proglacially disturbed sediments as a result of sagging over thawing dead ice. In the distal direction size and tilting of the thrusts decreases and folds are more open towards the undisturbed glacier forefield. Proglacial deformation can result in the formation of large, morphologically striking push moraines. They include the end-moraine system seen in front of the Holmströmbreen Glacier in Spitsbergen (Van der Wateren 1992).
2. *Subglacial structures.* Subglacial disturbances mainly result from shearing of older, overridden thrust structures or undisturbed sediments. Otherwise the glacial overburden can result in the formation diapirs and intrusive deformations, which result from sediment mobilisation and the extrusion of pore water. Subglacial deformation exerts a destructive role on older landforms and can in extreme cases level an undulating terrain to a flat till plain. Subglacial glaciotectonics includes the displacement of megablocks discussed above. The deformations on the north Norfolk coast of East Anglia have also been partly interpreted as subglacially formed diapirs by Banham (1975) and Ehlers *et al.* (1991).
3. *Supraglacial structures.* On stagnant and decaying ice, sediment is enriched also on the glacier surface. These deposits usually attain a texture disturbed by numerous sag features when they are lowered to the ground during further ice melt.

The formation of push moraines is often associated with the frontal oscillations of a glacier or ice sheet. However, even if the ice margin stagnates completely but the ice base moves, the ice-profile gradient causes differentiated stress in the bed sediments so that they can be pushed out in front of the ice margin (Piotrowski, personal communication). Eissmann (1987) has discussed the formation of glacigenic disturbances using the example of the Schmiedeberg end moraine in East Germany (Fig. 34). He assumes that the disturbances are mainly caused by an upward-directed vertical sediment injection during the thrusting. On the contrary, Van der Wateren (1992) emphasised the horizontal component of glacial thrusting (Figs 35 and 36). Field evidence seems to indicate that both processes can play a decisive role depending on local circumstances.

The large thrust moraines in the marginal areas of the Nordic glaciations represent major sediment accumulations. As a rule, they must be related to equivalent zones of glacial erosion ('exaration'). Aber & Bluemle (1991) speak of characteristic 'hill–hole pairs' (Fig. 37). In the Dutch literature the depressions are mostly referred to neutrally as **glacial basins** (e.g. De Gans *et al.* 1986), a typical example of which is the Amsterdam Basin (Fig. 38). In the German literature the basins are called **tongue basins**. Such basins, particularly in the area of the older glaciations, have subsequently been filled with sediments and are no longer visible at the surface. Older investigations have mostly concentrated on the thrust moraines and neglected the adjoining tongue basin. However, for an accurate understanding of the formation processes a combined investigation of both features is required. Amongst the modern works which attempt this more comprehensive approach are, for instance, Eybergen (1987), Van der Wateren (1987, 1992), Croot (1988) and Aber *et al.* (1989).

For the process of ice thrust, two different scenarios have been considered: frozen or unfrozen ground.

If a glacier advances over frozen ground and its sole is not at the pressure-melting point, the ground will freeze to the glacier base. As the ice continues moving, the resulting stress may

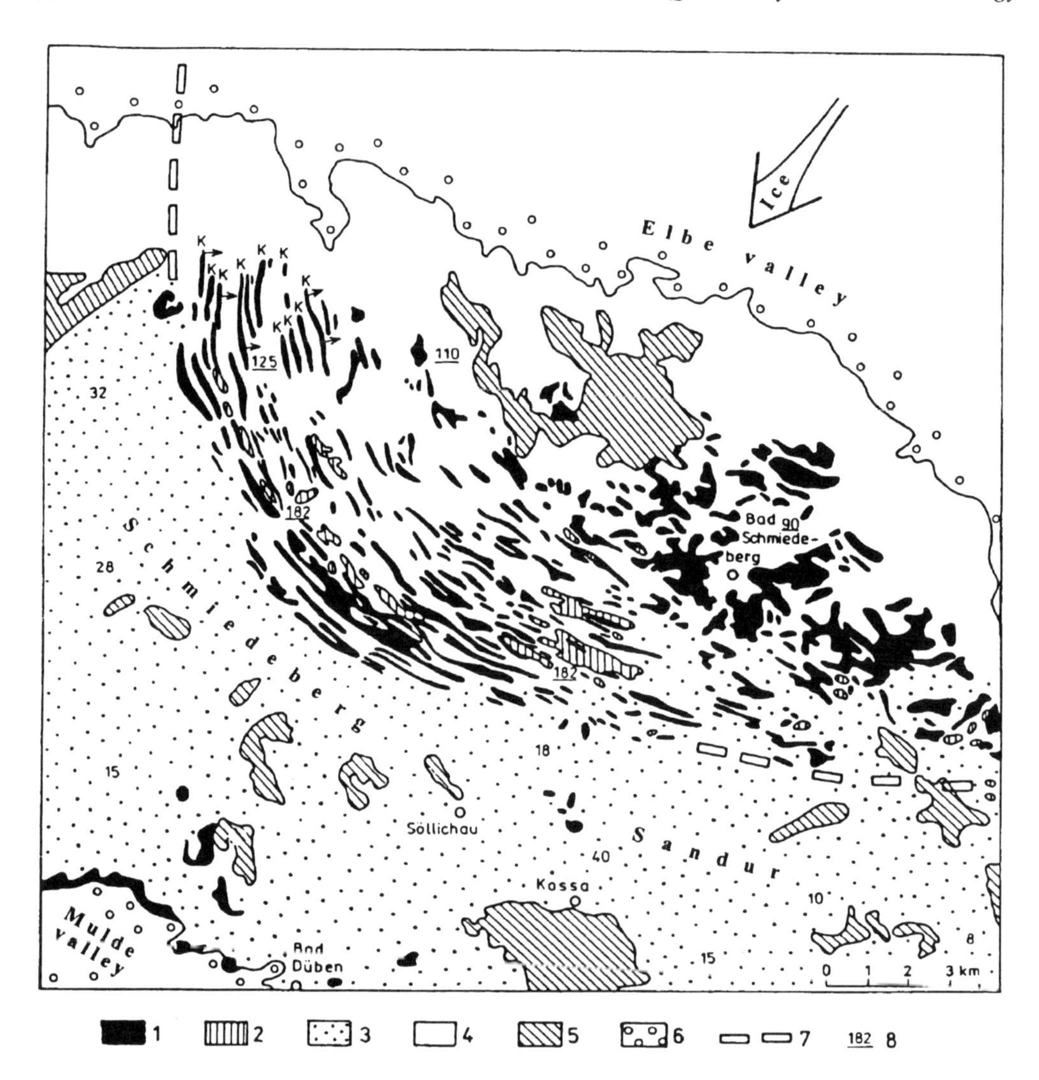

FIGURE 34 Sketch map of the Schmiedeberg push moraine. 1 = outcropping or near-surface Tertiary rocks (K = lignite seam); 2 = dragged-up gravel and sand, mainly of Early Pleistocene age (gravels of Elbe and Oschatz Mulde rivers); 3 = sands and gravels mainly of the Schmiedeberg sandur (numbers; thickness in m); 4 = undifferentiated meltwater deposits beyond the Elbe valley; 5 = till, mostly Saalian; 6 = floodplain and lower terrace of the Elbe and Mulde rivers; 7 = probable continuation of the outer margin of the glaciotectonically disturbed zone; 8 = elevation above sea level (from Eissmann 1987)

exceed the shear strength of the substratum. It is sheared off. In areas of deep-reaching permafrost shearing will occur in layers that are most susceptible to gliding. As clays freeze much later than sands, clay layers are most susceptible to shearing under this process. Consequently, clays have been frequently observed at the base of thrust sheets (Mathews & Mackay 1960).

However, no freezing-on is required to create thrust moraines. If a water-saturated clay is overridden by a fast-moving glacier, the sudden overburden results in a strong increase in pore-

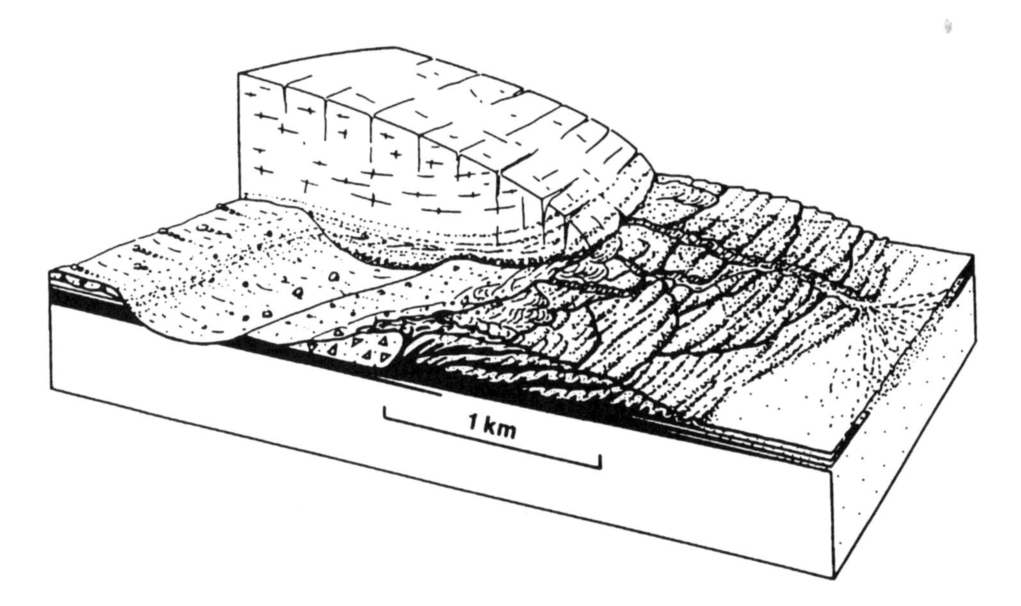

FIGURE 35 Simplified block diagram of the Holmströmbreen push moraine (Spitsbergen), an example of a push moraine built of folds and fold nappes mainly composed of fine-grained sediments (black). Vertical scale exaggerated. Triangles: till (from Van der Wateren 1992)

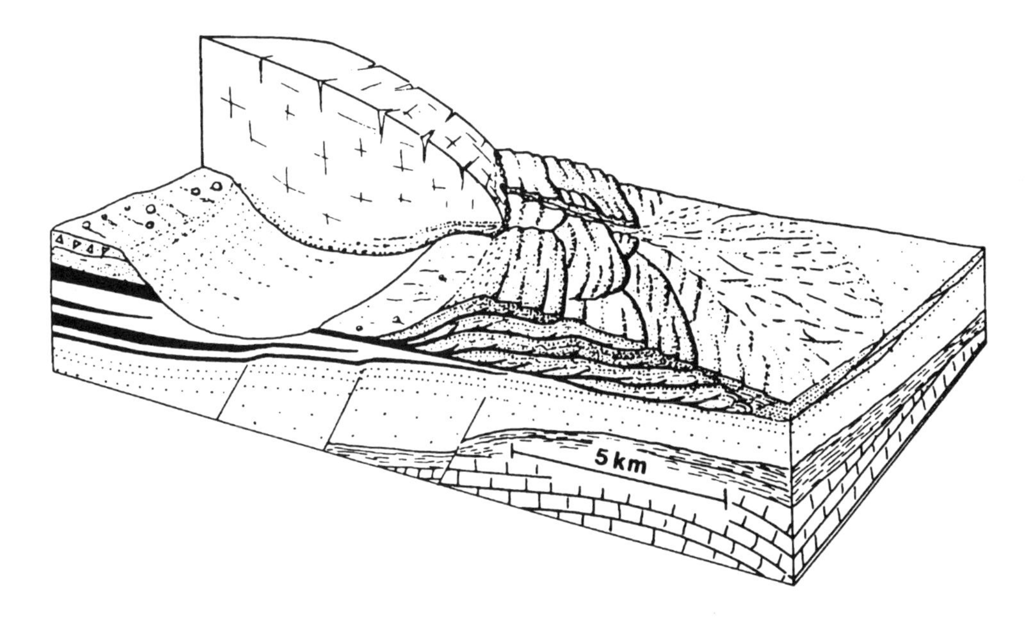

FIGURE 36 Simplified block diagram of the Dammer Berge push moraine (Niedersachsen, North Germany), an example of a push moraine built of large nappes composed of coarse-grained sediments on a thin basal shear zone of fine-grained sediments (black). Vertical scale exaggerated (from Van der Wateren 1992)

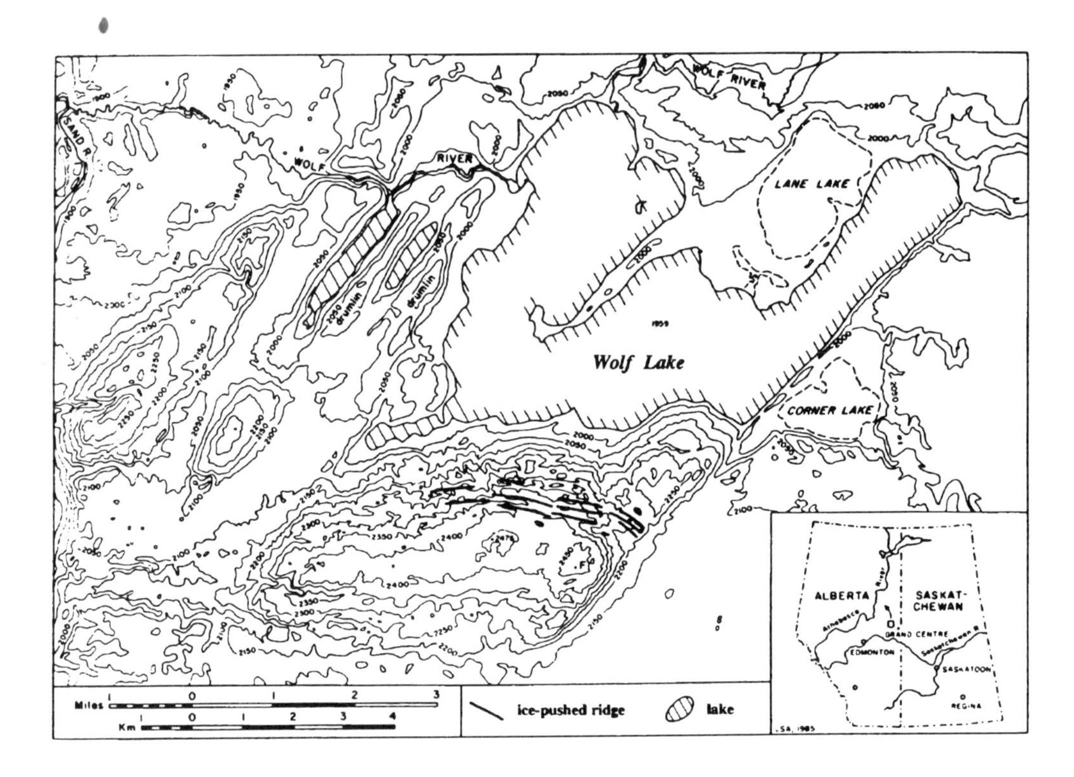

FIGURE 37 Ice-pushed ridge and adjoining lake (hill–hole pair) at Wolf Lake, Alberta. Contour interval 50 feet (ca. 15 m) (from Aber 1989)

water pressure. Effective pressure is lowered and shear strength minimised. Because of the low permeability of fine-grained sediments it takes a relatively long time until sufficient pore water can be squeezed out to relieve the pressure (Moran 1971). Thus two factors facilitate glaciotectonism at the permafrost wedge: (i) the rheological boundary between unfrozen and frozen ground, and (ii) high pore pressures built up on the proximal side of the permafrost wedge (Piotrowski 1993 and personal communication). For details of the mechanics of frozen ground, see also Williams & Smith (1989).

Investigations in the Uelsen and Dammer Berge end moraines in Germany have revealed that partially subhorizontal nappe-like thrust sheets are the dominant features, which in small exposures are difficult to identify as thrust phenomena (Van der Wateren 1992, Kluiving 1994). As in orogeny, in glaciotectonics it is not easy to explain the mechanism behind the formation of large nappes moved over major distances. A prerequisite is the formation of a glide plane on which the nappes can move with minimal friction. Fine-grained sediments (silts and clays), as they are found at the base of many thrust zones, are ideal in this context. The Dammer Berge and the Uelsen end moraines lie a short distance from the central German uplands. Here Tertiary and older clays come close enough to the land surface to have been susceptible to glaciotectonics (W. Richter *et al.* 1951, Van der Wateren 1987, 1992).

In the North German Saalian thrust moraines generally Tertiary or Cretaceous clays were primarily deformed. In contrast, within the Weichselian Glaciation area this role is partly taken

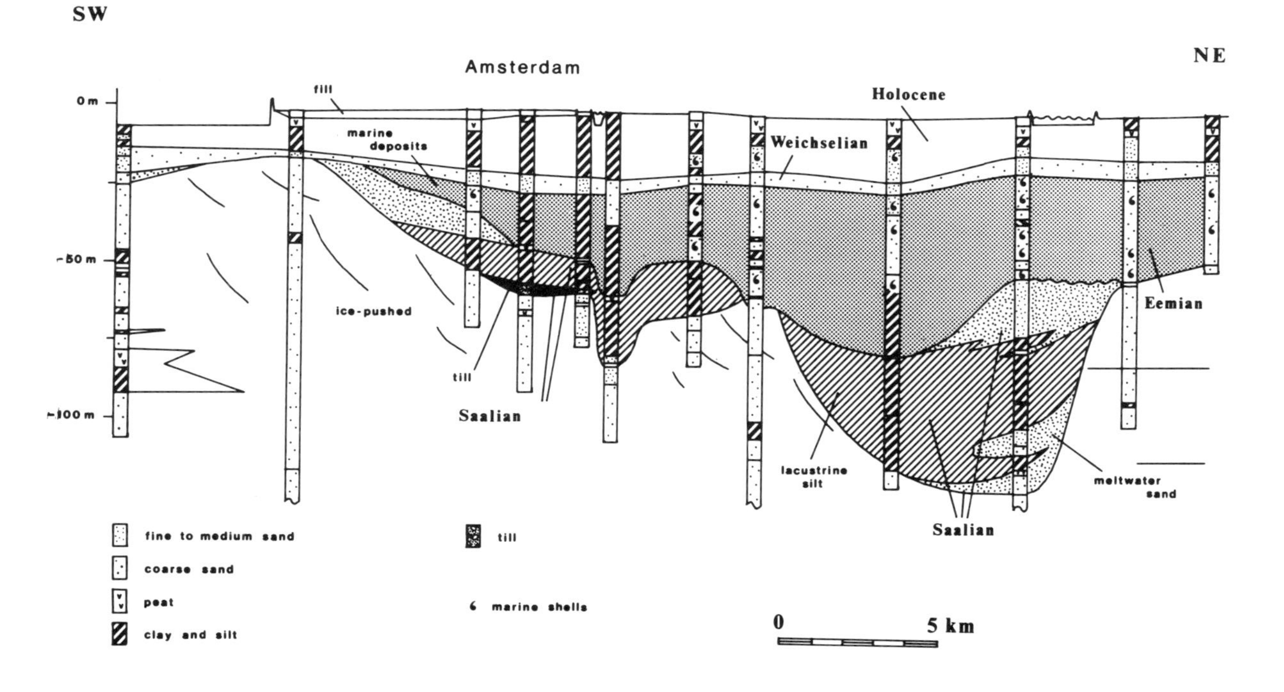

FIGURE 38 Longitudinal section through the Saalian Amsterdam glacigenic basin, ×100 vertical exaggeration (from De Gans *et al*. 1986)

by older till strata. This is the case even more so in North America, where some Wisconsinan end moraines consist entirely of till. In the large Danish push moraines of Ristinge on Langeland (Sjørring *et al.* 1982) or Lønstrup Klint in northern Jutland (A. Jessen 1931), planes of décollement developed within Quaternary marine clays.

Apparently not all end moraines formed during oscillations in the course of lateglacial ice retreat. Older thrust moraines overridden by a later ice advance may be preserved as landforms. For instance the Dammer Berge end moraine in Niedersachsen is assumed to have formed during the advance phase of the Older Saalian Glaciation (Van der Wateren 1987). Were it formed during the retreat phase, the till sheet of the Older Saalian should have been included in the thrusting. However, this is not the case. Instead, the thrust deposits are overlain by patches of till (K.-D. Meyer 1987). Overridden end moraines are also known from the Great Lakes region of North America (Dreimanis, personal communication). They are also referred to as 'palimpsest moraines' (Mickelson *et al.* 1983). However, in the case of the Dammer Berge the landforms appear very fresh. Neither the melting glaciers of Iceland nor of the Alps so far seem to have released any comparable major formerly overridden end moraines during their retreat.

With regard to their position relative to the ice margin, **interlobate moraines** are a special case. In the suture areas between ice lobes, such as for instance in the Great Lakes region of North America, in Latvia or Finland, till as well as meltwater sediments are concentrated to some extent (Punkari 1984, J. Lundqvist 1987). Often they comprise kame-like ice-contact features. Large examples include the Harricana, Bedford-Belair and Burntwood-Knife interlobate moraines in Canada (Dredge & Cowan 1989). Where deposition occurred into standing water, as in parts of Sweden or Finland, ice-contact deltas are characteristic features (Lundqvist 1989). Rapid advances by individual ice lobes led in Latvia to the formation of a special type of interlobate moraine. These so-called island-shaped moraines comprise highlands composed of ice-thrust sequences that were completely surrounded by ice lobes (Āboltiņš & Dreimanis 1995).

Where the ice sheet terminates in a lake, moraines do not necessarily form at the ice margin. If the outer margin of the ice floats on the water, glacial debris will accumulate at the grounding line. The resulting landforms are referred to as **equilibrium end moraines** in North America, from where several cases have been described. A well-investigated example is the Sakami Moraine in Québec (Hillaire-Marcel *et al.* 1981). Equilibrium end moraines also occur in the eastern Arctic, and many moraines around glacial Lake Agassiz may be of this type (Dredge, personal communication).

In the Alpine glaciation area, thrust moraines are rare. In Hantke's compilation (1978, 1980, 1983) the term occurs only twice, and in both cases without reference to concrete examples. Haeberli (1979) has described a minor Holocene push moraine from the marginal area of the Gruben Glacier (Saas valley, Valais Alps). Thrusts reaching several tens of metres deep, like those described from the marginal regions of the Nordic glaciations, seem to be lacking in the inner Alpine region. Instead, widespread accumulative end moraines mark the former ice-marginal positions. Parts of these consist of elongate ice-marginal alluvial cones formed by successive deposition of sediment layers. This is especially true for moraine ramparts in higher topographic positions. Thrust sediments can be partly incorporated in the landforms. However, tilting of the sediments alone is no proof of glaciotectonics. Particularly in small exposures, sag features that formed over melting dead ice may be easily mistaken for glaciotectonics. The outer form of the deposits does not in all cases reflect the internal structure and genesis of the landform (Schlüchter 1980a) (Fig. 39).

FIGURE 39 Strongly glaciotectonically disturbed proximal meltwater sediments in the Stucki gravel pit in Linden-Jassbach (Switzerland). Exposure parallel to glacier flow direction (from right to left). SCH = zone with distinct shear planes; LF = completely overridden recumbent fold; TF = 'funnel fold' (partly collapse structure over decaying dead ice or incomplete recumbent fold) (from Schlüchter 1980a)

3.4.3 Ice-decay landscapes

During each glaciation much sediment accumulates immediately on or adjacent to the ice margin. Sand and gravel accumulate in crevasses and in depressions on the glacier surface. In front of the ice margin isolated dead-ice blocks are covered by meltwater deposits. After the melting of the ice, kame terraces, kames and remnants of crevasse fillings remain. Where these features occur in major quantities, melting of the dead ice may lead to the formation of a characteristic ice disintegration landscape, which is often referred to as **hummocky moraine** (Hambrey 1994) (Figs 40 and 41). Ice-decay landscapes are met in the Nordic and in the Alpine glaciation areas alike.

The importance of ice-decay landscapes has often been underestimated. The characteristic landform assemblages have been described by Gravenor & Kupsch (1959) from Canada, by Aartolahti (1974) from Finland and by Marks (1994) from Poland. They include forms which are superimposed on older landform elements (e.g. end moraines or drumlins), and much of what Salisbury (1888) regarded as a uniform end-morainic landscape is covered by ice-decay landforms. However, it should be noted that ice decay can also include saturated sediments squeezing-up between ice blocks (Hoppe 1952). Flint (1971) has shown a good example of a reticulated pattern of 3-m-high disintegration ridges from Nashua, Montana.

In North America hummocky moraine is widespread in the Plains region. The landform assemblage includes regularly spaced knobs and kettles as well as irregularly spaced features of

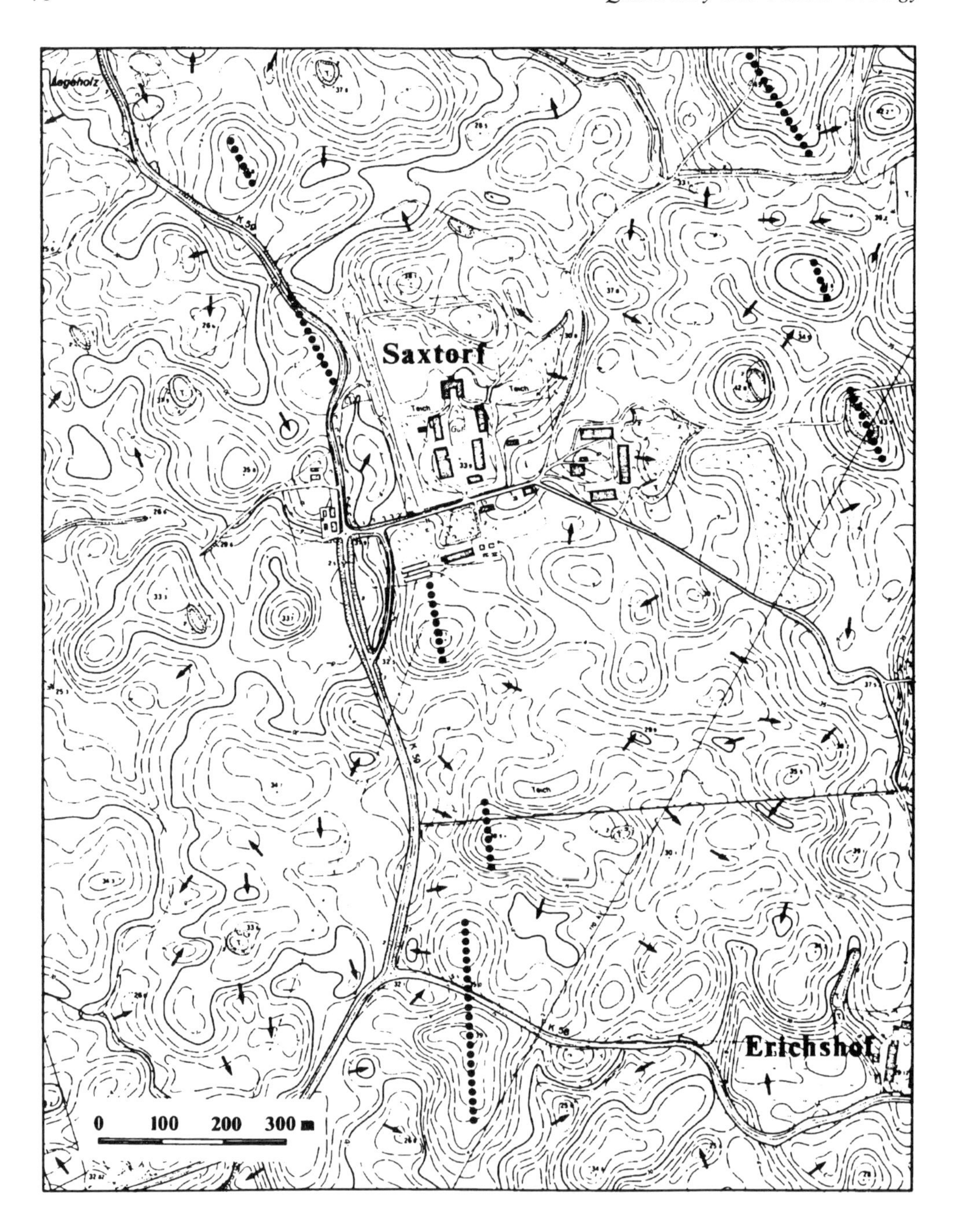

FIGURE 40 Ice-decay landscape at Saxtorf, Schwansen. The arrows mark closed depressions; the dotted lines indicate assumed ice-marginal positions. Topographic map 1:5000, Sheet Saxtorf (from Ehlers 1990a)

FIGURE 41 Ice-decay landscape at Saxtorf, Schwansen (from Ehlers 1990a). Bildflug: TK1/87, Str. 22, Bild Nr. 61. Vervielfältigt mit Genehmigung des Landesvermessungsamtes Schleswig-Holstein vom 16.12.92

various shapes including sinuous and nearly circular ridges. The dominant material of the features is till, but other ice-contact deposits also occur. Their morphological characteristics have been described by Gravenor (1955), Gravenor & Kupsch (1959), Bik (1969) and Parizek (1969).

Whilst the characteristic landforms of ice-decay landscapes have been described repeatedly (e.g. Sugden & John 1976), the processes which lead to the formation of those forms have gained little attention. Gripp (1929) reported the formation of a hummocky till plain by decay of loam walls 10–15 m high consisting of the till fill of former crevasses from the ice-marginal area of the Nathorst Glacier on Spitsbergen. Todtmann (1960) noted similar features from the Brúarjökull on Iceland. There, however, the crevasse fillings were only 1–2 m high. Such features seem to be relatively rare. Woldstedt (1939), in the course of his glacier investigations on Iceland, came to the conclusion that the hummocky till plain was mainly formed by thawing of buried dead ice. In this context Gripp (1974) speaks of a **meltdown landscape**.

Ice-disintegration features are found in vast areas in the prairies of Canada and the United States. Because of its position at the southern edge of the Baltic Sea, the North German glaciation area also provided favourable conditions for the separation of major dead-ice fields from the active ice sheet. However, the presence of dead-ice landforms should also not be underestimated in the Alpine region (Gareis 1978). In many cases, however, their occurrence is limited to smaller areas, such as the Bachhauser Filz and Buchsee areas, east of the Starnberger See (Schumacher 1981, Jerz 1987). The landforms resemble those found in North Germany. Again, it is not always easy to differentiate between ice-pushed strata and sediments distorted by dead-ice decay in exposures (Van Husen 1985a). Extensive dead-ice landscapes are found in front of the Outer Würm end moraine of the northern Rhine Glacier (e.g. at Rohrsee north of Ravensburg, Schreiner 1992).

3.4.4 Kettle holes

Chamberlin (1891) and Salisbury (1892) were the first to describe extensive pitted outwash plains from North America. Closed depressions (kettle holes) are often regarded as a characteristic of the young morainic landscape. It has been assumed that the irregular topography left behind after thawing of the buried dead ice can only survive one warm stage. The enhanced erosion and redeposition of a subsequent cold stage would either fill the old kettles or connect them to form continuous valleys (Marcinek *et al.* 1970). On the basis of these assumptions, the separation of older morainic landscape from younger morainic landscape has partly been mapped on the basis of the presence or absence of closed depressions (Werth 1912). Today it is a well-established fact that kettle holes in many cases also occur in old morainic landscapes (e.g. Garleff 1968). A closed depression may originate by a number of different processes:

1. Most kettle holes are best explained by **sagging over melting dead ice**. This applies to most of the kettle holes which either formed after the melting of overridden and till-covered dead-ice blocks or where ice got covered by meltwater sands and gravels (Woldstedt 1961). The buried ice need not necessarily have been glacier ice. Also frozen meltwater (Galon 1965) or icings (Kozarski 1975, Coxon 1978) can be sediment-covered to the same effect, as has been reported from Spitsbergen (Olszewski 1982). Buried dead ice can persist beneath the surface for rather long periods. French & Harry (1988) have thus demonstrated by fabric measurements that the ice core of the Sandhills moraine on Banks Island, Canada, was in fact a remnant of Late Wisconsinan ice. A thin cover of ablation till protected the underlying ice from further melting until today.
2. K.-D. Meyer (1973) suggested that some of the closed depressions may have formed by

meltwater scouring in former **moulins**. That such scour processes can create even deep hollows in bedrock can be shown from examples in Norway.

3. **Pingos** have also been discussed as possible causes for kettle-hole formation. Pingos are common features in recent periglacial areas around the Arctic (see chapter 5.1.6). When the ice core melts, a circular lake with a rim ridge remains. Maarleveld & Van den Toorn (1955) described such features from the Netherlands and Sparks *et al.* (1972) from East Anglia, Britain. However, most of the kettle holes without rim ridges are unlikely to have a pingo genesis.
4. Permafrost decay also leads to the formation of numerous mainly water-filled depressions (alasses). The process of permafrost decay is termed **thermokarst** or cryokarst.

A number of closed depressions in Europe contain deposits of last interglacial (Eemian) age; there can be no doubt at least that these survived the periglacial reshaping of the landscape as kettle holes. In North Germany such examples are exclusively found on terrain sloping less than 2°, so that solifluction was minimal (Ehlers 1978). Dead ice seems to have been the cause for the origin of most kettle holes. The time of dead-ice thaw largely depends on the thickness of the sediment cover. In some cases in North Germany and Poland dead ice has survived well into the Allerød and beyond.

4

Meltwater Deposits and Landforms

4.1 DISCHARGE OF MODERN GLACIAL STREAMS

The hydrology of recent glaciers has been described in detail by Röthlisberger & Lang (1987) and Lawson (1993). The following overview is based mainly on these sources. The discharge behaviour of glacial streams is mostly determined by the thermal conduction and energy balance of the surface terrain. It differs fundamentally from the discharge behaviour of other streams. Precipitation falling as snow generally has a negative effect on meltwater discharge, because radiation is reduced during a precipitation event, and afterwards the high albedo of the fresh snow lowers the temperature on the glacier surface.

The discharge of glacial meltwater is subject to daily and annual cycles, both of which are controlled primarily by solar radiation and the resulting air-temperature fluctuations. Maximum outflow lags slightly behind the daily radiation maximum. Consequently, the daily fluctuations are most pronounced during the summer months. Immediately after the end of the summer ablation period, an exponential decline of meltwater outflow is observed. In spring, discharge beginning once again is retarded because of the glacier's snow cover. Moreover, the englacial and subglacial drainage systems show a considerable retention capacity. Ninety per cent of the discharge may occur within just a few weeks!

In the Alps, above the altitude of 3500–4000 m, nearly 100% of the precipitation falls as snow and is initially stored. The hydrological mass balance of the Aletsch Glacier (Fig. 42) shows that meltwater discharge is largely restricted to the period between May and October. Monitoring of the water balance is usually based on the period of a hydrological year – i.e. from the beginning of October to the end of September in the following year. Beginning at the end of May, melting dominates over precipitation which, in unfavourable years (e.g. 1975/76), results in a negative mass balance. On the other hand, there are years with abundant precipitation and a positive mass balance. In the long term, years with a positive mass balance will result in a glacial advance.

Apart from the regular discharge fluctuations, aperiodic fluctuations and extreme flood events occur. Three processes result in maximum discharge – extreme melt rates, heavy rain and sudden outbursts of accumulated meltwater stored englacially or in glacial lakes. In the Alps and in the Scandinavian mountains, extreme melt rates mainly result from summer high-pressure weather conditions. During such events, the lowermost part of the glacier tongue melts fastest, because its low albedo promotes heating-up. Heavy precipitation falling as rain, even at

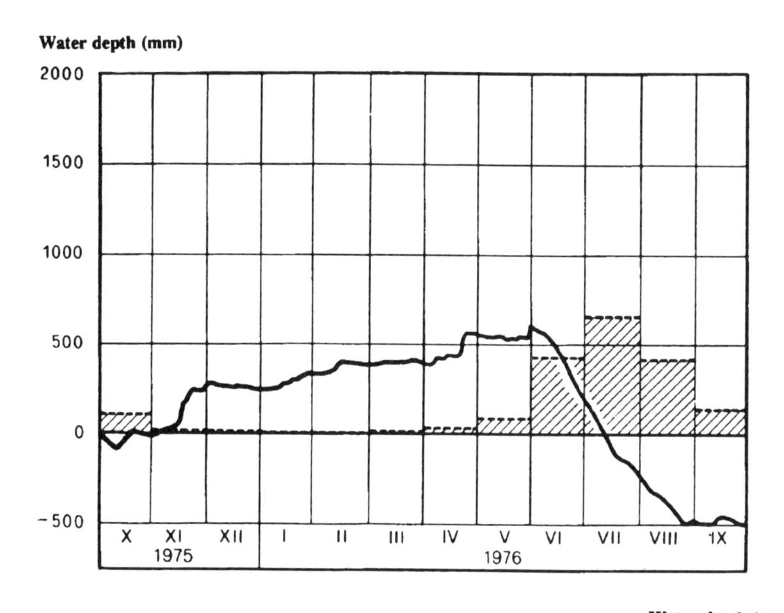

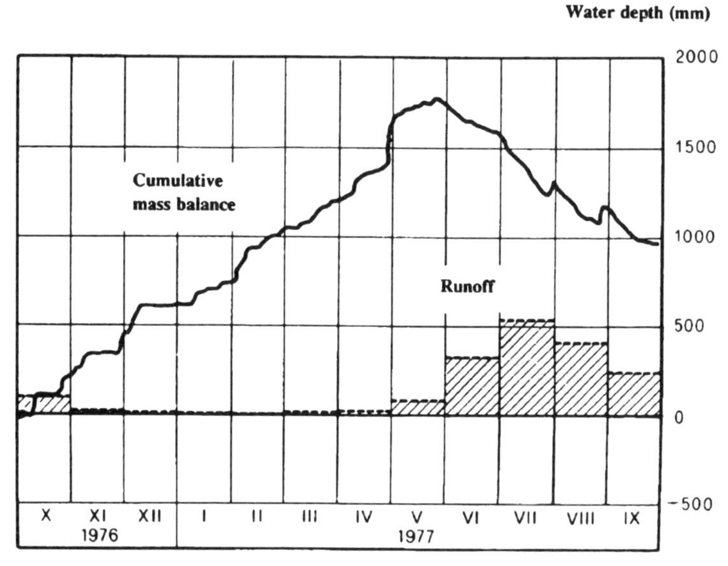

FIGURE 42 Annual accumulation (storage) and ablation (drainage) in the catchment area of the meltwater stream of the Aletsch Glacier, depicted as cumulative hydrological mass balance $P - R - E$ in mm water column. P = precipitation, R = runoff, E = evaporation. The vertical columns give the runoff measured at the Massa/Blatten gauge in the same scale. The hydrological year 1975/76 was characterised by low precipitation, resulting in low accumulation. Accordingly, the mass balance turned already negative in July. The hydrological year 1976/77, in contrast, was characterised by long and partly strong precipitation from October to the end of May. The ablation period was also interrupted by two accumulation events. The result was one of the highest positive mass balances since 1922 (from Röthlisberger & Lang 1987)

high altitudes, also leads to increased discharge, particularly when it occurs in combination with high melting rates (e.g. during summer thunderstorms in the late afternoon or in the evening). Sudden outbursts of accumulated meltwater may be triggered by changes within the unstable subglacial meltwater drainage pattern.

The major part of the meltwater is produced on the glacier's surface. Surface melt rates of the European Alpine glaciers range between approximately 0.1 and 10 m water head per year, depending on elevation. The basal melt rate is only about 0.01 m per year, depending on friction and geothermal heat flux. Very little water is produced by glacier flow itself. At a starting temperature of 0°C, an ice block would have to descend 34 km in order to melt, due to transformation of potential energy into heat. In the case of valley glaciers, the lateral supply of surface and ground water from the valley sides adds to the total drainage.

The drainage type depends to a high degree on the shape of the glacier surface. Snow or firn act similarly to an unsaturated pore-water aquifer. The meltwater seeps into this layer until it reaches a layer of lower permeability – usually the glacial ice. The water reacts like ordinary ground water – a groundwater table is formed, and the discharge takes place under gravity conditions according to Darcy's Law. Darcy's Law states that the flow in a permeable medium follows the equation:

$$Q = k \times i \times A$$

where A = cross-sectional area of discharge, i = hydraulic gradient and k = coefficient of permeability. In contrast to an ordinary pore-water aquifer, the permeability of snow is not constant but changes with time during the process of transformation into firn and ice. The aquifer drains in three directions – discharge on the firn line in channels on the glacier surface, discharge into glacial crevasses or seepage through the ice (Fig. 43).

In the ablation zone of the glacier, meltwater drainage can be observed on the glacier surface. Yet in most cases this water vanishes into the glacier after a short distance. Major

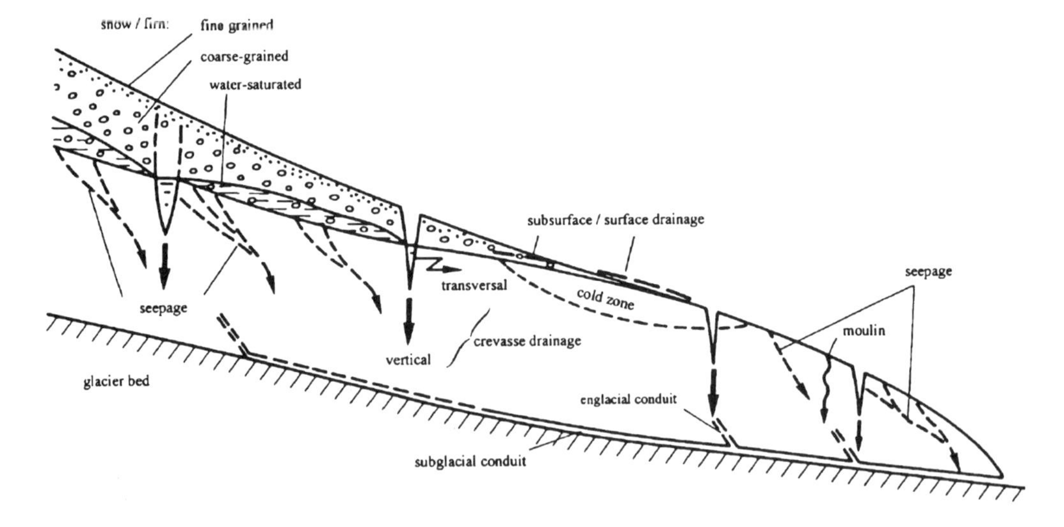

FIGURE 43 Schematic diagram of glacial meltwater runoff through supraglacial, englacial and subglacial drainage (from Röthlisberger & Lang 1987)

supraglacial streams flowing to the edge or terminus of the glacier are exceptional. Most of the water disappears into moulins or crevasses, and infiltration through the ice plays an important role. In the transitional region, between the accumulation and ablation zone, a shallow permanently frozen layer occurs. Thickness and extension of this layer are determined by climatic factors. During winter and early spring this layer may expand over the entire ablation area, if the latter is not covered with wet snow. The latter is often the case on glaciers in maritime humid-temperate zones (Röthlisberger & Lang 1987).

Seepage in temperate glaciers occurs through the interstices between the ice crystals. It is disputed, however, to what extent this permeability really has an effect. Nye & Frank (1973) believe that infiltration is possible through a series of such interstices between three ice crystals. Lliboutry (1971) has questioned this view. He reckons that the frictional heat generated in this process would lead to melting of the glacier. Nye (1976), however, assumes that the amount of water necessary to trigger such a process is not available. Trapped air bubbles in the voids, as well as deformation and recrystallisation of the ice, however, limit this mode of drainage. An important part of the downward-directed drainage from the glacier surface towards the base occurs via moulins, in which the water initially pours down in free fall. Moulins usually form in the upper marginal part of a crevasse zone, and it is assumed that cracks in the ice give rise to their formation. The water does not fall directly down to the glacier base. Vertical shafts alternate repeatedly with slightly inclined or horizontal tunnel sections. Sediments carried along are deposited in basins forming at the bottom of the first fall segment within the glacier. In the ablation zone, these sediment accumulations finally emerge at the glacier surface where they appear as isolated dirt cones (Röthlisberger & Lang 1987).

Discharge at the glacier base is carried out by channels underneath the ice. These form cavities either in the ice ('R channels' or 'Röthlisberger channels') or in the rock ('N channels' or 'Nye channels'), or a combination of both (Röthlisberger 1972, Nye 1973). Cavities in the bed may be linked by conduits in the ice (Walder 1986). In the absence of tunnels or conduits, water may also drain as a thin film between ice and glacier bed (Weertman 1972). Different drainage systems may coexist in different parts of a glacier. Lliboutry (1983) and Hooke (1984) both state that in many cases discharge may occur under atmospheric pressure. However, especially in the case of large glaciers with substantial ice movement at the glacier base, there is a strong tendency towards the closing of cavities, leading to meltwater discharge under pressure (Röthlisberger & Lang 1987). Such conditions result in increased meltwater erosion.

A special case of meltwater discharge is the drainage of glaciers in volcanic areas. Here, eruptions under the ice may result in catastrophic outbursts of meltwater (so-called **jökulhlaups**), well known, for instance, from the Icelandic glaciers (Björnsson 1988). The largest historically documented jökulhlaup (caused by an eruption of the Katla volcano on Iceland) is calculated to have had an outflow of 100,000 to 300,000 m^3 per second. Intense meltwater discharge may also result from the sudden drainage of ice-dammed lakes – in this context outflows of up to 3000 m^3/s have been measured on Iceland (Björnsson 1992).

In the proglacial zone of recent glaciers only a few traces of deep meltwater erosion are found. For example, Sharp *et al.* (1989) describe a subglacially generated groove pattern in front of the Glacier de Transfleuron (Switzerland). The configuration of this pattern resembles the subglacial Pleistocene channels of northern Germany, except the dimensions are 10,000 times smaller. The depth of the grooves is approximately 0.1 m , the width approximately 0.2 m and the lateral distance between the grooves about 1.5 m. The grooves run subparallel and point towards the former ice margin. These features formed in Cretaceous limestone.

In front of the ice margin, sand and gravel are deposited as the velocity of the meltwater decreases. The processes governing the formation of glacial outwash plains formed in this way have been analysed by Krigström (1962) on Iceland, and Church (1972) on Baffin Island (Canada).

Proglacial meltwater activity is not restricted to the accretion of vast sandur plains and distal lacustrine sediments. Substantial meltwater erosion may also occur. Church (1972) found that in the case of the Baffin Island valley sandurs, the actual sediment supply from the glacier was low. The present meltwater streams of the sandurs are mainly reworking proglacial outwash sediments dating from the last glaciation. Increased erosion occurs particularly in the proximal zone of the sandur, close to the ice margin. This situation is comparable to the final melting phase of the large continental ice sheets of northern Europe, where the supply of sediments ceased after the inactivation of the ice masses, so that the meltwater streams began to incise their respective sandurs.

A substantial portion of the sediment transport on recent sandur plains occurs during extreme highwater events, as a result of summer heavy rainfall or of 'jökulhlaups'. During drainage of a minor ice-dammed lake on Baffin Island, the discharge of the meltwater stream rose from less than 5 m^3/s to 200 m^3/s within 24 hours and then decreased to 20 m^3/s within two hours. Such discharge behaviour is regarded as typical for meltwater outbursts. During the drainage event, the ice channel gradually widens as the result of lateral melting of its walls, reaching its maximum diameter shortly before the impounded water completely drained (Church 1972).

Krigström (1962) distinguished three major morphogenetic zones on Icelandic sandur plains, each characterised by different landforms. In the proximal zone (close to the ice margin) there are only a few narrow, deep meltwater streams occupying relatively stable, slightly incised beds. Downstream in the central part of the sandur, the streams bifurcate into numerous branches, which show strong lateral erosion. This central part of the proglacial drainage system is the typical zone of the sandur plain. Streams are both wide and shallow. Sandur plains draining into lakes or the sea tend to develop a third, distal zone in which the meltwater streams join together to form a sheet-like drainage system and accumulate a delta. Here, in addition, the water is shallow in most places; although deeper channels can also occur.

The longitudinal profile of a sandur plain is usually slightly concave. Sandurs can have a rather irregular cross profile, and differences in height of several metres have been observed. When major floods occur in the accumulation area of the sandur plain, the surroundings of the main stream undergo strong deposition. Consequently, in the following stages of development there are always radical path changes. Sedimentation on recent sandur plains occurs as channel fill and deposition of sediment sheets and levees. Often, both the Baffin Island and Iceland sandur plains show no more than rudimentary bedding. Wherever individual layers can be distinguished, planar bedding prevails. The longitudinal axes of elongate pebbles are preferentially orientated at right angles to the flow direction. This, however, is only true for beds deposited by strong floods. Clasts at the base of shallow runnels are aligned much less regularly (Church 1972).

4.2 TRACES OF PLEISTOCENE MELTWATER ACTIVITY

Meltwater extends the effects of glaciation far beyond the limits of the actual ice advance. Meltwater activity has played a considerable role in overall cold-stage erosion and sediment transport. This becomes obvious when the depth and density of buried channels in areas like northern Germany are compared to the overall till thickness in the same areas. The meltwater sediments in these areas also contain far more local, i.e. freshly eroded material, than the tills. In hardrock areas where this is not so evident, traces of strong meltwater erosion have been found. For instance, in East Anglia the Chalk surface is not directly overlain by till but separated from it by a thin sheet of meltwater deposits, and the Chalk surface shows clear signs not only of plucking but also of meltwater erosion (Ehlers 1990a). Fluvial transport has been found responsible for about 90% of the sediments accumulated at active glacier margins in Alaska (Evenson & Clinch 1987).

4.2.1 Tunnel channels

Weichselian/Devensian incisions in northern Europe are characterised by series of elongated lake basins, which have been interpreted as kettle holes. In some cases river valleys cross Weichselian/Devensian incisions without deflection of the stream course: an example of this is the double channel in the Tuchola Heath in Poland, described by Galon *et al.* (1983). Figure 44 shows the extension of this channel system to the south, where the incision is crossed by the Brda River. Galon *et al.* (1983) have shown that channels formed during the Pomeranian Substage of the Weichselian run beneath the terrace deposits of the River Brda. Sedimentological studies indicate the incisions had an original depth of over 35 m (below ground surface), and that sandur sediments occur within them. This implies that the incisions were formed earlier than the sandur, but were blocked by dead ice. Since the oldest organic deposits discovered within the incisions date from the Allerød Interstadial, it seems it was during this period that the kettle holes formed, following thawing of the buried ice.

In North Germany, a large number of linear incisions have been identified in the area covered by the Weichselian ice sheet, and the overall pattern strongly suggests that the features originated as drainage channels within this ice sheet. Some of them terminate at the outermost marginal position; others, especially in the eastern part of the area, terminate short of the margin and were probably formed during a recessional phase. However, none of the incisions extend beyond the outermost ice-marginal position. Where other ice-marginal features are missing, it is possible to reconstruct former ice margins from tunnel channel distribution. Although a number of well-studied examples of Weichselian linear incisions are known from continental Europe, until recently little or no information was available for the Laurentide Ice Sheet area. However, following the pioneering work of Wright (1973), the widespread occurrence of 'tunnel channels' has been reported (e.g. Nelson & Mickelson 1977, Attig *et al.* 1989, Mooers 1989). The dimensions are only slightly smaller than those in Europe; Wright (1973, p. 254) describes the Grindstone tunnel valley as being one of the largest, with a maximum depth of at least 70 m and an average width of ca. 300 m. In Schleswig-Holstein, the maximum depth of a Weichselian incision would be at least 80 m and the average width ca. 500 m. The latest work on the New York Finger Lakes (Mullins & Hinchey 1989) shows that they are also underlain by major incisions, forming enclosed depressions similar to the largest features in the North Sea, as is the case on the Scotian Shelf (Boyd *et al.* 1988).

4.2.2 Buried channels

Analysing meltwater activity one must distinguish between drainage beneath the ice (subglacial area) and drainage in front of the ice margin (proglacial area). Whereas accretion predominates

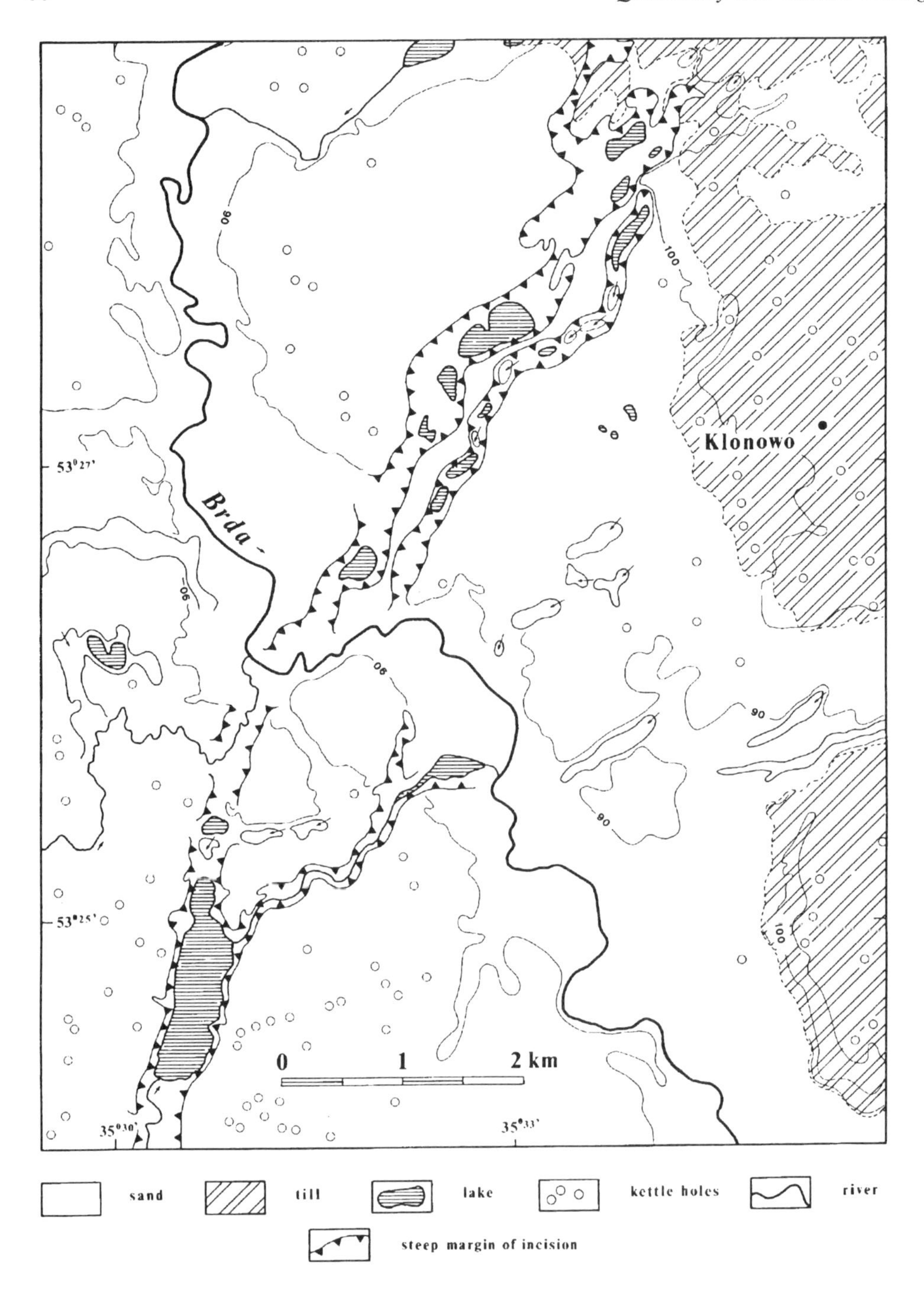

FIGURE 44 Double channel Weichselian incision in Pomerania, Poland, crossing the Brda sandur plain and crossed by the Brda River valley (from Ehlers & Wingfield 1991, after Galon *et al.* 1983)

in the proglacial area, subglacial drainage is dominated by erosion. Consequently, the southern peripheral zone of the North European Ice Sheet is interpenetrated by network-like systems of channels cut into the surface down to several hundred metres, running mostly in a radial pattern from the centre of glaciation to the former ice margin. The older channels are mostly filled with sediments to the level of the surrounding area. In the majority of the cases their course can only be detected by drilling. Only the most recent channels formed during the last ice age are still clearly visible at the land surface (Fig. 44). In the literature these forms are often referred to as tunnel valleys, because they were formed by meltwater flowing through tunnels underneath the ice. **Tunnel channels** might be a more appropriate term, as in contrast to normal valleys their thalweg has no continuous unidirectional gradient.

Modern geological investigation of the channels in Europe was initiated in East Germany. The first detailed investigations of Ice Age channel patterns were made by Eissmann (1967, 1975). In some cases the course of the channels could be traced for many kilometres in the opencast lignite mines in the Leipzig area. A summary map of the Elsterian age channels in northwest Germany was provided by Ehlers *et al.* (1984) (Fig. 45), and a similar map was compiled for East Anglia by Woodland (1970).

Until the early 1970s, only minor segments of the buried channels were known in sufficient detail. Since then systematic mapping, mainly for groundwater exploration in the marginal regions of the Nordic glaciation, has resulted in publication of numerous detailed maps (Netherlands: Ter Wee 1983a; Lower Saxony: Kuster & Meyer 1979; Hamburg: Ehlers & Linke 1989; Schleswig-Holstein: Hinsch 1979; Mecklenburg: Von Bülow 1967; Saxony: Eissmann 1975, Kupetz *et al.* 1989; Denmark: Houmark-Nielsen & Sjørring 1991). Also further east, in Poland, Belarus, Latvia, Lithuania and Estonia, corresponding incisions are known, though lower drilling density often does not allow detailed reconstruction of the channel patterns. Regionally the incisions have been interpreted differently (Mojski 1982; Tavast & Raukas 1982). The channels in Poland (Mojski 1982, Ber 1990) so closely resemble the forms existing in northern Germany that it is reasonable to assume that they formed under the same conditions.

The depths of the channels vary greatly. In northern Germany, a maximum depth of 434 m b.s.l. was found for the Elsterian Reesselner Rinne. Equally, in the southern North Sea, maximum depths of channels of more than 400 m b.s.l. have been measured. The distinctly shallower depth of the channels in Estonia, for example (maximum approximately 60 m b.s.l.; Tavast & Raukas 1982), may be the product of the ground consisting of hard rock (Palaeozoic limestone, dolomite and sandstone), whereas in northern Germany the Quaternary deposits are underlain by easily erodible soft rocks. Channel courses can be locally influenced by deep-seated tectonics. This is most obvious in the case of the so-called Alnarp valley, a trough following the margin of the Fennoscandian Shield through Skåne (southern Sweden) and northern Denmark (Houmark-Nielsen & Sjørring 1991). In addition, channel courses can be influenced by the strike directions of salt domes, as demonstrated by Piotrowski (1994a) in Schleswig-Holstein.

The intensity of channel formation varies through space and time during all ice ages.

- In the Weichselian glaciation in northern Germany only small channels were formed (to approximately 100 m deep); instead, accretion of sandur plains prevailed.
- In northwest Germany, during the Saalian glaciation, there was no significant formation of channels, but wide sandur plains were deposited.
- In the Elsterian glaciation, however, incision of channels of more than 400 m deep occurred.

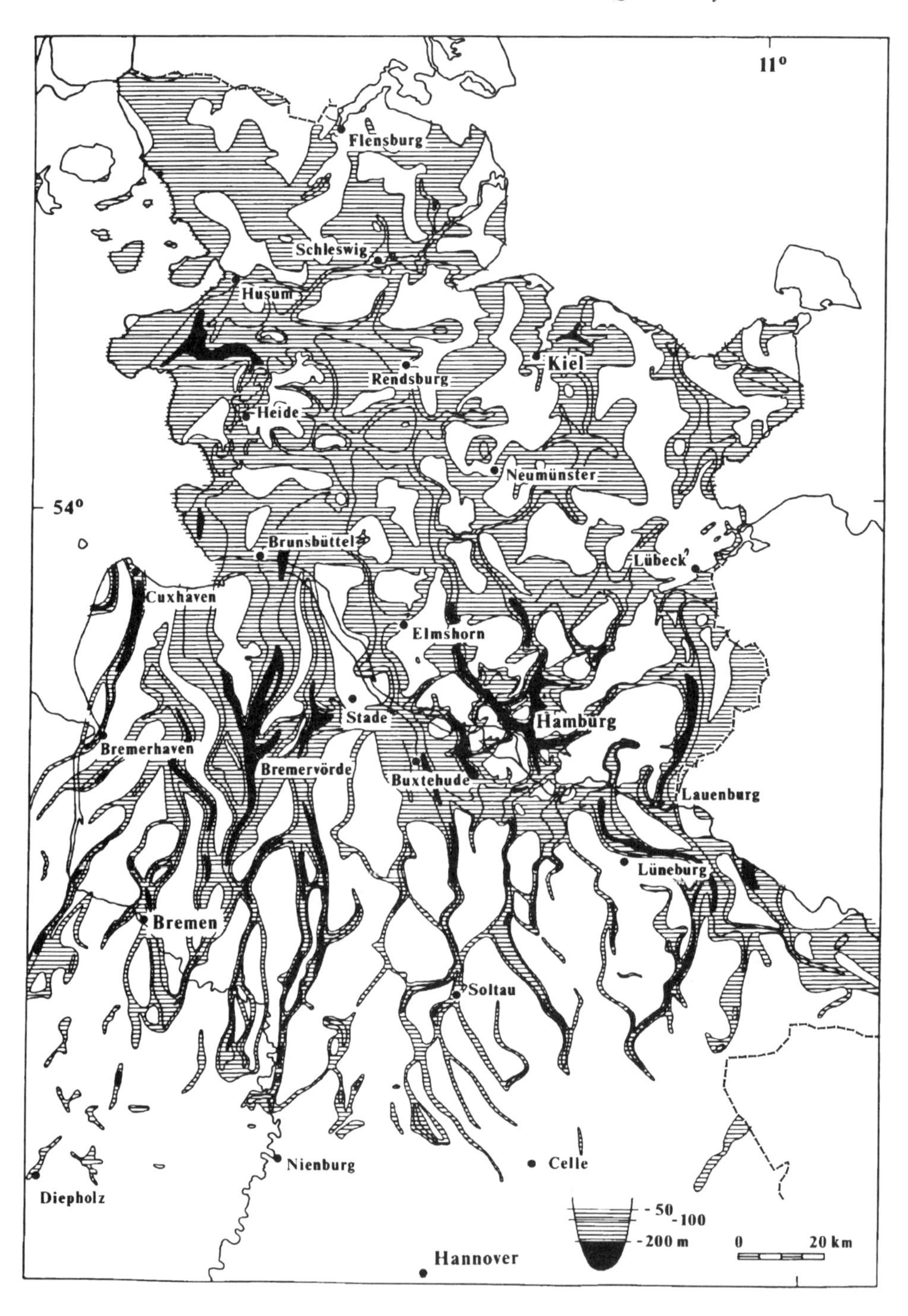

FIGURE 45 The course of Elsterian buried channels in northwest Germany (from Ehlers *et al.* 1984)

Yet these differences do not apply everywhere. Seismic investigations in the North Sea have revealed that channels formed during the last three glaciations (Cameron *et al.* 1987, Ehlers & Wingfield 1991). Only in the southern part of the North Sea is the oldest channel system most strongly pronounced. In the central North Sea, the deepest forms relate to the last, Weichselian Stage. Therefore, the findings from one region, like northern Germany, cannot automatically be transferred to other geographical regions.

Although the age and distribution of Pleistocene incisions on land and in the North Sea are largely known, their origin is still a matter of discussion. Four different hypotheses have been offered:

1. *Fluvial incision.* In the glaciated areas of North Germany, the suggestion that the incisions were of extraglacial, fluvial origin (Wolff 1907) is no longer accepted. The overdeepening, in some areas strongly pronounced, and the network-like interconnection of the channels, contradict a fluvial origin (Kuster & Meyer 1979). In Denmark, Lykke-Andersen (1986) interpreted some channels as Late Tertiary to Early Pleistocene river valleys overdeepened locally by glacial erosion. The widespread lack of till in the central and northern North Sea led Long & Stoker (1986) to propose that the incisions in these areas were formed by extensions of the European river system, modified by floods from 'jökulhlaups' as well as by later tidal influences (cf. Cameron *et al.* 1987). However, the distribution pattern does not resemble any fluvial system.
2. *Incision by subglacial meltwater.* Ussing (1903) interpreted the Weichselian channels in Denmark and North Germany as tunnel valleys eroded by subglacial meltwater, since they all end at former ice margins, leading into relatively high-level sandur. His assumption was that meltwater under high hydrostatic pressure eroded tunnels beneath the ice and lost its erosional capacity at the tunnel mouth with the decrease of velocity. Meltwater sands form the basal sediments of most of the linear incisions, whereas diamictons are rare and occur in discontinuous patches, which may be explained as slumped till or mudflows (Eissmann 1987).
3. *Incision by glacial scour.* Although Woldstedt (1913) originally followed Ussing (1903) in ascribing the tunnel channels to erosion by subglacial meltwater, he subsequently concluded that glacial scour was the principal erosional agent and that meltwater activity played only a minor role (Woldstedt 1952). An important argument in support of glaciogenic incision is the great similarity of the large Elsterian features to the overdeepened troughs of the Alpine foreland or the Norwegian fjords, which are mostly interpreted as formed by glacial erosion. Some of the alleged 'tunnel valleys', such as the 2-km-wide Kieler Förde or the 3-km-wide Eckernförder Bucht, are extremely difficult to explain by subglacial meltwater action. Most probably they are tongue basins rather than true channels. The proportion that different processes participate in channel formation is a matter of interpretation. Adjacent to some incisions, thrust features occasionally have been observed, which might have resulted from ice pressure (Bruns 1989, Piotrowski 1993).
4. *Liquefaction of sediments.* Boulton & Hindmarsh (1987) proposed that the water pressure below the sole of a growing ice sheet increases to exceed the overburden pressure of the ice. This leads to liquefaction of the unconsolidated sediments, which drain through major tunnels towards the ice margin. One would expect this process to cause the formation of dipping strata adjacent to the incisions. So far no traces of this have been found in densely drilled areas, like Hamburg, or on the seismic profiles of the North Sea. If the process

envisaged by Boulton & Hindmarsh (1987) has contributed to incision formation, it was probably at a much smaller scale than assumed by the authors. Another problem is that the incisions are also found cut into hard rock ranging in age from Precambrian to Mesozoic (Wingfield, 1990), which are not susceptible to liquefaction.

By using borehole information alone, the precise shape of channels can only be reconstructed with difficulty. Although, in northern Germany, a regular U-shaped profile has generally been suggested for the reconstruction of Elsterian channels, evaluation of seismic profiles from the North Sea shows this form is rather the exception. Only about one third of the deep incisions have a simple U-profile, whereas the rest show much more complicated cross profiles (Ehlers 1990a). In a few cases alone, can this be attributed to a meandering channel being repeatedly cut by the same profile. Detailed evaluations of seismic profiles of segments just 1 km apart have shown channels often consisting of a system of parallel incisions. Corresponding features have been observed for many Weichselian channels on the continent. This indicates that the channels were not generated in a single process, but through a sequence of several similar events. Occasional reactivations of old, abandoned channels during later glaciations are known (Piotrowski 1994a). The shape of the channels, with their local overdeepenings and irregularities of profile, definitely excludes extra-glacial formation, such as river valleys. In most examples subglacial meltwater erosion was the main force behind the channel formation.

The channels thus far have largely been described from Europe. However, there are also traces of major subglacial drainage channels in North America (see chapter 13.8).

4.2.3 Eskers

Subglacial meltwater activity not only led to the erosion of deep channels, but also to the accretion of elongate ridges, known as eskers (Swedish: åsar) (Fig. 46). Sefström (1836) originally considered the eskers as proof of a large boulder flood (cf. Rainio & Kukkonen 1985), whereas the glacialists later suggested that they were morainic ridges. Eventually De Geer (1897) was able to demonstrate that they were deposited by meltwater streams.

Today, it is known that eskers were in most cases accreted by subglacial meltwater streams, although occasionally englacial or supraglacial eskers occur (R.J. Price 1966). Eskers often consist of coarse-grained sand and gravel of varying grain sizes, together with boulders. The debris is often well rounded. Eskers form continuous ridges often of great length (in some places up to tens of kilometres), sometimes resembling railway embankments, yet often changing in width and height (Fig. 47). Branches may also occur (esker networks). Generally esker paths follow the thalweg of valleys or channel systems, but they also run across ridges. The subglacial drainage direction depends not on bed topography but on the inclination of the ice surface (Flint 1971). Eskers also form in active ice, though they can only be preserved where the surrounding ice has largely ceased to move. Therefore their course always reflects the orientation of the last ice movement. Thus esker orientation generally corresponds to the strike direction of the most recent glacial striae.

A special type of esker is the so-called engorged esker, first described by Mannerfelt (1945). These emerge during an advanced stage of ice disintegration in tributary channels draining lateral meltwater downhill to the central drainage channel. Quite often they run slightly diagonally, since the channels tend to shift downglacier due to the ice movement.

Whereas the large eskers in Scandinavia can be traced over tens of kilometres, only minor

FIGURE 46 Freshly formed esker at the terminus of Omnsbreen, Norway (Photograph: Ehlers 1980)

forms are found south of the Baltic Sea. In the area of the Laurentide Ice Sheet of North America, eskers are also largely restricted to the inner parts of the glaciation. Although some smaller eskers occur elsewhere, some of them in tunnel channels, a map of esker distribution largely corresponds to a map of crystalline bedrock distribution. Clark & Walder (1994) conclude that subglacial drainage was dependent on bedrock. In areas of a deforming bed, substrate channels would form instead of an arborescent drainage network of tunnels, leading to the formation of eskers (Walder & Fowler 1994).

Only a few eskers are known from the Alpine region. For example, Troll (1924) describes some examples from the marginal zone of the Inn–Chiemsee Glacier. The origins of other supposed eskers have been contested. Hantke (1978) shows geological cross sections through eskers in the Glatt valley (north of Zürich), but also calls attention to a series of misinterpreted features from other parts of Switzerland. Van Husen (1968, 1977) describes some small eskers from the Enns valley and Traun valley region of Austria.

4.2.4 Sandur plains, delta plains and gravel terraces

The widespread proglacial accretion of meltwater sands and gravels at the margins of recent glaciers attracted the interest of geologists quite early. After an excursion through Iceland, Torell (1858) came to the conclusion that the heathlands of northern Germany were formed during 'the great Ice Age' similar to the recent sandur plains at the margin of Vatnajökull (Fig. 48). This notion was not accepted in Germany until Torell held his lecture in Berlin on

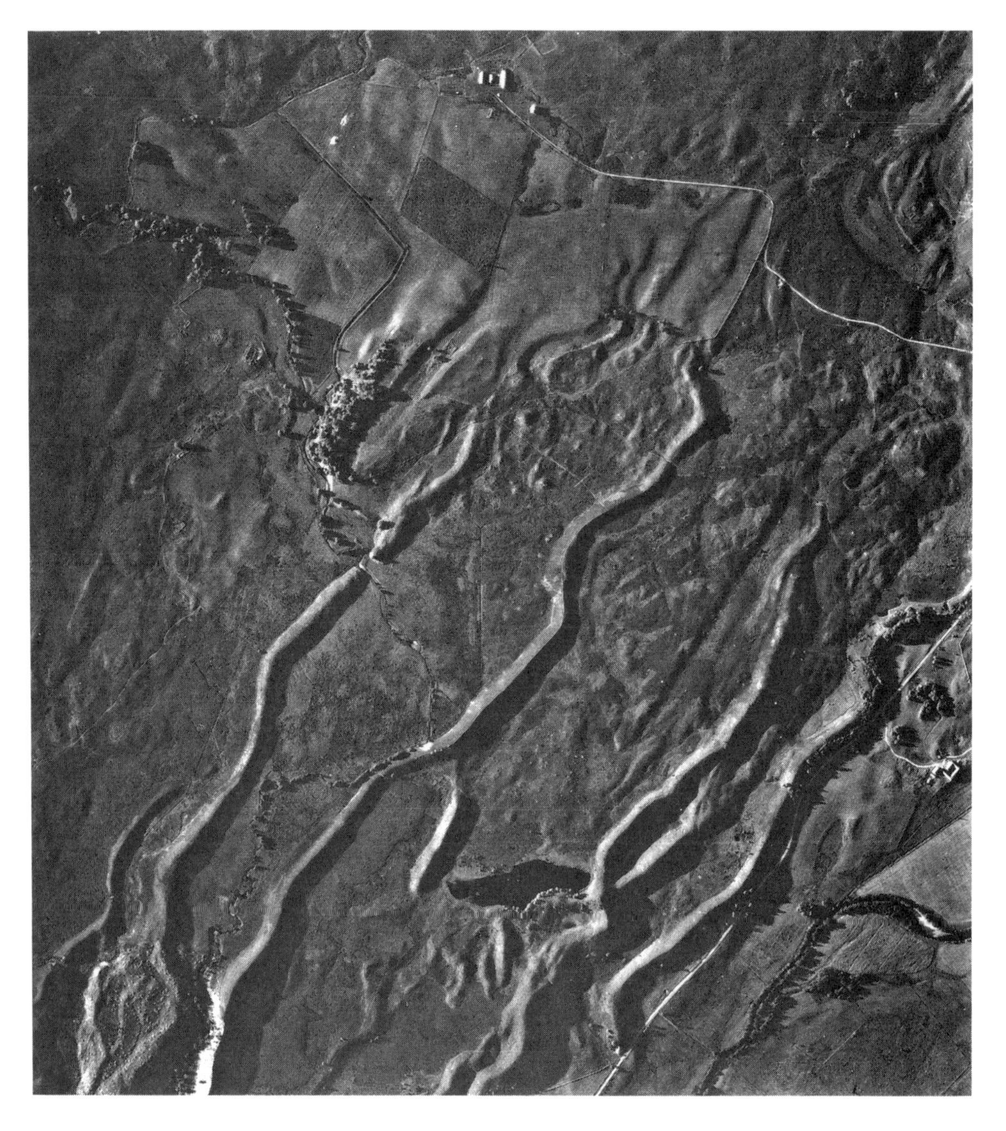

FIGURE 47 Vertical aerial photograph of eskers at Dell Farm, about 10 km SSE of Inverness (Scotland); width of the picture: 1.5 km (from Gray 1991; British Crown Copyright 1993/MOD reproduced with permission by Controller of Her Majesty's Stationery Office)

3 November 1875, on 'the polished surfaces and striae on the Muschelkalk surface at Rüdersdorf'. The term 'sandur' is used in the international Quaternary literature as an equivalent for outwash plain. The plural in Icelandic would be 'sandar', but often they are referred to as 'sandurs' instead.

Unlike the recent sandur plains of Iceland and Canada (Fig. 49), cross-bedding prevails in the Pleistocene sandur deposits in northern Europe (and North America) (Fig. 50). The difference may result from the different grain-size composition.

FIGURE 48 View across Skeiðara Sandur towards Vatnajökull (in the background, right), Iceland (Photograph: Ehlers 1990)

Meltwater deposits are widespread in northern Germany. With regard to their stratigraphical position in relation to the respective till sheet they can be subdivided into advance and retreat sandurs. Whereas advance sandurs mostly form thick, widespread sedimentary complexes deposited at the ice margin, retreat sandurs are mostly thin and largely restricted to single channels or valley fills. The sandurs of the Nordic glaciations correspond to the gravel trains of the Alpine glaciation area (cf. chapter 12).

In North America, meltwater deposits are mostly referred to as outwash or stratified drift. They occur as outwash fans or valley trains. The Mississippi River transported meltwater sediments from the Laurentide Ice Sheet down to the Gulf of Mexico. Whilst the proximal end of the outwash deposits is relatively coarse-grained, near the Mississippi delta the meltwater deposits comprise mainly fine sand and silt. However, cobbles and boulders are found as far as 300 km downstream from the nearest ice margin. They are interpreted as ice-rafted (Flint 1971).

Like the meltwater deposits of the Nordic ice sheets, the gravels of the Alpine glaciers largely predate the maximum ice advance. But in contrast to the Nordic ice sheets, gravel supply in the Alpine regions was provided by frost weathering in a period of climatic deterioration. This is reflected by the fact that tributaries in the lower reaches of the Alps started to contribute to the gravel aggradation at a relatively late phase, when frost weathering had also reached the foothills. These climatically controlled sediment bodies find their continuation in the proglacial river terraces (Van Husen 1981).

FIGURE 49 Meltwater sediments in the forefield of Kverkjökull, Iceland; height of the face ca. 6 m (Photograph: Ehlers 1990)

The grain size of meltwater deposits depends on current velocity, transport distance and composition of the source material. Gravels of the Alpine glaciation have been carried over much shorter distances than the meltwater sands and gravels of northern Germany. Therefore they are generally much coarser. Weiss (1958) and Hölting (1958) demonstrated that the grain size of sandur sediments decreases with growing distance from the ice margin and that sorting increases simultaneously. Meltwater deposits often show more than one grain-size maximum. This may result from the interaction of different modes of transportation (traction and saltation).

Glaciofluvial **delta plains** are common features not only in former ice-dammed lakes (see chapter 4.2.6), but also along the highest shorelines of the different stages of the Baltic, for instance along the Swedish east coast. The deltas were formed during deglaciation and generally have a facies architecture typical of the classic 'Gilbert-type delta' with bottomset,

FIGURE 50 Cross-bedded meltwater sands in a sandur at Daerstorf (Niedersachsen); height of the face 9 m (Photograph: Ehlers 1990)

foreset and topset beds (J. Lundqvist 1979, Ringberg 1991), i.e. a delta of the type G.K. Gilbert (1890) described from Lake Bonneville in Utah. The topset/foreset contact is taken to be the former sea level. The Salpausselkä ridges in Finland are large glaciofluvial accumulations at the contact between the Late Weichselian ice sheet and the Baltic Sea and its predecessor. According to different morphologies, several types of deltas may be distinguished (Glückert 1995). Where these ice-marginal formations are not in contact with feeder eskers, they are also referred to as 'marginal terraces' (Rainio 1995).

Whereas normally the drainage systems of the Weichselian ice sheet can be easily reconstructed based on morphology, this is more difficult with the older glaciations. In these cases only a thorough examination of the sediments can provide information about the former drainage pattern. Brinkmann (1933) was the first to successfully reconstruct former drainage systems from the internal structure of their deposits, of which the best suited for this purpose are cross-bedding measurements. The individual laminae of the cross-sets usually dip in the former flow direction (lee-side deposition). Thus, it is sufficient to measure the dip of a larger number of such cross-sets in an exposure in order to reconstruct the palaeoflow direction. The meltwater streams, with their heavy sediment load and intermittent flow, were not restricted to narrow stream channels. On the sandur plains near the ice margin, sheet-like sand and gravel deposits accumulated, similar to the modern sandur plains on Iceland. In braided river systems like these, a wide range of discharge directions can be measured. The flow directions within one stratigraphic sequence characteristically vary over a range of more than 180°. Nevertheless, the mean drainage direction can be reconstructed by averaging an adequate number of measurements. Experience shows that about 30–50 measurements are sufficient for this purpose.

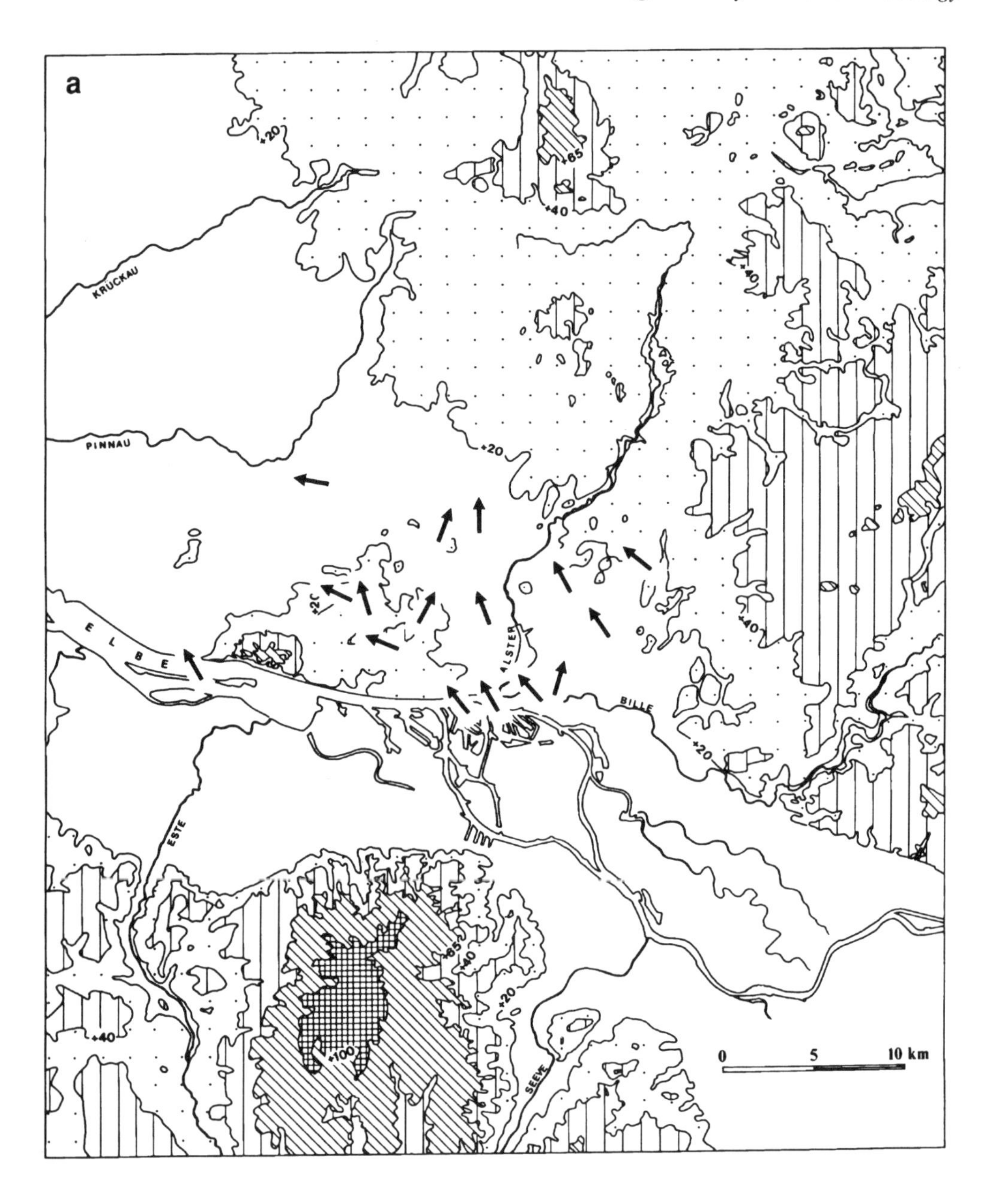

FIGURE 51 Reconstruction of the Saalian drainage systems in Hamburg based on palaeocurrent measurements. (a) Middle Saalian Glaciation, older phase; (b) Middle Saalian Glaciation, younger phase (from Ehlers 1990a)

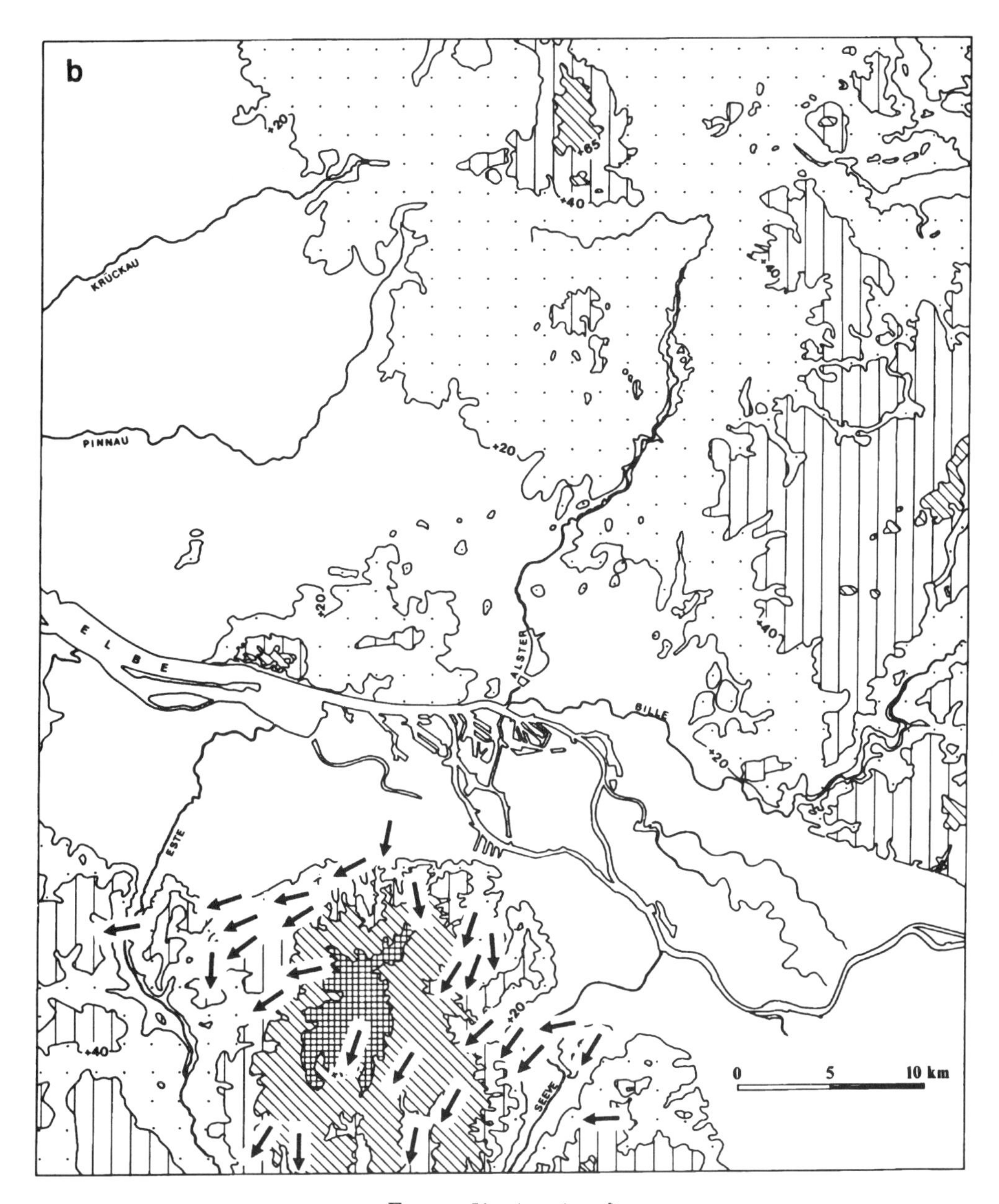

FIGURE 51 *(continued)*

Illies (1952) used this method for reconstructing the Pleistocene history of the lower Elbe River. The results of his measurements show that the Saalian meltwater streams were not directed towards the Elbe valley. Since his study, measurements of cross stratification have gained an important role whenever the palaeogeographic reconstruction of drainage systems is concerned.

However, the potential of this method had been overestimated initially. The measurements depend on the availability of good exposures. Normally, the information density is not sufficient to show, for example, the influence of salt structures on the development of Pleistocene drainage patterns. In addition, there is always a natural variability of drainage directions on a single sandur plain.

Where datable material is lacking, stratigraphical correlation of meltwater sediments is often quite difficult. Each additional measurement, however, adds to the better understanding of the regional drainage history. New outcrops can lead to a new assessment of former interpretations. Thus, the current state of knowledge of the development of the Saalian drainage systems in the Hamburg area has been gathered over the course of several decades (Illies 1952, Grube 1967, Ehlers 1978). When the ice advanced into the Hamburg area from a northeasterly direction, its meltwater initially drained northwestwards towards the Pinnau River via the lowland area northwest of Hamburg (Fig. 51a). In the process, sandurs were deposited north of the Elbe River in a level between 0 and 20 m. Only when this drainage outlet was blocked by the advancing ice did the meltwater find a new path to the southwest, towards the Weser–Aller ice-marginal valley. On this occasion the higher sandurs south of the Elbe were formed (Fig. 51b) (Ehlers 1990a).

In the early days of Quaternary research in northern Germany, sandur plains were regularly correlated with the nearest end moraine (e.g. Woldstedt 1925), though it is known today that this correlation was not always correct. Numerous sandur plains were formed during the advance phase of a glaciation and later overrun by the ice sheet. This, for example, applies to most of the large sandur plains in the Lüneburger Heide area. A distinction between advance and retreat sandurs is often only possible under favourable exposure conditions.

With respect to the stratigraphical correlation of meltwater deposits, the same petrographical methods may be used as those that are applied to till stratigraphy. It must be borne in mind that the composition of the meltwater sands and the corresponding glacigenic deposits are not entirely identical, at least in the north German glaciated region. As a rule, the respective meltwater deposits contain a higher proportion of local material. Petrographic investigations can be used to distinguish outwash units, especially if the material comes from different source regions. Examples of clearly discernible gravel accumulations are those from Schlüchter's (1976) investigations in the Aare valley in Switzerland.

In the Alpine glaciation area the thickest meltwater deposits were also laid down during the ice-advance phase. Gravel accretions of the last ice age overlain by Würm till have been described from the Inn valley (Fliri *et al.* 1970, Fliri 1973), the Enns valley (Draxler & Van Husen 1978), and the Gail valley (Van Husen & Draxler 1980). Corresponding advance gravels have been detected for the lateglacial Gschnitz stage (Mayr & Heuberger 1968, Van Husen 1977).

Outside each respective glaciation area, the Alpine meltwater gravels grade into the terraces of the proglacial landscape. Only at the end of the accretion period do tributaries from unglaciated catchment areas gain sufficient transport capacity of their own to contribute significantly to the accumulation process. Investigations in the Enns valley (Van Husen 1971) and in the Vienna region (e.g. Fink 1973) show that in those cases, only the upper part of the terrace gravel contains a notable admixture of local material.

Whereas the North German sandur plains are completely unconsolidated except for very local secondary cementation by iron or carbonate, carbonate-rich gravels in the Alpine piedmont region are quite often cemented to form the so-called 'nagelfluh' (calcrete). The degree of cementation in older deposits is more complete than in more recent ones; the increase, however, is not linear. Whilst Würmian gravels are largely unconsolidated, layers of 'nagelfluh' are often found in Riss Stage gravels. Older gravels are often entirely cemented to form a concrete-like hard rock, especially where they are covered by till and on valley flanks (Schreiner 1992).

4.2.5 Kames

Like so many glacial–geological terms, the word kame comes from the Gaelic. In Scotland and northern England the word 'kaim' describes irregularly shaped ridges (Francis 1975). The glacial morphological usage of the term dates from Jamieson (1874), but the current definition comes from Holmes (1947). Kames are landforms consisting of meltwater deposits accumulated in the ice or adjacent to an ice margin, with the ice playing merely a passive role, by providing one or more boundaries to deposition.

Kames are a characteristic feature of ice-decay landscapes. During ice melt great amounts of sediment are released, taken up by the meltwater and often redeposited adjacent to dead ice. The resulting landforms vary in shape. There are conical (Fig. 52), flattened, elongated or completely irregularly shaped kames appearing either singly or in groups (Gray 1991). Plateau-like kames occur, consisting of glaciolacustrine sediments. The latter include, for example, the plateau-like hills of Latvia, that are composed of a glaciotectonised core that is capped by glaciolacustrine deposits (Dreimanis, personal communication) (Fig. 53). Most other kames consist of sand and gravel of variable grain sizes. As a rule, they are internally well bedded and

FIGURE 52 Kame in the Kame-and-Kettle Moraine region, Wisconsin (USA) (Photograph: Ehlers 1993)

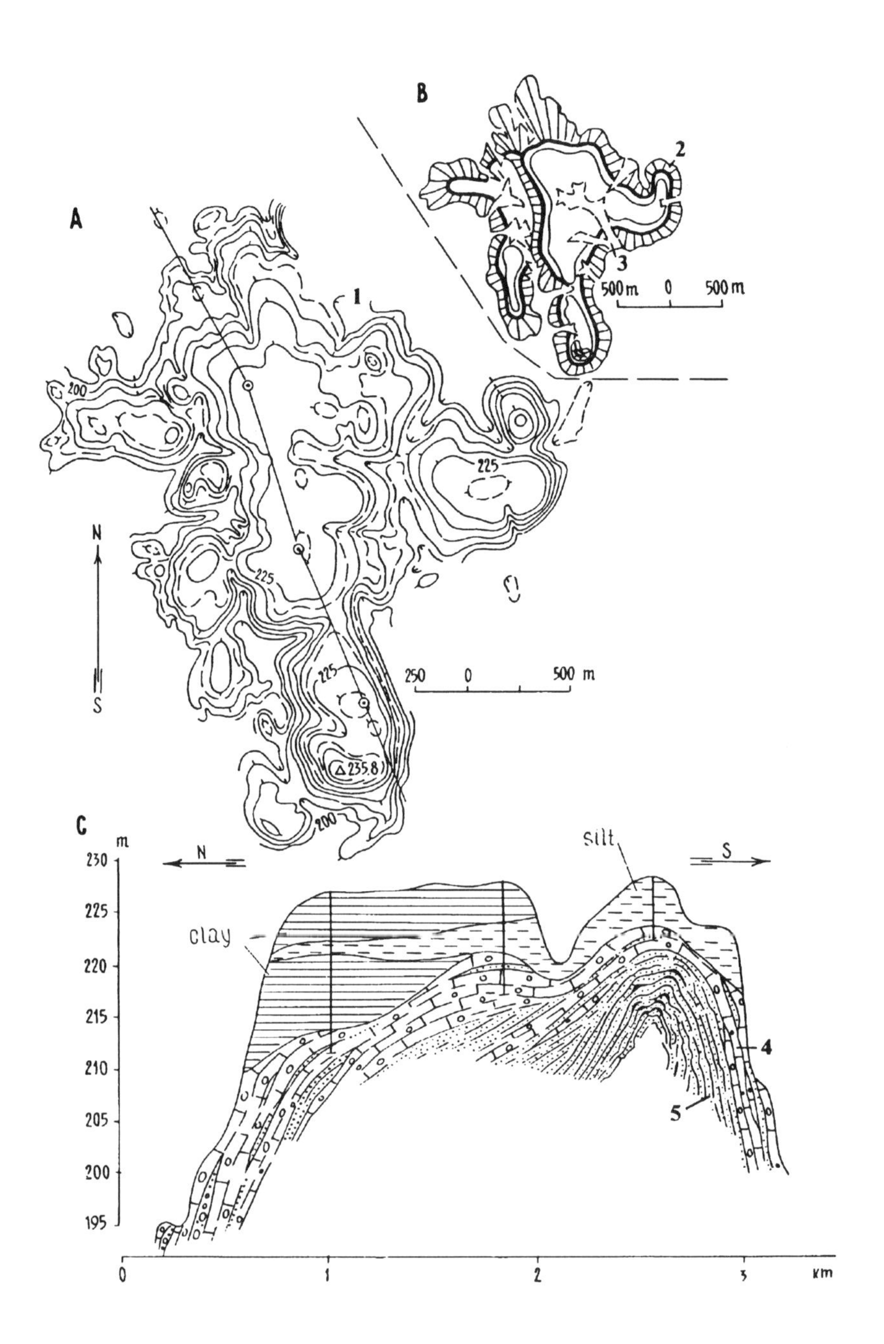

FIGURE 53 Skujene plateau-like hills in Latvia; hypsometric (A) and morphological (B) sketch, and cross section (C), of a plateau-like kame hill with glaciolacustrine clay cover. 1 = contour interval 5 m; 2 = steep slopes; 3 = ravines; 4 = reddish-brown laminated sandy, clayey basal till with thin sand and gravel interlayers; 5 = dislocated fine sand with thin intercalated silt layers (from Āboltiņš & Markots 1995)

often include cross-bedded units. The distinction of kames from landforms produced by meltwater erosion is sometimes difficult and requires good exposures.

The primary sedimentary structures of the glaciofluvial deposits show no difference from proglacial sandur deposits, but palaeocurrent measurements may reveal a more unidirectional drainage than in the latter. Numerous secondary structures like downthrow faults indicate sagging over decaying ice and collapse of former ice-contact margins (Shaw 1972).

The term **kame terrace** describes such deposits that have developed between the glacier and a neighbouring valley slope. It was originally used by Salisbury (1894). Most kame terraces were aggraded by glaciofluvial activity, but glaciolacustrine forms also occur. Kame terraces mostly form at the edge of valley glaciers, but sometimes also at the margins of larger ice masses. Very fine examples of valley kame terraces are found around Loch Etive and Loch Etteridge in Scotland (Gray 1991) (Fig. 54). In the Alps region, kame terraces are primarily known from the moraine amphitheatres of the southern side of the Alps (e.g. Habbe 1969); in the northern Alpine piedmont region they have been described from the region of the Iller Glacier (Habbe 1986a, b) and the Traun Glacier (Van Husen 1977), amongst others (Fig. 55). The terraced gravel trains of the Western Rhine Glacier area are also kame terraces in principle. Additionally, kames are found in areas formerly covered by the Scandinavian Ice Sheet (Niewiarowski 1965, Krüger 1983). The most famous kame terrace in northern Germany is the

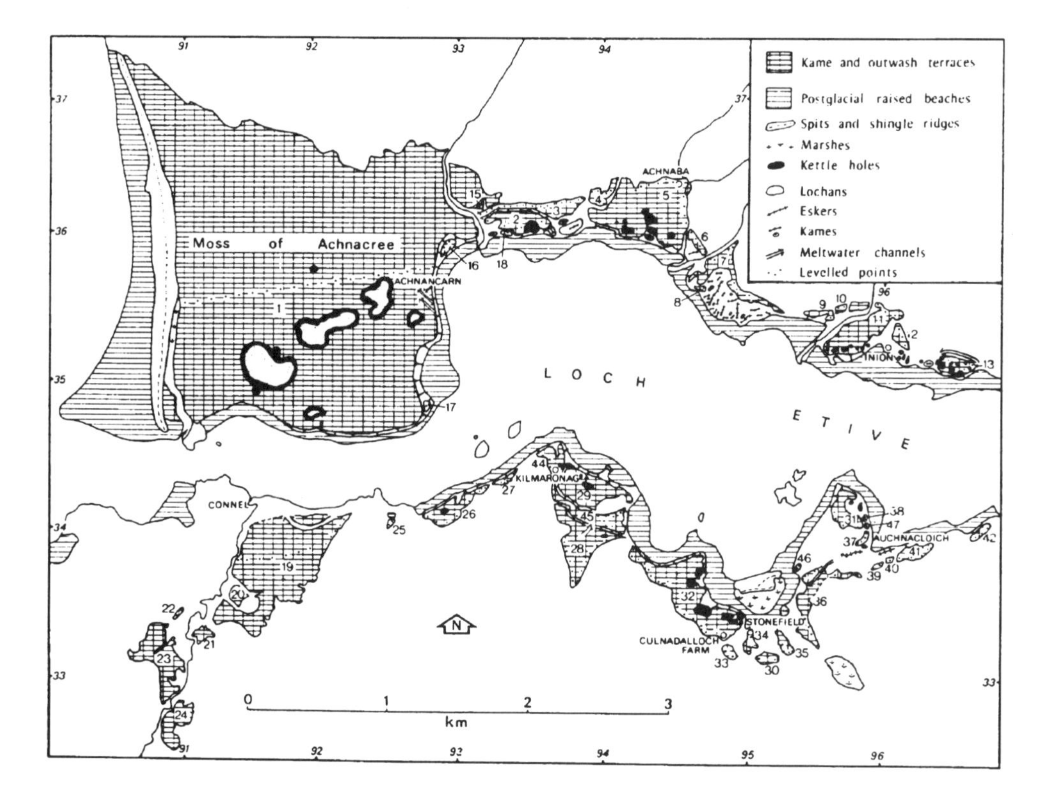

FIGURE 54 Kame terraces at the margin of Loch Etive near Oban, Scotland. They accumulated when the decaying glacier lay in what is now Loch Etive (from Gray 1991)

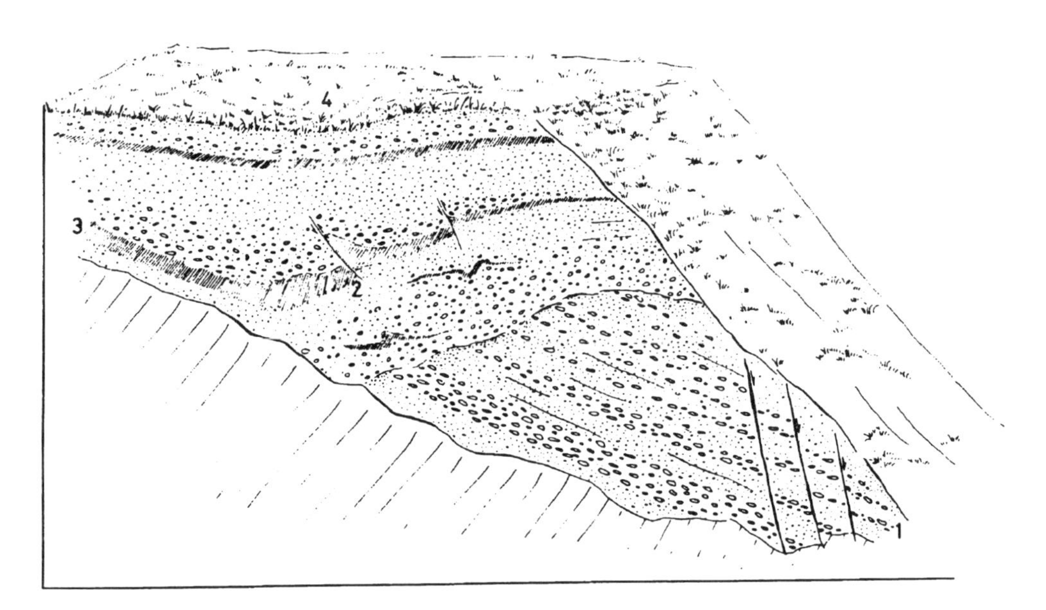

FIGURE 55 Internal structure of an ice-marginal terrace (kame terrace) from Wolfgangsee, Austria. 1 = delta bedding with syngenetic disturbances; 2 = sediments with synsedimentary disturbances; 3 = silt; 4 = dead-ice hollow (from Van Husen 1981)

so-called Porta Kame, an accretion of sand and gravel several kilometres long and up to two kilometres wide, situated at the southern edge of the Porta Westfalica, where the Weser River has cut through the Weser–Wiehengebirge Mountains (Woldstedt 1961). The largest kame deposit so far identified in northern Germany (11 km long, several hundred metres wide) is situated south of Kiel between Einfeld and Blumenthal (Stephan *et al.* 1983).

4.2.6 Ice-dammed lakes

The advancing ice sheets of the major continental glaciations blocked the drainage of numerous streams so that extensive ice-dammed lakes developed in front of the ice margin. Under the impact of a periglacial climate rock terraces formed along the lake shores. Dawson *et al.* (1987) reconstructed this process with the example of a former ice-dammed lake in southern Norway. Frost weathering initiates the process of terrace formation. At lake level, segregation ice forms in fissures and bedrock weakness zones. As a consequence of the higher thermal conductivity of the rock, the frost penetrates more deeply through the rock than through the lake; thus the frost surface attracts additional water from the lake (Fig. 56). Above the lake level, this process is limited by the lack of water. Therefore, frost weathering is mainly restricted to the lake level, i.e. to the cliff foot and the newly forming terrace. The resulting rock debris is partly carried away by winter lake ice and drift ice (Dionne 1979), and partly removed by wave erosion. Dawson *et al.* (1987) reconstructed a mean expansion in terrace width of 3 to 4 cm per annum (maximum value 7.07 cm per annum) for the lake they investigated. The same rates have to be taken into account for the formation of the rock terraces in Glen Roy, Scotland, during the Loch Lomond Stadial (Younger Dryas Chron). The three so-called 'parallel roads' in this valley

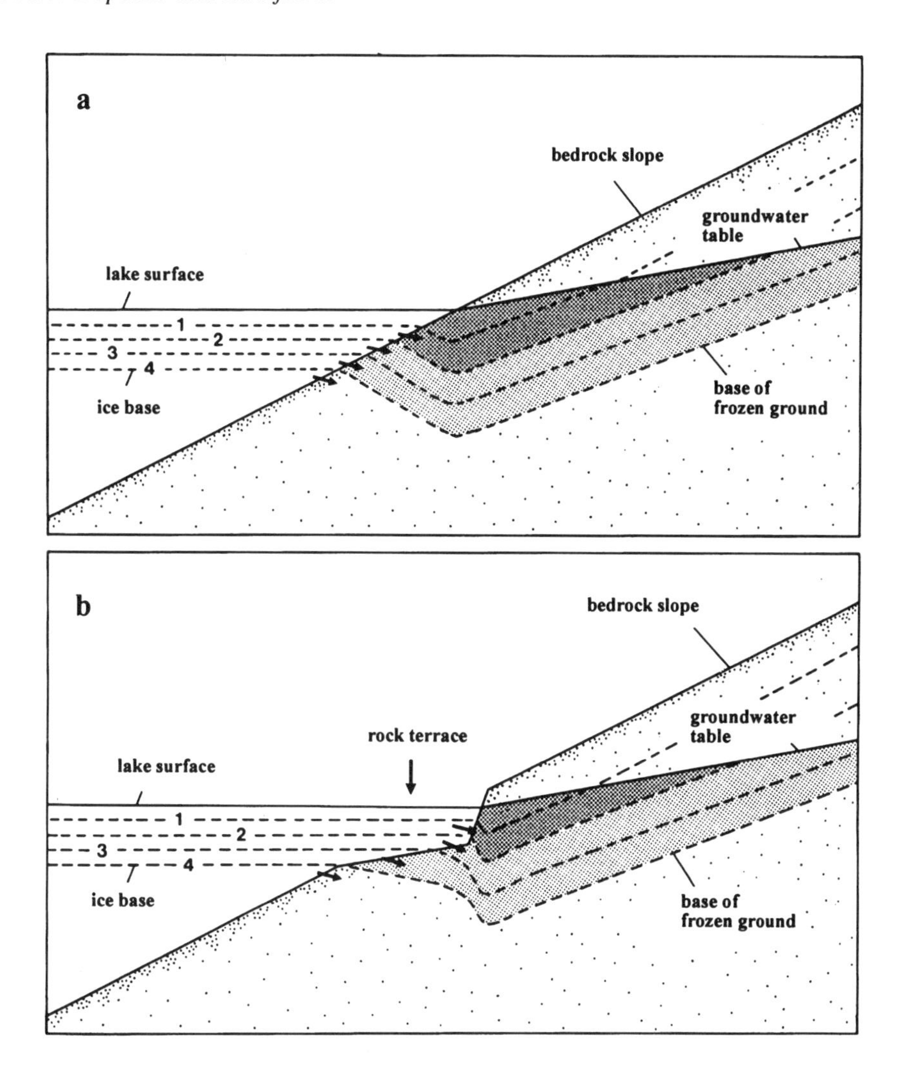

FIGURE 56 Formation of rock terraces by frost weathering at the rock/lake-ice interface (from Dawson *et al.* 1987)

represent three different levels of a former ice-dammed lake. These terraces are up to 12 m wide above which a rock cliff of up to 5 m in height occurs (Sissons 1978). In major lakes, drift ice can scour the bed. Under favourable conditions, traces of this can be seen on aerial photographs of lake plains (Dredge 1982).

Immense amounts of sediment were carried into the impounded lakes by the meltwater tributaries. At the mouths, sand and pebble deltas built up, whereas in the inner parts of the lake basin, clay and silt were deposited. The delta strata can be subdivided into 'foresets' dipping steeply into the basin at an angle of 30°, which are covered by thin, almost horizontal 'topsets'. The delta surface lies approximately at the former lake level (Fig. 57). Ice-dammed lake sediments have been studied in detail, for instance, by Shaw (1977) and Krzyszkowski (1993).

FIGURE 57 Grimsmoen, a giant delta of a former ice-dammed lake in Folldal, Norway. The surface altitude of the delta approximately reflects the former lake level (Photograph: Ehlers 1980)

The still-water sediments in the inner parts of the basin often show a sequence of thin, regularly spaced strata. This phenomenon is referred to as rhythmic bedding and the sediments are called rhythmites. Quite often, those layers represent annual stratification, so-called **varves** (De Geer 1884). The bright summer layers are sandy and silty and consist mainly of inorganic meltwater sediments, whereas the dark winter layers are marked by a higher content of organic compounds and clay (Fig. 58). As opposed to the gradual transition from summer to winter layer, the change from winter to summer layer is sharply delimited. Varve thickness and granulometry represent the drainage conditions of the respective year. As these are similar within major drainage basins, varve sequences can be correlated in a way similar to tree rings (see chapter 10.10). In some cases the lake sediments contain fossil tracks which may allow ecological interpretation (e.g. Gibbard & Stuart 1974).

Enormous ice-dammed lakes formed in North America at the end of the last glaciation. Lake Agassiz, northwest of the present Great Lakes, reached a maximum areal extent of almost one million square kilometres (Teller 1985); thus it reached about the same size as the Baltic Ice Lake, the largest European example of a proglacial lake formed during the retreat of the Weichselian ice in the Baltic Sea region. In tongue basins and in subglacial meltwater channels, a variety of overdeepened and partly closed depressions emerged that were quickly infilled when a meltwater influx occurred (Walther 1990).

Ice-dammed lakes were not restricted to the Nordic glaciations. As a consequence of different advance velocities of the Alpine glaciers, valleys were dammed and large proglacial

FIGURE 58 Varved clays overlying Saalian till in the Neumark–Nord opencast mine, southwest of Merseburg (Photograph: Ehlers 1992)

lakes formed especially in the eastern Alps. As a rule, a sequence of coarse, locally sourced basal material was deposited, overlain by fine-grained beds, locally including thick varved silts (Van Husen 1977, 1980). When the glaciers melted, ice-dammed lakes also formed at various places in the Alpine valleys. In most cases, the immense influx of sediments led to a rapid filling of these basins. Where the water level remained constant over a longer period of time, extensive deltas accumulated. A prominent example is the terrace near St Jakob in Südkärnten (South Carinthia, Austria) (Penck & Brückner 1901/09, Van Husen 1981).

In the lakes of the Alpine foothill region, rhythmites were deposited after the lake basins had become ice-free. In Lake Zürich, Kelts (1978) detected a pronounced annual stratification. The dominant chemical precipitation during summer resulted in bright layers and the deposition of organic detritus in autumn produced dark layers. However, not all rhythmic strata can be classified as varves. The Walensee in Switzerland, for instance, contains rhythmite sediments that do not originate from seasonal sedimentation cycles, and the sequence can be well dated. The sedimentation pattern changed dramatically in 1811 when the lake's catchment area was doubled resulting from the diversion of the Linth River into the Walensee. This event is easily identified in the sedimentary sequence. Examinations by Lambert & Hsü (1979) showed that from this date on, an average of two 'varves' per annum were deposited. These may result from suspension currents connected to high waters of the Linth that occur about twice every year, as indicated by gauge measurements.

4.2.7 Ice-marginal valleys

Girard (1855) was the first to point out that northern Germany is transversed by valleys that can be traced over long distances but are no longer occupied by a single through-flowing river. He distinguished three such palaeovalleys. Berendt (1879) realised that the origin of these depressions he described as 'main valleys' was related to major halts of the Pleistocene ice sheets. In addition, to the three valleys already mentioned by Girard, Berendt identified another further south that could be followed from Wrocław to Hannover. He realised that these valleys could not have formed simultaneously, but one after the other. The term '**urstromtal**' (plural 'urstromtäler') for these features was coined by Keilhack (1898a).

An ice-marginal valley (urstromtal) is the main drainage path of Ice Age meltwaters running subparallel to the ice margin; its headwaters are found at the present (European) main water divide, and when it was formed it drained the entire adjoining sector of the continental ice sheet (Liedtke 1962). In the English language 'urstromtäler' are often referred to as 'ice-marginal valleys' and in the Slavic literature they are called 'pradolinas'. They occur only in the northern European glaciation area. In the Alpine foreland, a sequence of successive drainage routes parallel to the ice margin was only formed in the Western Rhine Glacier region. In North America, the principal proglacial drainage remained directed south towards the Mississippi River, so that no major ice-marginal valleys could form. However, some proglacial spillways draining proglacial lakes, like the Illinois River valley, served a similar function (Dreimanis, personal communication).

By the end of the 19th century four large ice-marginal valleys had been identified in northern Germany:

Wrocław–(Magdeburg–)Bremen Urstromtal
Glogów–Baruth Urstromtal
Warsaw–Berlin Urstromtal
Toruń–Eberswalde Urstromtal

Whereas the Wrocław–(Magdeburg–)Bremen Urstromtal drained into the Weser River, the remaining three courses drained into the lower Elbe. Keilhack (1898b) assumed that each of the ice-marginal valleys should be correlated with one of the main Weichselian end moraines.

Occasionally additional ice-marginal valleys have been postulated like the Pomeranian Urstromtal by Keilhack (1898b), or the Reda–Łeba and Płutnica ice-marginal valleys of Augustowski (1965). Liedtke (1981) rejects the latter because of their insignificant size. However, in addition to the ice-marginal valleys draining towards the west, it is also important to mention the Pilica–Pripjat ice-marginal valley draining eastward in the direction of the Dniepr River.

Although their active period was only of short duration, the ice-marginal valleys were not formed in one but several steps. Therefore the eastern parts of the ice-marginal valleys are clearly terraced in some places (Liedtke 1957, Marcinek 1961). Today, the ice-marginal valley floors no longer have a persistent gradient. Where they are intersected by modern rivers (e.g. the Oder River or the Wisła) inversion of the gradient occurs. Thus, the Oder, east of Eberswalde, lies 34 to 36 m below the ice-marginal valley floor, and near Müllrose (south of Frankfurt on Oder) at 21 to 23 m. Similarly, the Wisła, near Fordon, lies 30 m lower than the level of the former ice-marginal valley. Thorough stratigraphic investigations have shown, however, that the ice-marginal valleys were originally undoubtedly continuous drainage paths for glacial meltwater. Their low gradient (in the case of the Toruń–Eberwalde Urstromtal, approximately 1:13,000 to 1:16,000) was sufficient. Liedtke (1981) refers to the example of the lower Volga River, the gradient of which is only 1:34,000.

The ice-marginal valleys are relatively recent landforms. For the older glaciations, no ice-marginal valleys have yet been reconstructed (K.-D. Meyer 1983c). Liedtke (1981) assumes that the presence of bedrock in the south prevented glaciofluvial erosion. The short duration of the maximum ice advances may also explain the lack of ice-marginal valleys. By counting the varves formed during the Elsterian and Early Saalian maximum ice advances, durations of only a few hundred years have been estimated. Apparently, the meltwater was dammed in front of the ice within the mountain valleys during these short periods.

Drainage parallel to the ice margin at the northern edge of the central German uplands cannot be demonstrated morphologically, although sediments prove the former existence of such drainage systems. Geological investigations have revealed that during the Early Pleistocene Menapian Stage, discharge from the central German rivers (Elbe, Saale, Weser) was for the first time redirected to the west through the Netherlands and into the North Sea (Zandstra 1983). For the Saalian Glaciation, a correlation between westward-directed sediment aggradation and the advance of the ice sheet is well established. Thus it has been demonstrated through geological mapping that the sediment body of the Weser Middle Terrace, which is up to 30 m thick and several kilometres wide, can be traced along the northern edge of the Teutoburger Wald Mountains in a westward direction.

5

Periglacial Deposits and Landforms

5.1 RECENT PERIGLACIAL PROCESSES

The term 'periglacial' dates from Łoziński (1909) and comprises the domain of frost climate. Used in the strict sense it means the effects of ground frost but in a broader sense it also includes other processes influenced by frost-dominated climate, such as the formation of loess – a definition favoured, for instance, by Washburn (1979).

Where frozen soil does not completely thaw for a period longer than one year, it is termed permafrost. Freezing by itself does not greatly affect the land surface. So-called 'dry permafrost' that occurs regionally in solid rock is basically nothing but a cold rock. The intense modification of the land surface, typical of large parts of the sub-Polar regions, only occurs where repeated freezing and thawing of water is involved.

At present, global permafrost covers an area of 25.4 million km^2, of which 24.9 million km^2 are situated in the northern hemisphere. The largest permafrost regions today are found in northern Asia (Fig. 59), such as Russia, where almost all of Siberia east of the Ob River is affected, except for some areas along the Pacific coast. The frost layer in northern Siberia is several hundred metres thick; maximum depths in central Siberia range from 1000 to 1600 m (Kondratjeva *et al.* 1993). In North America, continuous permafrost extends from Alaska to the southern rim of Hudson Bay. In the east, it is restricted to the northern part of the Ungava Peninsula, whereas the southern margin of the discontinuous permafrost lies at the southern rim of James Bay. In the North American Cordillera, Alpine permafrost reaches much further south, with isolated occurrences even in Mexico (Péwé 1983a). The greatest permafrost thickness in America was measured at Prudhoe Bay (Alaska) and on the North Slope of Alaska where it is 610 m thick (Ferrians 1994). As for Greenland, the third great sub-Polar region of the northern hemisphere, continuous permafrost is restricted to areas north of 66°N, reaching far less to the south than in America and Asia. The ice-free regions in these areas lie in the vicinity of the coast, where oceanic influences alleviate the frost climate.

Distribution of permafrost in Europe today is concentrated in northern Scandinavia and small parts of the high-mountain regions, including the Alps, Pyrenees and smaller parts of the Tatry, Carpathians, Abruzzi and the Scottish Highlands (King & Åkerman 1993). Local distribution of periglacial conditions outside the sub-Polar regions today is determined by altitude, lithology, vegetation cover, exposure and slope gradient. Altitude and vegetation cover have a significant influence on the intensity of the frost climate. In Scotland, the periglacial

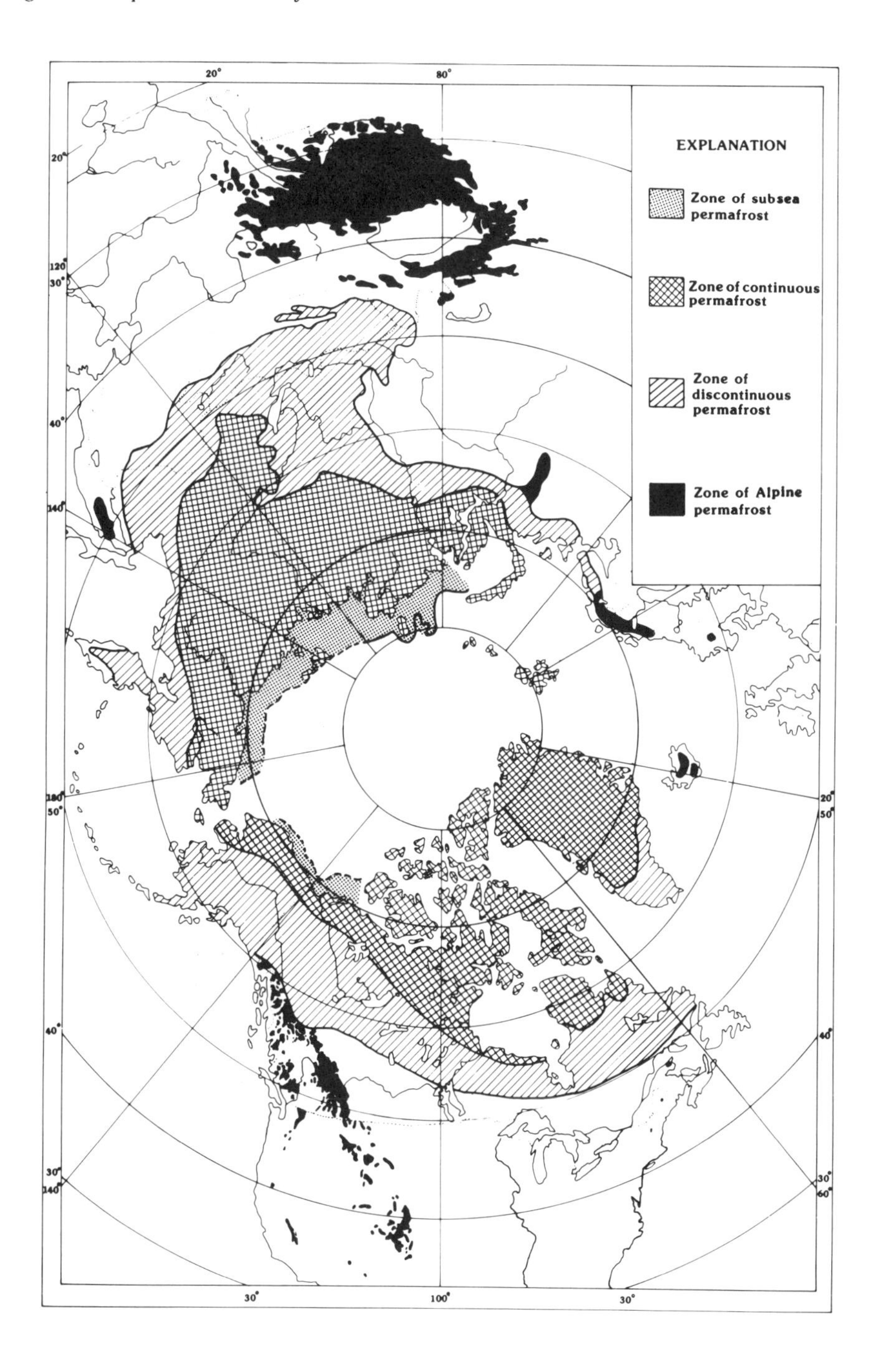

FIGURE 59 Modern distribution of permafrost in the northern hemisphere (from Péwé 1983b)

FIGURE 60 A–D: Four examples of modern frost weathering on Iceland (Photographs: Ehlers 1990)

FIGURE 60 *(continued)*

region today lies above ca. 600 m, but may descend to altitudes below 400 m where the natural vegetation cover is impaired. Differences in rock type play an important role in the production of different periglacial features. Quartzite, microgranite and granulite generate open boulder fields; sandstone and granite produce a coarse, cohesionless sandy matrix with dispersed stones; mica schist, clay, siltstones and lavas usually form a fine-grained diamicton. The extent of aeolian modification depends on exposure to the prevailing wind direction. On steeper slopes, the influence of mass movement is significant resulting in the formation of stone stripes, solifluction sheets and turf terraces (Ballantyne 1987).

Unlike glaciers and their deposits, periglacial processes have only relatively recently attracted the attention of scientific research. During the last few decades, exploration of the sub-polar regions has greatly increased because of growing economic interest. Overviews of the current state of research are provided by French (1976), Washburn (1979) and Ballantyne & Harris (1994), and permafrost dynamics forms the central theme of the book by Williams & Smith (1989).

5.1.1 Cryoplanation

Formation of flat land surfaces under the influence of periglacial processes is called cryoplanation. Frequent freeze–thaw cycles result in rapid disintegration of solid rock into coarse congelifracts (Fig. 60). In this process, boulder fields and boulder slopes are formed on flat and gently inclined terrain. Investigations within and outside the reach of the last ice advance in Scotland, the 'Loch Lomond Readvance' of the Younger Dryas period, have shown that congelifraction and formation of boulder fields took place primarily during the harsh permafrost conditions of the Late Weichselian glacial maximum. During the Younger Dryas period there was merely a slight periglacial modification of the land surface. In the Scottish Highlands, most of the slopes and plateaux are covered with a mantle of rock 0.5 to 1 m thick. The lack of erratic material indicates that this does not represent degraded morainic deposits, but results from *in situ* frost weathering. The frost heaving of rocks often leads to a concentration of blocks on the land surface, whilst the fine components are accumulated beneath (Fig. 61; Ballantyne 1984).

On inclined terrain, sliding of snow and firn may occur triggering abrasion and formation of cirque-like features. This process is called **nivation**. In the reach of elongate snow patches it may induce the formation of terraces. Demek (1969) and Czudek (1995) distinguish three morphographic stages in the development of cryoplanation terraces, all of which can be observed under favourable conditions on the same slope (1) The first is marked by nivation generating a small basin or a step. (2) In the second stage, a terrace forms that is dissected by frost wedges. On the terrace surface, inclined at approximately 2–7°, increased solifluction and slopewash prevail. (3) By further recession, planation of the entire summit is achieved ultimately when the slope gradients become so low that no further transport of material is possible. Frost heaving, frost sorting and deflation generate a boulder-strewn land surface. Exact measurements of terrace retreat and lowering of the land surface are not yet available. Cryoplanation terraces are typically 20–80 m wide and up to a few hundred metres long. Features similar in shape and origin, formed at the foot of the slope, are termed cryopediments (e.g. Demek 1968, Czudek 1995).

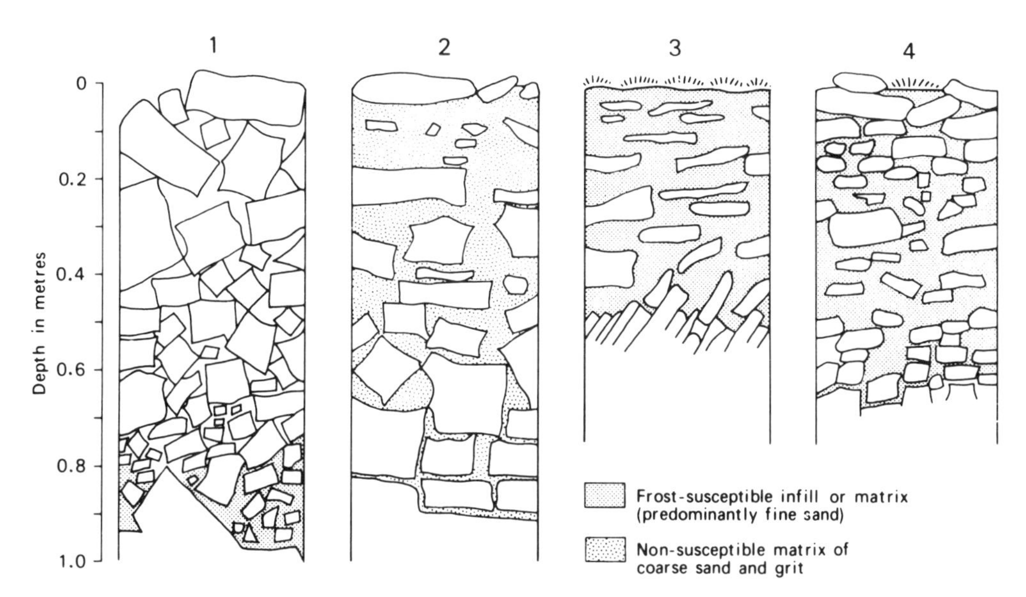

FIGURE 61 Schematic sections through mountain-top detritus on different lithologies. (1) Blockfield deposit on Cambrian quartzite, An Teallach; (2) regolith on Torridonian Sandstone; (3) regolith without vertical sorting, developed on fissile mica schists, Ben Wyvis; (4) regolith with pronounced vertical sorting, developed on siliceous schists, Ben Wyvis (Scotland) (from Ballantyne 1984)

5.1.2 Rock glaciers

In the periglacial talus zones of numerous high mountains, tongue-shaped frozen debris masses are found where the surface configuration is reminiscent of glaciers. These rock glaciers are usually several hundred metres long, 100 to 150 m wide, and 40 to 50 m thick. Genuine rock glaciers emerge from block-strewn slopes with a matrix of ice, unconnected to any glacier, and are therefore genuine permafrost formations (Barsch 1969, Haeberli 1975a). Similar forms emerge from glaciers covered with debris. Active rock glaciers move slightly down the valley. Measurements in the Alps have yielded maximum velocities of more than three metres per year (Haeberli *et al.* 1979). Inactive rock glaciers still possess an ice core but have ceased moving. Relic rock glaciers, however, are ice-free witnesses of former climatic conditions.

Active rock glaciers occur in the frost detritus zones of all major high mountains, including the North American Cordillera (S.E. White 1971, Luckman & Crocket 1979), the Alps (Barsch 1969, Haeberli 1985) and the Scandinavian mountains (Barsch & Treter 1976, Sollid & Sørbel 1992). Rock glacier formation is discussed by Whalley & Martin (1992). Rock glaciers develop within almost every rock type. The lowest margin for the occurrence of rock glaciers almost coincides with the 'climatic snow line', i.e. the theoretical boundary above which in flat areas the snow does not completely melt in the course of a year on a long-standing average. With increasing continentality, however, the distance between the snow line and rock glacier terminus increases. Thus, rock glaciers form more easily in continental comparatively drier climates with cold winters, because in more maritime areas, sites favourable for rock glacier formation are already occupied by glaciers (Barsch 1992). However, inactive rock glaciers have also been reported from Britain (e.g. Dawson 1977), but in most cases the precise mode of origin is still disputed (see, for instance, discussion of the feature at Beinn Alligin by Gordon 1993).

5.1.3 Cryoturbation

Irregular turbations and involutions of near-surface layers under the influence of ground frost are termed 'cryoturbations', after Edelman *et al.* (1936). They mostly occur in layered sediments of different grain sizes and, according to the shape of the respective patterns, descriptive terms range from patterned ground, turbated ground, involutions, drop soil, to convoluted soil (Fig. 62). For the formation of turbations three different explanations are offered (French 1976):

1. Some researchers assume that turbations are formed in the basal parts of the active layer, which starts re-freezing from above at the beginning of winter. Differences in grain size and water content result in differential rates of congelation so that pockets of unfrozen material can be trapped within already frozen ground. The volume increase of the frozen soil causes vertical deformation of strata.

   Although laboratory investigations show that this is possible, no direct evidence of the required cryostatic pressures has been found in the field (e.g. Mackay & MacKay 1976). Instead, dewatering and overconsolidation take place in the unfrozen zone, enclosed by the permafrost and the newly frozen layer, under water transfer to the adjacent frost layers (Mackay 1979, 1980). Therefore, apparently no viscous semi-liquid mass is formed by freezing from above and below.
2. Cryoturbation has frequently been interpreted as an effect of density differences between strata of different water content and grain-size composition. Thus, they genetically resemble load casts. Particularly favourable formation conditions might occur during the

FIGURE 62 Convolutions in Elsterian deposits. Peres opencast mine at Borna (Photograph: Eissmann 1977)

thawing of ice-rich permafrost (Eissmann 1981), but also late summer oversaturation of the active layer from heavy rain or thawing of ice-rich sediments may have a similar effect. In these cases, liquefaction may cause sediments of the subjacent bed to rise to the surface due to inversion of the density gradient.

3. Involution layers may also result from sediment turbation due to frost heaving and subsequent formation of segregation ice. This process has been ascertained by laboratory investigations and field measurements (Washburn *et al.* 1978, Mackay 1980). The widespread occurrence of sorted polygons is attributed to this process (cf. chapter 5.1.5).

In many cases differential frost heave and load casting commonly seem to work together in forming cryoturbations (Van Vliet-Lanoë 1991). Investigations on Banks Island (Canada) allowed French (1986) to demonstrate that cryoturbation and load casting result in strikingly similar features. Consequently, a palaeoclimatic interpretation of conturbations is not always possible.

5.1.4 Solifluction

Intense sediment movements take place on slopes in periglacial regions. In conjunction with slopewash, solifluction plays a decisive role. Even without a periglacial climate, soil may creep downslope under certain conditions (high water saturation, steep slopes, expandable clay minerals etc.). Mostly, however, the term solifluction is used in connection with the periglacial creep of water-saturated disaggregated rock particles. The term was first used by Andersson (1906). Other terms such as congelifluction (Dylik 1967) or gelifluction (Hamelin 1961) are used as synonyms. Because of areal erosion caused by this process, Liedtke (1975) also refers to it as cryodenudation.

Periglacial solifluction is triggered by high water saturation of the active layer, leading to a reduction in shear strength. When the **liquid limit** is exceeded the material starts moving slowly downhill. On slopes of approximately 2° the thawed ground begins to creep down the terrain surface, but the threshold value depends on the substratum. According to Ballantyne (1987), solifluction in recent periglacial regions of Scotland appears only on slopes with a gradient of 5 to 7.5°. Where vegetation cover is lacking or impaired, solifluction may lead to rapid redeposition of sediment. In this process, sorted polygons are distorted into stripes in the downslope direction. In extreme cases, water-saturation can result in a very mobile mass of sediment flowing off rapidly (Fig. 63). According to how much the process is influenced by vegetation, a distinction can be made between free and restricted solifluction (Büdel 1948). Consequently, a variety of characteristic micro-patterns can develop, such as solifluction tongues and loop soils. A review of the features produced is given by Karte (1979) (Table 3).

Büdel (1959) also makes a distinction between periodic and episodic solifluction. Whereas periodic solifluction occurs every year during early summer, episodic solifluction is thought to happen only rarely, under particularly favourable conditions. However, this discrimination is more or less of only theoretical value, because the resulting sediments show the same characteristics (cf. Washburn 1979).

5.1.5 Frost fissures and ice wedges

A characteristic of periglacial areas is patterned ground. Whereas stripes of sorted sediment prevail on slopes, on flat land polygonal patterns are formed. A distinction is made between

FIGURE 63 (a) and (b): Active layer on the banks of the Aldan River in Siberia (Photographs: Ehlers 1982)

TABLE 3 Types of gelisolifluidal small-scale landforms (from Karte 1979)

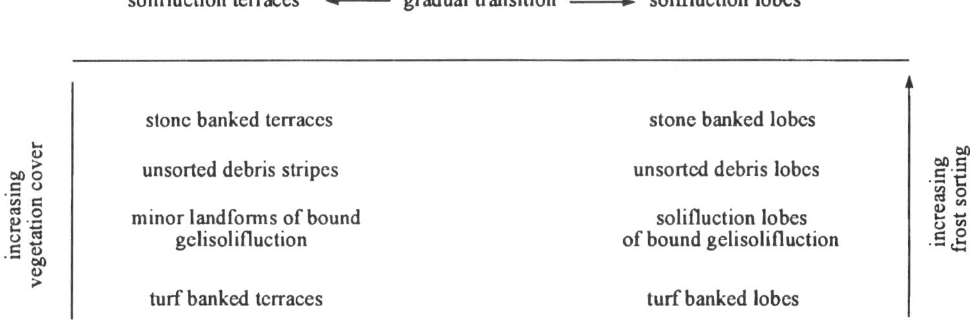

solifluction terraces ← gradual transition → solifluction lobes

solifluction terraces	solifluction lobes
stone banked terraces	stone banked lobes
unsorted debris stripes	unsorted debris lobes
minor landforms of bound gelisolifluction	solifluction lobes of bound gelisolifluction
turf banked terraces	turf banked lobes

increasing vegetation cover (downward) — increasing frost sorting (upward)

sorted and non-sorted polygons. Non-sorted polygons usually result from drying and cracking of the soil and/or the formation of ice wedges, whereas sorted polygons are generated by dynamic frost movements in the ground resulting in upwelling of fine material in boulder fields (Fig. 64).

Wedge-shaped ice veins that interpenetrate the ground down to several metres were first described from Alaska (Leffingwell 1915). These are known as ice wedges. Sudden cooling and rapid freezing of the ice leads to contraction and formation of cracks in a polygonal pattern. At the beginning of the thaw period of the following summer, water may enter the fissures and freeze. Thus, a complete closure of the cracks is made impossible and ice wedges are formed.

FIGURE 64 Sorted polygons on Dovrefjell in Norway (Photograph: Ehlers 1980)

The formation of ice wedges begins in the form of thin veins that constantly increase in size over the years (Fig. 65). Large ice wedges are formed preferentially in fine-grained or peaty substrate. In pure sand they are seldom found to exceed 0.1 to 0.5 m in width. Large ice wedges can build up to a depth of more than 30 m, expanding laterally until the intervening soil is compressed into columns, and the ice volume clearly exceeds the volume of the soil. Eventually further ice growth may lead to the formation of continuous ice masses with a thickness of more than 80 m, as reported from the New Siberian Islands (Washburn 1979).

The formation of ice wedges depends considerably on the substratum. Investigations in modern periglacial regions in Siberia led Romanovsky (1985, quoted in Pissart 1987) to state that an annual average temperature of −5.5°C is required for the formation of ice wedges in sand and gravel, −2.5°C in clay and −2°C in peat. Thus, ice wedges require higher temperatures to form in loess than sand. Reconstructing the climatic conditions that led to the formation of periglacial phenomena is problematic, since there is a difficulty in converting the

FIGURE 65 Modern ice wedge in Siberia (Photograph: Ehlers 1982)

initial soil temperature into air temperature. The differences may range between 1 and 6°C, depending on the winter snow cover (Gold & Lachenbruch 1973, quoted in Pissart 1987).

Low temperatures are a basic requirement for the formation of ice wedges, but they are not the only decisive factor. Particularly critical above all else is a sudden drop in temperature within the ground to a depth of 5 to 10 m. This is only possible where there is hardly any insulating cover of the ground either by vegetation or snow, as well as a thin summer active layer (Black 1976).

Commonly, ice wedges are categorised into two groups according to their mode of formation. **Epigenetic ice wedges** grow in already existing permafrost, and usually they are much younger than the surrounding material. The evolution of an epigenetic ice wedge is demonstrated in Figure 66a. Since ice wedges usually crack in the centre during successive winters, this is the position where new ice may be added during the transitional season. Because ice wedges are V-shaped, an epigenetic ice wedge grows in width, but insignificantly in depth (Mackay 1974). The progressive increase in ice-wedge thickness causes adjoining sediment layers to bend upwards.

Syngenetic ice wedges form in permafrost areas where recent sediment accretion is occurring. They grow as the surrounding terrain is being raised (Fig. 66b). They emerge in fluvial sediments, peat and solifluction deposits. If sediment supply and the degree of fissuring are in balance, then the ice wedge continues to grow upwards simultaneously with the sedimentation. Syngenetic ice wedges often appear as if they were tucked one into another (Fig. 66b), the oldest ice being found in the deepest part of the wedge.

Mackay (1990) introduced so-called **anti-syngenetic ice wedges** as a third group (Fig. 66c). These form when a terrain surface is lowered by erosion under continuing frost fissuring. In this case the ice wedge keeps growing downwards and the oldest ice is in the upper exterior margin.

When the ice thaws, the cracks are often filled by material slumping from the flanks, resulting in a downward bending of the strata. Often, normal faulting is observed in the sediments adjacent to ice-wedge pseudomorphs (ice-wedge casts) (Goździk 1973, Kolstrup 1980; see Figs 71a and b).

Ice wedges are truly permafrost features. However, they are easily confused with **sand wedges** (Péwé 1959), which can form under cold, arid conditions. These cracks contain no ice but are filled with non-ice materials from their initiation (usually wind-blown sand). Sand-wedge casts can be distinguished from ice-wedge casts by their near-vertical infilling with sand. The adjacent sediment is slightly upturned, and sometimes smaller wedges split off and extend beyond the wedge proper (Goździk 1973; see Fig. 71c).

Many forms observed in the field show characteristics of both ice wedges and sand wedges. Goździk (1973) refers to them as composite wedges or sand–ice wedges.

5.1.6 Pingos

The term 'pingo' comes from the Eskimos of the Mackenzie delta (Canada) and means 'small hillock' (Washburn 1979). The expression was introduced into scientific literature by Porsild (1938). In the Russian language the term 'bulgunnyakh' is often used (Jahn 1975). Pingos are usually single frost or ice hillocks with a round to egg-shaped base in plan (Fig. 67), with diameters reaching from just a few metres up to a maximum 1200 m and heights up to a maximum 100 m. In most cases, however, the diameter measures 20 to 300 m and the height from 5 to 70 m. The flanks are relatively steep (up to 35°). These obvious forms grade into

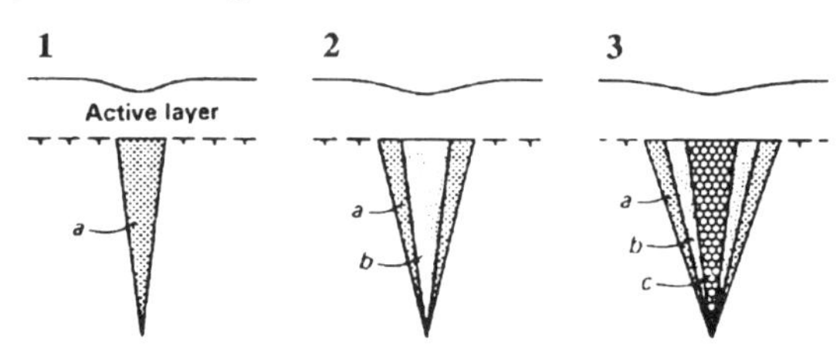

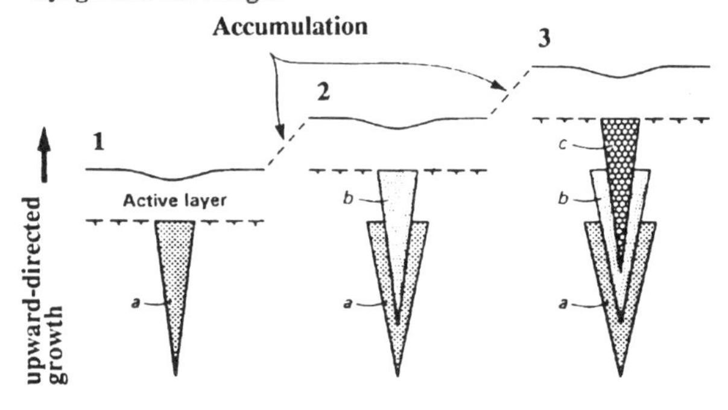

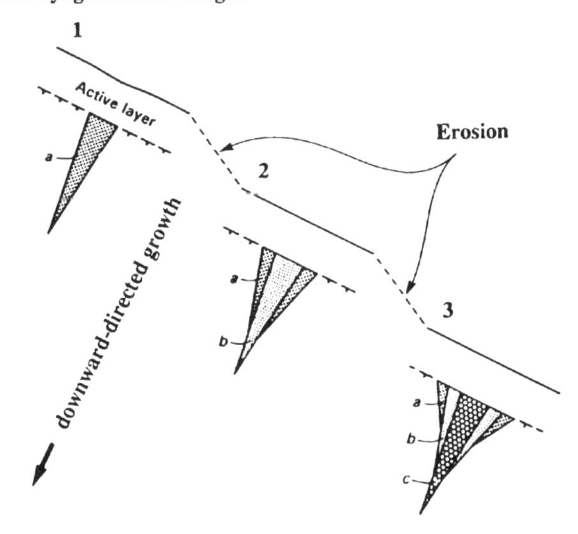

FIGURE 66 Schematic diagram showing the growth of epigenetic, syngenetic and anti-syngenetic ice wedges. Three growth stages are shown for each type. The diagrams of the epigenetic and syngenetic forms are based on N.N. Romanovsky (1978, Fig. 123). In the epigenetic ice wedge the oldest ice is on the flanks, in a syngenetic ice wedge in the outermost, lower parts. In the anti-syngenetic ice wedge the ice on the sides increases in age from bottom to top (from Mackay 1990)

smaller frost blisters or upward domings with different shapes. Pingos consist of an ice core with a soil cover 1 to 10 m thick (French 1976) protecting the core from thawing in summer. In contrast to short-lived **frost blisters** (Pollard & French 1985), pingos are multi-annual forms.

Two genetic types of pingos can be distinguished, the Mackenzie type and the East Greenland type.

1. Pingos of the Mackenzie type are generated in a closed hydrological system. They emerge in shallow lakes where unfrozen ground (talik) occurs like an island within the permafrost due to the insolating effect of the water. As the water table is lowered, or as the lake silts up, the insolation ceases, and the unfrozen ground underneath the lake is gradually narrowed down from all sides. The hydrostatic pressure of the trapped water eventually increases so much that a breakthrough occurs at the base of the permafrost, leading to freezing of the water and doming-up of the overlying ground (Fig. 68). The growth rate of a pingo may reach 1.5 m per annum in the initial state, but then decreases continually until the talik is entirely replaced by permafrost (Mackay 1973).
2. Pingos of the East Greenland type are formed in an open hydrological system in regions of discontinuous permafrost, where a lateral flux of percolating or ground water is possible. At the valley floor or on lower slopes, this water comes under hydrostatic and cryostatic pressure, so that a body of injection ice or segregation ice is formed. This type of pingo often occurs in small groups or swarms. In arctic sub-Polar regions, it is often found on slopes exposed to the south or southeast (French 1976).

FIGURE 67 Pingo (bulgunnyakh) in Siberia, east of the Lena River (Photograph: Ehlers 1982)

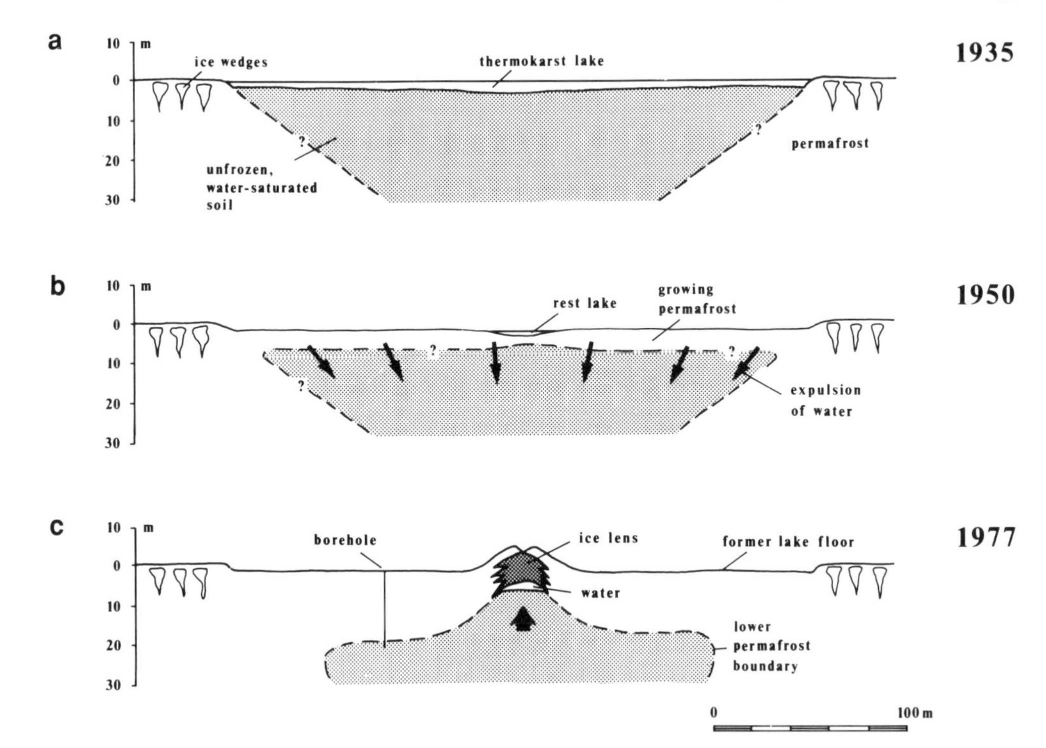

FIGURE 68 Growth stages of a pingo of the Mackenzie type (closed hydraulic system) in the Mackenzie Delta, Canada. (a) situation in 1935: thermokarst lake with low permafrost table. (b) situation in 1950: the lake is drained; permafrost is spreading from the surface. (c) situation in 1977: the water underground has come under pressure from the downward-advancing permafrost front, so that it rises in the centre of the former lake. The cover is uplifted. A pingo forms (after Washburn 1979)

Karte (1979) has compared the main characteristics of both pingo types (Table 4).

Some of the pingos are very recent features. In Siberia and North America, pingos are known to have formed only during the last few decades; others have been radiocarbon-dated to 4500–7000 BP. It is assumed that many pingos were generated during the cooling at the end of the Atlantic period (Washburn 1979).

When the ice lens grows too excessively, the protective sediment layer at the top of the pingo may break along star-shaped cracks and eventually start sliding down the flanks. In this case, the insulating effect of the cover is lost; the ice core melts and a small lake is left that might be surrounded by a low minerogenic ring-like ridge.

5.1.7 Palsas

Local frost heaving also occurs within the mires of periglacial regions (Fig. 69). The resulting features are called palsas. The term 'palsa' was originally used by the Lapps and Finns to describe a peat hummock with a frozen core (Seppälä 1988). Palsas are smaller than pingos, rarely exceeding 10 m in height and 50 m in width. They are often more flat-topped than pingos and tend to be more irregular in shape (Mackay 1978). Åhman (1976) and Seppälä (1976) assume that their formation requires a discontinuous winter snow cover, where patches with

TABLE 4 Differences between pingos with open and closed hydrological systems (from Karte 1979)

Closed system pingos	Open system pingos
occur in areas with thick, continuous permafrost	occur in areas with thin, discontinuous permafrost
mean annual temperatures below −6°C	mean annual temperatures between −1 and −6°C
round to oval shape	oval to oblong shape
occur mostly as singular hills, not in groups	occur mostly in small groups
in flat terrain, associated with former lake beds or river courses	at the foot of gentle slopes and on valley flanks, associated with seepage of sub-permafrost ground water
	in sub-Polar regions of the northern hemisphere mostly on S–SE-facing slopes

FIGURE 69 Cross section through a minerogenic palsa on a raised marine terrace at Varangerbotn, northern Norway (Photograph: Meier 1985)

thin snow cover allow the frost to penetrate deeply into the subsoil. This leads to a concentration of ground ice that can also involve the minerogenic base of the mire. Forms where the subsoil is also affected are called minerogenic palsas as opposed to organogenic palsas. Unlike most pingos, palsas contain various ice lenses 2 to 15 cm thick, separated by strata of frozen peat or mineral soil that dome up the peat cover. This heaving in turn allows less snow to be accumulated than in the surrounding area, hence enhancing the process further. In summer, the raised peat layer dries out, tensional cracks may form, and thawing of the core starts. The unfrozen remnants of palsas are either low (0.5–2 m high), circular rim ridges, rounded open ponds or pond groups, or open, vegetation-free peat surfaces (Seppälä 1988).

Palsas typically occur in the discontinuous permafrost region. They are found in Fennoscandia, Iceland, Canada and Alaska and are also widespread in Siberia. Whereas on Iceland the southern margin of palsa distribution almost corresponds with the 0°C annual isotherm, in Sweden it is situated at about the −2 to −3°C annual isotherm (Washburn 1979). The occurrence of palsas in Scandinavia today is largely restricted to areas north of the tree limit, with the southernmost site in Europe being found at 62°N on Dovre Fjell in Norway (Sollid & Sørbel 1974). In North America, the southernmost palsas have been reported from the Beartooth Mountains of Wyoming (Collins *et al.* 1984).

Palsas are comparatively young features, and can only emerge after the peat required for the isolation of the ice laminae has formed to a sufficient thickness. Bog formation in Finnish Lapland was only initiated about 8000 years BP (Seppälä 1971). Palsa formation began as late as at the turn of the Subboreal to Subatlantic period (about 2400 years BP), i.e. clearly after the postglacial temperature maximum. A large proportion of the Canadian features are thought to have formed within the last 200 years, and Priesnitz & Schunke (1978) even report many Icelandic palsas originating since 1965. Seppälä (1988) regards palsa growth and decay as a cyclic process, but many details are still unknown.

5.1.8 String mires

String mires (aapa mires) are swampy areas subdivided by strings of peat up to 2 m high and several metres wide, separated in most cases by approximately 10 to 50 m. String formation results from vegetational differences (Ruuhijärvi 1960). The irregular surface of the aapa mires causes uneven snow covering, thus allowing the ground frost to preferentially penetrate into the strings, resulting in further heaving. Seasonally the strings contain ice lenses. Through frost heaving the strings finally grow out of reach of the ground water, producing a vegetation typical of an extreme raised bog. Eventually peat formation comes to a standstill and lichens spread. The strings often form subparallel to the slope contours or are gently inclined downhill. On steeper slopes the aapa mires can attain a terraced character. On very steep slopes the strings are arranged at right angles to the contours, and distance between the individual strings decreases down to 3 to 4 m (Hallik 1975).

A correlation between string mires and periglacial conditions is disputed. So far, it is certain that their distribution clearly exceeds the present range of permafrost (e.g. Ruuhijärvi 1983). Probably their formation requires only harsh winter frost.

5.1.9 Thufur

A special sort of patterned ground is represented by the thufur. These are small earth hummocks, reaching a height of 0.1 to 0.5 m. They are formed primarily in a fine-grained substratum at least 0.3 to 0.4 m thick. The name thufur comes from the Icelandic language. It has not been possible to attribute their formation to a specific frost-dynamic process. However, the internal structure of the earth hummocks indicates that cryoturbation plays a role in their formation. Also differential penetration of ground frost, depending on the vegetation cover, is a precondition for thufur formation. The vegetation is important in creating insulation differences, protecting the frozen cores of the hummocks, and thus contributes to the further development of the thufur (Schunke 1975). Thufur may occur on slopes of up to 25° (Stingl 1969), though from a gradient of 5° upwards, transitional forms are found towards solifluction (Embleton & King 1975). With cessation of the necessary conditions thufur decay rapidly (Karte 1979). Some resemblance to thufur is found in the so-called 'buckelwiesen', a form of periglacial micro-relief described from European Alpine regions (e.g. Jerz 1993).

5.2 TRACES OF PLEISTOCENE PERIGLACIAL PROCESSES

During the Quaternary ice ages, permafrost increased in thickness and extended considerably further to the south in the non-glaciated regions (Fig. 70). In the North American Cordillera, the permafrost limit was lowered considerably. During the Wisconsinan maximum the 0°C mean annual air-temperature isotherm was about 1000 m lower than at present (Péwé 1983a). From the Appalachian Mountains numerous finds of relict permafrost features have been described (G.M. Clark 1968). As a consequence of the lowering of sea level, permafrost was also able to spread into areas that are today beneath the sea, for instance, in the Beaufort Sea and in northern Siberia. In the latter area relict permafrost is found from the present coastline north beyond the New Siberian Islands. During the major Quaternary ice ages, Central Europe, as far as the southern margin of the Alps, lay in the permafrost region. Its boundary can be followed eastward to the northern edge of the Black Sea and westwards to the French Atlantic coast (Poser 1948). A review of the development of permafrost in Europe during the last cold stage is given by Vandenberghe & Pissart (1993). In the glaciated areas, however, permafrost growth ceased. Beneath the base of a glacier where the temperature is at the pressure-melting point, permafrost is gradually degraded under the influence of geothermal heat.

The former permafrost distribution is well known from the occurrence of ice-wedge casts and other periglacial indicators; its thickness, however, can only be assessed in very few cases. The summer-thawed active layer above the frozen ground reached a maximum depth of about 2 m.

In the periglacial areas widespread solifluction occurred during the Pleistocene ice ages. Many of the corresponding micro-structures were later destroyed, but the sediments of the solifluction layer are still detectable at many places. In the Alpine region, where periods of increased solifluction occurred repeatedly during the Holocene, the soliflucted strata and their intercalated fossil soils can also be used for the reconstruction of postglacial climatic development (Furrer & Bachmann 1972, Furrer *et al.* 1975).

The numerous relic and fossil ground-frost features in Central Europe have escaped scientific attention until relatively recently. Gripp (1929) reported the occurrence of fossil frost cracks in northern Germany and presented photographs of occurrences in East Friesland. A few years earlier Kessler (1925) had identified wedge-like features from Thuringia to be frost cracks, and Soergel (1936) presented a comprehensive documentation of these features. In Britain, Hollingworth (1934) described active periglacial processes and landforms from the Lake District, and T.T. Paterson (1940) compared periglacial features around Cambridge with similar phenomena from Baffin Bay. In the United States, Schafer (1949) described relict ice-wedge casts from Montana.

Ice-wedge casts are to be found in numerous exposures in former periglacial regions. An overview of 22 localities with ice-wedge casts in Wisconsin was presented by Black (1965). Finds of fossil ice wedges are not restricted to the present land surface. In many cases

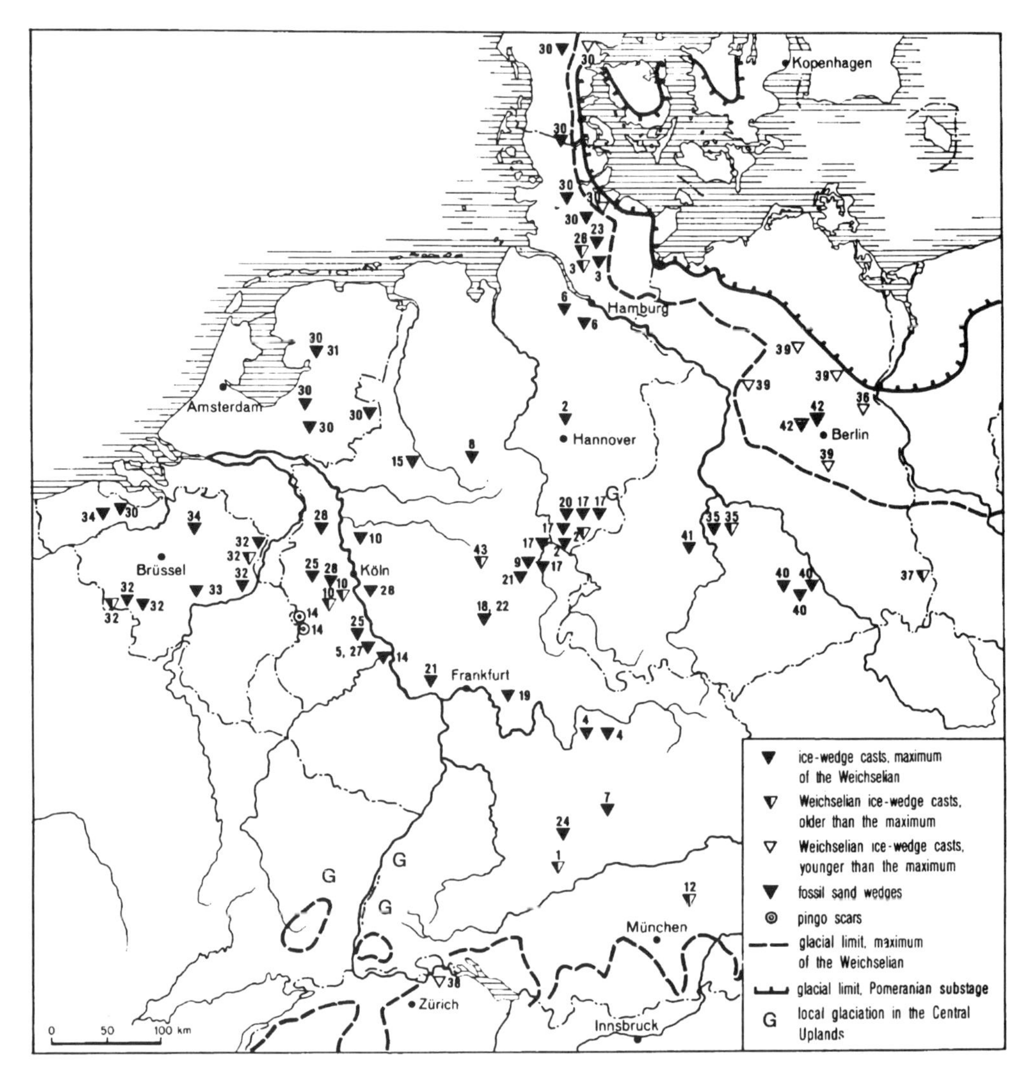

FIGURE 70 Traces of periglacial permafrost in Central Europe during the Weichselian glacial maximum (from Karte 1987)

ice-wedge casts of differing age have been identified, covered by younger deposits (e.g. Goździk 1973, 1995). More recent systematic investigations were made by Eissmann (1981, 1987) in the Leipzig area (Fig. 71).

Under favourable conditions, former polygon networks of ice wedges in till surfaces may also be identified on aerial photographs where there are slight moisture differences between the joint filling and the surrounding ground. Often they are even perceptible through different vegetation development. Examples from North Germany have been described by Svensson (1976), from Britain by Worsley (1966) and Morgan (1971), from Illinois by Johnson (1990) and from Canada by Morgan (1972, 1982). They have also been seen on air photographs of Ohio (H. Johnson, personal communication).

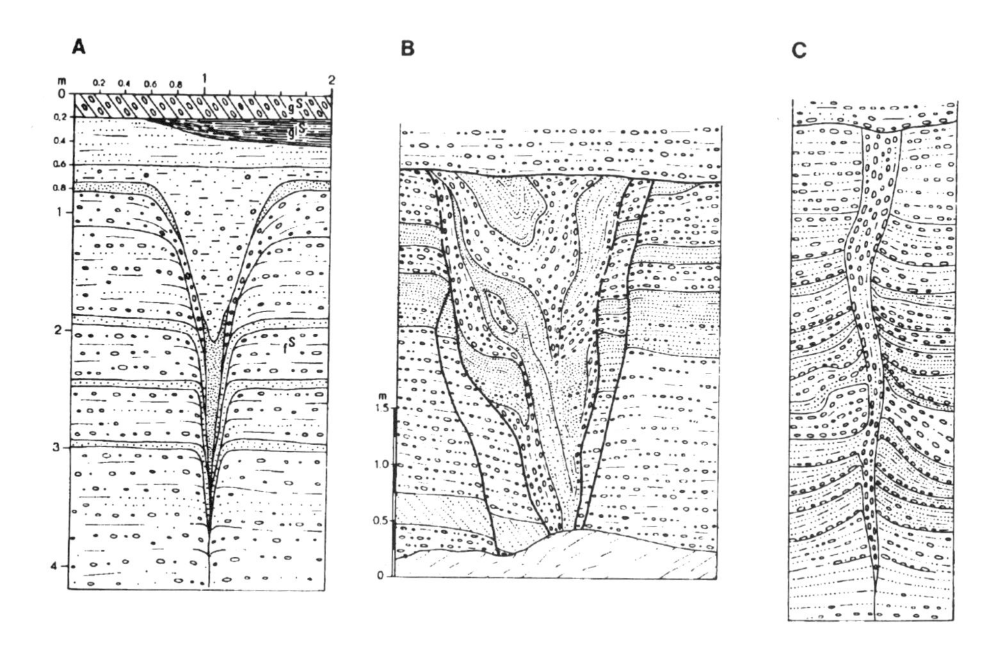

FIGURE 71 Typical ice-wedge and frost-wedge pseudomorphs from the Leipzig Lowland area, Saxony. A = ice-wedge pseudomorph with draw-down of adjoining layers, g^S = Saalian till, gl^S = Saalian Böhlen–Lochau Varved Clay, f^S = Early Saalian *Hauptterrassenschotter* ('Main Terrace gravels'); B = ice-wedge pseudomorph accompanied by graben-like downthrow features; C = narrow frost wedge with clearly upturned adjoining strata (from Eissmann 1981)

Under particularly favourable conditions, former ice-wedge networks may even be traced on sandur plains. As the moisture differences here are very small indeed, it is not expected that vast areas are perceptible from the air. In most cases only minor sections can be identified. The partial reconstruction of the polygons on Harksheide sandur plain (Fig. 72) is based on the evaluation of nine monitoring flights (Ehlers 1990a). The ice-wedge network shows throughout the same range of features. Pentagonal and hexagonal forms with a maximum diameter of about 20 m prevail, whilst rectangular forms occur locally. Polygon shape and size were found to be the same as those observed on till plains (e.g. Höfle 1983a).

Weichselian ice wedges are not entirely restricted to the unglaciated area. However, they are mostly small forms (Kolstrup 1980, H. Svensson 1984). The rapid warming at the end of the Weichselian Glaciation largely prevented the formation of ice wedges in western Europe; only when the cold returned during the Younger Dryas period could smaller fissures be formed once again (Böse 1991). In Poland, however, periglacial climate seems to have persisted after deglaciation, and full-sized ice wedges formed (Kozarski 1993).

From the Central European brown-coal districts immense deformations of strata have been known for a long time, including large-scale upturning of entire sediment series to form diapirs (Fig. 73). In the formerly glaciated area these features were originally related to the glaciations. However, because similar forms also occur outside the glacial limit, it soon became obvious that the diapirs are large periglacial structures (mollisol diapirs) (K. Kaiser 1958, Eissmann

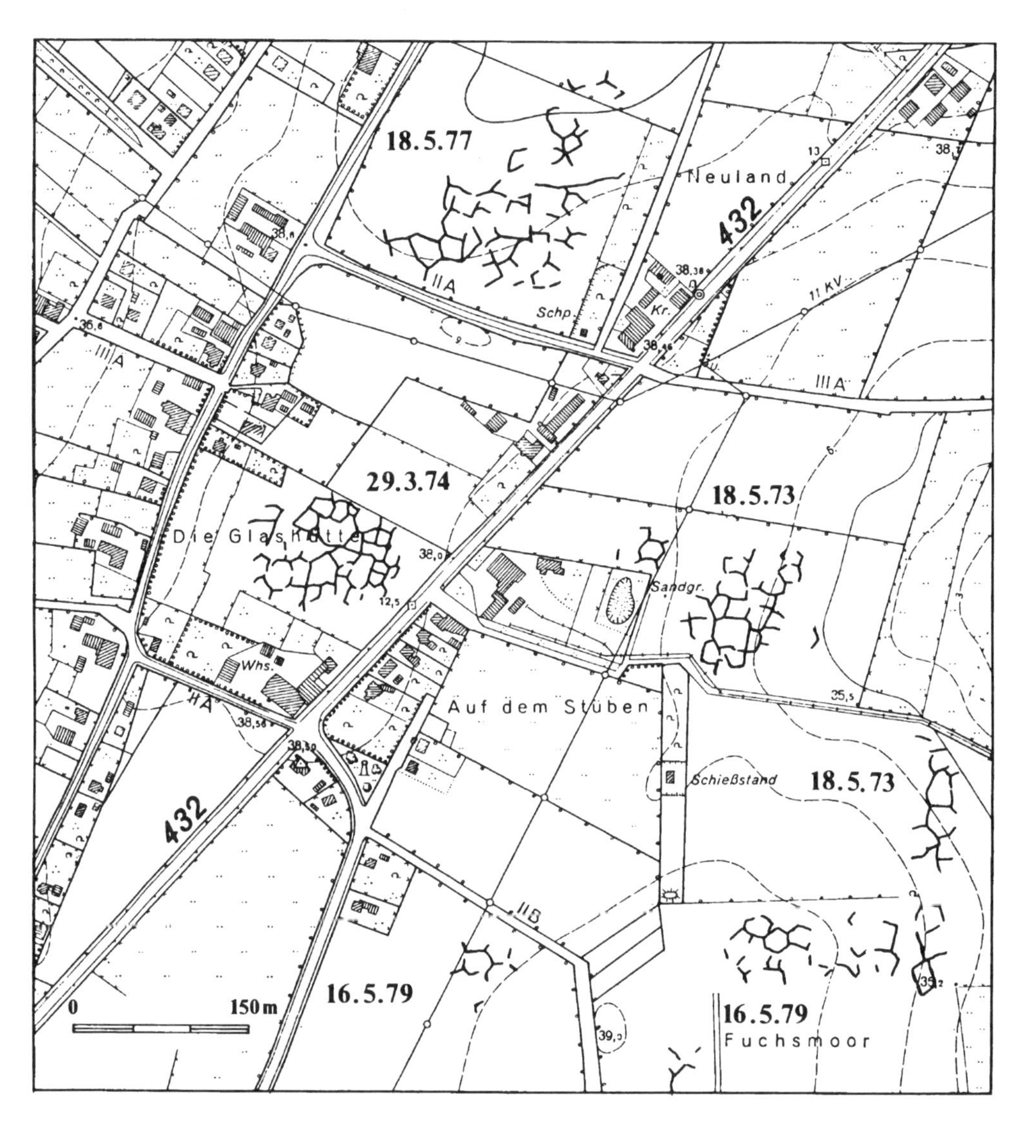

FIGURE 72 Ice-wedge polygon nets in the Harksheide sandur region north of Hamburg, reconstructed from several series of aerial photographs provided by Vermessungsamt Hamburg. Base map: Deutsche Grundkarte 1:5000, Sheet No. 6852 Glashütte-Nord (from Ehlers 1990a)

1975, 1978, Eissmann *et al.* 1995). Several generations of diapirs can be distinguished (Fig. 74). Indeed, Eissmann (1981) was able to demonstrate that diapirs in the Leipzig lowland formed during all cold stages of the later Pleistocene. However, the pre-Elsterian structures are distinctly smaller than those of the Elsterian, Saalian and Weichselian Stages. Based on the amplitude of the forms, assessments can be made of the thickness of each respective permafrost layer. Permafrost depth in central Germany during the early Elsterian Glaciation measured 18 m, in the early Saalian 30 m, during the Saalian maximum 40 m and in the Weichselian Glaciation 50 m (Eissmann 1981).

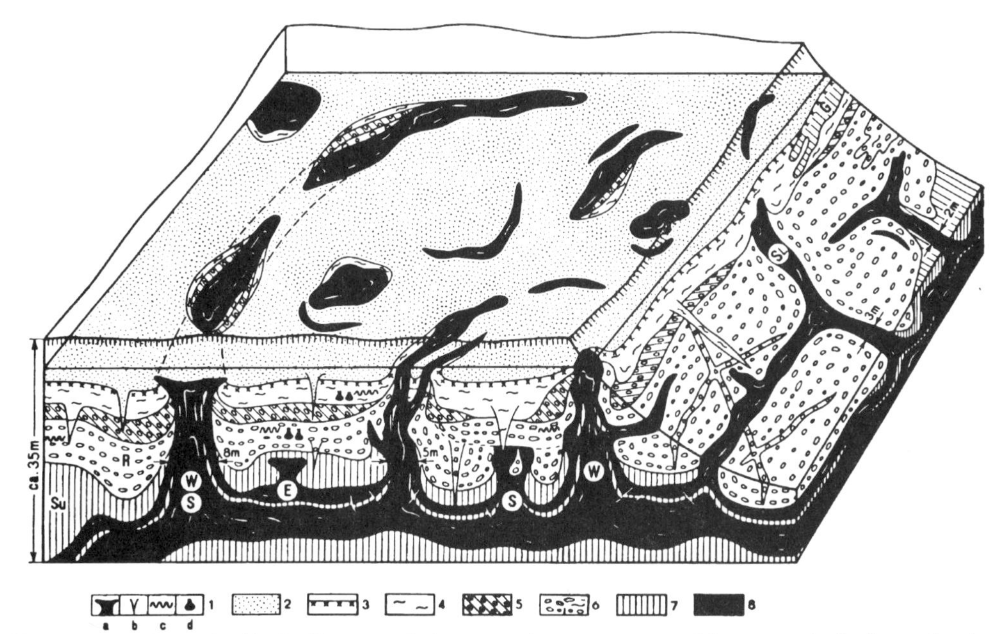

FIGURE 73 Synoptic block diagram of the most frequent types of brown-coal diapirs and other cryogenic features south of Leipzig. 1 = features of the permafrost and active layer: a = brown-coal diapirs, b = ice-wedge pseudomorphs, c = cryoturbations *sensu stricto*, d = drop soils; 2 = Weichselian loess complex; 3 = palaeosol of the Eemian Interglacial and Early Weichselian steppe soil; 4 = (late) Saalian solifluction and sediment flow complex; 5 = Saalian till; 6 = early Saalian sands and gravels ('Hauptterrasse') with intercalated solifluction layers; 7 = Tertiary clastics; 8 = brown coal with flow structures. Age of the brown-coal diapirs: E = late Saalian to early Saalian Cold Stage; S = predominantly early Saalian Cold Stage; SL = late Elsterian Cold Stage; W = Weichselian Cold Stage. Other abbreviations: G = slope failure; R = marginal trough; Su = karstic depression (from Eissmann 1981)

Remnants of ice-age pingos have been reported from the Netherlands (Maarleveld & Van den Toorn 1955), Britain (Sparks *et al.* 1972), various places in Germany (e.g. Garleff 1968) and Illinois (Flemal 1976). A number of these forms, however, may be traced back to other factors. Although organogenic palsas are not usually preservable, traces of minerogenic palsas can be found in the region of Pleistocene permafrost, e.g. on the Hohes Venn in Belgium (Vandenberghe & Pissart 1993).

Periglacial erosion in the North German lowlands has been described comprehensively by Lembke *et al.* (1970) (Fig. 75). Valley formation took place under the influence of permafrost, when the frozen subsoil prevented the water from seeping into the sandy substrate, so that it had to drain on the surface. On the slopes, solifluction occurred where the gradient exceeded 2°. The water-saturated active layer slid downhill like a mudflow, dragging along parts of the underlying substratum. Usually the finer material was washed out at the valley bottom, whilst sand and stones remained. This is how widespread stone layers formed at the bottoms of many recent dry valleys. Where steep slopes were undercut by the periglacial streams, slope failure and landslides developed. In the old morainic landscape, sediment reworking occurred repeatedly during several ice ages, and naturally only traces of the last (Weichselian) formation period are to be found on slope positions. In a number of valley floors, however, multilayered

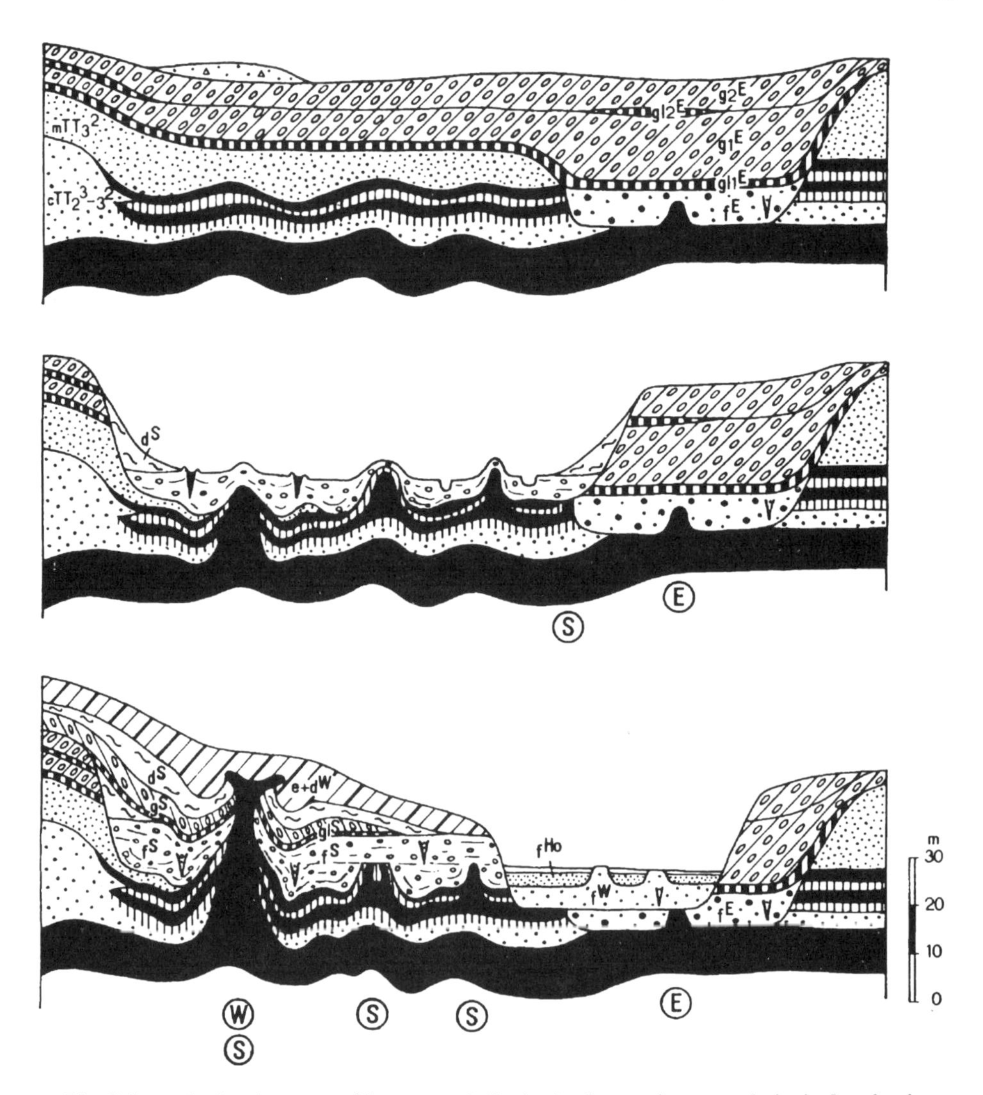

FIGURE 74 Schematic development of brown-coal diapirs in the southwestern Leipzig Lowland area, f^{Ho} = Holocene floodplain sediments; f^{W} = Weichselian Cold Stage 'Niederterrasse' (Lower Terrace); $e + d^{W}$ = Weichselian loess, solifluction and sediment flow deposits; d^{S} = Saalian solifluction and sediment flow deposits; g^{S} = first Saalian till; gl^{S} = Böhlen–Lochau Varved Clay; f^{S} = early Saalian 'Hauptterrasse' (Main Terrace); $g2^{E}$ = upper Elsterian till; $gl2^{E}$ = Miltitz–Pirkau Varved Clay; $g1^{E}$ = lower Elsterian till; $gl1^{E}$ = Dehlitz–Leipzig Varved Clay; f^{E} = early Elsterian terrace; $mTT_3{}^2$ = marine fine sands and silts; $cTT_2{}^3–_3{}^2$ = Tertiary terrestrial sediments with lignite seams; Diapirs: W = Weichselian, S = Saalian, E = Elsterian (from Eissmann 1981)

fills containing several gravel lags bear witness to repeated periglacial reworking (e.g. Fig. 76).

In the upland regions, the periglacial climate of the Quaternary cold stages also resulted in significant sediment movement. Central European upland slopes are, as a rule, covered by a sequence of several solifluction sheets that conceal the underlying rocks and form the substratum on which the recent soils developed (e.g. Schilling & Wiefel 1962, Semmel 1964,

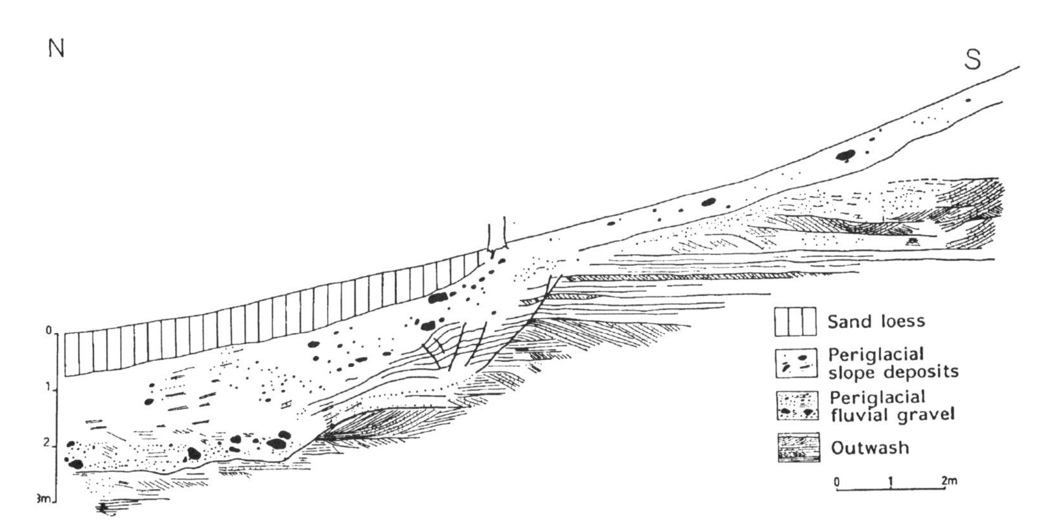

FIGURE 75 Profile from slope to bottom of a periglacial valley of the Frankfurt ice-marginal position at Buckow (Märkische Schweiz) (from Lembke *et al.* 1970)

1968, 1972, Stahr 1979). The overall extent of sediment reworking can only be assessed approximately, though Bremer (1989) reports that the solifluction layer of the last ice age, found at the foot of the south German escarpments, measures 3 to 5 m on average. Brunotte (1978) found Quaternary slope sediments to have an average thickness of 5 to 10 m at the foot of the Ahlsburg, an escarpment in the Niedersachsen upland region, and in some places periglacial deposits more than 20 m thick were found. In Niedersachsen and Hessen, Garleff *et al.* (1988) distinguished several generations of Quaternary pediments, which could partly be correlated with corresponding fluvial terraces. In these examples, the most recent (Weichselian) cover only forms a thin blanket, since the major part of the periglacial modification in the uplands took place during the earlier cold stages.

5.2.1 Fluvial processes

The fluvial processes in the periglacial region are often neglected in the available literature on periglacial morphology (e.g. Washburn 1979). However, that part of the Quaternary fluvial history that is actually represented by sediments and river terraces was largely deposited under periglacial conditions. The drainage behaviour of periglacial rivers resembles that of glacial meltwater streams. During the winter, frost brings drainage almost to a complete standstill; but in spring, when snowmelt sets in, there is a pronounced outflow maximum (the so-called nival flood). A crucial event in this process is the breaking-up of the winter ice, especially in those streams flowing from warmer to colder regions (e.g. the great Siberian rivers), which tend to be impounded for some time when thawing sets in. When the downstream ice barrier finally breaks up, this produces a flash flood with large floes of ice being carried away. The ice floes can transport major pieces of rock – so-called 'drift blocks'. Such flood events may result in enormous redeposition of material (Washburn 1979). Characteristic of the periglacial fluvial regime are braided rivers with wide floodplains, which give rise to the formation of valley trains (Fig. 77).

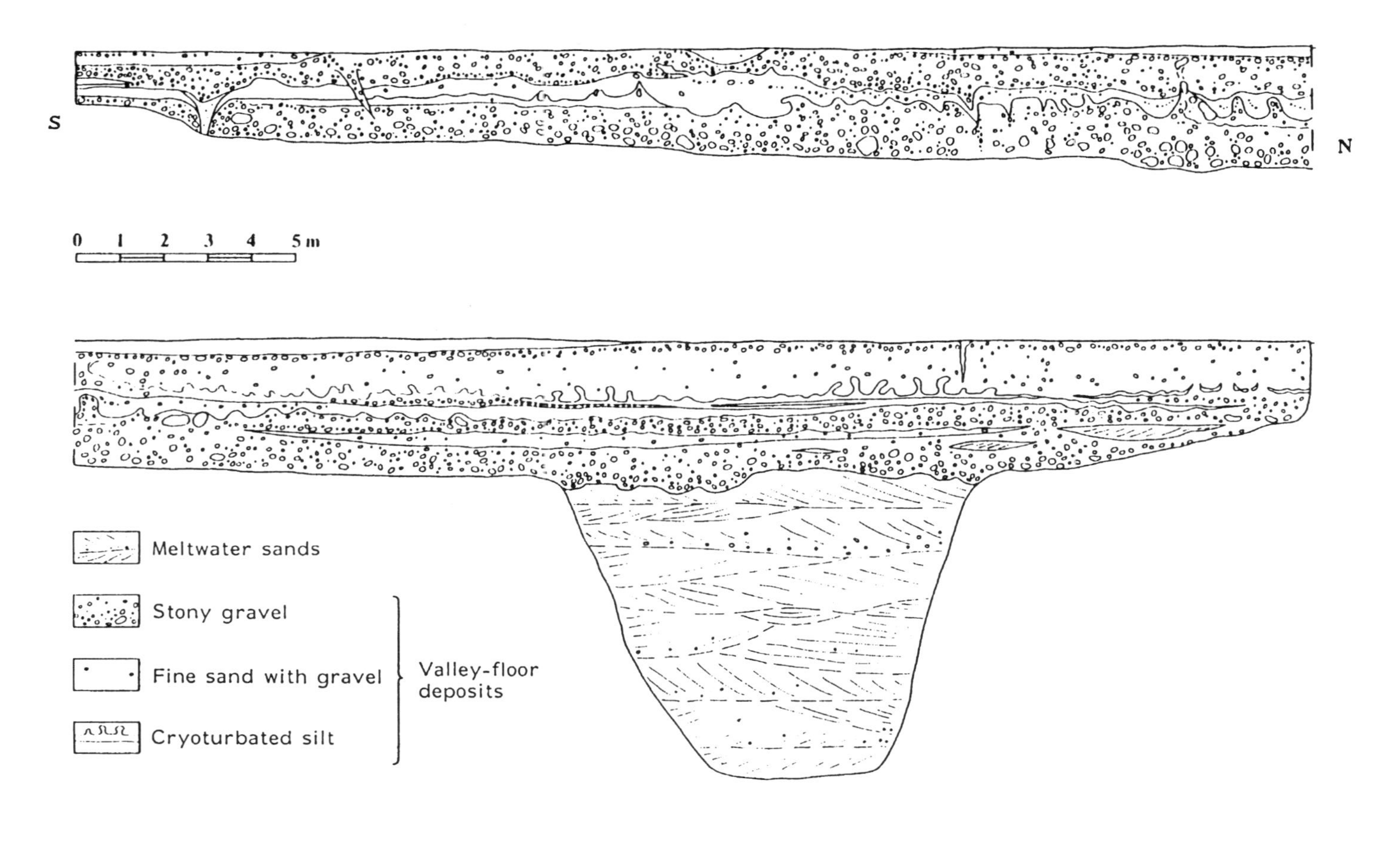

FIGURE 76 Longitudinal section through the floor of the Nenndorf dry valley at Eckel, south of Hamburg; vertical scale not exaggerated (from Ehlers 1978)

The mechanism of valley formation under periglacial climatic conditions has been long disputed. The old concept that ice ages were periods of accretion and interglacial intervals were periods of incision was questioned by Büdel (1969). His Spitsbergen explorations led him to the conviction that the periglacial region was an area of 'excessive valley formation', due to the effects of an 'ice crust'. As for Central Europe, Büdel assumed that during the wet and cold early glacial stages fluvial incision occurred, whilst during the cold and dry period of the glacial maximum aggradation took place. However, it is clear that during the cold stages repeated changes occurred between accumulation and incision. Thus, for instance, the last ice age in the Middle Rhine region is represented by two aggradational terraces (Bibus 1980). In addition, the style of sediments and their combination with cold-stage indicators give unambiguous proof that the great terraces of the Central European rivers were accreted under periglacial conditions (Semmel 1990).

Many of the valleys in Central Europe show an asymmetric cross section attributed to periglacial processes (Poser & Müller 1951, French 1972). Washburn (1979) points out that a series of factors may lead to the formation of asymmetrically shaped valleys. Amongst these are radiation, vegetation, wind direction, snow drift etc.

5.2.1.1 Terraces

Phases of erosion and accumulation alternate in every river as a result of short-term fluctuations in water supply. In the longer term, however, one of the two processes usually dominates. In the

FIGURE 77 Morsárdalur, a periglacial river valley in Iceland (Photograph: Ehlers 1990)

course of the fluvial history, nevertheless, there may be changes of the major processes. Whether a river changes from erosion to accumulation depends mainly on two factors – changes of water supply (climate) or changes of the gradient (tectonics). In the lower reaches of rivers, changes in sea level also exert a strong influence on fluvial dynamics. Multiple alternations between erosion and accumulation occur. During the accumulative phases, sediment is deposited on the valley floor, whilst during the following erosive episodes, these accumulations are cut through by the deepening river and partly removed, leaving behind a terrace.

The term terrace is used for both landforms and sediment bodies. In order to avoid confusion, the 'Commission on Terraces and Erosion Surfaces' decided that the term terrace should be used only to describe a landform but not the sedimentary body (Howard *et al.* 1968). However, this concept is not always applied. Geological mapping in Germany, for instance, uses nomenclature such as 'main terrace', 'middle terrace', and 'lower terrace' not only for landforms, but also as lithostratigraphic terms. This is problematic, since use of the word terrace for both landform and sediments is based on the concept that both can be attributed to the same aggradation cycle. However, terrace bodies can be built up in a very complex manner (Fig. 78), and, as a rule, geological investigation methods need to be employed to reconstruct their genesis.

Frequently terrace surfaces are correlated on the basis of their altitudes. This approach is again problematic because old terrace surfaces are seldom if ever unmodified river floodplain surfaces. This is because of later deposition or modification by periglacial or colluvial processes. Thus, the surface represents a hiatus in the sequence of strata, often comprising a longer period of time than the accumulated layers. The terrace base has also been used for correlation, but this is equally problematic because its formation may have been unrelated to the subsequent fluvial accumulation.

Information on the climatic conditions of the time in question are not available, but the sediments preserved contain clear indications of the conditions at the time of their deposition, and, therefore, the basis of terrace investigations should always involve analysis of their sediments. For distinguishing different sediment bodies, petrographical and sedimentological parameters should be utilised, such as grain size, fabrics, lithological composition and degree of weathering of gravel and sand fractions or heavy-mineral analyses. As an aid to the genetic interpretation of sediments, the degree of roundness (e.g. after Pettijohn 1975) or flattening (after Cailleux 1952) of clasts may be determined. The position of the individual sediment bodies relative to each other may be used to reconfirm the lithostratigraphy and to support correlations. Important clues are provided by investigation of internal bedding and position of the terrace base. An age-related classification of the strata can be achieved through analysis of intercalated organic deposits so that correlation can be achieved bio- or chronostratigraphically with other sequences. Furthermore, remains of molluscs or mammals, and palaeomagnetic measurements can provide clues for the sediment age. In addition, in younger terraces, archaeological finds are helpful for age-related classification. By these techniques a terrace stratigraphy can also be established in places where the sequence of terraces does not have a clear morphographic expression due to thick sedimentary cover, e.g. by loess (Gibbard 1985).

In places where the sediment body cannot be sampled because exposures are not available, the gradient of the terrace sediments or the landform surface can be used with caution. This is particularly possible in regions like the Alpine foreland, where terrace gravels were accumulated close to the ice front. The proximal parts of the terraces, i.e. the parts closest to the former ice margin, show the largest differences in altitude between the terraces of different age. When correlations are being made over greater distances it should not be

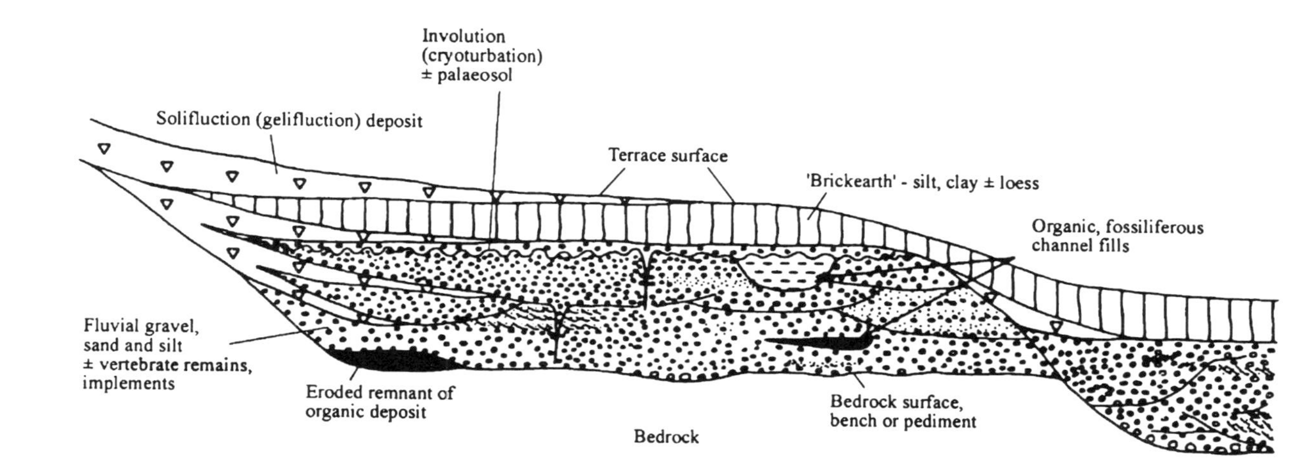

FIGURE 78 Section through an idealised terrace in the middle reaches of a river in southeast England (from Gibbard 1985)

forgotten that the gradient of the terrace body may differ from that of the recent valley floor and that differential uplift or downwarping may have resulted in tilting of the older levels of aggradation. For this reason, it is always advisable to relate the level of the terraces to sea level and not to the present valley floor (Tricart 1948).

Unfortunately, no universal system has yet been agreed for the nomenclature of terraces, so almost every author uses a nomenclature of their own, impairing extraregional correlation. In Germany, for instance, distinctions were initially made of a 'main terrace', a 'middle terrace' and a 'lower terrace'. This basic subdivision still exists for most of the rivers, but the meaning of the terms varies from place to place. Thus, for instance, the 'main terrace' of the Saale River was laid down in the Saalian, whereas the Rhine 'main terrace' was deposited as early as the Cromerian. After a short time, the subdivision was no longer sufficient to describe all the different terrace levels. This led to a subdivision of the middle terrace into an upper and a lower middle terrace and so forth. The swelling number of terraces eventually resulted in these notations being supplemented or replaced by a numbering system starting from the valley floor (e.g. Heine 1970). This succession, however, contradicts the numbering from older to younger units as is common practice in geology. Therefore an inversion of the numbering has been performed in more recent papers (e.g. Bibus 1980). Neither of these procedures is well-suited to describe complicated stratigraphic conditions. Difficulties regularly occur when further research reveals new members that must be added to the system. Therefore, it is highly recommended to work with local names or with type localities (Gibbard 1985).

The cold-stage terraces of the great rivers do not directly reflect the glaciations, though they are climate controlled. This applies to rivers in the periglacial region, as well as those originating in glaciated areas. In the Alps, all major Würm Stage gravel bodies predate the glacial maximum, and in formerly glaciated areas they are till-covered. Gravel supply began before the onset of glaciation, and virtually ceased at the Würm maximum. Drift blocks testify to deposition under periglacial conditions in braided river systems (Eppensteiner *et al.* 1973). Gravel supply was controlled by frost weathering. As soon as the climate ameliorated, the supply ceased (Van Husen 1981).

5.2.2 Aeolian processes

In periglacial regions, only a sparse vegetation generally develops. Along braided rivers the vegetation cover is often entirely lacking. Since a great part of the river bed is only filled with water during the springtime drainage peaks, vast areas are totally dry for long periods of the year, allowing large-scale deflation. Consequently, wind-blown accumulation of aeolian sediments is found all along the great Pleistocene river valleys – preferentially on the leeward sides. On the basis of their grain-size composition, a distinction is generally made between sandy and silty sediments (cover sand and loess), which largely appear separated geographically, but not always. Whereas the wind-blown sand may accumulate to form dunes, loess usually forms level layers with the surface only modified by subsequent redeposition.

Sand drifting with the wind can influence the subsoil, causing larger clasts as well as the bedrock to be gradually abraded. Not only ventifacts, but also numerous micro-patterns and facets on rock surfaces can be shown to originate from aeolian activity (Fig. 79). The high proportion of ventifacts in the former periglacial regions, such as the northern Great Plains of North America (Wayne & Aber 1991), reflect the strong influence of deflation in these barren or sparsely vegetated regions.

5.2.2.1 Cover sands

The largest Pleistocene dune regions of the western hemisphere are found in North America (H.T.U. Smith 1965). The Sandhills of Nebraska comprise an area of about 50,000 km^2 of stabilised dunes. Lugn (1935, 1968) introduced the term 'Sandhills Formation' for these aeolian deposits. The precise age of the dunes is still a matter of debate, and also their origin has not yet been adequately explained (Wayne & Aber 1991). However, it seems clear that a major period of dune formation occurred during the last 10,000 years (e.g. Ahlbrandt *et al.* 1983). Large dune fields are also found in parts of Canada, including the 'Great Sand Hills' of Saskatchewan as well as dune fields in the lower Churchill River valley (David 1977), or south of Harp River in east-central Labrador (Vincent 1989). One of the largest, still partially active, cold-climate sand seas is located on the coastal plain of northern Alaska, at the foot of the Brooks Range. It covers an area of about 11,600 km^2. Recent investigations indicate that the area has undergone stabilisation and later reactivation around 12 ka to 11 ka BP (Galloway & Carter 1993).

Weichselian to early Holocene wind-blown sands are also widespread in the western European lowland regions, such as in Great Britain (Catt 1977), Belgium (Pissart 1976), the Netherlands (Koster 1978, Ruegg 1983), Germany (H.-H. Meyer 1981), Denmark (Kolstrup & Jørgensen 1982) and Poland (Nowaczyk 1976); a collection of more recent works is found in Kozarski (1991) and a survey of the current state of research is offered by Pye & Tsoar (1990) and Koster *et al.* (1993). Chiefly, the wind-blown sands form a sheet-like deposit and, therefore, they are often referred to as cover sands (Koster 1982). True dunes with clearly shaped leeward slopes are rare, occurring preferentially in the vicinity of great river valleys and at the edges of the ice-marginal valleys. An overview of dune distribution and orientation in the European Lowlands and on the Russian Platform is provided by Zeeberg (1995). Whereas cross-stratified deposits dominate within the dune areas, horizontally bedded sequences prevail within the cover sands; they can also be distinguished from dune sand using other textural characteristics (Schwan 1988) (Fig. 80).

Within the Weichselian cover sand belt, a regional differentiation can be identified. Towards the east, the cover sands are replaced by dune fields mostly of the same age, which may be due to eastward-increasing aridity (Schwan 1988, Böse 1991).

Whether aeolian sand is concentrated to form dunes depends on the availability of a sufficient amount of sand and the moisture conditions. Climbing dunes, transverse dunes and/or barchanoid forms develop where abundant sand is available. A restricted sand supply with a more humid, vegetation-rich environment conversely result in the emergence of parabolic dunes. The latter may also develop into very elongated forms, finally producing longitudinal dunes. There are no specific shapes of dunes that can be linked to periglacial conditions (Koster 1988). Formation of sand patches instead of dunes occurs mostly where the amount of sand is not sufficient for dune formation. When coarse-grained residual sediments cover the terrain surface, deflation is impaired. Depending on the intensity of aeolian reworking, vegetation leads to the accumulation of either small isolated dunes or the deposition of irregularly shaped sand patches. There are no indications that the formation of either dunes or sand patches depends on the wind velocity (Dijkmans 1990).

The influence of snow on the deposition of aeolian sands was disputed for a long time (Pye & Tsoar 1990). The frequently observed stratification of sands had been connected with the alternating deposition of sand and snow. In other cases, snow was believed to be the reason for the lack of stratification. Koster & Dijkmans (1988) were able to show through field

FIGURE 79 Corrosion by wind-blown sand results in faceting of boulders (ventifacts): (a) faceted Rapakivi Granite, (b) preferential corrosion of kalifeldspar grains results in pitted granite surface (Photographs: Ehlers 1980)

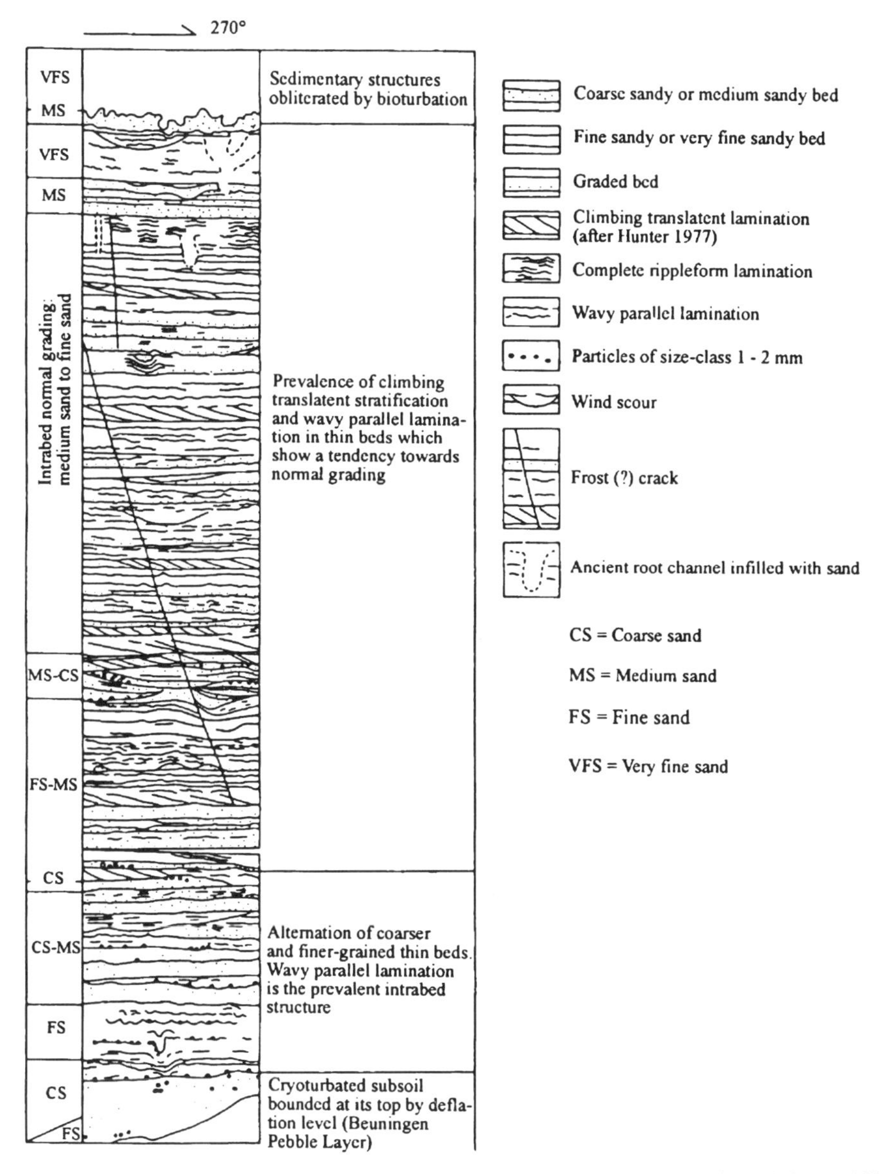

FIGURE 80 Typical aeolian cover sand profile from Twente, the Netherlands (from Schwan 1988)

investigations in the Great Kobuk Sand Dunes in northwest Alaska that specific thaw features may form on the leeward slopes of snow-covered dunes, primarily when an immense supply of sand exists. These structures result in various small-scale disruptions of the original aeolian stratification, so-called denivation features (Koster 1995). The sorting and almost horizontal stratification of the cover sands, however, can probably not be traced back to niveo-aeolian processes, and may result at least partly from periglacial slopewash (Habbe, personal communication).

Sometimes it is hard to distinguish whether sandy deposits that do not appear as dunes are of fluvial or aeolian origin. Cailleux (1942) showed that wind-transported sand grains are usually characterised by a matt ('frosted') surface, and Elzenga *et al.* (1987) distinguished aeolian and fluvial sediments based on this criterion. However, it should be taken into account that chemical processes may also produce matted grain surfaces (Washburn 1979).

Since the beginning of the Neolithic and the onset of deforestation, wind-blown sands in northwest Europe have been remobilised. These processes are partly documented by historical sources, but they can also be dated by pollen analysis and ^{14}C dating of sand-covered peat layers. A first great phase of aeolian sand drift commenced in Europe when sod fertilizing expanded (about AD 750 to 1200). For these more recent aeolian sands the term 'drift sands' is often used (Koster 1978, Castel *et al.* 1989, Castel 1991). The dune fields that arose from this process are called 'young dunes', in contrast to the older dunes of the Weichselian Stage and the early Holocene (Pyritz 1972).

5.2.2.2 Loess

The term 'loess' originally was used to denote the widespread fine soil in the Upper Rhine Graben area. It found its way into the scientific literature when it was used by Von Leonhard (1823/24). Lyell (1834) introduced the term into the English language literature. Subsequently, he also compared Rhenian and North American loesses (Lyell 1847). The origin of loess was long disputed. Lyell and many of his contemporaries regarded it as a fluvial deposit. Only after the investigations of Von Richthofen (1877) in China did it become clear that the European loess was an aeolian sediment originally deposited in a steppe-like landscape. In North America, the Mississippi valley loess was for a long time thought to be largely fluvial in origin (Chamberlin 1897, Russell 1944), but this concept was finally rejected by Holmes (1944), Leighton & Willman (1949) and Doeglas (1949). An overview of the state of knowledge on loess was provided by Pye (1987).

Loess is very well sorted, with a pronounced grain-size maximum in the coarse-silt fraction. Finer fractions cannot normally be taken up by the wind. In Australia, clay dunes ('parna') occur, but the clay has been transported as aggregates (Dare-Edwards 1984). Loess consists mainly of quartz, but to a lesser extent carbonate grains as well as feldspars, clay minerals and mica also occur. In addition, loess may contain heavy minerals which can be important for the relation of deposits to their source regions. Many loess deposits are found adjacent to major river valleys, in a downwind direction. This indicates that much of the loess was formed by deflation from poorly sorted outwash, accumulated in those valleys during cold stages. In summer, major parts of the valleys fell dry, where deflation occurred (Smalley 1972). Consequently, the greatest loess thicknesses are often found next to these rivers (e.g. Waggoner & Bingham 1961, Snowden & Priddy 1968).

There are two basic mechanisms that account for the generation of coarse-silt-sized quartz grains that form the bulk of loess material: glacial grinding (Smalley 1966) and cold-climate weathering (Zeuner 1949). Other mechanisms like salt weathering (Goudie *et al.* 1979, Pye & Sperling 1983), tropical chemical weathering (Nahon & Trompette 1982) or sandstone disintegration under hot, dry climates (B.J. Smith *et al.* 1987) may result in the same sort of grain size, but the contributions of these processes to global loess formation must be regarded as minor. Glacial grinding was mostly the dominant factor in North American and North European coarse-silt formation, whereas in Asia frost weathering in the mountains produced most of the desert loess (Smalley 1990).

Loess in pre-Quaternary deposits has rarely been identified (e.g. M.B. Edwards 1979). Quaternary loess is widespread. Indeed, Pécsi (1968) claims that it covers 10% of the global land surface, but 5% may be a more realistic estimate (Pye 1987). Loess is largely a cold-stage sediment deposited where the appropriate sediments were available for deflation. The correlation between loess formation and ice ages was pointed out by Soergel (1919), who found that loess was lacking on the young morainic plains (they were still ice-covered when loess was deposited), as well as on the most recent lower terraces (they accumulated during the last phase of loess formation). Thus, loess was identified as a sediment of the dry and cold glacial maximum.

In Central Europe, the northern limit of loess distribution lies approximately along the edge of the uplands. Thus a loess-free strip about 150 km wide remained in northern Germany and Poland between the margin of the Weichselian Glaciation and the loess belt. The clear northern boundary of the European loess has been interpreted as a result of different modes of transport. Whereas aeolian silt was transported by suspension, sand moved largely by saltation. However, there is no such clear distinction. In fact, a transitional sediment type is also found between sand and loess, that is referred to as **sand loess** (Vierhuff 1967). In his detailed investigations in the Lower Rhine region, Siebertz (1990) even distinguished six different types of aeolian sediments, which he grouped in a wind-blown sand series (grain-size maximum in the middle and fine-sand range) and in a loess series (grain-size maximum in the coarse-silt range), representing a more or less gradual transition. In Nebraska, the sand content of the Late Wisconsinan Peoria loess decreases from 60% near the ice margin to less than 5% in the more distal parts of the loess sheet; at the same time its thickness decreases from 28 m to less than 6 m (Lugn 1968).

In contrast to the northern Central European loess belt, the Alpine loess region reaches directly up to the margin of the glaciated area of the Würm Ice Age in some places (Grahmann 1932) (Fig. 81). The reason for this discrepancy lies in vegetation differences. Büdel (1951) assumed that the northern loess boundary must have been the borderline between vegetation-free congelifraction tundra and steppe-like areas covered by grass and herbs. The loess distribution also has an upper altitudinal boundary. Mountains like the Harz, the Rheinisches Schiefergebirge, or the Schwarzwald (Black Forest) are free of loess. Haase *et al.* (1970) give the upper margin of loess as at about 400 to 500 m. The altitudinal limit is interpreted as a vegetation limit, as is the northern limit.

In North America, loess deposits are widespread. They are found not only south of the glacial limit, but also cover the older part of the Wisconsinan glacial deposits meaning that loess deposition continued until the ice retreat. The loess belt reaches from the northwestern Rocky Mountains all across the United States. Its southernmost extension reaches along the Mississippi valley, where it is also thickest. In the east, a near continuous, though thin loess sheet reaches into eastern Ohio. Further to the east, only isolated small patches of loess are found (Ruhe 1983, Pye 1984).

The thickest loess deposits are found in China, where a maximum thickness of more than 300 m has been recorded near Lanzhou (Derbyshire 1983). On the Chinese loess plateau, the common thickness is about 80–120 m, whilst in Central Asia (Tajikistan and Uzbekistan) up to 200-m-thick loess occurs. In North America loess thickness rarely exceeds 60 m, and the same is true of the loesses of Argentina and of the Danube Basin. In most parts of Europe loesses are normally less than 20–30 m thick. Loess in Germany mostly reaches just a few metres in

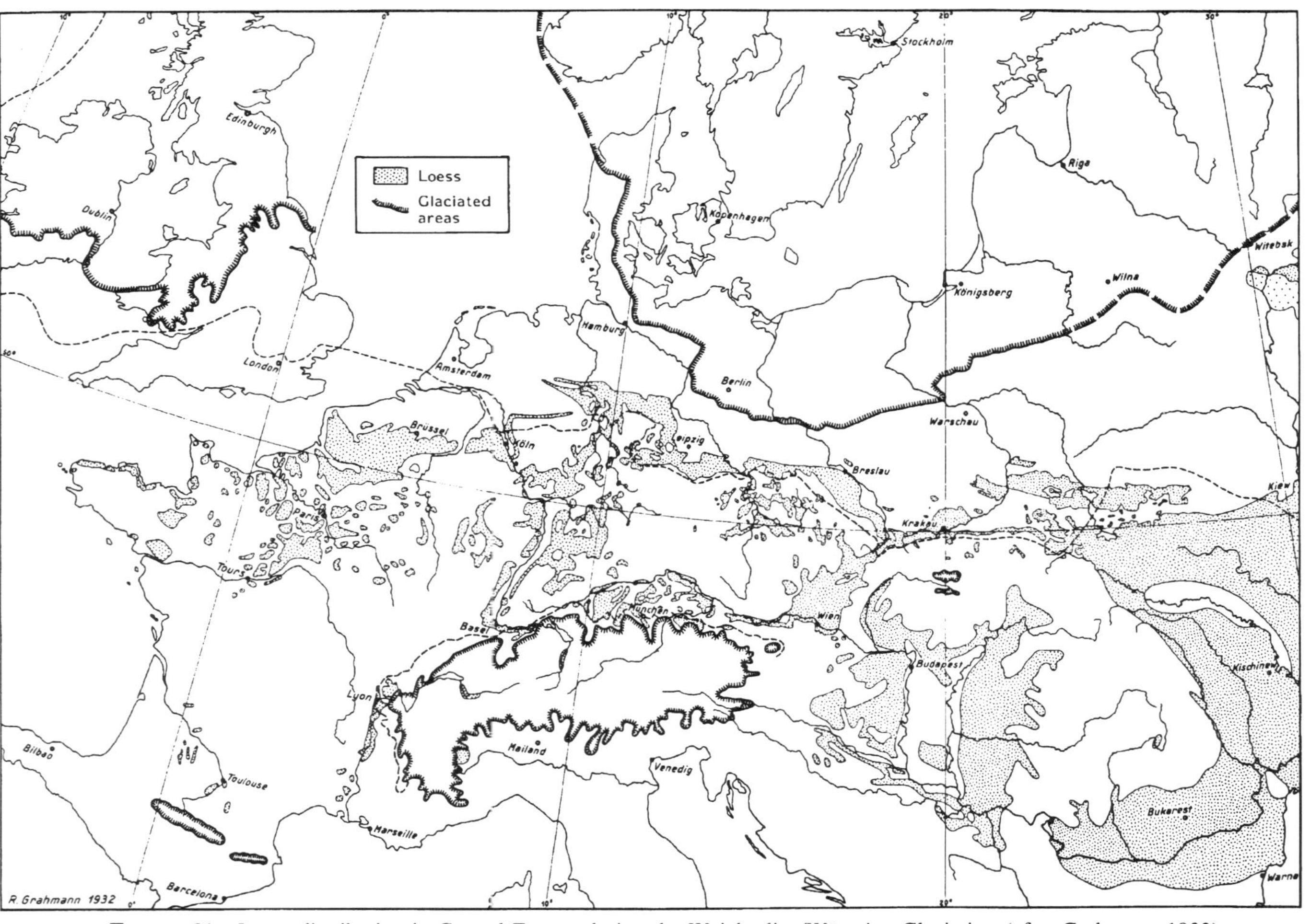

FIGURE 81 Loess distribution in Central Europe during the Weichselian/Würmian Glaciation (after Grahmann 1932)

thickness, although in the Kraichgau area loess is over 20 m thick and on the Kaiserstuhl volcano (Upper Rhine Graben) even up to 40 m. Significant loess deposits also occur in parts of Siberia, northern parts of India, New Zealand and Israel (Pye 1987).

In contrast to the wind-blown sands, loess normally forms a blanket covering older landforms. In mountainous areas it is thin on steep upper mountain slopes and thickens downslope. Gullying and river erosion can lead to formation of steep bluffs with an angle of 70–85°. Loess has a high permeability. Consequently in areas with a thick continuous loess cover, major parts of the drainage occurs subterraneously. Enlargement of drainage pipes can result in collapse of the roof and gully formation (Pye 1987).

6

Terrestrial Interglacial Environments

6.1 VEGETATION

The cold phases of the Pleistocene are interrupted by periods of more or less strong warming. These are subdivided into **interglacials** or **interglaciations** (warm stages) and **interstadials** or **interstades**. During the interglacials, climatic conditions were similar to or even warmer than today, whereas the interstadials were periods of less intensive warming. These climatic fluctuations were not felt worldwide with the same intensity, but differ gradually from one region to another. Whereas the polar regions were largely spared from climatic fluctuations, the middle latitudes experienced the most pronounced climatic changes. In contrast, in the lowlands of the inner tropics, constant climatic conditions largely prevailed throughout the Quaternary. In the mountainous areas, however, the altitudinal range of the different climates varied considerably. In addition, the coastal lowlands expanded considerably during periods of low sea level. As a result of the combined effects of these factors, the climatic changes of the Quaternary can also be reconstructed from many sedimentary sequences of the tropics. One example is the palynological investigation of sediment cores from the eastern Indonesian Seas (Van der Kaars 1991).

Climatic changes cause alteration of vegetation cover, the fauna, weathering and soil formation. These processes are interactive, causing climatic effects to be possibly modified by other factors, as is the case with soil acidification. Soil development during warm stages is an irreversible process leading to an increasing impoverishment of the substrate. The process can only be reversed if the substrate is replaced by fresh material.

The climatic changes of the past cannot be measured directly, but only reconstructed from preserved deposits. A particularly important indicator in this respect is the vegetation. In addition to floral macro-remains, pollen and spores provide ample and durable material, which can be used to reconstruct the vegetational history of warm and cold stages. However, in most substrates, nearly all organic remains are destroyed by oxidation. Favourable conditions for preservation are best provided by peat and gyttja, i.e. in mires and lacustrine deposits.

During the cold stages, the boundary of continuous forest cover shifted hundreds of kilometres towards the equator, and many areas close to the ice margins were effectively barren. The return of fauna and flora into these areas after the cold stages occurred surprisingly rapidly. Consequently, there has been much discussion about possible **refuges**, i.e. areas that remained ice-free and where at least part of the plant communities may have survived the cold stages. Of particular relevance to the concept of refuges are examples provided by islands or major peninsulas, such as Iceland or Scandinavia. The idea of refuges was first discussed at the end of the last century for Scandinavia. Warming (1888), for instance, thought that the flora of Greenland had survived in ice-free areas. Since then, some scientists have favoured the existence of refuges, whereas others supported a '*tabula rasa*' situation (Ives 1974). It seems that this question must be considered for each area separately. The discussion for the case of Iceland is summarised by P.C. Buckland & Dugmore (1991), who conclude that most species have re-immigrated by driftwood transport.

6.2 FAUNA

Whereas the faunal subdivision of the older geological periods is largely based on evolution, the Quaternary is too short for major evolutionary changes. Most species have remained largely the same throughout the last 2.4 million years. Faunal history of the Pleistocene in the middle latitudes largely reflects environmental changes, which are dominated by the climatic oscillations of the extremes between warm and cold stages. In Europe, during the cold stages, warmth-loving species either took refuge south of the Alps or died out. In the cold stages steppe faunas migrated into Europe from areas in northern and northeastern Siberia, Alaska and northwestern Canada. In the interglacials these faunas were replaced again by southern European–Mediterranean forms. In North America during periods of cold-stage low sea levels, immigration of species from Asia was possible via the Bering land bridge, provided that the passage was not blocked by glaciers. Each change from a warm stage to a cold stage and back triggered enormous changes in the faunal assemblages, not all of which were reversible.

In Europe, towards the end of the Cromerian Complex, the straight-tusked elephant *Palaeoloxodon antiquus* (formerly *Elephas antiquus*) replaced *Archidiskodon meridionalis* (formerly *Elephas meridionalis*), the southern elephant. Elephants (excluding mammoths and mastodons) are indicator fossils of the interglacials. The straight-tusked elephant became extinct after the Eemian. Other indicators of temperate conditions are red deer, roe deer and wild boar. A review of the climatic indicator value of large mammals has been given by Von Koenigswald (1988) (Fig. 82).

On the other hand, mammoth and woolly rhinoceros are generally regarded as indicators of cold stages. The mammoth developed from the steppe elephant (*Mammuthus trogontherii*), which appeared first during the Elsterian Cold Stage in central Europe. The woolly rhinoceros (*Coelodonta antiquitatis*) also occurred in the Elsterian for the first time. It became common within the cold-stage faunas after the Holsteinian. In combination with the true mammoth (*Mammuthus primigenius*), which occurred first in the Saalian Cold Stage, and the lemming (*Dicrostonyx*), these species can be regarded as typical elements of cold-stage faunas.

In contrast, wild horses characterise the transition between glacial and interglacial stages. They are elements of the steppe or savanna faunas. However, they are also present in open biotopes of the interglacials (Kahlke 1981).

6.3 WEATHERING AND SOIL FORMATION

The fundamentals of weathering and soil formation are discussed in detail in textbooks such as Catt (1988a), Buol *et al.* (1989) and Scheffer & Schachtschabel (1989). To provide an overview of soil formation for an international readership is difficult, because there is no internationally

FIGURE 82 Large mammals from Late Pleistocene deposits in the northern Upper Rhine Graben and their climatic demands (from Von Königswald 1988; drawings from Thenius 1962)

agreed soil classification system. The soil taxonomies used in various countries differ both in concept and terminology. Modern soil science goes back to concepts developed by Dokuchaev (1883), but they have diversified since then. The USA, Canada, Britain and Germany now all use different soil classifications, so the following text tries to avoid regional terminology as far as possible. For further soil studies the reader is advised to use local textbooks and local classification systems, such as Canada Soil Survey Committee (1978).

During the cold stages, soil formation (pedogenesis) was limited in the ice-free areas close to the ice margins. Chemical weathering and leaching were active in wet oxidising environments; clay translocation was inhibited as it is today in most arctic environments. The periglacial areas were sparsely vegetated, but where the decomposition of organic matter was restricted, it accumulated, in wetland areas for example. Arctic soils (Entisols of the US classification) developed in periglacial landscapes characterised by patterned ground. Both in the glaciated areas and in the periglacial zone, rejuvenation of soils was the dominant process, either by glacial deposition, aeolian processes or periglacial reworking of the active layer (see chapter 5). With the onset of warming, differentiated soil profiles develop (Fig. 83). As the pedogenic process proceeds, a succession of soil characteristics appear that provide a basis for classifying the soil at different stages. A common progression of soils on a typical till in Europe and North America develops from an Arctic soil (Entisol) to argillic brown earths or gleys (oxidised or wet Alfisols) (Fig. 83b). In general the stages of soil genesis are thought to be as follows:

1. During the early stages of the Holocene, the spreading vegetation provides litter of leaves and plant remains. Action of soil organisms mixes the organic matter with the soil substrate, forming a relatively thick topsoil rich in both humus and carbonate, and with a granular structure. The raw (parent) material thus develops into a soil profile with a humus-rich A horizon, overlying unweathered parent material, the C horizon. Further development of this A/C soil profile largely depends on its position in the landscape, which controls Eh (oxidation or reduction) conditions.
2. Leaching of carbonates can start before the ground is influenced by organisms. Leaching leads to a continuous deepening of the soil profile. The rate of leaching decreases with increasing depth because a portion of the percolating water never reaches the lower parts of the soil profile. Decalcification is followed by iron oxidation and formation of clay minerals, largely as a result of mica weathering. A new soil horizon is formed between the

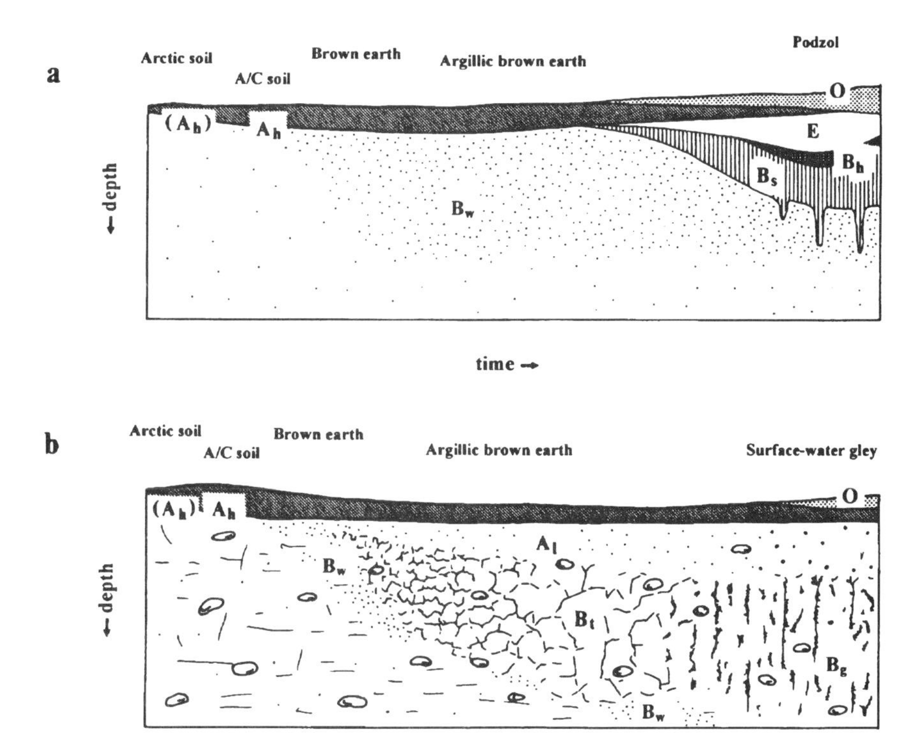

FIGURE 83 Possible postglacial soil profile differentiation on different substrates: (a) on sand, and (b) on till. Vertical axis: depth, horizontal axis: time (after Scheffer & Schachtschabel 1989)

A and C horizon, enriched in sesquioxides (iron, manganese) and clay. An A/B/C horizon profile develops, which in common terms is called a **brown earth**.

3. In contrast to the topsoil, the lower part of the soil profile contains few living organisms. Decalcification here results in a relative enrichment of clay and other minerals. By periodic expansion and shrinking of clay minerals in the B horizon, the soil attains a more blocky structure below the surficial horizons in fine-grained material. In the Holocene, intensive clay translocation occurred especially during the climatic optimum. This resulted in the formation of clay-rich (argillic) Bt horizons, which characterise the **argillic brown earths**.
4. Increasing clay translocation into the subsoil results in enhanced shrink–swell processes. Especially under the cooler and moister climate conditions after the Holocene climatic optimum, this resulted in seasonally poor drainage conditions. Shrink–swell in the subsoil results in formation of a prismatic structure. The argillic brown soils develop into **surface-water gleys**, provided there is enough rainfall. In Britain, interglacial soils are greatly enriched in illuvial clay, but drainage is not impeded because of frost disturbance during the Devensian Stage (Catt 1992).
5. Leaching of bases and nutrients leads to further acidification of the topsoil and consequently to a reduction in activity of living organisms. Reduced organic matter decomposition results in the accumulation of a thick litter layer. It overlies a bleached Ae horizon, stripped of sesquioxides by leaching under acid conditions. This is a characteristic of a **podzolic soil**. In the northern part of North America this stage has been reached only in the Great Lakes region and eastern Canada.

This sequence of soil formation has been repeated during all Quaternary interglacials in a similar way. Often, this cycle of soil formation was not completed in the relatively short periods of time between two cold stages. Also, soil formation has not always proceeded unhampered. Climatic deterioration (e.g. during the Younger Dryas) resulted in widespread interruption of soil formation and renewed periglacial mixing with the substrate.

The pedogenic pathway on sandy parent material produces a different sequence of soils. Because of the greater permeability of meltwater sands, the formation of a thick topsoil on these takes much less time than on till. As carbonates can also be leached more rapidly, brown earths can form after a short period of time. During this process, under pH values of about 5–7, the clay minerals formed by mica weathering are partially translocated and accumulated in the C horizon as clay bands. The cool and moist climate after the climatic optimum favoured acidification of the soil and loss of nutrients. This resulted in strong podzolisation, during which humus, iron and aluminium from the topsoil accumulated in the B horizon to form continuous horizons (Fig. 83a).

On loess, soil development in Europe during the early Holocene proceeded from incipient soils (Entisols) to chernozems (Inceptisols, or, if the dark A horizon was more than 25 cm thick, Mollisols). This process came to an end at the beginning of the Atlantic period, when a more humid oceanic climate encouraged forest expansion. The German chernozems are regarded as relics of early Holocene pedogenesis. The typical warm-stage soil on loess in western and Central Europe is an argillic brown soil (Alfisol). In eastern and southern Europe, chernozem formation continued far longer, in some areas lasting until the onset of cultivation of the steppe areas.

A special case of soil formation not restricted to interglacial climate are the **bogs**, because they consist of horizons with more than 30 weight-percent organic matter. Bogs form in areas of high water tables, or where surface water accumulates periodically and organic litter cannot be decomposed due to lack of oxygen. Two types of bogs are distinguished; the topogenic bogs formed on seasonally inundated land surfaces, and the ombrogenic bogs, formed because of an oversupply of rain water, independent of the groundwater table. A perhumid climate, with very high precipitation, results in increased accumulation of organic matter and finally formation of peat. Bogs develop best in a cool climate with acidic water, because under these conditions the activities of humus-decomposing soil organisms are decreased.

Wetlands are formed by the silting up of standing water, such as ponds, in the backwater of a river, around springs. The basic material is mainly reed (*Phragmites*), cat-tail (*Typha*), and/or sedges (*Carex*). When peat growth has reached the average level of the water table, the swamp plants are supplemented by alder (*Alnus*) and willow (*Salix*) (Overbeck 1975).

As soon as a freshwater bog grows above the water table, birch and pine start to spread on nutrient-poor substrates. This type of bog is an intermediate form between a freshwater bog and a raised bog and is therefore called a **transitional bog**. With ongoing bog growth, the plant community changes. Its nutrition relies increasingly on precipitation, which is poor in nutrients. Alder and willow disappear as rush, cottongrass, and peat moss (*Sphagnum*) spread. Some *Sphagnum* communities form a **raised bog**, which is often raised in its centre like the glass of a watch. Wetland development can lead from the formation of a mud via a freshwater bog and transitional bog to a raised bog. However, there are also raised bogs that have developed directly on minerogenic soils, e.g. podzols (Spodosols) or groundwater gleys (Entisols). The muskeg of Canada is a bog of either type (Pielou 1991).

Development of many recent bogs started in the lateglacial and postglacial with the formation of lacustrine deposits. The gyttjas of some dead-ice depressions can reach thicknesses up to 20 m. Under the warmer climate of the early Holocene, the ponds were silted up and freshwater bogs started to form. In Central Europe, many raised bogs formed during the hypsithermal Atlantic period (8000–5000 BP) with its humid and warm climate. In North America, where increased humidity started after the hypsithermal period, raised bogs began to develop between 4000 and 5000 BP.

Peat growth in coastal lowlands was partly favoured by the postglacial rise in sea level, which also caused a considerable rise of the groundwater table. Such was the case in the coastal areas of Germany, Denmark and the Netherlands (Scheffer & Schachtschabel 1989).

Soils on glacigenic deposits in the areas of the last and preceding glaciations may differ in thickness and intensity of development. On tills of the last glaciation in North and South Germany, argillic brown soils strongly influenced by surface water predominate. However, the prevailing soils on older tills of comparable composition are podzolic surface-water gleys, due to greater weathering and clay translocation (Stremme 1981). The difference is because for soil formation on Weichselian till, only the Holocene was available. In comparison, the older tills have undergone weathering and pedogenesis during two interglacials and a number of interstadials, leading to the formation of rubefied argillic brown earths in southern Britain (Catt 1988a). These intensive soil-forming processes could not be fully compensated by periglacial mixing and rejuvenation of the profiles (Stremme 1981).

Soils outside the limits of the last glaciation in many cases exhibit a number of features that developed before the Holocene. In some places traces of old warm-stage clay-illuviation horizons are found, the clay cutans of which were often fragmented during the Weichselian permafrost period. Such remains of older soil-forming processes are called **relict features**, and the soil is referred to as a **relict soil**. Periglacial aeolian processes have caused an enrichment of

FIGURE 84 Fossil soil of the Brørup Interstadial at Sonnenberg, Niedersachsen, North Germany. The soil profile is disturbed by periglacial involutions (Photograph: Ehlers 1980)

silt (loess) or sand in some soil profiles. The material may have been displaced by periglacial slope processes or mixed *in situ* by cryoturbation, so that in the upper part of the profile a new substrate was formed over old, autochthonous soil relics. In the Holocene, as a result of agriculture, humic material has often been washed downhill from the upper slopes and accumulated in the lower slope areas as colluvium (Catt 1987). Redeposited soil material such as colluvium is referred to as **soil sediment**.

Soils can also be used for stratigraphic correlations. In a number of places within the formerly glaciated area, **palaeosols** (fossil soils) are found under a cover of cold-stage deposits (Fig. 84). In soil stratigraphy, especially in North America, the individual units are referred to as **geosols**. The methods of palaeosol investigation have been outlined in a manual by Catt (1990).

Although in North America soil stratigraphy plays a major role in subdividing Quaternary deposits, in Europe it is largely restricted to loess sequences (see below). In Britain, palaeosols were first introduced as stratigraphic units by Rose *et al.* (1976), and a few more palaeosols have been identified since then (Whiteman 1981, Rose *et al.* 1985a, b, Kemp 1987). However, relatively few palaeopedological studies are available from the glaciated parts of continental Europe. The potential contribution of soil science to terrestrial Quaternary stratigraphy has not so far been fully exploited. Stephan (1981) has listed 28 occurrences of Eemian palaeosols in Schleswig-Holstein alone. He suggested that the type of soil formation during the last interglacial was similar to that of the Holocene. For the fossil, polygenetic Eemian soil in the

Schalkholz exposure in Schleswig-Holstein, for example, the following sequence of pedogenic events could be reconstructed: fresh till – A/C soil – argillic brown earth – argillic brown earth/surface-water gley – surface-water gley/podzolic soil – podzolic soil/gley. This should come as no surprise, because the similar climates of the Eemian and Holocene should result in similar pedogenetic pathways (Felix-Henningsen 1979). However, most Eemian soils have been strongly disturbed during the early Weichselian by periglacial processes. For example, cryoturbations and traces of solifluction are frequently observed. Typically, intrusions or incorporations of often white, gravel-free, fine to medium sands are discovered, which were originally probably deposited as cover on the Eemian land surface. Indeed, Stephan (1981) assumes these are the remnants of an aeolian cover sand.

Macroscopic identification of a palaeosol within a sequence of strata may be difficult. Most of the above-mentioned occurrences of palaeosols in Schleswig-Holstein were found in exposures, not in boreholes. Therefore, it is not surprising that older terrestrial soils (e.g. of the earlier Holsteinian Interglacial) have been observed only rarely (e.g. Felix-Henningsen 1979, Stremme 1981).

Beyond the maximum extent of the glaciations, Quaternary soil formation resulted in the alteration of older weathering profiles, already partially formed before the Quaternary. In the Rheinisches Schiefergebirge of central Germany, for instance, a weathering crust up to 150 m deep developed in the warm and humid climates of the younger Mesozoic and Tertiary. This often consists of a thin solum (distinctive soil horizons with structural characteristics) and a thick saprolite layer (deep weathering zone preserving original rock structure) (Felix-Henningsen 1990).

In contrast to Europe, the interglacials in North America are not defined by palynological investigations, but primarily by palaeosols. The Sangamonian Interglacial, which is the North American last interglacial, can be traced by a strongly developed soil from the Great Lakes southwards into western Texas. This soil varies according to substrate and climatic conditions. It is strongly developed both in loess and glacigenic deposits and can be used as a stratigraphic marker horizon. In the western United States, where no Illinoian tills have been found beyond the Wisconsinan maximum, the Sangamonian soil has often overprinted the remnants of soils formed during the preceding interglacials (Flint 1971).

However, as tools for stratigraphic correlation, palaeosols should be used with care. Not only is the precise onset and termination of soil-forming periods difficult to determine, but also the influence of parent material and relief may have been stronger than climatic factors. Diagenetic changes in buried soils have to be considered, and, of course, many former soils have been lost by subsequent erosion. Nevertheless, where they have been preserved, they can provide valuable information about the palaeoenvironment (Catt 1988b).

6.4 HUMAN ACTIVITIES

Man entered the history of the earth in the Pliocene. The oldest hominid finds (*Australopithecus afarensis*) in Ethiopia have been dated to ca. 3.8 Ma (C.M. Hall *et al.* 1984) and in Tanzania to 3.46–3.76 Ma (Drake & Curtis 1987). 'Lucy', the youngest specimen of *Australopithecus afarensis* known so far, from the Ethiopian Hadar Formation, has been $^{40}Ar/^{39}Ar$ dated to 3.18 Ma (Walter 1994).

The earliest traces of human activities in Europe are much younger. It is estimated that settlement started roughly about one million years ago. The few archaeological finds in most cases cannot be precisely dated (Ullrich 1989). Finds of early man (*Homo erectus*) from Europe are only known from seven places: Petralona in Greece (age disputed), Arago in France (about 70 remains), Ranuccio, Pofi and Castel di Guido in Italy, Vérteszöllös in Hungary, Mauer and Bilzingsleben in Germany. Damage of skull fragments found has often been interpreted as resulting from dismembering of corpses and special death rites (e.g. Ullrich 1989). However, most of the defects might be also easily explained by natural causes (Czarnetzki 1983). The oldest human find in Germany, the lower jaw fragment from the Grafenrain sand pit at Mauer (near Heidelberg) was found in 1907. However, the precise stratigraphic position of the find horizon of this *Homo erectus heidelbergensis* was much disputed. Whilst some workers put it into the Mosbachian (Cromerian Complex), others regarded an interval within the Mindel (i.e. Elsterian) Cold Stage as possible. Today it is correlated with the upper Cromerian Complex (Roebroeks *et al.* 1992, Roebroeks & Van Kolfschoten 1994, Müller-Beck 1995). The recent find of *Homo erectus heidelbergensis* from Boxgrove in southern England from coastal sediments also of Late Cromerian Complex age (Roberts *et al.* 1994) reinforce the view that humans first colonised northern Europe about half a million years ago.

Subdivision of archaeological finds is achieved largely on the basis of artefacts. The whole Quaternary until the end of the Weichselian glaciation was summarised under the Palaeolithic Period, the Old Stone Age. Further subdivision into an Early, Middle and Late Palaeolithic was based on changes in the tool-making techniques (Bosinski 1974), not on geochronological periods (e.g. Toepfer 1970), because the various techniques have been in part applied simultaneously. It is often impossible, especially for the pre-Weichselian, to provide a reliable age estimate, whereas classification within a certain industry should always be possible (Bosinski 1974).

With regard to the early settlement of Europe, some additional knowledge has been obtained in recent years. A key section for understanding the early settlement of Germany was found in the Kärlich clay pit (Neuwied Basin). Here, 12 archaeological horizons were identified, including the oldest evidence of man in Central Europe (Horizon Kärlich A; see Fig. 210) (Würges 1986). The more comprehensive finds, however, were detected in a higher sediment layer, deposited during the Kärlich Interglacial. On a peninsula within a small lake a group of *Homo erectus* settled during this interglacial, which had been dated to ca. 400,000 years BP (Van den Bogaard *et al.* 1989). Apart from numerous animal remains (straight-tusked elephant, cattle, horse, wild boar, red deer), stone and bone tools, hazel shells and fruit remains were found, indicating gathering activities of early man (Kröger *et al.* 1988). The interglacial palynologically resembles the Cromerian at the Bilshausen site (Urban in Bosinski *et al.* 1980) but might also be younger than Elsterian.

Early Palaeolithic artefacts are relatively rarely found. In Germany, the finds include the post-Elsterian sites of Wangen, Wallendorf, Markkleeberg (Grahmann 1955, Baumann & Mania 1983), Hundisburg (Fig. 85), Wolmirstadt and Gerwisch (Toepfer 1970). Middle Palaeolithic hunting places are more frequent. They include the famous Lehringen site (near Bremen), where a straight-tusked elephant was found with a yew spear 2.5 m long (Adam 1951, Thieme & Veil 1985). In 1985 a Middle Palaeolithic hunting place was found in the Neumark-Nord opencast lignite mine in the Geiseltal valley (Mania 1990). Two years later a second (Eemian) slaughter place of straight-tusked elephant was discovered at Gröbern, Kreis Gräfenhainichen, near Leipzig (Litt 1990).

At the transition from the Early Palaeolithic to Middle Palaeolithic, at the beginning of the Saalian Cold Stage, *Homo erectus* is replaced by early *Homo sapiens* (found, for instance, at

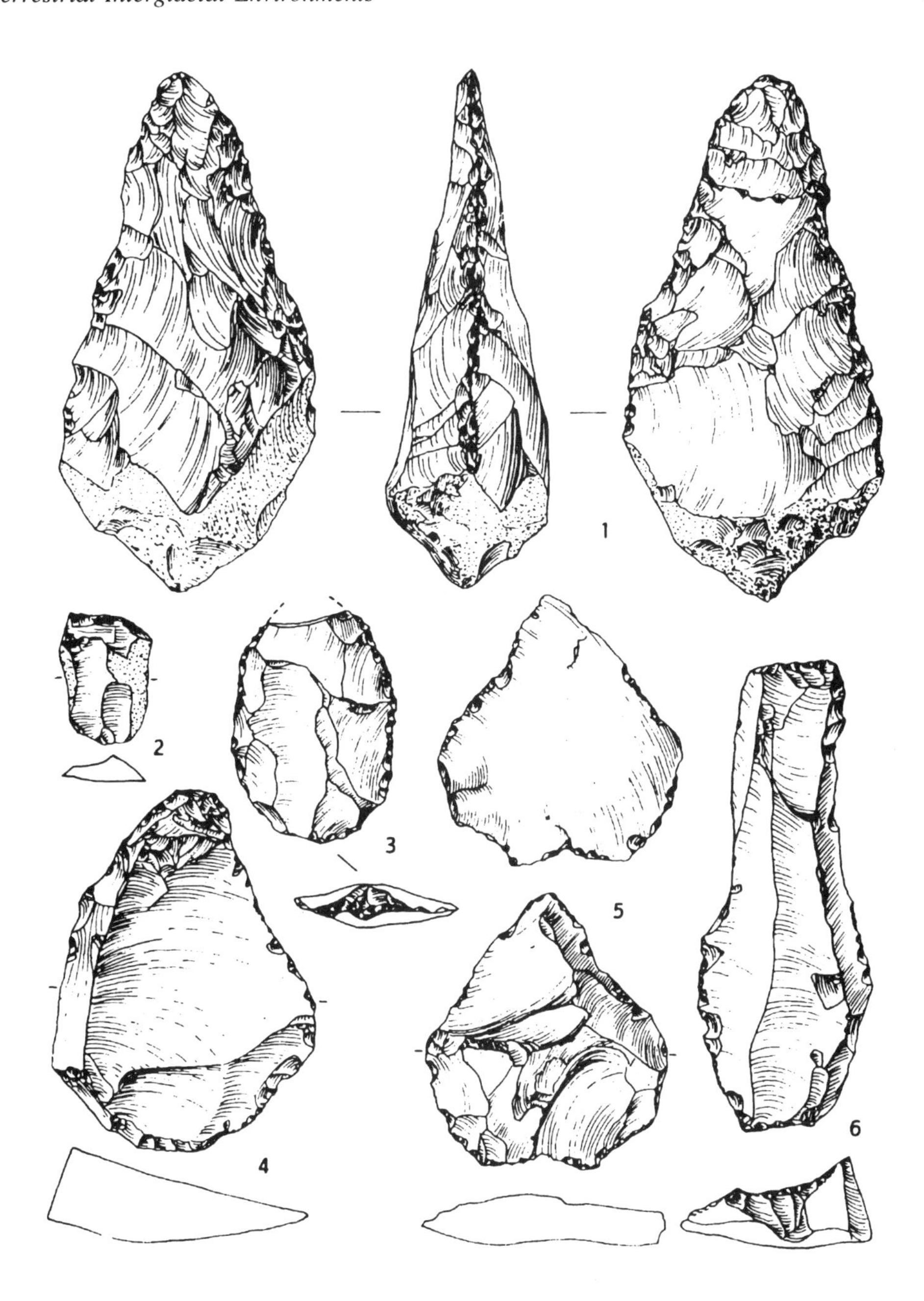

FIGURE 85 Acheulian artefacts (Early Palaeolithic) from Hundisburg, Kreis Haldensleben. 1 = handaxe, 2 = scraper, retouched on both sides, 3 = leaf-shaped point with retouched surfaces, 4 = flake, 5 = flake, 6 = Levalloisian blade (from Toepfer 1970)

Steinheim, Ehringsdorf and Swanscombe). The peak period of the subsequent *Homo neanderthalensis* falls into the second half of the Weichselian. It is assumed today that recent man (*Homo sapiens sapiens*) developed from Neanderthal man. Excavations at Mugharet Es-Skhul and Qafzeh in Palestine have detected unequivocal evidence of a transition between these two types (Bosinski 1985). Caves and rock protrusions formed preferred dwelling places, but, where these were missing, humans also settled in the open landscape. The best-known examples of the latter were excavated at Molodva on the Dniestr River (Ukraine). They consist of rounded buildings, about 8 m in diameter, the base of the wall being formed from mammoth bones (skulls, tusks, shoulder-blades, hips and long bones). On this base apparently a cupola-like, fir-covered construction was placed. Other sites of this type were identified at Ketrosy on the Dniestr and at Ripiceni-Izvor on the Pruth River (Romania). In Germany, a dwelling-place in the open landscape was found dating back to the Late Saalian (at Mönchengladbach-Rheindalen). The settlements apparently shifted seasonally in order to provide favourable conditions for hunting, or gathering of wild fruits (Bosinski 1985).

In contrast to Europe, North America was settled relatively late. The earliest claimed traces of human presence date back only to about 40,000 BP. However, the oldest indisputable settlement reconstructed so far was occupied by the Clovis palaeo-Indians in the western United States in the period about 11,500–11,000 BP, at the end of the last glaciation (Aikens 1984).

In the Early Palaeolithic, man was not yet able to interfere massively with nature and should perhaps be regarded as simply a member of the vertebrate fauna. He lived from hunting animals and gathering wild foods. Settlement density was extremely sparse. Only in the Holocene, in the early Atlantic period (about 6700–6400 BP), with the beginning of the Neolithic period (Late Stone Age), did agriculture begin. This led to extensive clearing of the forests and gradual restructuring of the natural landscape into the cultural landscape of today (Behre 1988).

In North America, agriculture may have started at about 4500 BP. However, it did not become a significant factor until much later. In Ontario, Indian farming started about AD 500 and ended as a result of tribal warfare about 1660 (McAndrews & Boyko-Diakonow 1989). Only after large-scale immigration of European settlers, beginning in the late 18th century, did farming start to dominate the landscape.

7

Marine Environment

7.1 SEA-LEVEL CHANGES

Evidence for former sea-level changes were first reported from the Mediterranean at the turn of the century. Early reports of raised shorelines in Algeria, southern Italy and Sicily were published before the First World War (De Lamothe 1899; Gignoux 1913). In this area Depéret (1918) distinguished five different levels of marine terraces:

Name	Height above sea level
Sicilian	90–100 m
Milazzian	55–60 m
Tyrrhenian	28–32 m
Monastirian	18–10 m
(lowest level)	7–8 m

He correlated these marine terraces with fluvial terraces in the Alps and in doing so equated the high sea-level stands with the then-established four ice age model. It was only when the principle of eustatic sea-level changes was established more than ten years later that it became clear that marine high stands were correlated not with cold stages but with warm stages instead.

At first many scientists assumed that the high stands described from the Mediterranean could be traced worldwide (e.g. Zeuner 1945, 1952). However, doubts began to emerge. In particular the Sicilian terrace was so high that even with total global ice melt, sea level would only rise to ca. 65 m. Zeuner (1945) attempted to explain this discrepancy by his concept of a gradual lowering of global sea level since the Tertiary. Recent investigations and critical evaluations of the available data refute this assumption. It is now well established that the marine terraces of the Mediterranean have been uplifted and displaced by recent tectonic movements, therefore removing the basis for extra-regional correlations (Hey 1978; H. Brückner 1980, 1986; Radtke 1983).

Sea-level changes are events that should be felt simultaneously across the globe. However, recent evidence has emerged that sea level may not form the universal reference level once supposed. When the surface of the earth is assumed to be at mean sea level (geoid), this plain in all places runs at right angles to gravity and forms an approximate rotational ellipsoid. Apart from the flattening of about 0.3% in the polar regions, which had already been detected by Newton (1687), there are additional secondary deviations from this ideal shape that are related to gravity differences in the earth's interior. The first suggestions of these deviations were detected from gravity measurements (cf. Daly 1940). High-precision satellite altimetry demonstrates that the real sea level is characterised by altimetric differences of about 180 m. The largest deviations from the geoid were registered near Sri Lanka (−104 m) and New Guinea (+74 m). In the Atlantic alone height differences amount to 118 m. The greatest positive anomaly is found in the central North Atlantic between Ireland and the Mid-Atlantic Ridge (+64 m), the largest negative anomaly (−54 m) along the deep-sea trench at the northern margin of the Minor Antilles (Fig. 86). Negative gravity anomalies cause negative deviations of sea level. In general, gravity anomalies situated close to the crust are felt more strongly than those deeper in the earth's interior. Therefore the influence of the earth's mantle is more strongly felt than the core. The influences of the crust and the sea-floor topography are especially pronounced (Emery & Uchupi 1984). The positions of the anomalies appear to be constant, at least over short periods of time. Whether major shifts have occurred during the Quaternary is a matter of debate (Devoy 1987). However, isostatic adjustment to Pleistocene ice-sheet growth will have had some influence on the geoid shape. Mörner (1976) even detects the influence of geoid changes on sea level during the Holocene. If this is the case then sea-level changes can only be interpreted in a very limited regional context (Shennan 1987).

Apart from subsurface influences temperature differences of sea water also have consequences for the shape of the sea surface. Even marine currents are depicted in the relief of the sea surface, displacements of such currents can cause sea-level changes of decimetres to metres. Cheney *et al.* (1984) demonstrated that the migration of a Gulf Stream meander north of Bermuda caused a sea-level change of 120 cm within five months.

Sea-level changes also result from the fact that water expands when heated. According to Pirazzoli (1989), the influence of this so-called **steric effect** on sea-level rise within the last hundred years is about 2–5 cm.

The mass balance of glaciers has remained largely constant over this period. The mass balance of the Antarctica Ice Sheet is regarded as slightly positive, and that of the Greenland Ice Sheet also roughly constant. Meier (1984) therefore concludes that in recent years the mass balance of the minor mountain glaciers worldwide has made a major contribution to sea-level rise, possibly as much as 1.4–5 cm over the last hundred years.

However, sea-level changes have not been restricted to the Quaternary. Vail *et al.* (1977) constructed a sea-level curve from the Jurassic to the present. A global lowering of sea level of about 200–300 m is assumed since the Late Cretaceous (cf. Emery & Uchupi 1984). Increases in sea-floor spreading and mountain building are regarded as responsible for this fall. If sea-floor spreading increases, ocean basins become larger and sea level falls. If a mountain range is formed, the concentration of lithospheric material in one area causes the depletion and lowering of the sea floor in others (Seibold & Berger 1993).

Sea level depends to a large degree on how much water is bound worldwide in glacier ice. The formation of extensive terrestrial ice sheets results in a lowering of sea level, their melting in a sea-level rise. These fluctuations are called **eustatic sea-level changes**. Recent evidence suggests that for the last glacial maximum a lowering of global sea level of about 130 m must be assumed (Shackleton 1987).

The loading of glaciated areas by ice up to several thousand metres thick results in a gradual sinking of the earth's crust. These movements are called **isostatic movements**. This term was

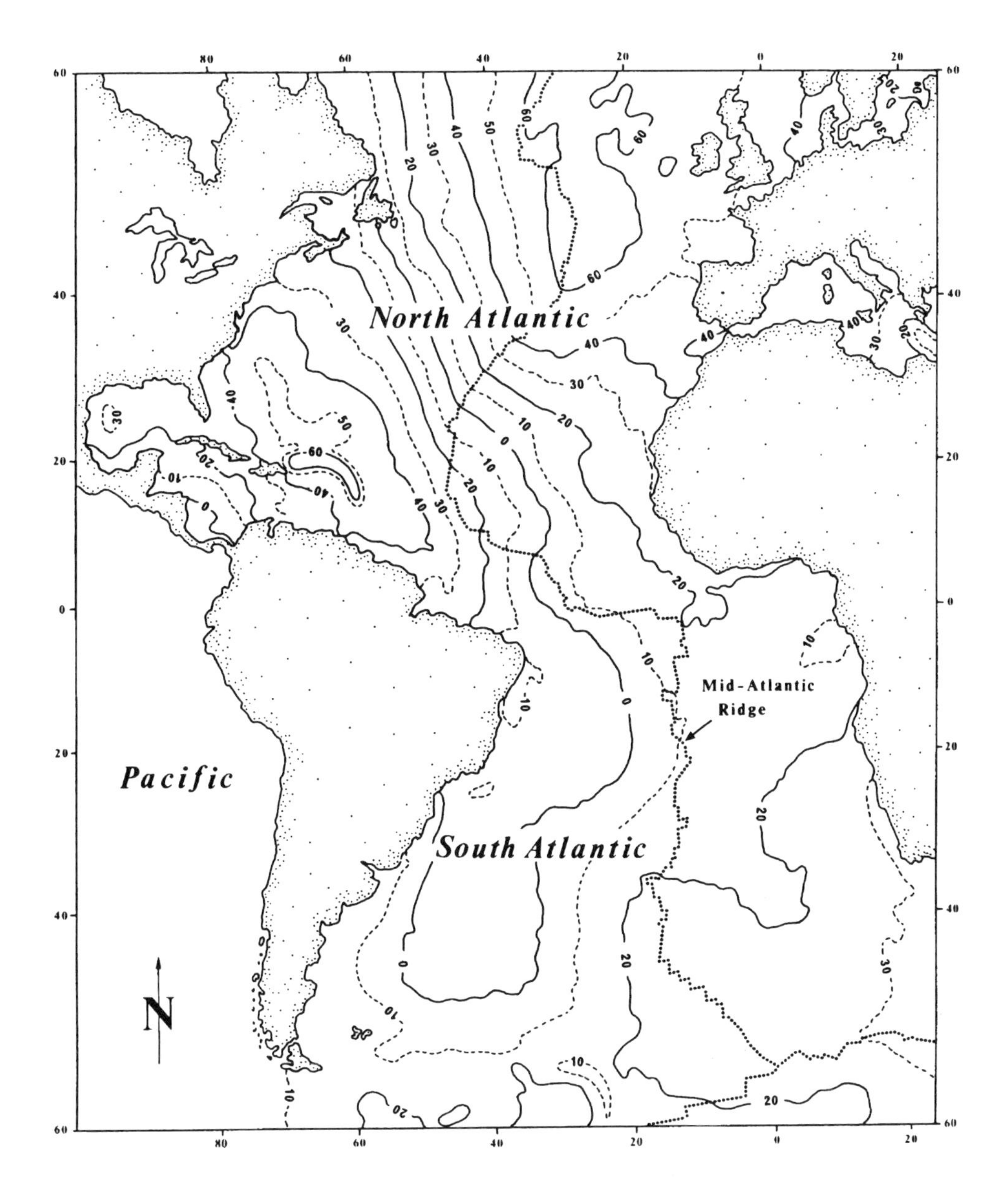

FIGURE 86 Surface of the Atlantic Ocean based on high-precision surveys by satellites GEOS-3 and SEASAT, shown as positive and negative deviations from the geoid. Contour intervals: 10 m (after Marsh & Martin 1982)

first introduced by Jamieson (1865). The continental plates of the earth's crust are on average 30 km thick and 'float' on the earth's mantle. This consists of three layers that differ with regard to their capacity to deform (rheology). The uppermost layer is the relatively rigid lithosphere, about 200 km thick. This is underlain by the soft asthenosphere, about 500 km thick, which is in turn underlain by the mesosphere, about 2200 km thick. Under the weight of the ice load, the earth's crust is depressed; when the ice sheet melts, the crust rebounds. Geophysical models demonstrate that the balance between the loaded, sinking part of the crust and its surroundings occurs in the asthenosphere (Officer *et al.* 1988).

The extent of the isostatic crustal movements in formerly glaciated areas has not been completely established. In Scandinavia and North America postglacial uplift has been reconstructed by evaluation of raised shorelines (Fig. 87). This method, however, can only provide information about the period after the onset of the postglacial transgression, not about the period when the area was still ice-covered. More recent investigations in Scandinavia have shown that about 10,300 BP central Scandinavia was depressed by about 450 m (Svendsen & Mangerud 1987). By extrapolating this value back into the period of maximum depression about 13,000 BP, considerably higher values have been estimated. Mörner (1980) considers the maximum depression of Scandinavia to be about 800 m.

Whereas glaciated areas were depressed under the overburden of ice, the adjoining unglaciated areas were simultaneously slightly uplifted. The **forebulge** resulting from this process must be regarded as the result of horizontal mass movements within the asthenosphere (Mörner 1980). That the build-up of the north European ice sheet resulted in the formation of such a marginal forebulge is indicated by investigations in western Norway. Svendsen & Mangerud (1987) demonstrated that Sunnmøre experienced a postglacial downwarping of about 20 m, which is interpreted as forebulge collapse after Weichselian uplift. Mörner (1980) regards

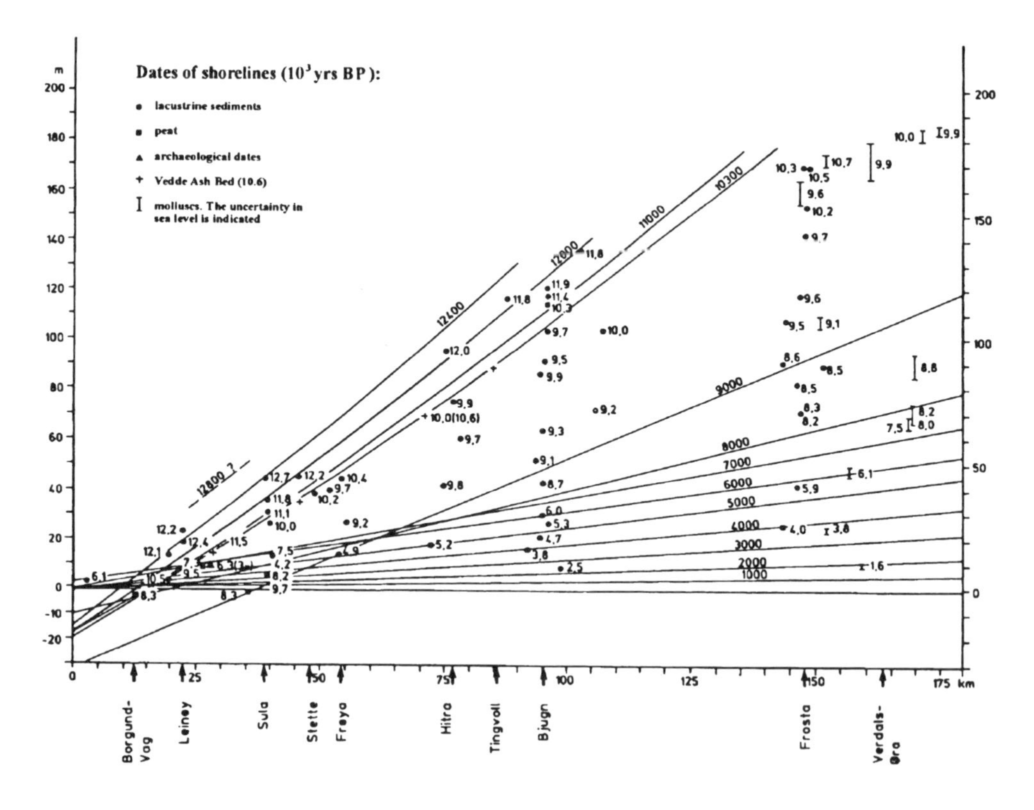

FIGURE 87 Postglacial uplift in Scandinavia (from Svendsen & Mangerud 1987)

an uplift of the central North Sea area of about 170 m as possible, but in his discussion of Weichselian sea-level changes, Boulton (1990) considers much lower values (about 30 m). Along the North American East Coast, in the forefield of the Laurentide Ice Sheet, values of slightly over 100 m have been measured (Pardi & Newman 1987). As the ice-load history and the viscosity–depth structure of the earth's mantle are not sufficiently known, geophysical modelling of isostasy and related sea-level changes is still in its infancy. The available geophysical models of sea-level changes, such as Lambeck (1993a, b) and Tushingham & Peltier (1991), are all based on the assumption that the earth is composed of laterally homogenous layers, and these models include only a small number of (up to five) such isoviscous units. The resulting models can only reproduce realistic sea-level histories for areas a considerable distance from the ice-sheet margins ('forefield') (Plag, personal communication).

Isostatic deformation of the earth's crust must have influenced ice movement and the development of drainage patterns. Because of the lack of shoreline evidence, the extent of isostatic movements in continental areas is very difficult to assess. It would be a mistake, however, to assume that no such isostatic adjustments have taken place.

Postglacial uplift of formerly glaciated areas continues to the present day. Values of about 1.5 mm/year have been measured in the area of the Late Weichselian glaciation centre in western Scotland (Shennan 1989). In the centre of the Scandinavian glacial loading the uplift rate is about 9.3 mm/year (Donner 1980). A component of recent Scandinavian uplift may, however, be the result of tectonics rather than isostasy (Nesje & Dahl 1990). From the distribution of postglacial uplift it can be deduced that practically the entire glaciated area has been affected by isostatic depression and subsequent uplift. It is therefore reasonable to assume much larger depression and uplift rates for the far more extensive Elsterian and Saalian ice sheets in Europe.

Isostatic downwarping of glaciated areas in many cases promoted lateglacial marine transgressions, which in turn influenced the flow pattern of the ice sheets. Where an ice margin floats in the sea under the influence of tides, major icebergs break loose. Marginal mass losses by calving are compensated by increased upstream glacier flow. This results in lowering of the glacier surface and ice extension within the catchment area. Extensive berg production results in the formation of an embayment in the ice margin, a **calving bay**. Rapid ice decay by calving of fast-flowing ice streams can trigger a chain reaction that may result in a rapid collapse of an ice sheet (Hughes 1987). Such a positive feedback system has been postulated by Eyles & McCabe (1989) for the termination of the Weichselian glaciation of the Irish Sea Basin. A similar effect can be assumed for the intrusion of sea water through the Hudson Strait into Hudson Bay (about 8500 BP) resulting in the separation of the ice domes in Keewatin and Labrador and subsequent rapid ice decay (Andrews 1987).

Because of the lowering of sea level during glacial stages, the contact between continental ice sheets and sea water during these stages occurred in areas which are now submerged and therefore difficult to investigate. As a result the significance of glaciomarine processes in the development of the Pleistocene ice sheets has emerged only very recently. In Germany, for instance, glaciomarine deposits had not been identified until Hinsch (1993) proved that the basal part of the Late Elsterian Lauenburg Clay in the Eggstedterholz borehole (Schleswig-Holstein) contains a marine fauna. Further identification of glaciomarine sediments can be anticipated in the future at least in the coastal areas. There can be no doubt that both the extent of isostatic downwarping and the importance of glaciomarine processes have been underestimated in many glaciated areas in the past.

Whereas eustatic sea-level changes had a globally synchronous expression, changes in ice volume have not occurred simultaneously in the different glaciated areas. The eustatic sea-level rise at the end of the last glaciation began at about 14,000 BP. This is also approximately when melting of the European and Laurentide ice sheets began. Whilst in Great Britain deglaciation had started 1000 years earlier, on West Spitsbergen it only started after the major ice sheets disappeared, ending no earlier than 11,000 BP (Boulton 1990). As a consequence, lateglacial and postglacial eustatic sea-level changes affected different areas with varying intensity:

1. *Great Britain.* Isostatic uplift was felt much earlier than sea-level rise in glaciated Britain as a result of early deglaciation. The extent of the associated regression is unknown. Eyles & McCabe (1989) interpret a sea-level fall of up to 100 m for the Irish Sea Basin. With the onset of global ice melt, however, Britain experienced a transgression almost everywhere with synchronous land uplift. Only in Scotland did uplift outstrip sea-level rise. In other parts of Britain major areas were inundated. Around the Wash, the 9000 BP coastline is situated about 50 m below sea level, and in the Channel area it is at a depth of 40 m. The minimum in the Wash area is the result of the collapse of the forebulge.
2. *Northwest Continental Europe.* In glaciated northwest Europe there is a clear distinction between zones of predominantly isostatic and predominantly eustatic influence. The central part of the glaciated area has only been affected by isostatic uplift. After deglaciation the marginal parts of the glaciated area (northern Jutland, western Norway) experienced a brief period of transgression, soon replaced by uplift. The dominance of eustatic sea-level rise is restricted to the extraglacial areas and is most strongly felt in the areas of the former forebulge.
3. *Spitsbergen.* A major component of postglacial sea-level rise was completed before deglaciation started in the western part of glaciated Spitsbergen. As a consequence, isostatic uplift was dominant throughout, and deglaciation was followed by a period of continuous regression. At present, however, transgression prevails (H. Brückner & Halfar 1994).

These examples demonstrate that there can be no general globally applicable model for the course of marine trans- and regressions and that strong regional differences prevail. It is therefore important to realise that most published sea-level curves are only regionally valid (Pirazzoli 1991).

7.2 MARINE CIRCULATION

Like the air masses of the atmosphere, the waters of the sea possess a global circulation pattern. Surface heat, salinity differences and wind drive this circulation, but ice cover and freshwater fluxes are also important. This circulation pattern is therefore referred to as a thermohaline system. This system has undergone considerable changes through the Quaternary and has had a strong influence on global climatic development.

Oceanic circulation can be compared with a large-scale conveyor belt driven by the polar Arctic water masses. The conveyor transports salt- and oxygen-rich waters through the depths of the oceans until they upwell in the northern parts of the Indian Ocean and the Pacific Ocean (Fig. 88) (Broecker & Denton 1989). As the North Pacific is effectively barred from the cold waters of the Arctic Ocean by the shallow Bering Strait, Arctic polar water is largely supplied from the northern parts of the North Atlantic. Thus the production of North Atlantic Deep Water greatly affects global oceanic circulation. Cold deep water is also produced around

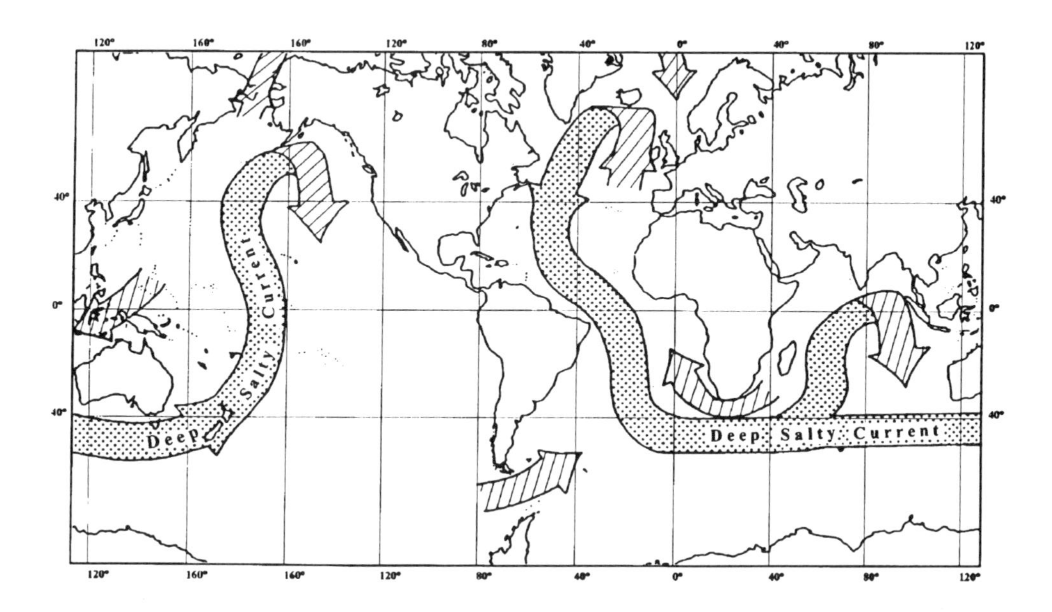

FIGURE 88 The great ocean conveyor belt is driven by the polar Arctic water masses. It transports salt- and oxygen-rich waters through the depths of the world ocean until they upwell in the northern Indian and Pacific Oceans (from Broecker & Denton 1989)

Antarctica, but the existence of a circumpolar ocean in the southern hemisphere inhibits exchange between Antarctic Deep Water and lower latitudes (Whitworth 1988).

The pattern of ocean circulation has a strong influence on climate. Ocean currents, such as the North Atlantic Drift, transport heat from low latitudes to high latitudes. This heat is released as North Atlantic Deep Water is formed. In winter the magnitude of the release of heat from deep-water formation in the North Atlantic is comparable with the amount of heat received through insolation (Broecker & Denton 1990).

Oceanic circulation depends not only on temperature but also on salinity. If major quantities of water vapour are removed from the sea, as occurs during major glaciations, the density of the sea water increases. Abundant transport of fresh water into polar regions reduces salinity and thus also reduces deep-water formation. Broecker & Denton (1990) assume that this happened in the course of ice-sheet melting in the North Atlantic about 11,000 BP. Glacial meltwater prevented the formation of North Atlantic Deep Water and thus disrupted the oceanic conveyor-belt system, giving rise to the Younger Dryas cooling event, which was felt most strongly in the circum-North Atlantic region.

Currents are largely driven by horizontal temperature gradients. During the cold stages global surface temperature difference between ice (0°C or less) and the tropics (ca. 25°C) was compressed into a much reduced latitudinal gradient than at present. The temperature gradient thus was greater, giving rise to stronger winds and stronger ocean currents. Equatorial and coastal upwelling were intensified. Thus, whilst fertility decreased in high latitudes due to ice cover, it increased in middle latitudes because of intensified mixing and in the subtropics due to upwelling (Seibold & Berger 1993).

The changes in oceanic circulation had major consequences on sedimentation. This in turn permits reconstruction of palaeoceanographic conditions from the sedimentary record. For instance, the stable-isotope composition of planktonic foraminifera can be used to calculate former surface-water temperature and salinity (Jansen & Erlenkeuser 1985). Using a number of sedimentary parameters, Henrich *et al.* (1995) were able to reconstruct the depositional regime of the Norwegian–Greenland Sea since the Late Saalian. They have shown that only during the warm phases of oxygen-isotope stage 5 did Atlantic water extend far to the north as at present. However, repeated ingressions of Atlantic water were also recorded during oxygen-isotope stage 6.

In the North Atlantic deposits of the last glacial stage are characterised by a number of layers with a high percentage of coarse-grained lithic material. These must have been deposited during phases of intense iceberg rafting (H. Heinrich 1988). The youngest of these so-called Heinrich Events occurred about 14,000 and 20,000 BP. They have been correlated with major meltwater outbursts from the North American ice sheet (Andrews & Tedesco 1992, Andrews *et al.* 1994). Clark (1994) assumes that the ice sheet had become unstable, resulting from a warming up of the glacier sole and the formation of a deformable bed. The glaciomarine Weichselian sedimentary sequence in the Norwegian–Greenland Sea is also characterised by varying amounts of ice-rafted detritus (Fig. 89; Baumann *et al.* 1995). It contains distinct layers of diamicton that represent short-term depositional events indicating instability of the continental ice sheets, similar to the Heinrich layers of the northwest Atlantic (Henrich *et al.* 1995).

7.3 MARINE DEPOSITS

7.3.1 Deep sea

The sea floor can roughly be subdivided into three major elements: shallow shelf seas, steeply dipping continental slopes and abyssal plains. Only deep sea trenches are lower in elevation than the abyssal plains. The abyssal plains are therefore areas of nearly continuous sedimentation providing a stratigraphical sequence allowing the reconstruction of a major part of the climatic history of the earth. Until the 1950s it was thought that this sedimentary record might extend back perhaps billions of years until seismic investigations revealed that the sediment cover of the Atlantic floor was about 500 m thick, and in the Pacific only about 300 m. It only became clear in the mid-1960s that the reason for this thin cover was the mechanism of plate tectonics and sea-floor spreading. The basic principles of deep-sea sedimentary distribution are discussed by Seibold & Berger (1993).

Under ideal conditions, the **deep-sea sediments** result from slow, vertical deposition on a relatively planar surface. Parts of the sea floor, however, have strong relief. Sediment slumps down steep slopes, bottom currents can redistribute it and powerful turbidity currents can accumulate extensive sheet-like deposits. The sedimentary sequence of the deep sea is therefore not uniform and discontinuities must be taken into consideration. In order to acquire relatively complete sequences, boreholes are preferably positioned on plateaux or on gently sloping plains where the danger of redeposition is small. The V28-238 and V28-239 boreholes drilled by the research vessel *Vema*, on which the fundamentals of the deep-sea oxygen-isotope stratigraphy were founded, come from water depths of 3120 and 3490 m in the Salomon Plateau region of the West Pacific (Shackleton & Opdyke 1973). Disconformities representing periods of non-

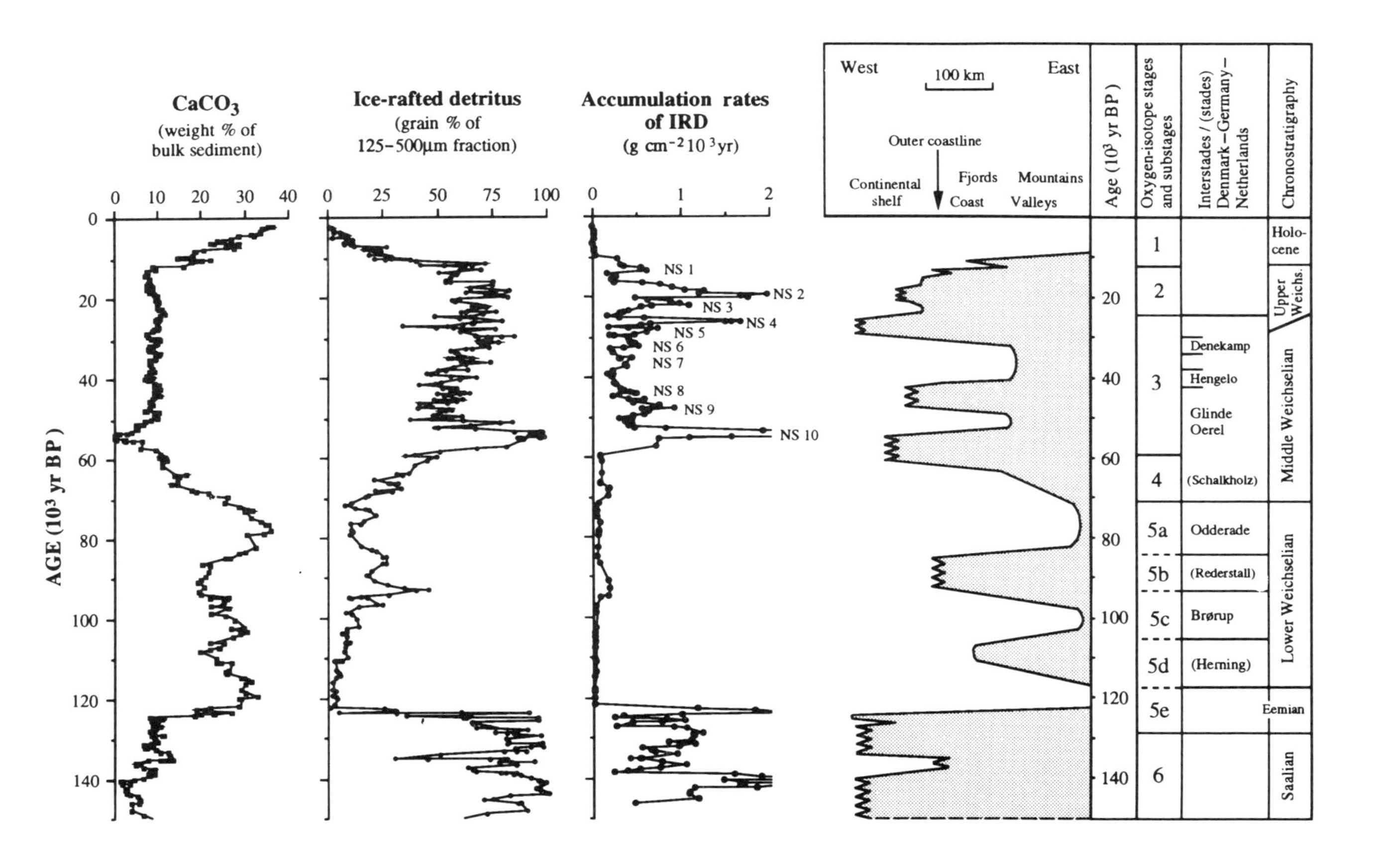

FIGURE 89 Marine stacked records of $CaCO_3$ content, amount of ice-rafted detritus (IRD) and accumulation rates of ice-rafted detritus of three sediment cores from the Norwegian Sea. Glaciation curve of western Scandinavia (right) as reconstructed from terrestrial data (from Baumann *et al.* 1995)

deposition are difficult to identify in sediment cores. The stratigraphical record of each borehole is therefore correlated with other boreholes, and radiometric dating and palaeomagnetic measurements provide a basic framework of fixed points into which the stratigraphic sequence is fitted.

Deep-sea deposits represent a mix of fine sediments derived from the land as well as the remains of marine microorganisms. The terrigeneous component comprises mainly clay but also silt and fine sand blown into the sea by strong winds. Coarse detritic layers are also occasionally found in deep-sea sediments. Only a very few mechanisms can explain their deposition. The most important of these is **drift-ice** transport. Additionally some coarse sediments are introduced by marine algae, driftwood (mainly roots of tree trunks), and mammals. Human activities have resulted in the input of some recent coarse sediment, mainly in the North Atlantic (especially cinder and ash from the steamship era). Whilst identification of the human input is relatively unproblematic, the reliable identification of drift-ice deposits is far more difficult. This is a major problem in the mapping of former drift-ice limits. Algal transport is largely restricted to warm seas, but in the northern hemisphere there is some overlap with the subpolar drift-ice zone so that interpretations should be made with caution (Emery 1963).

Drift-ice transport is not restricted to the area beyond the pack-ice boundary (Fig. 90). During the Pleistocene cold stages, cold currents could transport icebergs far to the south and north. In the North Atlantic, cold-stage drift-ice transport has been proved as far south as Morocco, in the South Atlantic from Antarctic waters to north of Cape Town (Emery & Uchupi 1984). Sediment redistribution is by no means limited to icebergs since sea ice can contribute considerably to sediment transport (Dionne 1993).

Wind-blown dust from the Sahara is also of major importance off the West African coast. Intercalations of fine-grained volcanic tephra are also occasionally found within deep-sea sediments. The major component of deep-sea sediments, however, consists of organic particles, mainly the tests and hard parts of foraminifera, radiolaria, diatoms and coccolithoporids. In the central Atlantic the Pleistocene sediments largely consist of *Globigerina* ooze. Because carbonate is progressively dissolved with increasing depth, a broad depth zonation of sediments results, starting with pteropod ooze (about 90% $CaCO_3$) down to a depth of about 2000 m, followed by *Globigerina* ooze down to 4500 m. At greater depths carbonate-depleted red deep-sea clay dominates (Seibold & Berger 1993).

The proportion of the different components varies regionally, and depends critically on the admixture of terrigeneous components. As two thirds of the continental rivers of the world drain into the Atlantic, the proportion of terrigenous material here is much higher than in the Pacific. As a rule, therefore, a sedimentary core of a given length from the Pacific will cover a longer period of time than one from the Atlantic. Sedimentation rates vary from less than 1 cm per millennium in some areas to 100 cm per millennium in upwelling areas. Using ^{14}C dating Broecker *et al.* (1958) demonstrated that deposition of terrigenous components during the last glacial maximum was three to four times higher than under interglacial conditions. Increased physical weathering, glacial erosion, fluvial erosion by meltwater activity, and the exposure of vast shelf areas to aeolian deflation due to lowered sea level are amongst the possible causes. The sedimentary sequence of the deep sea therefore reflects global climatic changes and can be used to reconstruct the climatic history of the Quaternary. Oxygen-isotope ratios ($^{16}O/^{18}O$) can be used to reconstruct the climatic history of the entire Quaternary (see chapter 10.1).

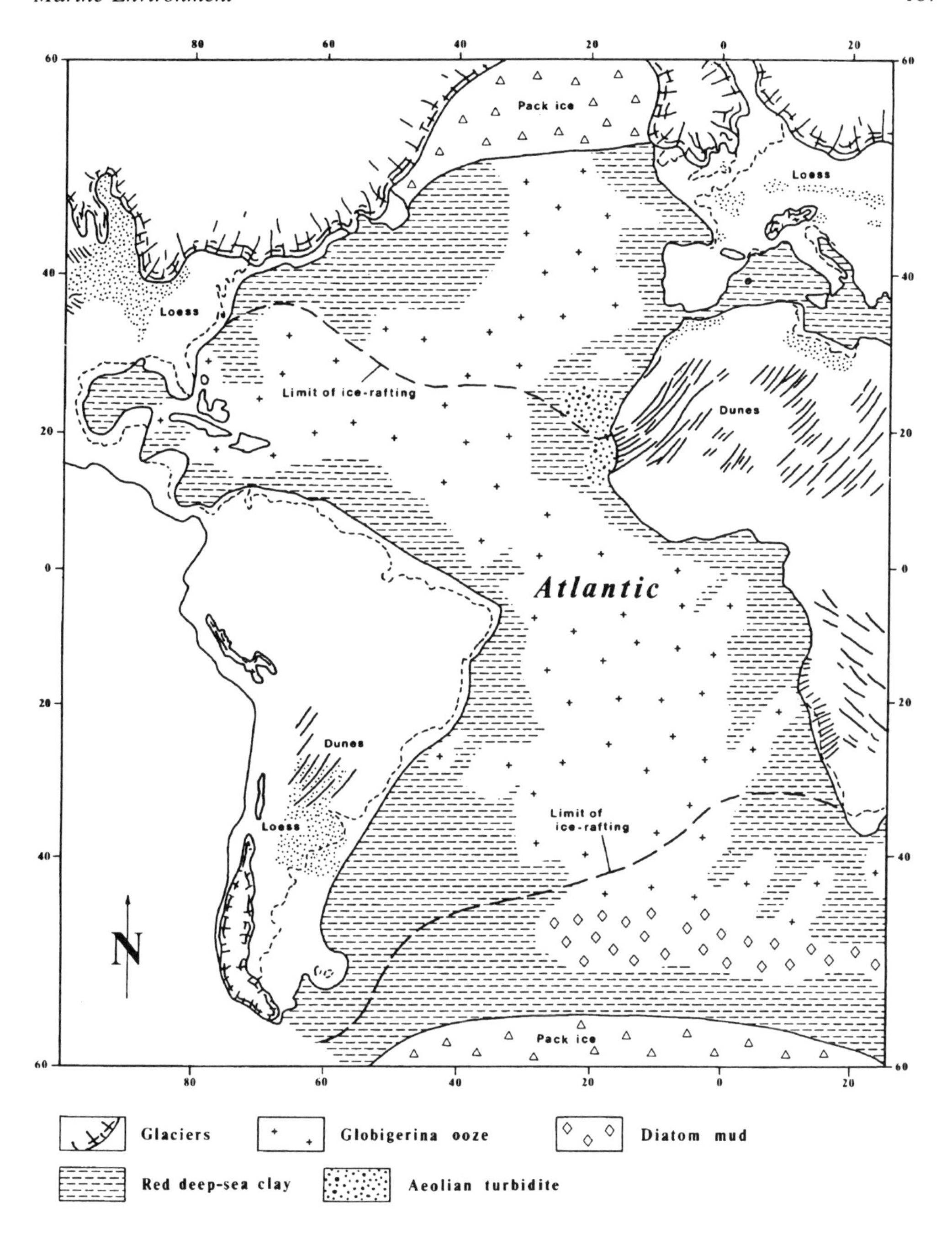

FIGURE 90 Sediment distribution in the Atlantic during the Weichselian Glaciation (from Emery & Uchupi 1984)

7.3.2 Shelf seas

Whilst deep-sea sediments allow reconstructions of the climatic changes of the Pleistocene, the margins of the continental shelf provide evidence for sea-level changes during the recent geological past. This is especially true in areas where sedimentation prevails causing prograding shelf margins. During periods of lowered eustatic sea level, sedimentation is interrupted and processes of reworking and erosion dominate the shelf areas. During these periods the coastline migrates seawards (regression), possibly into the area of or close to the shelf break. As a result, enhanced accumulation occurs on the continental slope enabling the pattern of sea-level changes to be reconstructed from slope sequences.

During a sea-level rise the coastline shifts landwards (transgression) and sedimentation on the shelf surface is resumed. Accumulation continues on the continental slope, but at a lower rate. If during the course of a transgression the amount of accumulation on the shelf surface is less than the erosion during the subsequent regression, it follows that only one sediment layer will be found on the shelf surface, representing the current sea-level high stand. If, however, accumulation prevails, the sequence of strata will contain the deposits of each transgression, separated from one another by an erosional hiatus (Mougenot *et al.* 1983) (Fig. 91). Under favourable conditions, for example along rapidly subsiding continental margins, it may even be possible to predict the sedimentary sequences. Comparisons with the oxygen-isotope curve can sometimes be used to date the sedimentary units, as was demonstrated in the case of the Seyhan-Ceyhan delta in the eastern Mediterranean (Piper & Aksu 1992).

Boulton (1990) described the sequence of processes that occur on a glaciated shelf during one glacial–interglacial cycle (Fig. 92): (1) During an interglacial the formerly glaciated land area undergoes glacio-isostatic uplift, giving rise to the formation of marine terraces. In the inner shelf area and within fjords marine deposition occurs. (2) During the early glacial stage, glaciers advance into the marine area. Older strata are partly tectonised and reworked. Till is deposited in subglacial situations and glaciomarine sediments accumulate in front of the ice margin. (3) During the glacial maximum the ice sheet advances as far as the shelf margin, resulting in enhanced deposition on the continental slope. Simultaneously, older strata are reworked in the inner shelf area and in fjords. (4) During ice decay deposition occurs in proximal areas near the glacier terminus whilst erosion prevails in the distal areas further away from the ice. On the continental slope slumping leads to enhanced deposition in the lower slope area. (5) As (1); deposition occurs mainly in the overdeepened parts of the inner shelf and in fjords.

In shelf seas the input of terrigenous matter is much more significant than in the deep sea. A significant amount of terrigenous material is deposited seaward of river mouths, whilst drift ice is able to transport much larger components (Dionne 1993). The active entrainment of sediment by sea ice on arctic and subarctic coasts can occur in three ways:

1. Sediment is frozen to the underside of grounding sea ice, e.g. during low tide. This process occurs preferentially in extensive tidal-flat areas. Repeated freezing-on of sediment results in bedded ice, where sediment-rich layers alternate with pure ice which forms when the floe refloats during high tide. In this way major quantities of sediment can be incorporated. On Baffin Island R. Gilbert (1990) found drift ice containing up to 670 g/l sediment 200 m distant from the shore. The grain-size composition of this material largely reflects the composition of the local sea floor. Even large pebbles and boulders can be incorporated (Dionne 1981).
2. Major amounts of suspended sediment can be incorporated especially during the early

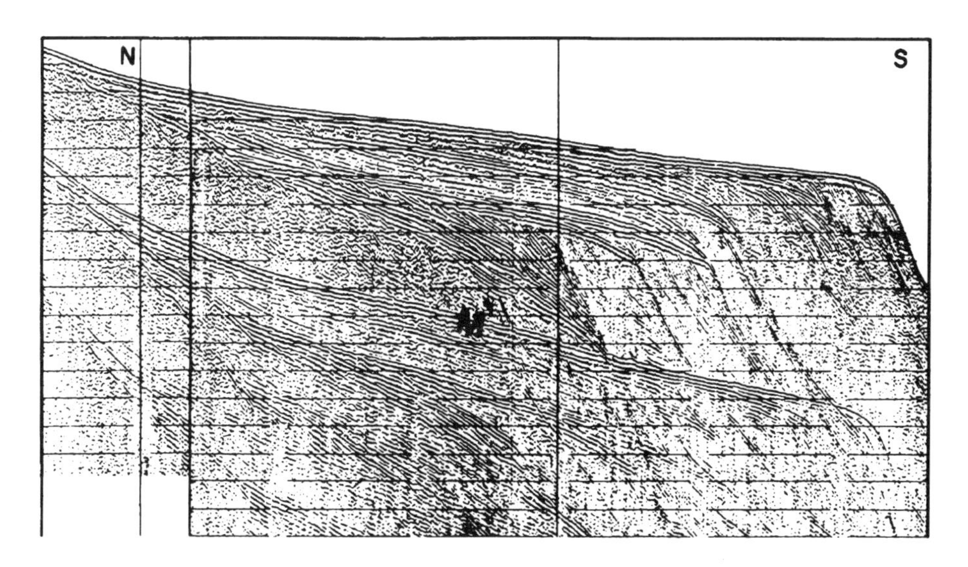

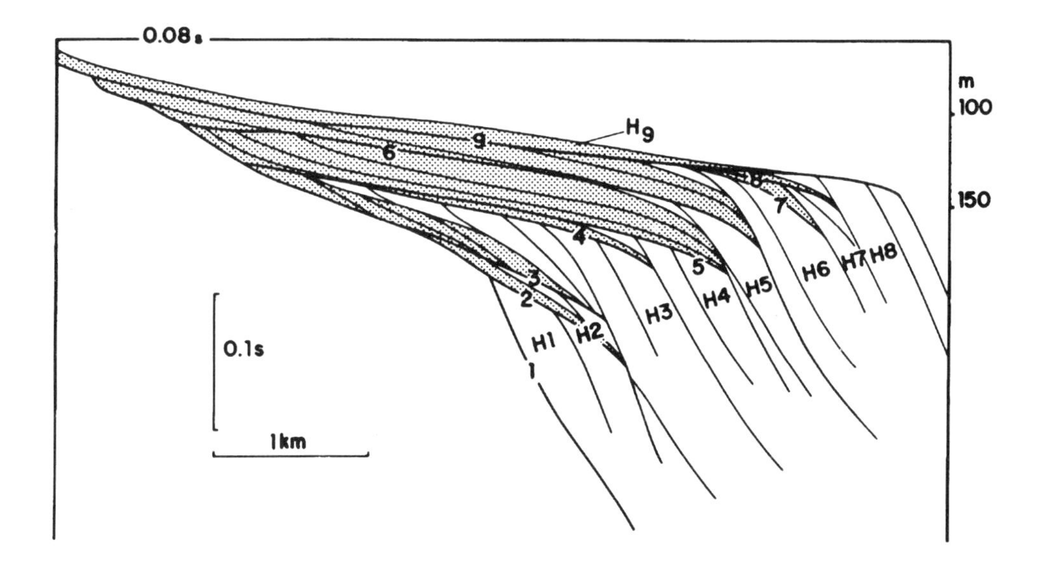

FIGURE 91 Example of a complex, prograding shelf with sigmoidal deposition. Above: air-gun profile from the Gulf of Genoa (M = multiple); below: interpretation. Vertical scale: 0.1 seconds. The sediment bodies with the letters H1–H9 were deposited during high stands of the sea (H9 during the Holocene), the erosional disconformities 1–9 represent low sea levels during cold stages. Vertical exaggeration about 11 times (from Mougenot *et al.* 1983)

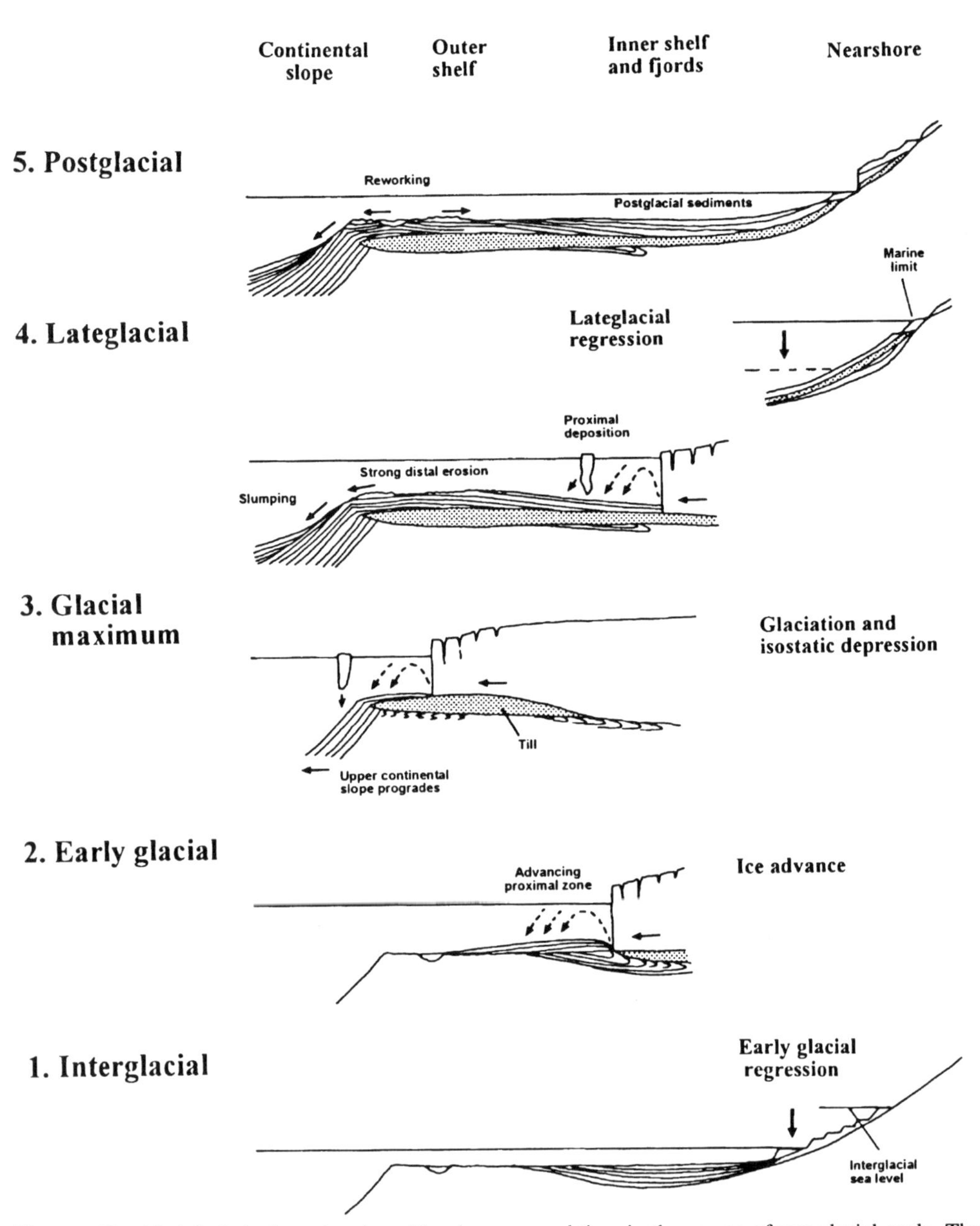

FIGURE 92 Model of glaciomarine deposition in space and time in the course of one glacial cycle. The relative changes in sea level are indicated (from Boulton 1990)

stages of ice formation. In this situation uniformly dark-coloured ice is formed. This process is significant along the margins of arctic and subarctic tidal-flat areas, but sediment content is comparatively low. Along the north coast of Alaska, Barnes *et al.* (1982) measured values of only up to 1.6 g/l.

3. In shallow water areas, deep cooling can lead to the formation of ice on the sea floor

(anchor ice). This also occurs in intertidal areas, i.e. areas that are dry during low tide. The exposed flats can cool so thoroughly that during reflooding ground ice forms. Under favourable conditions this ice can grow to such a size that buoyancy becomes dominant and blocks of sediment-laden ice start to float (Reimnitz *et al.* 1987). There are no data on the intensity of this process but it is assumed to be of relatively little importance.

Drift ice can form ice-pushed ridges several metres high and hundreds of metres long along arctic coasts (Dionne 1992).

Passive uptake of sediment also plays a significant role in beach areas. Waves and tides can transport sediment up to gravel size onto the ice, especially during break-up of the ice cover. In addition, avalanches, solifluction and meltwater runoff from the land can cross the pack ice and add to the sediment load. In many cases, however, sediment-laden ice is unable to float and the floes melt *in situ* (R. Gilbert 1990). Sediment can also be blown onto the pack ice by wind. Since a considerable proportion of the arctic land surface has only sparse vegetation cover, deflation is considerable. This mode of transport, however, is largely restricted to the fine-sand and coarse-silt fractions.

Most of the sediment entrained by these mechanisms into the pack ice is transported only short distances within the coastal zone. Only a minor proportion is carried to the deep sea.

Drifting icebergs not only cause considerable sediment transport in the glaciomarine environment (Fig. 93), but they also leave characteristic marks by ploughing the sea floor. Iceberg-induced furrows have, on average, a depth of about 2–5 m and a width of 20–25 m. The largest iceberg furrow so far observed, recorded on the Labrador Bank, had a depth of 17 m (Barrie 1980). The main control on iceberg ploughing is iceberg thickness. Dietrich *et al.* (1979) and Brooks (1979) measured maximum drafts of 187 m (113 measurements) and 220 m (221 measurements) off Greenland. Free-floating ice shelves have a physically constrained marginal thickness of about 250 m (Paterson 1994). Ancient iceberg scours formed during periods of lowered sea level and possibly greater iceberg thickness have been found to maximum depths of 715 m in Baffin Bay (Praeg *et al.* 1987) and 850 m on the Yermak Plateau between Greenland and Spitsbergen (Vogt *et al.* 1994). Iceberg ploughing creates a diamicton that is slightly coarser-grained than normal glaciomarine deposits as a result of winnowing and erosion of fines and in which fossils in life position are rare. Vorren *et al.* (1983) have termed these deposits **iceberg turbates**.

The precise extent of the Pleistocene glaciations in shelf areas is still incompletely known. L.H. King *et al.* (1972) have found submerged end moraines off Nova Scotia and B.G. Andersen (1979) mapped a series of Weichselian ice-marginal positions on the Norwegian shelf. The question of a possible glaciation of the Barents Sea has been discussed since the work of Schytt *et al.* (1968). In particular Grosswald (e.g. 1980) has strongly advocated arguments in favour of glaciation. Major parts of the area have now been systematically mapped (Vorren *et al.* 1988, 1990; Solheim *et al.* 1990) and it has now been demonstrated beyond doubt that the entire Barents Sea area was glaciated by a fully developed terrestrial ice sheet in the period of eustatically lowered sea level between 23,000 and 15,000 BP (Elverhøi *et al.* 1993). The sequence of events has been reconstructed in some detail. Since the Eemian, Spitsbergen has experienced three major glaciations; around 110 ka BP, 75–50 ka BP and 25–10 ka BP (Mangerud & Svendsen 1992). However, there are still many problems concerning the correspondence of the Barents Sea Ice Sheet expansions with the glacial phases of the Scandinavian mainland, and the determination of the eastward extent of the ice sheet.

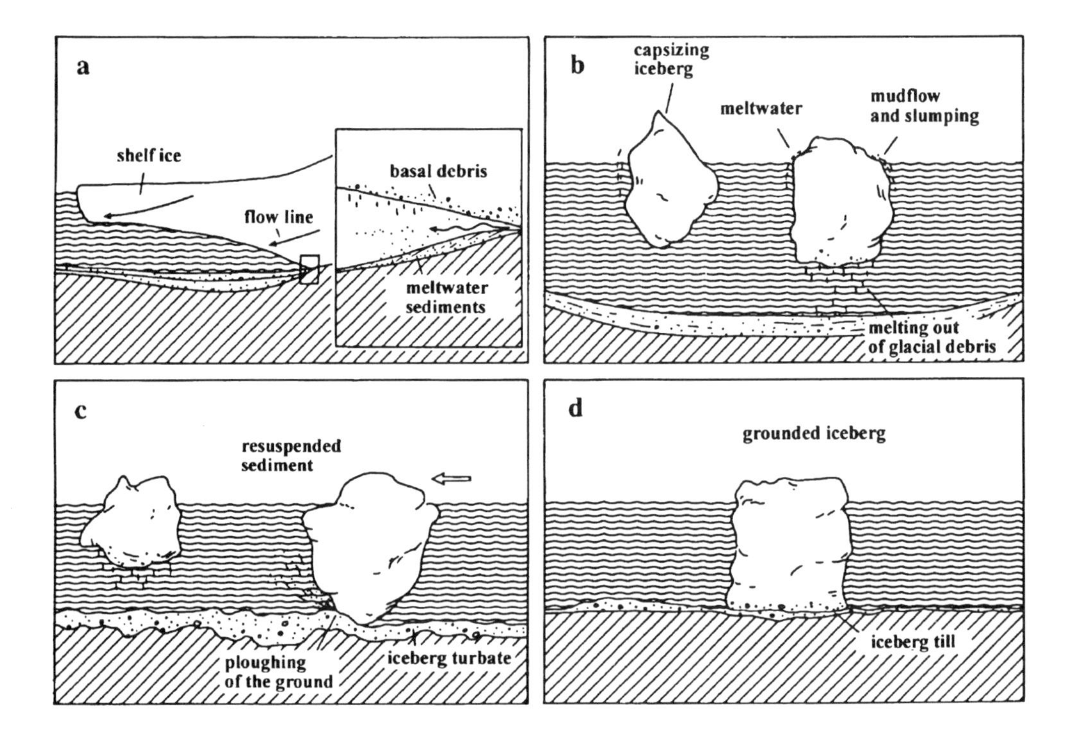

FIGURE 93 Sediment redistribution by shelf ice and floating icebergs. a = shelf ice; sedimentation by melting-out from the base and from subglacial meltwater streams; b = sedimentation by melting icebergs (rollover, dumping, mudflow, slumping, basal melt); c = sediment resuspension and ploughing of the sea bed by icebergs; d = melting of grounded icebergs (from Vorren *et al.* 1983)

7.4 COASTS

In subsiding areas such as the Netherlands or North Germany the distribution of former seas can only be reconstructed from their deposits. Evaluation of numerous boreholes has allowed the coastlines of the last (Eemian) and second-but-last (Holsteinian) interglacial high sea stand to be reconstructed. Additional, Early and Middle Pleistocene transgressions have been recorded in the Netherlands (cf. chapter 11).

Not every warm-stage transgression reached the present mainland areas. More-complete sedimentary sequences, which also contain evidence for minor transgressions, can be found in the shelf seas such as the North Sea Basin. So far this stratigraphical record has been largely investigated by seismic methods and stratigraphical correlation with the Pleistocene time scale has been only partly accomplished.

The combined evaluation of boreholes and seismic records off the North Carolina coast (USA) has allowed the identification of seven marine stratigraphical units separated by hiatuses. Biostratigraphical investigations and amino-acid dating have provided a broad chronostratigraphic subdivision of trans- and regressions. Sea-level high stands representing stages 23–29, 13 or 11, 9 and 5 have been identified (Riggs *et al.* 1992). Evaluation of the oxygen-isotope record in addition to recent geological investigations in North America indicate no intra-Weichselian high sea-level stand. Indeed, Wellner *et al.* (1993) have shown the Oxygen Isotope Stage 3 sea level off New Jersey to lie at a depth of about 20 m below present sea level.

Whilst in subsiding areas the reconstruction of former transgressions must be based on the sedimentary record, in areas of uplift former cliffs and raised beaches provide information on the exact position of the coast at a given time (Fig. 94). In glaciated districts former coastline features may be concealed by younger glacial sediments. A fossil cliff of the former coastline of the Eemian North Sea, for example, is exposed at Sewerby (Yorkshire, England) (Catt 1991b).

The east coast of the United States, south of Delaware Bay, is an area that has undergone continuous coastal tectonic uplift since the Pliocene. During the 1960s several ancient coastlines had been reconstructed (Flint 1971; Fig. 95). These raised shorelines are only separated by small differences in altitude, and because of the complex relationship between the different lithofacies involved they are difficult to date (Fletcher & Wehmiller 1992).

In parts of California, on the American Pacific coast, coastal uplift has led to the preservation of a sequence of up to 12 terraces, the highest of which now lies about 250 m above sea level. The lowest two of those terraces have been dated by uranium-series and amino-acid racemisation techniques to ca. 80 ka and ca. 120 ka BP; the ages of the older terraces can only be estimated (Hanson *et al.* 1992).

Some coral reefs provide more potential for accurate sea-level reconstructions. By $^{230}Th/^{234}U$ dating relict coral reefs of the Eniwetok atoll (Pacific) in combination with dates from the Bahamas and from Florida (Broecker & Thurber 1965), Broecker (1965) determined

FIGURE 94 Raised beach (42 m) at Squally Point, Nova Scotia, consisting of a wave-cut rock platform overlain by beach gravel (Photograph: R. Morrison)

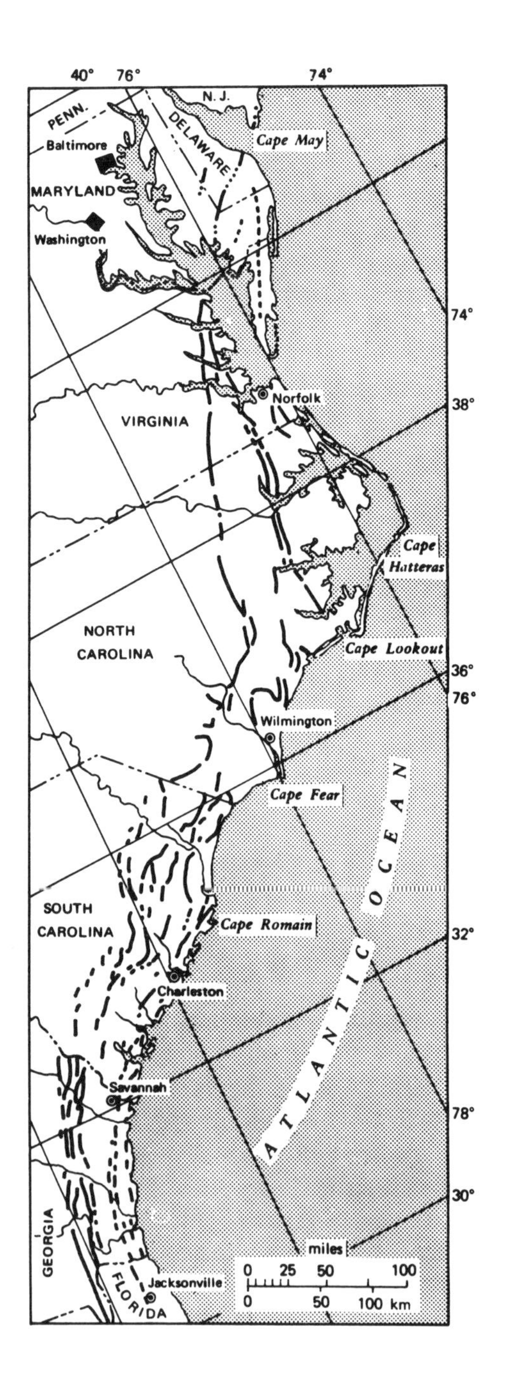

FIGURE 95 Raised shorelines along the east coast of the United States (inactive cliffs, beach ridges etc.). The stratigraphical interpretation still remains uncertain (from Flint 1971)

the ages of two sea-level high stands at 80,000 and 120,000 years BP. Subsequent investigations of terraces in Barbados have shown that not two but three high stands occurred during this time, dated to about 87,700 (Stage 5a), 112,000 (Stage 5c) and 125,000 BP (Stage 5e) (Edwards *et al.* 1987, Radtke *et al.* 1988). These high sea-level stands correlate well with the insolation maxima determined by Milankovitch. This provided the first evidence that the radiation curve was principally correct (Broecker *et al.* 1968). These dates were confirmed by dating raised coral reefs on New Guinea (Bloom *et al.* 1974). However, dating marine terraces and other sea-level indicators must be done with utmost scrutiny (Radtke 1989), because it appears that $^{230}Th/^{234}U$ dates often give minimum ages (Brückner, personal communication).

Elevated marine terraces can be used for reconstructing former sea-level changes under the assumption that the uplift rate has remained constant, rapid, and that the terraces contain datable material. The best sites for such investigations are raised coral terraces. A chronostratigraphy correlated with the deep-sea oxygen-isotope curve covering the last 250,000 years has been established on the Huon Peninsula, New Guinea (to Stage 7; Chappell & Shackleton 1986). On Barbados, Radtke *et al.* (1988) extended the chronostratigraphy to Stage 15, but a recently dated flight of terraces on Sumba (Indonesia) allows a reconstruction of sea-level changes during the last million years. Age determinations by $^{230}Th/^{234}U$ and electron spin resonance (ESR) indicate that Sumba has undergone nearly constant uplift of about 0.5 mm/year allowing simple correlation of the terraces with the oxygen-isotope curve. High sea stands, approximately at the same level as the modern sea ($\pm$5 m), occurred on Sumba during stages 5, 9, 11 and 25. During all other interglacials the sea level was lower than today (Pirazzoli *et al.* 1993). However, the results from one region alone cannot easily be transferred to others. In southeastern India, for instance, the invasion of the Eemian sea was the first transgression since the Pliocene to leave an onshore record (Brückner 1988). Comparisons of Quaternary sea levels on a global scale are needed before a clearer picture of the Pleistocene trans- and regressions can be developed.

8

Methods of Investigating Glacial and Interglacial Deposits

8.1 GRAIN SIZE

Grain-size analyses are standard procedures for the characterisation of unconsolidated rocks. They can be applied to aeolian, fluvial, lacustrine, marine and glacigenic deposits alike. In most of these environments sorting plays a major role, either by wind, flowing water or tidal currents. In these cases, grain-size analyses provide information relating to the transport conditions. Glacigenic deposits are different. In a glacier, sediment is not sorted but, quite the contrary, thoroughly mixed. Where grain-size curves from till show a certain degree of sorting, the reason normally results from the incorporation of major quantities of foreign materials, e.g. overridden meltwater sands.

Mixing of materials also leads to a certain homogenisation of the material. Therefore, in contrast to glaciofluvial deposits, till units can often be correlated over major distances. This relatively uniform composition caused Karrow (1976) in such cases to speak of 'mature till'. However, this state of maturity can only be reached where the ice has advanced over a relatively homogeneous substratum. The grain-size composition of basal tills, to a large degree, is controlled by the composition of the local bedrock (De Ridder & Wiggers 1956).

Grain-size composition can be used as a criterion for distinguishing different till units (e.g. Karrow 1976, Höfle 1980). The Older Saalian till of Niedersachsen (Lower Saxony, Germany), for instance, is normally relatively sandy, whilst the Middle Saalian till is often clay-rich. The Younger Saalian till – at least in the Hamburg area – occupies an intermediate position. The suitability of grain-size analyses for stratigraphical correlations is, however, rather limited (cf. Stephan 1987b). Apart from one relatively widespread till facies, each ice advance has also deposited one or more other lithofacies, the composition of which was partially controlled by changes in glacier dynamics. For example, the Older Saalian till of North Germany, apart from the sandy variety mentioned above, occurs as a clay-rich, reddish facies largely composed of far-travelled material.

Thick till sequences have to date rarely been studied in detail. Vertical and lateral variations in grain-size composition have been observed occasionally, but rarely quantified; a few examples are given by May & Dreimanis (1976). Often the basal part of a till sheet is found to be strongly enriched in local components (e.g. Broster & Dreimanis 1981).

The tills of the Alpine glaciations are characterised by relatively high proportions of coarse components. Whilst the gravel content of the Swiss and Austrian tills and the tills of the south

German Alpine foreland is generally about 20–50% (e.g. Van Husen 1977, Schlüchter 1981b, Grottenthaler 1989), in the North German tills it rarely exceeds 5%. In the literature little quantitative data on clast contents can be found, because grain-size analyses for practical reasons of sample size are often restricted to fractions smaller than 2 mm.

When grain-size analyses from different regions are compared, it must be remembered that the boundaries between particle sizes vary considerably. A survey conducted by the INQUA Commission on Genesis and Lithology of Quaternary Deposits in 1974/75 revealed the extent of the problem. Even within North America, two different scales (Wentworth and USDA) are in use, and accordingly the silt/clay boundary may be either at 0.002 or 0.004 mm. Most North American workers place the sand/silt boundary at 0.062 mm, though 0.074 and 0.050 mm are also used. In Russia, some workers place the upper boundary of silt at 0.100 mm (Raukas *et al.* 1978).

8.2 ERRATICS

Till is composed of two components: rock fragments and single minerals. Rock fragments that are not native to the area are referred to as erratics. Rock fragments dominate the major fractions, whereas in the 0.5–1.0 mm range most components are mono-mineralic. Each rock type that is represented in a till shows a bimodal distribution with one maximum in the gravel fraction and a second in the matrix. After short transport distances the gravel component dominates, whereas after longer transport grinding down of the clasts leads to a dominance of the matrix component (Dreimanis & Vagners 1971). In the course of glacier transport, weak rocks are more vulnerable to crushing than hard rocks. This results in the relative enrichment of resistant rock types in the gravel fraction, and a relative impoverishment in non-durable components like limestones and sedimentary rocks (Saarnisto 1990).

When the compositions of various grain-size fractions are compared it has been found that the fine material has been transported over longer distances than the coarse material (cf. Lindén 1975). According to Lillieskӧld (1990) transport distances in meltwater-transported debris tend to be longer than in glacially-transported deposits. This, however, is in contrast to findings from Canada and the marginal areas of the former ice sheets. Shilts & McDonald (1975) found, for instance, that in the Windsor esker, downstream fluvial transport was largely restricted to between 3 and 4 km. Transport distance in eskers is controlled by the length of the tunnel under the ice. In North Germany the meltwater sands always contain higher proportions of local material than the corresponding tills. The reason for this is that the North German pre-Quaternary bedrock consists almost entirely of Tertiary sands and clays with very few particles larger than 2 mm (Ehlers 1990b).

Torell (1865) was the first to construct an ice-stream model based on the distribution fans of erratics. J. Petersen (1899, 1900) concluded that the individual source areas were of varying importance during the various glaciations. Subsequently, clast-lithological investigations played a major role in stratigraphical investigations especially in Denmark, Germany and the Netherlands. The approach was mainly based on the identification of **indicator clasts**. The rocks best suited for this purpose are those from limited and well-known source areas. After the promising results of early clast-statistical investigations by Milthers (1934) and Hesemann (1939), Lüttig (1958) developed a method in which geographical longitude and latitude of the centres of the source areas of the determined indicators are summed and divided by the number of clasts. By this method a **theoretical indicator centre** (in German: Theoretisches

Geschiebezentrum, TGZ) is obtained, which frequently allows the stratigraphical classification of a till. When restricted to the most common lithofacies types, the following principles apply, for instance, to the Niedersachsen tills (cf. Table 5):

1. Elsterian tills in northern Niedersachsen are characterised by a relatively high percentage of western Fennoscandian material.
2. The Older Saalian (Drenthe) tills contain mainly material from central and south Sweden.
3. The Younger Saalian (Warthe) tills are characterised by East Baltic indicators.

The ideas about which rock types should be used as indicators differ from area to area and from one scientist to another. For instance, for the German geologists the Kinne Diabase is regarded as an indicator, but this does not apply to their Dutch colleagues (e.g. Zandstra 1988). From North Germany, Denmark and the Netherlands a great number of clast-lithological investigations have been published (e.g. K.-D. Meyer 1983b, Zandstra 1988, Lüttig 1991, Smed 1993). Smed (1993) has also developed a new method of presenting results in a far more graphical way than simply listing the coefficients (Fig. 96).

Quantitative indicator investigations have also been successfully applied in the Baltic States. Clast-lithological investigations are in addition successfully utilised in the central areas of the glaciations. Examples of modern clast-lithological investigations aimed at reconstructing ice-movement directions in the Fennoscandian hardrock area have been presented by Salonen (1986, 1987, 1991). The method can furthermore be used to locate ore deposits buried beneath a till cover (e.g. K. Eriksson 1983). This method is particularly useful for short- and medium-range distances. In long-distance transport of clasts, because of mixing of materials of different ages and repeated reworking of older materials the reconstruction of regional ice-movement direction may be difficult.

Britain, with its small-scale variation of rock types, would be a promising area for quantitative indicator-clast analyses and, to a certain extent, also some marginal areas of the North American Pleistocene ice sheets. For instance, Dionne (1994) described the characteristic erratics found at the mouth of the Saguenay River in Québec, discussed their possible source areas and the implications for Wisconsinan ice-movement directions. Prest & Nielsen (1987) provide an overview of long-distance transported erratics in the Laurentide Ice Sheet region (see also chapter 13.8).

A single glaciation can result in the deposition of very differently composed indicator assemblages, as has been demonstrated in the Netherlands (Zandstra 1988). There is no doubt that during the Saalian the Netherlands were only covered by one ice sheet, but it deposited more or less simultaneously tills from east-central Swedish, south Swedish, west-central Swedish and east Fennoscandian source areas. However, this does not mean that clast investigations are unsuited for correlations. If all available data regarding stratigraphy and texture of the deposits are considered, it can provide a valuable contribution to the solution of stratigraphical questions.

Because the composition of the indicator assemblages depends on grain size (Fig. 97), it is important to compare only counts of the same fractions. In North Germany, clast-lithological investigations concentrate on the 2–6 cm fraction, but in the Netherlands the fraction >2.7 cm is investigated. In each example the vast majority of erratics cannot be precisely determined, so that the stones evaluated only make up a minor percentage of the original material. However, according to K.-D. Meyer (1983b) about 10% of the clasts can be fully identified. In order to achieve reliable results, a large number of erratics must be examined (about 700–1000

TABLE 5 Indicator clast counts from northwest Germany, 2–6 cm fraction. 1 = Elsterian till from Hamburg, Elbtunnel; 2 = Older Saalian till from Wehden (Niedersachsen); 3 = red–brown, east Baltic facies of the Older Saalian till from Schneiderkrug (Niedersachsen); 4 = Middle Saalian till from Grauen (Niedersachsen); 5 = east Baltic Younger Saalian till (Vastorf type) from Emmendorf (Niedersachsen); 6 = Weichselian till from Groß Weeden (Schleswig-Holstein) (from K.-D. Meyer 1983b)

Sample No.		1	2	3	4	5	6
Nordic crystalline rocks		962	416	684	533	810	486
Local erratics (Tertiary)		77	15		10		8
Nordic crystalline rocks	%	40	34	18	27	21	27
Nordic sedimentary rocks	%	28	14	4	11	4	17
Nordic Palaeozoic limestones	%	14	29	63	18	52	18
Dolomites	%		1	11		19	
Old Red Sandstone	%			4		1.4	
Flint	%	12	22	0.3	34	2	23
Upper Cretaceous Conglomerate	%	4			10		15
Quartz	%	4					
Dala Sandstone		7	5	10	2	8	3
Hälleflinta		3	5	1	7	3	4
Tiger Sandstone					1		
Uralit Porphyrite				1			
Number of indicator pebbles		52	30	17	30	31	27
Åland Aplite Granite			1				
Åland Granite		4		6		10	
Åland Granite Porphyry						2	
Åland Quartz Porphyry			1				
Åland Rapakivi				1		1	
Alminding Granite				1			
Blue Quartz Granite			1				
Bornholm Granite					1	1	1
Bothnian Sea Quartz Porphyry						2	
Bredvads Porphyry		5	1				
Brown Baltic Quartz Porphyry					1		
Colonus Slate			1		3		4
Dala Porphyry		6			1		1
Digerbergs Sandstone		2			1		
Flivik Granite		1					
Garbergs Granite Porphyry		1					
Grey Växjö Granite		1		1			2
Grönklitt Porphyrite		2	1				
Halen Granite			1	1			
Hardeberga Sandstone		15	5		5	3	8
Höör Sandstone							1
Kalmarsund Sandstone			2				1
Karlshamn Granite						2	
Kinne Diabase					1		1
Nexö Sandstone				1			
Red Baltic Quartz Porphyry			1				
Red Växjö Granite		4	5	4	7	4	6
Rhomb Porphyry		3					
Skåne Basalt						1	
Småland Granite		3		1	3		
Småland Quartz Porphyry		1	1				
Spinkamåla Granite				1		2	
Stockholm Granite		1	3			1	
Tessini Sandstone		3	5		4		
Uppsala Granite						1	
Västerviks Quartzite			1				2
Virbo Granite					3	1	
Flint/crystalline ratio		0.3	0.64	0.016	1.23	0.08	0.84
TGZ	λ	14.63	15.89	17.03	15.36	17.38	14.44
	γ	58.32	57.45	57.91	57.17	58.40	56.56

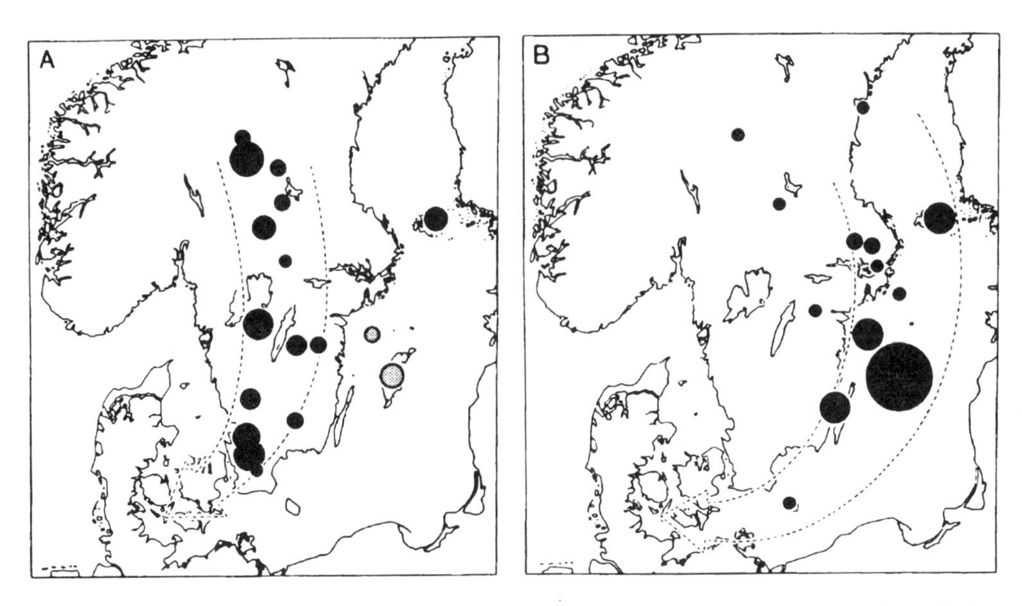

FIGURE 96 Erratic content of two Danish tills, the 'thick till' and 'discordant till' from Ristinge, Langeland, drawn after Per Smed's method (from Smed 1994)

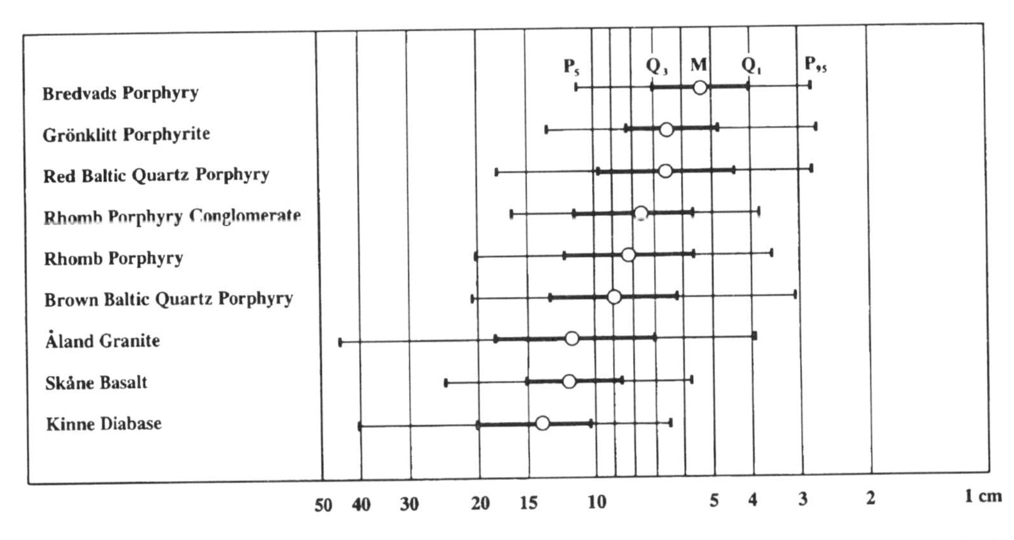

FIGURE 97 Grain-size distribution of different crystalline indicator rocks from meltwater sands in Denmark. M = median; Q_3, Q_1 = quartile; P_5 = 5%; P_{95} = 95% (after Gry 1974)

specimens). As the samples have to be picked directly from the exposure in order to avoid mixing of different stratigraphic units, sampling of a single till unit for one analysis may take as long as a day. The determination of the rocks is difficult and requires considerable experience and good knowledge of the source areas. The Scandinavian erratics and their source areas were first described in the books of Korn (1927, crystalline rocks) and Hucke & Voigt (1967, sedimentary rocks). More recent handbooks have been published by Gudelis (1971), Hesemann (1975), Zandstra (1988) and Smed (1989, 1994).

In the former German Democratic Republic, clast-lithological analyses were formalised in a 'Fachbereichsstandard' (Zentrales Geologisches Institut 1972). Analyses were performed in two fractions, 4–10 mm and >10 mm, using a till sample of 5 kg. For the evaluation of the 4–10 mm fraction, five rock types were distinguished: Nordic crystalline rocks (NK) or crystalline rocks (K), Palaeozoic limestones (PK), Palaeozoic schists (PS), dolostones (D) and flint (F). The analysis of the fraction >10 mm was used only to supplement the count of the 4–10 mm fraction. The stratigraphic interpretation was also included in the official 'standard', if only as a recommendation. This method takes an intermediate position between clast analysis and fine-gravel analysis (see below) and has been applied successfully as demonstrated from investigations in the Leipzig Lowland area. The two Elsterian tills there can be clearly distinguished on the basis of their gravel composition (Fig. 98). The lower Elsterian till contains a much higher percentage of quartz than the upper unit, which can be attributed to local reworking of older terrace gravels. In the upper part of the till unit (samples 6–8), the quartz content decreases progessively and the overall composition comes close to that of the overlying upper Elsterian till (Eissmann 1982).

As in glacigenic deposits and glacial outwash, the terrace deposits of the great rivers can also be correlated on the basis of their petrographic composition. The methods are discussed, for instance, by Bridgland (1986). Clast composition of fluvial deposits depends on the rocks in the catchment area and on weathering. Changes of the catchment area and climatic changes are reflected in the gravel spectrum, with most examples dominated by local material. This is especially true in places where older gravel bodies are reworked. In her investigation of terrace gravels of the Lahn River, Lipps (1985) (Fig. 99) categorised the rocks involved into three groups according to their hardness: (1) quartz, (2) chert, quartzite and ferrigenous gravel, (3) other, less resistant rock types (such as greywacke and Bunter Sandstone). As might be expected, the older gravels contain mainly hard rocks, whereas the more recent, less strongly weathered terrace gravels are characterised by higher proportions of easily weathered components.

8.3 FINE GRAVEL AND COARSE SAND

In all cases where major exposures are not available, or where more closely-spaced sampling is desirable, clast-lithological investigations based on indicator erratics can provide only a coarse basic framework for the stratigraphical interpretation, and need to be supplemented by other methods.

Since the mid-1950s, fine-gravel analyses of the fraction 3–5 mm have been applied with great success in the Netherlands (Maarleveld 1956). The method has been successfully applied in the Netherlands (Zandstra 1983), in North Germany (Ehlers 1990a, Panzig 1995) and Denmark (Sjørring & Frederiksen 1980, Houmark-Nielsen 1987).

With clast-statistical methods the *size* of the sample determines the degree of detail that can be achieved. A larger sample allows greater accuracy, although there is not a linear increase in the level of confidence (Fig. 100). In order to identify a component that occurs in a sample with a percentage of 1% at a probability of 95%, more than 4000 clasts must be identified. As this far exceeds any reasonable effort, a compromise between effort and reliability must be found (Bridgland 1986). A reasonable sample size would be about 300 grains.

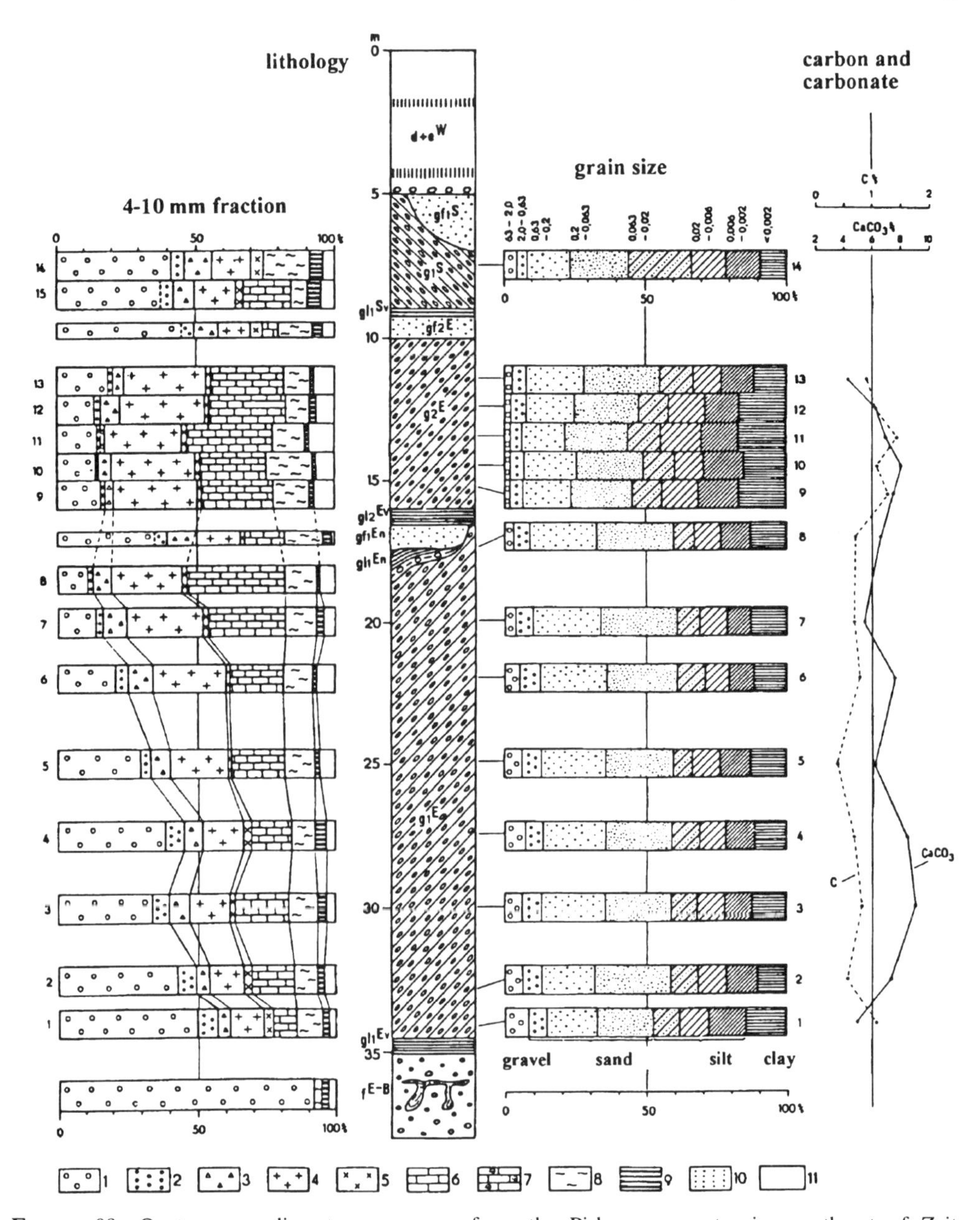

FIGURE 98 Quaternary sedimentary sequence from the Pirkau opencast mine northeast of Zeitz (Leipziger Bucht) with diagrams of clast and grain-size composition and contents of carbonate and free carbon. 1 = quartz, 2 = residual quartz, 3 = flint, 4 = crystalline rocks (apart from eruptive rocks), 5 = eruptive rocks (mainly porphyries), 6 = limestones, general, 7 = Muschelkalk, 8 = schists (schist, quartzite, greywacke), 9 = lydite, 10 = sandstone, 11 = rest. Abbreviations: B = 'Brüggen Cold Stage' (Praetiglian); E = Elsterian Cold Stage; S = Saalian Cold Stage; W = Weichselian Cold Stage; $_{g1}E$ = first Elsterian till; $_{g2}E$ = second Elsterian till; $_{g1}S$ = first Saalian till; $_{d+c}W$ = slopewash and aeolian deposits of the Weichselian Cold Stage; f = fluvial deposits; gf = meltwater sands and gravels; gl = glaciolacustrine silt; v = advance outwash; n = retreat outwash (from Eissmann 1982)

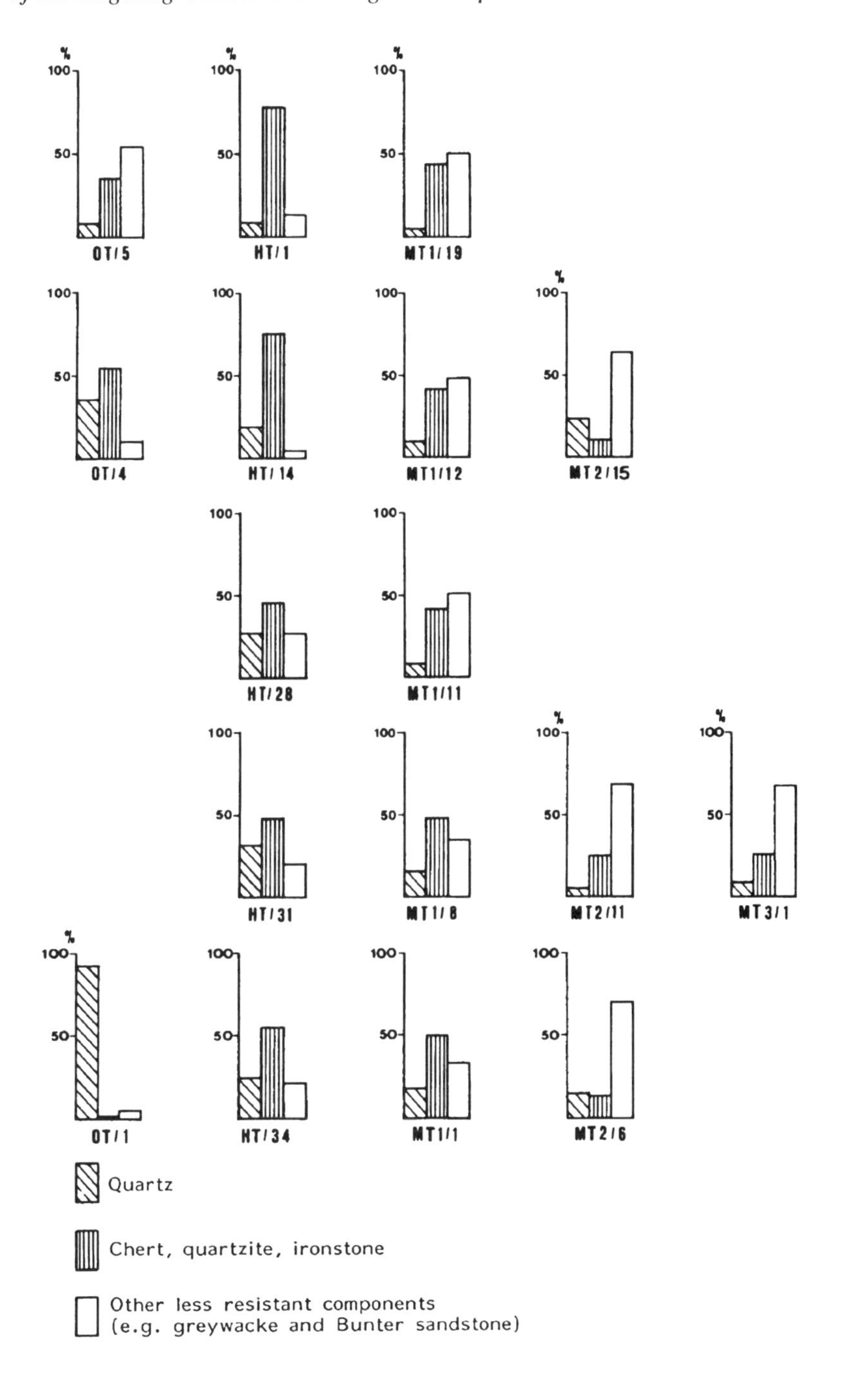

FIGURE 99 Petrographic analyses of terrace gravels from the Lahn River, western Germany. OT = Upper Terrace; HT = 'Hauptterrasse' (Main Terrace); MT = Middle Terrace (from Lipps 1985)

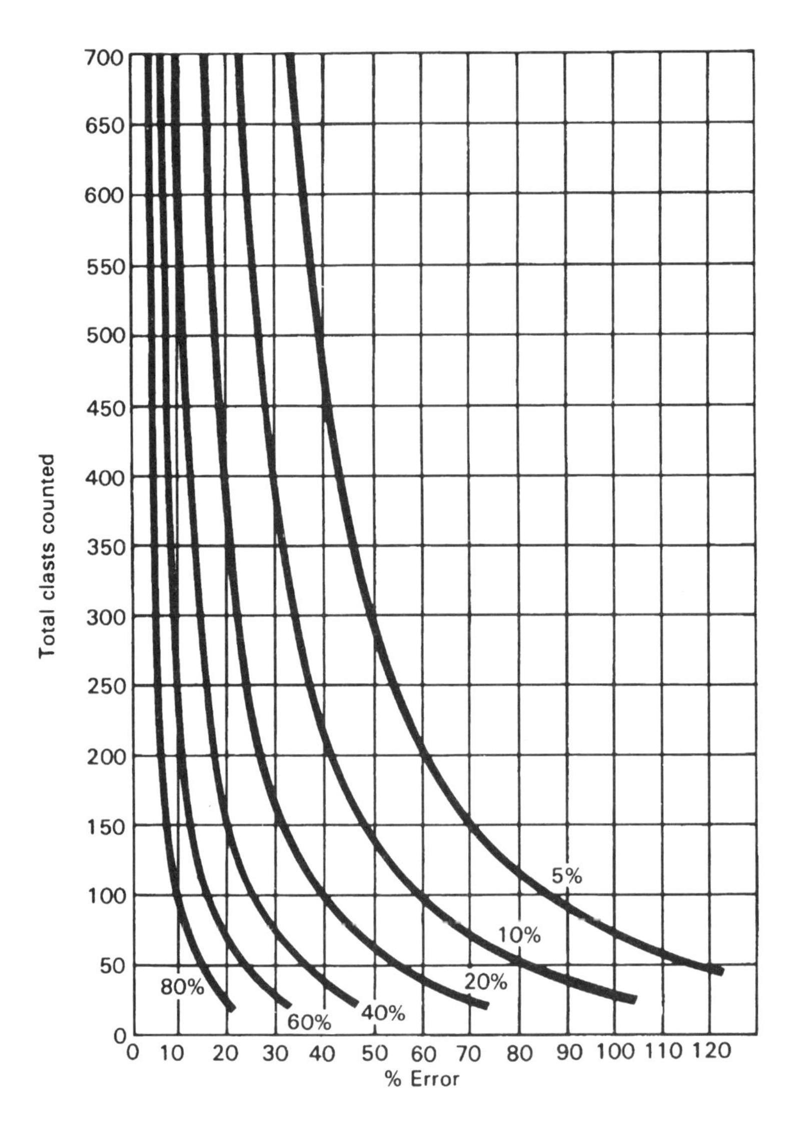

FIGURE 100 Statistical validity of clast-lithological analyses depending on sample size, standard error 95% confidence level. Curves for components with a content of 5, 10, 20, 40, 60 and 80% of the overall sample. Example: In a sample of 300 clasts there is 95% confidence that a component which is represented with 5% in the sample will be represented at values between 2.5 and 7.5% in the count (from Bridgland 1986)

In counts using the Dutch method about 30 different rock types are distinguished. Most of these are represented in such small percentages that, for statistical evaluation, several rock types must be grouped to form a major rock group. In the first subdivision, limestones and non-limestones are distinguished, and subsequently the non-limestones are further subdivided into quartz, flint, crystalline and sedimentary rocks (e.g. Ehlers 1990a).

Fine-gravel analysis provides no results that could not in principle be achieved by evaluating indicator erratics. Its great advantage, however, is that because of the small quantities needed for analysis, samples from boreholes can also be evaluated. The source area of the individual components cannot be determined with the same degree of precision as with indicator analysis, but fine-gravel analysis also allows a rough differentiation of major source areas.

As mentioned above, the petrographic composition of Quaternary sediments generally depends on grain size. For example, a typical Elsterian till sample from the Hamburg area (high quartz content, relatively low content of limestone), demonstrates that quartz clearly dominates the fractions smaller than 2 mm (Fig. 101). With grain sizes larger than 2 mm the quartz content tends towards zero.

The grain size best suited for stratigraphic differentiation depends on the lithology of the deposits. It is also possible to investigate fractions smaller than fine gravel lithologically. Dreimanis (1939, 1947) has demonstrated that different Pleistocene sediments can also be distinguished using their coarse-sand composition, and has differentiated sandstone, limestone, dolomite, melanocratic (dark) minerals, feldspars and quartz. The 0.5–1 mm fraction was found to be best suited for distinguishing the different tills in Latvia (Danilans 1970) using this method.

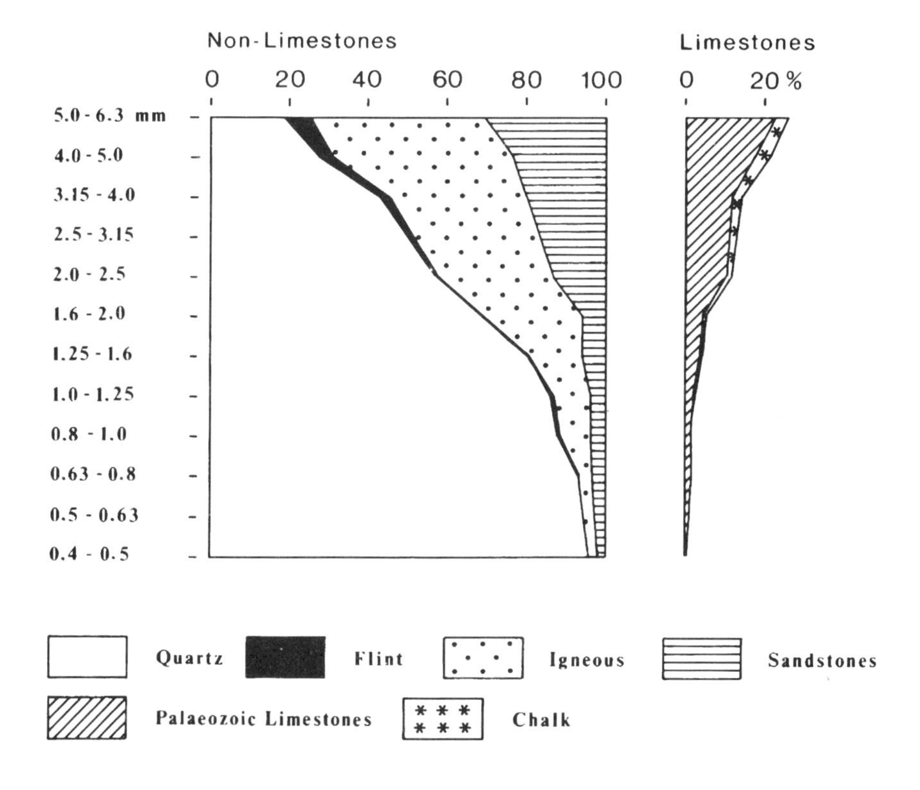

FIGURE 101 Composition of a till sample according to its grain size (Elsterian till from a depth of 34–35 m from core drilling Dradenau KB 42 in Hamburg-Waltershof) (from Ehlers 1990a)

8.4 CLAST MORPHOLOGY

Clast roundness provides information about the mode of transport a rock has undergone. The term was originally used by Wentworth (1919). Roundness deals with the sharpness of the edges and corners of a clast and is independent of the shape of the rock fragments (discoidal, spherical, bladed, rod-shaped) (Pettijohn 1975). Following Cailleux (1952) the degree of roundness R can be expressed in the following formula:

$$R = \frac{2r1}{L} \times 1000$$

where L is the length of the clast and $r1$ the radius of the smallest curvature. Since the curvature radius of the sharpest edge can only very vaguely be determined (through comparison with a chart), other procedures are often employed for the determination of the curvature. On the international level, a classification scheme is frequently used that provides an optical distinction into five categories (well rounded, rounded, subrounded, subangular and angular); cf. Pettijohn (1975) (Fig. 102).

Periglacial congelifraction creates predominantly angular gravel, till is characterised by subangular clasts, and fluvial pebbles in contrast are mostly subrounded or well rounded. As roundness depends largely on the type of rock, it only makes sense to compare equal rock types.

In a situation where deposits consist largely of gravel, the study of clast shape can be of great importance. This applies, for instance, to the flint gravels of southeast England. Gibbard (1986) developed a key for the distinction of flint pebbles from various environments. Complete nodular specimens with a porous surface and nodular protuberances are likely to be derived directly from chalk bedrock. Rounded flints with high sphericity, a thin patinated surface, black–grey–white in colour and with small crescentic cracks are most likely derived from Tertiary deposits. Rounded flints with lower sphericity and a tendency towards flattened faces in one plane, with frosted surfaces and patina brown–grey–white in colour, and crescentic cracks are probably derived from beach gravel. Broken flints with fresh corners are the product of frost shattering.

The forms of major clasts are not only useful to obtain information about transport and depositional environments. Krygowski & Krygowski (1965) have developed a method of measuring the mechanical 'rollability' of sand grains. This method which provides an estimate of the degree of abrasion sand grains have undergone, has also been applied to tills (Krygowski 1969). Without additional information, however, the results are difficult to interpret.

Since the introduction of the scanning electron microscope (SEM), it has been possible to reconstruct some additional aspects of the sedimentological history of sand grains. The concept of environmental interpretation of surface textures was first published by Krinsley & Donahue (1968). Following the publication of an atlas of quartz sand surface textures (Krinsley & Doornkamp 1973), the method has been applied in various investigations. In North Germany,

FIGURE 102 Roundness classes according to Pettijohn (1975). From left to right: well rounded, rounded, subrounded, subangular, angular

for instance, it is possible to differentiate between quartz grains originating from glacially crushed fresh Scandinavian rocks and others reworked from Tertiary sands. Mechanical wear of quartz grains in a high-energy environment leaves conchoidal fractures and cleavage plates (Figs 103 b, d), whereas chemical weathering results in rough, pitted surfaces (Figs 103 a, c) (Ehlers 1990a). Comparisons of subglacially and supraglacially transported quartz grains indicate that subparallel and conchoidal fractures result from subglacial crushing and not from mechanical weathering of the source rocks (Mahaney *et al.* 1991).

8.5 HEAVY MINERALS

Heavy-mineral analysis as a tool for stratigraphical investigations was developed in the 1930s. In the Netherlands, Edelman (1933) applied the method successfully to distinguishing fluvial and glaciofluvial deposition of different ages. At the same time, Derry (1933) was studying heavy minerals as indicators of till provenance in the Toronto area of Canada. Boenigk (1983) comprehensively describes the technique of heavy-mineral analysis and the range of its applications. For the composition of a heavy-mineral spectrum three aspects are of crucial importance:

1. The source area determines the mineral spectrum available for deposition.
2. Transport results in an alteration of the original mineral composition by means of reworking, mechanical wear and sorting.
3. Weathering and diagenesis may further change the heavy-mineral spectrum after deposition.

Under favourable conditions it is possible to reach palaeogeographical conclusions based on heavy-mineral analyses. A classic example is the work of Boenigk (1970, 1978, 1981) on the Lower Rhine (Fig. 104). He reconstructed the gradual development of the catchment area of the River Rhine in the transition period from the Tertiary to the Quaternary using the heavy-mineral composition of the sediments.

Most heavy minerals are relatively resistant under recent climatic conditions. Conversely, extensive weathering of heavy minerals occurred during the Tertiary. The epidote group has proved to be relatively resistant to strong chemical weathering; garnet and amphibole disintegrate much more readily. In general, the following succession of weathering can be established under neutral to low-acid conditions (Boenigk 1983):

Extremely stable: zircon, tourmaline, rutile, anatase, brookite, topaz, spinel, zincblende, corundum.
Very stable: disthene, andalusite, sillimanite, titanite, staurolite, epidote, monacite.
Fairly stable: monacite, mica (biotite), pyroxene, hornblende, garnet, olivine.
Very unstable: fayalite (olivine), apatite, carbonate, false galena.

In addition, the formation of new heavy minerals can change the composition of the heavy-mineral spectrum considerably. Formation of zircon, tourmaline, garnet and apatite have been described, yet they are usually restricted to additional growth on detrital grains hardly changing the overall spectrum. Marked alteration, however, can be caused by the appearance of newly formed iron oxides and hydroxides, as well as by pyrite, barite, anhydrite, celestine, carbonate, false galena and anatase. These young minerals, which do not belong to the original heavy-mineral spectrum, must be removed in the laboratory prior to counting (Boenigk 1983).

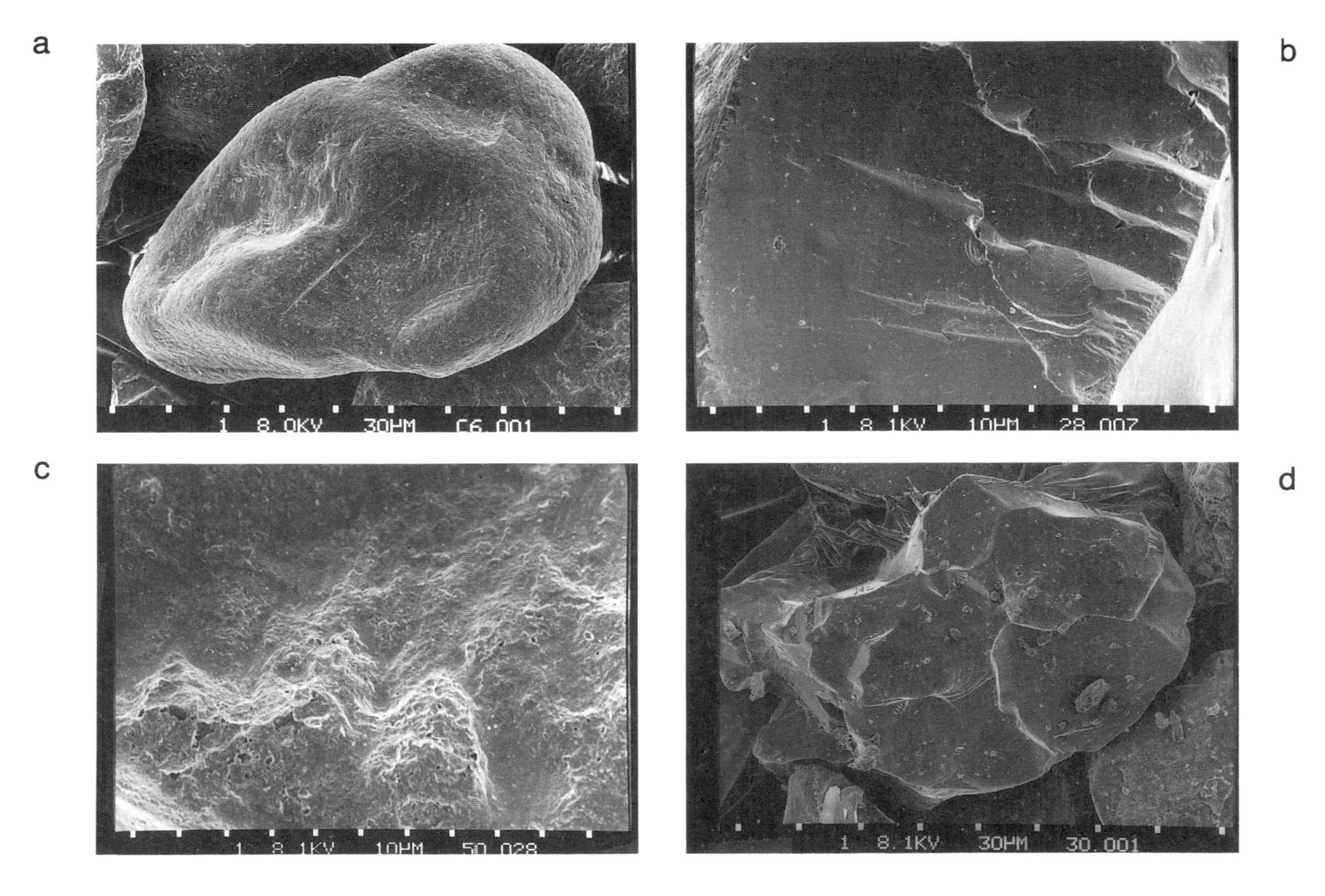

FIGURE 103 Scanning electron micrographs of sand grains from North German tills. (a) and (c): rounded quartz grains reworked from the underlying Miocene sands; (b) and (d): grains worn by glacial transport with characteristic fractures (Photographs: Ehlers 1990a)

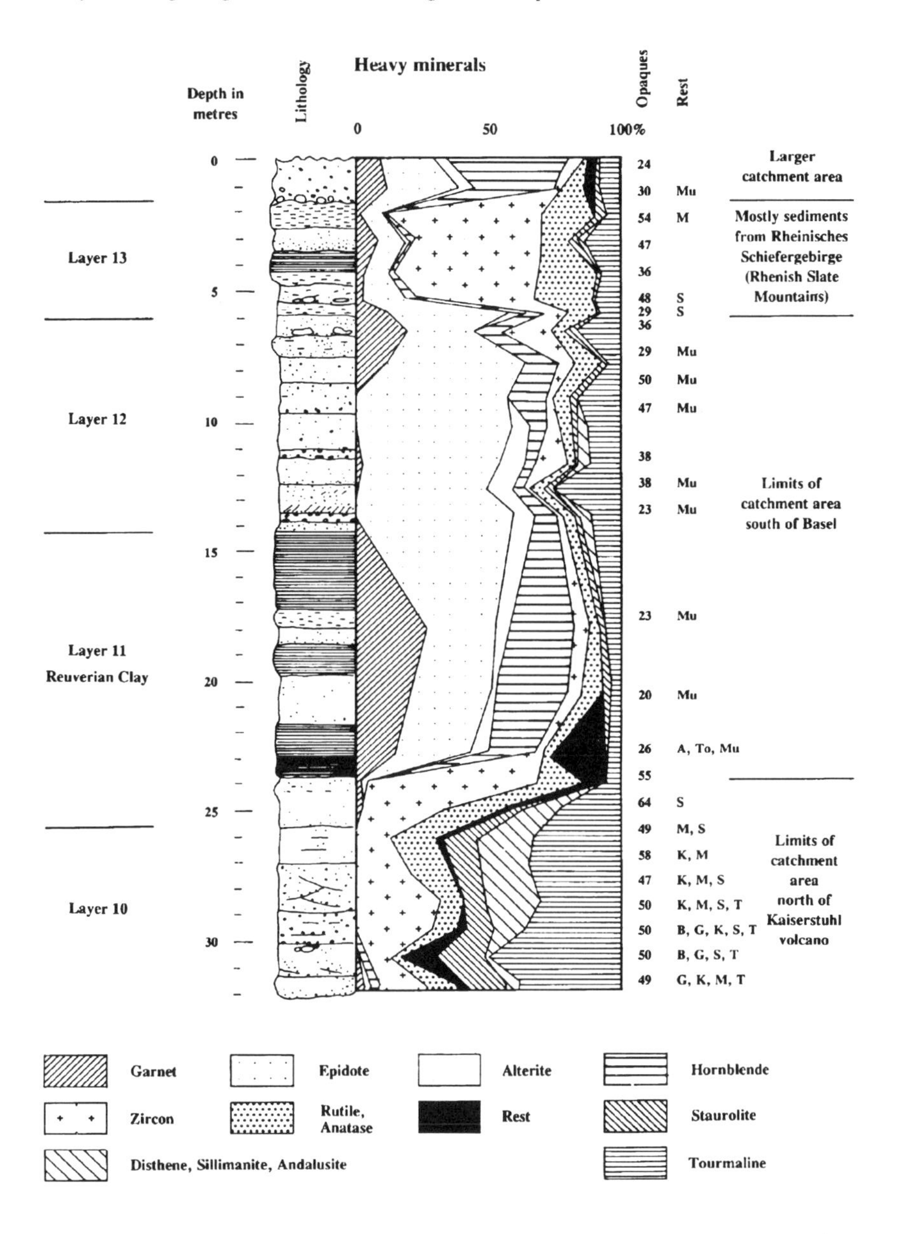

FIGURE 104 Heavy-mineral diagram of the Plio-Pleistocene deposits in the Fortuna opencast mine, Lower Rhine Embayment. Key to 'Rest' column: A = apatite; B = biotite; C = chlorite; E = epidote; Gl = glaucophane; H = hornblende; K = corundum; M = monazite; Mu = muscovite; P = pyroxene; S = spinel; Si = siderite; T = titanite; To = topaz. Layer 10: deposits of the Rhine when its source was north of the Kaiserstuhl mountain. Layers 11 and 12: The catchment area now reaches into the south German molasse region; the garnet–epidote–green hornblende association is characteristic. Layer 13: renewed reduction of the catchment area. Deposits from the Slate Mountains (with zircon) dominate. They are overlain by the Rhine spectrum as in layers 11 and 12 (from Boenigk 1983)

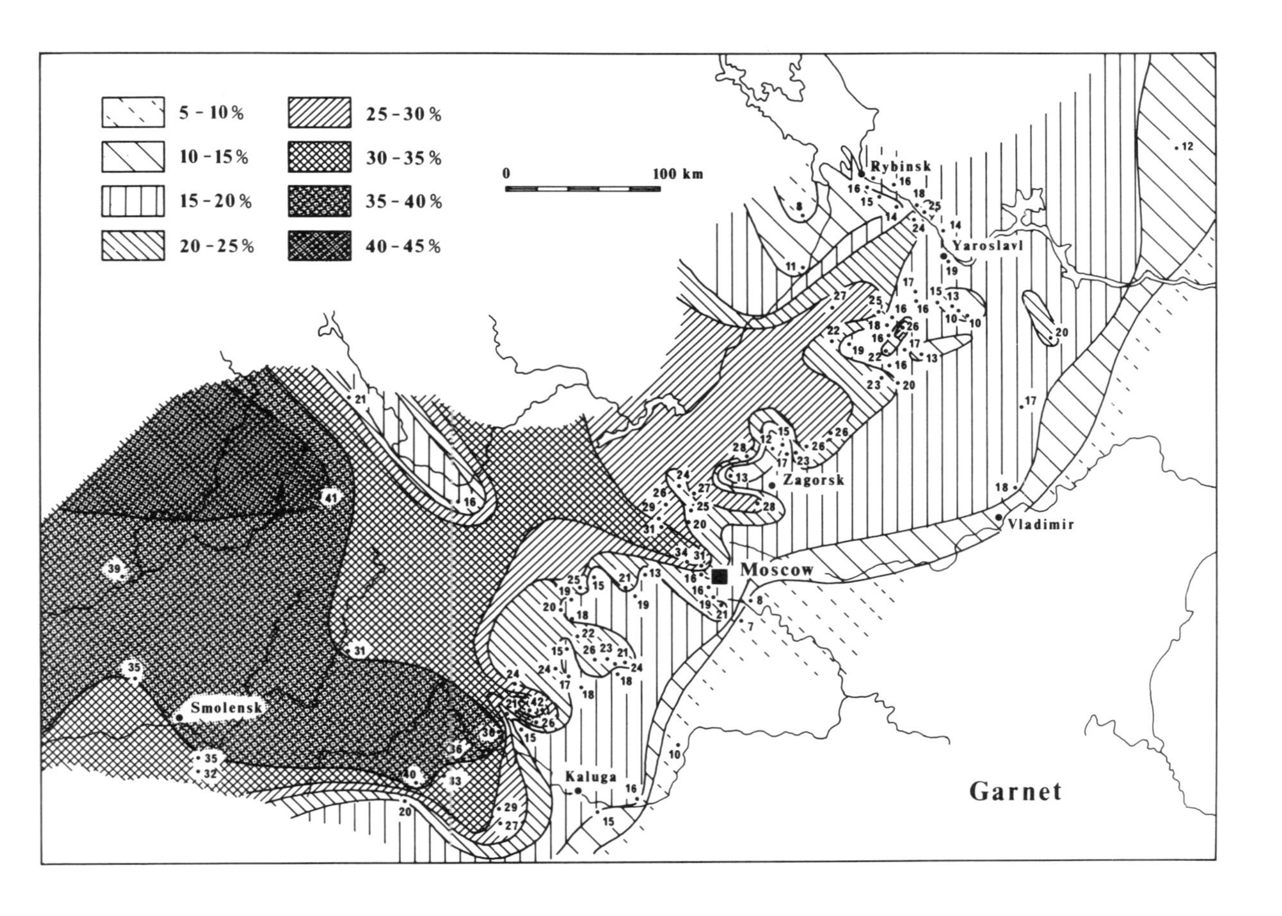

FIGURE 105 Garnet content of the Moscow Till, central Russia (from Sudakova 1995)

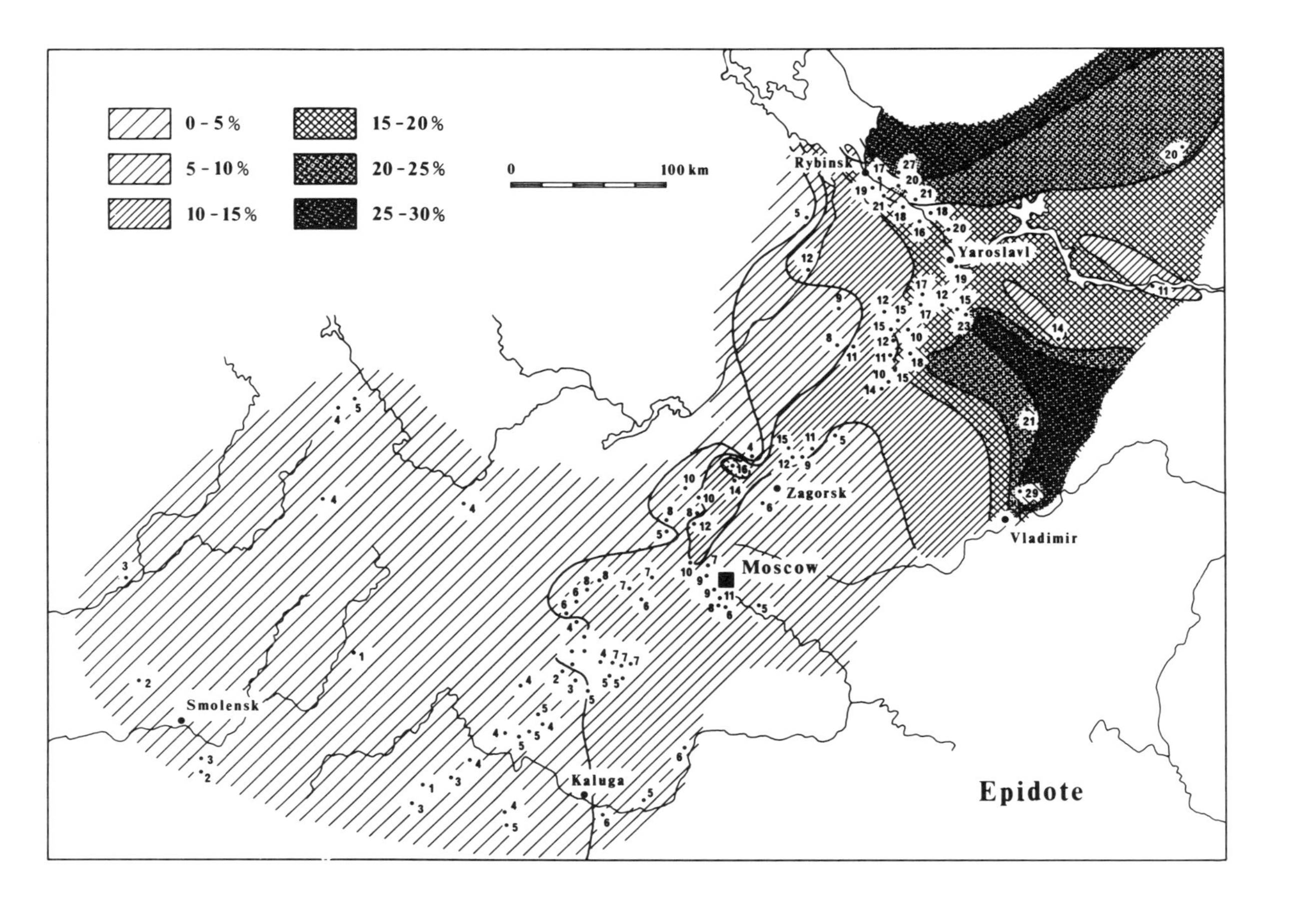

FIGURE 106 Epidote content of the Moscow Till, central Russia (from Sudakova 1995)

Wearing and breaking-up of heavy-mineral grains during transport are of only minor significance. Alternatively, sorting effects can have a considerable influence on the composition of the heavy-mineral spectrum. Certain minerals that predominantly occur as small grains (e.g. zircon and rutile) may be found preferentially in fine-grained sediments, whereas other minerals like staurolite are coarse-grained and thus more frequent in coarser sediments. Therefore, grain-size effects should always be taken into account when assessing counts (Boenigk 1983).

The depositional environment is of immense significance for the heavy-mineral composition. Thus, in dune sands easily rollable round grains tend to be enriched; in lake deposits lamellar minerals like biotite, chlorite, muscovite, and relatively light heavy minerals predominate; whereas under the influence of waves heavy grains are concentrated in placers. Consequently, the proportion of heavy minerals increases compared to the overall mineral spectrum, and indeed, recent heavy-mineral concentrations can be observed on most beaches.

Heavy-mineral analyses have been most successful in the investigation of fluvial sediments, but they can also be applied to glacigenic deposits. Since very small samples can be used, the method allows investigation of fine-grained sediments and gives the opportunity to differentiate within a single sedimentary unit.

In North America, heavy-mineral analyses have been extensively used for till stratigraphic investigations in Illinois (Willman *et al.* 1963). In the Great Lakes region, two major source areas have been distinguished by their heavy-mineral suites: the Grenville Province in the east is characterised by a high heavy-mineral content, with garnet and tremolite being the main components, whereas the westerly Superior/Southern Province has low heavy-mineral contents, dominated by epidote (Dreimanis *et al.* 1957, Gwyn & Dreimanis 1979). In the British Quaternary sequence, heavy-mineral analyses have been successfully applied to the Middle Pleistocene deposits of East Anglia to distinguish between British and North Sea source areas (Perrin *et al.* 1979, Bridge & Hopson 1985).

Heavy-mineral analyses have been used with great success for provenance studies in Russia. In the Moscow till sheet (Saalian), material from Fennoscandian source areas with their high garnet content can be clearly distinguished from epidote-rich tills from the Timan–Urals region (Sudakova 1995) (Figs 105 and 106). In principle, the same type of analysis can be applied to light minerals, e.g. by identifying different feldspar types. However, where this has been done no stratigraphic distinctions have been found (Willman *et al.* 1963, Hüser 1982).

In the Alpine region Dreesbach (1985) has undertaken heavy-mineral analyses in the Iller–Loisach Glacier region. She demonstrated that heavy-mineral composition provided no means of stratigraphical differentiation. Both the deposits from the Würm Glaciation and older glacigenic sediments are similarly characterised by central Alpine heavy-mineral assemblages (amphibole, garnet, epidote, staurolite). Only when distinct changes in the catchment areas occurred could significant results be anticipated. However, it is important to remember that older sediments may have been reworked and incorporated in younger tills. The heavy-mineral composition of Quaternary sediments in the Swiss Mittelland region has shown that the composition of the mineral spectrum there is largely controlled by the degree of reworking from the local substratum (Molasse rocks) (Gasser & Nabholz 1969).

8.6 CLAY MINERALS

As with heavy minerals, the clay-mineral composition of glacigenic deposits also can be used to a certain degree. Whilst the heavy minerals largely reflect the composition of volcanic and metamorphic rocks, clay minerals can be used to distinguish sedimentary source areas. In Illinois, for instance, clay minerals are used to determine the influence of Cretaceous sediments from western source areas (high montmorillonite content). In combination with heavy- and light-mineral analyses, clay mineralogy has been investigated in great detail in Illinois (Willman *et al.* 1963, Willman & Frye 1970), with more recent studies including, for example, Stewart & Mickelson (1976). Vortisch (1982) has also produced a comprehensive clay-mineralogical study of North German till. He concluded that the clay mineralogy of tills over longer distances (several tens of kilometres) can rarely be used for stratigraphical correlation. However, it is very homogeneous over short distances. In the case of the Elbe-Seitenkanal investigations by Vortisch (1982), three different till units were distinguished, the most distinctive being the Elsterian till (Fig. 107). This may result from the fact that in contrast to the other two (Saalian) tills, the dark grey Elsterian diamicton contains a high proportion of reworked Miocene mica clay, producing its dark colour. The partial homogeneity seems to be a characteristic of tills from the marginal areas of a glaciation. Closer to the glacial centre, e.g. in Scandinavia, the variability is far greater. For clay-mineralogical investigations the same restrictions apply as for heavy mineralogy; the lengthy preparation procedure required renders them useful only when very small samples of fine-grained material are available (e.g. from boreholes).

With regard to the transport mechanism and source areas of the till matrix, clay mineralogy has proved to be a very useful tool. In most cases, the material has only been transported over short distances, as demonstrated in both the Alpine area (Peters 1969) and the area of the Nordic glaciations. However, there are exceptions. Haldorsen *et al.* (1989) showed clearly that the matrix of the so-called 'red tills' in the Netherlands, in contrast to the matrix of the other Dutch till types, contained little or no smectite. This indicates that it must have been transported long distances almost exclusively. The Scandinavian hardrock areas are virtually free of smectite, whereas the Mesozoic and Tertiary deposits of the Dutch–German–Polish lowlands contain abundant smectite.

8.7 GEOCHEMISTRY

Geochemical investigations of glacigenic deposits play an important role in mineral prospecting in areas like the Canadian or Fennoscandian shields (e.g. Shilts 1975, 1984b, K. Eriksson 1983). Like boulder trains, geochemical dispersal fans can be traced back under favourable conditions to their source areas. Glacial dispersal trains may be crudely described as consisting of two parts: a 'head' or area of high element concentration, which quickly decreases down-ice to a ribbon-shaped 'tail' of dispersal (Figs 108 and 109). The aim of geochemical prospecting is to find, identify and trace this tail back to the source. Such studies must be accompanied by investigations of ice-movement directions, as indicated by striae, fabrics and indicator clasts (Shilts 1976).

Besides economic significance, geochemistry can also be used for stratigraphical correlation. The geochemistry of tills primarily depends on bedrock source areas, although it is also strongly influenced by grain size, carbonate content and type/degree of weathering. Clay-rich tills are characterised by lower SiO_2 content and by higher proportions of elements bound to clay minerals, such as K, Al, Fe, Ti and Mg (Cornelius 1984).

The trace elements show the same dependencies in principle. Ba, Co, Cu, Ga, Pb, Rb, U and Zn are found to be enriched in the clay fraction, and naturally their percentage depends on the

A

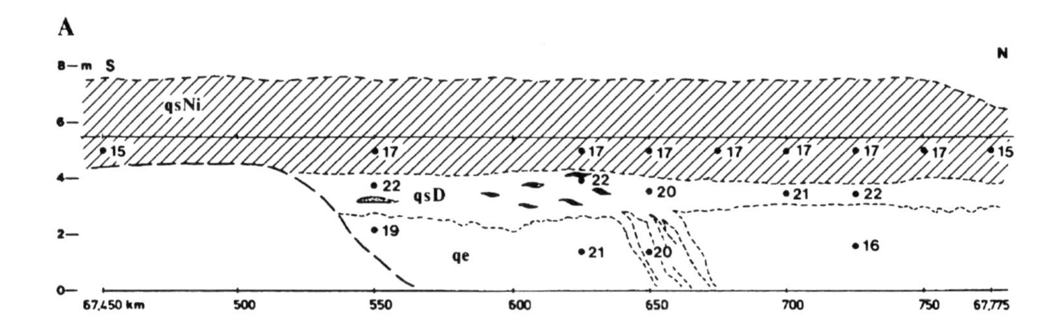

B

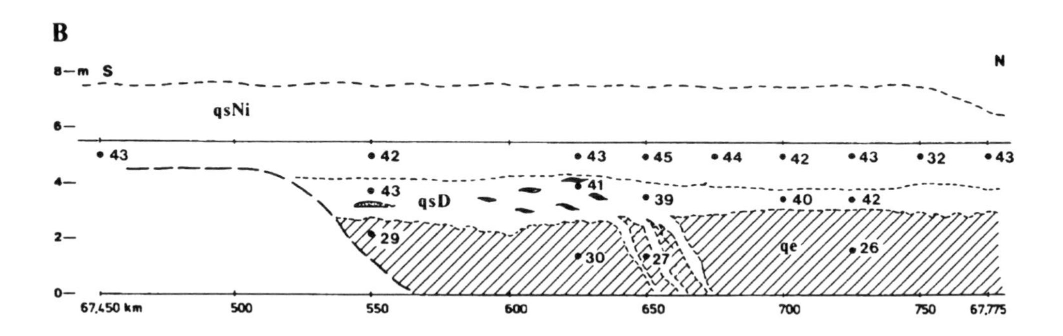

C

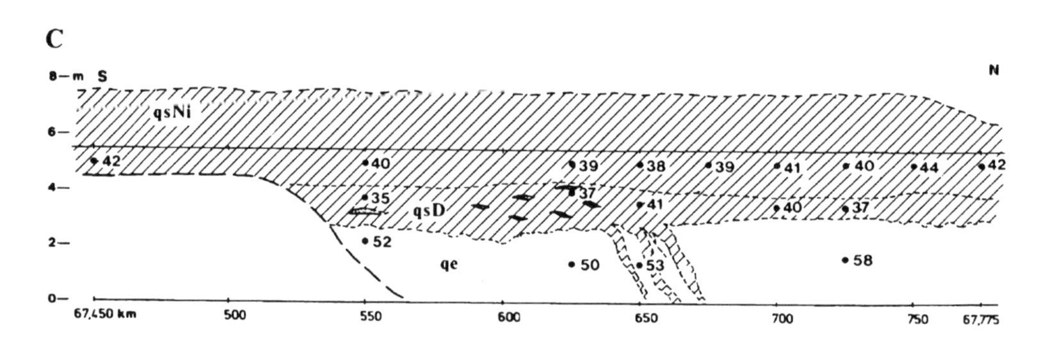

FIGURE 107 Clay-mineral composition of different till units from the Elbe-Seitenkanal construction site, contents given in percentages; qsNi = Middle Saalian till, qsD = Older Saalian till, qe = Elsterian till: (A) kaolinite and chlorite, (B) illite, (C) mixed-layer minerals (from Vortisch 1982)

clay content of the overall sample. Conversely, Sr behaves very much like Ca, with its distribution depending largely on the carbonate content. Accordingly, it decreases in the upper, weathered parts of the profile along with the depletion of carbonate. Other trace elements are not susceptible to weathering. Cornelius (1984) discovered that the decalcified parts of the Saalian tills in Hamburg were relatively enriched in Pb and Zn and to a lesser degree also in V, Th, Rb and Ga (Fig. 110).

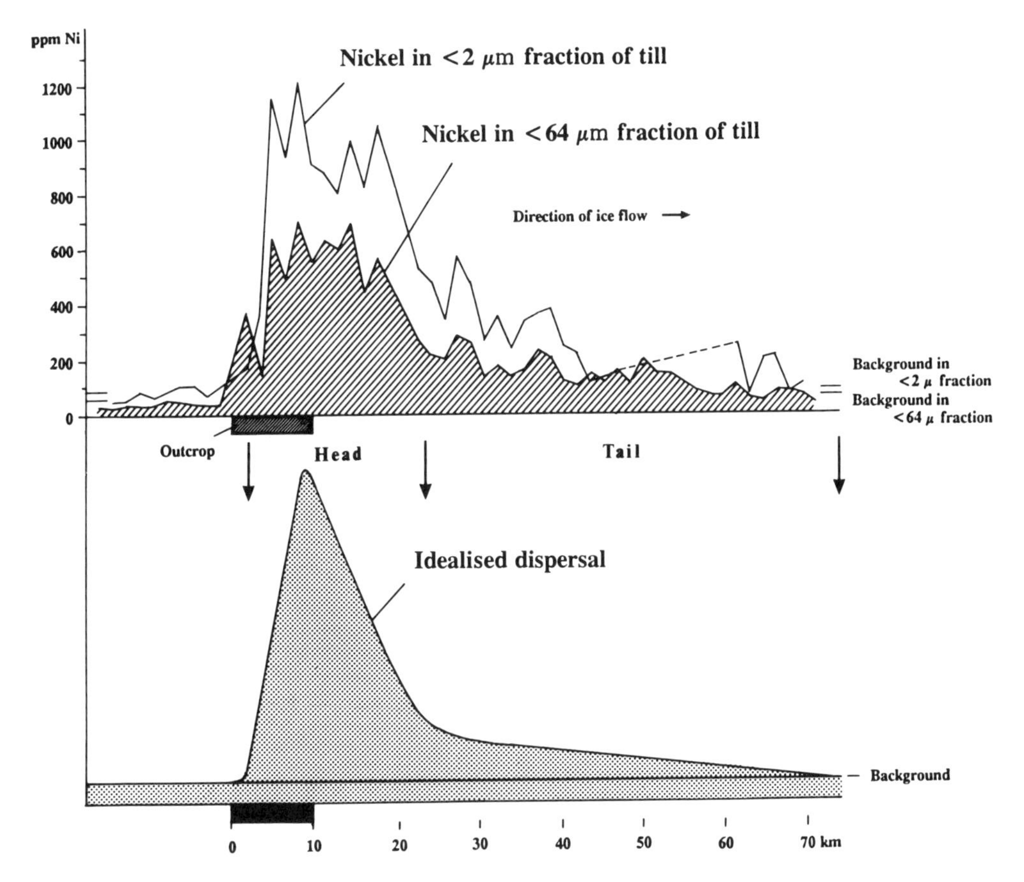

FIGURE 108 Actual (top) and idealised (bottom) dispersal curves showing the relationship of the head and tail of a Ni concentration (after Shilts 1976)

General rules with regard to the effects of weathering have not yet been established. Geochemical alteration of the clay fraction within Quaternary weathering profiles largely depends on the type of processes taking place. For instance, in the decalcified upper parts of the Middle and Older Saalian tills Cornelius (1984) found the following alterations:

- Middle Saalian till: Rb and K decreased, Fe, Cr, Ni and V enriched
- Older Saalian till: Rb and K enriched, Fe, Cr, Ni and V decreased.

In the upper part of the Middle Saalian till, long-term weathering and pedogenesis has altered the clay-mineral composition during two interglacials (Eemian and Holocene). During this process easily soluble potassium was removed, whereas iron and its companion ions chromium, nickel and vanadium were enriched as free or pedogenic oxides. The potassium enrichment in the Older Saalian decalcified till resulted from a strong increase of the illite content due to intensive mica weathering (Cornelius 1984).

Sulphides and carbonates in coarser fractions (>4 mm) are less affected by weathering than finer components. This is why limestone clasts may still occur within the decalcified matrix of a weathered till profile. As a single limestone clast would heavily distort the results from a

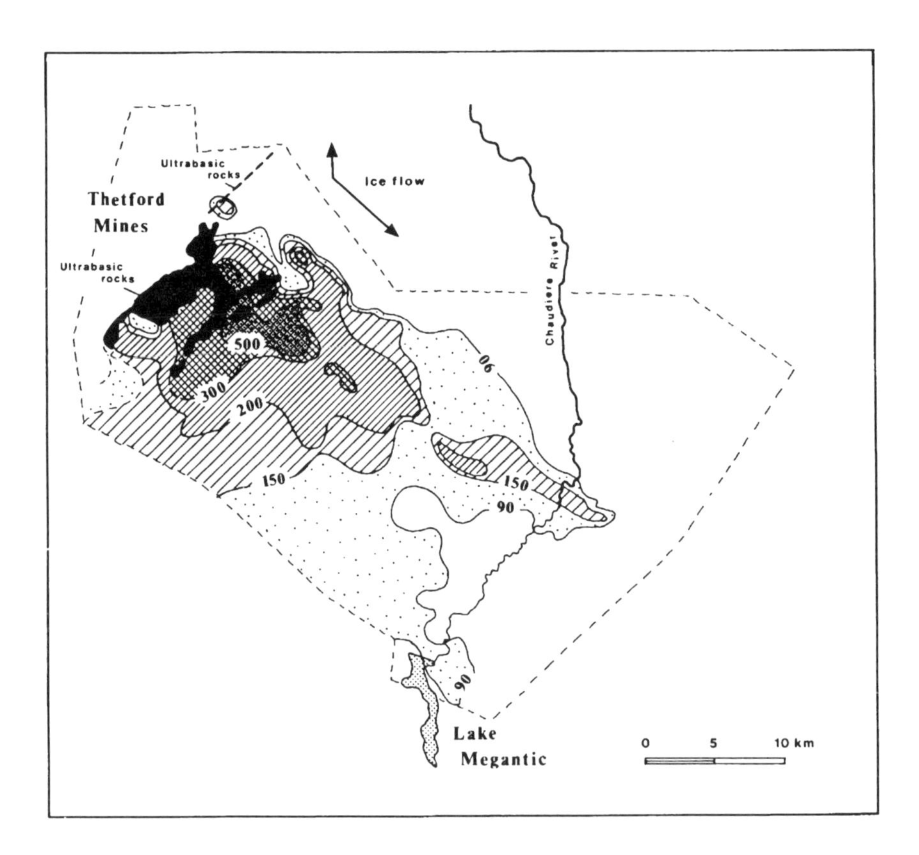

FIGURE 109 Nickel dispersal from ultrabasic outcrops in the Thetford Mines area, Canada (after Shilts 1976)

small sample used for geochemical analysis, the analyses of several parallel samples of the same unit may yield strongly differing results. This 'nugget effect' can be avoided if, instead of bulk samples, the investigation is restricted to a relatively narrow grain-size fraction (Shilts & Kettles 1990).

It seems that geochemistry is only of limited use for routine stratigraphical correlation in areas like Britain or North Germany. However, the method is a valuable tool in prospecting and for investigating weathering and pedogenetic processes. Besides this, an assessment of geochemical background values like the natural content of arsenic and heavy metals is important to detect and evaluate possible recent soil contamination.

8.8 MICROMORPHOLOGY

The investigation of soil structures under the microscope with thin sections has been applied in pedology since the pioneering works of Kubiena (1931, 1938). The preparation of samples and their interpretation has been described by Catt (1990). It is invaluable, for instance, not only to clarify the pedogenetic processes that have occurred but also to determine relict soil characteristics of older interglacials that have been preserved in recent soils. Studies of this type

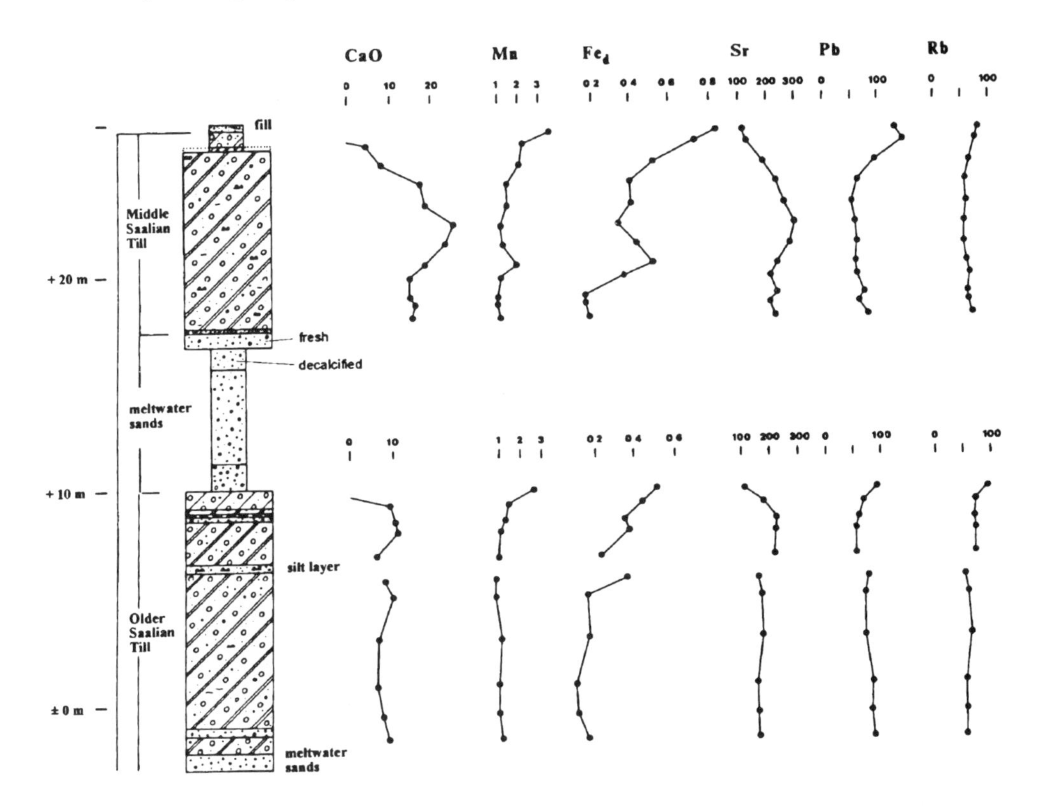

FIGURE 110 Geochemical composition of two Saalian tills in a core drilling in the Wittenbergener Heide (Hamburg). Main elements as percentages: CaO = calcium oxide; Mn = manganese; Fe_d = dithionite-soluble iron; trace elements in ppm: Sr = strontium; Pb = lead; Rb = rubidium (after Cornelius 1984)

include Felix-Henningsen (1979), Whiteman & Kemp (1990), Xiaomin *et al.* (1994) and Bronger *et al.* (1995).

However, micromorphology can also be used for detailed sedimentological investigations. Van der Meer (1982) has refined this method into a very effective tool for distinguishing diamictons of different origin (Fig. 111). Many processes like lodgement, shearing, or sediment flow leave characteristic features which can be detected in micromorphological investigations (Van der Meer 1993). For example, it was possible to show that sediments in a certain end moraine in Patagonia (Argentina) were glaciolacustrine, rather than subglacial, in origin (Van der Meer *et al.* 1992), that diamictons in the southern North Sea include flow and lodgement till facies (Van der Meer & Laban 1990), and that the Irish Sea Till is a basal till and not a glaciomarine deposit (Van der Meer *et al.* 1994).

8.9 MAGNETIC SUSCEPTIBILITY

As a rule, rocks contain a certain percentage of magnetic minerals like magnetite and maghemite. In glacigenic deposits, this proportion depends on the composition of the source rocks that have been overridden and incorporated by the glacier. Like clast composition, heavy

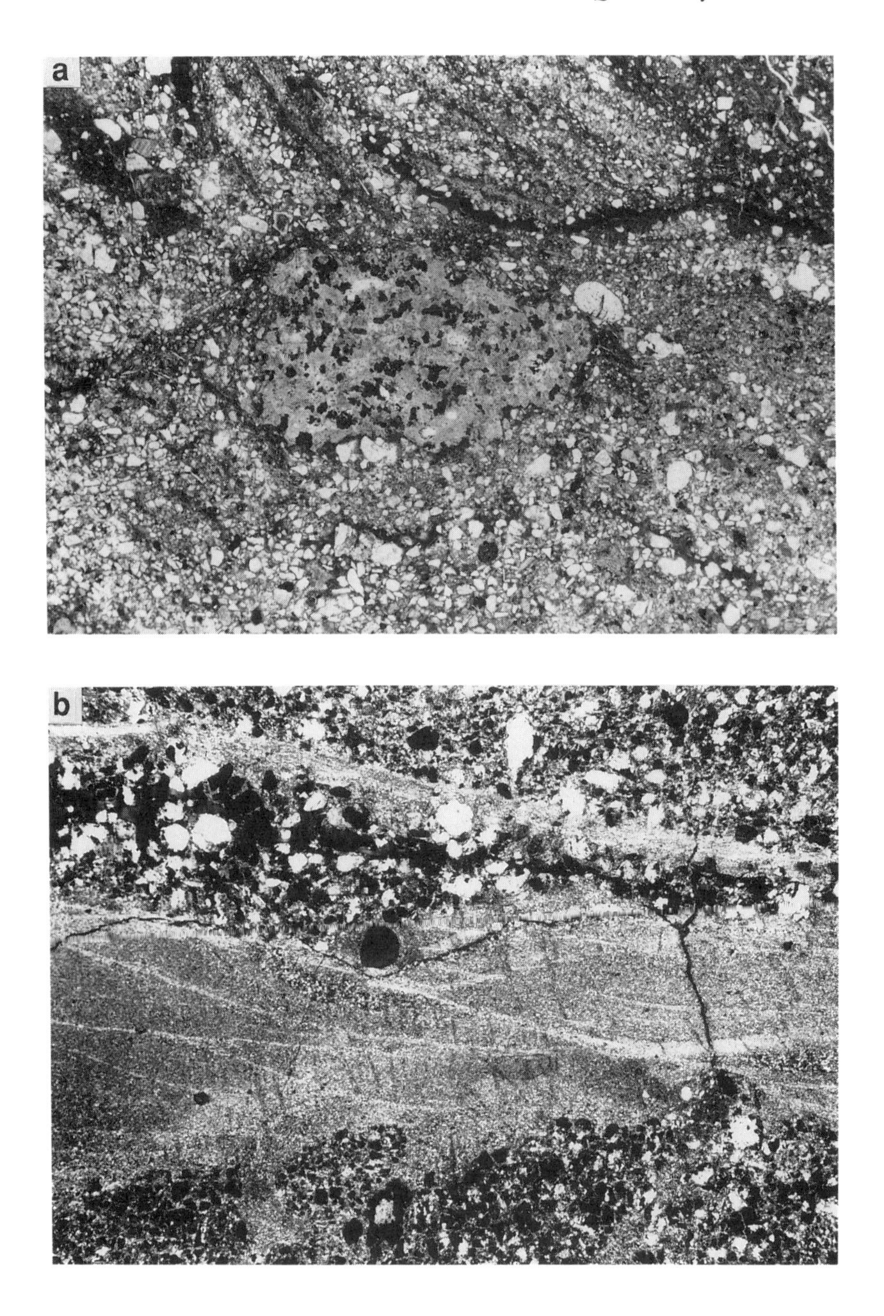

FIGURE 111 Photomicrographs of thin sections of till samples studied for micromorphological characteristics; a = Irish Sea Till; small gravel particle with triangular extensions. This structure is known as a pressure shadow and clearly related to shear; b = anastomosing system of discrete shears in a Saalian till from Lunteren, the Netherlands (from Van der Meer 1993)

minerals and clay minerals, the content of magnetic minerals can be used to characterise a deposit. For this purpose, it is not necessary to count the individual grains but simply to measure the relative magnetisation of the sample, i.e. the strength of attraction exerted towards a known magnetic source. This is termed magnetic susceptibility. It depends not only on the mass of magnetic particles but also on size and shape of the individual grains. Because magnetic susceptibility is easily measured (required sample size: 30 g), it can be used conveniently for the correlation of Quaternary deposits.

To a major degree, magnetic susceptibility has as yet only been used to correlate North American till samples. In particular, long-distance transported tills with material from the Canadian Shield were found to be rich in magnetic minerals. Thus, the three tills of the eastern Green Bay Lowland (Wisconsin) can be distinguished by their sand content and magnetic susceptibility (Fig. 112), with the oldest till in this case having the lowest susceptibility (McCartney & Mickelson 1982). Magnetic susceptibility has been found to be useful in addition for the stratigraphical correlation of loesses and lake sediments. In deep-sea sediments it has proved possible to correlate with the changes in $\delta^{18}O$ (Steens *et al.* 1990). Magnetic susceptibility parallels the glacial–interglacial cycles in loess sequences (Kukla *et al.* 1988; see chapter 15), and it has recently been found to correlate with the pollen and organic carbon contents of European crater-lake sediments (Thouveny *et al.* 1994).

8.10 SEISMIC INVESTIGATIONS

In many regions, the sea floor is extremely flat. In contrast to terrestrial investigations it is therefore rarely possible to use recent landforms to draw conclusions about the Quaternary

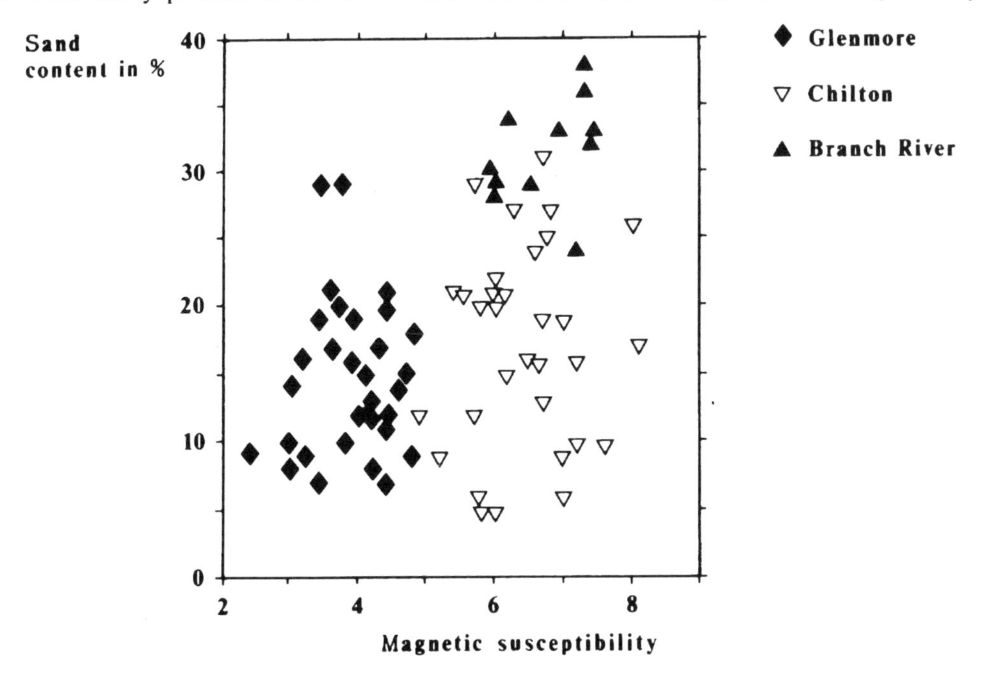

FIGURE 112 Sand content and magnetic susceptibility of the <2 mm fraction of Wisconsinan tills from the eastern part of the Green Bay Lowland (Wisconsin) (from McCartney & Mickelson 1982)

development. For this purpose one is forced to rely almost entirely on the results of geological and geophysical investigations, which has advantages and disadvantages. Whereas in continental areas the structure of the subsurface must be reconstructed from point data (boreholes), in marine studies modern methods of shallow seismics allow the continuous recording of long profiles. Unfortunately, the seismostratigraphic units recorded offshore cannot be directly correlated with sequences onshore. This can only be achieved via the evaluation of intentionally planned boreholes, that provide orientation points within the seismostratigraphic network.

With seismic profiles, a choice must be made between deep penetration and high resolution. Where both objectives need to be met, a combination of several instruments must be applied. The British Geological Survey (BGS) in its offshore mapping programme usually applies three different systems: boomer, sparker and air gun.

	Boomer	Sparker	Air gun
Frequency (in Hertz)	700–8000	200–600	15–200
Depth of disturbance below sea floor (m)	1.5–2.5	8.5–12	25–30
Penetration depth (m)	20–60	300–450	850–1200
Resolution (m)	0.5–2	5 –10	15–20
Vertical exaggeration	25–30-fold	9–12-fold	6–7.5-fold

The three systems are used simultaneously and at the same time the sea-floor morphology is recorded with side-scan sonar. In the British sector of the North Sea, for instance, such profiles were recorded on average every 5–10 km. The results of the investigations are presented on 1:250,000 maps. The records contain an enormous wealth of information; only a small part of which has been analysed so far.

A technical problem is the transformation of seismic profiles into real cross sections. Seismic velocities to a large degree depend on the composition of the penetrated medium. Therefore, it is only to a limited extent possible to deduce sediment thickness from seismic measurements. As crude estimates the following factors may be used: in salt water 100 msec are equivalent to a depth of 75 m, whereas in sediment they are equivalent to a depth of 85 m. In salt water the influence of salinity and temperature can be measured and calculated with relatively little effort, but the true seismic velocities in sediment can only be determined on core material. In Cretaceous chalk, for instance, seismic velocities of ca. 2.0–5.0 km/s have been measured (Wingfield, personal communication).

Figures 113 a–d give examples of shallow seismic measurements from the North Sea. In the boomer records shown, the vertical distance between the depth contours is 25 msec (21 m in sediment), in the sparker records 50 msec (43 m) and in the air-gun records 100 msec (85 m). The horizontal distance between the vertical lines (fixes) on average is about 1.75 km. In reality on board the research vessel it is not distance that is measured but time intervals (of ten minutes), the real distances ranging between about 1.3 and 2.1 km. At every fix, the exact position of the research vessel is determined via a satellite navigation system (NNSS satellite 1102 system), so that the precise position of the fixes can be reconstructed.

A technical problem results from the limited resolution of the profiles. The sparker records are used as standard profiles, having a horizontal resolution of 10 m. The theoretical vertical resolution is 1 m but, in addition to the downward deteriorating quality of the recordings, because of the echoes, a much lower resolution must be accepted. Each reflector is depicted repeatedly because of the echoes. Strong reflectors, like the sea floor or stone layers, are depicted as a series of two or three parallel reflectors of about equal strength, the uppermost of which is the true reflector. Within the band of echoes all other signals are wiped out, causing vertical resolution to be restricted to about 4 m.

Furthermore, the evaluation of lower parts of the profiles is hampered by the occurrence of multiples of the higher profile parts. The multiples result from the reflection of the signals from the sea surface and from the sea floor, so that all signals are recorded several times. The multiples are characterised by being a vertically exaggerated true reflection of higher parts of the same profile.

The example shown is taken from a BGS survey of the central North Sea (Ehlers & Wingfield 1991). For ease of comparison, all three records have been transferred to the same horizontal scale, but note that the height exaggeration differs considerably. The clearest information about the internal structure of the subsurface is provided by the sparker profile (Fig. 113a) which shows three major and two smaller Weichselian/Devensian channels under the sea-floor surface, incised through relatively transparent marine deposits. Both the double channel at fixes 58/59 and the shallow depression between fixes 60 and 61 display very irregular bases. In both cases the channels are infilled with well-bedded sediments.

In the boomer profile (Fig. 113c), it can be seen that the channel fill was not completed in one sedimentation event, but was interrupted by periods of erosion. The discordance in the upper part of the channel fill is most striking and marks a transition from draped bedding in the lower part of the channel to subhorizontally bedded fill in the upper part. Whereas the boomer profile depicts the fill of those two channels in great detail, the fill of the channel left of fix 58 is nearly invisible, most likely indicating the presence of sand.

In the boomer record, not only the sedimentary structures, but also the bedforms of the sea floor are shown in great detail. Thus, the small slope west of the double channel and the hill east of fix 61 are clearly visible, but also the depression that has formed above a pipeline on the sea floor (east of fix 60). One should not, however, be confused by the strong vertical exaggeration: the step in the western part of the profile is only about 8 m high.

Within the sparker profile, channels of an older erosional phase are also seen, which may be attributed to the Saalian/Wolstonian Cold Stage. Their channel fill displays numerous point reflectors, probably caused by stones. In contrast to the Weichselian/Devensian channel fill, there can be no doubt about the glacial or at least ice-proximal infill of the older channels. The Boomer profile does not penetrate sufficiently deep, so that the interpretation must rely on the sparker and air-gun profiles. The air-gun profile (Fig. 113d) gives the least resolution, but is invaluable for the determination of the depth of deeper horizons and channels.

8.11 GEOPHYSICAL LOGGING

For economic reasons, most well boreholes today are rotary counterflush drillings. The poor sample recovery using this method can partly be balanced by comparison with geophysical well logs. The method, originally used in the oil industry, has been applied to hydrogeological exploration since the early 1960s. Under favourable conditions, the logs allow characteristic till

a

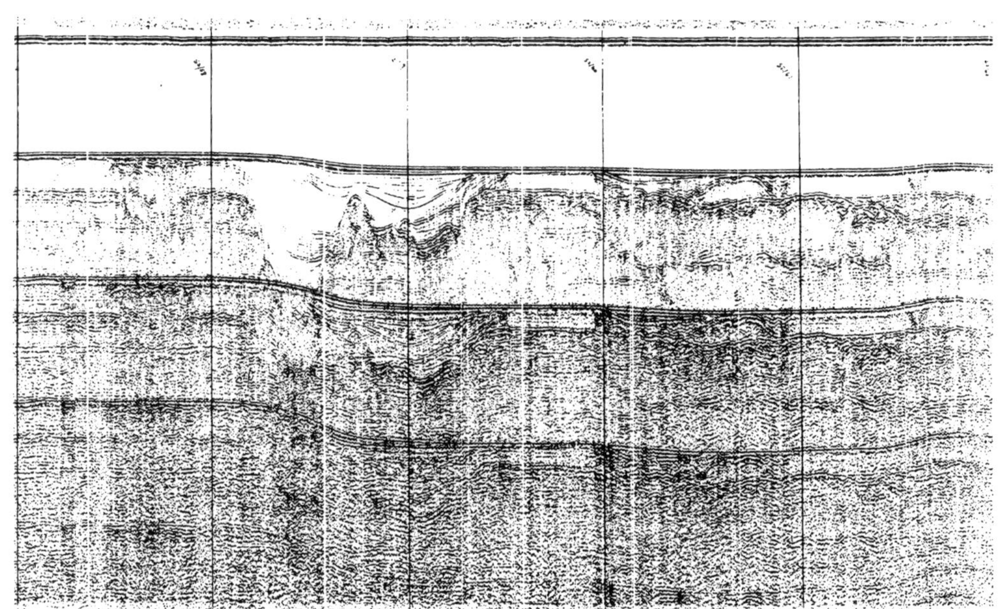

b

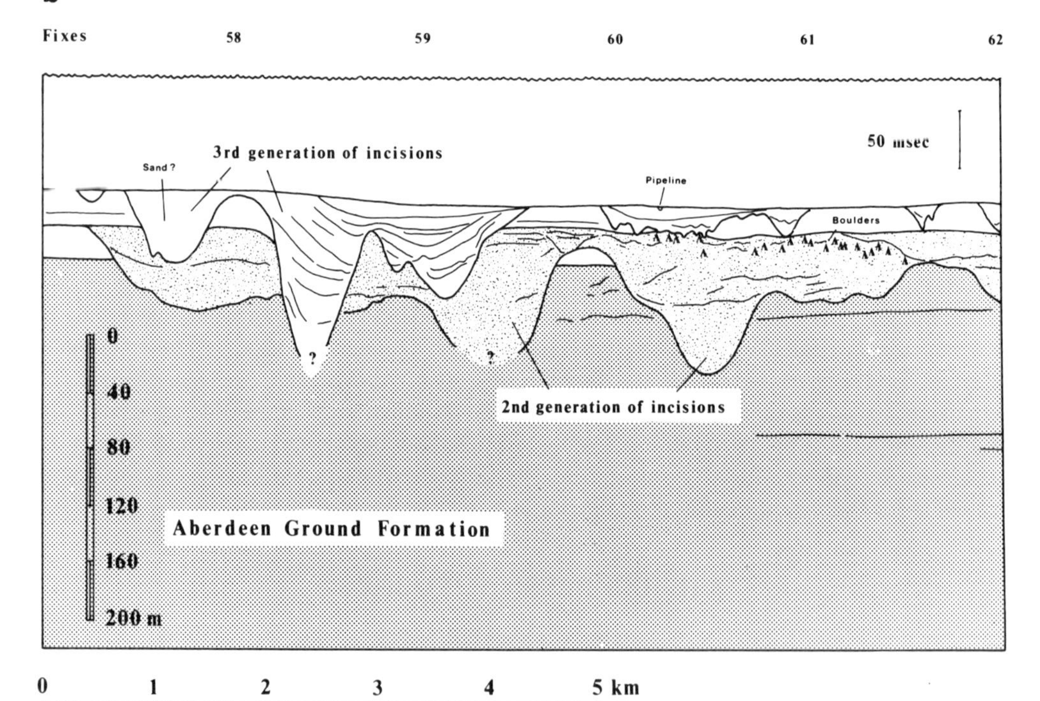
Fixes
58
59
60
61
62
50 msec
Sand ?
3rd generation of incisions
Pipeline
Boulders
2nd generation of incisions
0
40
80
120
160
200 m
Aberdeen Ground Formation
0
1
2
3
4
5 km

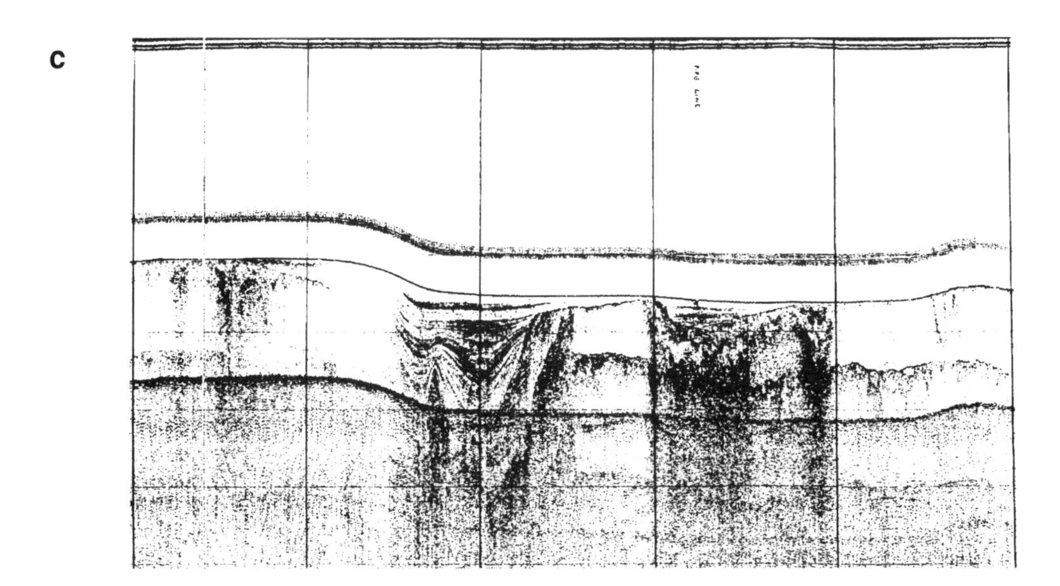

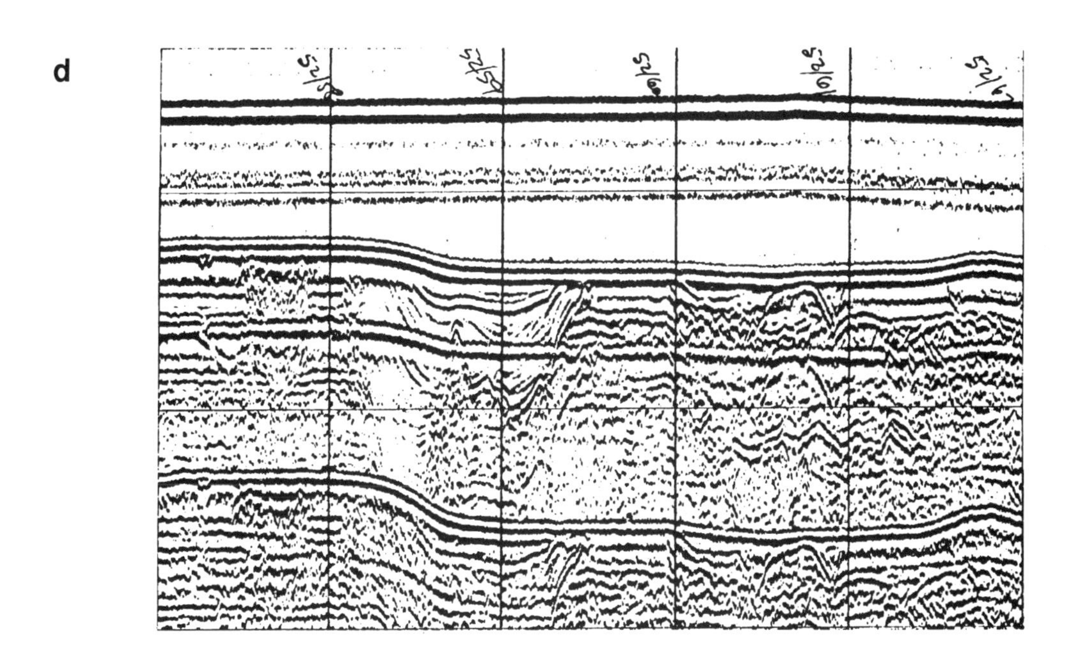

FIGURE 113 Shallow seismic profile from the central North Sea: (a) sparker profile, (b) interpretation, (c) boomer profile, (d) air-gun profile (from Ehlers & Wingfield 1991)

layers to be identified and followed over distances of several kilometres (Ehlers & Iwanoff 1983). Two standard methods employed in groundwater drillings are resistivity and gamma ray logs. Figure 114 gives a good example of the results obtained by both of these methods (cf. Ehlers & Linke 1989).

In a **resistivity log** an electrical current is measured which is passed through the sediment. The current is emitted by an electrode at one end of a sonde and measured at the other end. The passage of electricity is largely determined by pore volume. Coarse-grained sands yield a high resistivity, silt and clay a low resistivity, and in dry sand the resistivity is extremely high (normally not measured). Resistivity measurements are strongly influenced by borehole diameter, the type of mud fluid and the geometry of the zone around the borehole invaded by the mud fluid. If only the so-called 'normal device' is used, the current-emitting electrode is positioned at the base of the sonde and the measuring electrode on the upper end. Two different spacings of electrodes are applied, the so-called Short Normal (16 inch) and Medium Normal (64 inch). The Medium Normal gives a larger penetration depth, so the influence of mud fluid contamination is largely reduced. However, because of the wide spacing of the electrodes, thin beds cannot be recorded. The Short Normal is better suited to determining the exact boundaries between strata.

In the case of drilling 7040 B 520 (Fig. 114), two measurements have been made, because the first survey failed to reach the bottom of the borehole. The curves were recorded at different horizontal scales, but comparison of the two measurements shows that the results are exactly equal.

The **gamma-ray log** is used to measure the natural gamma radiation of the sediment. In contrast to the resistivity log, measurement is point-specific, which gives the gamma curves a much greater resolution than those obtained from resistivity measurements. Because radioactive decay is a random process, no two gamma-ray logs from the same drilling will look completely the same. However, comparison of the two measurements in well 7040 B 520 reveals that the differences are small compared to those that result from the different lithologies. Natural gamma radiation in Quaternary deposits is largely a result of radioactive ^{40}K, which occurs in feldspars and, more importantly, in clay minerals. This means that the gamma-ray log provides an opportunity to determine the grain-size composition of the drilled strata. In the gamma-ray log, clays are represented by high values, sands by low values and glacial diamictons appear somewhere in the mid-range, as do glaciolacustrine silts.

Geophysical well logs have, for instance, been used to trace individual till units over hundreds of kilometres in southern Saskatchewan (Christiansen 1971a). The evaluation of geophysical well logs reveals that the internal composition of thick till sequences may vary considerably (Ehlers & Iwanoff 1983, Ehlers 1990a). For example, the Older Saalian till in northern Hamburg consists of a relatively clay-rich basal unit, a sandy middle layer, followed by a more clay-rich unit and topped by a very clay-rich upper layer (Fig. 115). Marked differentiation can also be observed in the other tills encountered in the profiles illustrated. Since well logs are routinely recorded for every major borehole, they provide a valuable tool for stratigraphical correlation, especially in densely drilled urban areas.

8.12 BIOLOGICAL REMAINS

During the Quaternary the fauna and flora were forced to adjust to dramatic climatic and environmental changes. The methods briefly outlined below are used to reconstruct the

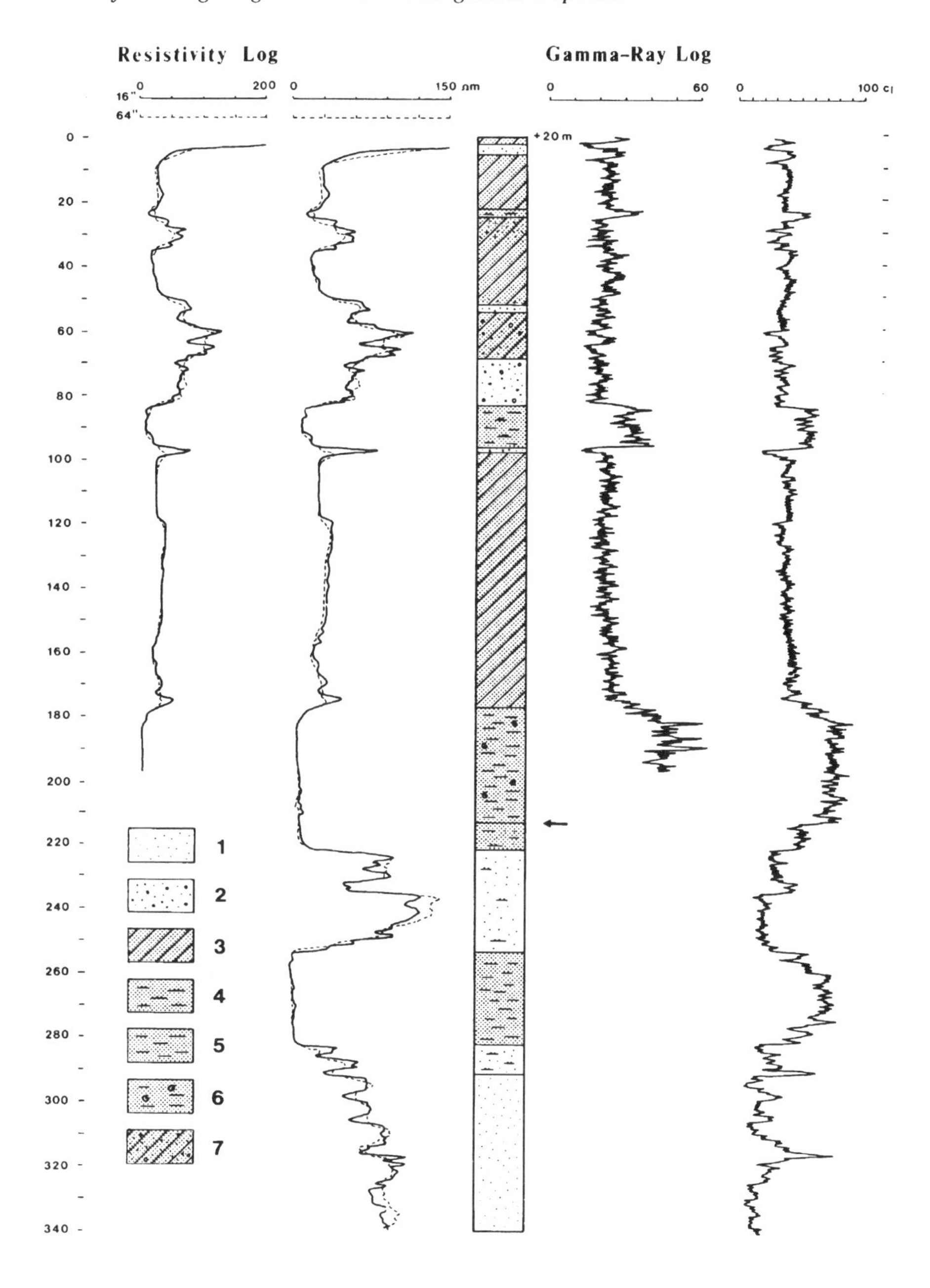

FIGURE 114 Two geophysical logs of a well drilling in Hamburg. Whilst the resistivity logs are quite identical (differences due to scale), the gamma-ray logs differ in detail. 1 = fine sand; 2 = gravelly sand; 3 = till; 4 = glaciolacustrine silt; 5 = glaciolacustrine clay; 6 = marine, shelly clay; 7 = clast-rich till; arrow: local base of Quaternary (from Ehlers & Linke 1989)

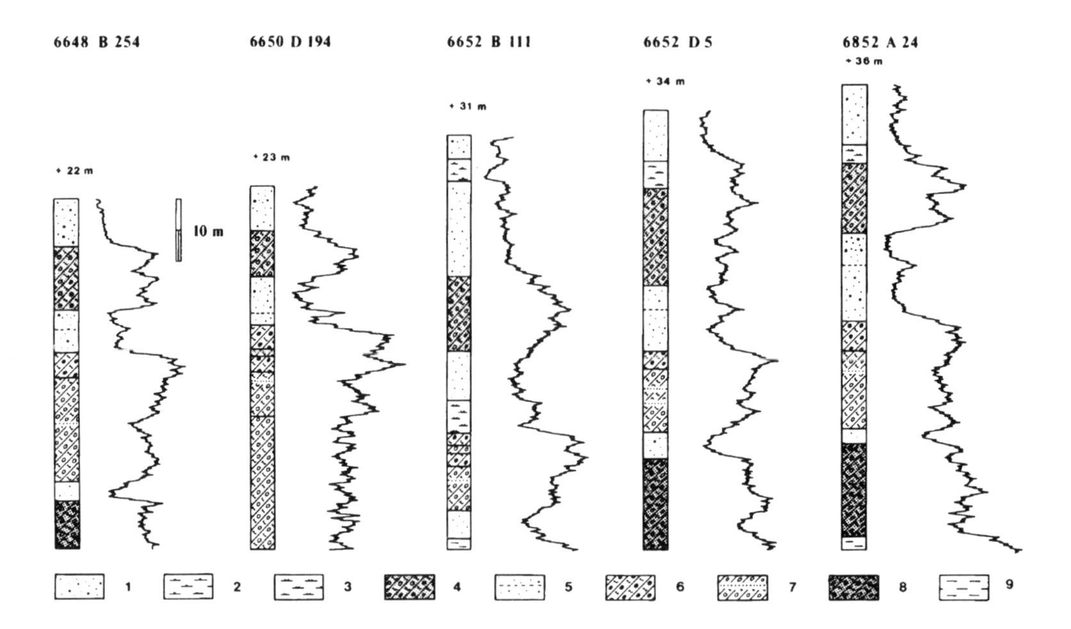

FIGURE 115 Correlation of till units in five boreholes from north of Hamburg, based on gamma-ray logs; 1 = fine-grained meltwater sands, dry; 2 = muds of the Eemian Interglacial; 3 = glaciolacustrine silt; 4 = till of the Middle (and Younger?) Saalian Glaciation; 5 = fine-grained meltwater sands with silt layer (stippled); 6 = Older Saalian till, red facies; 7 = Older Saalian till, normal facies (grey) with sand layer; 8 = till of the Elsterian Glaciation; 9 = Miocene mica clay (from Ehlers 1990a)

biological response. They provide information about the ecological conditions of the period in question, and can also provide palaeoclimatic information. However, primarily they are not dating methods, although under favourable conditions they may indicate the age of a certain deposit.

8.12.1 Macroscopic plant remains

The investigation of fossil Quaternary floras began with the analyses of plant macrofossils in the last century. The method of investigation was developed by Nathorst (1873) in Sweden and was soon applied throughout Europe and America. Macroscopic plant fossils are parts which are visible to the naked eye, in contrast to microfossils like pollen and spores. They include such diverse remains as seeds, cones, twigs, leaves, moss remains (bryophytes), algae and fungi. In contrast to microfossils they can often be determined to species level (Warner 1990). Recent work in North America includes papers by N.G. Miller (1973, 1987, 1989), N.G. Miller & Calkin (1992) and N.G. Miller & Thompson (1979).

Macroscopic plant remains are most useful in reconstructing local ecological conditions. For

---

FIGURE 116 (Opposite) Macroscopic plant remains from the Ipswichian interglacial deposits at Bobbitshole near Ipswich. Aquatics to the left, reedswamp species in the centre and marsh plants to the right. The changes in plant remains reflect environmental changes from open water to reedswamp to marsh (after West 1957)

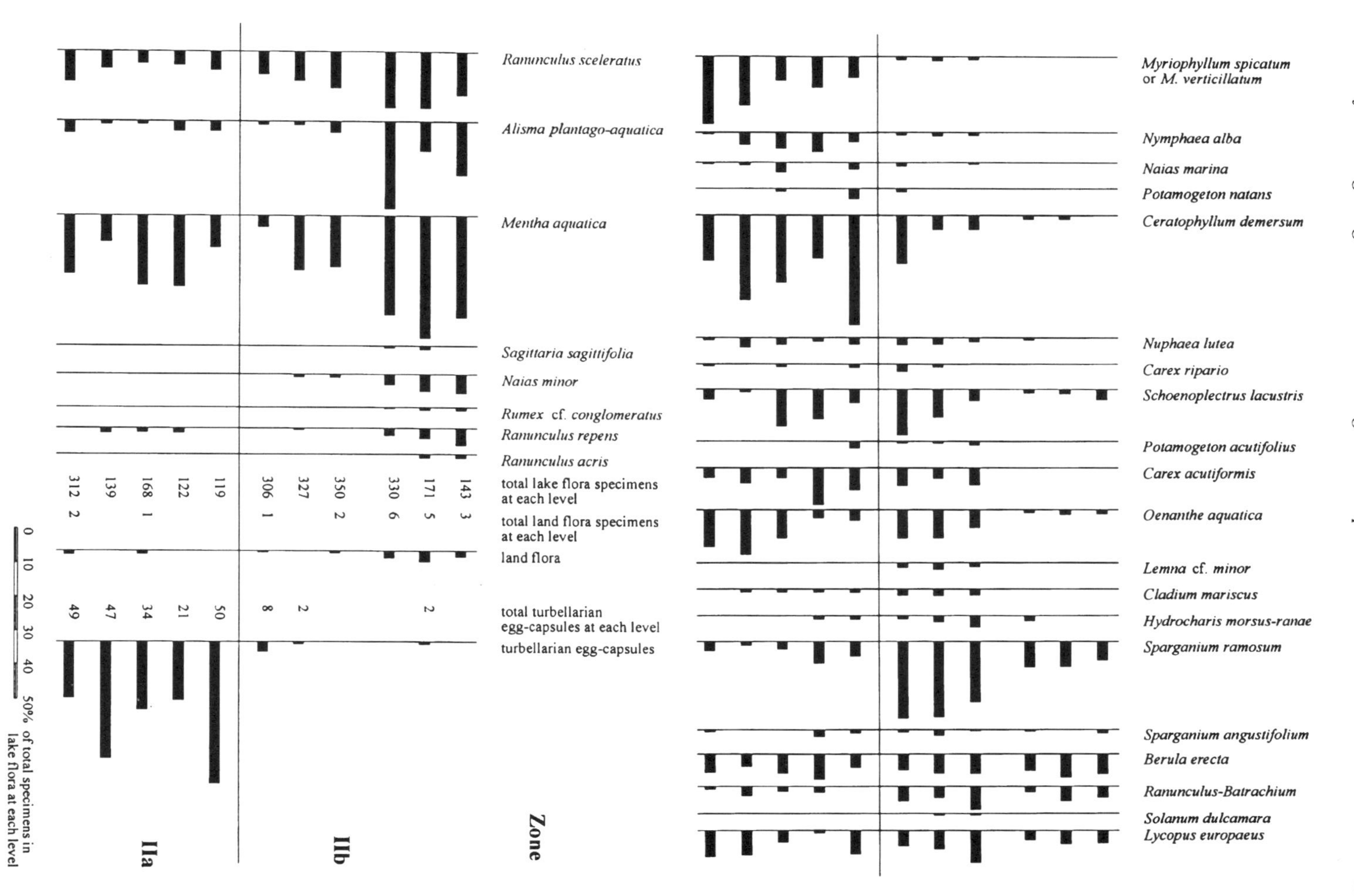
Ranunculus sceleratus
Alisma plantago-aquatica
Mentha aquatica
Sagittaria sagittifolia
Naias minor
Rumex cf. conglomeratus
Ranunculus repens
Ranunculus acris
total lake flora specimens at each level
total land flora specimens at each level
land flora
total turbellarian egg-capsules at each level
turbellarian egg-capsules
143 3
171 5 2
330 6
350 2
327 2
306 1 8
119 50
122 21
168 1 34
139 47
312 2 49
Zone
IIb
IIa
0 10 20 30 40 50% of total specimens in lake flora at each level
Myriophyllum spicatum or M. verticillatum
Nymphaea alba
Naias marina
Potamogeton natans
Ceratophyllum demersum
Nuphaea lutea
Carex riparia
Schoenoplectrus lacustris
Potamogeton acutifolius
Carex acutiformis
Oenanthe aquatica
Lemna cf. minor
Cladium mariscus
Hydrocharis morsus-ranae
Sparganium ramosum
Sparganium angustifolium
Berula erecta
Ranunculus-Batrachium
Solanum dulcamara
Lycopus europaeus

example, in the pollen diagram from the last interglacial (Ipswichian) site at Bobbitshole, England, a rise in Cyperaceae has been recorded towards the top of the sequence. Investigation of the plant macrofossil remains revealed that this does not reflect a regional vegetational change, but results from the local development of the pond shifting from open water to reedswamp to marsh (West 1957; Fig. 116).

8.12.2 Pollen and spores

C.A. Weber, who was also one of the pioneers of plant macrofossil investigations, began in the 19th century to investigate the pollen content of mires (Weber 1893, 1896). The Swede Lennart Von Post (1916), however, was the first to reconstruct the vegetational history of a bog using quantitative pollen analysis (palynology). Originally this new method was only applied to postglacial peats, but Firbas (1927) and Jessen & Milthers (1928) extended their investigations to older deposits. For palynological investigations, as a rule, a sequence of samples is analysed that represents a complete sediment profile. Thus, not only are the different genera recorded that have occurred in a specific area at a given time, but also their pattern of immigration and disappearance is seen. Under ideal conditions, the resulting pollen diagram covers the whole period from the treeless tundra at the end of a cold stage via the gradual immigration of forest trees and the optimal distribution of a thermophilous flora to the late interglacial coniferous forest and back to tundra conditions at the onset of the following cold stage. The successive steps of this vegetational development are normally characterised by the dominance of certain plant communities, which can be utilised to subdivide the vegetational development into distinct pollen zones.

The pollen flora is strongly influenced by local conditions. A small pond in the middle of an oak forest will naturally contain large quantities of local oak pollen, whereas a pond in an open grassland savanna landscape will contain mostly long-distance-transported pollen and spores. For regional interpretations therefore deposits of large lakes or bogs are preferred. Different pollen types travel over different distances, since pollen grains vary enormously both in size and shape (Fig. 117). Some (e.g. *Pinus sylvestris*) have special air bags which help them to stay in the air.

The vegetational development as reconstructed from palynological investigations is to a large degree controlled by climatic factors. However, a pollen diagram is not a composite climate curve. Because the individual plants are immobile, vegetation requires a long time to adjust to positive climatic changes. Whilst a sudden intrusion of harsh climate can eliminate sensitive plants immediately over vast areas, re-immigration occurs at a much slower pace, because seeds and fruits of many species (like fir or beech) are transported over only small distances, and because the individual plant needs decades before being able to reproduce. The re-immigration of spruce into western Canada after the end of the last ice age is shown in Figure 118.

Gradual re-immigration of plants has resulted in a sequence of plant communities. For the Holocene in Europe, this has been thoroughly documented by Lang (1994); for North America a similar documentation has been published by Wright *et al.* (1993). Differences from one interglacial to another can be used for stratigraphical correlation of the individual deposits. However, many pollen sequences are incomplete, either because sedimentation or peat growth started too late in the interglacial, or because silting up of the water or the end of peat growth put an early stop to deposition, so that the last part of the interglacial is unrepresented. In such cases stratigraphical correlation is seriously impaired.

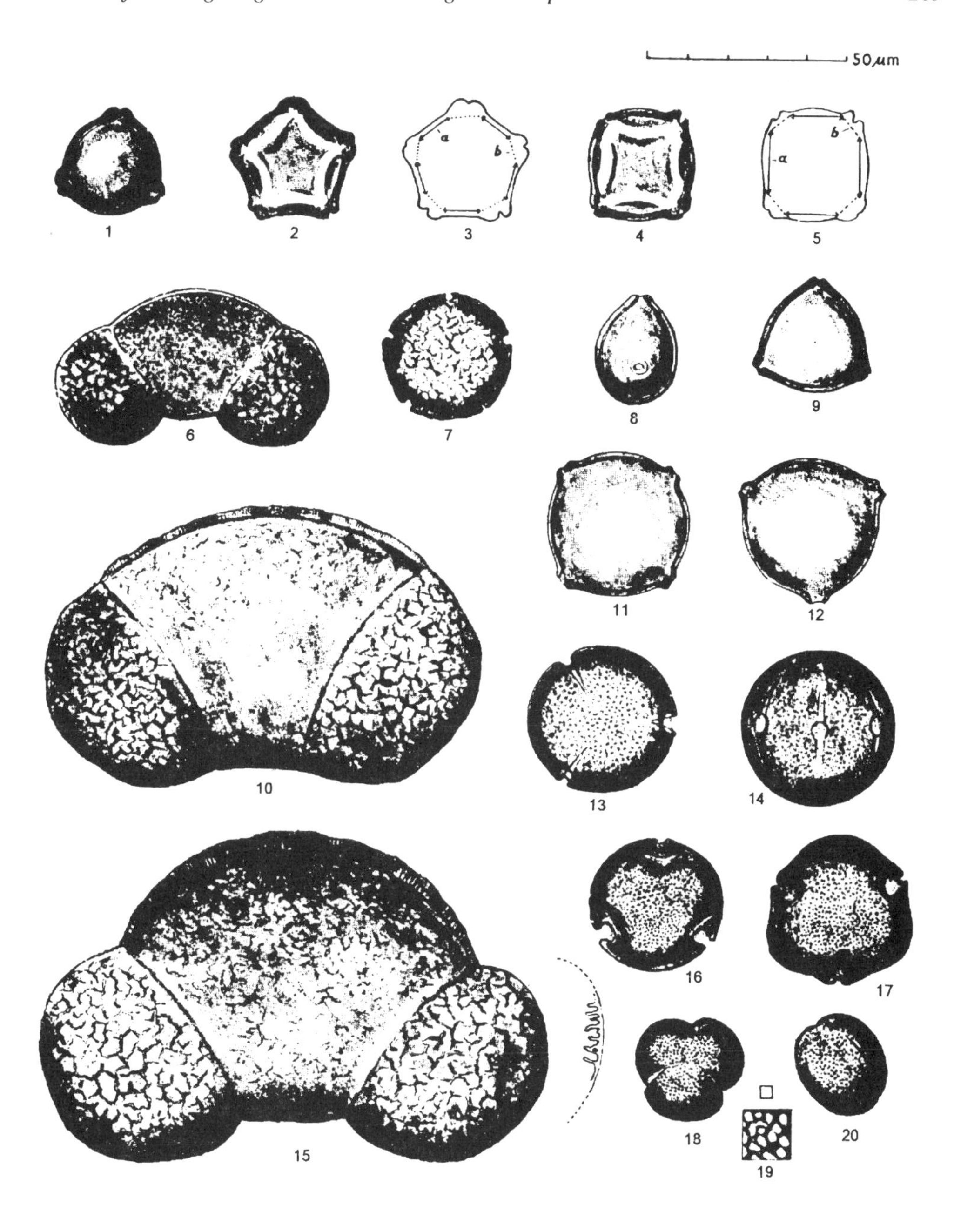

FIGURE 117 Pollen grains of some important plants of the Quaternary interglacials in central Europe. 1 = birch (*Betula*); 2–5 = alder (*Alnus*); 6 = pine (*Pinus*); 7 = elm (*Ulmus*); 8–9 = hazel (*Corylus*); polar and equatorial view; 10 = spruce (*Picea*); 11–12 = hornbeam (*Carpinus*); 13–14 = beech (*Fagus*); 15 = fir (*Abies*); 16–17 = lime (*Tilia*); 18–20 = oak (*Quercus*) (from Overbeck 1975)

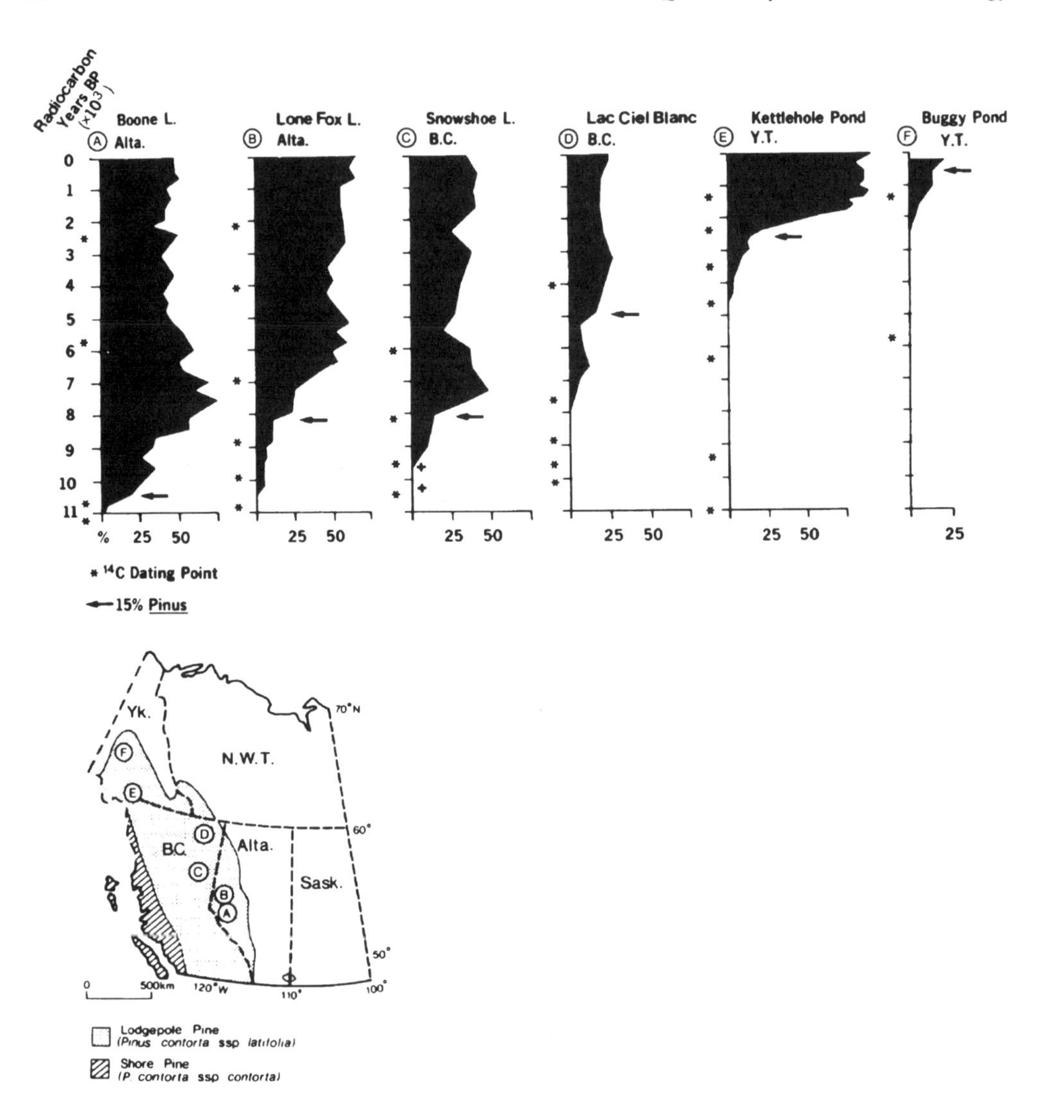

FIGURE 118 *Pinus* pollen in percentage of total pollen for sites from western Canada. The diagrams indicate the gradual re-immigration of pine in the postglacial period from southerly refuges. B.C. = British Columbia; Yk., Y.T. = Yukon Territory; N.W.T. = North-West Territories; Alta. = Alberta; Sask. = Saskatchewan (after MacDonald 1990)

Comparison of the last three interglacials in North Germany may serve as an example of their stratigraphical differences. In the Holsteinian Interglacial, fir and spruce immigrated rather early. In contrast, fir occurs much later in the Eemian and north of the Elbe River spruce is found only in low quantities and is restricted to the end of the interglacial. In the Holocene, spruce occurs in the Thuringian Forest under natural conditions (Overbeck 1975), fir occurs additionally in the Harz Mountains and their foothills, as well as in the Lüneburger Heide area; in all other areas both have only been introduced by recent forestry (Firbas 1949). On the other hand, beech (*Fagus sylvatica*) is restricted in the Holsteinian to very small quantities (H. Müller 1974b) and is absent from the Eemian, whereas, in the second half of the Holocene, it is widespread throughout North Germany.

Tropical and polar regions were hardly affected by the Quaternary climatic fluctuations, whereas between a marked shifting of climatic zones has taken place. It is suggested for central Europe that such temperate periods be regarded as interstadials (in contrast to interglacials) during which the vegetational and climatic conditions did not reach a state comparable to the postglacial optimum. According to this definition, the relatively warm periods of the Early Weichselian must be also regarded as interstadials (Behre 1989). Some interstadials are characterised by tundra-like conditions (e.g. the northwest European Glinde and Denekamp Interstadials), others are characterised by coniferous forest mixed with some birch in its early phase (e.g. Brørup Interstadial), or even by thermophilous forests (St Germain). The occurrence of interstadial forests with an 'interglacial'-type vegetation demonstrate that the term 'interstadial' must be defined on a regional basis.

The weakly developed interstadials at the beginning of the Middle Weichselian (Oerel, Glinde) are in part rich in heaths (Ericaceae) (Behre & Lade 1986). This indicates that until then only minor rejuvenation of the soils had occurred, with the fossil soils observed at the Keller and Schalkholz sites (Schleswig-Holstein) pointing towards similar conditions. In both cases, strong periglacial activity only began after the last soil had formed. In contrast, the organic deposits of the Dutch Middle Weichselian interstadials apparently formed after the onset of periglacial processes. The pollen diagrams from the Moershoofd, Hengelo and Denekamp interstadials are characterised by a lack of Ericales (Behre 1989).

The typical vegetational development of the north German Eemian is briefly discussed as an example of an interglacial vegetational development. The palynological subdivision of the Eemian in northwestern central Europe was originally based on the subdivision of Jessen & Milthers (1928), which is rarely used today. It has been replaced by more modern subdivisions of Selle (1962), Behre (1962) and H. Müller (1974a). The most recent proposal was presented by Menke & Tynni (1984) who distinguish seven pollen zones (Fig. 119):

Zone I: Birch zone. The zone is characterised by a dominance of birch pollen (*Betula*). Typical light-loving plants (heliophytes) of the Late Saalian, like *Helianthenum* and *Hippophae*, no longer occur. *Juniperus* and *Artemisia* are still present, however, with low, steadily decreasing percentages. The *Pinus* content increases gradually. The boundary with Zone II is at the decline from the *Betula* maximum and at the beginning of the *Ulmus* rise. Zone I is missing in many pollen diagrams.

Zone II: Pine–birch zone. This zone is characterised by a relative *Pinus* maximum and a simultaneous steady decline of *Betula*. Pollen of thermophilous plants occurs regularly, but only at low percentages. The initial decline from the *Pinus* maximum forms the most suitable boundary with the subsequent Zone III.

Zone III: Pine–mixed-oak-forest zone. *Pinus* still dominates. *Betula* decreases further, and the percentage of mixed-oak-forest pollen (especially *Quercus*) increases. *Acer* (maple), *Hedera* (ivy) and *Viscum* (mistletoe) are present at low quantities. Roughly at the beginning of Zone III *Fraxinus* (ash) also appears for the first time. The upper boundary of Zone III is at the first appearance of *Corylus* (hazel). Zone III is often weakly developed and difficult to separate from Zone II.

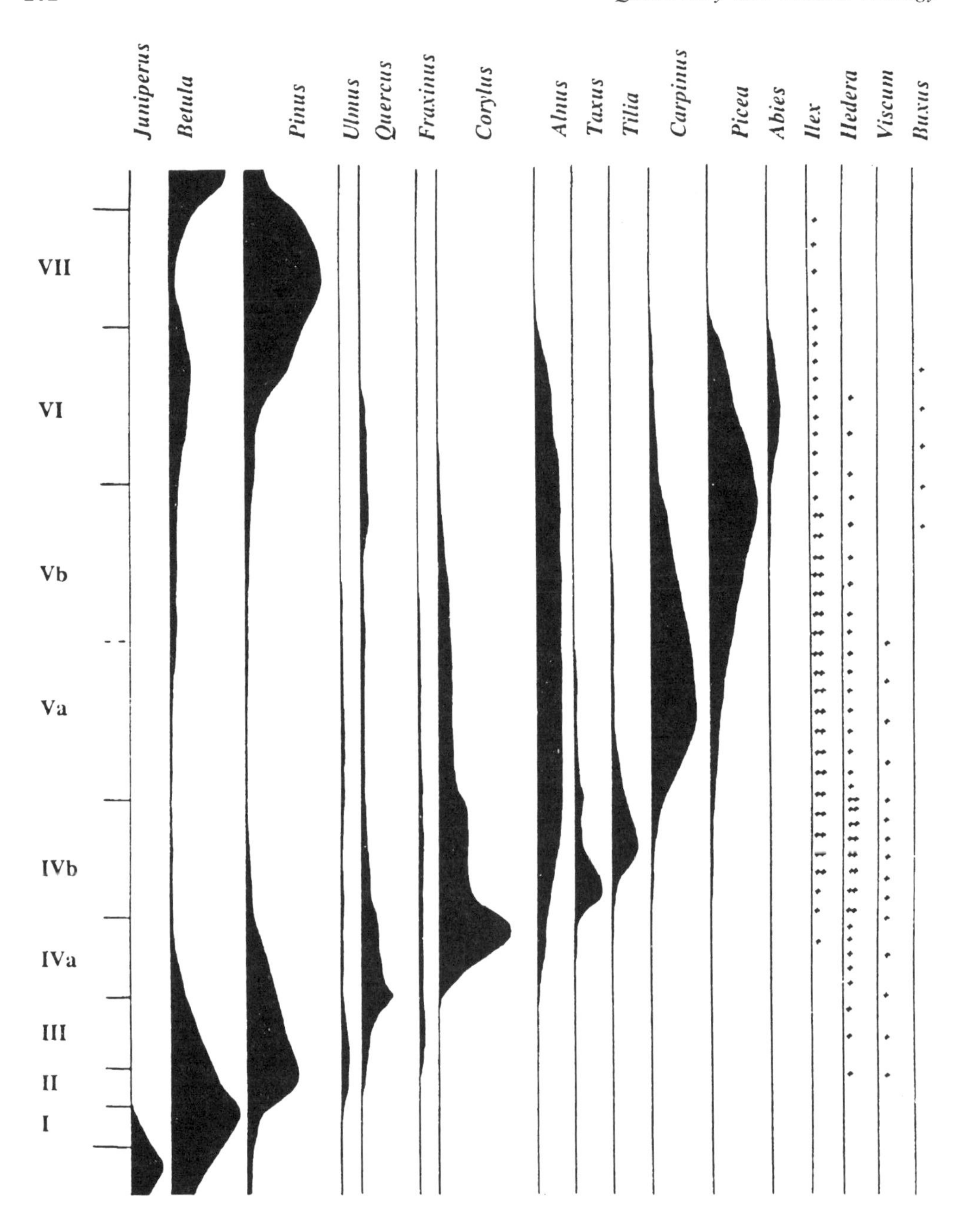

FIGURE 119 Schematic and simplified pollen diagram of the Eemian Interglacial from central and western Holstein, Germany (from Menke & Tynni 1984)

Zone IVa: Mixed-oak-forest–hazel zone. This zone is particularly characterised by the rapid increase of *Corylus* and *Alnus* (alder), as *Pinus* decreases. The first *Ilex* pollen occurs, and the pollen curves of *Taxus* and *Tilia* start from this zone. The upper boundary is defined at the first clear decline from the *Corylus* maximum and the rapid increase of *Tilia*, which may occur considerably later than the first occurrence of the species (the rational *Tilia* boundary).

Zone IVb: Hazel–yew–lime zone. After the *Corylus* maximum, the percentage of hazel drops to about 50% and remains at that frequency for some time. The *Taxus* and *Tilia* curves reach their maxima. In the upper part of Zone IVb, the curves of *Carpinus* (hornbeam) and *Picea* (fir) start to rise. The boundary with Zone V is placed at the intersection between the *Tilia* decline and the *Carpinus* rise.

Zone V: Hornbeam–fir zone. The earlier part of Zone V is mostly characterised by a more or less pronounced dominance of *Carpinus*. The percentage of *Corylus* drops again markedly. *Picea* increases gradually. As *Carpinus* declines in the upper part of Zone V, *Quercus* and *Pinus* start to increase again. The boundary with Zone VI is either the rational *Abies* boundary, the upper rational *Corylus* boundary or the beginning of the *Pinus* rise coinciding with the middle of the younger *Quercus* maximum.

Zone VI: Pine–spruce–fir zone. *Pinus* increases as the frequency of thermophilous trees decreases further. *Abies* (fir) gains some importance in the lower part of Zone VI in western Schleswig-Holstein (up to a maximum of 10%) but decreases thereafter. The percentage of *Picea* decreases in favour of *Pinus*. The boundary with Zone VII is placed by H. Müller (1974a) at the decrease of *Abies* below 1%. Behre (1962) used instead the upper empirical *Abies* boundary (i.e. the upper appearance of *Abies*) and the rise of non-arboreal pollen.

Zone VII: Pine zone. This zone is characterised by a clear *Pinus* dominance. *Picea* fades out either at the boundary VI/VII or shortly thereafter. After a marked *Pinus* maximum, in the upper part of the zone, *Betula* starts to increase.

The end of the Eemian is defined as the end of a continuous forest cover with its transition to a subarctic vegetation. This transition has also been used by Grüger (1979) to define the Eemian/Weichselian boundary in his Samerberg profile. This boundary is used because it represents the change from a warm-stage dense vegetation cover and its well-protected soils, to cold-stage conditions with their stronger sediment redeposition.

Any interglacial deposit with the pollen sequence presented above in northern Germany might easily be classified as 'Eemian', but in many cases the situation is not so straightforward. In the past, additional interglacials were postulated on the basis of deviating pollen diagrams, but this dating was later revised.

Investigations of the vegetational history do not always allow a safe stratigraphic interpretation. In each case, the stratigraphical position of the deposit and its palaeogeographical context must be considered. Occasionally, different methods of investigation can lead to different results. Although the interglacial sites of Gröbern and Grabschütz (Germany) can be easily correlated palynologically with the other North German Eemian sites (Litt 1990), the investigation of plant macrofossils by Mai (1990) led to a different conclusion. According to him only the Gröbern site is Eemian, whereas Grabschütz is regarded as a subcontinental, thermophilous interglacial of different age. This, however, seems hardly compatible with the present knowledge of the stratigraphical sequence between Saalian and Weichselian.

Sometimes interglacial or interstadial deposits cannot be easily separated from stadial, cold-climate deposits in long pollen sequences. Redeposited and far-travelled arboreal pollen may have contaminated deposits of a treeless period, especially in the case of lake deposits. For a clear subdivision therefore, the inclusion of non-arboreal pollen is crucial. Apart from the general rise in non-arboral pollen, the increase of light-loving plants (heliophytes) such as *Artemisia, Plantago,* and *Armeria* can provide clear evidence of cold-stage conditions (Menke 1975; Grüger 1989). Good examples are provided by the cold-stage sections of the pollen diagrams of Rederstall in Schleswig-Holstein (Menke & Tynni 1984) and Oerel in Niedersachsen (Behre & Lade 1986).

As a rule, palynological investigations are undertaken in undisturbed sequences like mires or lake deposits, which guarantee largely continuous sedimentation. Pollen analysis of fluvial deposits and also of cave sediments is possible, but reworking and discontinuous sedimentation may be a problem. Pollen analyses of marine deposits are also possible, but under shallow-water conditions a considerably higher rate of redeposition and biases caused by sorting must be taken into account, so that only limited information can be obtained about vegetational development. However, under favourable conditions pollen assemblages can be used for direct correlation of terrestrial vegetational sequences with the oxygen-isotope record (Heusser & Shackleton 1979).

Redeposited pollen and spores can also be utilised for stratigraphical interpretations. Indeed, redeposited microfossils occur within the matrix of most tills, and Thomson (1941) in his early investigations in the Quaternary deposits of Estonia, even spoke of a 'wealth of tree pollen'. Redeposited microfossils in tills and other glacial deposits have later been utilised both as indicators of ice-movement directions and as a tool in lithostratigraphical work (Ringberg & U. Miller 1992).

Although long-distance transported pollen might be entrained into a glacier (McAndrews 1984), the pollen content of tills largely results from the incorporation of older sediments, especially those deposited in the last temperate period before the ice sheet growth. The Late Weichselian tills of Finland contain pollen from intra-Weichselian interstadials, which are characterised by high proportions of birch (*Betula*) pollen as well as abundant non-arboreal pollen and spores (20–40%). Where Eemian deposits have been incorporated, the *Betula* pollen is accompanied by *Pinus* and *Alnus*, whereas non-arboreal pollen occurs in lower percentages (e.g. Donner & Gardemeister 1971).

In Estonia, it has been possible to distinguish reworked Eemian and Holsteinian pollen assemblages in tills. Thus, palynological investigations have been used to distinguish tills of different ages (Liivrand 1991), although comparable investigations in Ontario, Canada (Dreimanis *et al.* 1989), have been less successful. Of the 13 investigated samples, only two yielded significant proportions of reworked warm-stage pollen. The other samples were characterised by high *Pinus* values. This was interpreted as reflecting the reworking of Early Wisconsinan or Early Illinoian interstadial deposits. In Denmark also, palynological investigations of till samples have been carried out (cf. review by K.S. Petersen 1983). The material was found to have been derived from local sediments (Hansen 1979).

Pollen is not only incorporated into tills by reworking of older sediments, but also by later soil formation. Felix-Henningsen & Urban (1982) compared decalcified and fresh till samples from the Rotes Kliff section on the island of Sylt (North Sea). In addition to *Pinus, Picea,*

Betula, Alnus and *Salix* pollen, the fresh till contained pollen of a number of Early Pleistocene and Pliocene species, almost certainly incorporated from the underlying substratum, a short distance from the place of deposition. In the pedologically altered upper, decalcified part of the till the situation was completely different, the pollen almost exclusively containing Gramineae and Ericaceae. In this case apparently, the original pollen content of the till had been destroyed by pedogenesis and replaced by in-washed grass and heather pollen. Since no tree pollen was found, this process was thought to have occurred during a Saalian interstadial.

8.12.3 Diatoms

Diatoms (siliceous algae) are well suited for palaeoecological investigations. Diatomaceous deposits provided the basis for diatom systematics both in Europe and America (e.g. Ehrenberg 1854). A recent review of diatom research in the United States has been provided by Patrick (1984), and Smol (1990) discussed the palaeoecological investigations of algae in Canada. The potential value of diatoms as ecological indicators was recognised long ago (e.g. Hustedt 1953). They have been used with great success for the reconstruction of Holocene environmental changes. For instance, the influence of human settlement and agriculture in Northern Ireland is well reflected in the diatom assemblages of Lough Neagh lake sediments (Battarbee 1973). At the Kirchner Marsh site in Minnesota, pollen and diatom analyses have been used together in a detailed analysis of Holocene vegetational and limnological development (Brugam 1980). They are well suited, for instance, to determine transgression or regression contacts, which is important for determining sea-level changes or rates of isostatic uplift (e.g. Lie *et al.* 1983, Grönlund 1991; Fig. 120). However, very few pre-Holocene sites have been systematically sampled either in America or Europe. Diatoms have not undergone evolutionary changes within the short period of the Quaternary, but react sensitively to changes in water depth, nutrients and salinity, making them ideal indicators of palaeo-ecological conditions (e.g. Smol *et al.* 1986). In contrast to palynology, no modern extra-regional diatom biostratigraphy is available.

As with pollen, redeposition of diatoms can be a problem. In marine deposits freshwater diatoms are commonly found, brought in by the rivers or even by the wind. Under highly alkaline or very acid conditions, diatoms dissolve, which can result in assemblages dominated by stronger forms. In the deep sea diatoms like other organisms also dissolve under high pressure at great water depths (Lowe & Walker 1984).

8.12.4 Mammals

Animals can adjust more rapidly to climatic change than vegetation; however, for their diet they mostly depend on plants. Early investigations of Quaternary fauna largely focused on the large mammals; a review of the evolution of Quaternary mammals and their ecology is provided by Thenius (1962, 1972), Stuart (1974, 1982) and Lundelius *et al.* (1983). In particular, **large mammals** were regarded as indicator fossils of the Quaternary.

Mammal remains (bones, teeth and antlers) are far more perishable than pollen or even wood and are best preserved in carbonate-bearing deposits such as fresh loess or travertine. Fresh glaciofluvial or river deposits can also contain bones and teeth. For instance, dredging in the Weichselian meltwater gravel of the Elbe ice-marginal valley has retrieved numerous mammal remains. However, the faunal assemblage is mixed, i.e. reworked from other sites, and includes

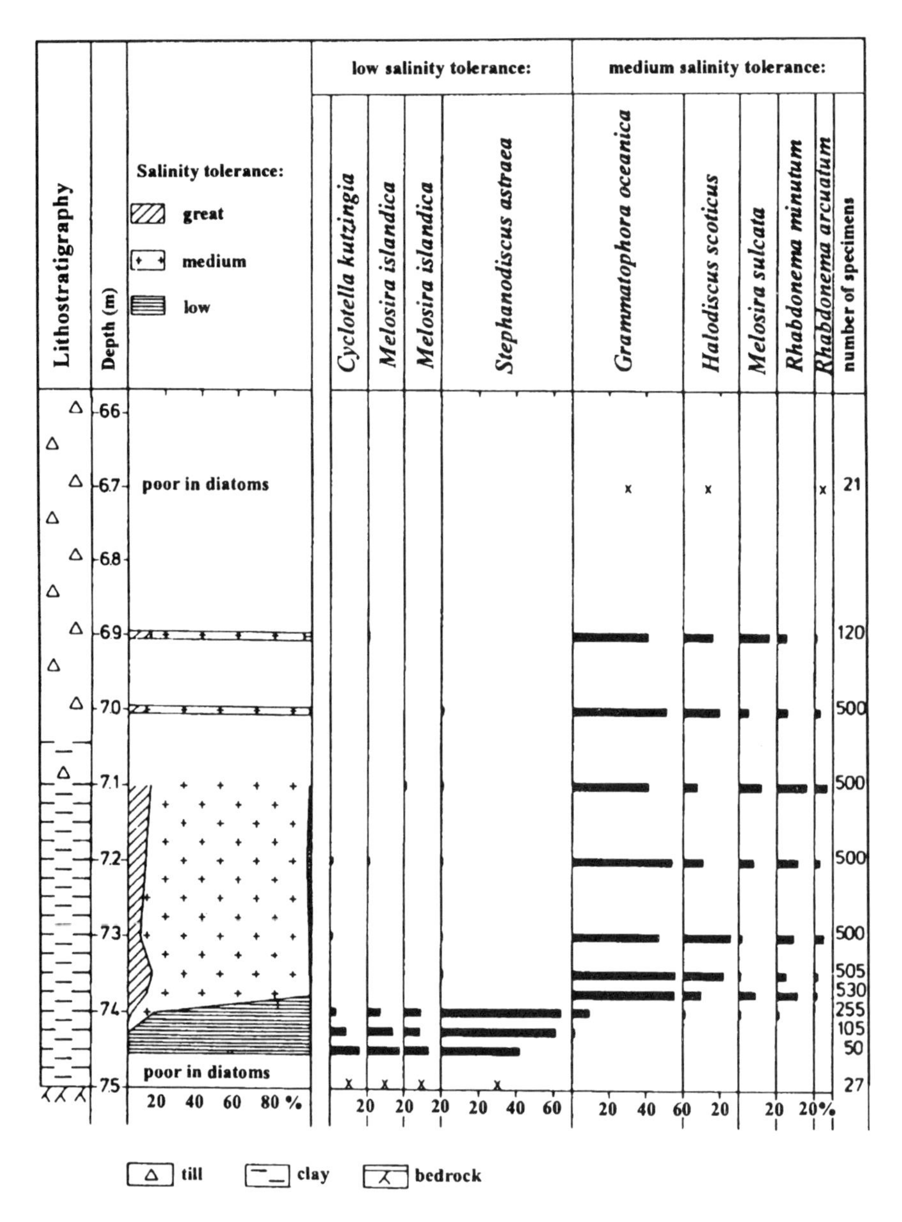

FIGURE 120 Diatom diagram from Peräseinäjoki, Finland, showing diatoms as halobian groups and according to their salinity preference. Freshwater lake deposits are overlain by marine sediments of the Eemian Baltic Sea (after Nenonen *et al.* 1991)

both temperate and cold-stage elements, so that they are not useful for environmental reconstruction.

Large mammal remains in most cases have been preserved only as single specimens or in small numbers, so that statistical analyses are limited. **Small mammals** are more suitable for such investigations, and basic investigations using assemblages of these creatures have been undertaken, for instance, by Jánossy (1969) and Von Koenigswald (1973). During the development of *Arvicola* (a vole species), the teeth (molars) have undergone a gradual evolution. This can be evaluated statistically and used for stratigraphic interpretations (W.-D. Heinrich 1987).

Some groups of small mammals have undergone comparatively rapid evolution during the Quaternary. With the voles this has proceeded sufficiently rapidly to allow at least a rough age estimate based on the respective faunas. *Mimomys* occurs from the Early Pleistocene to the Cromerian, and is replaced during the Cromerian Complex (after the Matuyama/Brunhes boundary but prior to the Cromerian IV) by *Arvicola terrestris cantiana*. After the Holsteinian Interglacial, this species is followed by *Arvicola terrestris* ssp. A and B (Van Kolfschoten 1990; Von Koenigswald & Van Kolfschoten 1993).

In addition to floral investigations, the faunas also provide numerous clues to the cold-stage environmental conditions. For the stratigraphical interpretation, however, only rough estimates can be provided, and therefore a large number of specimens are desirable to furnish the best possible interpretations. For example, at the Belvédère site near Maastricht (the Netherlands), in terraces of the River Maas, five separate mammal horizons have been identified (Van Kolfschoten 1985). The oldest fauna cannot be correlated with any specific stage because of insufficient material, but the mammal remains of the Maas terrace gravels (Belvédère 2) exhibit characteristic tundra elements of the Early Saalian. The overlying faunas 3 and 4, in the sandy Maas deposits, represent a transition to warm-temperate climatic conditions. The presence of *Arvicola terrestris* ssp. A indicates that the deposits still predate the Saalian ice advance into the Netherlands. In support, Van Kolfschoten (1981) demonstrated that the more progressive form *Arvicola terrestris* ssp. B occurs in ice-pushed deposits at Rhenen, which were undoubtedly thrust by the Saalian ice. The transition from *Arvicola terrestris* ssp. A to *Arvicola terrestris* ssp. B therefore must have taken place during the Saalian Cold Stage. The interpretation of Belvédère horizon 3 is based on the evaluation of 105 specimens, that of Belvédère 4 on 638 specimens of small mammals, in comparison with only 23 large mammal remains at the same site (Van Kolfschoten 1985). The youngest mammal level at Maastricht, Belvédère 5, according to its stratigraphical position and the exclusive occurrence of *Arvicola terrestris* ssp. B, can be assigned with the Early Weichselian (cf. Van Kolfschoten 1990).

8.12.5 Molluscs

Both marine and non-marine molluscs have been used in the reconstruction of Quaternary environments. The study of marine Quaternary molluscs dates from J. Smith (1838), who identified fossil cold-water species in raised marine deposits near Glasgow. In North America, a benchmark paper on biometrical analysis of molluscan assemblages was published by Schenck (1945), and recent investigations include, for instance, studies by Emery *et al.* (1988), Meijer (1990), Kennedy *et al.* (1992) and Hinsch (1993). Research nowadays has largely concentrated on microfossils. In many cases molluscs have only been used for radiometric or amino-acid dating, although molluscs provide valuable palaeoenvironmental evidence (Peacock 1993).

Non-marine molluscs are found in a wide range of deposits, including loess, cave sediments, fluvial, lacustrine and glaciolacustrine deposits, and travertine (B.B. Miller & Bajc 1990). In the course of his loess investigations in the former Czechoslovakia, Ložek (1964) was the first to recognise the fundamental importance of **loess molluscs**. With the help of malacological investigations it is possible to reconstruct on a large scale the environmental conditions during the deposition of the central European loesses. The basic type of association is represented by the *Pupilla* faunas. Although *Pupilla* is found today in open landscapes such as moist grassland, steppe or warm rocks, in the loess sequences they are representative of a cold and dry steppe climate. Apart from *Pupilla muscorum* and *Pupilla loessica*, also characteristic of these faunas is *Vallonia tenuilabris*, a species that is found at present in the cold regions of northern Asia. A different environment is represented by the *Columella* fauna, which indicates a cold but moist climate. This, together with the *Striata* fauna, represents the transition towards interglacial conditions, as well as the early glacial environment of dry steppes (Fig. 121). Because of continental conditions having drier climates and better preservation of carbonate shells, as well as thicker loess sequences, malacological investigations have been undertaken most successfully in the Czech Republic, Slovakia and Hungary. In Germany, most of the loess profiles are decalcified throughout. In France, however, Rousseau & Puisségur (1990) have found a well-preserved mollusc record in the loess sequence of Achenheim, Alsace, spanning the last 350,000 years, and Rousseau (1991) has used this record to construct a rough estimate of palaeotemperature and precipitation. In North America, extensive investigations of Pleistocene molluscs in Ohio have been published by LaRocque (1966, 1967, 1968, 1970).

Mollusc faunas can also provide valuable environmental information outside the loess areas. In Britain, detailed malacological investigations of interglacial deposits were conducted by Sparks (1956, 1957, 1980). An early review was presented by Kerney (1977), but, more recently, Keen (1982) has investigated fossil molluscs on the Channel Islands and Preece (1990) studied the molluscan fauna of the Middle Pleistocene interglacial deposits of Little Oakley, Essex. In Germany, for example, the Eemian lake deposit at Gröbern (Fuhrmann 1990) and the travertine at Weimar-Ehringsdorf (Mania 1973) have been studied. Mollusc faunas cannot only be used for palaeoecological studies, but also for the reconstruction of migration patterns, such as Late Wisconsinan molluscan migration in the Great Lakes region of North America studied in detail by Bajc (1986).

8.12.6 Beetles

Amongst the animals that have been locally preserved in large quantities are also the beetles (Coleoptera). Until a few decades ago, Pleistocene fossil insects were all regarded as belonging to extinct species. However, Russell Coope eventually demonstrated that this was not the case (Coope 1959). It could even be shown that the ecological requirements of the various species and genera underwent little change during the Quaternary. Because of their high mobility, beetles can react much more quickly to climatic change than plants, and they may even expand into regions where soil development does not yet allow the growth of higher plants (Morgan & Morgan 1990). Therefore, it can be no surprise that the results of investigations of fossil Coleopteran faunas do not always coincide with those of palynological investigations (Coope 1977). Coope *et al.* (1961) came to the conclusion that one of the Middle Weichselian warming periods in Britain, the Upton Warren Interstadial, was considerably warmer than indicated by its vegetational development. Whereas for plant growth the annual climatic pattern and length of the growing season are decisive, the beetle faunas largely reflect July temperatures. Therefore, considerable differences between maritime and continental regions must be taken into consideration. The area of the cold pole of the earth in eastern Siberia is also a region of high July temperatures, and this aspect has to be considered when the lateglacial climatic

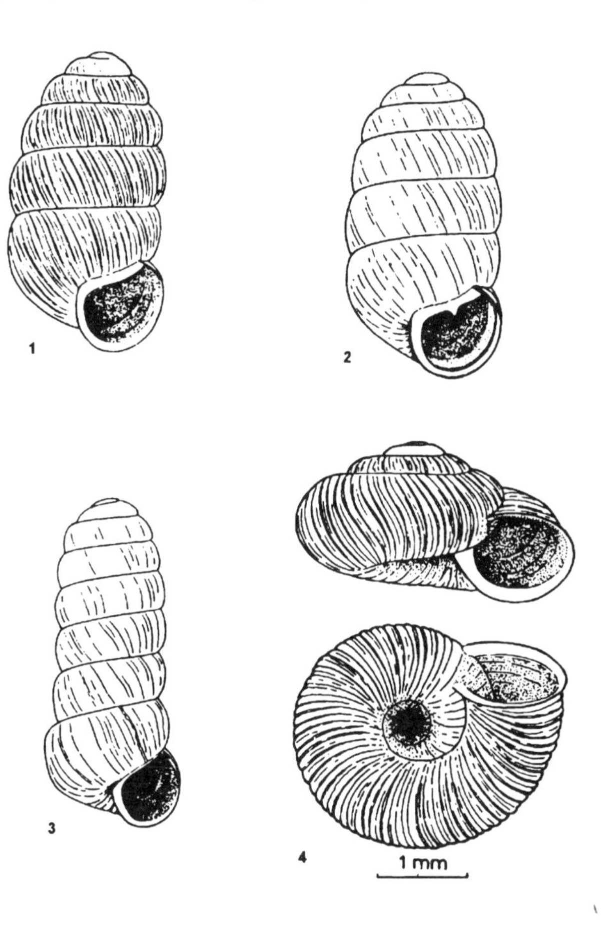

FIGURE 121 Characteristic loess molluscs. 1 = *Pupilla loessica* Lozek, 2 = *Pupilla muscorum* (Linné), 3 = *Columella columella columella* (Martens), 4 = *Vallonia tenuilabris* (A. Braun) (from Ložek 1965)

development is reconstructed. During the Older Dryas, for instance, relatively high July temperatures occurred together with very cold winters (Lemdahl 1988). The investigation of beetle faunas has developed into a standard method of reconstructing Quaternary environments. An overview of the state of the subject is presented by Elias (1994).

8.12.7 Foraminifera

An important method for palaeontological investigation of marine Quaternary deposits is foraminifera analysis. That foraminifera in deep-sea cores could provide information about past climate was first detected by Schott (1935), and Phleger *et al.* (1953) found evidence for nine ice ages in sediment cores from the Atlantic Ocean. Since the work of Emiliani (1955) foraminiferal assemblages have not only been used to reconstruct former climates but also the oxygen-isotope composition of the foraminiferan shells (see chapter 10.1). This method has become the most important branch of Quaternary research, allowing the reconstruction of the complete climatic history of the Quaternary.

The first quantitative analyses of shallow-marine foraminifera faunas were published by Feyling-Hanssen (1954), and numerous studies have since been made, especially from Denmark and North Germany. With the help of foraminifera analyses, it was possible to reconstruct some aspects of the ecological conditions at the Saalian/Eemian/Weichselian transition in northern Jutland, where this period is represented entirely by marine deposits (Knudsen & Lykke-Andersen 1982) (Fig. 122). According to changes in the faunal composition – as in pollen analysis – marine deposits can be subdivided into a sequence of foraminiferal zones. These biozones allow correlation even between incomplete sequences. Because of the very small size and high abundance of foraminifera, only small samples are required, so that the method is well suited for the correlation of boreholes (Knudsen 1985). With the help of foraminiferan analysis it can be demonstrated, for example, that the marine transgression at the end of the Elsterian Cold Stage had already invaded Denmark and North Germany under cold climatic conditions, before the onset of the Holsteinian Interglacial (Knudsen 1980, 1987). Foraminiferal investigations have also been successfully applied in North America, for instance in reconstructing the palaeoenvironments of the Late Wisconsinan Champlain Sea (Guilbault 1993) and of areas on the Canadian continental shelf (Osterman & Nelson 1989).

The locally high content of marine microfossils in Danish tills has made it possible in many cases to use the foraminifera content for stratigraphic correlation. A basic publication in this respect is that of K.S. Petersen & Konradi (1974), and the application of similar investigations for stratigraphic questions is demonstrated, for example, in Sjørring *et al.* (1982) (Fig. 123). Houmark-Nielsen (1987) uses foraminifera analyses as an important criterion to distinguish Weichselian from Saalian tills.

8.12.8 Ostracods

Ostracods have two calcitic valves, hinged dorsally, most shells being between 0.5 and 2 mm in diameter. Their fossils are preserved in well-buffered sediments of oceans, lakes, ponds and streams. Quaternary freshwater ostracods cannot be used in stratigraphy, but they are good ecological indicators. An overview of the analytical procedures is provided by Delorme (1990).

After the first investigations by Woszidlo (1962), Mania (1967) used ostracod faunas for his detailed investigations of the Ascherslebener See deposits. With this method, he was able to establish a detailed stratigraphic subdivision of the Weichselian. The first studies in North America were undertaken by Klassen *et al.* (1967) and Delorme (1968). More recently, Penney (1989) has investigated ostracods and foraminifera from Eemian tidal flat deposits in southwestern Denmark and Pietrzeniuk (1991) has evaluated the ostracod fauna from the Schönfeld Eemian site in the Lausitz area, East Germany (Fig. 124). The ostracod faunas perfectly reflect climatic events. One advantage in comparison with pollen analysis is that autochthonous sequences can also be recovered from cold-stage lake deposits. For example, in an investigation of the Eurach 1 core drilling (at Penzberg, Oberbayern), about 20,000 ostracods were sampled. The faunas clearly reflect the climatic changes from full-glacial conditions of the Riss Cold Stage to the Eemian Interglacial (Ohmert 1979). Further investigations have been reported from Britain (Lord & Robinson 1978, Horne *et al.* 1990) and

FIGURE 122 (Opposite) Foraminiferal diagram, temperature curve and chronostratigraphy of the Late Quaternary marine deposits from Vendsyssel, Denmark. A = arctic forms, B = boreal forms, L = lusitanian forms; d = deep-water forms; l = shallow-water forms (from Lykke-Andersen & Knudsen 1991)

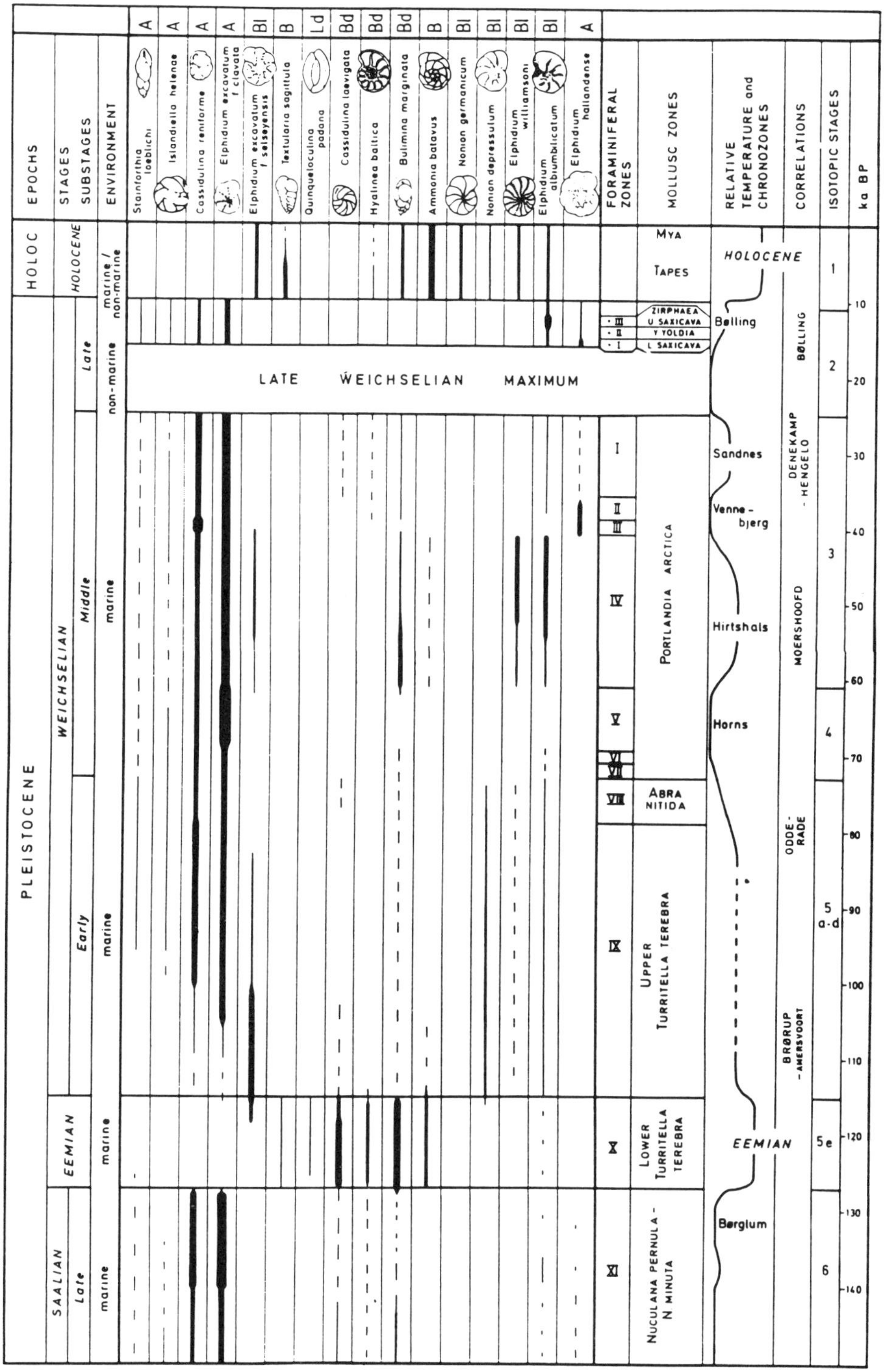

EPOCHS: PLEISTOCENE | HOLOC

STAGES: *SAALIAN* | *EEMIAN* | *WEICHSELIAN* | *HOLOCENE*

SUBSTAGES: *Late* (Saalian) | *Early* | *Middle* | *Late* (Weichselian)

ENVIRONMENT: marine | marine | marine | marine | non-marine | marine / non-marine

Stainforthia loeblichi — A
Islandiella helenae — A
Cassidulina reniforme — A
Elphidium excavatum f clavata — A
Elphidium excavatum f selseyensis — Bl
Textularia sagittula — B
Quinqueloculina padana — Ld
Cassidulina laevigata — Bd
Hyalinea baltica — Bd
Bulimina marginata — Bd
Ammonia batavus — B
Nonion germanicum — Bl
Nonion depressulum — Bl
Elphidium williamsoni — Bl
Elphidium albiumbilicatum — Bl
Elphidium hallandense — A

LATE WEICHSELIAN MAXIMUM

FORAMINIFERAL ZONES: XI | X | IX | VIII | VII | VI | V | IV | III | II | I | • I | • II | • III

MOLLUSC ZONES: NUCULANA PERNULA - N MINUTA | LOWER TURRITELLA TEREBRA | UPPER TURRITELLA TEREBRA | ABRA NITIDA | PORTLANDIA ARCTICA | L SAXICAVA | Y YOLDIA | U SAXICAVA | ZIRPHAEA | TAPES | MYA

RELATIVE TEMPERATURE and CHRONOZONES: Børglum | EEMIAN | Horns | Hirtshals | Venne-bjerg | Sandnes | Bølling | HOLOCENE

CORRELATIONS: BRØRUP - AMERSVOORT | ODDE-RADE | MOERSHOOFD | DENEKAMP - HENGELO | BØLLING

ISOTOPIC STAGES: 6 | 5e | 5 a-d | 4 | 3 | 2 | 1

ka BP: 140 | 130 | 120 | 110 | 100 | 90 | 80 | 70 | 60 | 50 | 40 | 30 | 20 | 10

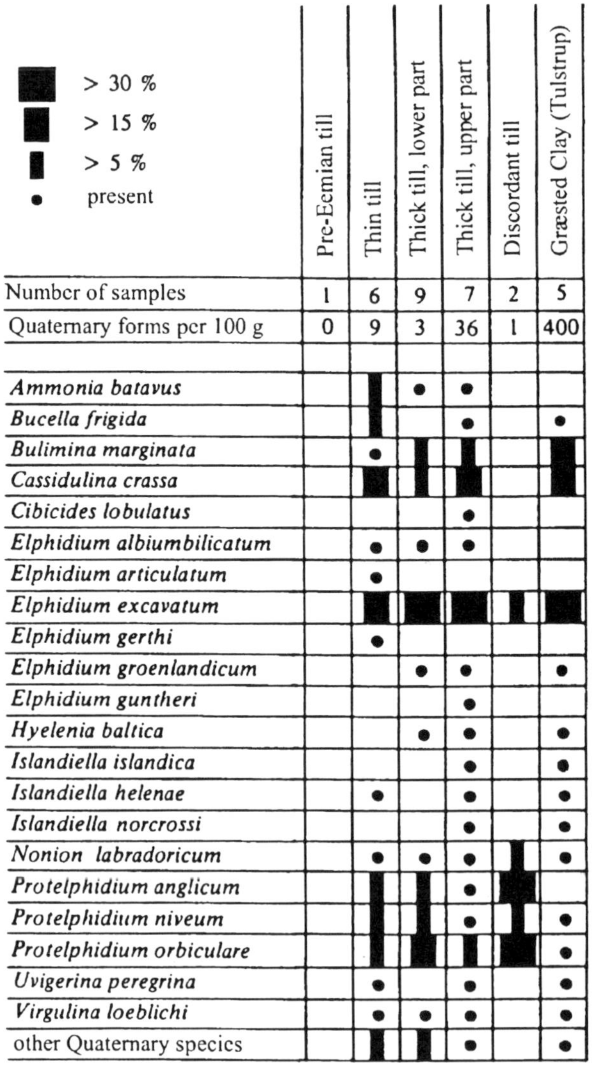

Legend: ■ > 30 %; ■ > 15 %; ▮ > 5 %; • present

	Pre-Eemian till	Thin till	Thick till, lower part	Thick till, upper part	Discordant till	Græsted Clay (Tulstrup)
Number of samples	1	6	9	7	2	5
Quaternary forms per 100 g	0	9	3	36	1	400
Ammonia batavus		> 5 %	•	•		
Bucella frigida		> 5 %		•		•
Bulimina marginata		•	> 5 %	> 5 %		> 15 %
Cassidulina crassa		> 15 %	> 5 %	> 15 %		> 15 %
Cibicides lobulatus				•		
Elphidium albiumbilicatum		•	•	•		
Elphidium articulatum		•				
Elphidium excavatum		> 15 %	> 30 %	> 30 %	> 5 %	> 30 %
Elphidium gerthi		•				
Elphidium groenlandicum			•	•		•
Elphidium guntheri				•		
Hyelenia baltica			•	•		•
Islandiella islandica				•		•
Islandiella helenae		•		•		•
Islandiella norcrossi				•		•
Nonion labradoricum		•	•	•	> 5 %	•
Protelphidium anglicum		> 5 %	> 5 %	•	> 30 %	
Protelphidium niveum		> 5 %	> 5 %	•	> 5 %	•
Protelphidium orbiculare		> 5 %	> 15 %	> 5 %	> 30 %	•
Uvigerina peregrina		•		•		•
Virgulina loeblichi		•	•	•		•
other Quaternary species		> 5 %	> 5 %	•		•

FIGURE 123 Redeposited Quaternary foraminifera in tills from Ristinge Klint, Denmark. For comparison: foraminifera from the marine Middle Weichselian Græsted Clay from Tulstrup (from Sjørring *et al.* 1982)

Denmark (Penney 1987). The investigations have confirmed the high palaeoecological indicator value of the ostracods, but at the same time they seem to be of little or no stratigraphic indicator value (Birks & Birks 1980). The differences between ostracod faunas of Gröbern and Grabschütz (Fuhrmann & Pietrzeniuk 1990a, b) does not indicate different interglacials. According to the geological situation both sites must be regarded as Eemian (Eissmann & Litt 1995).

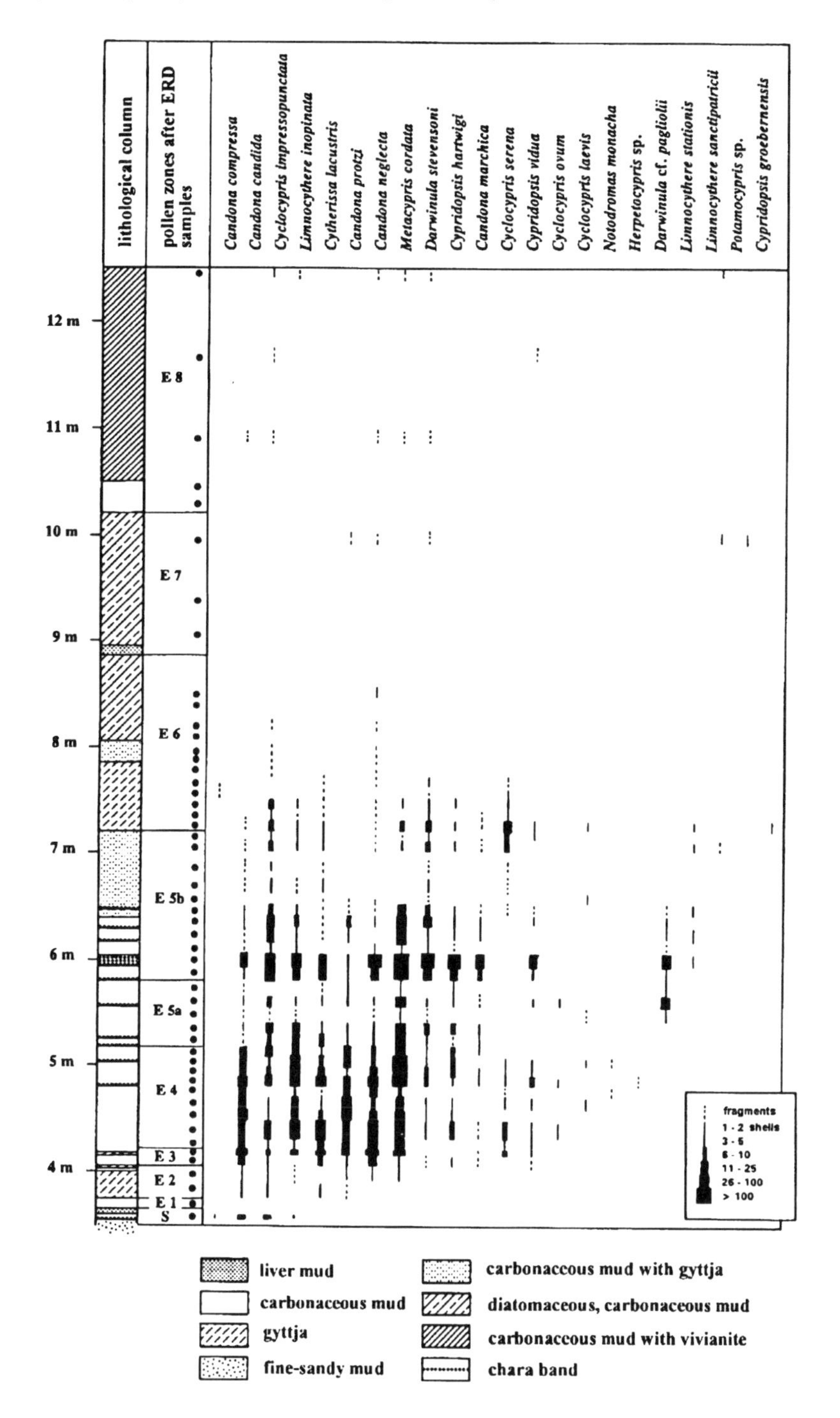

FIGURE 124 Ostracod diagram of the Eemian Interglacial from Schönfeld (Niederlausitz, East Germany) (after Pietrzeniuk 1991)

SECTION III

QUATERNARY STRATIGRAPHY

9

Principles of Stratigraphy

Stratigraphy (from Latin 'stratus' = layer, and Greek 'graphia' = description) deals with the subdivision of rocks into layers and formations, their lithology and fossil content. Stratigraphy does not represent a single, rigid scheme into which the individual strata and events can be fitted. Stratigraphic tables often comprise a number of different stratigraphies which, because of their different methodological approach, are neither identical nor in every respect comparable. This should be borne in mind when stratigraphy is discussed. The rules of stratigraphical work have been established by the 'International Subcommission on Stratigraphic Classification of the IUGS Commission on Stratigraphy', published by Hedberg (1976) and Salvador (1993). The following overview is largely based on these sources.

Any formal stratigraphical unit must be defined properly. For each unit a type section (or type region) is required, in which both the upper and lower boundary of the unit against the preceding and following units are exposed. The type section must be described, and the definition must be published in a widely circulated scientific journal. Formal stratigraphical units are identified by capitalised first letters of unit and rank terms (e.g. Brown Bank Formation), informal names should not be capitalised (e.g. Saalian till).

9.1 CHRONOSTRATIGRAPHY

One aim of stratigraphical investigations is to put the observed strata into chronological order. This is done in chronostratigraphy. The individual chronostratigraphical systems (Quaternary, Tertiary, Cretaceous etc.) form the basic framework of any stratigraphy. The chronostratigraphical units and their boundaries are by definition isochronous. This means that they are the same age wherever they occur. However, they do not necessarily coincide with lithostratigraphical or biostratigraphical units or boundaries (see below). Therefore their definition must be agreed on amongst scientists. In the case of the Quaternary, the lower boundary is not generally agreed. The problem is that the onset of cold, ice-age conditions was not simultaneous on a global scale. Therefore at the moment two different lower boundaries of the Quaternary are in use (1.6 million and 2.4 million years; see chapter 1).

9.2 LITHOSTRATIGRAPHY

The basis of practical stratigraphical work is lithostratigraphy. In lithostratigraphy the rock strata are distinguished according to their composition. Tills are distinguished from sands or organic strata. The principal lithostratigraphical unit is the **formation**. A formation can be subdivided into **members** and **beds**. Where necessary, two or more formations may be collected into a larger unit, termed a **group**, but its usage in the Quaternary is quite limited.

9.3 BIOSTRATIGRAPHY

In biostratigraphy rock strata are classified according to their fossil content. Biostratigraphy is distinct from litho- and chronostratigraphy in that it is based on the presence or absence of certain components (the fossils) in the rock record. Thus whilst every rock has a certain age and a certain lithology, it does not necessarily contain fossils. Therefore a biostratigraphical classification can only be applied to certain types of strata. In the Quaternary fossil-free units like tills or meltwater deposits are excluded.

The principal biostratigraphical unit is the **biozone**. Several biozones with common biostratigraphical features may be combined to superzones, and where more detailed subdivision is needed, subzones may be introduced. However, in contrast to lithostratigraphy and chronostratigraphy, biostratigraphy is mostly non-hierarchical. This is true for instance for the pollen zones of an interglacial, the sequence of which represents the vegetational development of that respective chronostratigraphical unit.

It is evident that lithologically defined boundaries are not necessarily everywhere of the same age. For instance, the Weichselian ice sheet advanced earlier into central Sweden than into North Germany, and re-immigration of oak after the last glaciation occurred much earlier into southern Germany than into Denmark. The resulting units and their boundaries are time-transgressive. Therefore there is seldom a direct coincidence between litho-, bio- and chronostratigraphical boundaries.

9.4 OTHER STRATIGRAPHIES

Apart from the above stratigraphies several other approaches can be used for stratigraphical subdivision, including **oxygen-isotope stratigraphy** (chapter 10.1) and **magnetostratigraphy** (chapter 10.2). Others would be **seismostratigraphy** (as applied in seismic investigations for instance at the bottom of the North Sea or on the continental shelf regions) or **pedostratigraphy** (as used, for instance, for the correlation of fossil soils in loess).

Landforms can also be used for stratigraphical correlations in what is referred to as **morphostratigraphy**. In the 1950s in particular, correlations of end moraines in a so-called 'end-moraine stratigraphy' played a major role. However, landforms are not stratigraphical units in the true sense, and it has proved difficult to match end moraines with respective lithostratographical units. Neither do the end moraines necessarily mark the outermost ice-marginal position of a certain ice advance, nor does every end moraine correlate with an individual till sheet.

The sedimentary record, especially of cold stages, never represents complete periods of time but only certain short events within that period. According to this pattern, in recent years, especially in North America, a new understanding of the stratigraphic sequence has developed that is summarised under the term **event stratigraphy**. This concept, for instance, largely underlies the overview of the Canadian Quaternary by Fulton (1989).

It is always unfortunate when terms from different stratigraphies are mixed. This problem often arises from the fact that many poorly defined traditional terms are in general use and hard to replace. These terms may appear convenient because everybody is accustomed to them, but tradition becomes a burden where it results in hazy terminology and incorrect usage of undefined terms. More effort is needed to overcome those obstacles. Examples of this approach are given in the new North American Stratigraphic Code (1983) which includes diachronic and allostratigraphic units.

10

Dating Quaternary Deposits

The basic law of stratigraphy, as defined by the Danish medical doctor Nicolaus Steno (= Niels Stensen) in 1669, states that in an undisturbed position lower layers are older than those above, the so-called 'law of superimposition'. This rule also applies to Quaternary stratigraphy. There are means of correlating strata that are not in direct contact with each other, and many are briefly discussed in the methodological chapters of this book. However, even the most careful application of these methods has resulted in a stratigraphical framework that was later found to be incomplete. Periods of deposition have alternated with non-deposition, and often erosional events have removed major parts of the stratigraphic sequences.

All terrestrial Quaternary records are incomplete. This applies even for the long loess records of China and central Asia, which may span the whole Pleistocene but include numerous gaps caused by non-deposition. There is only one environment on earth that in many cases guarantees continuous deposition, that is the deep sea. Unfortunately, the deep-sea record cannot directly be correlated with the terrestrial stratigraphy. Therefore indirect means of correlation must be found. The only criterion that can be applied to all deposits is age. A layer deposited in Ohio 15,000 years ago can be safely correlated with a sediment from the deep-sea floor of the same age. Unfortunately, age determination of Quaternary deposits in many cases is far from being as precise as desirable.

Colman *et al.* (1987) published a classification of dating methods. They differentiate six groups:

1. *Sidereal* (calendar or annual) methods which determine calendar dates or count annual events
2. *Isotopic* methods, which measure changes in isotopic composition due to radioactive decay
3. *Radiogenic* methods, which measure cumulative non-isotopic effects of radioactive decay, such as crystal damage and electron energy traps
4. *Chemical and biological methods*, which measure the results of time-dependent chemical or biological processes
5. *Geomorphic methods*, which measure the results of complex interrelated, time-dependent geomorphic processes
6. *Correlation* methods, which establish age equivalence using time-independent properties.

Ten of the methods listed by Colman *et al.* are briefly outlined below.

10.1 OXYGEN ISOTOPES

Ocean-water oxygen is mostly made up of two isotopes: ^{16}O and ^{18}O (about 0.2%). During evaporation, the heavier isotope ^{18}O is preferentially enriched in the liquid phase and depleted in the vapour phase. On the other hand, ^{18}O condenses preferentially from the vapour. Evaporation and condensation are dependent on temperature. Because ocean temperatures are far more stable than air temperatures, the latter have a strong bearing on the isotopic composition of the precipitation that finally adds to the mass balance of a glacier or ice sheet. Thus in ice cores isotopically lighter winter layers can be distinguished from heavier summer layers (Paterson 1994).

Under unchanging climatic conditions ^{16}O is transported back into the sea at the same rate by precipitation and fluvial drainage, resulting in an annual balance. During the cold stages, a major proportion of the precipitation is not returned to the sea for a prolonged period but is stored in the glaciers and ice sheets of the continents. As a consequence, sea water is gradually depleted of ^{16}O. Marine organisms with calcareous shells incorporate both oxygen isotopes in a proportion that varies with that of sea water. In this way it is possible to reconstruct the former composition of the sea water from the oxygen-isotope ratio of fossil shells and, indirectly, to estimate the extent of glaciation during the time of deposition.

At lower temperatures foraminifera also incorporate a higher proportion of ^{18}O into their shells. After investigating the oxygen-isotope composition of several deep-sea cores from the Caribbean and the North Atlantic, Emiliani (1955) suggested that the changes he had found primarily reflected glacial/interglacial temperature changes. Later Shackleton (1967) found that oxygen-isotope composition was more strongly influenced by ice volume than Emiliani had assumed. After publication of the oxygen-isotope curves of two Pacific deep-sea cores (Shackleton & Opdyke 1973, 1976) it became clear that the deep-sea oxygen-isotope record reflected the entire climatic history of the Quaternary.

The world seas have a clearly developed temperature stratification. Whereas the near-surface sea water (down to a depth of about 300 m) with a certain delay follows the seasonal temperature changes of the atmosphere, the temperature of the oceanic deep water (below 1000 m depth) has most likely remained almost unchanged throughout the entire Pleistocene. The shells of bottom-dwelling foraminifera, so-called benthonic forms (Table 6), therefore show a

TABLE 6 Depth range, temperature regime and temperature range of various foraminiferan species (after Hemleben & Spindler 1983)

Foraminiferan species	Approximate depth range (m)	Temperature regime/range (°C)
Globigerinoides ruber	0– 50	Subtropical (20–29)
Globigerinoides sacculifer	0–100	Tropical (15–30)
Globigerina bulloides	0–100	Subarctic/subantarctic (0–25)
Globoquadrina pachyderma	50–300	Arctic/Antarctic (2–7)
Neogloboquadrina dutertrei	50–300	Temperate/subtropical (14–25)
Globorotalia crassiformis (left coiling)	100–300	Subtropical (0–25)
Uvigerina spp.	100–2000	(0–20)
Cibicoides spp.	>30	(0–20)

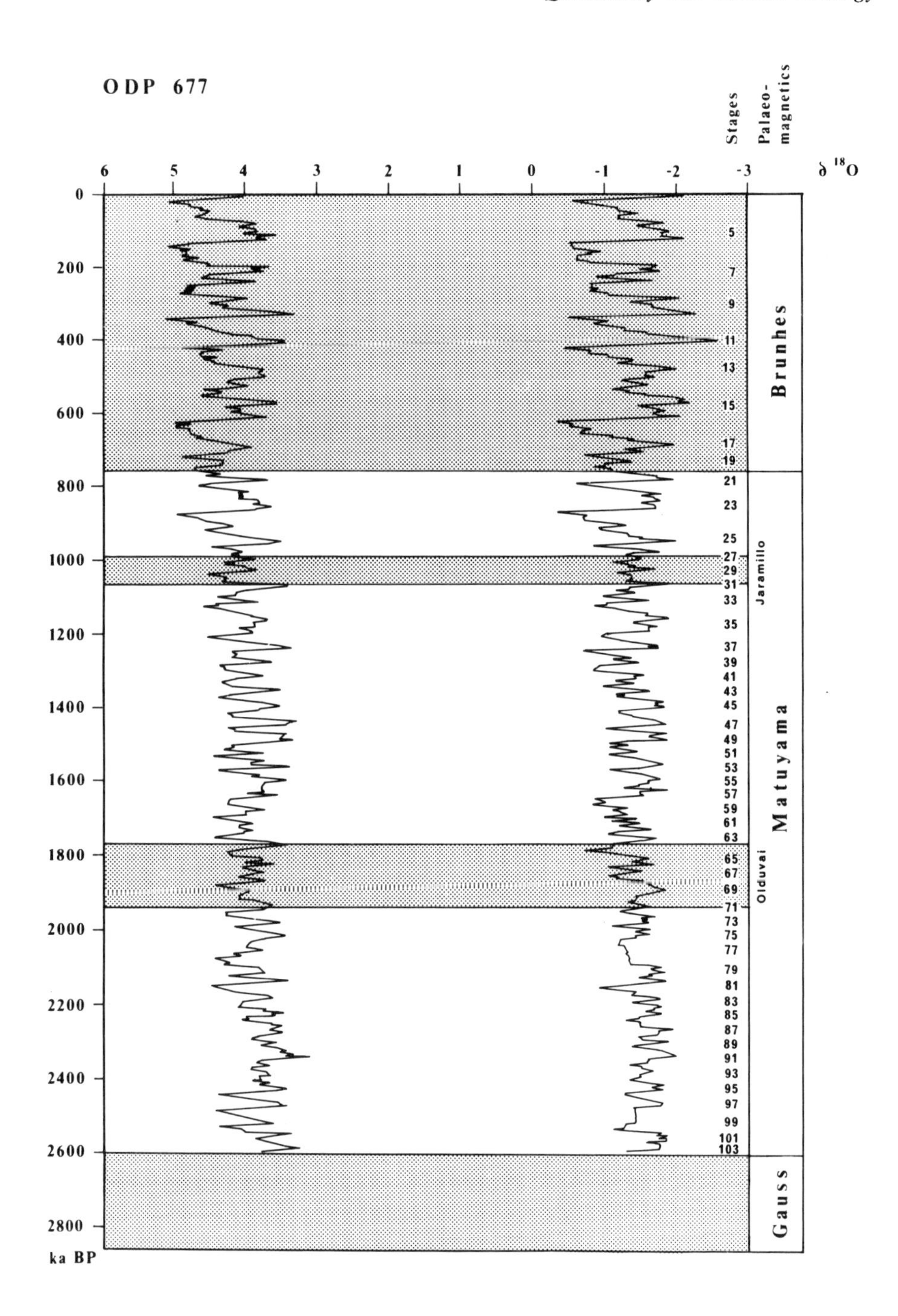

FIGURE 125 Oxygen-isotope curves for benthonic (mainly *Uvigerina senticosa*, left) and planktonic foraminifera (mainly *Globigerinoides ruber*, right) together with warm-stage numbers and magnetostratigraphy; deep-sea drilling ODP 677. Age in ka before present; $\delta^{18}O$ in per mille, related to the PDB (Peedee Belemnite) standard (after Shackleton *et al.* 1990)

curve of oxygen-isotope composition that is largely independent of temperature changes and only reflects alterations in global ice volume. It has been shown that the foraminifera *Uvigerina* is best suited to determine the oxygen-isotope ratio. However, benthonic foraminifera are often represented in the deep-sea sediments in much lower quantities than planktonic forms, so that often both forms must be used (Fig. 125). Apart from generalised curves that represent major periods of time, today also highly detailed oxygen-isotope curves are available. Comparison of different deep-sea cores demonstrates that oxygen-isotope curves can be correlated worldwide with a high degree of reliability (Fig. 126).

In areas with coastal upwelling, such as the west coasts of the southern continents, the nutrient-rich upwelling water causes high sedimentation rates. Upwelling depends on wind conditions which have undergone considerable shifts during the climatic changes of the Pleistocene, resulting in higher sedimentation rates during glacial stages. In areas with low sedimentation rates the resolution of the oxygen-isotope curve is considerably reduced by

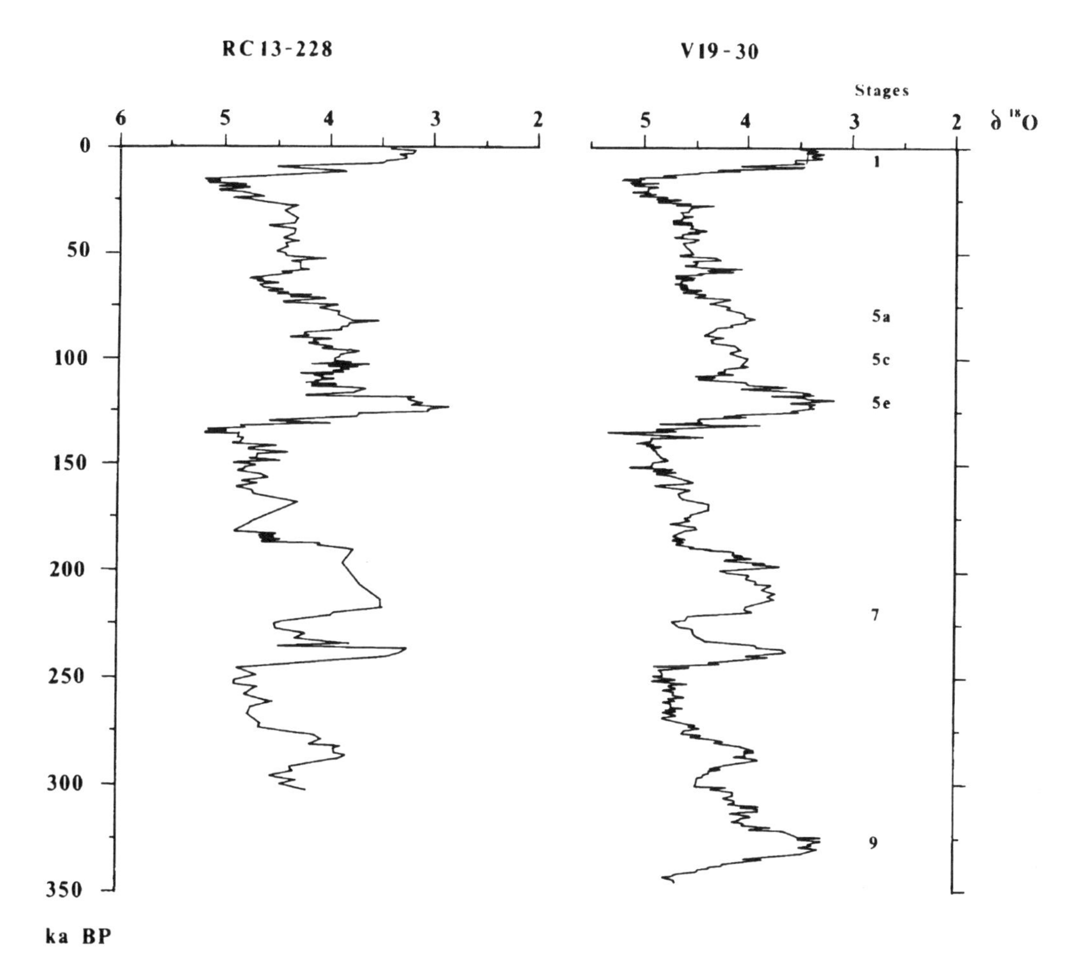

FIGURE 126 Oxygen-isotope curves for benthonic foraminifera of deep-sea drillings RC13-228 (South Atlantic) and V19-30 (Equatorial Pacific). Age in ka before present; $\delta^{18}O$ in per mille, related to the PDB standard (after Shackleton 1989)

bioturbation. A certain limitation to precision also results from the fact that the bottom water needs some time to react to climatic changes. A time lag of up to 1000 years must therefore be considered – depending on the site of the borehole (Shackleton 1977).

The oxygen-isotope ratio is normally expressed as the deviation in parts per mille of the ^{18}O ratio from the $^{18}O/^{16}O$ ratio of a standard, which was originally a belemnite from the Peedee Formation. This value $\delta^{18}O$ is calculated using the formula

$$\delta^{18}O = 1000 \cdot \left[\frac{^{18}O/^{16}O \text{ sample} - ^{18}O/^{16}O \text{ reference}}{^{18}O/^{16}O \text{ reference}}\right]$$

Under the overburden of fresh deposits, older strata become increasingly compressed. However, the compaction of deep-sea sediments plays only a minor role. The increase in density from 1.5 to 1.8 g/cm^3 with increasing depth occurs over several hundred metres (Garrison 1981, Emery & Uchupi 1984). It results less from settling than from diagenetic processes. Overall the consequences are so minimal that with regard to the short Quaternary cores used for oxygen-isotope determinations it can be neglected.

Therefore Shackleton & Opdyke (1973) used a linear interpolation between the Matuyama/ Brunhes boundary and the present. In core V28-238 this period covers a core length of 12 m, in core V28-239 about 7.25 m. Linear interpolation gives an age of 123,000 BP for Oxygen Isotope Stage 5e. This value has in the meantime been ascertained independently by U/Th dating both of continental deposits and coral terraces (on Barbados). Deeper cores allow the correlation of the complete Quaternary sequence with a well-defined time scale. Fixed points are provided by the magnetic reversals which have been dated elsewhere. Additionally, an improved correlation of the oxygen-isotope curve with changes of the earth's orbital geometry (global forcing) has been made, so that the sequence of events and their ages can be regarded as relatively well established (e.g. Shackleton *et al.* 1990).

Oxygen-isotope measurements have not only been used in deep-sea research, they are also applicable to ice cores (e.g. Anklin *et al.* 1993, Dansgaard *et al.* 1993, Grootes *et al.* 1993) and to speleothems. Lake sediments can also be analysed. For instance, Siegenthaler *et al.* (1984) have obtained oxygen-isotope curves from lake sediments in Switzerland which compare excellently with the Dye 3 Greenland ice core (Dansgaard *et al.* 1982).

Whilst Alpine glaciers and mountain ice caps usually only contain relatively young ice, the large ice sheets of Greenland and Antarctica have preserved much older ice. Ice cores from these areas thus allow the reconstruction of past climatic conditions. The first deep ice core on Greenland was recovered at Camp Century from 1963–66 (Dansgaard *et al.* 1969, 1971). In Antarctica, the Vostok ice core of 1983 (Jouzel *et al.* 1987) provided a record that reached slightly beyond the last interglacial. A different approach was used by Reeh *et al.* (1991) who evaluated the isotopic record of the Greenland ice margin at Pakitsoq, which also provided a record dating back to the last interglacial (Fig. 127).

Two cores were drilled at the top of the Greenland Ice Sheet. The core of the European Greenland Ice-core Project (GRIP) was drilled at Summit, where today no horizontal ice movement occurs. This was the locality where the longest record could be expected, and the cores were drilled almost to the base of the ice. The core drilled by the American Greenland Ice Sheet Project (GISP2) was drilled 28 km further to the west. Whilst the upper 2750 m of the cores compare relatively well, there are major discrepancies in the lowermost parts (Johnsen *et al.* 1995) (Figs 128 and 129).

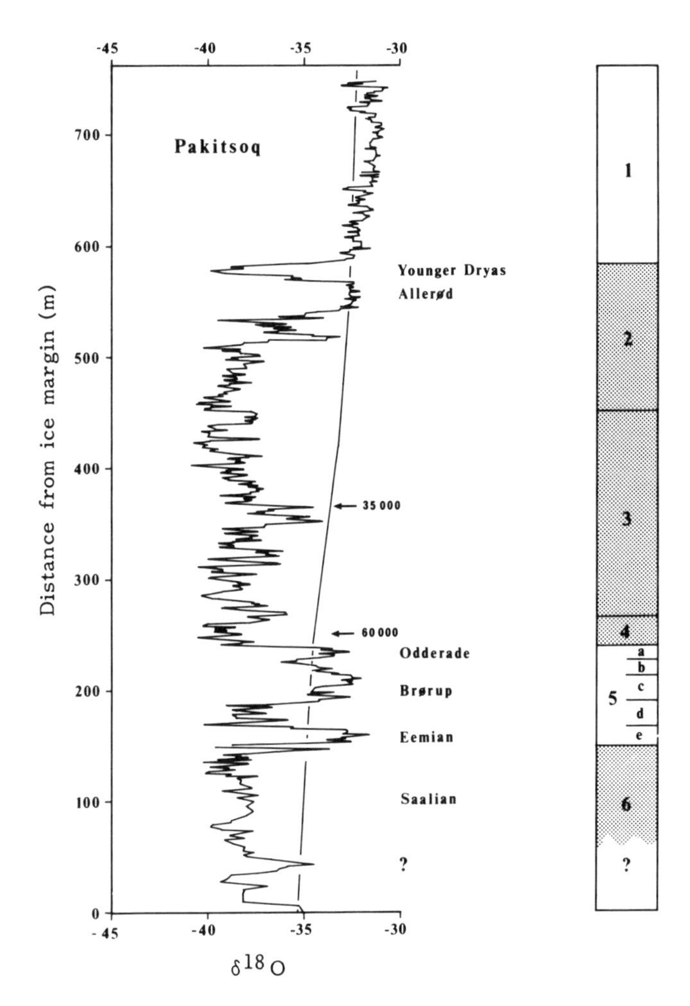

FIGURE 127 $^{18}O/^{16}O$ measurements of ice samples from the margin of the Greenland Ice Sheet at Pakitsoq, as compared with the oxygen-isotope deep-sea stratigraphy. The high resolution of the ice samples allows the identification of even minor climatic oscillations (after Reeh *et al.* 1991)

The most spectacular result of the two ice cores was that the last interglacial apparently had been very different from the Holocene. Whilst the Holocene oxygen-isotope record of the two cores shows no climatic oscillations, the Eemian part is characterised by a series of extreme fluctuations in $\delta^{18}O$ values. If the latter reflect climatic changes, they seem to indicate alternation from cold-stage to warm-stage climate and back within decades (Anklin *et al.* 1993, Dansgaard *et al.* 1993). This seems to be in contrast to the pollen record from Europe, in which the Eemian appears as a period of uninterrupted interglacial vegetation (e.g. Woillard 1978, Menke & Tynni 1984, Litt 1994). The continuous presence of frost-susceptible plants like holly (*Ilex*) and ivy (*Hedera*) seems to exclude the occurrence of mid-Eemian periods of severe winter cold. However, it must be borne in mind that climatic oscillations are felt more strongly in higher latitudes (Johnsen *et al.* 1995). Moreover, the resolution of the terrestrial record is

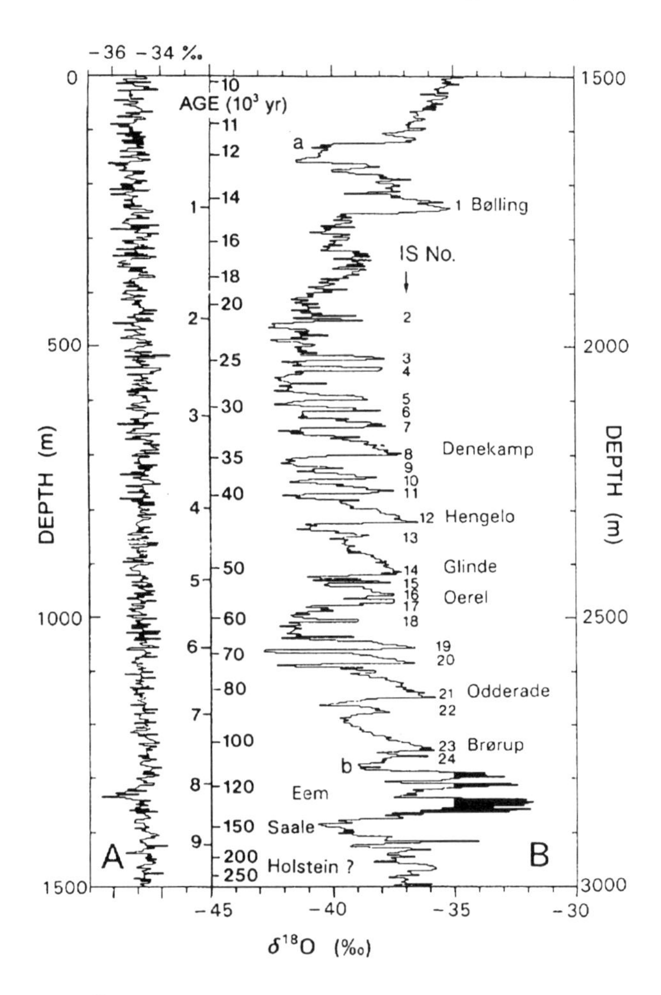

FIGURE 128 The GRIP $\delta^{18}O$ profile plotted on a linear depth scale (to the left and right). The two sections (A and B) are separated by a time scale that was established by counting annual layers back to 14,500 years BP and, beyond that, by ice-flow modelling (from Dansgaard *et al.* 1993)

lower than that of the ice cores, and biotic changes have a certain response time that may mask sudden climatic events (E. Larsen *et al.* 1995). Oscillations in magnetic susceptibility measured in sediment cores through palynologically dated Weichselian/Eemian sequences have yielded a pattern comparable to the GRIP ice core (Thouveny *et al.* 1994). New studies of Eemian sequences in central and southern Europe suggest that slightly cooler conditions occurred in the middle and late parts of the interglacial (Guiot *et al.* 1993, Tzedakis *et al.* 1994). That the climatic oscillations deduced from the GRIP ice core are real, is also suggested by palaeotemperature reconstructions from Norwegian speleothems, that suggest cold periods around 139,000 ±5000 BP, 129,000 ±5000 BP, 114,000 ±5000 BP and 100,000 ±10,000 BP (Lauritzen 1995). That the climatic oscillations of the Eemian show up so dramatically in the GRIP core, whilst the known climatic changes of the Holocene are very little represented, may

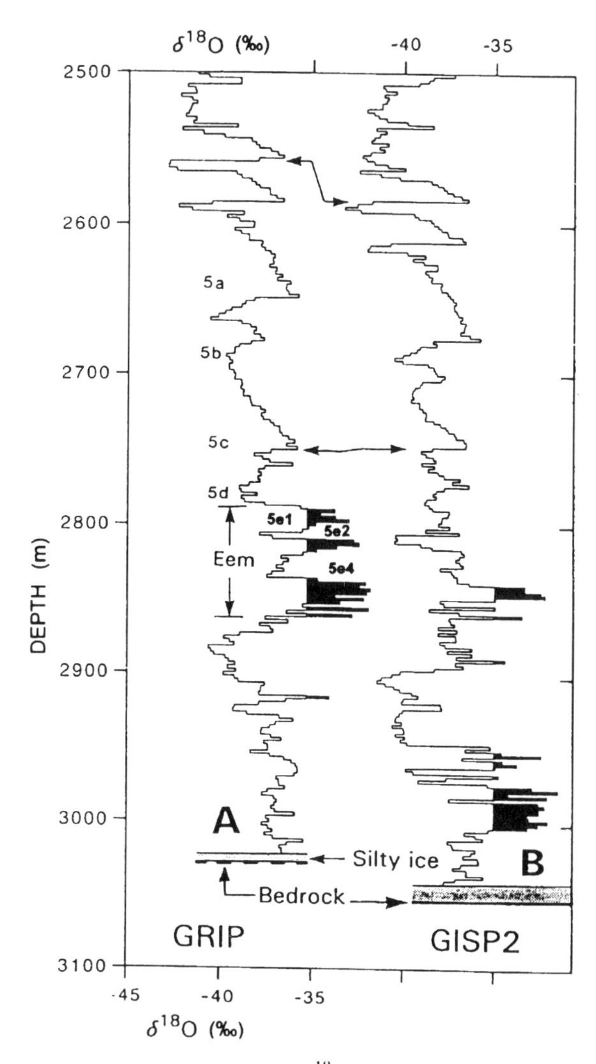

FIGURE 129 The deepest parts of the GRIP and GISP2 $\delta^{18}O$ records plotted on a linear depth scale. The $\delta^{18}O$ values higher than present are emphasised in black. The GRIP Eemian of this figure, as defined by Dansgaard *et al.* (1993), may not correspond exactly with the Eemian of the European pollen diagrams (from Johnsen *et al.* 1995)

reflect changes in oceanic circulation rather than principal climatic difference (Larsen *et al.* 1995).

10.2 MAGNETOSTRATIGRAPHY

When lava cools to below ca. 500°C, the newly formed crystals of magnetic minerals are aligned in accordance with the earth's magnetic field (Brunhes 1906, Matuyama 1929). The same applies to the settling of magnetic particles in sediments, though to a lesser degree. In the laboratory this orientation can be measured. As the magnetic field of the earth has changed repeatedly, these changes can be used to provide a basic stratigraphic framework, the so-called magnetostratigraphy. The method has been described, for instance, by Løvlie (1989) and R. Thompson (1991).

The magnetic history of the earth is subdivided into major polarity epochs. The recent (normal) Brunhes Epoch started about 780,000 years ago. It was preceded by the reversely polarized Matuyama Epoch, which began about 2.6 million years ago. So the Matuyama and Brunhes Epochs cover the entire Quaternary. The Matuyama Epoch was interrupted by two shorter periods of normal polarity. Such short-term magnetic reversals are called events. Those events in the Matuyama Epoch are the Jaramillo and the Olduvai Events. All these can be used to help in dating Quaternary sequences.

Magnetostratigraphy is not a dating method in the true sense. In fact, the magnetostratigraphic boundaries must be dated by other means (e.g. by the K/Ar method). As the dating methods have improved gradually, so have the age estimates of the palaeomagnetic reversals. Therefore in the older literature different ages for the boundaries can be found than in current publications. The most recent subdivision according to Shackleton *et al.* (1990) is:

Brunhes Epoch	780,000	to	present
Matuyama Epoch	2,600,000	to	780,000
Jaramillo Event	1,070,000	to	990,000
Olduvai Event	1,950,000	to	1,770,000

In addition to these major units, some short-term excursions from the dominant polarity regime have been noted, most of which are not yet very well established. One of the best-documented cases is the Blake Excursion, which occurred during the last interglacial about 117,000 BP, but even in this case there is not a consistent global pattern.

Palaeomagnetic investigations have helped to subdivide the long loess profiles of China. In combination with magnetic susceptibility, they allow a close correlation with the deep-sea stratigraphy (Heller *et al.* 1991).

Not only can reversals of the earth's magnetic field be used for dating, but also to a certain extent can changes in inclination and declination (secular change). Continuous records for the last 10,000 years have, for instance, have been established for the deposits of Lake Michigan (Creer *et al.* 1976) and Lake Windermere, Britain (Thompson & Turner 1979), and for the period 10,000–16,000 years BP in Russian Karelia (Ekman & Iljin 1995). These variations are cyclic, with wavelengths of 2000 to 3000 years. Consequently they require long sequences spanning several cycles for good correlations.

10.3 RADIOCARBON DATING

Radiocarbon dating is the most widely used method of radiometric age determination in Quaternary research. The method was developed by Libby (1952), for which he was awarded a Nobel Prize. Radioactive carbon (^{14}C) forms from nitrogen (N) in the stratosphere as a result of cosmic radiation. It oxidises to CO_2 and mixes with the 'normal' CO_2 of the atmosphere which contains the stable isotopes ^{12}C (ca. 99%) and ^{13}C (ca. 1%). ^{14}C occurs at an abundance of 10^{-12}. Atmospheric carbon is dissolved in fresh water and salt water and absorbed by plants and animals. Through the permanent exchange with atmospheric carbon, the ^{14}C content of living plants and animals is at balance with the atmospheric composition. After death, however, the exchange ceases and the radioactive carbon within the body decays at a half life of 5730

±30 years, until it finally vanishes completely. By measuring the content of ^{14}C as compared to the content of stable carbon isotopes, it is possible to determine when the organism died.

The method can be applied to wood, peat, seeds, charcoal, bones, teeth, shells, corals, speleothems, soil humous, leather and a number of other substances. It covers the age range of 300 to 70,000 years and the dates are given in radiocarbon years BP (before present). 'Present' by definition means AD 1950, which must be taken into consideration especially with very young ages. If only the notion BP appears after the number, the date is a conventional radiocarbon date, i.e. the age is given as it was measured. However, as the ^{14}C content of the atmosphere has varied slightly in the course of time, radiocarbon dates can be calibrated by comparison with a standard curve based on ^{14}C-dated tree-ring records. Such calibrated radiocarbon dates normally include the notation 'cal BP'.

Dates are given as value $X \pm Y$ BP. Y in this expression is usually a one standard deviation estimate of the error of measurement. However, some laboratories (e.g. the Geological Survey of Canada) report to 2σ. It refers to the technical quality of the measurement only and says nothing about the overall reliability of the date. The highest degree of precision of radiocarbon dating is in the order of ±20 to 50 years (Geyh, personal communication).

Calibration has shown that in certain age intervals radiocarbon dates are equivocal. This applies, for instance, to the periods 8850–8650, 6150–5950, 5300–5000, 4800–4550 and 2750–2350 BP (Stuiver & Pearson 1993). Additionally, plateaux of radiocarbon ages have been found around 12,700 and 10,000 BP (Becker & Kromer 1986, Ammann & Lotter 1989). The latter is especially unfortunate, because it covers the period of the Pleistocene/Holocene transition and the Younger Dryas.

The biggest problem with radiocarbon dating is contamination. Because the samples used are very small, even minute additions of modern carbon to the sample will result in a much too young date. This is especially problematic where very old samples which are near the maximum limit of the method are to be dated. The slightest contamination in that case will yield dates around 30,000–50,000 BP – 50,000 years is still the maximum age that can be dated by the method (Geyh, personal communication).

Whereas in conventional radiocarbon dating the radioactive decay is measured, newly developed **accelerator mass spectrometry** (AMS) is able to measure the atom ratios (e.g. of ^{14}C, ^{13}C and ^{12}C). The technique has the advantage that much smaller samples can be analysed than with conventional radiocarbon dating (1 mg or less as compared to several grams or more), and that dating is much faster. The method is not restricted to radiocarbon dating but can be applied to other long-lived cosmogenic radionuclides, such as, for instance, ^{10}Be. The disadvantage is that the equipment is expensive and therefore only available at relatively few laboratories (Geyh & Schleicher 1990).

10.4 POTASSIUM/ARGON DATING

Whereas radiocarbon is used to date organic matter, the potassium/argon method is applied to mineral constituents of volcanic rocks. The scientific background of the method was developed by Aldrich & Nier (1948). In nature potassium occurs in the form of two stable isotopes, ^{39}K (93.3%) and ^{41}K (6.7%), as well as the radioactive ^{40}K (0.01%), which decays simultaneously to ^{40}Ar and ^{40}Ca. As ^{40}Ca cannot be distinguished from any other ordinary Ca, the decay must be quantified by measuring ^{40}Ar. Because ^{40}Ar is a gas, it vanishes unless it is trapped in a stable crystal lattice. Therefore the potassium/argon method is applied to minerals like feldspars and micas. A starting event is required that has released all possibly occurring older ^{40}Ar to the atmosphere. Therefore the method is restricted to volcanic deposits. Even in those cases old ^{40}Ar may remain trapped within parts of newly formed crystals and, if undetected, it may result in excessive ages. Originally, the method was applied to relatively old rocks; Geyh & Schleicher (1990) list it as a standard method for ages greater than 3–5 million years.

Instead of measurements of ^{40}Ar, it has become possible with the **argon–argon method** (Merrihue & Turner 1976) to determine directly the content of the radioactive parent element ^{40}K. Under irradiation with fast neutrons the main isotope ^{39}K turns into ^{39}Ar. The quantity of this artificially produced element can be determined together with the content of natural radioactive ^{40}Ar. As this can be achieved in one single sample, any errors resulting from inhomogeneities of the material can be avoided. The very short half life of 269 years for ^{39}Ar means that no natural ^{39}Ar occurs in the sample. Thus the measured ^{39}Ar is proportional to the ^{39}K content of the rock, which in turn is proportional to ^{40}K (Richards & Smart 1991).

A modern variant of the $^{40}Ar/^{39}Ar$ method is the **laser technique** in which the samples are heated by a focused laser beam (York *et al.* 1981). Very small samples can be analysed, so it is possible to obtain dates for individual mineral grains. According to Hu *et al.* (1994) the laser technique allows dating of samples even as young as Holocene (Geyh & Schleicher 1990).

10.5 THORIUM/URANIUM DATING

Natural radioactive uranium occurs in small quantities in most rock types. Several dating methods are based on the radioactive decay of uranium; they are grouped under the term uranium-series dating. They are all based on the fact that if the parent isotopes remain undisturbed in a rock for over ca. 2 million years, they will develop a state of equilibrium with their daughter isotopes. This equilibrium is disturbed when chemical elements are leached from their source rock and redeposited. Uranium is very susceptible to leaching, whilst its daughter isotopes, like ^{230}Th, are not. Redeposited uranium thus is in disequilibrium with its daughter isotopes; it can be measured and used for dating.

The most widely used method of disequilibrium dating from the uranium-series group is the $^{230}Th/^{234}U$ method, developed by Barnes *et al.* (1956). It determines the deficiency of ^{234}U. With mass-spectrometric techniques it can be used to date materials between several thousand and 500,000 years (Edwards *et al.* 1986/87). The method is based on the fact that uranium is readily soluble in carbonate-containing water in the form of UO_2. Those ions are incorporated into carbonate precipitate from the water.

Subsequently, at a half life of 248,000 years, ^{234}U decays into ^{230}Th. Suitable materials for dating thus include speleothem, coral, mollusc shell, travertine, calcareous tufa, peat, inorganic marl and lacustrine sediments. A uranium concentration above 50 ppb is required (Geyh & Schleicher 1990). The method is based on the assumption that the sample contained radioactive uranium but no ^{230}Th at the time of formation. The latter can be checked by measuring the content of stable ^{232}Th, which occurs together with non-authigenic ^{230}Th.

10.6 FISSION-TRACK DATING

The method goes back to Price & Walker (1962, 1963). Of the two naturally occurring uranium isotopes ^{238}U and ^{235}U, only ^{238}U decays in a dual way. Apart from emitting alpha radiation, it also decays by spontaneous fission. This nuclear fission leaves minute tracks in minerals and glasses, which can be made visible under the microscope after etching. Track density is set in relation to the uranium content of the sample. In most cases dating is restricted to materials at least a few hundred thousand years in age. Under favourable conditions, younger ages can be determined (Wagner & Van den Haute 1992). For dating Quaternary deposits, the preferred mineral is zircon, which is found in volcanic deposits. Fission-track dating has been successfully applied, for instance, in correlating the Pearlette Ash layers of the Unites States with their sources in the Yellowstone Park region (Naeser & Naeser 1991).

One of the great advantages of the method is that single mineral grains can be analysed. This allows for an easy detection of contamination. On the other hand, like all defects caused by radiation, fission tracks tend to heal with time, especially under high temperatures. This results in ages that are too young. However, it is possible to test the annealing behaviour of the material in question under variable heat under laboratory conditions and to recalculate their long-term behaviour. Zircon has proved to be most stable against healing, titanite is slightly less stable and apatite and volcanic glasses are relatively unstable.

10.7 THERMOLUMINESCENCE, OPTICAL STIMULATED LUMINESCENCE AND ELECTRON SPIN RESONANCE

Three methods of age determination are based on the fact that radiation damage to minerals increases with mineral age. Thermoluminescence dating (TL), optical stimulated luminescence dating (OSL) and electron spin resonance (ESR) mainly differ in the method used to measure this damage. The methods are described in detail by Wagner (1995).

In nature, the internal crystal structure of minerals deviates from the ideal textbook lattice structure. Two types of deficiencies occur: primary deficiencies were built in during mineral formation, and secondary deficiencies have been acquired in the course of time, as a consequence of alpha, beta or gamma radiation. The defects act as traps for electrons released from minerals as under nuclear radiation. When the mineral is heated, the electrons are released. Some of them recombine at so-called luminescence centres, which are another type of defect in the crystal lattice, under emission of photons. The release of large numbers of photons results in a measurable light effect, the thermoluminescence (Wintle 1991). Thermoluminescence was already observed by Robert Boyle some 300 years ago, and Wiedemann & Schmidt (1895) traced it back to the effects of ionising radiation. The concept of utilising this phenomenon for dating was first proposed by Daniels *et al.* (1953).

The TL signal is proportional to the number of trapped electrons. Therefore it is also proportional to the time of exposure to radiation. Ideally, therefore, a linear correlation should exist between exposure and thermoluminescence. However, because with increasing age more and more electron traps are already occupied, saturation is reached after some time.

In order to allow dating, there must have been an event in the past that has emptied the electron traps. As heating releases the electrons, exposure to fire or high temperatures are ways of zeroing the signal. Therefore the TL method was first applied to date pottery and burnt flint in archaeology. Prolonged exposure to sunlight also empties most of the electron traps in quartz and feldspar crystals, though not all. As this is a gradual process, a bleaching curve results. If dating is to be successful, the sample should have been exposed to sunlight long enough to empty most of the electron traps. In the laboratory the natural thermoluminescence of the sample (or equivalent dose, ED) is measured and compared to an artificial TL signal induced in the same sample by exposure to a calibrated radiation source (Wintle 1991).

TL dating was originally used in archaeology to determine the age of ceramics. Only later it was learnt that it could also be applied to sediments, quartz and feldspars. The method has been used to date Pleistocene sediments in the former Soviet Union since about 1965. In Britain, Wintle (e.g. 1981, 1986) has proved that the method can yield reliable results. The state of research has been presented, for instance, by Aitken (1985), Wintle (1990) and Forman & Nachette (1991). However, there are great controversies about the question to which type of deposit and to what age range this method can be applied. The fundamental precondition is that a sediment must have been exposed to sunlight long enough to eliminate all older TL signals, and then buried until being sampled. Thus loess and aeolian sands are suitable materials. The method, however, has also been applied to glaciofluvial and glaciolacustrine deposits, which may not have received sufficient exposure to light before deposition. Even tills have been dated. Berger *et al.* (1992) applied TL for dating loess up to 800,000 years old, and Wagner (1995) even regards TL as suitable for an age range between 100 and 1 million years. Many others, however, put the upper age limit closer to 100,000 years.

The OSL dating technique was introduced by Huntley *et al.* (1985). OSL dating utilises the same sort of defects as TL, but only the light-sensitive traps. A low-power laser is used instead of heat to free trapped, light-sensitive electrons (Geyh & Schleicher 1990). As total bleaching under sunlight is only a matter of seconds, the method has the advantage that it can also be applied to materials that were exposed for a very short time, making it appear suitable for a wider range of sediments (Aitken 1994).

Mejdahl & Funder (1994) applied both TL and OSL dating to a variety of sediments from East Greenland. They found that some samples showed a young OSL age but had retained an older TL age from an earlier bleaching event. Dating of sediments from the last glacial/interglacial cycle gave reasonable results, whereas deposits from older glaciations showed a large spread of ages. Shallow-marine and fluvial sediments fell within $\pm 15\%$ of the expected ages, whereas ice-proximal glaciolacustrine and glaciofluvial sediments were found to be unsuitable for luminescence dating.

Whereas TL and OSL determinations include the damage of the original luminescence centres, the ESR method (Zeller *et al.* 1967) uses microwave energy to determine the centres *in situ*. This has the advantage that the unaltered sample remains available for further studies. The age range is assumed to cover at least the whole Pleistocene. The method has been applied to carbonates like speleothem, travertine and mollusc shells. However, the method is not unproblematic. It is assumed that the centres remained stable over a long period of time. But, like other radiation damages (see fission-track method), the centres undergo annealing processes. Molluscs have been found to be unsuitable (Katzenberger 1989), and speleothems and travertine have not correlated well with uranium/thorium dates (Henning & Grün 1983). A review of the possibilities of ESR and its limitations are given by Schwarcz (1994). The age range covered by ESR includes the entire Quaternary, but the method will not come close to the precision of some other dating methods, like ^{14}C or $^{40}Ar/^{39}Ar$.

TL, OSL and ESR methods have in common that they provide the means of determining ages of deposits which otherwise could not be dated.

10.8 BERYLLIUM DATING

The ^{10}Be method was developed by Arnold (1956). It is not yet regarded as a standard method, but it has great potential for applications to Quaternary science. For instance, it can be applied to carbonate-free pelagic sediments. The dating range extends from 100,000 years to about 15 million years. Radioactive ^{10}Be is produced in the atmosphere by a reaction between cosmic radiation and nitrogen and oxygen. ^{10}Be becomes attached to aerosols and with precipitation is brought to the earth's surface. At the same time, ^{10}Be is also formed by reactions between that proportion of secondary cosmic radiation that reaches the earth's surface and oxygen and silicon of silicates. This ^{10}Be is several orders of magnitude rarer than the cosmogenic ^{10}Be. The two thirds of cosmogenic ^{10}Be that falls into the oceans is bound to solid matter because of its strong affinity to surfaces, and after a relatively short period of about 1000 years, it is deposited in deep-sea sediments. The ^{10}Be that falls on land is held back by fine-grained particles of soil and sediment.

The ^{10}Be content of loess/palaeosol sequences has been used for stratigraphical correlations. It gives a curve similar to the deep-sea oxygen-isotope curve and thus has been used as an indicator of palaeoclimate (Shen *et al.* 1992). The changes in ^{10}Be content are thought to reflect different sedimentation rates between rapidly deposited cold-stage loess and palaeosols that had been exposed to ^{10}Be influx for a long time. In combination with the $\delta^{18}O$ chronology the ^{10}Be method can achieve an accuracy up to 5–10 ka in dating continental loess profiles. In Louchuan, China, the method has also been successfully applied to the loess stratigraphy in combination with magnetic susceptibility (Beer *et al.* 1993).

The ^{10}Be produced *in situ* on the land surface can be used for dating how long the respective surface has been exposed (Monaghan *et al.* 1983). This has been successfully applied to dating when certain formerly glaciated terrain became ice-free (Nishiizumi *et al.* 1986). Exposure dating is never applied with ^{10}Be alone.

10.9 AMINO-ACID DATING

An introduction and review of the method that was developed by Abelson (1954, 1955) is given by G.H. Miller & Brigham-Grette (1989). The proteins within organisms are composed of about 20 different types of amino acid. Most of them consist of molecules centred by a single chiral carbon atom. Those occur in two different varieties, L-isomers which deflect polarised light to the left, and D-isomers which deflect it to the right. Proteins with a high molecular weight in living organisms contain only L-isomers. After death of the organism, those amino acids are gradually transformed into D-isomers, until an equilibrium ratio between the two isomers is reached. The D/L equilibrium ratio is 1.0–1.3, depending on the type of amino acid. This process of amino-acid transformation is called racemisation, or epimerisation for a related process. It can be used with caution for relative age determination (e.g. G.H. Miller & Mangerud 1985). It has been found that racemisation is not a linear process. After rapid racemisation during the first ca. 350,000 years, the process slows down considerably. Nevertheless, when calibrated, the method can yield age estimates.

The advantages of the method are that it is fast, requires little laboratory equipment and is therefore cheap, so that investigations of large numbers of samples are possible. One of the shortcomings, however, is that racemisation depends strongly on the diagenesis temperature of the sample. In addition, different mollusc species yield different results, so that only fossils of the same species should be compared. Under unfavourable conditions even parts of the same mollusc shell may yield variable results (Sejrup *et al.* 1994). Amino-acid dates have been used to distinguish deposits of unknown but potentially different ages (e.g. Bowen 1991). It has been applied to the correlation of interglacial sites in the North Sea (Fig. 130) (Knudsen & Sejrup

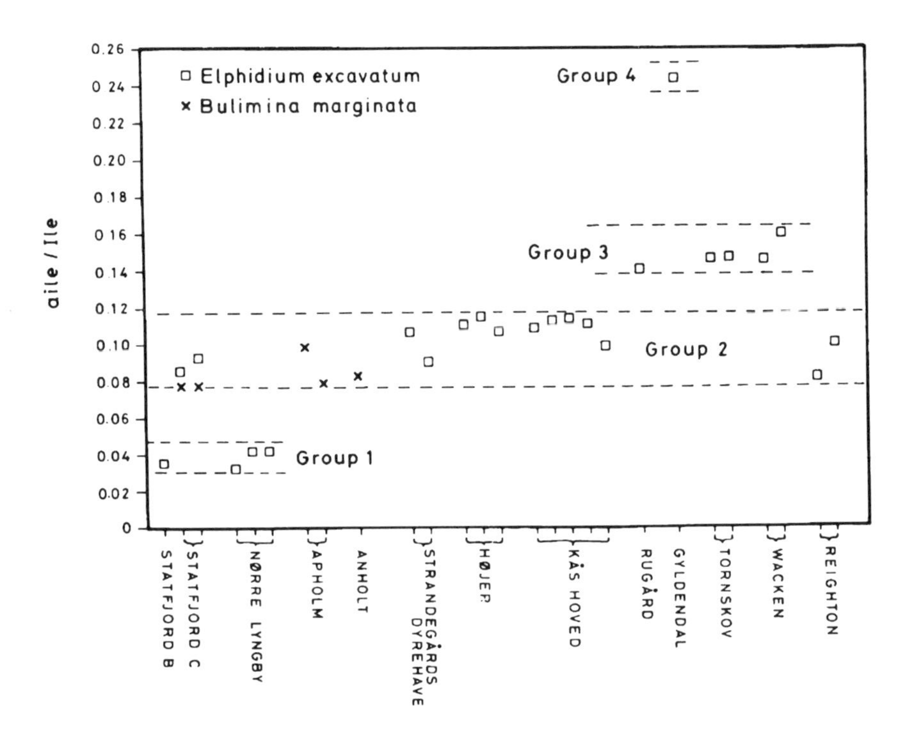

FIGURE 130 Isoleucine epimerisation in the benthonic foraminifera *Elphidium excavatum* and *Bulimina marginata* from sites in the Norwegian North Sea, Denmark, Germany and England (from Knudsen & Sejrup 1988)

1988), in the Mediterranean, coastal America, pluvial lakes, and in Australia. The method can be used on mollusc shells, ostracods and foraminifera but does not yield satisfactory results from bone or wood. The time range varies depending on the climatic history. It ranges from one million years in the arctic to 300,000 years in the US Gulf Coast region (Karrow, personal communication). In some cases, however, the dates obtained are in conflict with other evidence. For instance the D/L ratio of the marine deposits of Kås Hoved in Denmark indicates an Eemian age, whereas the foraminifera suggest that it is a Holsteinian deposit (Knudsen & Sejrup 1988).

Because the rate of racemisation depends on temperature, it is possible from the D/L ratios to derive palaeotemperatures for the period of diagenesis if the age of the sample is known (G.H. Miller & Brigham-Grette 1989). Mangerud *et al.* (1992), for example, used D/L ratios and a postulated basal temperature of 0°C, to calculate the duration of the last glaciation on Svalbard to be less than 10,000 years.

10.10 VARVES

That the rhythmic layering of deposits from glacial lakes reflects an annual alteration of drainage conditions was originally detected by De Geer (1884). In 1912 he was able to construct a varve chronology over the last 12,000 years. The technique as it is applied today has been described by Strömberg (1983).

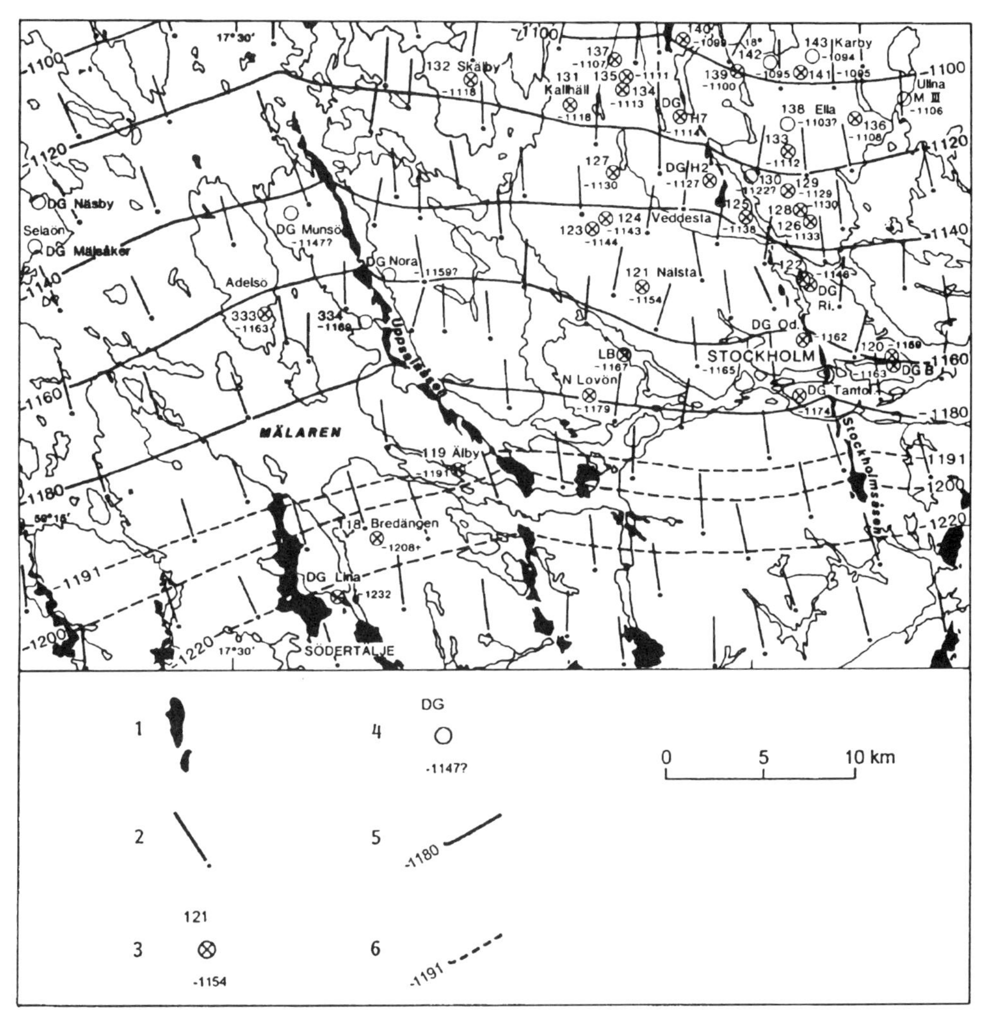

FIGURE 131 Striae, eskers and ice-marginal positions in central Sweden. The age of the individual ice-marginal positions is determined by varve chronology. 1 = esker; 2 = youngest set of striae; 3 = varve-measurement site 121, bottom varve dated to −1154 before varve zero (zero varve = 7288 BC); 4 = varve-measurement site of other authors (DG = De Geer); 5 = ice recession line according to varve data and youngest striae; 6 = ice recession line, uncertain position (from Strömberg 1989)

In order to draw a varve chronology numerous measurements are necessary, either in exposures or core drillings (Figs 131 and 132). Reproducibility of the results depends on the frequency of characteristic strata within the sequences. Correlations over a distance of more than 10 km are generally problematic. Annual stratification can only be preserved in places where the layers are undisturbed by benthonic faunal activities. Bioturbation destroys bedding. For this reason glacial lakes with their water temperature close to the freezing point provide a favourable environment. The best varves are formed in fresh water. In salt water flocculation of clay particles results in masking of the annual stratification so that measurements are impossible. The annual varves may be subdivided by fine, so-called 'daily' varves reflecting

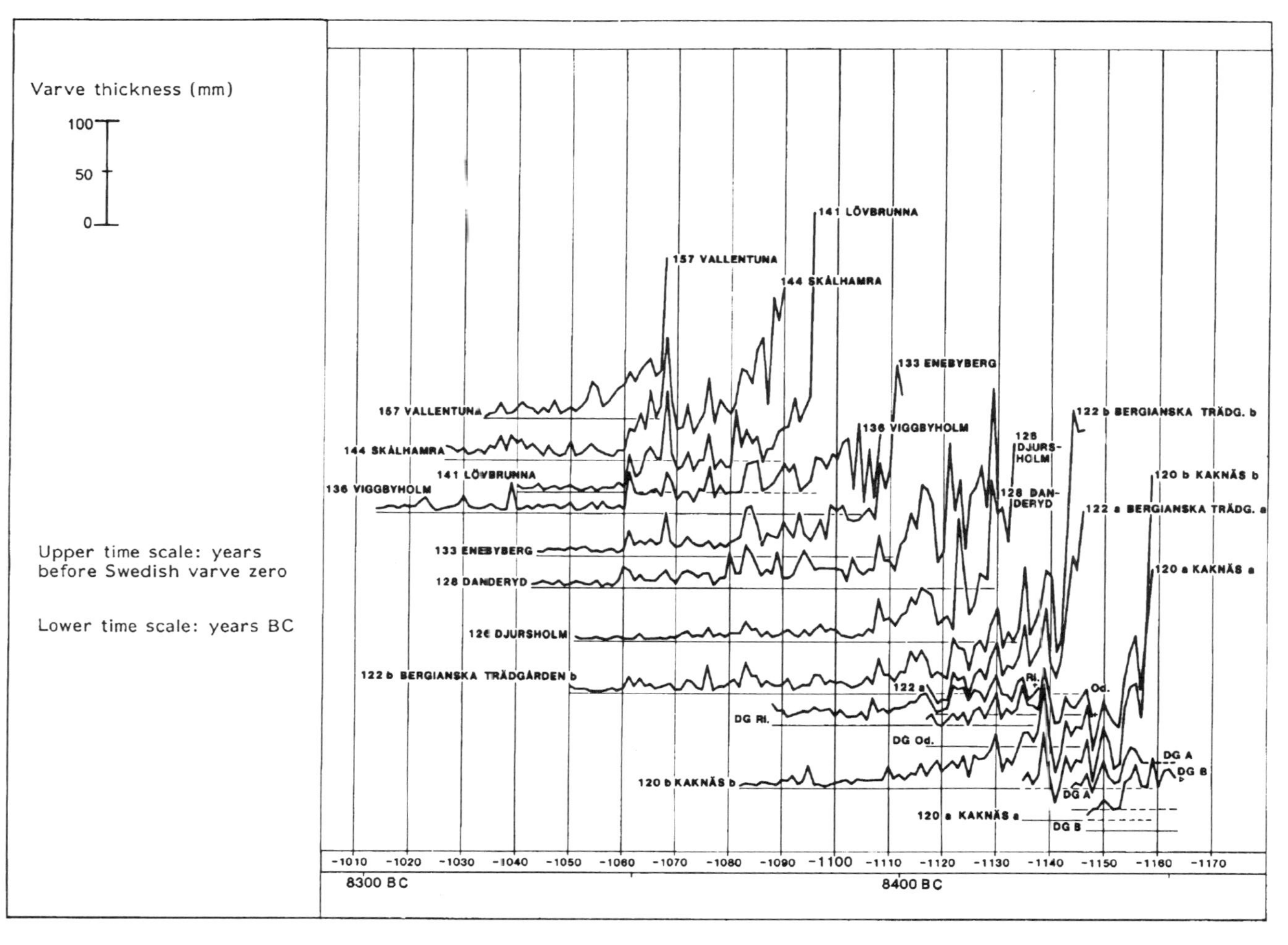

FIGURE 132 Correlation of varve diagrams from central Sweden (for position see Fig. 131) (from Strömberg 1989)

single outflow events within the course of the year. In the case of strongly developed 'daily varves' there is the danger of their misinterpretion as annual features. Measurements of varves can best be undertaken at exposures. If they are not available, closely spaced drillings are required so that recurrences of the stratification due to underwater slumping can be detected. Their identification is often difficult even if the distance between drillings is very short.

The Weichselian lateglacial varved clays in Sweden have allowed the complete reconstruction of the deglaciation chronology. The detected ice-marginal positions strike perpendicular to the orientation of the youngest glacial striae. Typical recession rates of the ice margin are 200 to 300 m per annum; maximum values lie at 1000 m, with minimum values at 50 m annually. As early as 1940, De Geer presented a complete varve chronology for Sweden, the so-called Swedish Time Scale. Since then, the time scale has been revised, and the age of the so-called zero varve is now assumed to be 118 years older than originally thought. Through measurements of the annual stratification in the estuary of the Ångerman River, the varve chronology could be extended beyond the date of complete deglaciation up to the present day (Lidén 1913, 1938). The annual layering there results from seasonal changes in stream discharge. The total Swedish varve chronology down to the present day comprises 12,850 annual varves; the margin of error is calculated at +35/−205 by Cato (1987) (Strömberg 1989, Ringberg 1991, 1994). In Finland, similar work undertaken by Sauramo (1918, 1923), has most recently been revised by Niemelä (1971). In the meantime it has also been possible to connect the Finnish varve chronology to the Swedish Time Scale (Strömberg 1990).

The classic studies of varved silts and clays were undertaken in Scandinavia. However, deposits of ice-dammed lakes are widespread in the marginal areas of all glaciations. In the Leipzig area of Saxony (Germany), varved clays occur at the base of the till units of each major ice advance. However, they are very thin and comprise only a few annual strata, a fact that indicates a rapid advance of the ice (Eissmann 1975). The Dehlitz–Leipzig varved clay at the base of the first Elsterian ice advance, for instance, reaches a thickness of just 2 m and comprises a maximum of 88 varves. It is assumed that part of the varves had been eroded when the glacier overrode the area. With respect to the advance velocity of the glacier, Grahmann's (1925) calculation is still valid that it must have been high but still under 110 to 120 m per annum. The glacial lake in which the banded clay was deposited covered a maximum area of about 750 km^2 (Eissmann 1975). These large lakes were rather short-lived. Moreover, when the glaciers of the Saalian Glaciation blocked the drainage of the Weser and Leine rivers in western North Germany, the impounded lake there existed only for a few decades (Gassert 1975). In this case the areal extent and the outlets of the proglacial lake are not well known. Thome (1980) assumed that the impounded Fulda–Werra fluvial system was dammed until drainage at an altitude of 300 m over the Neustädter Sattel and the Schwalm towards the Lahn River became possible. The corresponding sediments, however, have not yet be detected.

Of considerably longer existence were the numerous ice-dammed lakes of the disintegration phase of the North European Ice Sheet. For example, lacustrine sediments locally more than 100 m thick were deposited in the widespread lake system that existed in northern Germany at the end of the Elsterian Stage.

The large ice-dammed lakes in North America have also deposited varved clays. Pioneering investigations were conducted by Antevs (1925, 1928), but radiocarbon dating proved that the deposits in question were much younger than originally thought. With the exception of O.L. Hughes' (1965) investigations in northern Ontario, little recent work has been done with regard to varve chronology. Modern studies have mainly addressed the depositional conditions in the lakes (e.g. Ashley 1975, Gustavson 1975, Eyles & Clark 1988).

Varve studies are not necessarily restricted to glacial lakes. Laminated diatomite deposits in North Germany have allowed estimates of the duration of both the Eemian and Holsteinian interglacials (H. Müller 1974a, b). Investigation of laminated gyttja in Lake Gościąż in Poland has yielded more than 12,000 couplets of light, mostly calcitic and dark, organic-rich layers. Comparison with the pollen sequence and calibration with radiocarbon dates suggests that the Younger Dryas cooling here lasted 1600 years (Pazdur *et al.* 1987, Ralska-Jasiewiczowa *et al.* 1992). This, however, is in disagreement with most other assessments of the Younger Dryas length. Detailed varve-chronological investigations in Lake Soppensee in Switzerland have shown a Younger Dryas of about 700–900 years (Lotter 1991), which agrees better with radiocarbon-dated sequences. Maar sediments have also yielded long varved sequences (Zolitschka *et al.* 1992). Where the varved layers reflect the carbonate content of the sediment, carbonate solution may hamper the investigations (Geyh, personal communication).

11

Quaternary Stratigraphy of Northern Europe

11.1 PALAEOGEOGRAPHIC SITUATION AT THE BEGINNING OF THE PLEISTOCENE

Knowledge of the onset of the Pleistocene is limited. The most complete sequence from the marginal region of the North European glaciation is known from the Netherlands, where Tertiary and Quaternary deposits more than 500 m thick have been preserved in the Rhine and Meuse estuaries. The region is a young subsidence zone, a continuation of the North Sea central graben (P.A. Ziegler 1982) (Fig. 133). Even longer sedimentary sequences have been preserved in the troughs between salt domes of Schleswig-Holstein, e.g. at Oldenswort on the Eiderstedt peninsula (Menke 1975).

In the Late Tertiary the sea withdrew from the northwest European Basin, and fluvial deposition prevailed. Since the early Miocene the rivers of the Fennoscandian Shield and the Baltic Platform in the north, as well as of the Variscian Mountains in the south, transported clastic sediments into the northwest European and East German/Polish basins. Uplift of the surrounding areas led to increased sedimentation (P.A. Ziegler 1982). The deposits of this system are light grey to whitish quartz sands, into which layers of clay and brown coal are intercalated. The gravel fraction consists predominantly of well-rounded quartz grains as well as very rare quartzites and silicified sedimentary rocks, the latter consisting mainly of former limestones from the eastern Fennoscandian Shield and the Baltic Platform. The drainage system was generally aligned along the course of the depression occupied by the present Baltic Sea, and is therefore termed the **Baltic River System** (Bijlsma 1981). Figure 134 shows a reconstruction of this river system (Gibbard 1988).

The Baltic River System remained functioning during the Early Quaternary, and its deposits can be traced from the Bremen region of Germany well into the Netherlands. The deposits contain lydites from the southern central German uplands as well as large blocks, which must have been transported by drift ice from the eastern Baltic region (Gripp 1964). The 'Loosen Gravels' of Mecklenburg are regarded as part of this river system (Von Bülow 1969; see Fig. 152).

Evidence of periods with a cold climate and related frost phenomena can be traced back to the Late Tertiary in northwest Europe. Sharp-edged silificate clasts in the Miocene lignite sands of Besenhorst (near Geesthacht, east of Hamburg) might hint at ice transport (Ehlers *et al.* 1984). The 'lavender blue silificates' found between the Miocene lignite beds of the Lower

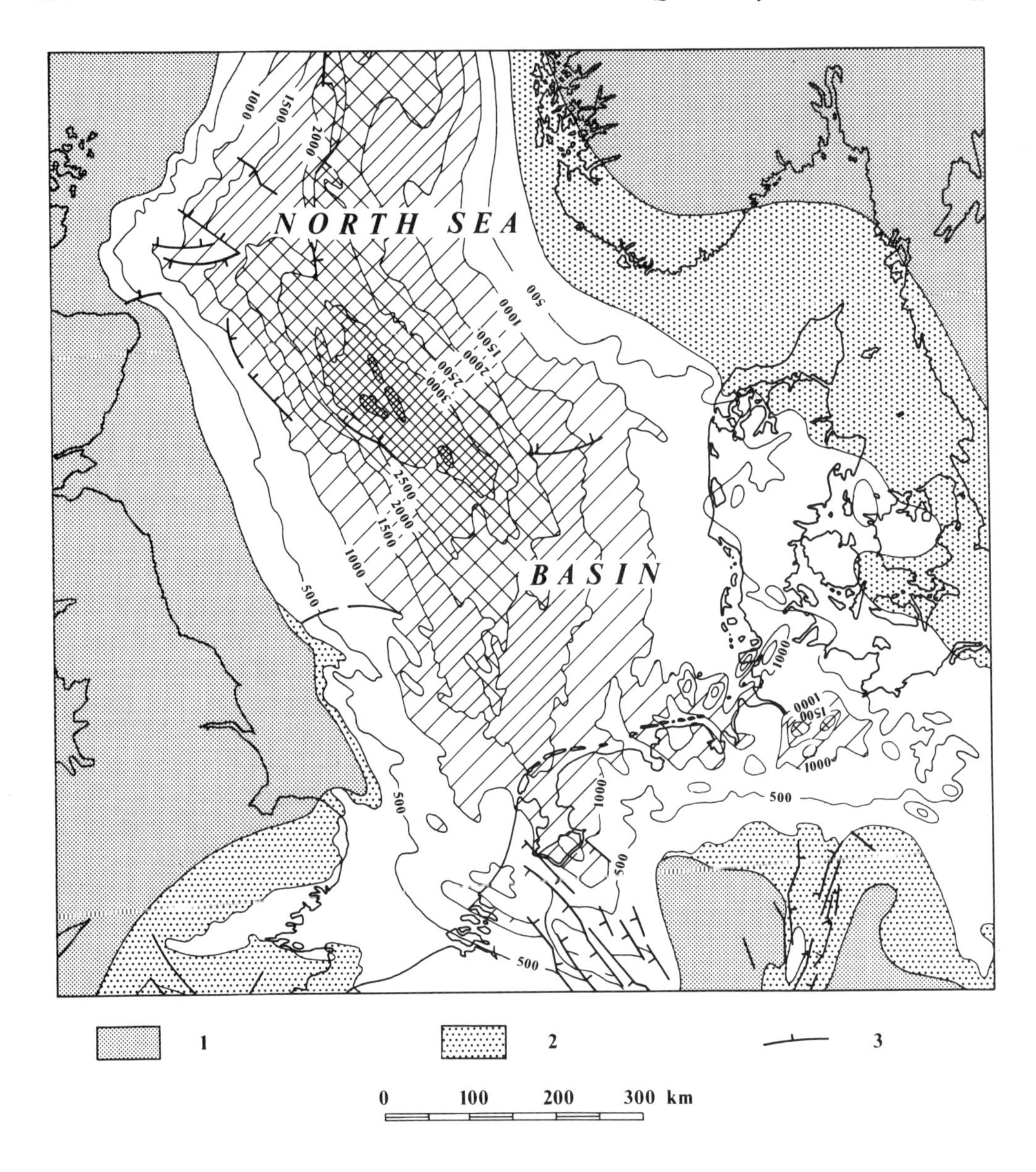

FIGURE 133 Distribution and thickness of Tertiary deposits (excluding the Danian) in the North Sea basin (after Ziegler 1982); thickness in metres. 1 = continental areas; 2 = no deposits; 3 = fault (from Ehlers 1988)

Lausitz in East Germany (H. Ahrens & Lotsch 1976) are similarly interpreted. Ice rafting has been postulated for the subangular Scandinavian sandstone blocks, up to 0.5 m in diameter, and hornstones found in the kaolin sands on the island of Sylt. Von Hacht (1979) first reported large blocks of sand that could only have been transported in a frozen state. It remains doubtful that the poorly orientated scratches on Danian-age flints and other hard rocks also noted by Von

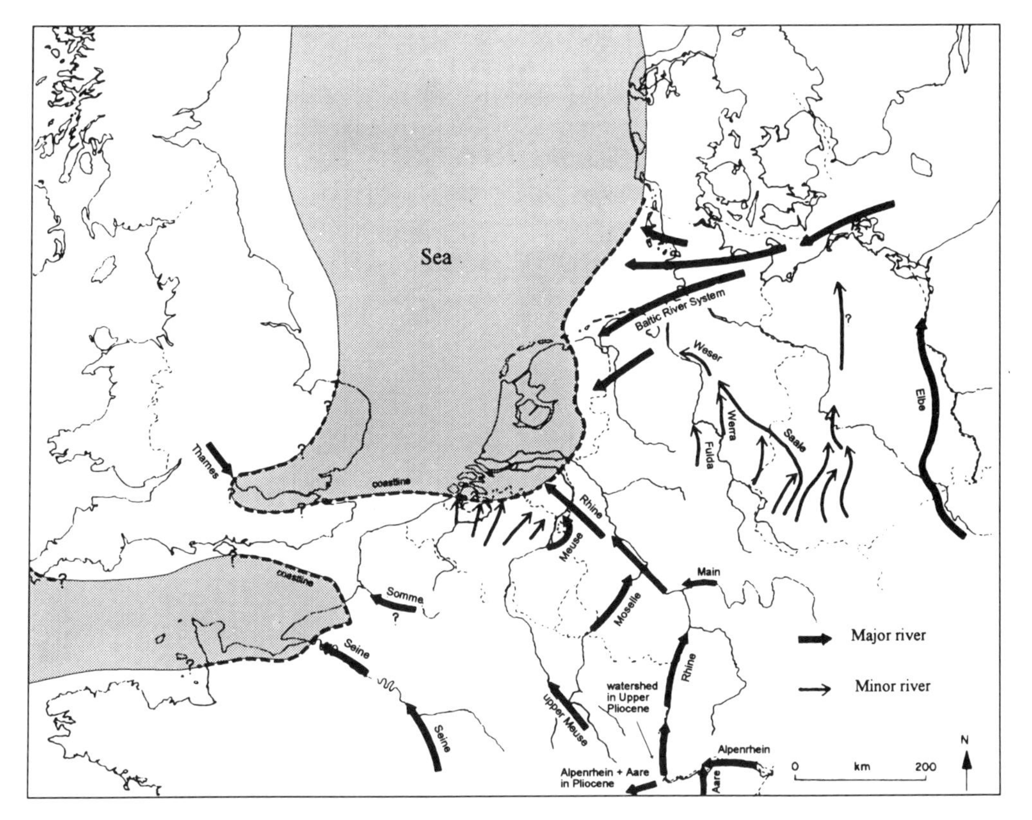

FIGURE 134 Schematic palaeogeographical reconstruction of the drainage systems for the period Reuverian to Tiglian (Late Pliocene–Early Pleistocene) (from Gibbard 1988). The Baltic Sea does not yet exist. The Weser and Elbe are tributaries of the Baltic River System. The Alpine Rhine (Alpenrhein) still drains into the Rhône, and becomes confluent with the Upper Rhine in the Latest Pliocene

Hacht (1987) should be interpreted as glacial striae. On the basis of heavy-mineral analyses by Burger (1986), the kaolin sands are of Lower Pliocene Brunssumian age.

11.2 SUBDIVISION OF THE PLEISTOCENE IN NORTH GERMANY AND THE NETHERLANDS

The basic subdivision of the North German Quaternary stratigraphy was established by the end of the last century. At that time the actual length of the Ice Age was unknown. It has been assumed since the late 19th century that North Germany, like the Alps, had been affected by three glaciations (Penck 1879). This concept, however, found general acceptance only after Keilhack (1896) in Berlin and Gottsche (1897a) in Hamburg had identified a third (oldest) till unit in North Germany. The names Elsterian, Saalian and Weichselian first appeared in 1911 on the 1:25,000 map sheets of the Prussian Geological Survey and they gained general acceptance in the 1920s. According to Woldstedt (1929) they were originally proposed by Keilhack.

After it had been discovered that the Alps had been glaciated (at least) four times instead of three (Penck & Brückner 1901/09), there have been no lack of attempts to identify corresponding numbers of glaciations elsewhere. However, this has not been possible in Britain or Ireland, nor in western Central Europe. In Germany, neither the subdivision of the Saalian into two independent glaciations nor the assumed detection of an additional older (Elbe) glaciation (Van Werveke 1927) could stand up to close inspection. Further to the east, however, traces of more glaciations have been identified. In Poland, for example, these include at least the pre-Elsterian Nidanian and Narevian glaciations (e.g. Lindner & Marks 1994, Lindner 1995).

Later, palaeovegetational and sedimentological investigations in the Netherlands showed that the known warm and cold stages of the Pleistocene had to be supplemented by some additional stages. In 1957, Zagwijn supplemented the stages known at that time as follows:

Weichselian Cold Stage
Eemian Warm Stage
Saalian Cold Stage
Holsteinian Warm Stage
Elsterian Cold Stage
Cromerian Warm Stage

by another five (older) stages:

Menapian Cold Stage
Waalian Warm Stage
Eburonian Cold Stage
Tiglian Warm Stage
Praetiglian Cold Stage

Originally it had been thought that each 'stage' of this subdivision had been either warm or cold. However, Zagwijn (1963) demonstrated that within each of the older six 'stages' repeated changes between boreal and warm-temperate climate had occurred (Fig. 135). This means that they are more complex units composed of several warm and cold events. During one of the cooler phases of the 'Tiglian Warm Stage', even permafrost occurred in northern Belgium (Kasse 1988). Palaeoclimatically the Tiglian thus forms a complex unit which, however, according to the international stratigraphic rules (Hedberg 1976), can be correlated from one place to the other on the basis of their biostratigraphical characteristics (Zagwijn in Gibbard *et al.* 1991). A similar definition applies to the 'Cromerian Complex Stage', which comprises at least four warm and three cold events (Gibbard *et al.* 1991).

Palynological interpetation of the Early Pleistocene sediments from the Netherlands is problematic. Many of the pollen analyses were undertaken from clastic deposits, or from thin peat or mud beds intercalated in gravelly deposits. In clastic deposits, however, a certain degree of reworking must always be taken into account. This is often indicated by pollen and spores derived from Tertiary or Mesozoic rocks. Furthermore, mechanical abrasion of the pollen grains can influence the composition, and some oscillations in pollen composition may reflect changes in depositional environment rather than vegetational changes (De Jong in Gibbard *et al.* 1991). The Dutch Quaternary stratigraphy is shown in Figure 135.

More favourable conditions for the preservation of good pollen sequences are found locally in the Early Pleistocene deposits of North Germany. In a karstic depression on the Lieth salt dome in Schleswig-Holstein, a sedimentary sequence with five Early Pleistocene warm stages has been preserved, which according to Menke (1975) represent the period from the Tiglian to

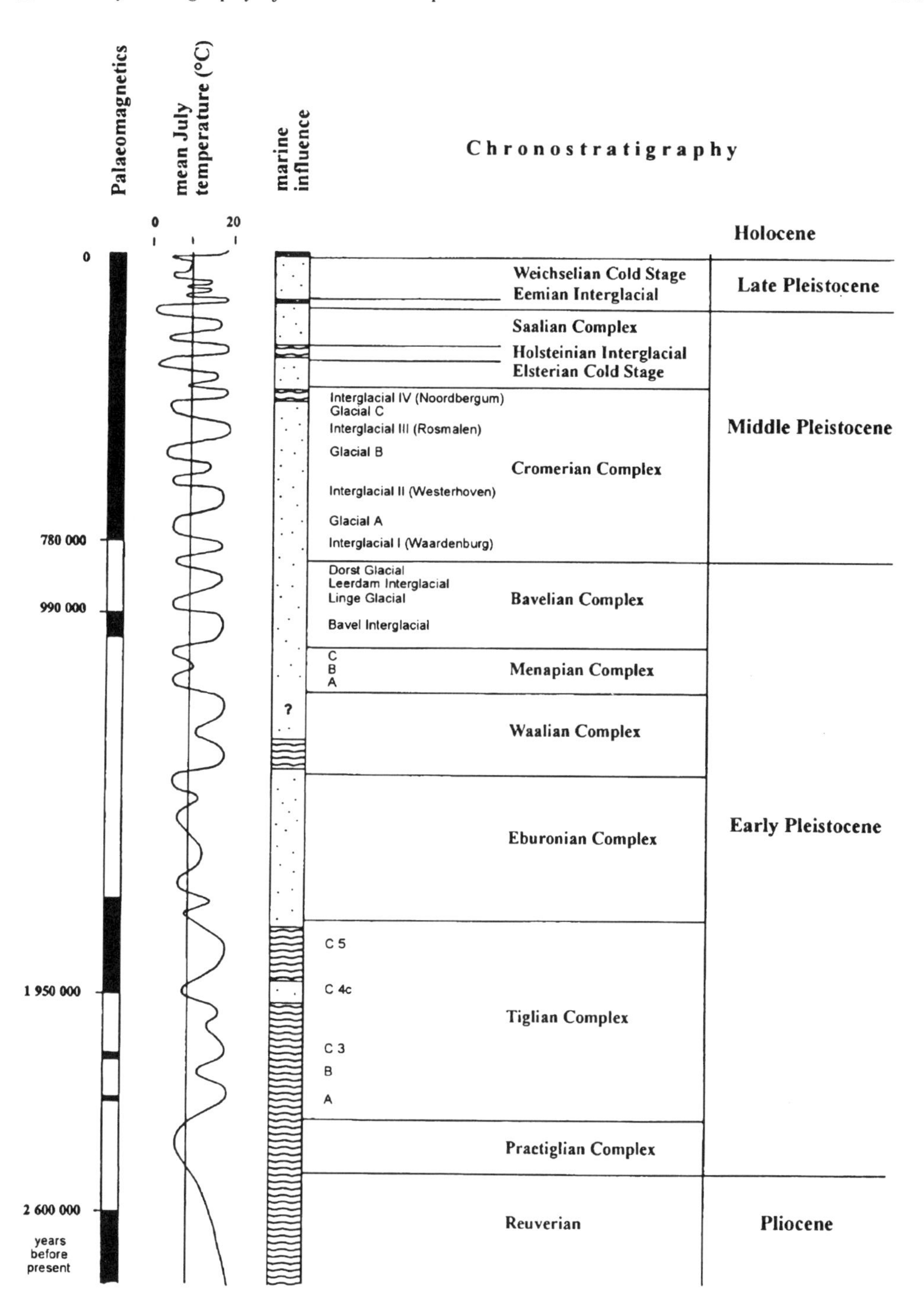

FIGURE 135 Climatic curve and chronostratigraphy of the Pleistocene in the Netherlands and adjoining regions (after Gibbard *et al.* 1991)

beyond the Menapian. The sediments are autochthonous peats and muds, which in contrast to their Dutch equivalents show a clear vegetational development for each event. Connection with the Pliocene vegetational records is possible via the Oldenswort borehole. Moreover, the younger part of the Early Pleistocene (probably the end of the Menapian to the early Elsterian) has been encountered in boreholes in a karstic solution hollow on top of the Gorleben salt dome; only a brief account of which so far has been published (H. Müller 1992).

11.3 PRAETIGLIAN

The Praetiglian is the first true cold stage in Europe. Approximately at the same time there was a strong increase in ice-rafted detritus in the ocean west of Norway, which probably resulted from more extensive glaciations of Scandinavia (Mangerud *et al.* 1996). Menke (1975), following the suggestion of Lüttig (1965), refers to the Early Pleistocene periods of cool or cold climate as **cryomers**, and to the periods of warm climate as **thermomers**. The pollen diagrams of the Praetiglian resemble those of the Weichselian Cold Stage (Zagwijn 1975). Pollen of thermophilous plants are almost completely lacking. The only trees present are *Betula, Pinus* and *Alnus*. Herb pollen (Gramineae, Cyperaceae, Ericaceae) dominate. Comparable pollen associations are found nowhere in the Pliocene or Miocene, although the Late Pliocene pollen assemblages indicate a certain degree of cooling (Menke 1975, Suc & Zagwijn 1983).

In the western Netherlands (e.g. at Brielle and Ockenburg), thick Praetiglian sequences have been preserved. The ratio arboreal/non-arboreal pollen varies. However, this may not result from climatic oscillations but reflect changing sedimentary conditions. The high percentage of Mesozoic pollen is especially striking (e.g. *Classopollis*). The source area of this pollen is unknown, but possibly it is derived from the Chalk areas of the English Channel.

The British sequence of 'preglacial' Pleistocene deposits is far from complete, and a satisfactory correlation with the Dutch sequence has not been possible until very recently, mainly because of the lack of precise dates and the identification of significant biostratigraphical events. Results from British Geological Survey work in Suffolk strongly suggest that the Red Crag Formation, contrary to earlier ideas (e.g. West 1980a), may be assigned to the Netherlands' Praetiglian or Late Reuverian Stages (Zalasiewicz *et al.* 1988).

11.4 TIGLIAN

Most warm stages of the Early Pleistocene cannot be recognised without additional stratigraphical information. The oldest warm stage of the Tiglian, Tiglian A, however, has a very distinct pollen assemblage, characterised by the simultaneous presence of *Fagus* (beech) and *Tsuga* (hemlock) (Zagwijn 1963). A similar pollen assemblage has been described from the Frechen I Interglacial of the Lower Rhine Bight by Urban (1978). Menke (1975) correlates his Meinweg Warm Stage and Nordende Warm Stage of the Lieth sequence with the Tiglian A. In doing this, regional floristic differences must be explained. The vegetational communities characteristic of the Lower Rhine region differ in their greater diversity of species from the communities on poorer soils, such as the kaolin sands at Lieth.

In the later substages of the Tiglian, as well as in the Waalian, beech pollen is missing, and *Tsuga* only occurs very irregularly. Apart from the other Tertiary relics, the Tiglian C also contains *Phellodendron, Actinidia* and *Magnolia* as macrofossils; their pollen, however, is rarely found. The Tiglian C, according to Menke (1975), can be correlated with the Ellerhoop

Warm Stage of Lieth. An interglacial that might be correlated with the Tiglian C5–6, has been found at Brüggen (Urban 1978).

All warm stages of the Tiglian are characterised by the presence of the water fern *Azolla tegeliensis*, which occurs only in this stage. Megaspores of this species have occasionally been found in younger deposits (e.g. of the Weichselian Cold Stage), but they could be clearly identified as reworked material. The more recent form *Azolla filiculoides* occurs for the first time in the Late Tiglian, after pollen zone TC4c. From then on it occurs in all warm stages until the Wacken Interglacial.

Within the Beerse Member of the Tiglian, which correlates with pollen zone TC4c, cryoturbations and ice-wedge casts have been observed in northern Belgium (Kasse 1988). During this period an important regression occurred, which in the Dutch literature is referred to as the 'Main Quaternary Regression'. It marks an important faunal change. When the North Sea transgressed again into parts of the southwestern Netherlands in the Late Tiglian, apart from *Ellobium pyramidale, Macoma praetenuis* and *Mya arenaria*, almost all warmth-loving mollusc species that had still been present in the Middle Tiglian were missing. They do not appear again in the deposits of any younger transgressions (Meijer in Gibbard *et al.* 1991).

The Tegelen Clay (Tiglian) is well known for its diversity in fauna and flora. Its small-mammal association includes *Mimomys pliocaenicus, Mimomys blanci, Mimomys reidi* and *Ungaromys* cf. *nanus*, but *Mimomys pitymoides* and *Microtus (Allophaiomys)* are absent. These vertebrate finds have been made in layers that are correlated with pollen zone TC5 (Van Kolfschoten in Gibbard *et al.* 1991).

In Britain the Ludhamian to early Beestonian stages are all correlated with the Tiglian of the Netherlands (Gibbard *et al.* 1991). True cold-stage sediments have been reported from the Baventian Stage, the British equivalent of the Dutch Tiglian C4c. They are thought by some (e.g. Solomon in Funnell & West 1962, Bowen 1991, Hart & Boulton 1991) to include the first evidence of possible ice-rafted detritus in the British North Sea sequence. However, the minerals found may simply record longshore drift derived from the North German river system (A. Burger, personal communication).

11.5 EBURONIAN

Not many details are known about the climatic and vegetational history of the Eburonian. It is subdivided into seven climatic units. During the warm stages, cool coniferous forests spread into the Netherlands, whilst during the cold stages an open, treeless vegetation prevailed. Ice-wedge casts and cryoturbations of this period have been described from the region around Tegelen (Zagwijn 1975, Gibbard *et al.* 1991). Within the Eburonian fauna a first indicator of a glacial climate, the lemming *(Dicrostonyx)*, has been recovered in a borehole at Brielle (Netherlands) (Van der Meulen & Zagwijn 1974, Van Kolfschoten & Van der Meulen 1986).

The Harderwijk Formation, a sequence of fluvial sands, was deposited in the Netherlands starting in the Late Tiglian and ending in the Waalian Stage. This formation contains clasts from Norway and Sweden, including Silurian fossils, but mostly restricted to the fine-gravel fraction (De Jong & Maarleveld 1983). The composition of these quartz-rich sands resembles that of the Pliocene to Early Tiglian kaolin sands of the Baltic River System. In the older parts of the sequence the quartz content in the fine-gravel fraction is about 90%, in the Waalian it decreases to 80–90%, and feldspar content rises to 6–15% (Zandstra 1993).

In the British Early Pleistocene sequence a large hiatus is present between the British

Pastonian and Cromerian Stages (Mayhew & Stuart 1986), so that no apparent equivalents of the Eburonian, Waalian, Menapian and Bavelian stages have yet been identified, although deposits of the Thames system may span this period. The Valley Farm Soil, a strongly developed palaeosol complex, formed in this depositional hiatus (Kemp 1985, Kemp *et al.* 1993).

According to the international stratigraphical subdivision, the Tertiary/Quaternary boundary (at 1.77 million years) would lie approximately at the boundary between the Tiglian and Eburonian. However, as pointed out in chapter 1, this is not feasible for the North European Quaternary stratigraphy.

11.6 WAALIAN

The warm periods of the Waalian are characterised by the occurrence of numerous Tertiary relics, including *Tsuga, Pterocarya, Carya* and *Eucommia*, as well as *Azolla filiculoides*. The Waalian can be subdivided into two warm events (Waalian A and C) and an intervening cold event (Waalian B) (Fig. 136) (Zagwijn *et al.* 1971). The cold event seems to have been rather mild. Apart from Tertiary relics, thermophilous plants have been found throughout, although at reduced numbers (Zagwijn 1975). A similar picture is presented by the Tornesch Warm Stage from Lieth, Schleswig-Holstein, the North German equivalent of the Waalian (Menke 1975).

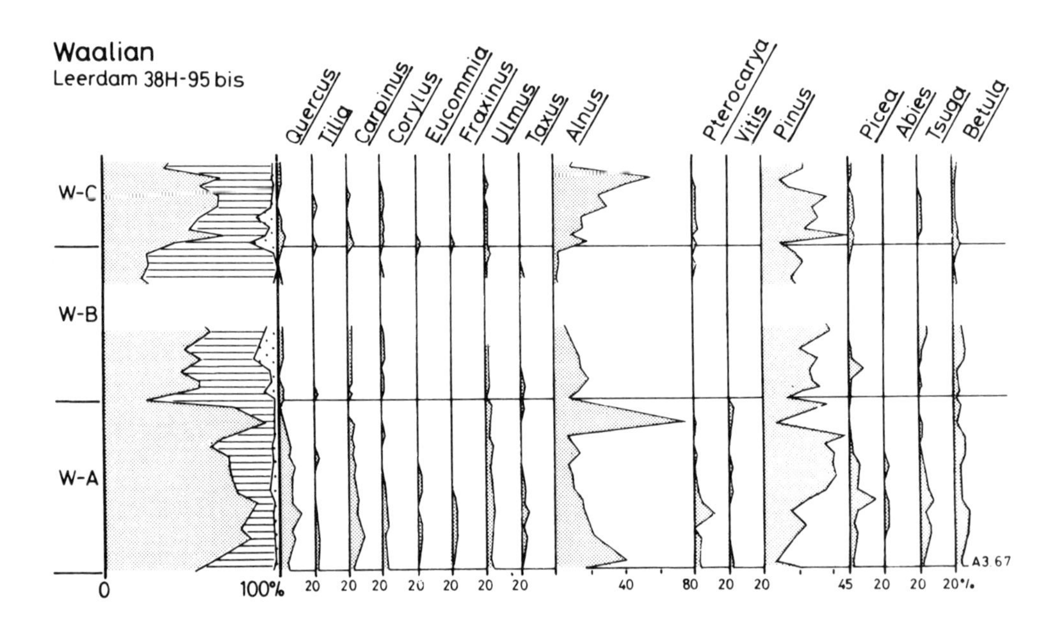

FIGURE 136 Pollen diagram of the Waalian Warm Stage from borehole Leerdam 38H-95bis. Typical profile of an Early Pleistocene warm stage from the Lower Rhine area; there are no distinct indications for any systematic vegetational development and maximum distribution of certain tree species (from De Jong 1988)

11.7 MENAPIAN

The Menapian was originally identified as a cold stage by Zagwijn (1957). At the type site (Veghel borehole in the Rur valley graben) there are indicators of two cold phases, separated by an interstadial with *Pinus* and *Betula*. At least temporarily the Lower Rhine region at that time formed part of the periglacial zone. In the Tegelen region ice-wedge casts and cryoturbations have been reported. The Hattem Beds of the Menapian are the oldest strata in the Netherlands that yield major quantities of Scandinavian erratics from eastern Fennoscandian and central Swedish (Dalarna) source areas (Zandstra 1983, 1993). Haflidason *et al.* (1991) and Sejrup *et al.* (1995) have described a till from the Norwegian shelf of probable Menapian age.

11.8 BAVELIAN

In his investigations of the Early Pleistocene sequence at Lieth, Menke (1975) identified two interglacials that were younger than Waalian and older than the Cromerian. New investigations have confirmed this finding. The newly defined unit has been called the Bavelian Complex by Zagwijn & De Jong (1984). The Leerdam Warm Substage of the Bavelian was originally placed within the Menapian Complex (as an interstadial; cf. Zagwijn 1975). The interglacial sequence of Bavel at that time was regarded as older than Menapian. The Bavelian comprises at least two warm and two cold substages:

Bavel Warm Substage
Linge Cold Substage
Leerdam Warm Substage
Dorst Cold Substage

The warm periods of the Bavelian Complex, also in the Netherlands, do show a clear vegetational succession that indicates a gradual spread of thermophilous species simultaneous with the rise of temperatures. In the Bavel Warm Substage five pollen zones have been distinguished so far:

Bv5: *Pinus–Picea* zone
Bv4: *Tsuga–Abies–Carpinus* zone
Bv3b: *Carpinus–Eucommia–Tsuga* zone
Bv3a: *Carpinus–Eucommia* zone
Bv2: *Alnus–Picea–Ulmus–Tilia* zone
Bv1: *Pinus–Alnus–Ulmus* zone

The Bavelian *sensu stricto* (i.e. the Bavel Warm Substage within the Bavelian Complex) coincides with the Jaramillo Event, a period of magnetic reversal within the Matuyama Epoch. The first magnetic reversal occurred within pollen zone Bv2, and the second after zone Bv4 (or at the beginning of the Linge Cold Stage). Thus the Bavel Warm Substage in contrast to most other stages can be precisely dated to the period 990,000–1,070,000 BP (according to Shackleton *et al.* 1990). However, climatic history may have been more complex than the pollen records indicate. The magnetically reversed Jaramillo Event spans oxygen-isotope stages 27 to 31. The Bavel Warm Substage is regarded as an equivalent of the Uetersen Warm Stage of Menke (1975).

The small-mammal fauna of the Bavel Warm Substage at the type locality contains

Mimomys savini and *Microtus arvalis*. For *Microtus* this is a very early occurrence. Recently remains of *Microtus (Allophaimys)* have also been found in the type deposits. This subspecies is missing in all deposits younger than the Matuyama/Brunhes magnetic reversal boundary. The transition from the *Allophaimys* fauna to the *Microtus* (*s.s.*) fauna seems to have occurred between the Bavelian and the early Cromerian (Van Kolfschoten, written communication).

During the Linge Cold Substage temperatures dropped approximately to the level of the Weichselian Cold Stage. The vegetation was largely restricted to herbs, as well as *Pinus, Betula* and *Juniperus*.

During the Leerdam Warm Substage again thermophilous vegetation spread out. Three pollen zones have been distinguished:

Ld3: *Pinus–Alnus–Betula* zone
Ld2b: *Pinus–Quercus–Carpinus–Ulmus* zone
Ld2a: *Pinus–Ulmus* zone
Ld1: *Betula zone*

In contrast to the Bavel Warm Substage, in the Leerdam Warm Substage neither *Carpinus* nor *Tsuga* is present in major quantities. *Pinus* dominates throughout. Zagwijn (1974) still regarded this sequence as an interstadial. New, better borehole information necessitated the modification of this concept. The final phase of this warm stage is especially comparable with the end of the preceding Bavel Warm Substage. The Leerdam Warm Substage is equated with the Pinneberg Warm Stage in Lieth (Menke 1975).

The Dorst Cold Substage, like the Linge Cold Substage, was a period of very low summer temperatures. In the pollen diagram this is reflected by a dominance of herb pollen (Zagwijn & De Jong 1984). Both the Bavel and Leerdam Warm Substages have recently also been identified on top of the Gorleben salt dome (H. Müller 1992).

Even with these new stages it has to be conceded that the Early Pleistocene stratigraphy of northwest Europe still shows considerable gaps. As the reversely magnetised Cromerian I must be correlated with Stage 21 of the deep-sea oxygen-isotope stratigraphy, between the Bavelian Complex and the Cromerian several additional cold and warm stages must have occurred that have not so far been identified.

11.9 CROMERIAN

The term 'Cromerian' for the following sequence of warm and cold events is not a very fortunate choice. It originally referred to the warm-stage deposits of the Cromer Forest Bed in East Anglia, which have been known since the 19th century (cf. West 1980a). When Zagwijn & Zonneveld (1956) encountered Middle Pleistocene warm-stage deposits in a clay lens within the Dutch Sterksel Formation at Westerhoven, they originally correlated them with the English Cromer site. However, later investigations have shown that the English Cromer Forest Bed Series comprises at least two complete interglacials (Pastonian and Cromerian, cf. West & Wilson 1966), neither of which could be safely correlated with the Westerhoven Interglacial. It now appears that the British Pastonian is an equivalent of the late Tiglian (Gibbard *et al.* 1991).

Further investigations in the Netherlands resulted in the detection of a second Cromerian Interglacial. This lies stratigraphically below the Westerhoven Interglacial. It is characterised by the occurrence of *Quercus, Ulmus* and *Carpinus* (up to 20%). The only Tertiary relics present are *Celtis* and *Parthenocissus*, as well as *Eucommia*, which is missing in the Westerhoven Interglacial (Cromerian II) and all younger interglacials. This oldest Cromerian interglacial of the Dutch stratigraphy is referred to as the Waardenburg Interglacial (Cromerian I). It also does not correlate with the English Cromerian. In contrast to the Cromerian II (Westerhoven) Interglacial its deposits are reversely magnetised. Thus the Matuyama/Brunhes reversal is situated between the Cromerian I and II interglacials, i.e. in Cromerian Glacial A (Zagwijn *et al.* 1971). This implies that the Cromerian I can be correlated with deep-sea oxygen-isotope stage 21, Glacial A with stage 20 and the Cromerian II either with stage 19 or 17 (see Figs 126 and 135).

In the course of further investigations in the Netherlands, two additional Cromerian interglacials were identified. Neither of these has been preserved completely at any site yet found, and the individual occurrences are difficult to correlate. Thus the Cromerian Complex at present includes four interglacial events:

Cromerian IV (Noordbergum)
Cromerian III (Rosmalen)
Cromerian II (Westerhoven)
Cromerian I (Waardenburg)

During the Cromerian IV a transgression occurred in parts of the Netherlands. In the terrestrial region in the Cromerian IV of the Netherlands the oldest occurrences of *Arvicola terrestris cantiana* have been recorded (Van Kolfschoten in Gibbard *et al.* 1991).

In various other places throughout Europe, Cromerian deposits have been found (for review see Turner 1996). The precise stratigraphic position of the individual sites, however, has remained rather unclear. The English Cromerian of **West Runton** in East Anglia, mentioned above, should be correlated either with the Dutch Cromerian II or with a so-far undetected interglacial between the Cromerian II and III. With regard to the mammal stratigraphy, the English Cromerian sites are also difficult to compare with those in the Netherlands. At West Runton, the stratotype of the Cromerian *sensu stricto*, and at several other British sites, including Little Oakley and Sugworth, *Mimomys savini* occurs, whilst other *Mimomys* species are missing. In the Netherlands there is no equivalent to this faunal assemblage. However, the English Cromerian *sensu stricto* does compare well with the Thuringian Voigtstedt Interglacial (see below; cf. Stuart & Lister in Gibbard *et al.* 1991).

The interglacial of **Osterholz** (near Elze, Niedersachsen, North Germany), described by Grüger (1968), should be older. In addition to high percentages of *Carpinus* it also contains up to 13% *Eucommia*. Thus a correlation with the Cromerian I seems most suitable (Zagwijn *et al.* 1971).

The interglacial deposits of **Bilshausen**, situated in North Germany between Göttingen and the Harz Mountains, were found in a dark bituminous clay, which is over- and underlain by loess. Palynological investigation by H. Müller (1965, 1992) has shown that it represents a fully developed warm-stage sequence sandwiched between two cold-stage deposits. Within the mixed-oak-forest zone, elm is strongly represented. The oak increase occurs rather late in the interglacial. *Picea, Pinus* and *Abies* play a major role. There is a well-pronounced *Carpinus* peak. *Eucommia* is missing. Varve counts have shown that the warm stage lasted about 27,000 years. It may be correlated with either the Cromerian II or IV Interglacial.

From Thuringia Erd (1970) described two Cromerian interglacials, an older **Artern Interglacial** and a younger **Voigtstedt Interglacial**. The vegetational development of the

Artern Interglacial resembles that of the Cromerian II of the Netherlands, but in contrast with the latter it is reversely magnetised (Erd 1978). The Voigtstedt Interglacial might possibly be an equivalent of the Bilshausen Interglacial. Its correlation with the Dutch Cromerian Interglacial Complex is still uncertain (Eissmann 1995).

In Denmark, the Harreskovian Interglacial (two occurrences at **Harreskov** and **Ølgod**) has been known for a long time (Jessen & Milthers 1928). S.Th. Andersen (1965) made clear that it represented a Cromerian interglacial. However, so far it is uncertain with which of the Dutch Cromerian events it might be correlated although it is generally equated with Cromerian II.

In Latvia the **Zidini Interglacial** is regarded as a warm stage of the late Cromerian Complex. It is correlated with the Belovezh Interglacial in Belarus. The deposits encountered in a borehole at 83.5–102.8 m below the surface contain a pollen flora that indicates a double climatic optimum. The first is characterised by *Ulmus, Tilia* and *Corylus* and the second by *Carpinus, Ulmus* and *Tilia*. The diatom flora of the deposit contains elements that were extinct by the late Middle Pleistocene (Dreimanis & Zelčs 1995).

The Cromerian Complex also includes the famous mammal site of **Tiraspol** of the lower Dniestr (Ukraine), which according to palaeomagnetic investigations must be placed around the Matuyama/Brunhes reversal. The fauna contains, amongst others, *Alces latifrons* and *Mimomys*, i.e. animals that became extinct after the Cromerian (Nilsson 1983).

The interglacials of the Cromerian Complex were separated from each other by severe cold substages. At least during the Cromerian A an extensive ice sheet seems to have developed in Scandinavia. A till has been reported from the central North Sea that was deposited at that stage (Sejrup *et al.* 1987) and, indeed, the Cromerian interglacial deposits at Harreskov and Ølgod overlie glacigenic deposits (Jessen & Milthers 1928). Drainage of the central German rivers during that period occurred via a far southern path through the Netherlands (Fig. 137). Whether deflection of the river courses by a Scandinavian ice sheet played a role in this cannot be said with certainty.

Meanwhile in southern **Britain** a series of fluvial aggradations comprising predominantly quartz-rich gravels of the Kesgrave Formation were deposited by the ancestral Thames (Hey 1991). These deposits range from the 'Pre-Pastonian a' Substage (Tiglian) to the early Anglian and contain ample evidence of cold-climate deposition. Finds of volcanic rocks, especially rhyolithic tuffs derived from North Wales (Whiteman 1983), within these gravels have been interpreted as indicating early glaciation in the Welsh Mountains, the so-called 'Berwyn Glaciation'. Since these erratics have been recovered from several members within the Kesgrave Formation, it has been thought that this glaciation may have been multiphased (Bowen *et al.* 1986). However, there is relatively little chronological control on these fluvial aggradations, and therefore any detailed interpretation must await further discoveries. Recent finds of organic temperate-stage sediments at Ardleigh, Broomfield, Wivenhoe and Little Oakley in Essex (Gibbard *et al.* unpublished; Bridgland 1994) indicate that the younger Kesgrave members span the late 'Cromerian Complex' (early Middle Pleistocene). They provide evidence for two or possibly three additional stages in the British sequence. The Kesgrave Formation can be traced into the offshore region, where it forms part of the fluvial–deltaic Yarmouth Roads Formation (Balson & Cameron 1985). The latter also includes material derived from the Rhine, Meuse and the North German river system (Zagwijn 1974, 1985).

In the central areas of the Nordic ice sheets most unconsolidated deposits older than the last glaciation have been removed. However, there are exceptions. Remnants of deep kaolin weathering have been observed at several places in Sweden (Lundqvist 1985). Also in parts of

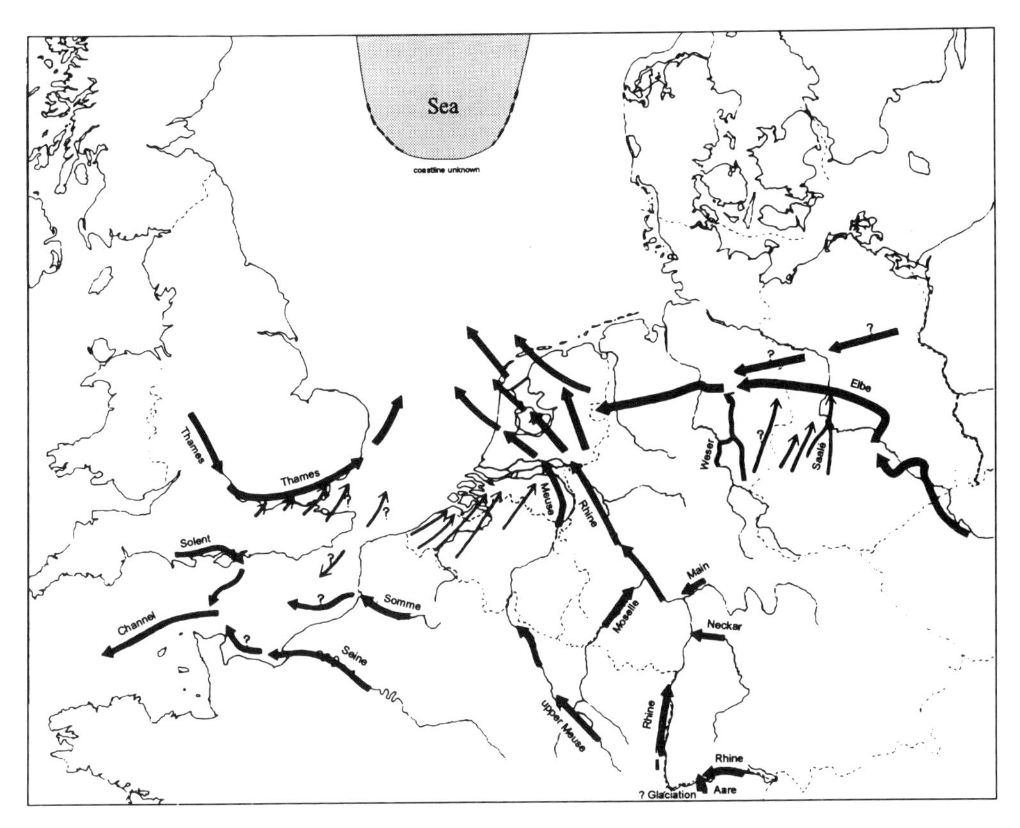

FIGURE 137 Schematic palaeogeographic reconstruction of the drainage system during a cold stage of the Cromerian Complex (early Middle Pleistocene) (after Gibbard 1988)

central Lapland (Finland), for instance, the metamorphic Precambrian rocks are covered by weathered regolith of Middle to Late Tertiary age (Hirvas 1991). Bedrock erosion thus seems to have been moderate (Nenonen 1995, Saarnisto & Salonen 1995).

11.10 ELSTERIAN COLD STAGE

The oldest glaciation that has been proved throughout the north European glaciated area is the Elsterian glaciation. Its deposits are especially well developed and well exposed in Saxony and adjoining areas and therefore Keilhack chose the term 'Elsterian' after one of the principal rivers in the region.

The advance of the Elsterian ice changed the regional drainage system. Rivers that hitherto had mostly drained towards the Baltic Sea were partly dammed by the advancing ice sheet and partly forced to flow west or east. For example, the Elbe River was dammed and formed a large lake south of Dresden. The Oder (Odra) River drained via the 'Moravian Gate' towards the Danube and Black Sea (Šibrava 1986).

The extent of Elsterian glaciation is not clear in all regions. Deep subglacially formed channels can be traced into the northern Netherlands. For example, near Noordbergum a channel over 300 m deep occurs (Ter Wee 1983a), but known occurrences of Elsterian till in

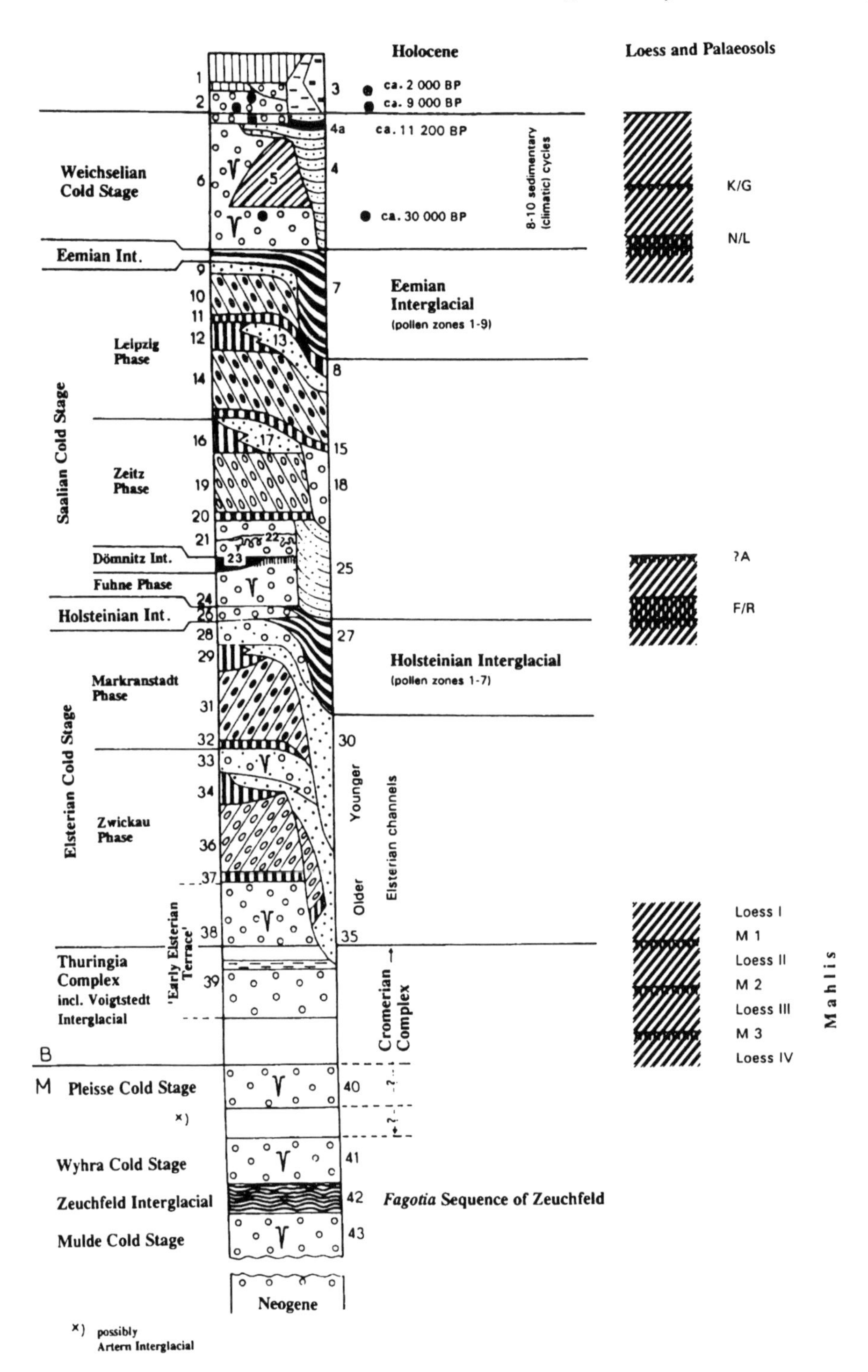

Holocene
Loess and Palaeosols
ca. 2 000 BP
ca. 9 000 BP
ca. 11 200 BP
Weichselian Cold Stage
ca. 30 000 BP
8-10 sedimentary (climatic) cycles
K/G
N/L
Eemian Int.
Eemian Interglacial
(pollen zones 1-9)
Leipzig Phase
Saalian Cold Stage
Zeitz Phase
Dömnitz Int.
Fuhne Phase
?A
F/R
Holsteinian Int.
Holsteinian Interglacial
(pollen zones 1-7)
Markranstadt Phase
Elsterian Cold Stage
Zwickau Phase
Younger
Older
Elsterian channels
'Early Elsterian Terrace'
Thuringia Complex incl. Voigtstedt Interglacial
Cromerian Complex
Loess I
M 1
Loess II
M 2
Loess III
M 3
Loess IV
Mahlis
B
M
Pleisse Cold Stage
Wyhra Cold Stage
Zeuchfeld Interglacial
Fagotia Sequence of Zeuchfeld
Mulde Cold Stage
Neogene
x) possibly Artern Interglacial

FIGURE 138 (opposite) Stratigraphy and facies of the Pleistocene in central Germany (Leipzig Lowland and surrounding area) (after Eissmann 1990).
Holocene:
1 = younger and older alluvial loam;
2 = younger and older fluvial sands and gravels with radiometrically dated fossil wood;
3 = silt, gyttja, lake marl, peat;
Weichselian Stage:
4 = slope wash, solifluction, lacustrine and fluvial sediments (sands, silts, marls and peat);
4a = volcanic tuff from the Laacher See eruption;
5 = loesses and loessic sediments;
6 = fluvial sands and gravels with radiometrically dated plant remains;
Eemian Stage:
7 = lacustrine silt, mud, clay, peat;
Saalian Stage:
8 = glaciolacustrine varved clay and silt;
9 = fluvial and glaciofluvial sand and gravel;
10 = Upper Saalian Till, upper unit;
11 = Upper Breitenfeld Varved Clay;
12 = Lower Breitenfeld Varved Clay;
13 = glaciofluvial sand and gravel;
14 = Upper Saalian Till, lower unit;
15 = Upper Bruckdorf Varved Clay;
16 = Lower Bruckdorf Varved Clay;
17 = glaciofluvial sand and gravel;
18 = fluvial and glaciofluvial Pomßen Gravel and Sand;
19 = Lower Saalian Till;
20 = Böhlen and Lochau Varved Clay;
21–24 = Main Terrace Complex gravels ('*Hauptterrassen-Komplex*');
  21 = upper gravel;
  22 = silt and fine sand, strongly cryoturbated Markkleeberg horizon;
  23 = mud, silt and fine sand;
  24 = lower gravel;
25 = slope wash, solifluction and lacustrine sediments (fine sand and silt);
Holsteinian Stage:
26 = fine-grained gravel and sand (*Corbicula* Gravel);
27 = lacustrine sediments (silt, fine sand, mud, peat, diatomite);
Elsterian Stage:
28 = fluvial and glaciofluvial gravel and sand;
29 = glaciolacustrine varved clay and silt;
30 = glaciofluvial sand and gravel (e.g. Krippehna Gravel);
31 = Upper Elsterian Till;
32 = Miltitz Varved Clay;
33 = fluvial and glaciofluvial gravel and sand (e.g. Möritzsch Gravel);
34 = Broesen Varved Clay and Silt;
35 = glaciofluvial sand and gravel;
36 = Lower Elsterian Till;
37 = Dehlitz–Leipzig Varved Clay;
38 = gravel and sand, interfingering with thick slopewash and solifluction sediments (fine sand and silt);
pre-Elsterian:
39 = fluvial gravel and sand with a silt horizon (Knautnaundorf Horizon);
40, 41 = fluvial gravels and sands (Lower and Middle Early Pleistocene terraces);
42 = lacustrine and fluvial silt and sand;
43 = fluvial gravel and sand (Upper Early Pleistocene terrace);
B/M = Brunhes/Matuyama boundary;
v = ice-wedge casts.
Loesses with important palaeosols: K/G = Kösen-Gleina Palaeosol; N/L = Naumburg-Lommatzsch Palaeosol Complex; A = Altenburg Palaeosol, F/R = Freyburg-Rittmitz Palaeosol; M1–3 = Mahlis Palaeosols.
(from Eissmann 1995)

the Netherlands are currently restricted to a few thin layers (Zandstra 1983). It must be remembered that during subsequent cold stages major parts of the Elsterian tills may have been eroded, although north of the Weser River continuous till sheets of this age occur (Höfle 1983b).

In the southern parts of Niedersachsen and in Nordrhein-Westfalen the former presence of Elsterian glaciation can in most cases only be inferred from redeposited Scandinavian erratics in contemporaneous terrace deposits. According to Thome (1980), however, there is evidence of an Elsterian ice advance into the Münsterland basin. Klostermann (1985, 1992) also regards an Elsterian ice advance into the Lower Rhine region as possible. However, the thrust sequences of the end moraines west of the Weser River contain no Elsterian till (Skupin *et al.* 1993).

In Saxony and Thuringia, where the Elsterian sediments were not overridden by the Saalian ice sheet, a more accurate reconstruction of the maximum extent of the Elsterian glaciation is possible. Here the maximum distribution of Nordic erratics, the so-called 'flint line', equals the maximum spread of the Elsterian ice sheet. It is found at an altitude of 300–480 m (Wagenbreth 1978). Towards the west it dips markedly. In the Wesergebirge Mountains of Niedersachsen (Lower Saxony) the flint line is found at an altitude of 200 m a.s.l. (Kaltwang 1992). There, however, it marks the upper limit of the Saalian ice sheet; the Elsterian limit being lower in this area.

In the Saxonian–Thuringian area, especially in the vast lignite opencast mines, two Elsterian tills have been found (Fig. 138). Both ice advances were apparently only separated by a short warming period, the Miltitz Interval (Eissmann 1975). During this interval the ice melted at least as far north as Mecklenburg, but no afforestation occurred.

The Elsterian tills of West Germany normally contain a relatively large proportion of western Scandinavian material. Rhomb Porphyries and other indicator pebbles from south Norwegian source areas are frequently found (K.-D. Meyer 1983b), together with flint conglomerate pebbles from the bottom of the Skagerrak. However, in none of the cases do Norwegian rocks predominate; at best they make up a few per cent of the (crystalline + sedimentary) indicator rocks. Rhomb Porphyries even occur in Elsterian tills around Leipzig. These occurrences are attributed by Eissmann to an early Norwegian–west Swedish ice stream, which may have extended as far as the Bornholm region in the east, but was later displaced by a north Swedish–Finnish ice stream. In eastern Germany and in Poland the 'Baltic' facies of Elsterian till represents the normal lithofacies. This means that the tills tend to be rich in Palaeozoic limestones and poor in flint (Eissmann *et al.* 1995).

The subsurface of the southern marginal areas of the north European glaciation is dissected by a network-like system of deep incisions, the direction of which trends radially away from the centre of the glaciation towards the ice margin. The incisions occur in two types. The deeply-incised **channels** (Fig. 139) have been discussed in chapter 4.2.2. The shallower, wider channels of the second type, are referred to as **troughs** ('wannen'), may have been formed during the first advance of the Elsterian ice. In the river valleys and under lakes the advancing ice would have met thinner permafrost and would have been able to cut deeply into the subsurface (cf. Maarleveld 1981), and in this way could have formed these shallow features.

The largest known 'wanne', the 'Elbtal-Wanne' (Elbe Valley Trough) between Coswig and Riesa (Saxony), is over 100 km long and 20 km wide. The age of the 'wanne' can be demonstrated because undisturbed Early Elsterian terrace gravels rest on an erosion surface beneath the basin infill. At the base of the 'wanne', till has been found. This indicates that the

FIGURE 139 Relief of the Quaternary base in the Cottbus region, East Germany; contours in metres above and below sea level (from Kupetz *et al.* 1989)

basin was formed by glacial erosion ('exaration') and only modified by meltwater action. The till, which is mostly characterised by a strong uptake of local material, is in turn overlain by the deposits of an ice-dammed lake, termed 'Prettiner Folge', which reach a maximum thickness of 160 m. Recent investigations have shown that the 'Elbtal-Wanne' formed in several phases mainly during the Elsterian II ice advance. Early infill of the basin occurred over dead ice. Several till bodies within the fill have been identified as relics of Elsterian II age. It was not until the Elsterian lateglacial that the dead ice finally melted completely. By the beginning of the Holsteinian Interglacial the 'wanne' was already almost completely filled, so that continuous sediment transport through the depression was possible (A. Müller 1973, 1988, Eissmann *et al.* 1995).

The deeper Elsterian channels in Niedersachsen end approximately at a line from Diepholz via Nienburg to Celle. Whilst the shallower channels as a rule are not infilled by Late Elsterian glaciolacustrine and marine Holsteinian sediments, both are often represented in the deep channels. This implies that the deep channels were either formed later (Eissmann & Müller 1979), or that they have remained active as meltwater drainage paths at least until the end of the glaciation.

When the Elsterian ice melted in the unfilled portions of the channels, large ice-dammed lakes formed in which silt and clay were deposited. In North Germany these glaciolacustrine deposits are called **'Lauenburger Ton'** (Lauenburg Clay). The distribution of this facies extends westwards into the Netherlands, where this sediment is referred to as 'potklei' (pottery clay) or, more properly, the Peelo Formation. It is found in the provinces of Friesland, Groningen and Drente. It consists of a complex of glaciolacustrine clays, silts and fine sands; in the channels it can reach a thickness of over 150 m. The composition of the Lauenburg Clay reflects the progressive decay of the Elsterian ice. Whereas the older layers are rich in dropstones and mixed with sand, the sorting increases towards the top where a varve-like sequence of laminated sediments is found. If they represent annual layering, deposition in the Hamburg region seems to have lasted for over 2000 years. The upper part of the Lauenburg Clay is often reddish in colour, a feature Hinsch (1993) assumes results from redeposition of the fine-grained fractions of a Late Elsterian East Baltic till facies.

In **Denmark** Elsterian tills are known from exposures near Esbjerg (no longer accessible), as well as from the Røgle cliff (Fyn) (Sjørring 1983). The Trelde Næs cliff section, which presumably includes Elsterian till overlain by Holsteinian deposits (Fig. 140) has been drawn by Houmark-Nielsen (1987). The results of drillings indeed indicate that a bipartite Elsterian till exists in Denmark. An earlier ice advance from the north, characterised by Rhomb Porphyries from the Oslo area, seems to have been followed by a Baltic ice advance, which transported greater amounts of flint and Palaeozoic limestones (Sjørring 1983).

In **Britain**, as on the continent, the Anglian (Elsterian) tills are the first widespread unequivocal evidence of large-scale Pleistocene glaciation. Arrival of the Anglian Ice Sheet caused burial of the Kesgrave Formation in all but the southernmost part of East Anglia. The Anglian Ice Sheet advanced into and overrode the contemporary valley of the River Thames in Hertfordshire and also advanced into its southern tributary valleys (Gibbard 1977, 1979). Throughout the region subglacial tunnel valleys were formed, radiating broadly parallel with the ice movement (Woodland 1970).

In northeastern Norfolk ice from both Scandinavia and the British glaciation centres interacted during the Anglian Stage. The ice sheet terminated in a large ice-dammed lake (Lunkka 1994). Although up to five different Anglian tills can be distinguished in parts of East Anglia, no evidence of a major intervening ice-free period has so far been discovered (cf. Hart & Boulton 1991, Ehlers *et al.* 1991). The previously identified Corton event is no longer interpreted as an interstadial, since it contains no evidence of marked climatic amelioration (West & Wilson 1968) but appears to represent a change in ice dynamics manifest as local ice-front retreat. Recent studies indicate deposition of these deltaic sands as a product of glaciofluvial activity (Pointon 1978, Hopson & Bridge 1987, Ehlers & Gibbard 1991).

It has been generally assumed that the end of the Anglian/Elsterian saw a rather abrupt climatic change to the temperate conditions of the succeeding Hoxnian/Holsteinian Stage.

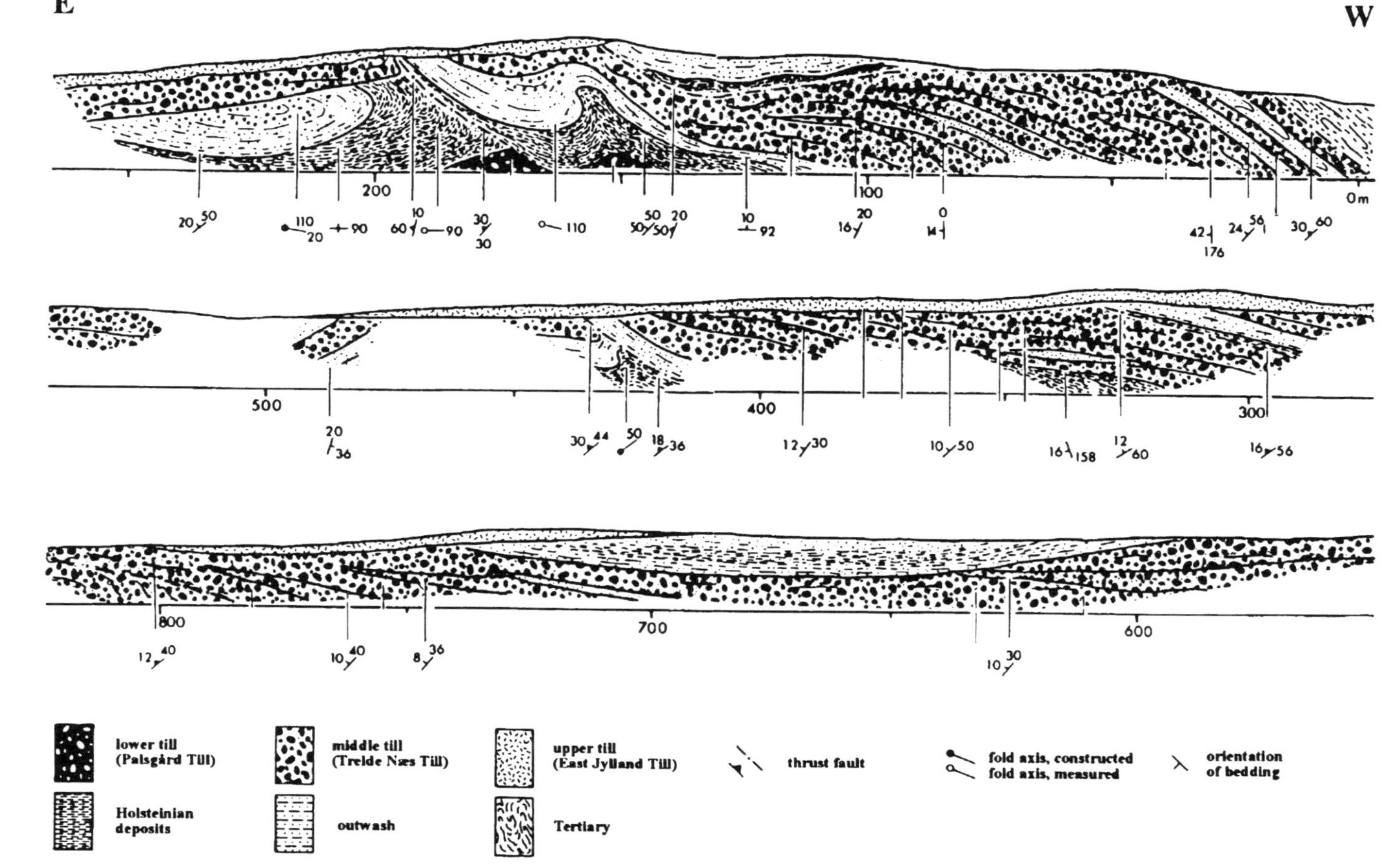

FIGURE 140 Cliff section at Trelde Næs north of Fredericia, Jutland. Elsterian till overlain by Holsteinian Interglacial lacustrine deposits (after Houmark-Nielsen 1987)

However, recent investigations in the Schöningen opencast mine in Niedersachsen, have suggested three Late Elsterian interstadials (Offleben I, II and Esbeck Interstadials) in which *Betula, Pinus* and *Picea* forests were established (Urban *et al.* 1988).

11.11 HOLSTEINIAN INTERGLACIAL

During the Holsteinian Warm Stage northwest Europe experienced a major marine transgression. For North Germany it was the first marine transgression since the Miocene. The extent of the Holsteinian Sea was not reached again in either the Eemian or the Holocene. Gottsche (1898), who was the first to examine the warm-stage deposits of North Germany, did not give them names but referred to the strata of the two warm stages as 'Interglacials I and II' (Gottsche 1898). After the term Eemian for the Interglacial II had been introduced and soon largely accepted (see below), Penck (1922a) introduced the term 'Holstein Interglacial'. However, he did not regard the sediments in question as equivalent to Gottsche's Interglacial I but thought them to represent a transgression that should be placed intermediate between the Eemian and the Holocene. This mistake was corrected by Grahle (1936).

On the basis of their contained faunas, the marine Holsteinian deposits can be clearly differentiated from Eemian marine strata. Unlike during the Eemian when the transgression occurred only after an early part of the interglacial, the sea had already invaded parts of Jutland and North Germany by the end of the Elsterian. In the Eggstedt borehole, Schleswig-Holstein, marine fauna has been encountered at a depth of −133.3 m under 48 m of Lauenburg Clay (Hinsch 1993, Knudsen 1993). In addition, the upper parts of the Lauenburg Clay sequence contains an arctic–boreal marine fauna in places. There can therefore be no doubt about the early onset of the transgression. Marine Late Elsterian deposits in Schleswig-Holstein can be traced as far inland as Kellinghusen, about 50 km from the present North Sea coast. The early transgression may have been a result of isostatic depression of the land surface. However, it must also be borne in mind that the base of the marine Elsterian sediments in Eggstedt lie at 83 m below present sea level; in Hamburg, where the marine Holsteinian strata overlie early Holsteinian warm stage lacustrine deposits, their base is found at 50 m below present sea level. Here the transgression did not occur before the *Abies* rise, i.e. in the second half of the interglacial.

Apart from glacio-isostatic adjustment, long-term crustal movements have occurred during the course of the Pleistocene. Whilst in North Germany the transgression maximum of the Holsteinian Interglacial (uncorrected for later compaction) is situated at −13.5 m, the same level in East Anglia in the Nar Valley is located at +23 m (Ventris 1985), so that a height difference of 36.5 m remains to be explained.

The transgression of the Holsteinian North Sea was at first restricted to the Elsterian channels, which were only incompletely filled with sediment at that time. Here cold-tolerant species may have persevered far into the interglacial. Although marine Holsteinian deposits occur in Essex (Clacton; Pike & Godwin 1953) and France and Belgium (Paepe & Baeteman 1979, Sommé 1979), there was no marine connection through the English Channel to the central North Sea, so that thermophilous (Lusitanian) faunal elements were forced to immigrate the long way around Scotland. Therefore the mollusc fauna of the marine Holsteinian in North Germany is characterised by boreal species even at the climatic optimum. Only with the occurrence of the *Ostrea* facies, which is widespread in Schleswig-Holstein, were major areas flooded by the Holsteinian Sea (Hinsch 1993) (Fig. 141).

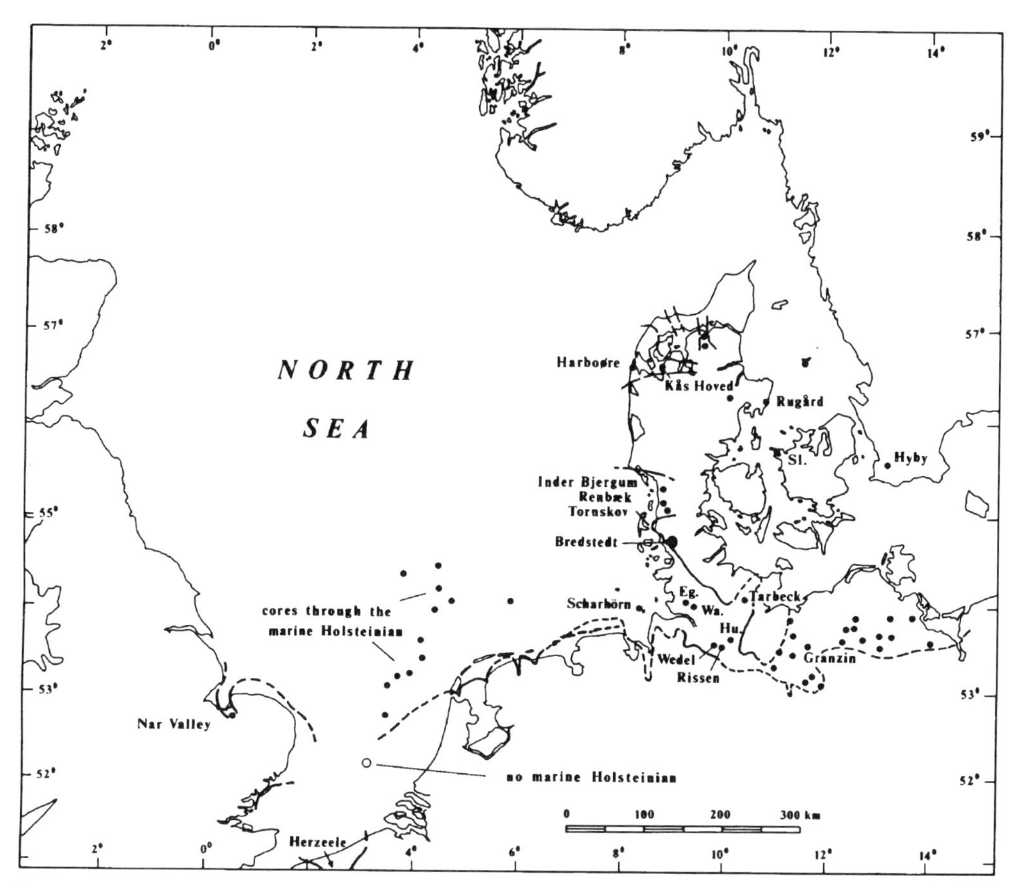

FIGURE 141 Extent of the Holsteinian Sea and important terrestrial Holsteinian sites; Eg. = Eggstedterhulz, Wa. = Wacken, Hu. = Hummelsbüttel (after Ehlers 1988; supplemented after U. Müller *et al.* 1995 and written communication by Knudsen)

Beyond the reaches of the Holsteinian transgression, the interglacial is represented by fossil soils, lacustrine and organic strata. In particular, the diatomite of the northern Lüneburger Heide area has provided valuable information about the course of the interglacial. K.-J. Meyer (1974) proved that the fine rhythmic lamination of the diatomite represents annual layers. The annually layered part of the sequence begins in pollen zone III (birch–pine period). It was possible by extrapolation to determine that the overall duration of the interglacial was about 15,000–16,000 years (H. Müller 1974b).

The vegetational development of the Holsteinian has been described by numerous authors (e.g. Hallik 1960, S.Th. Andersen 1965, Erd 1970, H. Müller 1974b, Linke & Hallik 1993) (Fig. 142). The interglacial starts with the expansion of plants of the subarctic to cool-temperate climatic zone. Pollen of this zone I includes much reworked Tertiary material. In contrast to the Eemian, during the Holsteinian the thermophilous trees in North Germany appeared almost simultaneously. The North German Holsteinian is characterised by the early appearance of *Abies* (fir) and the presence of *Celtis* and *Azolla*. Towards the end of the interglacial as the oak and elm decline, *Pterocarya* (wing nut) is regularly encountered. The occurrence of this tree is

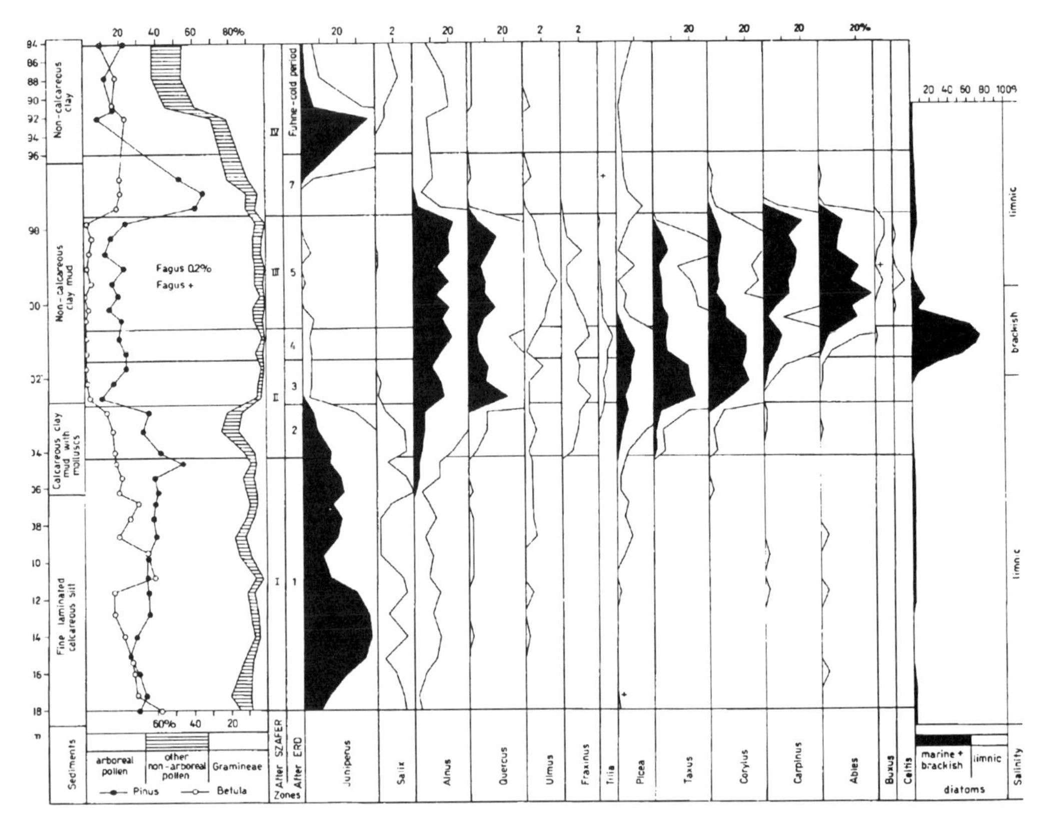

FIGURE 142 Pollen diagram from the Holsteinian Interglacial from Pritzwalk, East Germany (after Erd 1970)

also recorded in the Alpine region (see chapter 12.11) and in the Hoxnian deposits of Britain, which are the equivalent of the continental Holsteinian (West 1956, 1980b, Turner 1970).

Although palynological investigations are not always well-suited to dating sediments, as pointed out by Bowen (1991), many temperate events show a very characteristic vegetational sequence, which allows them to be correlated. In **Britain** this applies to the classic Hoxnian sequences at Hoxne, Marks Tey and elsewhere, which can be clearly distinguished from the classic Ipswichian deposits at Bobbitshole (Ipswich), Trafalgar Square etc. (West 1980b). In contrast to these there are other Middle or Upper Pleistocene sites that do not fit so readily into the scheme and that may either represent additional stages (like the continental Wacken/Dömnitzian) or are atypical equivalents of the Hoxnian and Ipswichian Stages.

In **Ireland** the situation is more complicated. Middle Pleistocene interglacial deposits there have been identified at Boleyneendorrish northeast of Gort (County Galway) and at several other sites (Mitchell 1948). Originally, this Gortian Interglacial (Watts 1964) had been correlated with the Hoxnian/Holsteinian. In this case, the Ipswichian/Eemian Interglacial would be missing in Ireland. This still is a widely accepted concept. However, it has been challenged by Warren (1979, 1985) and Synge (1980, 1981) who regard the Gortian as equivalent to the Ipswichian/Eemian Stage. Middle Pleistocene interglacial deposits seem to exist in Ireland (e.g. Coxon & Flegg 1985), but the exact stratigraphy is still under discussion.

Further east, in **Poland**, traditionally the Mazovian Interglacial is correlated with the Holsteinian. However, the Ferdynandovian Interglacial seems also to represent the same pre-Saalian, post-Elsterian period. Like the Holsteinian, it is characterised by the occurrence of *Pterocarya* and *Azolla filiculoides*. There are two climatic optima. The lower one is characterised by *Quercus, Ulmus, Abies, Tilia* and *Corylus*, the upper one by *Carpinus, Alnus, Corylus* and *Quercus*. They are separated by a cool interval with taiga-like vegetation, represented by *Pinus, Picea* and *Betula*. The Ferdynandovian was regarded as a more continental representation of the Mazovian (Janczyk-Kopikowa 1975). However, lately the Ferdynandovian has been thought to predate the latter and to be separated from it by a period of widespread glaciation. Whether this is correct or not remains unclear, but Lindner (1992), for example, correlates the Ferdynandovian with the Cromerian Voigtstedt Interglacial.

In **Russia** the equivalent of the Holsteinian is the Likhvin Interglacial (Sudakova & Faustova 1995), or the so-called 'Great Interglacial' of the older literature. It contains a characteristic mammal assemblage, the Singil Fauna, which includes *Arvicola mosbachensis, Lagurus transiens-lagurus* and voles of the genus *Microtus*. The stratigraphical details are still unclear. The Likhvin Interglacial *sensu lato* seems to comprise a complex of several warm and cold stages. According to some workers, the Likhvin *sensu stricto* may have had a double climatic optimum. It is overlain by periglacial deposits (loess) (Velichko & Faustova 1986).

In the **Baltic States** the Butėnai Interglacial of Lithuania, the Pulvernieki Interglacial of Latvia and the Karuküla Interglacial of Estonia are equivalents of the Holsteinian. Marine deposits of this stage have been found along the Baltic Sea coast of Latvia (Dreimanis & Zelčs 1995). A thorough modern palynological investigation of the interglacial deposits in the Baltic States has recently been undertaken by Liivrand (1991).

In **Finland** the peat bed found at Naakenavaara is thought to be equivalent to the Holsteinian Interglacial. However, it might also be older. The sequence includes wood of *Larix* (larch) and consists almost entirely of remains of *Aracites johnstrupii*, a now extinct plant, the remains of which have not been found elsewhere in Finland. The most recent occurrences of this plant are known from Belarus in Likhvin/Holsteinian age deposits. The peat at Naakenavaara is separated from the Eemian deposits by a single till bed, which may suggest that Finland was ice-covered throughout the Saalian Stage (Hirvas 1995).

In **Sweden** two sites in central and southern Sweden are correlated with the Holsteinian Interglacial. At Öje, a complex sequence of till-covered, silty, partly laminated sediments with organic material has been studied. Pollen and macrofossil analyses show that the vegetation was dominated by coniferous trees, including pine, spruce and larch. Because of the presence of exotic conifers not growing in Scandinavia today, the Öje sediments have been correlated with the Holsteinian (Robertsson & García Ambrosiani 1992). Pollen analysis of lacustrine silty sediments from a borehole at Hyby in the Alnarp valley, southernmost Sweden, show low pollen frequencies of *Corylus* and mixed-oak-forest, together with *Picea, Abies, Taxus, Frangula* and *Azolla* massulae, which suggests correlation with the Holsteinian Interglacial (U. Miller 1977).

In **Denmark** numerous sites with marine Holsteinian deposits have been identified, including Tornskov and Inder Bjergum in southern Jutland. A recent overview is provided by Knudsen (1994). Limnic Holsteinian deposits have been investigated at Vejlby in Jutland (S.Th. Andersen 1965). They correlate closely with the neighbouring North German sites.

Holsteinian mammal faunas are relatively rare in Central Europe. Mammalian remains have been described from Neede (Netherlands) by Van Kolfschoten (1988, 1990). Apart from horses and forest rhinoceros and red deer (*Cervus elaphus*), mainly rodents such as *Clethrionomys* and *Arvicola terrestris cantiana* have been found. *Sorex* and *Pliomys* are absent. The faunal assemblage indicates an open forest landscape – a picture that is supported by the palynological investigations (Van der Vlerk & Florschütz 1953).

The Holsteinian Interglacial also includes abundant evidence of human activities. The Bilzingsleben site is possibly of this age (cf. chapter 11.13) and in Steinheim an der Murr a skull fragment of early man was found in 1953. The site was correlated with the Holsteinian on the basis of its elephant fauna (Adam 1954). The warm climate is suggested by finds of water buffalo, forest elephant, aurochs, giant deer and forest rhinoceros (Czarnetzki 1983).

11.12 FUHNE COLD STAGE

At the end of the Holsteinian the climate grew considerably colder and permafrost developed in Central Europe. This period is referred to as the Fuhne Cold Stage, a term coined by Knoth (1964). During this period the profiles at Wacken, Dömnitz and Hoogeveen indicate that fine sands several metres thick were deposited. In Wacken the upper part of the sequence is characterised by a cryoturbated 'drop soil', indicative of cold-climatic influence. In the Netherlands cryoturbations, large ice-wedge casts and stone layers were found. The pollen diagrams indicate that deforestation occurred in places including Vejlby (Denmark; S.Th. Andersen 1965), Pritzwalk (Brandenburg) and Hoogeveen (Netherlands) at the end of the interglacial. Only a sparse vegetation cover remained, consisting largely of *Pinus, Betula, Salix, Artemisia, Juniperus* and partly Ericales (Erd 1970, Zagwijn 1973).

Indications have been found that the Fuhne Cold Stage may include minor climatic oscillations. For instance, Erd *et al.* (1987) found that the Holsteinian deposits at Rossendorf near Dresden are overlain by cold-stage sediments characterised by an increase in heliophytes, *Betula* and *Salix*, which were overlain by an interstadial deposit showing renewed immigration of birch and pine. After this interstadial subarctic conditions returned again. The interstadial may possibly be correlated with the Missaue Interstadial identified in the Schöningen brown-coal mine near Helmstedt (Niedersachsen) (Urban *et al.* 1988).

11.13 WACKEN INTERGLACIAL

The first hints of an additional interglacial following the Holsteinian were discovered almost simultaneously by Erd and Menke. Erd reported the principal results of his investigations at Pritzwalk on a DEUQUA meeting in Lüneburg 1964, whilst Menke presented the first complete pollen diagram of the new warm event at the annual meeting of the 'Arbeitsgemeinschaft Nordwestdeutscher Geologen' in Nienburg of the same year (Fig. 143). He found a layer of interglacial deposits about 1 m thick, separated by cryoturbated sands from the underlying 34 m of the Holsteinian sequence at Wacken, Schleswig-Holstein (cf. Menke 1968).

Evaluation of the palynological record provided the following picture. The climatic deterioration at the end of the Holsteinian had at least resulted in elimination of all thermophilous trees and also *Picea*. The Wacken Interglacial therefore began with a phase in which *Betula* and *Pinus* dominated, represented in the Wacken sequence by a swamp peat. The following development saw an immigration of thermophilous trees similar to the Holsteinian. However, there were a few differences. *Carpinus* immigrated relatively early. *Picea* appeared earlier than *Carpinus* and remained poorly represented throughout the interglacial. *Abies* was

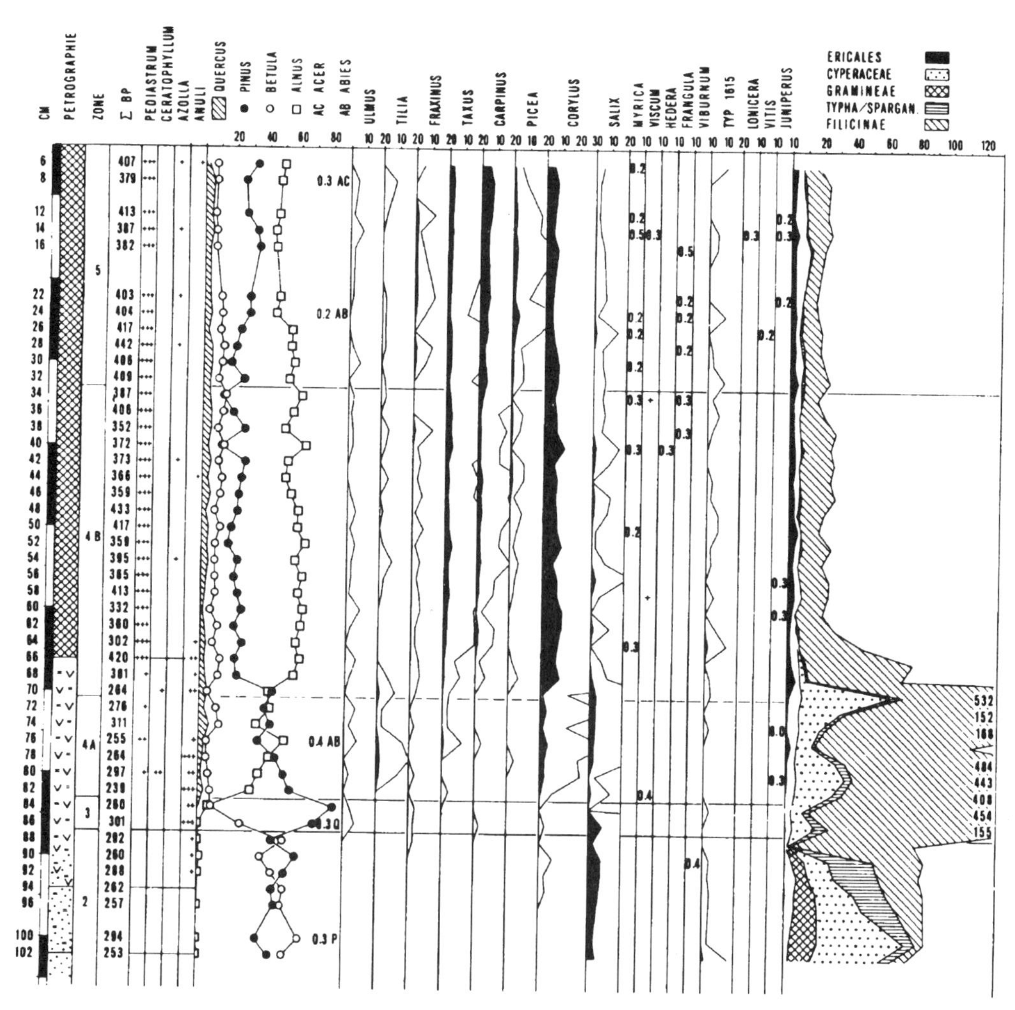

FIGURE 143 Pollen diagram from the Wacken Interglacial from Wacken, Schleswig-Holstein (from Menke 1980b)

not present. The new interglacial thus differed markedly from the Holsteinian and could also not be mistaken for the late climatic oscillations of the Holsteinian that had been found by H. Müller (1974b) in the diatomites of the Lüneburger Heide area, because the latter did not include any complete expulsion of the thermophilous flora from North Germany (Menke 1980b).

The vegetational history at Dömnitz closely resembles that of the Wacken site. Erd (1970, 1973) demonstrated that the event that he termed 'Dömnitz Interglacial' was not simply a warmer interstadial but a fully-developed interglacial, and that its vegetational history resembled neither that of the Holsteinian nor the Eemian. The Wacken Interglacial was found to be the last warm period in which the water fern, *Azolla filiculoides*, occurs (Erd 1978).

Almost at the same time Zagwijn (1973) found deposits indicating a temperate climate at the construction site of two new locks at Hoogeveen and Peelo in the Netherlands. These sediments rested on Holsteinian deposits and were covered by Saalian till. At this site the sequence indicated reforestation followed a period of forest retreat after the Holsteinian, leading to a mixed forest with *Pinus, Alnus, Carpinus* and *Picea*, in which *Pinus* (probably because of the short period of time available) had remained the dominant tree species. Zagwijn (1973) concedes that the deposits might be called interglacial, but he referred to them only as the 'Hoogeveen Interstadial'. The correlation of this deposit is still open. Zagwijn (1973) regards it as a Saalian interstadial, but, notwithstanding its weaker development, does not exclude a correlation with the Wacken/Dömnitz Warm Stage.

The introduction of a new interglacial was initially met with widespread scepticism. No marine transgression was known from the Wacken Warm Stage. The Wacken site was ice-pushed and the Pritzwalk sequence was only known from boreholes. Until that time Holsteinian and Eemian interglacial deposits had been found in many places, but never before had deposits of this new interglacial been identified. Criticism also focused on the point that the marine clays at Wacken only represented the early phase of the Holsteinian Interglacial (pollen zones I–III). The clays grade upwards into marine sands, which in turn are overlain by fossil-free 'flaser sands'. Consequently, Woldstedt & Duphorn (1974) initially interpreted the Wacken Warm Stage as nothing but the upper part of the Holsteinian sequence that was missing in the marine clays. The evidence against such an interpretation, however, is the fact that the regression of the Holsteinian Sea in Schleswig-Holstein did not occur before pollen zones X–XII (Menke 1980b), so that the vegetational development of the Wacken Warm Stage cannot be regarded as a continuation of that represented by the marine Early Holsteinian deposits. The Wacken Warm Stage deposits are ice-pushed, but they form part of a major raft within which they lie concordantly and undisturbed on the Holsteinian and Fuhne deposits. Palaeogeographical investigations in the surrounding area have shown that the likely source area for the raft was the 'Hadenfelder Rinne', an Elsterian channel east of the site (Menke, written communication 1992).

This concept has now become generally accepted. New sites of Wacken/Dömnitz Warm Stage deposits have been identified, including Maastricht Belvédère in the Netherlands (Van Kolfschoten 1988). At the meeting of the Subcommission for European Quaternary Stratigraphy in Hamburg 1986 on the 'Type Regions of the Holsteinian Interglacial', it was agreed that the Holsteinian Interglacial ends with the transition from boreal to subarctic climate both in Munster Breloh and at Pritzwalk (E1 borehole) (Jerz & Linke 1987). Consequently the Wacken Interglacial forms no part of the Holsteinian. At the moment it is included – together with the Fuhne Cold Stage and the Saalian Cold Stage *sensu stricto* – within the **Saalian Complex** (Litt & Turner 1993).

In the meantime another, more complete occurrence of this interglacial has been identified at the opencast brown-coal mine at Schöningen, Niedersachsen (Urban *et al.* 1991). Nevertheless, it cannot be denied that deposits of this interglacial are rather rare. The reason for this may be that at the onset of the Wacken event most of the kettle holes (dead-ice depressions) from the Elsterian Cold Stage that might have served as sediment traps had been completely filled.

In **Poland**, the Zbójno Interglacial was distinguished, based on palynological investigations of peat at Zbójno in the Holy Cross Mountains (Lindner & Brykczynska 1980). They are thought to be equivalent to the interglacial deposits at Schöningen. Some other interglacial sites in Poland have been reinterpreted lately and are now also thought to belong to the Zbójno Interglacial (Marks, personal communication).

In **Britain**, the Hoxnian (Holsteinian) deposits in East Anglia are overlain by inwashed fine sediments that filled the depositional basins. They are interpreted as indicating the establishment of periglacial conditions. At Hoxne, however, Wymer (1983) has identified interstadial deposits overlying the interglacial lacustrine and the fluvio-periglacial beds in his excavations. It would seem possible that this could be the equivalent of the continental Wacken/Hoogeveen/Dömnitz event (Gibbard & Turner 1988).

Human settlement during the Wacken Interglacial should have been similar to that during the Holsteinian. Between Holsteinian and Wacken Warm Stage no full-scale cold stage occurred during which glacigenic deposits were deposited, therefore clear stratigraphic distinction between both interglacials is not always possible. According to Mania (written communication), the Bilzingsleben site (Thüringen) which has been correlated with the Wacken Interglacial could also be of Holsteinian age. An ESR date from this site has yielded an age of 320,000–412,000 years BP (Schwarcz *et al.* 1988). In Bilzingsleben a human settlement was found beside a shallow lake near a karstic spring. The finds were embedded in travertine (carbonate precipitate). Apart from numerous flint artefacts, bone, horn and ivory tools were also recovered, as well as remains from meals (bones of prey animals). Typical *Homo erectus* skull fragments and teeth were also found (Mania 1989a).

Furthermore, the foundations of three houses were marked by large animal bones and stones arranged in ovals or circles 3–4 m in diameter. In front of the entrance, which opened to the southwest, a fireplace was found in each case. The houses probably consisted of structures of wooden poles covered with hides or fur, all of which had disappeared. Next to the living quarters workshops were also found. Apart from coarse, heavy gravel tools of quartzite, Muschelkalk limestone and crystalline rocks, numerous small flint tools were also in use. The latter were worked on one or both sides and retouched along the edges. Traces of wear indicate that the tools had been used for peeling or carving wood, and probably also for cutting and drilling (Mania 1989a).

The tools enabled the Early Palaeolithic hunters to go after large game. Amongst the Bilzingsleben finds forest elephant, forest and steppe rhinoceros, wild ox, wild horse and bear amounted to 60% of the prey. Medium-sized game (deer, roe and boar) form 17%, small game (mostly beaver) 20% of the prey. Additionally, remains of large fish were found (tench and catfish). These humans were hunter-gatherers. Consequently, the travertine of Bilzingsleben includes egg shells, mollusc shells, cherry stones, numerous fruits, nuts, seeds, and leaves of edible plants. In the surroundings of the site at that time a light *Buxus–Quercus* forest grew with numerous openings and shrubs. The depressions were covered by extensive reed areas overgrown by willow thickets and riverside forests (Mania 1989a).

11.14 SAALIAN COLD STAGE

The Saalian Cold Stage in Germany is traditionally subdivided into two major ice advances, the Drenthe and the Warthe advance. The term **Warthe** dates back to early works of Woldstedt (1927a, b). The **Drenthe** was introduced much later by Woldstedt (1958), following a suggestion by Van der Vlerk & Florschütz (1950).

The Saalian Cold Stage, like the Weichselian, started in North Germany, Denmark and the Netherlands with a prolonged non-glacial cold phase. This was interrupted by several interstadials. The deposits of that period identified so far are not sufficient to draw a clear picture of the climatic development. In the Netherlands the Saalian *sensu stricto* started with a period of considerable cooling. A treeless tundra prevailed, but frost intensity was slight. A weak amelioration subsequently led to the Bantega Interstadial (Zagwijn 1973), during which a birch–pine forest spread, with minor percentages of *Alnus, Corylus* and *Quercus*. Only after this and possibly other, weaker interstadials did the continental ice sheets finally invade North Germany and the Netherlands.

In **Denmark**, two interstadials are known that have been placed in the early Saalian, named Vejlby Interstadials I and II (S.Th. Andersen 1965). Because no Wacken Interglacial stratum is present at Vejlby, it is quite possible that the interstadials belong within the Fuhne Cold Stage and not the Saalian *sensu stricto*. Vegetational development in both cases was limited to a birch–pine forest. On the basis of the thickness of the deposits and the arboreal pollen frequency in the total pollen spectrum, it seems that the older (Vejlby I) interstadial was warmer than the younger event. A comparable event, the Dockenhuden Interstadial, has been described from North Germany (Linke & Hallik 1993), but in this case it might also be part of the Fuhne Cold Stage.

In **Britain**, the equivalent of the continental Saalian is the Wolstonian glaciation. The name is adopted from a village on the River Avon in northern Warwickshire, from which Shotton (1953) first described the deposits which were later designated as the stratotype (Mitchell *et al.* 1973). They comprise glacigenic deposits in the area around Coventry, Rugby and Leamington which have been correlated with those overlying Hoxnian temperate deposits in at least two places, Nechells and Quinton in Birmingham (Shotton 1983, 1986, Horton 1989). These glacigenic deposits must be pre-Devensian, because they are outside the limit of the last glaciation. In contrast to most areas in western continental Europe, in Britain the Wolstonian glaciation seems to have had a more limited extent than the preceding Anglian and the subsequent Devensian glaciations except in the English Midlands and eastern England. Even the existence of the Wolstonian Stage has been hotly debated for several decades. This dispute however, must be regarded as settled (see discussion in Ehlers & Gibbard 1991, Rice & Douglas 1991). Rose (1987, 1988) regards the Wolstonian of the Midlands as not sufficiently well established and instead has proposed the definition of a new stratotype in East Yorkshire. Meanwhile, Gibbard & Turner (1990) have proposed boundary stratotypes for the stage in East Anglia. The exact extent of the Wolstonian ice sheet is still not known. There is no question that the Welsh glacier reached the West Midlands (Maddy *et al.* 1991). Some argue that east coast ice reached as far south as Norfolk (Straw 1991), whilst others think that its outermost limit was further north, somewhere in Lincolnshire (cf. Catt 1991b). However, ice-marginal deposits of the Wolstonian have recently been detected in northwest Norfolk, south of the Wash (Gibbard *et al.* 1992).

Nevertheless, a striking discrepancy remains between the situation in the Netherlands with the extensive Saalian deposits and where the Elsterian is virtually missing (e.g. Ter Wee 1983a, b) and in Britain, where the Wolstonian glaciation is of minor importance but the Anglian ice reached as far south as the Thames valley. This anomaly stimulated Cox and others (Bristow & Cox 1973, Cox 1981) to argue that the supposed Anglian tills of areas like Norfolk might really be of Wolstonian age. However, in many places, such as at Hoxne and Marks Tey, Hoxnian Stage deposits fill kettle-hole depressions directly overlying Anglian glacial deposits, so that the stratigraphical correlation of the latter with the continental Elsterian is beyond doubt (cf. West & Gibbard 1995).

Because Wolstonian glacial deposits are seldom exposed, very little is known about the intra-Wolstonian stratigraphy. In Yorkshire, there seems to be just one Wolstonian till, the so-called 'Basement Till' in the coastal cliff sections of Holderness (Catt & Penny 1966, Catt

1991b). The age of this till has been questioned recently. N. Eyles *et al.* (1994) regard it as the deposit of a surging Late Devensian ice lobe in the western part of the North Sea basin.

The extent of the Saalian ice sheet in the North Sea is still an open question. In the Dutch North Sea sector, Saalian till can only be traced about 40 km seaward of the coast (Joon *et al.* 1990). Saalian till is missing in the central and northern North Sea. This seems to indicate that there was no contact between British and continental ice sheets at the time (Sejrup *et al.* 1987, Long *et al.* 1988). However, on the basis of ice-movement directions in the Netherlands, Rappol *et al.* (1989) concluded that British and continental ice must have been in contact to explain the flow directions they discovered. If this is correct, then the maximum glaciation in Britain would have coincided with that in the Netherlands and was therefore of Drenthe age (Gibbard 1991, Gibbard *et al.* 1992), i.e. Older Saalian Glaciation.

In **Poland** the Middle Polish Glaciation (Saalian) is subdivided into two major glaciations, the Maximum (Odra) and the Warta (Warthe) Advance. A third ice advance before the Odranian Maximum is referred to as 'Premaximum Stadial'. Lindner (1988) has suggested that this may be a correlative of the Fuhne Cold Stage, but the evidence is still equivocal (Mojski 1995). The maximum advance of the Odranian Glaciation reached the Sudeten Mountains and crossed the Moravian Gate, extending well into the Czech Republic, beyond the watershed separating the Baltic from the Black Sea catchment area. Till-petrographic studies in western and central Poland and in Silesia suggest an ice advance from the north and northeast (Czerwonka & Krzyszkowski 1992). The Warta ice sheet did not extend as far south as the Maximum advance. Its outermost position in the type area of central Poland was located around the upper reaches of the Warta River (Rzechowski 1986). However, recently it has been suggested that the Warta ice sheet in Poland may possibly have advanced much further south, beyond the limits of the Odranian Glaciation (Marks, personal communication).

As in North Germany, the zone of the Saalian Glaciation in western Poland is characterised by a number of prominent end moraines (Rotnicki 1974). Ruszczynska-Szenajch (1982), however, also describes depositional end moraines from the area of the Wartanian Glaciation around Warsaw.

In **Russia** the Saalian Cold Stage saw two glaciations, the Dniepr Glaciation and the Moscow Glaciation. During the older Dniepr Glaciation, the continental ice sheet advanced far to the south into the **Ukraine**. Its outermost ice margin was situated more than 100 km south of Kiev; in the Dnieper valley it advanced more than 50 km beyond Kremenchuk to a position south of 49°N (Matoshko & Chugunny 1995). This is the most southerly extent of all Pleistocene ice sheets in Europe. In Russia the deposits of the Dniepr Glaciation contain a distinct erratic assemblage, dominated by material from northeastern sources. This corresponds with the results of fabric measurements in the northern part of the Russian Plain and points towards a glaciation centre in the Pai Hoi and Novaya Zemlya region. In contrast, the Moscow Glaciation was centred on Fennoscandia. Its ice sheet coalesced with that from the Urals almost at the foot of the mountains. The southernmost boundary of this advance in Russia was situated less than 100 km south of Moscow (Sudakova & Faustova 1995, Sudakova *et al.* 1995).

In **Finland**, Saalian deposits have been preserved only in exceptional cases, such as the Lappajärvi meteorite crater (Niemelä & Tynni 1979, Donner 1988). Two Saalian tills have been identified, the older of which contains clasts derived from northern or even northeastern source areas. The younger Saalian till, like the Weichselian tills, was deposited by ice advancing from the northwest (Saarnisto & Salonen 1995).

FIGURE 144 Extent of the Elsterian and Saalian glaciations in the Netherlands, Germany, Czech Republic and Poland and important sites. 1 = unglaciated area; 2 = maximum extent of the Weichselian Glaciation; 3 = maximum extent of the Saalian Glaciation; 4 = maximum effect of the Elsterian Glaciation (after Ehlers 1990a)

11.14.1 Older Saalian Glaciation

Within the Saalian Complex in North Germany, three till sheets separated by meltwater deposits can be distinguished. Because the local stratigraphies still do not correspond, they are referred to here as 'Older', 'Middle' and 'Younger' Saalian till. The extent of the Older Saalian ice advance is shown in Figure 144.

The Older Saalian ice advance in **Lower Saxony** is called the 'Main Drenthe advance'. Its ice sheet covered almost all of Niedersachsen, it intruded into the Münsterland Bight and advanced up to its southern margin (Klostermann 1992, Skupin *et al.* 1993). In the west, it reached the Lower Rhine and left behind enormous pushed end moraines, the most prominent of which is the Rehburg end moraine. Previously interpreted as recessional end moraines (of the so-called 'Rehburger Phase'), their formation during the advance phase and subsequent overriding by the ice is documented by the occurrence of the same sandy basal till in the foreland and on top of the thrust ridges (K.-D. Meyer 1987). Later, ice masses from the eastern part of the Baltic Sea overrode the area and deposited a red–brown till in northwest Germany and the Netherlands, which is characterised by east Baltic indicators (much Palaeozoic limestone, little or no flint, usually some dolostones) (Fig. 145).

In the **Netherlands** Ter Wee (1962), Jelgersma & Breeuwer (1975) and Maarleveld (1981) originally subdivided the Saalian Glaciation into five different phases, all represented by push moraines. However, only a single Saalian till sheet has ever been found. The Older Saalian till occurs in a number of strikingly different lithofacies types. However, Zandstra (1983) has clearly demonstrated that the clast associations involved do not represent deposits of different glaciations but a sequence of different ice masses within a single major ice advance.

In **Denmark** Saalian tills have been found in various places. Houmark-Nielsen (1987) distinguishes three Saalian ice advances. The oldest of these deposited a sandy, quartz-rich till, the Trelde Næs Till that contains abundant Norwegian rocks and other stones characteristic of an ice advance from a northerly direction. The indicators include flint conglomerates and ash-bearing marine diatomite of the Palaeocene Fur Formation, formerly referred to as 'moler'. It is assumed that the ice crossed the border into Schleswig-Holstein. A second ice advance from a northeasterly direction followed, which – after accumulating fine-grained, coarsening-upwards meltwater sands – deposited another sandy, quartz-rich till. This Ashoved Till is characterised by a central Swedish indicator association (Kinne Diabase) and Jurassic rocks incorporated at the margin of the Kattegat (Katholm erratics). This ice advance is probably the equivalent of the North German Older Saalian ice advance (Houmark-Nielsen 1987).

Climatic deterioration during the Saalian led to a reduction but most likely not to a complete absence of **human settlements** in Germany. Palaeolithic artefacts have been found in Saalian deposits in various places. In many cases, however, they are reworked from sediments of the preceding Holsteinian or Wacken Interglacial or from one of the Early Saalian interstadials.

At Markkleeberg (near Leipzig) at the base of the Early Saalian terrace of the Pleisse and Gösel rivers 9000 Middle Palaeolithic Late Acheulian artefacts have been found. Groups of people at that time frequented the valleys in order to obtain fresh flint for tool making. On the river banks fresh flint of the right type was exposed in the Elsterian tills. Because the site represents a working place of early humans, complete tools are rare; the finds comprise 80% flakes and 17% cores. Only 3% of the finds were tools, most of which were apparently left behind because they had various faults (Mania 1989b).

At the end of the Older Saalian the ice possibly melted back as far as beyond the southern

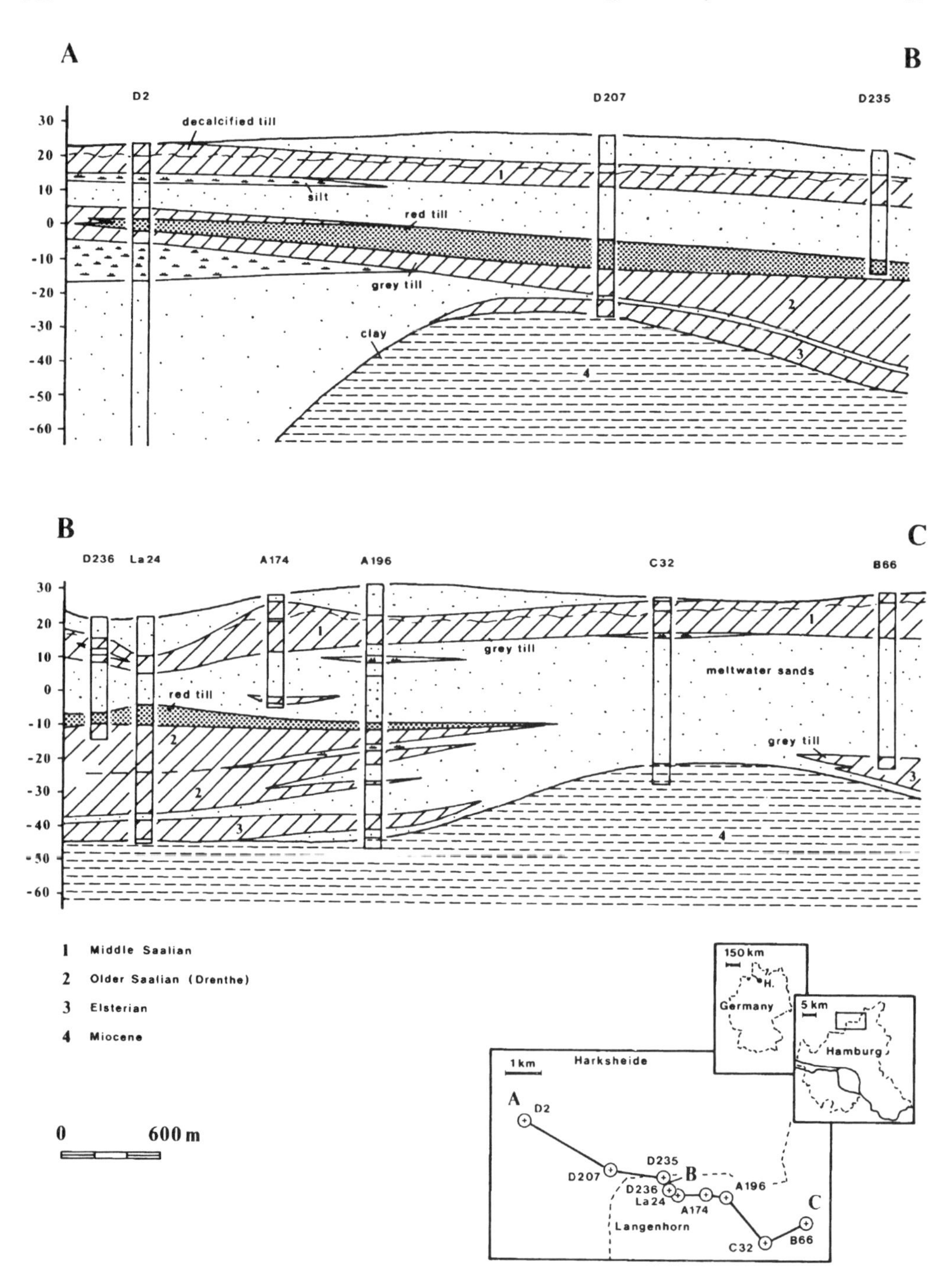

FIGURE 145 Distribution of the red facies of the Older Saalian till in a cross section at the northern margin of Hamburg. Vertical exaggeration 20-fold; depth in metres (from Ehlers 1992)

Baltic Sea coast. Denmark and Northern Germany became ice-free. Exposures in Dithmarschen and northern Niedersachsen indicate that the Older Saalian till at this time was exposed at the land surface. Before being covered with younger sediments, it was dissected by a polygonal ice-wedge network (Höfle 1983a). Under the influence of periglacial climate, a gravel lag formed on the Older Saalian till (Stephan *et al.* 1983).

The **warmer intervals** between the individual ice advances of the Saalian in North Germany and adjoining areas are characterised by a lack of organic deposits. Apart from the fragile bryophyte remains described by Grube (1967) (*Calliergon giganteum*), no intra-Saalian autochthonous organic deposits are known, although all three ice advances left behind kettle holes that would have formed ideal sediment traps for the preservation of such deposits. Only in the subsequent Eemian Interglacial were those depressions foci of peat growth and gyttja, diatomite or lake marl deposition. No intra-Saalian high sea-level stand has been reported from the North Sea. Nevertheless, repeatedly an interglacial status has been postulated for the interval between the Older and Middle Saalian ice advances. The supposed warm stages include the 'Uecker Warm Stage' of Röpersdorf (Erd 1987) and the 'Vorselaer Schichten' (Klostermann *et al.* 1988). In both cases, however, the unequivocal stratigraphic position within the glacial part of the Saalian (i.e. after the Wacken Interglacial) has not been determined.

As evidence for a 'Treene Interglacial' between Drenthe and Warthe Stages in North Germany palaeosols have mainly been used (particularly by Picard 1959 and Stremme 1960, 1981). The stratigraphical position of some of these fossil soils remains doubtful (see discussion in Ehlers *et al.* 1984). Menke (1985b) demonstrated that the fossil soil at the type locality of the 'Treene Warm Stage' (Picard 1959) was actually of Eemian age.

The potentially equivalent intra-Saalian bleached loams on the Isle of Sylt described by Felix-Henningsen (1983) most likely were not formed in an interglacial but in an interstadial. The deep-penetrating wet bleach mainly suggests pedogenesis under very moist conditions. The translocation processes that in part extend down to more than 10 m indeed indicate that for a longer period of time no permafrost existed in the area. Investigations of the pollen content of the bleached loam by Felix-Henningsen & Urban (1982) yielded evidence of an open vegetation. Grasses, *Calluna* and other Ericaceae seem to have been present during the formation of the bleached loam. Pollen of thermophilous trees found in limited numbers most likely cannot be interpreted as having been derived from local vegetation; it is either redeposited or might reflect long-distance transport.

Other indicators also suggest that the climate during this phase cannot have been particularly warm. There are strong indications that in protected positions dead ice from the Older Saalian advance survived in Schleswig-Holstein until the very end of the Saalian Stage (cf. Stephan 1974). The Eemian basin of Grabschütz in Saxony, described by Eissmann (1990), Wansa & Wimmer (1990) and Eissmann & Litt (1995) apparently was also already excavated by the first Saalian ice advance (Zeitz Phase) and filled by a mixture of till and ice. The buried dead ice persisted until the next ice sheet of the so-called Leipzig Phase overrode the area. Under the new ice cover or at least during the subsequent ice-decay phase, the dead-ice block melted and the till sank down as a more or less continuous sediment body. Only then was the 10–12-m-deep kettle hole formed, in which later the Eemian deposits accumulated (Wansa & Wimmer 1990).

In **Poland** detailed investigations in the Bełchatów opencast mine (near Łodż) yielded evidence for an intra-Saalian interstadial only, which stratigraphically might represent the interval between Older and Middle Saalian till. The sequence in the Bełchatów opencast is thought to comprise five Saalian tills and meltwater sands, which overlie pollen-bearing, supposedly Mazovian (Holsteinian) deposits. An organic silt has been found between the second and third till from the base, with a pollen content representing reforestation up to a coniferous forest including a few broad-leaved trees (Krzyszkowski 1991a, b). However, many stratigraphical problems remain unresolved.

The Odintsovo Interglacial in **Russia** is thought to be the equivalent of an intra-Saalian interglacial separating the deposits of the Dniepr Glaciation (Drenthe) from those of the Moscow Glaciation (Warthe). However, the stratigraphical position of the deposits in question is uncertain (Velichko & Faustova 1986). The same applies to the Snaigupėlė Interglacial of Lithuania (Kondratienė 1981), which may predate the glacial part of the Saalian.

11.14.2 Middle Saalian Glaciation

The subsequent ice advance began in North Germany with the deposition of meltwater sands. In contrast to the glaciofluvial deposits of the Elsterian, which in North Germany are largely concentrated in buried channels, the meltwaters of the Middle Saalian accumulated vast outwash fans, the deposits of which can be several tens of metres thick. During the maximum phase of this advance, when the ice lay south of the present Elbe valley, the meltwater drained southwards and then via the Aller-Weser ice-marginal valley ('urstromtal') towards the North Sea (K.-D. Meyer 1983c). It is thought that the Middle Saalian ice advanced at least as far west as the Altenwalder Geest hills, south of Cuxhaven (Höfle & Lade 1983, Van Gijssel 1987).

Within the area covered by the Middle Saalian Glaciation a number of marked end-moraine ridges are found. They include the hills of the Neuenkirchen and Falkenberg ice-marginal positions. The stratigraphical position of these end moraines is not well understood in all cases (K.-D. Meyer 1983a). The strike direction of the ridges seems to indicate that at that time the ice margin had formed a number of ice lobes and tongues, although perhaps not quite to the degree postulated by Hövermann (1956). Further north in the more central parts of the ice sheet, however, a radial ice movement was clearly developed. Fabric measurements from the area around Hamburg yield clear NE–SW maxima for this advance (Ehlers 1978, 1990a).

A clear correlation of the West German Saalian till stratigraphy with the respective sequences in **Mecklenburg–Vorpommern**, **Sachsen–Anhalt** and **Brandenburg** has not yet proved possible, although three Saalian tills have been found in those areas. However, they differ in composition from the West German tills. Eissmann (1986, 1990) has correlated the Saalian stratigraphy of the Saale–Elbe region with northwest Germany.

In **Denmark** the Trelde Næs Till and Ashoved Till (both Older Saalian) are overlain by a widespread chalk-rich Saalian till (Lillebælt Till), separated from the older tills by sands. Its area of distribution reaches from the German border north to northern Jutland (Skærumhede). Since this is the uppermost Saalian till, Houmark-Nielsen (1987) correlates the Lillebælt Till with the North German Younger Saalian Till.

The Warthe ice sheet at its maximum in western Europe terminated more than 100 km inside the Drenthe maximum. Meltwater passage towards the west, therefore, was open during this phase. The oldest ice-marginal valley identified in North Germany is the Aller-Weser Urstromtal which served as a drainageway for the meltwaters of the Middle Saalian Glaciation. Towards the east its catchment area can be traced to the Letzlinger Heide region (north of Haldensleben, Sachsen-Anhalt) (Glapa 1971). The main water divide during this phase was situated between the Warta and Pilica rivers; the latter drained eastwards via the Pripiat River to the Dnieper (Pilica–Pripiat Urstromtal; Różycki 1965).

In **Russia** the extent of the Moscow Ice Sheet (Warthe) was also less than that of the preceding Dniepr Glaciation (Drenthe). West of the Ural Mountains in most areas the ice advanced as far as the divide between northern and southern drainage systems. Fewer ice-dammed lakes were therefore formed than during the subsequent Weichselian glaciation. Increased meltwater drainage via the Volga into the Caspian Sea may have caused overflow to the Black Sea via the Manytsh Depression.

11.14.3 Younger Saalian Glaciation

After the Middle Saalian Glaciation, the ice margin retreated far to the northeast and parts of the ice sheet stagnated and melted *in situ*. The active ice margin was probably situated in the present Baltic Sea area. In the proglacial area, widespread dead-ice masses were covered by outwash during the subsequent readvance.

In North Germany drainage was directed largely towards the west; only limited outwash was deposited in elevated terrain (cf. in Hamburg–Borgfelde, Grube 1967), whilst in large areas the Younger Saalian till directly overlies Middle Saalian till. Further east, around Vastorf (southeast of Lüneburg), however, thick meltwater sands accumulated during the Younger Saalian ice advance.

Indicator counts have demonstrated that the composition of the Younger Saalian till varies locally. In northeastern Niedersachsen a red–brown, east Baltic facies of the Younger Saalian till is widespread (K.-D. Meyer 1983b). At the outermost ice margin in Schleswig-Holstein, tills have also been found that are rich in east Baltic indicator clasts (e.g. at Hohenwestedt and Osterrade; Stephan *et al.* 1983). In immediately adjoining areas, however, the east Baltic components are lacking. For example, in northwestern Schleswig-Holstein the flint content of this till is higher, so that it cannot be distinguished from Middle Saalian till (Stephan in Ehlers *et al.* 1984).

Fabric measurements in Hamburg and northeastern Niedersachsen show that ice movement during this last Saalian ice advance was mainly directed towards the west. In Schleswig-Holstein, however, southerly directions have also been observed (Stephan *et al.* 1983).

A line of marked end moraines in the northern Lüneburger Heide area has traditionally been associated with the Younger Saalian 'Warthe' ice advance. However, more recent investigations show that many assumed Younger Saalian end moraines in Niedersachsen often have a much older core. The Garlstorfer Wald–Toppenstedter Wald ridge, for instance, consists largely of thrust Tertiary strata. The internal structure shows that the thrust occurred from the east. The age of the push moraine can be deduced from the stratigraphy: because Elsterian deposits were incorporated in the thrusts, the ridge must postdate the Elsterian Glaciation. On the other hand, the Tertiary strata to the east of the ridge are covered by thick Older Saalian till, so they were no longer accessible for thrusting during the Middle and Younger Saalian. Therefore this ridge is almost certainly an Older Saalian push moraine.

The extent of outwash accumulation during the Younger Saalian seems to have been rather limited. Many areas that were previously interpreted as Younger Saalian sandur plains have been found to be covered by patches of Middle Saalian till (K.-D. Meyer 1983c).

The idea that during the Younger Saalian a continuous ice-marginal valley existed from Wrocław via Magdeburg to Bremen, as envisaged by Glapa (1971), has been questioned by

K.-D. Meyer (1983c). According to Liedtke (1981), however, this valley can indeed be traced eastwards beyond Wrocław. A much further eastward continuation towards the Warta River even seems possible. Liedtke (1981) considers that the Warta must have drained westwards, because all cols further east were too high to be crossed.

The climatic development at the end of the Saalian probably very much resembled that at the end of the Weichselian. In northern Denmark marine deposits are preserved from the transitional period between the Saalian and Eemian. By investigating benthonic foraminifera from a borehole on the Isle of Anholt (Kattegat), Seidenkrantz (1993) demonstrated that warming at the end of the Saalian was interrupted by a strong cooling event roughly comparable to the Younger Dryas phase at the end of the Weichselian (cf. chapter 11.16). At Brokenlande in Schleswig-Holstein Menke & Ross (1967) found a gradual transition from a cool phase characterised by *Artemisia* (Phase A) via the spread of *Hippophae* and several herbs and perennial plants (Phase B) and a period dominated by *Hippophae* and *Juniperus* shrubs (Phase C9) to the birch period of the early Eemian (pollen zone I). This sequence of strata seems to have formed after the Kattegat Cooling.

11.15 EEMIAN INTERGLACIAL

The term 'Eemian' (after the small Eem River in the central Netherlands) was coined by Harting (1874). It gained international acceptance more than 30 years later following a comprehensive investigation by Danish geologists. Madsen *et al.* (1908) were the first to compare marine deposits of the last interglacial in Denmark, North Germany and the Netherlands, and they used Harting's term. Later, the terrestrial deposits of the last interglacial were also included in the definition. The climatic development of the Eemian very much resembled that of the Holocene. Overall, however, the Eemian seems to have been slightly warmer. Therefore it is no surprise that the Eemian global sea level rose slightly higher than that of the Holocene. This, however, does not apply to North Germany (Fig. 146).

The oldest **marine Eemian** deposits in Schleswig-Holstein belong to the birch period (pollen zone I). The marine transgression thus occurred later than during the Holsteinian, but considerably earlier than in the Holocene. In contrast to the Holsteinian, a direct connection existed from the North Sea to the southwest, so that thermophilous Lusitanian faunal elements could easily immigrate. Accordingly, in the Dutch Eemian sequences they are even more strongly represented than in their German equivalents (Hinsch 1993).

The **vegetational development** of the Eemian in the northern Central European lowlands is very distinct from that in the Holsteinian. It can be summarised as follows (Fig. 119; Menke & Tynni 1984, Litt 1994). The early phase of the interglacial is characterised by re-immigration of birch and pine (zones I and II). In contrast to the Holocene this was followed – prior to the appearance of hazel (*Corylus*) – by immigration of oak (*Quercus*) and elm (*Ulmus*) (zone III). This light pine–mixed-oak-forest was rapidly joined by hazel. Hazel became the dominant tree type even in the upland sites (oak–hazel period, zone IVa and hazel–yew–lime period, zone IVb). A characteristic of the Eemian vegetational development is also the strong hornbeam (*Carpinus*) phase, succeeded by an equally strong spruce (*Picea*) peak. Subsequently fir (*Abies*) and pine (*Pinus*) immigrated (pine–spruce–fir period, zone VI). Finally, pine gained dominance (pine period, zone VII), until towards the end of the interglacial birch began again to spread.

The vegetational history of the Eemian is so well known that Menke (1984) was able to undertake a detailed comparison with the present interglacial for Schleswig-Holstein, including

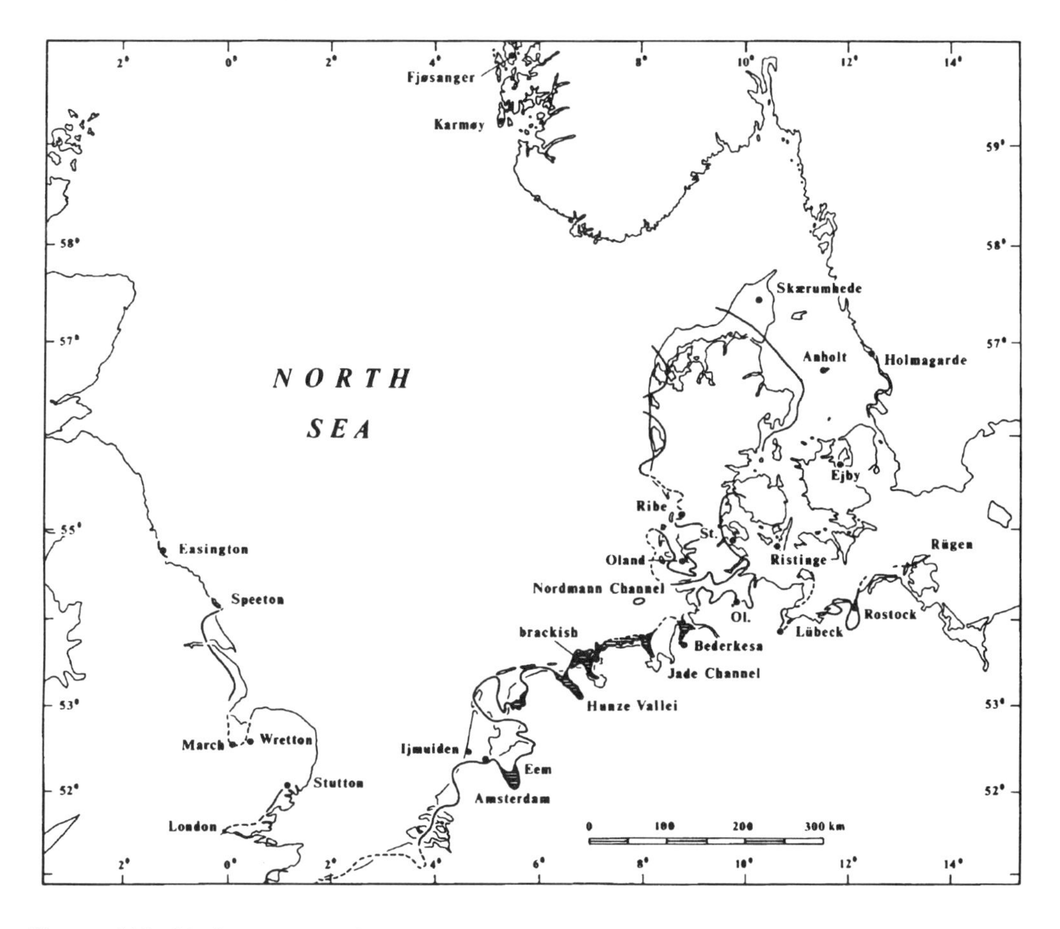

FIGURE 146 Maximum extent of the Eemian Sea (after Ehlers 1988; supplemented after U. Müller *et al.* 1995)

calculations of the **temperatures**. Both warm stages show strong resemblances. Thus in both interglacials the mistletoe (*Viscum*) immigrated about 2000–5000 years after the beginning of the interglacial and disappeared about 7000–8000 years after afforestation from the landscape of northern Holstein. Mistletoe is an indicator of warm summers. During their climatic optima both interglacials experienced July temperatures about 2°C warmer than at present (Menke 1984). In the Lausitz area of East Germany, Striegler (1991) came to the same conclusions. Two complete skeletons and more than 1000 isolated plates, as well as other bone parts, of the pond tortoise (*Emys orbicularis*) have been recovered from the Eemian sequence at Schönfeld (Niederlausitz). Today the northern limit of the area colonised by this tortoise is found in southern France, but further to the east, under more continental climate, it reaches into eastern Germany and most of Poland (Stuart 1979). According to the distribution of the pond tortoise in the Eemian climatic optimum, a July mean temperature of 19°C must be assumed for the Lausitz area. That is slightly above the present 18°C July temperatures. Comparison of the ostracod fauna with the pollen sequence indicates that the temperature optimum coincides with Menke's pollen zones IVb–V (see chapter 8.12.2) (Pietrzeniuk 1991).

Clear differences between the Eemian and the Holocene can be found with regard to the frost-susceptible genera. In this respect holly (*Ilex*) is an important climatic indicator. Its recent easternmost boundary roughly coincides with the 0°C January isotherm. In contrast, during the Eemian *Ilex* reached much further east than today (cf. Frenzel 1967). Ivy (*Hedera*) was also far more widely distributed in the Eemian than in the Holocene. This implies that the Eemian had much milder winters than the present interglacial. Distributions of the submediterranean box (*Buxus*) and the royal fern (*Osmunda*) point in the same direction (Menke 1984).

After the Eemian climatic optimum, *Abies* immigrated into North Germany. Simultaneously spruce increased significantly. This period can be roughly compared with the Subatlantic period of the Holocene. Until the middle of this period, well after the onset of podzol formation, North Germany formed part of the temperate zone. On the basis of the vegetational pattern, July temperatures in Schleswig-Holstein then were about 15°C. Towards the end of this period North Germany again was close to the northern limits of the more demanding deciduous trees and Abies, and by the end of zone VII (of Menke 1984) the subarctic forest boundary again passed through Schleswig-Holstein. The northern tree limit in Fennoscandia today follows July temperatures of about 10°C.

In contrast to the calculation of palaeotemperatures, it is much more difficult to determine **rainfall**. Menke (1984) assumes that Eemian precipitation was probably higher than today. This seems to be indicated by the occurrence of oceanic plant elements (e.g. *Ilex*) in regions of Central Europe, which today are regarded as subcontinental (central Germany, Poland). Although extreme wet bleaching of soils and diatomite formation are unknown from the Holocene of North Germany, they were widespread during the Eemian. Nevertheless, the climatic differences between the Eemian and the Holocene are minor rather than substantial (Menke 1984).

Palaeopedological investigations, especially in Schleswig-Holstein (e.g. Felix-Henningsen & Stephan 1982), have provided detailed information on **soil development** in the Eemian Interglacial. Before the onset of pedogenesis, the deposits of the Younger Saalian Glaciation had been covered by a widespread solifluction sheet or reworked periglacially. This provided a fresh, unweathered substratum for soil formation. The fossil soils of the Eemian as a rule are polygenetic, reflecting the climatic development and changing landscapes of the interglacial. After decalcification and browning, during the climatic optimum clay translocation occurred, and thus development of argillic brown earths. From the end of the *Carpinus* period onwards, decreasing summer temperatures, increasing depletion of nutrients, and the spread of fir and heather initiated the formation of podzolic soils. However, the preserved soil profiles are often eroded at the top, so that the traces of podzolisation are not preserved in all cases.

The three major pedogenic periods of the Eemian, according to Menke (1981), can be correlated with the major vegetational phases. Decalcification occurred during pollen zones I to IVa. It was followed by clay translocation in pollen zones IVb to V. Finally, podzolisation took over during pollen zones VI and VII. Strong weathering and high precipitation, especially during the warm-oceanic second phase, seem to have resulted in the migration of silicic acid and thus, via the strong proliferation of silicic algae, in the formation of the thick diatomite deposits of North Germany (Menke 1981).

In **Britain** the equivalent of the Eemian is the Ipswichian Stage. Interglacial deposits of last interglacial age have been found in numerous places, including Bobbitshole, south of Ipswich, which provided the stage name (West 1957) and was later designated as the stratotype (Mitchell *et al.* 1973). A problem has been that neither this nor any other Ipswichian deposit spanned the entire interglacial time, so that its vegetational history had to be reconstructed from numerous pieces (Jones & Keen 1993). On the basis of sites in East Anglia, West (1980b) provided a composite botanical record of the Ipswichian, which still remains valid after investigation of many additional localities. A relatively long Ipswichian sequence has been recovered at Beetley, East Anglia (Phillips 1976, West 1991), and at Wing, East Midlands (Hall 1980).

The Ipswichian pollen diagrams are characterised by high percentages of non-tree pollen associated with remains of large mammals. Grazing seems to have had a major influence on the vegetational development (Phillips 1974, Stuart 1976, 1986, West 1991).

Further to the east, numerous Eemian sites have been studied in **Poland**, e.g. Krupinski (1986), Tobolski (1986), Mamakowa (1989). During the Eemian, the lower Wisła region underwent a major transgression, the deposits of which have been investigated by Makowska (1979, 1986). The Polish Eemian floras show strong similarities with the Saxonian sites of Litt (1994) but reflect the regional slightly more continental drier conditions. In **Russia**, the equivalent of the Eemian is the Mikulino Interglacial, named after a site north of Smolensk. The Mikulino Warm Stage was also characterised by a major transgression along the arctic coasts of Russia and the Baltic Sea transgressed into parts of Russia and the Baltic States. This may be partly attributed to incomplete isostatic rebound after the extensive Saalian Glaciation and also may reflect higher global eustatic sea level during the period.

In **Finland** Eemian deposits have been identified in various places (Donner 1983). Marine Eemian deposits are found at relatively high altitudes, including Ollala, Haapavesi (116 m a.s.l.; Forsström *et al.* 1987), Norinkylä, Teuva (112 m a.s.l.; Grönlund 1991) and Ukonkangas, Kärsämäki (106 m a.s.l.; Grönlund 1991). The type locality in Finland for the Eemian is Evijärvi (B. Eriksson *et al.* 1980, Donner *et al.* 1986, B. Eriksson 1993). During the climatic maximum the region was vegetated with a mixed-oak-forest, including *Quercus* and *Carpinus* and up to 15% *Corylus*, indicating that the Eemian was warmer than the Holocene.

In **Sweden**, about ten sites are correlated with the Eemian Interglacial (Robertsson & García Ambrosiani 1992). The till-covered, organic brackish–marine sediments at Bollnäs are situated at a level which indicates that in central Sweden the Eemian transgression reached over 90 m above contemporary sea level (Fig. 157). The microfossils and macrofossils indicate a climate as least as warm as today. The sediments represent the middle part of the Eemian (García Ambrosiani 1990). Other sites with more or less complete Eemian terrestrial sequences are situated at Stenberget (Berglund & Lagerlund 1981), Margareteberg (Påsse *et al.* 1988) and Leveäniemi. The flora and beetle fauna from the Leveäniemi site in northern Lapland also indicates climatic conditions warmer than the Holocene (J. Lundqvist 1971, Robertsson & Rodhe 1988).

In **Norway**, in the zone of maximum glacial erosion, pre-Weichselian deposits are naturally extremely rare. Mangerud (1970) discovered the first Eemian site at Fjøsanger, near Bergen. Detailed descriptions of the *in situ* and complete Fjøsanger Interglacial are provided by Mangerud *et al.* (1981). Both palynologically and by means of amino-acid dating, it can be correlated with the Eemian of Central Europe (Sejrup 1987, Mangerud 1991b). This far north *Carpinus* and *Abies* are absent. Yet, both the vegetation and the marine fauna (molluscs and foraminifera) show summer temperatures ca. 2°C warmer than at present.

Numerous Eemian sites have been reported from **Denmark**. The southern parts of the country were partly covered by a shallow transgression of the Eemian Sea (Fig. 146), whilst northern Jutland and the Isle of Anholt were covered by deeper water. Marine Eemian deposits in the latter area reach a thickness of up to 65 m. The foraminiferal fauna suggests that water depth was over 100 m. Here the marine sequence starts in the Late Saalian and continues well into the Weichselian Stage (Knudsen 1994). Lacustrine Eemian deposits have been reported from various sites, including Hollerup in Jutland (S.Th. Andersen 1965).

The Eemian **fauna** of Central Europe differed markedly from that of the Holocene. *Hippopotamus amphibius incognitus* spread into central England, and the Upper Rhine graben was inhabited by water buffalo (*Bubalus murrensis*). Forest elephant and forest rhinoceros, as well as giant deer (*Megaloceros giganteus*) occurred together with deer, roe, fallow deer (*Dama dama*) and boar (Kahlke 1981, Von Koenigswald 1988). The reasons for this more diverse faunal composition are sought in climatic causes (stronger Atlantic influence) as well as a lack of concurrence of predators (Von Koenigswald 1991).

Traces of **human settlement** are relatively rare in Central Europe in the region north of the uplands. The same is true of Britain (Wymer 1981, 1988). The reason may be that the respective evidence is now concealed by younger sediments (e.g. loess), or scattered by periglacial reworking during the subsequent Weichselian Cold Stage. Further to the southeast, in the Czech Republic and in the Thuringian Basin, numerous artefacts have been discovered in freshwater carbonates, not all of which, however, can be dated to the Eemian. The Eemian age of the travertines from Weimar (Belvederer Allee) and Taubach, however, are undisputed. Artefacts have also been recovered from the deposits of several Eemian ponds in the Halle–Leipzig region at sites including Grabschütz, Gröbern and Rabutz. Th. Weber (1990) points out that a great coincidence exists between the morphological assemblages of these and other finds that can be safely dated to the Eemian. At the same time, clear differences from the Saalian Acheulian industries from the same region support the correct stratigraphical correlation.

In the travertine at Weimar–Ehringsdorf (Fig. 147) a resting place of hunter-gatherers who lived during an interglacial beside the springs in the Ilm valley has been found. An overview of the geological, palaeontological and archaeological investigation is provided by Steiner (1981). The contemporaneous vegetation consisted of an open mixed-oak-forest; a savanna-like landscape with extensive shrubs. The fire places of the camp are represented by extensive burnt layers. Artefacts were produced using the Levallois technique, i.e. they are blades chopped off from a prepared corestone. As raw material apart from flint, chert, porphyry, porphyrite and quartz were also used. In addition to marginal retouches, both surfaces were often worked. The lower travertine at Ehringsdorf has traditionally been interpreted as Eemian, and the upper part as Weichselian interstadial (Soergel 1926, Kahlke 1975). However, according to new palaeontological investigations, this interpretation has been questioned. A greater age seems possible, with regard to the mammal fauna (Musil 1975, Heinrich 1981, 1982, Jäger & Heinrich 1982).

Further finds equated with the Eemian have been reported from the Burgtonna travertine and from the lacustrine clays at Rabutz. The stone tools are of the Mousterian culture. This industry has been attributed to the 'classic' Neanderthal man (Mania 1989b).

Hunting for nutrition played an important role and it was still concentrated on large mammals. At Hollerup, in Jutland (in the vicinity of Randers), bones of *Dama dama* have been found that have been worked by humans within Eemian deposits. They represent the oldest traces of human activities yet found in Denmark (Møhl-Hansen 1955). At Lehringen on the Aller River, an elephant carcass with a wooden lance still in it provides evidence of the hunting methods of that time (Adam 1951). Hunting with wooden lances had already been practised in Spain (Torralba) in the Early Palaeolithic period (Bosinski 1974). In a brown-coal mine at Gröbern, northeast of Bitterfeld, Germany, the slaughterplace of a forest elephant was found.

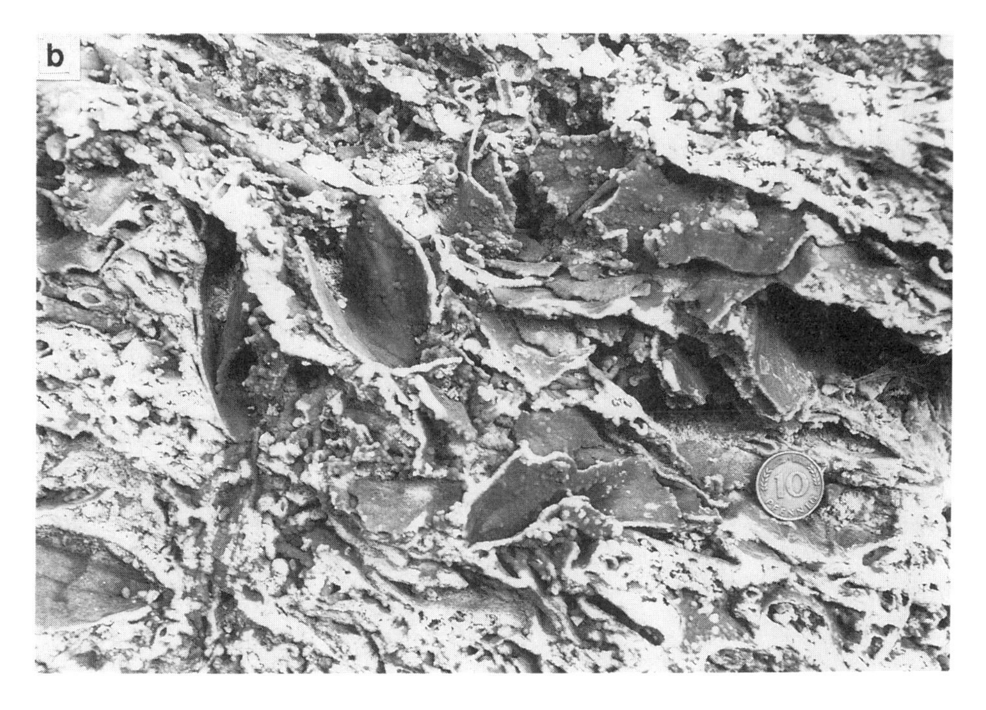

FIGURE 147 The travertine at Weimar–Ehringsdorf. (a) general view of the section; (b) leaves imprinted in the travertine (Photographs: Ehlers 1992)

The elephant died in the shallow water of a small lake and had been cut up by flint tools. Whether the animal died from hunting activities or of natural causes cannot be said with certainty (Th. Weber & Litt 1991). Apart from hunting, there are numerous traces of gathering activities. For instance, in Rabutz numerous burnt hazel shells were found (Toepfer 1958).

11.16 WEICHSELIAN COLD STAGE

The climatic history of the Weichselian Stage is well known today (Fig. 148). This has been largely furthered by investigations of deep-sea sediment cores (Shackleton 1987) and ice cores (e.g. Dansgaard *et al.* 1971, Jouzel *et al.* 1987). The first hints of a possible subdivision of the Early Weichselian Cold Stage came from palynological investigations in Denmark and in the Netherlands. S.Th. Andersen (1957, 1961) reinvestigated the deposits of what had been described by Jessen & Milthers (1928) as deposits of a 'younger temperate period' overlying Eemian peat at Brørup (between Esbjerg and Kolding) where he found a temperate interstadial with forest development. This event has been named after its stratotype locality, the Brørup Interstadial. More details of the climatic development at the beginning of the Weichselian have

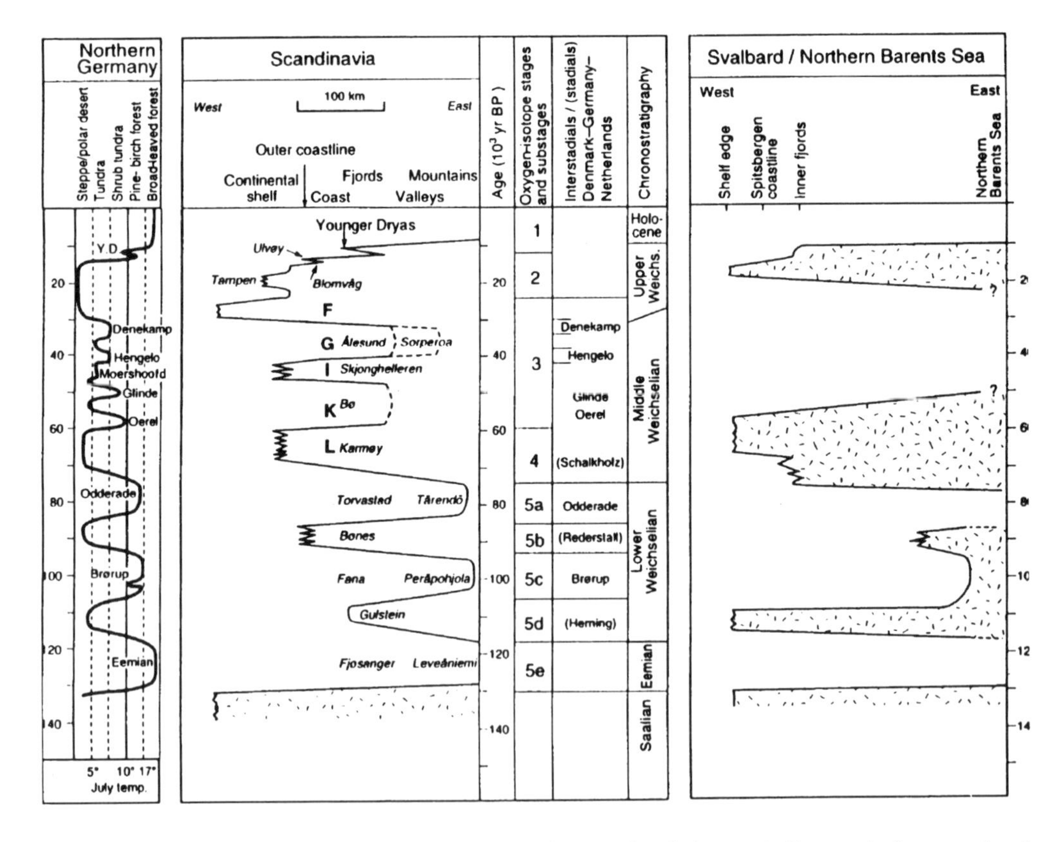

FIGURE 148 Late Pleistocene glacial fluctuations and vegetational changes in Europe. Left: vegetational changes in northern Germany; centre: time–distance diagram of the Scandinavian Ice Sheet development; right: time–distance diagram of the Barents Sea/Svalbard (Spitsbergen) Ice Sheet development (from Mangerud *et al.* 1996)

since been revealed at a number of sites in North Germany, where a series of fossil soils was found overlying the Eemian land surface. Amongst these are the sections of Rederstall (Menke & Tynni 1984), Schalkholz (Felix-Henningsen 1983) and Keller (Menke 1982) in Schleswig-Holstein, and Sonnenberg in Niedersachsen (Lade & Hagedorn 1982). At Odderade in Schleswig-Holstein the deposits of another relatively warm interstadial, the Odderade Interstadial, were identified (Averdieck 1967). The most complete sequence of Early Weichselian deposits in North Germany has been encountered in a closed depression at Oerel (Niedersachsen) (Behre & Lade 1986).

The Oerel profile, above the Eemian, shows an undisturbed sedimentary sequence with the deposits of two forested and two forest-free interstadials. The older of these is the Brørup Interstadial peat. Zagwijn (1961), working in the Netherlands, had previously detected an Early Weichselian interstadial in a stratigraphic position assumed to be between the Eemian and the Brørup, which he termed 'Amersfoort Interstadial'. According to more recent investigations, however, this is the lower part of Brørup, which includes a minor cold and dry interval, which Reille *et al.* (1992) refer to as their 'Montaigu Event'. The Brørup is correlated with deep-sea oxygen-isotope stage 5c (cf. Behre & Lade 1986, Tobolski 1986). The second major Early Weichselian interstadial is the Odderade Interstadial (Stage 5a).

The Oerel sequence correlates well with the long terrestrial pollen diagrams from southern Europe and with the deep-sea oxygen-isotope curves. The two warm periods in the Early Weichselian sequence at Grand Pile in the Vosges (St Germain I and II; Woillard 1975) are the equivalents of these interstadials (Grüger 1979, Menke 1980a).

Of the more weakly developed interstadials at the Oerel site only the first, the Oerel Interstadial, can be correlated with similar occurrences in southern Europe. For the second Middle Weichselian interstadial, the Glinde Interstadial, no counterparts have yet been identified. In the Middle Weichselian, climate remained much harsher. Van der Hammen *et al.* (1967) had identified three treeless interstadials of Middle Weichselian age, which they called the Moershoofd, Denekamp and Hengelo Interstadials. They are not very well defined and difficult to distinguish. The younger part of the Middle Weichselian possibly was a period of a relatively uniform climate, during which edaphic factors rather than climatic changes determined whether humic layers were deposited or not. The radiocarbon ages of the respective layers show a broad scatter (Behre 1989).

Cooling in the Early Weichselian led to the renewed build-up of ice sheets in the circumpolar region (Fig. 148) and to a lowering of sea level. The latter had a fundamental influence on the climatic development in Central Europe. During the Early Weichselian the sea level in the Brown Bank area of the southern North Sea was situated at a depth of −40 m (Zagwijn 1989). Throughout the remaining period of the Weichselian, sea level never rose above that depth. This meant that the present coastal areas of the Netherlands and North Germany were hundreds of kilometres from the sea. Consequently they came under the influence of a much more continental climate than today (colder winters, warmer summers). However, the occurrence of some suboceanic species (such as *Osmunda*) in the Early Weichselian interstadials indicates that the winters cannot have been too severe (Menke & Tynni 1984). However, north Denmark remained under marine influence; until just before the invasion of the Late Weichselian ice sheet, northern Jutland down to a line from Ålborg to Hanstholm was covered by the sea (Houmark-Nielsen 1989).

Whilst on the Continent there seem to have been at least six interstadials within the Early and Middle Weichselian, of which the earliest (Brørup and Odderade) were the warmest (Behre

& Lade 1986), the situation in **Britain** appears to be different. The Chelford Interstadial is most probably the equivalent of the Brørup. The Brimpton Interstadial of Bryant *et al.* (1983) is thought to be an equivalent of the Odderade, as well as possibly the Wretton Interstadial of West *et al.* (1974). However, for the Upton Warren Interstadial (dated at 42,000 BP) there seems to be no continental equivalent. It was too short for trees to invade Britian, but, according to the insect fauna, summers were very warm (Coope & Angus 1975).

In central Sweden till-covered pre-Late Weichselian sediments were first identified in Jämtland (J. Lundqvist 1967, 1969) and later in many other parts of the country. Most of the sites should probably be correlated with the Brørup Interstadial, although some may also be equivalent to the Odderade Interstadial (J. Lundqvist 1992). The key sites indicate that mean annual temperature during the Brørup was 2–3°C cooler than today in central Sweden (Robertsson & García Ambrosiani 1992). The Tärendö Interstadial in easternmost Norrbotten, northern Sweden, and its equivalents in central Sweden have been tentatively correlated with the Odderade Interstadial (Lagerbäck & Robertsson 1988). The climate was considerably colder than during the Brørup and there are signs of permafrost and strong wind erosion (Lagerbäck 1988a, b).

Whilst knowledge of the warm phases within the continental Early Weichselian has largely improved within the last 30 years, the stadial vegetation of the Middle Weichselian is relatively poorly known. At Kobbelgård on Møn, Denmark, Kolstrup & Houmark-Nielsen (1991) found a sequence of silt and fine sand mostly containing pollen of grasses, sedges and *Artemisia*. Thermoluminescence dates suggest that the strata were deposited around 20,000–24,000 BP. The vegetational pattern generally coincides with what Steinmüller (1967) had found in the Goldene Aue area, in the margins of the Harz Mountains.

Based on the intensity of periglacial processes (ice-wedge casts and cryoturbations), palynological investigations and the reconstruction of former snow-line altitudes, Kolstrup (1980) reconstructed the climatic conditions of the Weichselian maximum in the Netherlands. She calculates that mean January temperatures were about −8°C and mean July temperatures about +9 to +10°C. This represents a drop of mean temperatures by 7 or 10° as compared to the modern climate (cf. Van Vliet-Lanoë 1989). Under these conditions strong periglacial reworking occurred in the unglaciated areas, where in addition fluvial and aeolian processes played a major role.

The cold intervals of the beginning Weichselian Stage possibly resulted in a widespread retreat of **human settlement** from major areas. During most phases of the Weichselian, however, Germany was not completely deserted. Settlement density is difficult to calculate, but on the basis of the prey and nutrient potential, Müller-Beck (1983) estimates that the Neanderthal population of southwestern Germany (Baden-Württemberg) probably did not exceed 1000 people. Cold-stage settlements have been found, for instance, in the volcanic depressions in the eastern Eifel mountains (Early Weichselian/Middle Palaeolithic). Conard (1990) has demonstrated that in the open loess landscape of the region mainly horses, red deer and bison were hunted.

The youngest Middle Palaeolithic artefacts belong to the so-called 'Blattspitzenkultur'. Fabrication of flint artefacts had been refined to such a state that thin specimens, carefully retouched on both surfaces, the so-called 'blattspitzen', could be produced. The most important finds of this culture were made in the Ilsenhöhle, a cave at Ranis (Thuringia), where under the collapsed roof a campsite of Middle Palaeolithic humans has been preserved (Fig. 149). The age is estimated to be around 38,000–40,000 years BP. Above this horizon (Ranis 2 and 3)

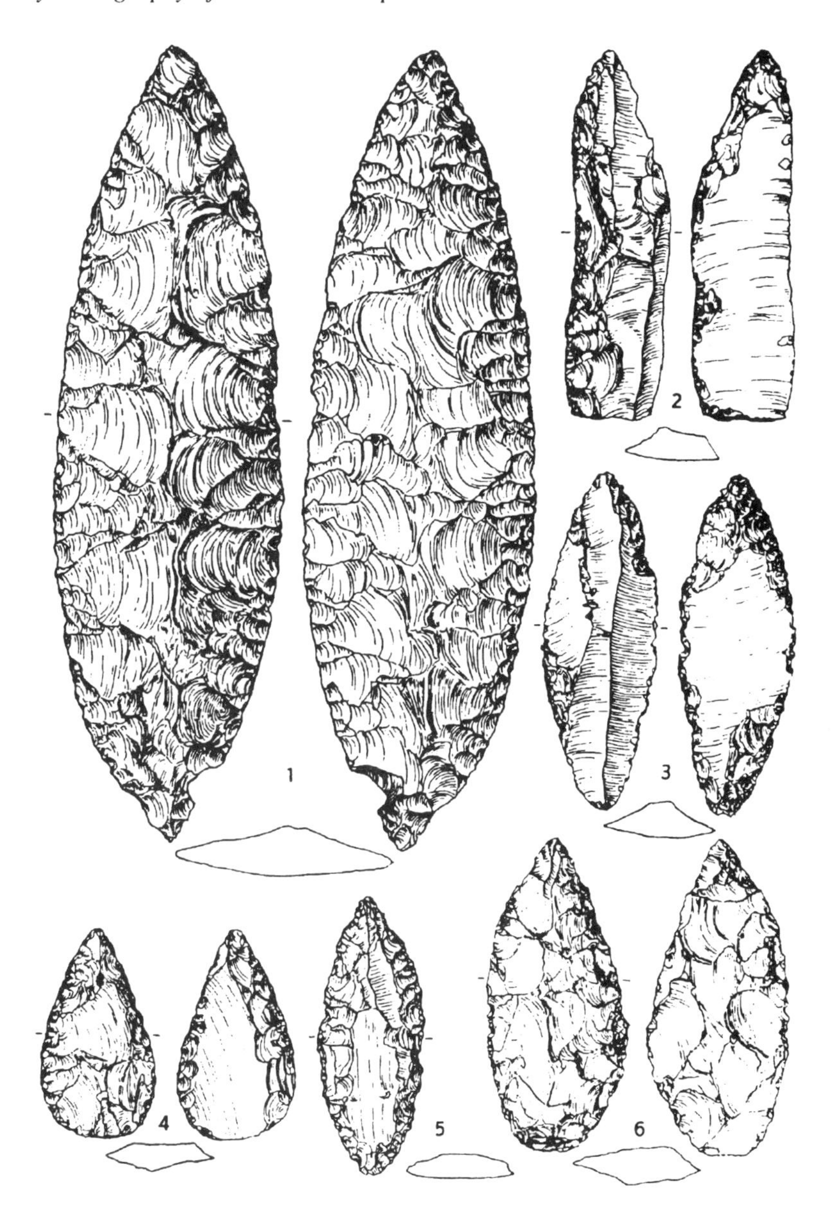

FIGURE 149 Artefacts of the 'Blattspitzenkultur' from the Ilsenhöhle site, find horizon Ranis 2 (Middle Palaeolithic). 1 = double point, 2 = blade with retouched end, 3 = double point with slight surface retouch, 4 = leaf point, 5 = double point, 6 = leaf point (from Toepfer 1970)

additional artefacts were found (Ranis 4), dated to about 25,000 BP. The artefacts include knives with a curved back of a type belonging to the Late Palaeolithic Gravettian industry. At the turn from Middle to Late Palaeolithic, anatomically modern man entered the stage (*Homo sapiens sapiens*), the so-called 'Cromagnon type'. Finds from this age are extremely rare in North Germany. Apparently at the Weichselian maximum only a few groups of hunters advanced into the northern tundra. Since about 30,000 BP hunting methods had been largely improved by the introduction of the bow and arrow (Feustel 1989).

In southern Germany traces of Middle Weichselian (Würmian) human cultures are also relatively rare. Their young age allows dating using the radiocarbon method and thus the finds of Aurignacian and Gravettian artefacts in various caves in southern Germany are relatively well-dated. They all fall in the period after about 30,000 BP. Between 20,000 and 17,000 BP is a marked gap of finds, indicating that climatic conditions prevented human settlement. The Solutréen industry occurred in France during this period, but has not yet been found in Germany (Hahn 1983).

The most detailed profile of Weichselian deposits found so far in Germany is in the deposits of the Ascherslebener See, a former lake at Königsaue in Sachsen-Anhalt. The profile contains traces of nine Weichselian interstadials. Unfortunately, the sequence has only been investigated for its ostracod and mollusc faunas, and not palynologically (Mania & Stechemesser 1970). From the banks of the Ascherslebener See artefacts of the Micoquo-Pradnikian industry (Königsaue A, C) and the Mousterian (Königsaue B) have been found in two Early Weichselian interstadials. Living quarters were not found; they seem to have been located beyond the narrow zone that has been preserved (Mania 1989b).

The maximum **ice advances** of the last cold stage both in Britain and continental Europe did not occur before the last part of the Weichselian (Worsley 1991). Nevertheless, there can be no doubt that extensive glaciers also existed in northern Europe in the Early Weichselian (Mangerud 1983, 1991a, b, B.G. Andersen & Mangerud 1989, J. Lundqvist 1992). The exact extent of the Early Weichselian glaciation in Scandinavia, the deposits of which have been identified in Norway, Finland and northern and central Sweden, so far is unknown. It is certain, however, that it was less extensive than the Late Weichselian glaciation. Finnish Lapland was already ice-covered between the Eemian and the Brørup Interstadial. This is suggested by the fact that deposits of the two periods are always found separated from each other by a till bed (Hirvas 1995). However, further to the south there is no evidence of Early Weichselian glaciation (Nenonen 1995). Hütt *et al.* (1993) have presented numerous thermoluminescence and optical stimulated luminescence dates of Early Weichselian buried podzols and overlying sands that seem to indicate that central and western Finland remained ice-free.

In Denmark, K.S. Petersen & Kronborg (1991) postulated an Early or Middle Weichselian ice advance based on thermoluminescence dating. Recent investigations of the Klintholm sequence on Møn have revealed that the Early Weichselian was characterised by periglacial conditions. Between 75 and 40 ka BP Baltic glaciers advanced twice into the region. The advances were interrupted by an ice-free, periglacial period around 50,000 BP. The first glaciation deposited a reddish-coloured till with East Baltic erratics, the second glaciation left a thick grey till of Baltic origin. After deglaciation, a lake basin developed in the western Baltic, increasingly affected by ice-rafting. Its deposits were later glaciotectonically deformed during the Late Weichselian maximum (Houmark-Nielsen 1994). In North Germany no unequivocal evidence for such an ice advance has yet been found. Similarly, in Skåne no traces of a glaciation prior to the Gärdslöv Interstadial have so far been identified (ca. 20,000–30,000 BP) (U. Miller 1977, Berglund & Lagerlund 1981).

According to current knowledge, the **Late Weichselian glaciation** started around 28,000 BP and reached its culmination at about 22,000–18,000 BP. Radiocarbon dates from Poland suggest that it was a very rapid, surge-like advance (Kozarski 1995).

The Lower Wisła region of **Poland** is the type region of the Vistulian (Weichselian)

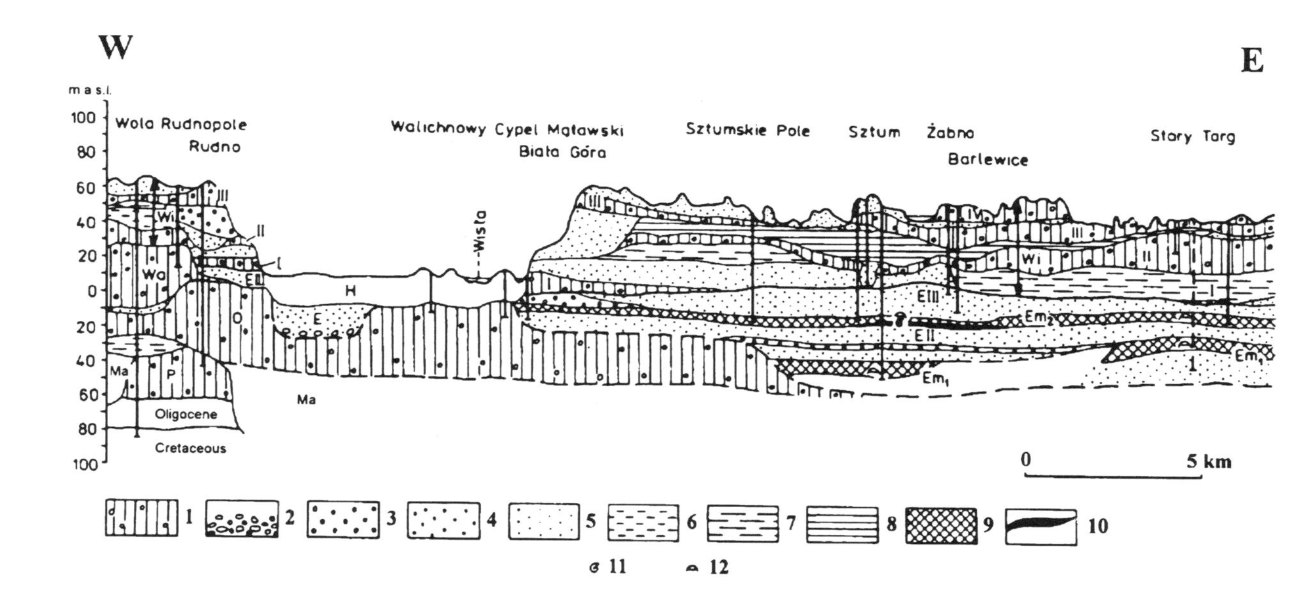

FIGURE 150 Cross section through the Quaternary deposits of the Lower Wisła valley. 1 = till; 2 = boulders and gravel; 3 = gravel; 4 = sand; 5 = fine sand; 6 = silt; 7 = clay; 8 = varved clay; 9 = marine silt and clay; 10 = silt and clay with organic remains; 11 = freshwater fauna; 12 = marine fauna; E = Eemian Interglacial; Ma = Mazovian Interglacial; O = Odra Glaciation; Wa = Warta Glaciation; W = Vistulian Glaciation; H = Holocene (after Makowska, from Lindner 1992)

Glaciation. Five Vistulian till units have been identified, overlying marine Eemian deposits (Fig. 150). In contrast to, for instance, the German sedimentary sequence, the tills include deposits of an Early Vistulian glaciation. In Poland the maximum advance of the Vistulian (Weichselian) Glaciation extended about 50 km south of Poznań. The ice sheet formed several more or less distinct lobes, the most conspicuous of which was the Wisła lobe, which extended upstream to the Płock region (Mojski 1995). The Vistulian stratigraphy has been summarised by Kozarski (1980). The ice advance was locally accompanied by glaciotectonic deformation, most intensive during the Chodzież Readvance (Kozarski 1992). In the Suwałki region advance of narrow ice tongues resulted in the formation of a number of valley-side tectonic features (Ber 1987). Subglacial drainage resulted in the formation of a dense network of meltwater channels (Galon 1965, Kozarski 1967, Pasierbski 1979, Lankauf 1982, Niewiarowski 1992). Proglacial outwash accumulated sandur plains, the sedimentary facies of which largely resemble the North German examples (Zielinski 1989). The age of the Vistulian ice advances is about 15–20 ka BP, according to thermoluminescence dating (Mojski 1992) or ca. 14.5–20 ka BP according to radiocarbon dates (Kozarski 1995).

The outermost limit of the Vistulian ice sheet is locally characterised by the occurrence of some major ice-marginal landforms, some of which represent well-developed push moraines (Kasprzak 1985), whilst others are depositional end moraines. Some of the latter are characterised by well-developed ice-proximal scarps (Kasprzak & Kozarski 1989). However, detailed geological investigations have shown that some of them have a core with a glaciotectonic structure dating from the previous glaciation (Rotnicki 1975). In the extra-glacial area, periglacial processes resulted in large-scale alteration of the original relief. Detailed investigations of the Ostrzeszów Hills thrust moraine show that disintegration of the original ridges from the previous glaciation was not always controlled by internal composition. Many crests today run at right angles to the original geological structure (Rotnicki 1974).

In the south, the Tatry Mountains were glaciated; ice tongues reached down to 914 m a.s.l. (Dzierżek *et al.* 1986).

The maximum extent of the ice sheet south of the Baltic Sea has generally been clarified. Gripp (1924) mapped the Weichselian ice margin in Schleswig-Holstein on the basis of morphological criteria and he concluded that the Weichselian ice had not crossed the Elbe River. Although his line was heavily criticised, it has remained largely valid until today. Older ideas that the Weichselian ice might have crossed the Elbe valley near Lüneburg have not been substantiated. For example, the Eemian deposits at Lauenburg on the Elbe River are not till-covered (Fig. 151; K.-D. Meyer 1965). The chalk-rich tills of the Lüneburg area are Saalian. The Middle Saalian tills in Niedersachsen generally are characterised by high contents of Cretaceous chalk, which results in low weathering depths and a fresh appearance of the substrate.

In Schleswig-Holstein, according to Gripp (1964) three major end-moraine belts can be distinguished, which he termed A (outermost), M (middle) and I (innermost) ice-marginal positions. The M moraine belt is especially well-developed. It forms the southern rim of the Lübeck basin. Stephan & Menke (1977) distinguished five Late Weichselian ice advances in the area around Kiel, which almost certainly can be combined into three major advances. At present it is not yet possible to distinguish them everywhere in Schleswig-Holstein. Petrographic investigations are of limited use in this respect, because most of the Weichselian tills have a very similar clast composition. Similarly, correlations with the Weichselian tills of Mecklenburg-Vorpommern and Brandenburg are hardly possible (Stephan *et al.* 1983). In

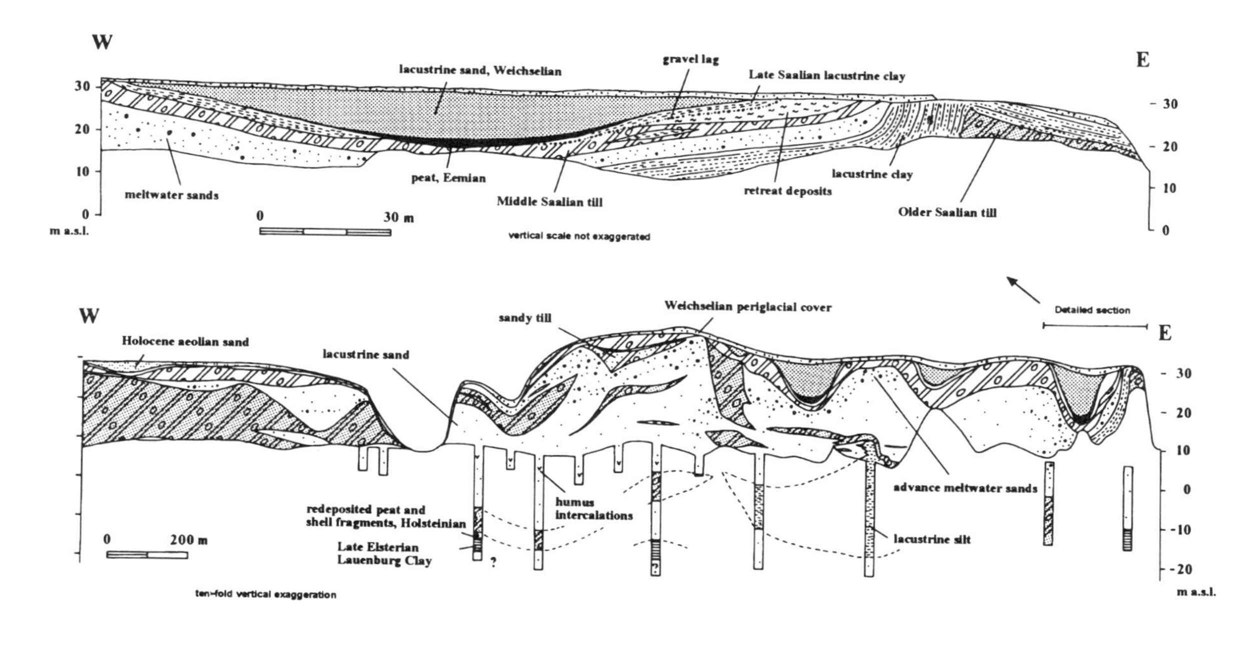

FIGURE 151 The Quaternary section from the Elbe River cliff at Lauenburg, Schleswig-Holstein (after Meyer 1965)

contrast to many of the older glaciations, the Weichselian ice margin was strongly divided into individual lobes and glacier tongues. The overall morphology indicates that, at least temporarily, the Baltic Sea depression played a decisive role in controlling ice-movement directions.

Whereas in West Germany the Late Weichselian glaciers only covered a small margin of the coastal zone south of the Baltic Sea, in eastern Germany the ice advanced more than 200 km inland. The classic morphostratigraphic subdivision of the Weichselian Glaciation was based on mapping of the various ice-marginal positions in Mecklenburg-Vorpommern and Brandenburg. Here Woldstedt (1925) distinguished a **Brandenburg**, **Frankfurt** and **Pomeranian 'Stage'**, each of which could be subdivided into several belts of end moraines. This morphological subdivision of the glaciation, however, is not reflected in the stratigraphy. According to clast-lithological investigations by Cepek (1962), only two Weichselian till units have been distinguished that could be correlated with the Brandenburg and Pomeranian Phases. The end moraines of the so-called 'Frankfurt Stage' thus only represent landforms created by an oscillation during the ice retreat from the maximum Brandenburg ice-marginal position (Cepek 1965). Minor oscillations of the ice margin have locally resulted in deposition of more than two till units (e.g. in Berlin; Böse 1979).

In Mecklenburg, as elsewhere in East Germany, only two tills of the Weichselian Glaciation could originally be recognised (Cepek 1972). However, the existence of a thin, third till unit was proved by Heerdt (1965), which he correlated with the 'Rosenthal end moraine'. This till, according to more recent investigations, can be easily distinguished from the two older Weichselian tills on the basis of the stratigraphy, but also on its clast composition. The distribution of this youngest Weichselian till is limited to the area north of the Rosenthal end moraine (gW3 in Fig. 152). The till sheet generally forms a cover on the pre-existing landforms; the greatest thicknesses are therefore found in depressions (Rühberg 1987).

This last ice advance, which is an equivalent of the Fehmarn Advance in Schleswig-Holstein, disturbed the subsurface very little. Equally in Schleswig-Holstein this ice advance has hardly altered the relief (Prange 1978). The ice in many cases overrode dead ice of the Pomeranian Phase. Only when the ice sheet encountered steep slopes were major push moraines formed, including the Kühlung in East Germany, as well as the Retzow-Gülitz heights and the Brohm-Jatznick ridge. Geological mapping has demonstrated that this ice advanced south to Malchin and Laage. Most eskers in Mecklenburg-Vorpommern were formed during the Rosenthal Advance, but in front of the ice margin no major sandurs were accumulated (Rühberg 1987).

Therefore three Late Weichselian ice advances can be distinguished in northeast Germany. The maximum Brandenburg Advance is assumed to have taken place about 18,000 BP but no exact dates are available. The Pomeranian Advance is older than 13,500 BP (Duphorn *et al.* 1979) whilst the Mecklenburg Advance, which is equivalent to the Fehmarn Advance of Schleswig-Holstein (Stephan & Menke 1977), is assigned to the period 13,200–13,000 BP. It thus coincides with the Oldest Dryas period. Its deposits are overlain by sediments of the Meiendorf Interstadial (Bock *et al.* 1985).

During the Weichselian Glaciation in Denmark, North Germany and Poland numerous subglacial meltwater channels were formed. Most of them are still visible on the present land surface. One well-investigated example is the Stellmoor tunnel valley, northeast of Hamburg (Homci 1974, Grube 1983). This valley has an average width of almost 500 m and a depth of ca. 15 m and is almost half filled with Late Weichselian and Holocene sediments, so that it has

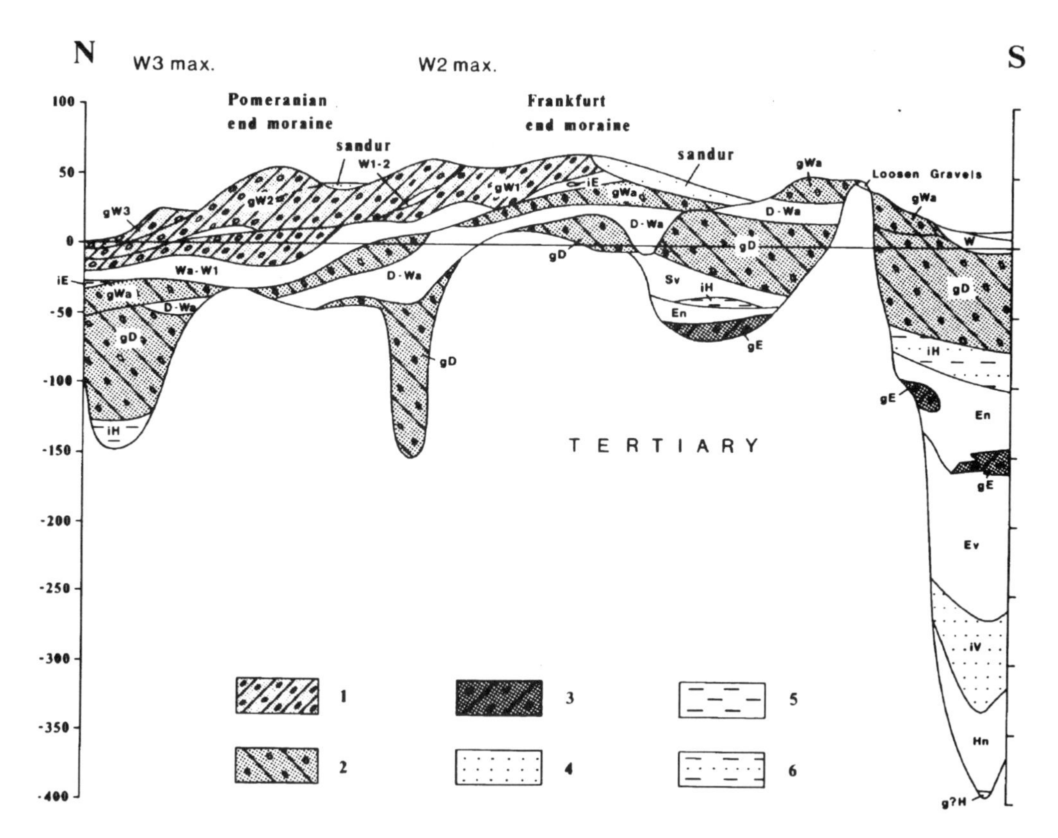

FIGURE 152 Schematic and idealised cross section through West Mecklenburg (between the Baltic Sea and Elbe River). 1–3 = tills (1 = Weichselian, 2 = Saalian, 3 = Elsterian), 4 = sand, 5 = silt to clay, 6 = interbedded silt and sand. gW3 = till of the Mecklenburg Advance; W2–3 = sands and silts between gW3 and gW2; gW2o = till of the Pomeranian Main Advance; W2u–W2o = sands and silts between the two Pomeranian Advance tills; W2u = till of the First Pomeranian Advance; W1–2 = sands and silts between W1 and W2 tills; gW1 = till of the Brandenburg/Frankfurt Advance; W = valley sands of Weichselian age; Wa–W1 = sands and silts between W1 and Warthe tills; iE = marine or lacustrine Eemian deposits; gWa = Saalian Warthe till; D–Wa = sands and silts between Warthe and Drenthe tills; gD = Saalian Drenthe till; Sv = Saalian advance sands; iH = marine Holsteinian deposits; En = Elsterian retreat sands and silts; gE = Elsterian tills; Ev = Elsterian advance sands and silts; iV = marine Voigtstedt deposits (?); Hn = Helme retreat sands (?); g?H = large erratics; remnants of the Helme Glaciation (?) (from U. Müller *et al.* 1995)

a flat floor today. However, detailed geological mapping by Homci (1974) has shown the base of the post-Weichselian sediments to be very irregular, including various basins and swales. Its topography leaves no doubt about its subglacial origin; well-preserved remnants of small eskers indicate its genesis by subglacial meltwater drainage. The genesis of the other Weichselian channels is also not in doubt; none of them extends beyond the outermost Weichselian limit.

During the Weichselian Glaciation, the lower Elbe valley between the Havel River mouth and the North Sea served continuously as the main drainage path parallel to the ice margin, so that no changes of river course took place. Further to the east, however, four to five major valley tracts can be distinguished, which as **urstromtäler** (pradolinas) one after the other

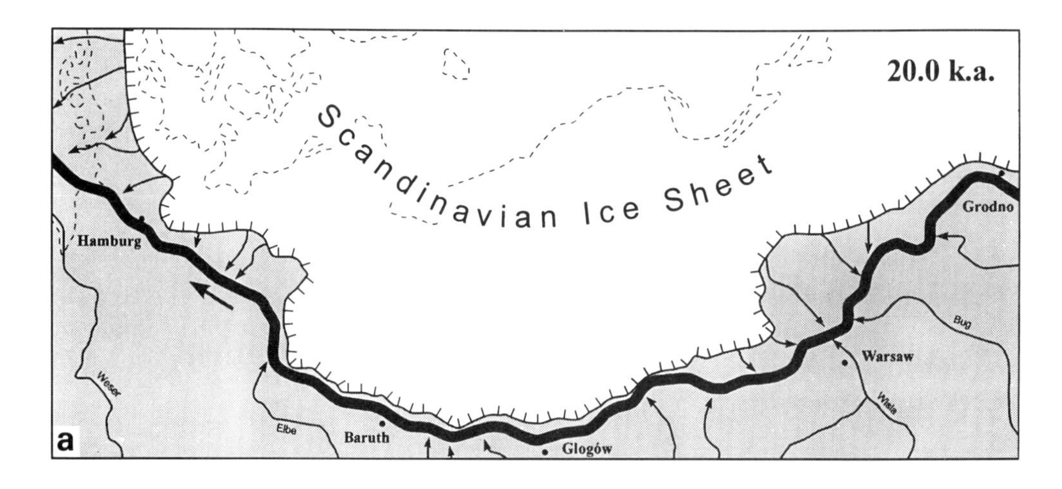
20.0 k.a.
Scandinavian Ice Sheet
Hamburg
Grodno
Bug
Warsaw
Weser
Elbe
Baruth
Glogów
Wisla
a

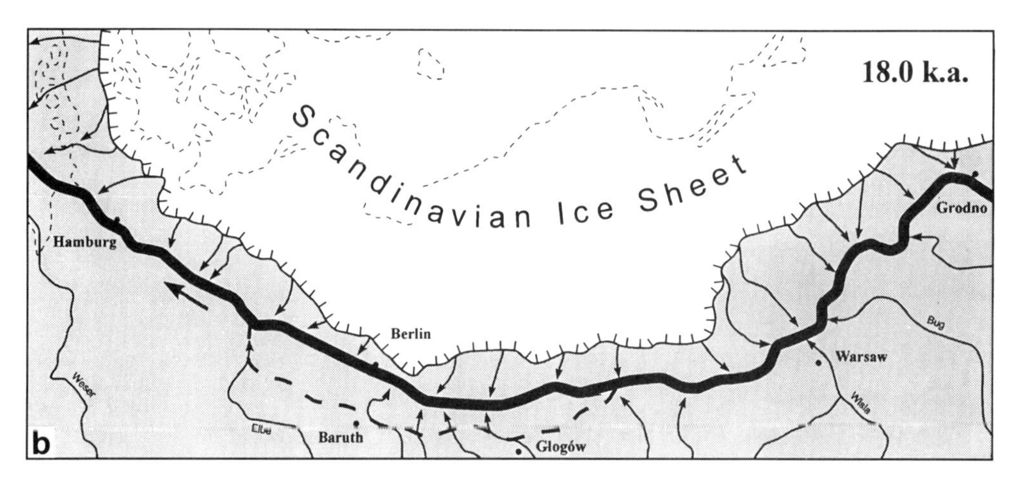
18.0 k.a.
Scandinavian Ice Sheet
Hamburg
Grodno
Bug
Berlin
Warsaw
Weser
Elbe
Baruth
Glogów
Wisla
b

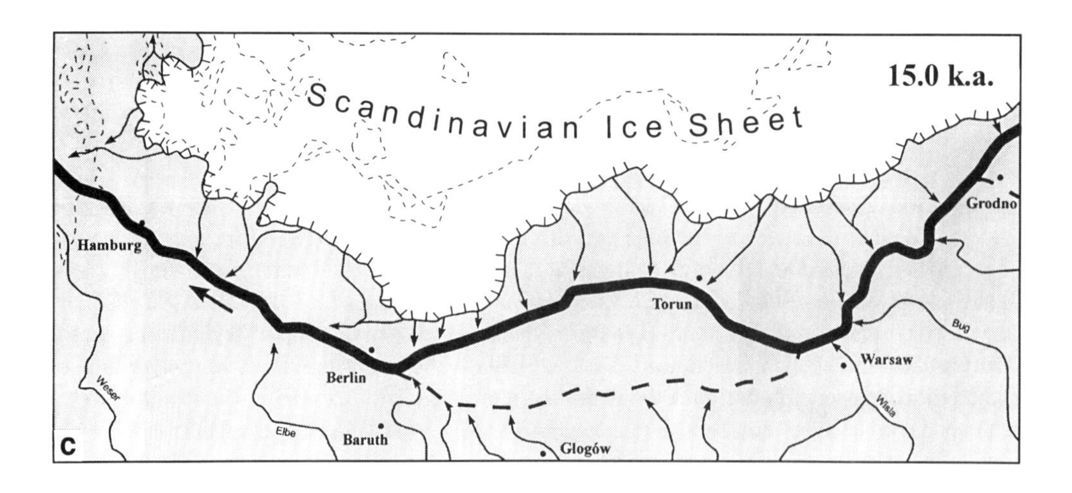
15.0 k.a.
Scandinavian Ice Sheet
Hamburg
Grodno
Torun
Bug
Warsaw
Berlin
Weser
Elbe
Baruth
Glogów
Wisla
c

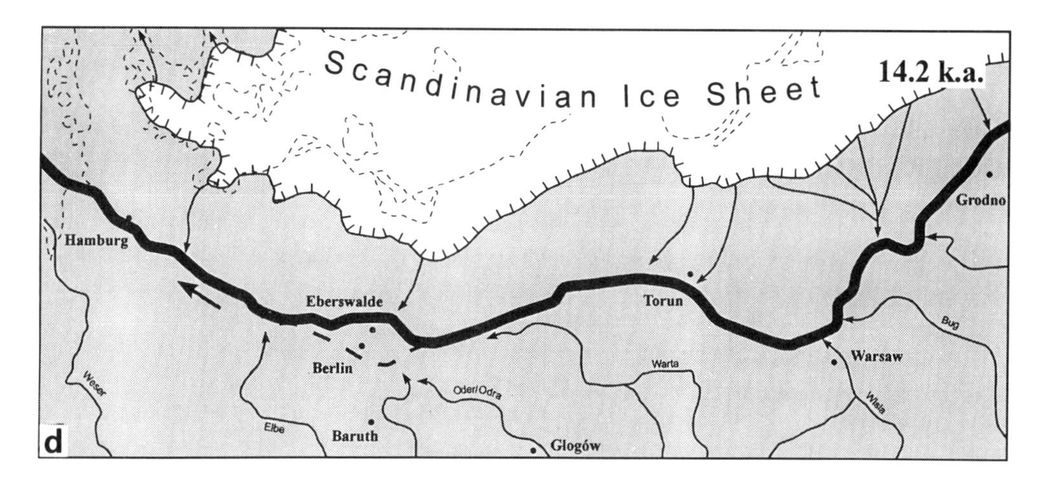

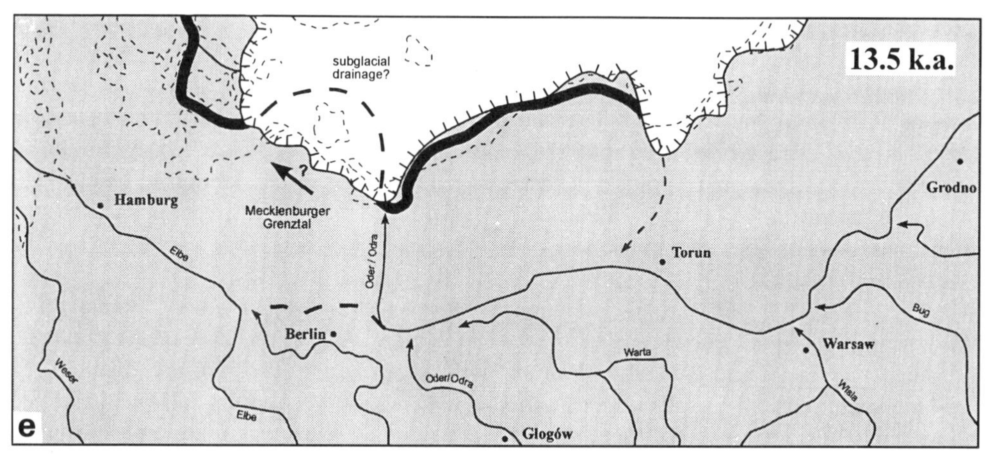

FIGURE 153 Urstromtäler (ice-marginal valleys) in North Germany. (a) Brandenburg ice-marginal position with Glogów–Baruth Urstromtal; (b) Frankfurt ice-marginal position with Warsaw–Berlin Urstromtal; (c) Main Pomeranian ice-marginal position with Toruń–Berlin Urstromtal (?); (d) Rosenthal ice-marginal position with Toruń–Eberswalde Urstromtal; (e) Velgaster ice-marginal position with Netze–Randow Urstromtal (after Liedtke 1981 and Marks and Stephen, personal communication)

drained the southern sector of the Weichselian ice sheet (Fig. 153). In the Weichselian the main watershed occurred considerably further to the east than during the Saalian Glaciation. The upper reaches of the Glogów–Baruth Urstromtal, the oldest of the four major Weichselian ice-marginal rivers, originated in the area around Minsk (Belarus). Further east the meltwaters drained through the tributaries of the Dniepr towards the east and south. The easternmost parts of the oldest Weichselian urstromtal, which drained the meltwaters of the Brandenburg Advance, are found at an altitude of about 190 m (Liedtke 1981).

The Frankfurt ice-marginal position is connected to the Warsaw–Berlin Urstromtal. This and the following ice-marginal streams are considerably shorter than the oldest one, but they still reach northeast into the area around Vilnius and Molodečno (Liedtke 1981). Liedtke (1957) assumes that during the Pomeranian Advance a Toruń–Berlin Urstromtal was formed initially (c in Fig. 153). At that time the drainage in the east via the Noteć and Warta rivers and the Oderbruch was already ice-free, whereas in the west the more southerly ice margin forced the meltwaters to drain southwards via the Buckow Gap into the Berlin Urstromtal. During the Weichselian ice-decay phase the drainage was first maintained via the Toruń–Eberswalde Urstromtal, which continued to function until the formation of the Rosenthal end moraine. According to Liedtke (1981), during the formation of the Velgast end moraine the ice had melted back far enough so that drainage in the west could be redirected via the Mecklenburger Grenztal into the western Baltic Sea depression and via the Belt to the North Sea. This youngest ice-marginal stream, Liedtke's Noteć–Randow Urstrom, ceased functioning finally when the Wisła River mouth into the Baltic became ice-free.

Further east on the **Russian Plain**, the Dniepr River also received meltwater from the Scandinavian Ice Sheet and flowed to the Black Sea. At this time the Black Sea served as a slightly brackish flow-through basin draining into the Mediterranean Sea, which had a lower level at this time. Still further east, ice-dammed lakes overflowed via the Volga River to the Caspian Sea, the level of which rose by 28 m (Grosswald 1980, Chepalyga 1984). This caused a major transgression but was insufficient to reopen the outlet to the Black Sea through the Manytsh Depression.

In **Denmark** the Mecklenburg Advance of North Germany is correlated with the Bælthav Advance (Houmark-Nielsen 1987). The end of the Weichselian in Denmark as much as in Germany, was characterised by the isolation of dead-ice masses. Their decay led to the formation of vast ice-disintegration landscapes with characteristic landform assemblages (Stephan & Menke 1977, Prange 1979, Stephan *et al.* 1983). The identification of individual retreat phases as had been attempted earlier by Gripp (e.g. 1964) and more recently by Walther (1990) is not possible.

In **Britain** the Late Devensian glaciation is referred to as the Dimlington Stadial (Rose 1985). Its glaciers advanced far south along the east coast of Britain, touching the north coast of Norfolk. An ice tongue, following the Vale of York, advanced into Lincolnshire, damming the Trent to form proglacial Lake Humber. In the Pennines ice flow generally followed the major valleys (Catt 1991a, b). The ice sheet also intruded into the Wash, damming up a lake in the southern Fenland (West 1993). In the North Sea, the Dogger Bank remained ice-free, but major parts of the central and northern North Sea were glaciated (Ehlers & Wingfield 1991).

In the Irish Sea, ice advanced far to the south, touching the South Wales coast and possibly reaching as far as the Isles of Scilly (Harris & Donnelly 1991, Scourse 1991). However, many details of the glacial history are still far from clear. Not only are there problems of connecting the British and Irish stratigraphic sequences with those of continental Europe, but also with one another. The last glaciation of **Ireland**, the Midlandian (Mitchell *et al.* 1973) or Fenitian Glaciation (Warren 1985) consisted of a major Irish ice sheet, and a smaller Kerry/Cork ice sheet in the southwest. The flow lines and boundaries of the ice sheets have been reconstructed from landforms, striae and indicator erratics. The distribution of Galway Granite erratics, for instance, clearly demonstrates the southern extent of an Irish ice sheet (Warren 1991a, b), but this seems to have been older than the last glaciation. The distribution of periglacial features suggests that a corridor more than 50 km wide between the two Irish ice sheets remained unglaciated (Coxon & O'Callaghan 1987, Coxon 1988). A comprehensive overview of the glacial landforms in Ireland has been provided by McCabe (1985).

The dynamics of the Irish ice sheets are still hotly debated. Landforms along the west coast seem to indicate discharge of a major ice stream via Sligo Bay into the North Atlantic, forming a calving bay (McCabe *et al.* 1993). Isostatic depression under the weight of the ice sheet is also demonstrated, for instance by the occurrence of an ice-contact glaciomarine delta in the Upper Carey valley, Northern Ireland (McCabe & Eyles 1988), and by shallow-marine deposits at Portballintrae (McCabe *et al.* 1994). Isostatic downwarping in the Irish Sea basin has also been thought to have resulted in marine contact and glaciomarine deposition. The widespread occurrence of what is referred to as 'Irish Sea Till', a shelly diamicton found on the western shores of the Irish Sea, provides the evidence for this advance. This was formerly interpreted as a product of shelf ice (Devoy 1983). In Counties Dublin and Wicklow, however, it extends to altitudes of 300 m OD, clearly beyond any marine limit (Warren 1991a), therefore terrestrial ice must have also been involved. As a result of glacio-isostatic depression, development of a calving bay may have resulted in rapid disintegration of the combined British/Irish Ice Sheet at the end of the glaciation (Eyles & McCabe 1989, 1991).

In Scotland in the north, the Hebrides were not completely crossed by mainland ice during the Devensian. There is evidence of older glaciation that twice covered the northwestern Hebridean shelf to the shelf break, but the age of these ice advances is still uncertain (Stoker & Holmes 1991). The Outer Hebrides supported a local ice cap during the Late Devensian (Peacock 1984, 1991). The Inner Hebrides and the Orkney Islands were overridden by Scottish ice during the Late Devensian (Peacock 1983, Sutherland 1991), whereas Shetland and the Færøer Islands supported local ice caps (Jørgensen & Rasmussen 1986, Sutherland 1991). It has been postulated that parts of Scotland, mainly Caithness and Buchan, remained ice-free during the last glaciation (Cameron *et al.* 1987). Buchan is an area where deep pre-Quaternary weathering products have been preserved (A.M. Hall 1991). However, more recent investigations seem to indicate that both areas were probably covered by Late Devensian ice (A.M. Hall & Bent 1990, Sutherland & Gordon 1993). As on the Continent, the unglaciated parts of Britain and Ireland underwent strong periglacial reworking (Ballantyne & Harris 1994).

The extent of the Late Weichselian ice sheet in the **North Sea** is still a matter of debate. On the basis of seismic investigations, Long *et al.* (1988) and Johnson *et al.* (1993) concluded that the last British and Scandinavian ice sheets did not meet. However, the distribution of buried subglacial channels on the North Sea floor suggests a much wider extent of glaciation than assumed (Ehlers & Wingfield 1991). Recently, the investigation of sediment cores has shown that both ice sheets were in contact in the northern North Sea. AMS radiocarbon dates show that this event took place rather early, between 29,400 and ca. 22,000 BP. This ice advance thus predates the Dimlington Stadial (Sejrup *et al.* 1994).

The present landscape of Scandinavia was largely formed during the last glaciation. However, in places pre-Late Weichselian sediments have been preserved, including, for instance, sediments underlying till in Gudbrandsdalen, Norway (Bergersen & Garnes 1983), or the Veiki moraines in northern Sweden (Lagerbäck 1988a). It is thought that cold-based ice in the central parts of the ice sheet helped to preserve both pre-Weichselian landforms and deposits (Sollid & Sørbel 1994).

In **Russia**, the Late Weichselian Valdai Glaciation only covered the northern part of the country; the outermost ice margin having remained about 250 km north of Moscow. In the northeast, the Scandinavian Ice Sheet interacted with the Urals/Novaya Zemlya glaciation centre. West of the Urals there is no indication of an extensive Early Weichselian ice sheet. In the Timan–Pechora–Vychegda region the Early Valdai is only represented by fluvioperiglacial gravels and sands (Faustova 1995).

During the Weichselian maximum, the ice margin passed across the southeastern part of

Lithuania, and the outermost limit was situated only a few kilometres southeast of Vilnius. Ice retreat began slowly, the Pomeranian ice-marginal position being only about 50 km behind the maximum position. During the later phases of the Weichselian, the ice-marginal landscape was sculptured by major ice streams, leading to the formation of numerous streamlined landforms in Latvia (Aboltinsh 1989, Zelčs 1993, Dreimanis & Zelčs 1995) and **Estonia** (Raukas & Tavast 1994) and to the deposition of wide lobate end-moraine arcs. They comprise the Luga (13,000 BP), Otepää (12,600 BP) and Sakala (12,250 BP) end moraines in Latvia and the Pandivere (12,000 BP) end moraine in Estonia (dates after Raukas 1992). The later Palivere ice-marginal formation in Estonia (11,200 BP) resembles the Finnish Salpausselkä ridges. Apart from end moraines on the island of Saaremaa, it mostly consists of ice-marginal meltwater accumulations, deposited into the Nomme Lake, an early development phase of the Baltic Ice Lake. Dating of the older end moraines is largely hypothetical, but the Pandivere ice-marginal position can be linked via St Petersburg with the Finnish varve chronology (Karukäpp *et al.* 1992).

The Weichselian ice sheet sculpted the present-day landscape of **Finland**. Apart from the large ice-marginal formations (Glückert 1995), the characteristic landforms include drumlins, flutes, rogen moraines, hummocky moraine, De Geer moraines, deltas, sandurs and kames (Aartolahti 1995). During deglaciation, the Scandinavian Ice Sheet in Finland was subdivided into four major ice lobes (Fig. 154). The extent of the individual lobes and ice movement directions have been reconstructed by detailed geological investigations.

The early history of **deglaciation** in continental Europe is not well dated. After the ice had reached its maximum extent in Denmark about 18,000 BP, and about 15,000 BP, no active ice of the main Weichselian ice stream (Northeast Ice) was left in Denmark. The advance of the Young Baltic ice from the south and southeast reached its maximum before 14,200 BP. Deglaciation of Kattegat and southern **Sweden** started about 14,000 BP (Lagerlund & Houmark-Nielsen 1993). According to Smed (1994), by about 13,000 BP the ice sheet in Germany and Poland had retreated to the coastal zone of the Baltic Sea, only covering Rügen and parts of the Polish coast. In east-central Sweden recession of the ice margin occurred at an average rate of 200–300 m per year, with maximum values of about 1000 m. Stockholm was ice-free by about 10,430 BP (Risberg *et al.* 1991). At about 8860 BP the ice margin had retreated from the present Baltic Sea area (Strömberg 1989).

In **Finland**, part of the ice recession can be dated by varve chronology which covers a period of 2800 years (Sauramo 1929). It is a floating chronology, i.e. it has not been extended up to the present, but correlation with the Swedish varve chronology (Cato 1987, Strömberg 1990) has allowed precise dating of Finnish deglaciation to the period between 12,100 and 9300 BP. The Salpausselkä Ice-Marginal Zones are included in the chronology. Salpausselkä I formed between 11,100 and 11,300 BP and Salpausselkä II between 10,600 and 10,800 BP. This varve dating also allowed Donner (1978, 1982) to demonstrate that the Salpausselkäs did form not synchronously throughout their length. For instance, formation of Salpausselkä I started 350 years earlier in the west than in the east, and the ice margin retreated 100 years earlier from the west than from the east. However, the Salpausselkäs all represent the same climatic oscillation, i.e. the Younger Dryas Stadial (Saarnisto & Salonen 1995).

At the end of the Allerød, Finland still seems to have been barren. The earliest vegetation proven for the end of the Younger Dryas Stadial indicates tundra conditions (*Artemisia* zone). After formation of the Salpausselkä III ridge, about 10,000–9950 BP, a birch-dominated vegetation spread into the ice-free areas of Finland (Glückert 1995).

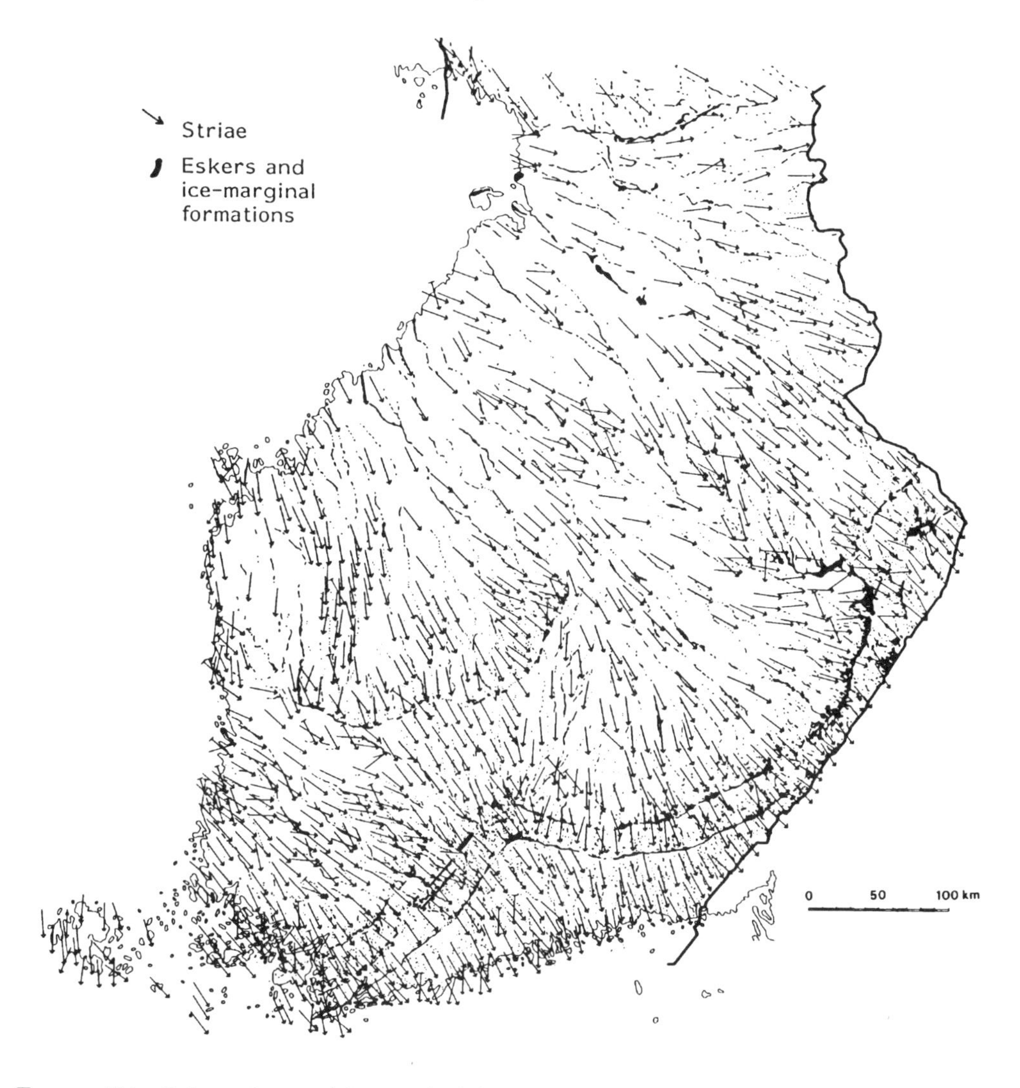

FIGURE 154 Striae, eskers and ice-marginal formations in southern Finland (Geological Survey of Finland, Department of Quaternary Geology 1986)

During the last cold stage, all of Europe except for Portugal and parts of Spain, as well as northern Asia, were inhabited by a fauna completely different from the present one, including mammoth (*Mammuthus primigenius*) and woolly rhinoceros (*Coelodonta antiquitatis*) (R.-D. Kahlke 1994). When the climate improved towards the end of the glaciation, the fauna adjusted to changes in a far more radical way than in any previous interglacial. The end of the Weichselian saw the extinction of a considerable proportion of the large mammals. In Europe these include mammoth (*Mammuthus primigenius*), bison (*Bison priscus*) and woolly rhinoceros (*Coelodonta antiquitatis*). In North America the loss of major terrestrial mammals is greater than the combined losses of the preceding three million years. Similar developments occurred worldwide, in some regions earlier (Australia), in others later (Madagascar, New

Zealand). A connection with increasing human settlement seems to be beyond doubt. The causes, however, are not only a result of hunting but also of increasing alteration of the natural environment (Martin 1990, Stuart 1993). The extinction of the giant deer (*Megaloceros giganteus*) in Ireland, for instance, occurred about 10,600 BP, which is about 1600 years prior to the archaeologically dated arrival of man on the island. Barnosky (1986) explains this specific extinction as a result of climatic deterioration in the Younger Dryas period. However, extinctions before the last cold stage had mostly been balanced by evolution or by immigration of new species. It seems that this adjustment was largely prevented by interference of man.

When in the lateglacial, the Weichselian ice started to melt about 13,000 BP, Central Europe was rapidly resettled. The Elster–Saale region of Germany seems to have had a relatively high population density during this period. The about 100 or so sites identified so far mainly comprise open-air settlements. Main immigration into Saxony and Thuringia seems to have occurred via Switzerland, along the Danube River and across the Frankenwald Mountains. Connections to the Middle Rhine region and to the North German lowlands (Hamburg culture) have also been demonstrated. The hunter-gatherers of that age belonged to the Magdalenian industry (Fig. 155). Tools, like arrowheads, were produced from the horns of reindeer as well as from flint (Feustel 1989).

The end of the Weichselian in Europe is characterised by a major period of climatic warming, which may be subdivided into a number of interstadials; Bock *et al.* (1985) distinguish a Meiendorf, Bølling and Allerød Interstadial. During the **Meiendorf Interstadial** climatic conditions already seem to have resembled those of the later Allerød. However, no afforestation occurred in Schleswig-Holstein. The archaeological finds of the 'Hamburger Stufe', the oldest of the so-called 'Federmesser' cultures, come from strata deposited after the *Hippophae* maximum of this interstadial. The **Bølling Interstadial** saw similar climatic conditions.

The strongest of the three Late Weichselian warm-climate oscillations is the **Allerød Interstadial**. Its type locality (north of Copenhagen) was described by Hartz & Milthers (1901). The British equivalent of the Allerød is the Windermere Interstadial. During the Allerød a birch–pine forest spread into North Germany and adjoining areas. When compared with the modern distribution of these species, a relatively low average summer temperature of about 12°C might be reconstructed. However, the presence of certain water plants indicates that in fact summers had been considerably warmer (about 15–16°C). This discrepancy can be explained by the fact that water plants could spread much faster than trees, so that the absence of species like oak, elm and alder, which could also grow under these conditions, was probably a reflection of delayed immigration (Zagwijn 1975). The end of the Allerød in major parts of Germany is easily identified because of the presence of an ash layer originating from an eruption of the Laacher See volcano in the Eifel Mountains about 11,000 BP.

Usinger (1985) is of the opinion that the Bølling Interstadial *sensu* Iversen (1954) cannot be separated from the classical Allerød. Also Walker *et al.* (1994) in their review of the climatic development at the end of the last glaciation also prefer to speak neutrally of a 'Lateglacial Interstadial'.

The **Younger Dryas** period, following the Allerød (about 10,800–10,000 BP) in northern Europe, saw the last prominent readvances of the Nordic ice sheets. The end moraines of this readvance include the Ra moraines of Norway (e.g. Sollid & Sørbel 1979, Sørensen 1979), the Middle Swedish end moraines (e.g. J. Lundqvist 1988, Bergström 1992), the Salpausselkä ridges of Finland (e.g. Glückert 1986, Rainio 1995) (Fig. 156) and the Kalevala Stage end

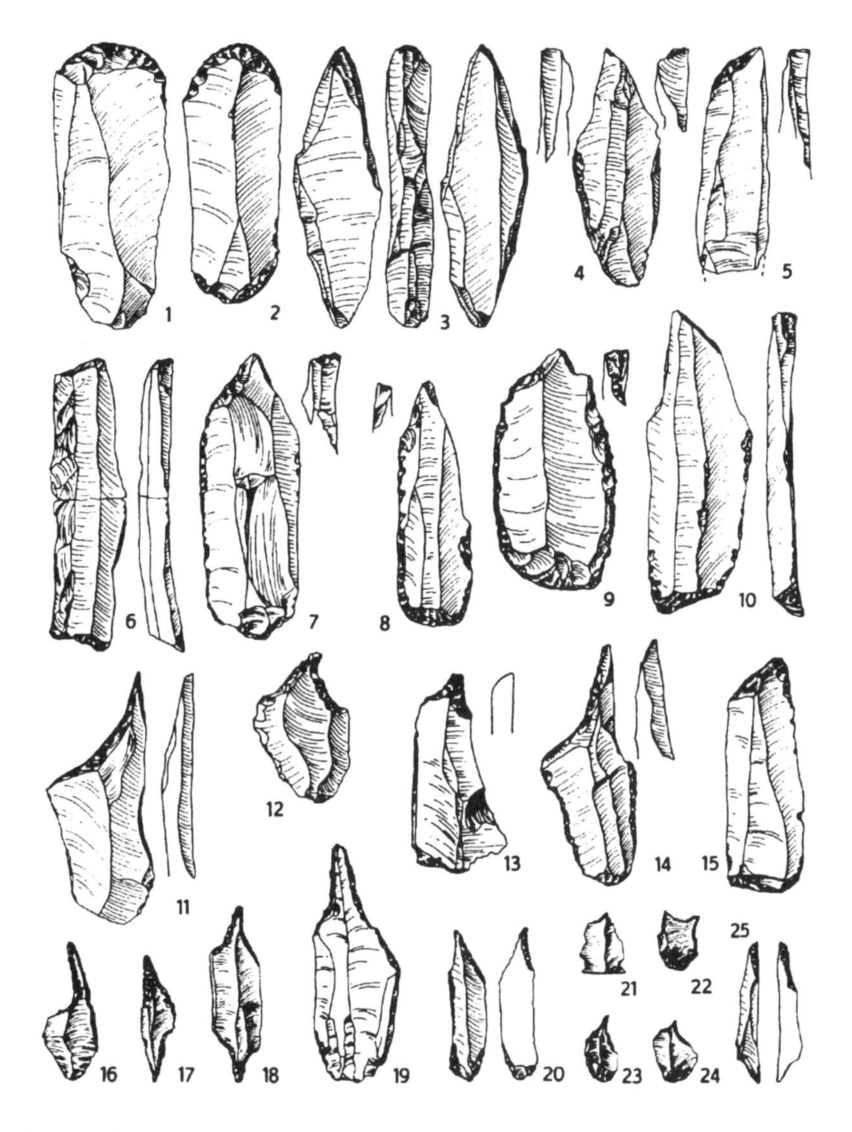

FIGURE 155 Artefacts of the Magdalenian (Late Palaeolothic) from Nebra (1–3, 6, 9, 11, 14, 20, 25), Groitzsch D (5), Groitzsch C (7, 8, 10, 13, 15), Groitzsch A (19), Halle, Galgenberg (12) and Saaleck (4, 16, 17, 18, 21–24). 1 = blade scraper, 2 = double scraper, 3 = double burin, 4, 5, 9, 11 = burin, 6, 7 = ridged knife, 8, 10 = burin scraper, 12, 13, 15 = prongs, 20–24 = fine drills, 17, 18 = double drills, 14, 16, 19, 25 = drills (from Toepfer 1970)

moraines of Russian Karelia (Rainio & Saarnisto 1991, Ekman & Iljin 1995). The climatic development in southern Sweden and Denmark has been reviewed by Berglund *et al.* (1994). In Scotland the equivalent of the Younger Dryas Stadial is the Loch Lomond Stadial, in Ireland the Nahanagan Stadial. Cirque glaciers formed in the mountains not only of Scotland but also of northern England, Ireland and Wales (e.g. Sissons 1980, 1981, Gray & Coxon 1991). The ice sheet in the west-central Grampians reached a maximum size of 200 km^2 and a thickness of

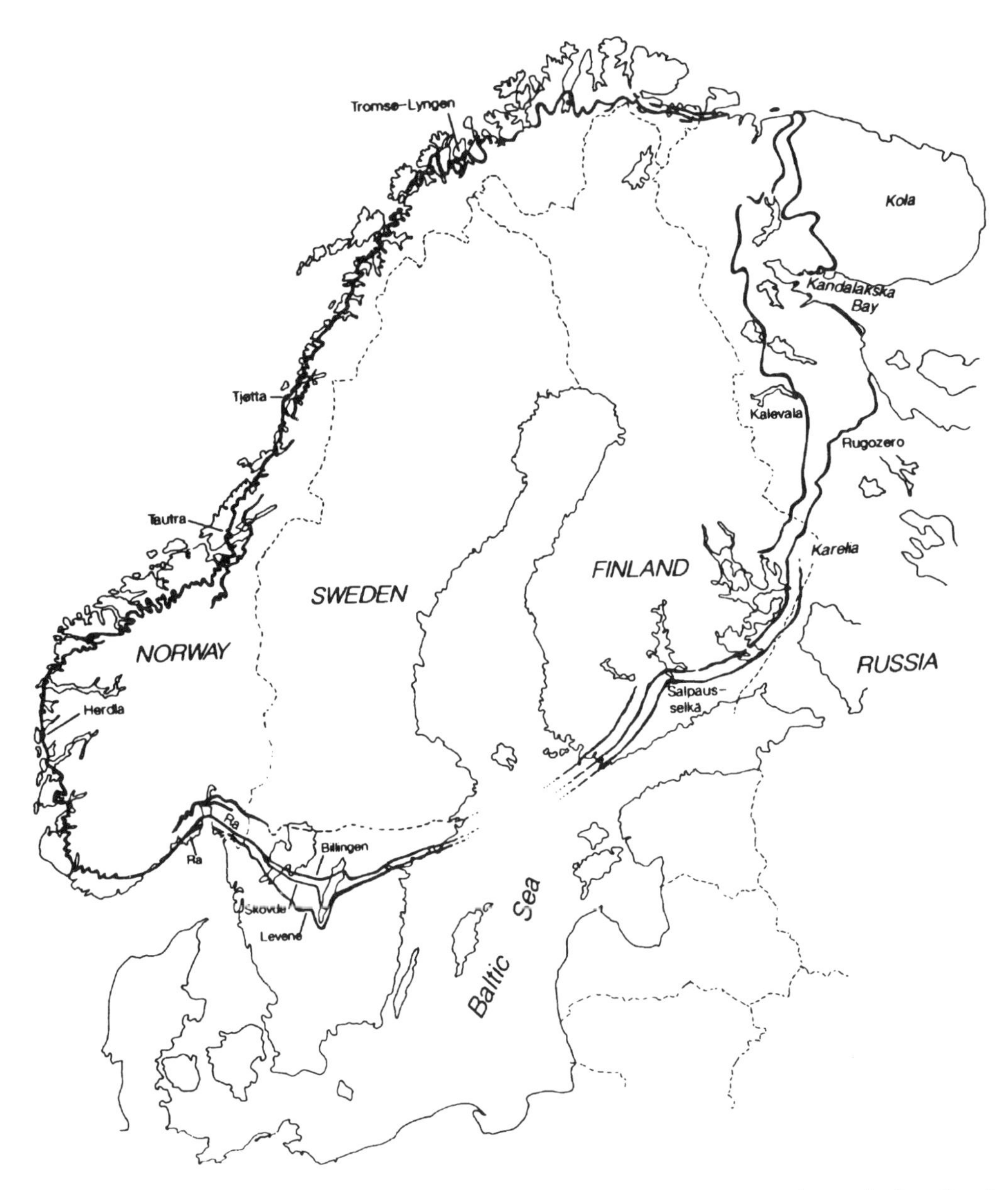

Figure 156 End morains and ice-marginal deposits of the Younger Dryas readvance in Scandinavia (from Andersen *et al.* 1995)

400 m (Walker *et al.* 1994). The Vedde Ash from the Icelandic volcanic region forms an important marker horizon within the deposits of this period; it was originally dated to 10,600 BP (Mangerud *et al.* 1984), but new AMS dates indicate an age of 10,300 BP (Bard *et al.* 1994). As yet it has been identified in deposits of Younger Dryas age from Scotland, Norway and the North Sea. During the later phase of the Younger Dryas Stadial aeolian activity was considerably revived (Bohncke *et al.* 1993).

The age and duration of the Younger Dryas period are difficult to determine, because it largely coincides with a radiocarbon age plateau around 10,000 BP. The plateau reflects a period during which the atmospheric $^{14}C/^{12}C$ ratio constantly decreased, which makes exact dating impossible (see chapter 10.3). Investigation of a laminated gyttja in Lake Gościąż in Poland has yielded more than 12,000 couplets of light, mostly calcitic and dark, organic-rich layers. Correlation with a pollen diagram and calibration with radiocarbon dates suggests that the Younger Dryas cooling lasted 1600 years (Ralska-Jasiewiczowa *et al.* 1992, Goslar *et al.* 1993). This is in disagreement with most older assessments of the Younger Dryas length (e.g. Zolitschka *et al.* 1992). The absolute time scale of the GRIP and GISP2 Greenland ice cores suggest that the Younger Dryas lasted for 1150 years (Johnsen *et al.* 1992) or 1300 years (Taylor *et al.* 1993). Detailed varve-chronological investigations in Lake Soppensee in Switzerland have also shown a Younger Dryas of about 1140 years (Hajdas *et al.* 1993).

The climatic changes of the lateglacial naturally caused considerable cultural adaptations. The faunal changes, like the disappearance of mammoth, were of major consequence. For instance, the Ahrensburg Group hominids in the Younger Dryas, which has been especially well investigated in the Hamburg region, had adapted their lifestyle to the seasonal migration of the reindeer. Gripp's (1943) investigation of the bones and horn finds demonstrated that the hunting sites represented summer camps; no remains of animals slain in winter were found.

11.17 HOLOCENE

The subdivision of the Holocene currently used in Europe dates from investigations made at the end of the last century in Norway and Sweden. The Norwegian Blytt originally introduced the terms Boreal, Atlantic, Subboreal and Subatlantic to characterise the recent Norwegian floral distribution (Blytt 1876). After further publications by Blytt and the Swede Sernander, these terms were used in Sernander's dissertation to characterise the postglacial vegetational development (Sernander 1894). The gradual development of the classification is reviewed by Mangerud *et al.* (1974).

Sernander (1894) was well aware of the problem that was inherent in his subdivision. He wrote: 'It is obvious that a truly temporal coincidence between e.g. an aspen–birch horizon in Skåne and a similar zone in Jämtland *a priori* has to be regarded as most doubtful.' (Sernander 1894: 4; quoted after Mangerud 1982). Immigration of plants over distances of thousands of kilometres must have taken considerable time. Sernander therefore suggested that age determinations had to rely on other methods, such as climatic oscillations or sea-level changes. After the 1910 Geological Congress in Stockholm, the terminology of Blytt and Sernander spread all across northern Europe. The system was originally based on macrofossils from peat. In the 1930s when Quaternary geologists turned their interest to lake deposits and when the analysis of macro-remains was supplemented by pollen analyses, the terms lost their original meaning and were largely applied to separate the newly created pollen zones (e.g. Nilsson 1935).

With the advent of radiocarbon dating in the 1960s, it became possible to date organic deposits directly. The contrast between the biostratigraphical analyses and their chronostratigraphical interpretations then became obvious. As a consequence, Mangerud *et al.* (1974) suggested a new definition for Sernander's old terms based on chronology, which on the one hand as closely as possible reflects the traditional usage of the terms but on the other hand avoids any geological definition. According to their chronology, the following periods were distinguished:

Subatlantic	2500 BP–present
Subboreal	5000–2500 BP
Atlantic	8000–5000 BP
Boreal	9000–8000 BP
Preboreal	10,000–9000 BP

This subdivision is generally applied today by most workers in northern Europe (Mangerud 1982).

The vegetational development of the postglacial period in Central Europe has been investigated by Firbas (1949) and Overbeck (1975). The following characterisation of the individual phases is largely based on Overbeck (1975).

The **Preboreal** was characterised by reforestation and a rapid rise in temperature. In the pollen associations this is expressed by the disappearance of the heliophytes and the predominance of widely distributed *Betula* and *Pinus*.

In the **Boreal** warming continued. Hazel (*Corylus*) spread. The distribution of mistletoe, ivy and holly (*Viscum, Hedera* and *Ilex*) indicate that summer temperatures were possibly already higher than at present in the Boreal.

In the **Atlantic** period the climatic optimum of the Holocene was reached. In North Germany summer temperatures were about 2–3° higher than at present. Since the beginning of the Atlantic period, the percentage of pine in the North German vegetation declined strongly. Mixed-oak-forest spread, containing *Alnus, Ulmus, Quercus* and *Tilia*. Later ash (*Fraxinus*) also immigrated. In the upland regions beech (*Fagus*) appeared; for a short time it advanced as far north as eastern Holstein. Spruce (*Picea*) spread in the Harz Mountains.

Man in the early Holocene still lived on hunting, gathering and fishing. About 10,000 BP agriculture developed in the Near East; about the same time the first animals were domesticated. Soon after, pottery making began. The new forms of economy required fixed settlements, which meant the end of nomadism. This cultural revolution characterised the shift from the Mesolithic to the Neolithic period. Because the introduction of agriculture and livestock breeding did not occur simultaneously, the opening of the Neolithic cannot be fixed to a certain date all over Europe. In Central Europe, the Neolithic Bandkeramik spread between 6700 and 6400 BP. Loess areas were settled first; between 5500 and 5000 BP agriculture also spread into the North German lowlands. The distribution of the 'Trichterbecherkultur' of that period, named after the shape of their ceramics, is characterised by a widespread severe deforestation (Behre 1988).

Since the onset of the **Subboreal** period, the influence of agriculture was felt in the vegetational associations. In the pollen diagrams this change is largely indicated by the appearance of the agricultural weeds, such as plantain (*Plantago*). Subsequently, extensive clearing of former forest areas occurred, leading to intensified erosion. In the river valleys flood loam began accumulating. In the course of the Subboreal temperatures dropped by about 2–3°. Amongst the trees, beech (*Fagus*) and hornbeam (*Carpinus*) spread.

In the **Subatlantic** period the climate finally became cooler and more humid. Beech spread further north. The increasing effects of human interference with the natural vegetation are shown in most pollen diagrams.

The temporal correlation of the various pollen zones of the North German subdivision largely correspond to the system suggested by Mangerud *et al.* (1974). Only the transition from Subboreal to Subatlantic has been placed by Overbeck (1975) at about 1100 to 900 BC (3100–

2900 BP) and thus considerably earlier than Mangerud *et al.* His pollen zone X (oak period) and the first part of his pollen zone XI (beech period) would, according to Mangerud *et al.* (1974), still form part of the Subboreal.

At the beginning of the Holocene the **glaciers** were melting worldwide as a result of global warming. In Norway, deglaciation of the last ice sheet took place almost exclusively in the early Holocene. The deglaciation pattern and the resulting landform assemblages have been described, for instance, by Carlson *et al.* (1979), Sollid & Sørbel (1979, 1981) and Sollid & Reite (1983). Numerous reconstructions have been published on the glacier retreat, particularly from the Alps (e.g. Patzelt 1975, Porter & Orombelli 1985, Furrer 1991; cf. chapter 12.15). The extent of this glacier retreat is hard to determine. For instance, Porter & Denton (1976) assume that by the Middle Holocene a large number of mountain glaciers had disappeared completely, and that they only started to reform in the later cooler and/or moister periods. They refer to this process as 'Neoglaciation'. There are various opinions about the extent of this process and the number and age of the Holocene ice advances. There also seem to have been major regional differences (F. Röthlisberger 1986).

By investigating peat and lake deposits from the marginal regions of Jostedalsbreen (Norway), Nesje *et al.* (1991) showed that dynamically active, extensive valley glaciers were only present in Scandinavia during the early Preboreal period. Later in the Preboreal, their volume decreased. However, at the Preboreal/Boreal boundary another major glacier advance occurred, the outermost boundary of which has been reconstructed at about 1 km beyond the limit of the 'Little Ice Age'. At about 9000 BP the Scandinavian Ice Sheet had vanished (Birks *et al.* 1994). Strong warming during the Atlantic period resulted in an uphill migration of vegetational zones. At this time, during the Holocene climatic optimum, elm trees (*Ulmus*) grew in southern Norway up to an altitude of 700 m a.s.l. This is the present upper boundary of birch and implies that the snow line at that time was raised by about 400 m, so that only small glacier remnants at high elevations would have survived on the Jostedals Plateau (Nesje *et al.* 1991). This situation is in contrast to the development in the Alps where the glaciers did not disappear (see chapter 12.15).

By radiocarbon dating of tree trunks, it has been shown that in Scandinavia spruce was also distributed up to much greater heights in the Boreal and Atlantic periods than at present (e.g. Karlén 1988). Towards the late Atlantic/Boreal transition, elm declined and birch spread again, reflecting renewed formation of glaciers in the Jostedals Plateau. Simultaneously, a short period of strong temperature decline has been recorded in European terrestrial mollusc assemblages (Rousseau *et al.* 1994). The first major ice advance in Scandinavia did not occur before 3700 and 3100 BP. Warming during the early Middle Ages again resulted in enhanced melting. The readvance of the Little Ice Age had already begun in Scandinavia around AD 1030–1080. From historical records it can be shown that about 1650–1680 another serious climatic deterioration was felt, finally leading to the neoglacial maximum extent of the glaciers by the mid-18th century (Nesje *et al.* 1991).

11.18 QUATERNARY DEVELOPMENT OF THE BALTIC SEA

11.18.1 Origin of the Baltic Sea

The early development of the Baltic Sea (prior to the Weichselian Stage) is largely unknown. At the time of the Baltic River System (from the Late Tertiary to the Early Pleistocene Waalian

Stage) fluvial deposits were transported from Scandinavian source areas into North Germany and the Netherlands. Remnants of these deposits include the Loosen Gravels in Mecklenburg and the Kaolin Sand on the Isle of Sylt. During their deposition the Baltic Sea depression cannot have existed. The further development from the Menapian to the Elsterian Cold Stage is unknown. If and how far earlier glaciations during this period were able to form a proto-Baltic Sea cannot be said at present. There can be no doubt, however, that glacial erosion played a prominent part in forming the Baltic Sea basin, and that the basin has increased in size and depth through the Pleistocene. The relatively low percentage of Cretaceous chalk and flint in the Elsterian till lithology in North Germany may be an indication that at that time the Tertiary cover overlying the Cretaceous in the western Baltic Sea region was still largely intact (K.-D. Meyer 1991).

The first clear evidence of a Baltic Sea stems from the Holsteinian Interglacial, several marine interglacial deposits of which have been described from the Baltic Republics (Cheremisinava 1970, Danilans 1973, Dreimanis & Zelčs 1995). No marine Holsteinian deposits are known from Poland, but in Germany the Holsteinian transgression from the North Sea can be followed across Schleswig-Holstein into eastern Mecklenburg–Vorpommern. Brackish facies are also found in northwestern Brandenburg (Cepek 1967). Several occurrences of marine Holsteinian deposits in western Mecklenburg and northeastern Niedersachsen seem to indicate a marine connection between the North Sea and the Baltic Sea via the lower Elbe region (Woldstedt & Duphorn 1974).

That the glaciers of the Older Saalian Glaciation crossed the Baltic Sea depression without apparently having greatly influenced their advance direction (K.-D. Meyer 1991) does not necessarily negate the possibility of a Holsteinian predecessor of the Baltic Sea. The same procedure was repeated during the Weichselian Glaciation, at a time when the Baltic Sea definitely existed. For the flow behaviour of the ice sheets glaciodynamic conditions seem to have been more important than bed morphology (Ehlers 1990a, b).

It is well established that the Eemian Sea covered major parts of the present Baltic Sea basin, marine Eemian deposits being known from Denmark, Schleswig-Holstein and Mecklenburg–Vorpommern. The occurrences of Eemian deposits along the route of the Nord-Ostsee Kanal identified in borings seem to indicate that a narrow connection between the North and Baltic seas existed across Schleswig-Holstein (Kosack & Lange 1985). In Poland two transgressions of the Eemian Sea can be distinguished. Here a wide bay in the lower Wisła valley reached inland beyond Gdańsk (Makowska 1979). However, the extent of the Eemian Sea northwards and eastwards is not completely known. Marine Eemian deposits have been identified on the Isle of Prangli in Estonia (Liivrand 1991). There can be no doubt that the Eemian Sea not only reached the Finnish coast but also affected parts of northern Russia, where marine Eemian deposits have been found at Mga and Petrozavodsk. That the influence of the Eemian Sea reached further east than the Holocene Baltic Sea has been explained by the major extent of the Saalian ice sheets and the different deglaciation history of the Saalian and Weichselian ice sheets (Forsström *et al.* 1988; Fig. 157). The centre of the glaciation and therefore the maximum isostatic depression may have been located in southern Finland, as suggested by Forsström & Eronen (1985), but it cannot be supported by any field evidence (Donner 1988). The Eemian transgression was probably the last marine transgression into the Baltic Sea basin before the Holocene.

During the Middle Weichselian those parts of the Baltic Sea not covered by ice formed a large glaciolacustrine basin. Ice-dammed lake sediments from this period on Møn have been

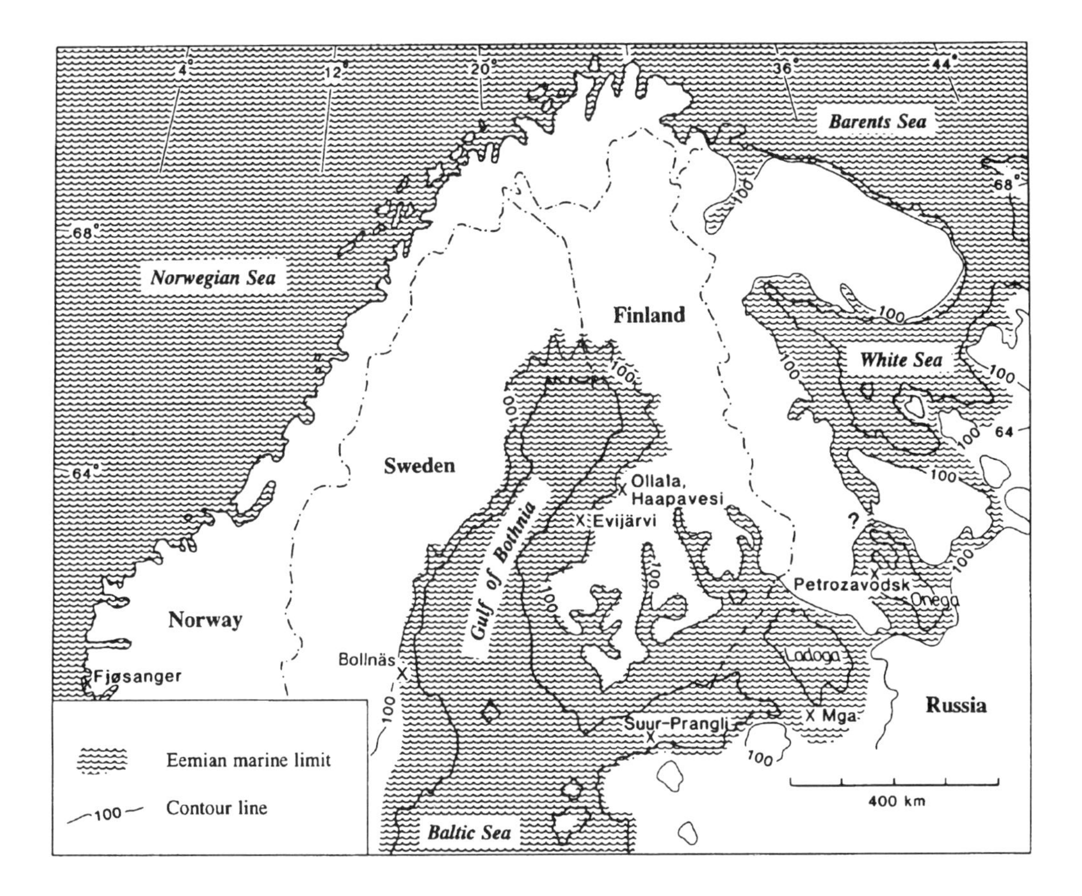

FIGURE 157 Maximum extent of the Eemian Sea in Fennoscandia, in the Baltic region and adjoining parts of Russia about 3000–4000 years BP early in the Eemian Interglacial (from Forsström *et al.* 1988)

found to contain increasing amounts of ice-rafted detritus, indicating the advance of the Late Weichselian ice sheet. From 15,000 to 25,000 BP the Baltic Sea was completely covered by the Scandinavian Ice Sheet (Houmark-Nielsen 1994).

11.18.2 Postglacial development of the Baltic Sea

That the Baltic Sea region after the last glaciation experienced important changes of sea level has been known since the early 18th century. De Geer (1888/90) was the first to draw a map of the highest marine levels so far proved in Sweden. A few years later (De Geer 1896) he presented the basic principles of the postglacial history of the Baltic Sea and convincingly demonstrated the effects of isostasy on Scandinavia. The influence of eustasy was still unknown at that time. It was proved only a quarter of a century later by Nansen (1922) and especially by Ramsay (1924).

The development of the Baltic Sea after the end of the last glaciation can be roughly subdivided into four major phases (Fig. 158):

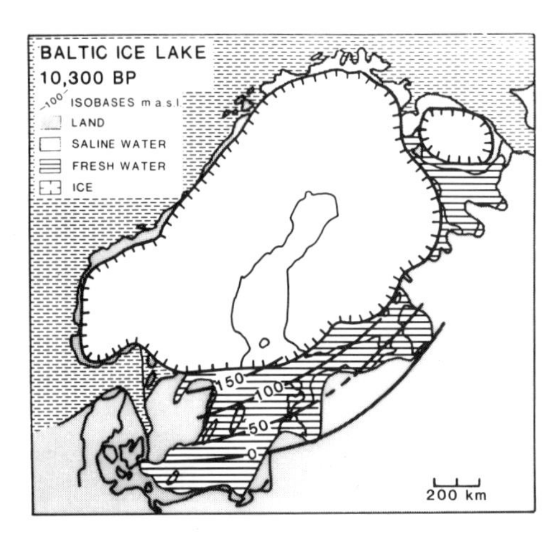

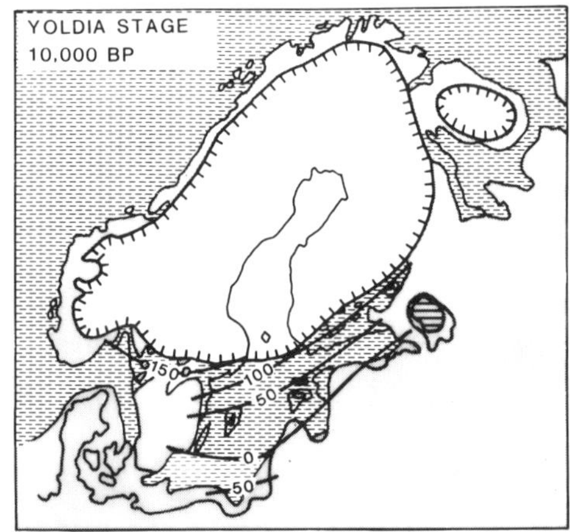

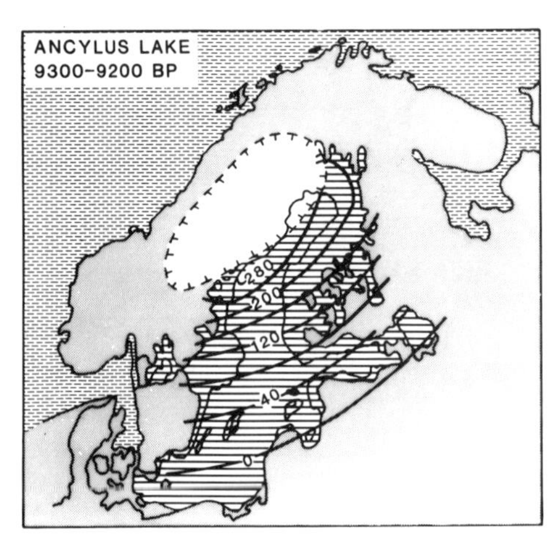

FIGURE 158 Postglacial development stages of the Baltic Sea. A = Baltic Ice Lake, B = Yoldia Sea, C = Ancylus Lake, D = Litorina Sea (from Eronen, unpublished)

1.	Baltic Ice Lake	from deglaciation to 10,300 BP
2.	Yoldia Sea	10,300–9500 BP
3.	Ancylus Lake	9500–8000 BP
4.	Litorina Sea	8000 BP–present

The **Baltic Ice Lake** was the last to be detected. At the turn of the century the opinion still prevailed that the Baltic Sea basin had remained ice-filled until finally the Yoldia Sea broke through the barrier. Munthe (1902) in his description of map sheet Kalmar for the first time mentioned that an ice-dammed lake preceded the Yoldia Sea phase. Today it is known that the Baltic Ice Lake developed gradually by merging of numerous local ice-dammed lakes, which only combined to form a single, large lake after the Baltic Sea basin became ice-free (Kvasov 1978). Most likely the Baltic Ice Lake did not extend into northeastern Finland, as had been assumed by Sauramo (1958). Moreover, the idea that the ice-lake phase might have been preceded by a lateglacial incursion of the sea (the lateglacial Yoldia Sea of Sauramo 1958) remains to be substantiated (Hyvärinen & Eronen 1979).

The first drainage of the Baltic Ice Lake was to the Kattegat through the Danish/Swedish straits, probably the Øresund Strait (Agrell 1976); the lake level lay above the global sea level. The western margin of the lake was formed by the Darss Sill, a threshold at the bottom of the Baltic Sea between the Darss Peninsula of Mecklenburg and the Danish Isle of Falster. The shape and size of the lake changed rapidly, depending on further ice melt and on the progressive isostatic uplift. The varved clays from the Baltic Ice Lake in the Bornholm Basin and Hanö Bight have been connected with the south Swedish varve chronology. They were shown to have been formed during the Bølling Interstadial (Duphorn *et al.* 1981, Ringberg 1991). Björck & Digerfeldt (1984, 1989) demonstrated that final drainage of the Baltic Ice Lake occurred in two phases. An interruption of several hundred years was imposed when the glaciers in the Younger Dryas phase readvanced strongly, blocking the outlet. The final drainage of the Baltic Ice Lake occurred ca. 10,300 BP when the ice margin in Sweden retreated north of Mount Billingen. This event was more or less simultaneous with the transition from the Younger Dryas phase to the Preboreal, i.e. with the Pleistocene/Holocene boundary. The zero varve of the Finnish varve chronology refers to the point when the level of the Baltic Ice Lake dropped to global sea level (Fredén 1979).

The traces of this event around Mount Billingen at Skövde, in the area between Lakes Vänern and Vättern in Sweden, were not found until recently. Högbom (1912) had interpreted a small erosional channel cut into Cambrian sandstone at St Stolan (15 km north of Skövde) as the outlet of the Baltic Ice Lake. When G. Lundqvist (1958) reinvestigated the area around Mount Billingen, he was more sceptical. The relatively small landforms contrast with the water masses which should have been released by the outburst of the gigantic Baltic Ice Lake. The difference in level between the Baltic Ice Lake and global sea at Mount Billingen at that time amounted to 26 m (Fredén 1979). Strömberg (1992) described coarse and huge bouldery glaciofluvial sediments at Klyftamon, ca. 10 km west of Mount Billingen. He interpreted the sediments as having been formed during roughly 90 years of gradual drainage of the Baltic Ice Lake.

The high-Arctic salt-water mollusc *Yoldia arctica* (today: *Portlandia arctica*) characterises the next stage of the development of the Baltic Sea, the **Yoldia Sea**. It is thought that during this phase marine salt water intruded into the Baltic Sea in an eastward direction when the ice margin had retreated from Mount Billingen and the level of the Baltic Sea had adjusted to world sea level. However, the influence of the salt water may have been considerably less than originally thought. Already De Geer (1940) had assumed that the environment in the Stockholm region was favourable for *Yoldia arctica* only for a very short time, hardly more than a hundred years. Precisely how far to the east the salt-water influence actually reached is debatable. The salt-water influence during the Yoldia Phase has been traced by foraminifera, diatoms and various molluscs found from Dalsland–Värmland in the west to the Stockholm area in the east (e.g. Brunnberg & Possnert 1992, Wastegård 1995). The short brackish phase can be followed in the sediments of the Baltic from southeast of Stockholm to Gotland (N.-O. Svensson 1989, Wastegård *et al.* 1995). Abelmann (1985) only found brackish diatoms of the Yoldia Phase in cores from the Karlsö Basin (near Gotland) on the Baltic Sea floor. In contrast, only lacustrine diatoms have been preserved in the Gotland Basin, Bornholm Basin and Gdańsk Bight from the Yoldia Phase.

The end of the Yoldia Sea was reached when, as a consequence of continuing land uplift, the connection to the open sea, the so-called Närkesund, became increasingly shallower. Finally the threshold between Kilsbergen and Tiveden was uplifted sufficiently to prevent any saline deep currents from entering the Baltic basin. The name **Ancylus Lake** for the subsequent freshwater phase was introduced by De Geer (1890). It refers to the freshwater mollusc *Ancylus fluviatilis*. The onset of the Ancylus Lake Phase predates the first appearance of *Alnus* in Sweden, which means it occurred around 9200–9300 BP. The drainage of such a large lake to the west theoretically should have resulted in the incision of a major canyon. After a lengthy search this outlet was finally identified by Munthe (1927) and Von Post (1928) at Degerfors in Sweden. The traces in the field, however, are relatively scarce, and the spread of major boulders could just as easily have resulted from boulder-rich till (G. Lundqvist 1958). A renewed investigation of the area by Fredén (1979) provided no convincing evidence of an outlet of the Ancylus Lake. Probably global sea level at this time was approximately equal to the water level in the Baltic Sea, and inflow of salt water was only prevented by the high sill between the two basins (Fredén 1979). Because of the differential isostatic uplift, the Ancylus Lake reached its maximum extent in the south later than in the north. The height of the water level in the Baltic Sea basin caused drainage to occur via the Darss Sill threshold. During the course of this event a drainage channel 10–20 m deep was cut into the subsurface. This catastrophic drainage resulted in an overall drop of the lake level of about 20 m (Kolp 1986). The westward extension of the Ancylus Lake beyond the Darss Sill finally led to renewed connection of the basin to the sea.

The **Litorina Phase** represents the last phase of Baltic Sea development during which the present connection with the sea was established via the Øresund, Lillebælt and Storebælt straits. The term Litorina Phase was originally used by Lindström (1886). The phase is named after the snail *Littorina littorea*. The change from the Ancylus Lake to the Litorina Sea did not occur abruptly but rather in a gradual transition. In Blekinge, southern Sweden, the salt-water diatom *Mastogloia smithii* was already found in the Late Boreal, about 8500 BP (Berglund 1964). In Ångermanland, further to the north, the transition occurred about 1000 years later (Fromm 1938). The Litorina Phase can be subdivided into a number of subphases: the Mastogloia Sea (low salt content), the Litorina Sea *sensu stricto* (highest salt content), the Limnea Sea (brackish influence, starting about 4000 BP) and the Mya Sea (present Baltic Sea, the deposits of which are characterised by the mollusc *Mya arenaria* which had immigrated in the 16th/17th century.

The coast of Mecklenburg-Vorpommern was affected by the Litorina transgression about 7800 BP. In Schleswig-Holstein and Mecklenburg signs of a regression at about 1000 BP have been observed, the effect of which had been limited to the western Baltic Sea because of the different isostatic uplift pattern. The alternation between transgressive and regressive phases seems to have had a decisive influence on the formation of the straightened coastline and the systems of beach ridges and recurved spits along the southern Baltic Sea coast (Kolp 1982).

12

Quaternary History of the Alps

12.1 TECTONIC FRAMEWORK

The Alps are a young mountain range, the orogenesis having begun in the Cretaceous (Austrian and Vorgosauian orogenic phases). Since then the central Alps have remained above sea level, though large-scale uplift did not occur before the Late Tertiary. However, orogenesis was still incomplete at the beginning of the Quaternary, and uplift continues up to the present day, at least in major parts of the Alps. In the Hohe Tauern range in the central eastern Alps, uplift of up to about 1 mm/year has been demonstrated by high-precision levelling (Senftl & Exner 1973). In addition, uplift rates in the Swiss Alps of over 1 mm/year have been measured (Gubler 1976). A profile from the Schwarzwald (Black Forest) to the Nufenenpass (St Gotthard) illustrates this extent of the recent uplift (Fig. 159).

Most tectonic features of the Alpine bedrock predate the glaciations. Large-scale tectonic movements, however, seem to have modified the extent of Pleistocene erosion. Nevertheless, there is no evidence of any Late Pleistocene displacement of hardrock units. Similarly, evidence is lacking for the older gravel trains having been displaced by tectonics, as was assumed by Penck (1922b). Nevertheless, Schlüchter (1981a) points out that the Early Quaternary 'Deckenschotter', found in the highest positions of northeastern Switzerland, are either missing or covered by younger deposits in the central Swiss Mittelland region. In this area the former valley floors have been buried under sedimentary sequences several hundred metres thick, whilst similar deposits are lacking in northeastern Switzerland. This would seem to indicate that the eastern parts of the central Swiss Mittelland have been uplifted since the Middle Pleistocene, whilst at the same time the central parts have subsided. This tendency is most obvious in the western Mittelland area in the Lac Leman (Lake Geneva) area (Gubler 1976).

The southwestern corner of the Schwarzwald (Black Forest) subsided during the Early Pleistocene, resulting in the diversion of the Alpine Rhine to its present westerly course. This area is still subsiding today. The same is true of the Rhine graben, which is still tectonically active (Hantke 1978). In contrast, the south German Alpine foreland was largely uplifted so that the Quaternary gravel trains form a sequence of terraces, in which the oldest gravels are found in the highest positions. The only exception is the Munich gravel plain, where subsidence has prevailed and the oldest gravels are covered by younger deposits (Knauer 1931, Jerz 1981).

Apart from true tectonic movements in the Alps, a certain isostatic component must be taken

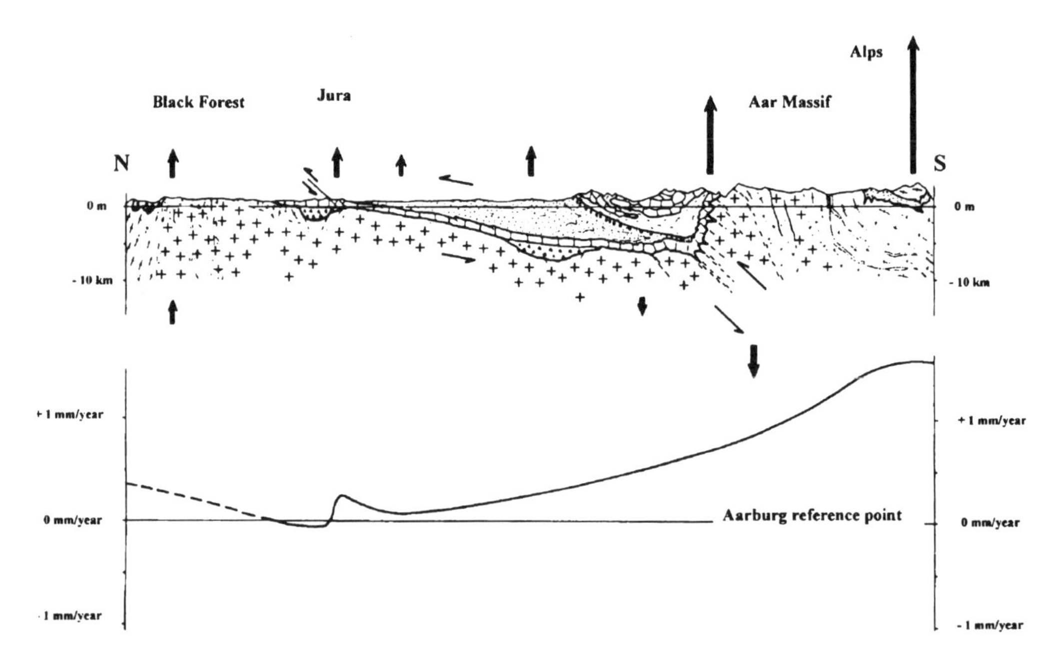

FIGURE 159 Movement directions and rates of uplift along a cross section from the Schwarzwald (Black Forest) to Nufenenpass (Swiss Alps) (from Müller *et al.* 1984)

into account, in accordance with other formerly glaciated areas. By using a viscoelastic model of the crust and mantle, Gudmundsson (1994) showed that the recent Alpine uplift rate lies within the boundary of an assumed glacio-isostatic rebound. Jäckli (1965) assumed that at least some of the frequent rock falls result from this effect. However, this opinion has been disputed, since the extent of the glaciated area and ice volume were relatively small. Whilst the North European glaciation (without Siberia and the Barents Sea) covered an area of 3,300,000 km^2, only 126,000 km^2 were ice-covered in the Alps. If the ice volumes are compared, the ratio is about 40:1 (Schneider 1976).

12.2 ORIGIN OF THE VALLEYS AND BASINS

The present drainage pattern of the Alps is controlled by tectonics and lithology. The valley courses generally follow earlier tectonic lineaments. At least the major Alpine valley system was outlined in the Tertiary. In the eastern Alps this can be demonstrated by the distribution pattern of inner Alpine Tertiary deposits and by preserved palaeolandforms. In the southern Alps canyons several hundred metres deep formed when the Mediterranean fell dry during the Messinian salinity crisis (Felber *et al.* 1991), and also the southern Alpine lakes Lago Maggiore, Lago di Lugano, Lago di Como, Lago d'Iseo and Lago di Garda may have first formed at Messinian time (Bini *et al.* 1978, Finckh 1978). In the glaciated areas most major valley courses have remained largely unchanged since preglacial times.

The ice streams of the Alps have reshaped the pre-existing relief to a glacially moulded landscape. One element of this relief are the U-shaped valley troughs. A characteristic trough valley has a wide, flat valley floor and steep to vertical flanks, which higher up grade into more

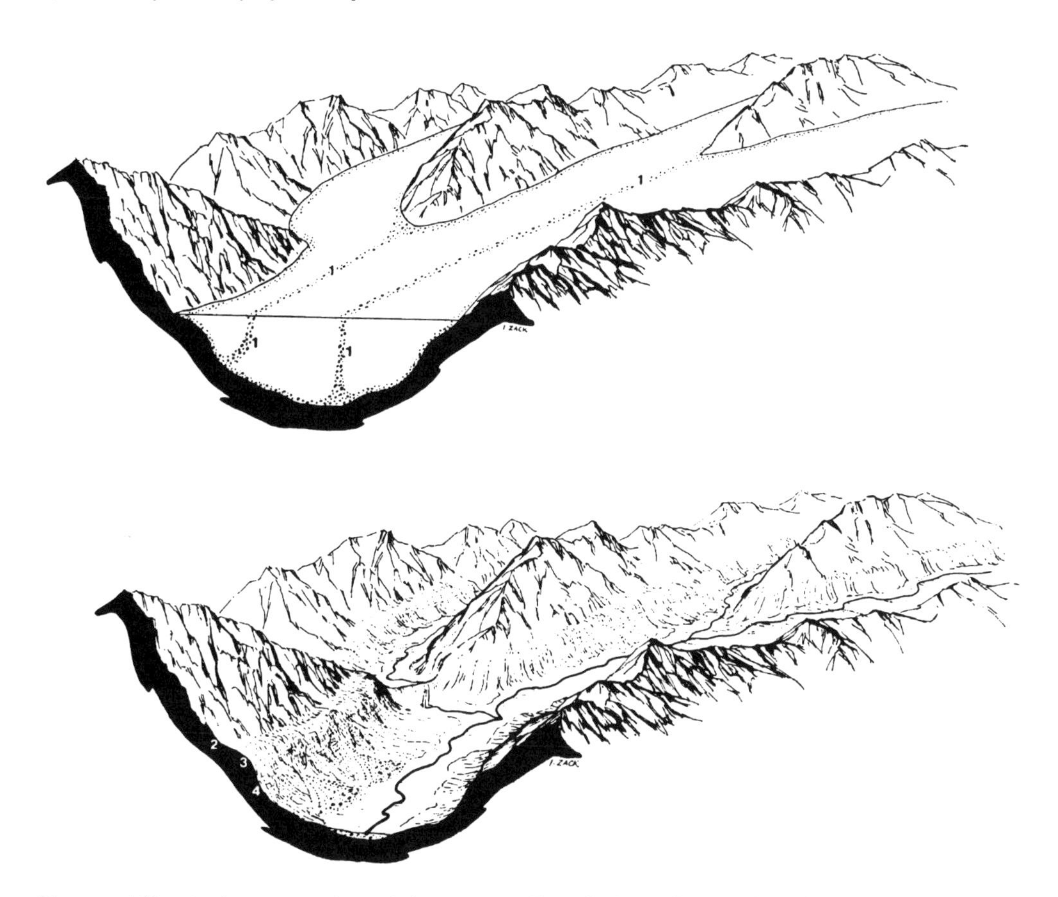

FIGURE 160 An ice stream formed from two valley glaciers of approximately equal thickness. The individual ice masses are separated by a medial moraine (1). The ice stream alters the shape of the valley to a trough (U-shaped valley). Above the ice stream frost weathering results in sharply modelled landforms. After ice decay the worn landforms of the trough valley are clearly visible: Schliffkehle (2), Trogschulter (3), Trogwand (4). The minor tributary valley that was occupied by a much smaller glacier now ends as a hanging valley high above the main valley floor. The step is not yet cut through by the creek that drops as a waterfall into the main valley (from Van Husen 1987)

gently sloping parts. Broad trough shoulders *sensu* Louis (1952), which might be interpreted as relics of former valley floors, have been comparatively rarely preserved. There can be no doubt that even during the last glaciation the ice cover rose above the trough shoulders. Above the glacially sculpted parts a few peaks are found that have been affected neither by local nor long-distance glaciation. They are called nunataks. The upper limit of glacial erosion ('schliffgrenze') is well developed, especially in the central Alps. Since glacial erosion was much stronger in the main valleys than in the tributaries, many of the latter today finish as hanging valleys, high above the modern main valley floor (Fig. 160).

Glacial polish and roches moutonnées indicate strong glacial erosion. Additionally, glacial erosion has largely modified the longitudinal profile (thalweg) of the valleys. Thus numerous overdeepened basins were formed. In the eastern Alps comparison between glaciated and unglaciated valleys indicates overdeepening has in some cases been favoured by pre-existing, tectonically controlled steeper parts of the preglacial valleys.

That the Alpine valleys are overdeepened has been known since the last century (e.g. Penck 1901). The causes of this were a matter of debate. Some authors (e.g. Heim 1885, 1919) at first assumed tectonic or isostatic reasons. Today, it is well established that the overdeepening was largely caused by glacier and meltwater erosion. Until recently the extent of this process has been largely unknown, because most of the basins have been refilled with sediments. Only numerous drillings and geophysical investigations during recent years have allowed a rough assessment of the extent of glacigenic overdeepening.

Strong glacial incision reflects the erosional capacity of the glacial debris enclosed in the basal parts of the ice, as well as the high ice-flow velocities. New seismic investigations in the Inn valley have revealed that there Quaternary sediments reach a thickness of about 900–1000 m. Between Wattens and Schaftenau, the base of the Quaternary lies at a depth of 300–400 m below sea level (Weber *et al.* 1993). Overdeepened rock basins (Seiler 1979) are found especially in narrow sections of the Alpine valleys, where the ice had to pass at particularly high velocities (Frank 1979, Van Husen 1979). At the confluence of valley glaciers pronounced excavation resulted in the formation of confluence basins (e.g. near Innertkirchen in the Aare valley). These forms originated before the Würmian Stage, most of them presumably much earlier. The most noteworthy overdeepenings still visible in the field are found near the outer margins of glaciation. Glacial tongue basins are occupied by the great lakes of the southern and northern Alpine margins (e.g. Lake Constance, Lake Garda). Overdeepening of the lakes at the southern Alpine margin reaches maximum values of 660 m at Lago di Como and 680 m at Lago Maggiore (Schlüchter 1981a).

The Alpine valleys have not been preserved as they were left by the glaciers. After deglaciation they were still subject to considerable erosion and reworking. These processes were particularly effective during the lateglacial but persisted in a markedly reduced way throughout the Holocene. In oversteepened regions great sections of rock were detached, and pressure release triggered rock falls (Abele 1994). Slabs of rock (especially clay schist, phyllite and mica schist) have moved downhill only gradually (valley thrust; Stiny 1941), whereby in extreme cases open clefts were left gaping in the upper slopes ('mountain tear-up' *sensu* Ampferer 1939). At the entrances of branch valleys, large alluvial cones accumulated. They frequently dammed the drainage, initiating the formation of lakes and swampy areas (Van Husen 1987) (Fig. 161).

The sequence of events causing the overdeepening may partly be reconstructed for the tongue basins of the Alpine foothills. It has turned out that valley incision set in rather early (Eberl 1930). In the Traun–Enns Platte area in Upper Austria, Kohl (1974) demonstrated that the so-called 'Ältere Deckenschotter' ('Older Cover Gravels') of Penck had not been deposited on an almost horizontal base, but infilled channels of the Tertiary land surface. Jerz reached the same conclusion for the Wolfratshausen basin (south of Munich). However, glacial basins from such an early stage have not been found. Jerz (1979) assumed that the glaciers of the early cold stages spread out in a fan-like pattern and were restricted to the area close to the Alpine margin. Grimm *et al.* (1979), however, found that in the Salzach Glacier region a pre-Mindel glaciation advanced in channels far into the Alpine foreland.

For the Mindel Glaciation a proto-Wolfratshausen tongue basin has been identified (Fig. 162). This first incision reached only 20 to 30 m below the present land surface (down to a depth of 560 to 550 m a.s.l.). At that time the basin had a slighty different shape; for instance at

FIGURE 161 Pleistocene valley development in the eastern Alps. The sketch diagrams show an Alpine valley, occupied by a glacier, the margin of which is marked by moraines (1) and kames (2). After ice decay in the massive hard rocks in the background a rockfall has formed (3), whilst in the weaker, less solid rocks in the foreground, mountain tear-up (4) and valley thrust (5) occur. The kame terraces on the opposite slope are rapidly eroded, forming a large alluvial cone (6). By this process the main stream is blocked and swampy regions are formed (7) in which fine-grained sediments and peat occur (Van Husen 1987)

Wolfratshausen it extended 2 km further to the west. During the Riss Glaciation, the Wolfratshausen basin was distinctly overdeepened. The incision reached down to more than 130 m below the present valley floor (deeper than 450 m a.s.l.). This basin was then largely filled up towards the end of the Riss Glaciation, and possibly a residual lake remained. During the Würmian Stage, the glacier advanced several kilometres beyond the Wolfratshausen basin. In this process part of the old basin filling was cleaned out and replaced by younger sediments (Jerz 1979, 1987).

The Wolfratshausen basin is an example that some landforms developed early in the Quaternary and were merely overprinted during the last glaciation. This, however, cannot be generalised. For example, in the Salzach Glacier region, Weinberger (1955) showed that

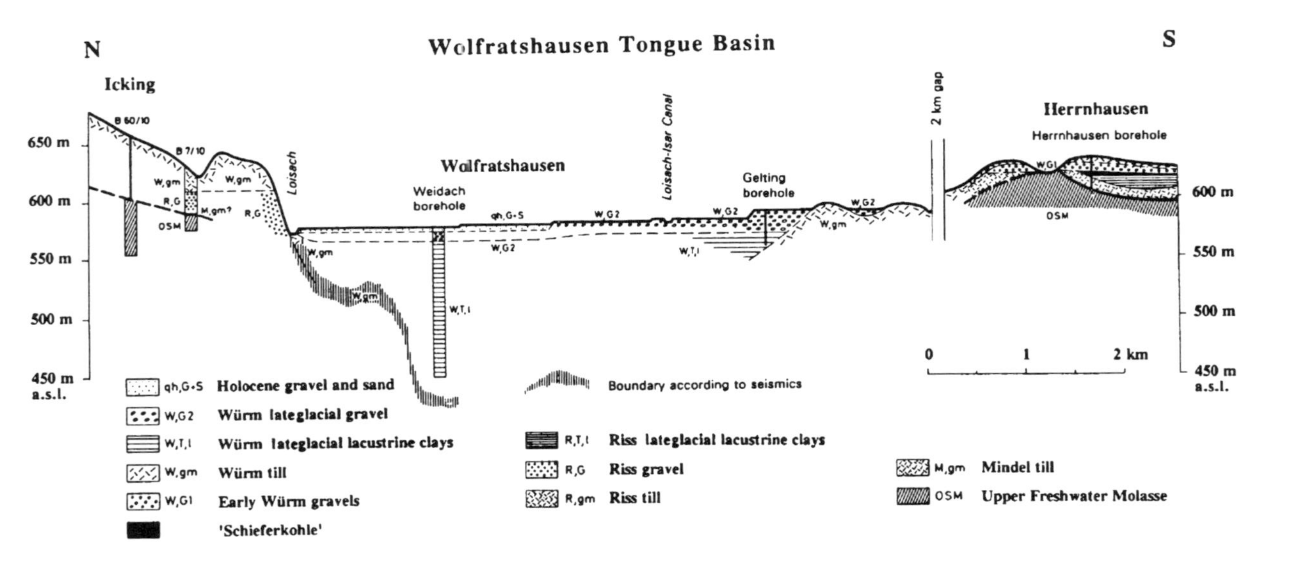

FIGURE 162 N–S profile through the Wolfratshausen basin (from Jerz 1979)

cold-stage valley fills had repeatedly initiated shifting of the valleys, with the glaciers of the next ice advance following the new course into the Alpine foreland. Kohl (1974) found corresponding conditions for the Traun–Enns Platte. The present drainage network there had only developed during the Würmian.

12.3 CLASSIC SUBDIVISION OF THE ALPINE QUATERNARY STRATIGRAPHY

In the Alpine region the term 'glaciation' in a climatostratigraphic sense refers to a period during which glaciers reached down into the foothills. Periods during which this was not the case should be referred to as 'cold stages'.

The classic Alpine Quaternary stratigraphy is largely a morphostratigraphy. It is based upon the concept that the so-called **glacial series** consists of a sequence of tongue basins with drumlins, moraine belts and gravel spreads. The term 'glacial series' was coined by Penck & Brückner (1901/09). Penck and his student Brückner laid much emphasis on demonstrating the direct connection between glacial gravel spreads, terraces and moraines of the same age. In the northern foothills of the Alps comprehensive morphological sequences of this type were initially found only for three glaciations; for the fourth this evidence was produced much later through investigations in Upper Austria (Weinberger 1950, Kohl 1958). Rögner (1979), however, considers he has detected intercalated meltwater deposits and morainic material of a fifth glaciation. In the cases of the other older glaciations, no such complete landform assemblages have been preserved.

The morphostratigraphic method was developed in the northern Alpine foothills; it is there where it is best applied. In the southern and French margin of the western Alps, it is much more difficult to use this method. These restrictions also apply to central Switzerland, where the Jurassic mountains barred a free outflow of glaciers and meltwaters.

With regard to glacial morphology, the southern German Alpine foothills may be roughly subdivided into three large areas:

1. The west is dominated by the former Rhine Glacier. The ice here advanced far into the Alpine foothills. During the penultimate glaciation it extended across both main water courses, the River Rhine in the west and the Danube in the east. Almost everywhere in this region older glacial deposits (pre-Riss) are covered by younger strata.
2. In the east, the Danube is separated from the formerly glaciated region by a wide belt of Tertiary uplands ('Tertiäres Hügelland'). Glaciofluvial aggradation in this region was restricted to comparatively narrow river valleys, in which only deposits of the more recent glaciations have been preserved.
3. In the central area between the Riss and Lech rivers on the other hand, Early Pleistocene gravel was deposited. Multiple changes of drainage directions resulted in the formation of gravel spreads of several cold stages, which have been preserved at different altitudes. Here Penck (1882) was able to demonstrate that the Alps had experienced multiple glaciations during the Quaternary. It was also in the Riss–Iller–Mindel region where the classic subdivision into the four Alpine glaciations was established (Günz, Mindel, Riss, and Würm) (Penck 1899, Penck & Brückner 1901/09):

Hochfeld (Böhener Feld)	Günz Ice Age
Grönenbacher Feld	Mindel Ice Age

Hitzenhofer Feld	Riss Ice Age
Memminger Feld	Würm Ice Age

Subsequently, two additional, earlier cold stages were also identified in this region; the Donau and Biber cold stages (Eberl 1930, Schaefer 1956). Only the Würm Cold Stage has been formally defined; it should therefore be referred to as the 'Würmian'. To avoid confusion, the other, informal stage names of the Alpine Quaternary stratigraphy are given here in the traditional German form, e.g. 'Riss' instead of 'Rissian'. A major revision of the Alpine Quaternary stratigraphy is presently under way (see, for instance, Ellwanger *et al.* 1995). However, the surroundings of Memmingen still remain as a key region for the Alpine Quaternary stratigraphy today.

Where the Alpine Quaternary stratigraphy extends beyond the limits of the glaciated region, it becomes largely a fluvial gravel stratigraphy (Schaefer 1951), and the morphostratigraphic approach has also been applied. In some cases the assignment of the individual gravel spreads to the respective ice advances has posed a major problem. Weidenbach (1937) subdivided the gravel spreads into 'advance gravels' and 'retreat gravels', depending on their relations to the appropriate end moraine. Van Husen (1983) has emphasised that the gravel spreads are not strictly 'glacial' in origin, since aggradation had set in long before the glaciation proper, during the period of deteriorating climate. The accompanying thinning of the vegetation led to increased erosion and rivers overloaded with sediment laid down gravel in their courses, thus raising the entire valley floor. This is supported by the fact that simultaneous accumulations of gravel terraces in the non-glaciated Alpine valleys had also occurred (Draxler & Van Husen 1989). Among these are the enormously thick terrace sediments of the Inn valley that reach far into the tributary valleys (Paschinger 1957) (Fig. 163).

The Alpine Quaternary sequence is highly incomplete, and quite often it is not easy to establish a relationship between a gravel spread and a terminal moraine. A well-known example of a long-disputed stratigraphy is represented by the Inn Valley Terrace near Innsbruck. Radiocarbon dating of the varved clays at Baumkirchen finally proved that the terrace accumulated before the Würmian glacial maximum (Fliri 1973).

The Inn Valley Terrace overlies the so-called 'Höttinger Brekzie' (Fig. 164), a brecciated deposit of an ancient debris cone, strongly carbonate-cemented. Besides fragments of red sandstone, limestone and dolomite, the remains of a rich flora were discovered, from which Von Wettstein (1892) recorded *Rhododendron sordellii*, the Pontian Alpine rose, the distribution of which is today reduced to the region south of the Alps. The central Alpine situation of the breccia provided proof of a warmer interglacial climate between periods of till deposition. During its formation the region must have been ice-free up to the high mountain crests and this evidence finally settled the dispute between mono- and polyglacialists at the 'Deutscher Geographentag' congress in Innsbruck (1912). However, the age of the interglacial deposits has remained disputed to this day. The succession of strata is as follows:

Upper till
Inn Valley Terrace sediments
Lower till
Höttinger Breccia
Basal till

Originally the 'basal till' was assigned to the Mindel Glaciation, and the Breccia to the

FIGURE 163 Inn Valley Terrace at Innsbruck; view from Seegrube towards the south (Silltal) (Photograph: Heuberger 1959)

Mindel/Riss Interglacial (Paschinger 1950). However, it has since been demonstrated that the overlying Inn Valley Terrace sediments should be assigned to the Würmian (and not to the Riss/Würm Interglacial). Thus the lower till could have been deposited by Early Würmian glaciers that did not reach the Alpine foothills (Heuberger 1974). This would imply that the 'Höttinger Breccia' may have formed in the Riss/Würm Interglacial, but it could be older.

Morphostratigraphical correlations are not without problems. Since the different zones of a glaciated area do not always act simultaneously, the outermost end moraines of different regions may well belong to different stages, i.e. they are not coeval. In addition, landforms of older glaciations were often subsequently eroded and overprinted. As long as it was believed that the Alpine region was only affected by four great glaciations, morphostratigraphy seemed adequate for the tasks at hand. However, today litho- and biostratigraphical methods are increasingly used to update the Alpine Quaternary stratigraphy, and palaeomagnetic investigations contribute to an improved dating of the sedimentary record. Unfortunately, the latter method is applicable only to very few places because suitable fine-grained, sufficiently magnetised sediments are absent from many key sites.

12.4 TRACES OF OLD GLACIATIONS

Nowhere in the Alps have Quaternary deposits been completely preserved. Therefore the basic stratigraphical subdivision is combined from various regional stratigraphies, though in many cases the determination is difficult because age control is lacking. Indications of Early

FIGURE 164 Höttinger Breccia, lower part, in which tunnelling provided the decisive evidence that the breccia was underlain by till. Exposure in an abandoned quarry on the Innsbruck–Hungerburg road (Photograph: Heuberger 1993)

Pleistocene cold stages have long since been recognised in the Italian and French Alps. In both areas an abrupt change from fine-grained sedimentation to the deposition of coarse gravel and conglomerates took place during what was regarded as the 'Upper Pliocene' (which would be included in the Pleistocene according to the definition adopted here). However, the evidence is ambiguous. The change in sedimentary environment might either be explained by tectonic causes or by climatic changes (Billard & Orombelli 1986).

In the intra-montane basin at Leffe, above Bergamo in Lombardy (Italy), Early Pleistocene 'schieferkohle' was described by Taramelli (1898). It represents moderately altered, heavily compressed peat. The strata, in excess of 100 m thick, contain mammal remains and have been thoroughly examined by Lona (1950) and Venzo (1950). The results of long-standing investigations were finally compiled by Lona & Bertoldi (1973). The sequence of sediments begins with Early Pleistocene lake deposits belonging to the Matuyama Epoch, according to palaeomagnetic investigations carried out by Billard *et al.* (1983), and thus must be older than 780,000 years. Although the palynological investigations offer evidence for clear glacial–interglacial cycles, no direct correlation to glacial deposits has so far been established. Since the age of the sedimentary cover has not been determined ('Mindel' according to Venzo 1952), the stratigraphical position of the Leffe sequence within the Pleistocene has not been defined. Venzo's (1952) attempt to correlate with the North Alpine stratigraphy is no longer valid; a modern investigation of the profile is still awaited.

In France, ancient traces of glaciation have also been identified on the Chambaran Plateau, west of Grenoble. Here in a sequence of strata thought to be older than 1.6 million years, a clayey sediment with striated boulders has been found. However, Bourdier (1963) could not determine with certainty whether this diamicton represented a slope deposit or a till. On the western Chambaran Plateau the coarse gravels are overlain by strongly cemented loess beds that contain, apart from a fossil soil, a rich mammal fauna (Viret 1954). The faunal assemblage suggests that the age of the loess is ca. 2.2 million years (Guérin 1980).

The oldest traces of glaciation in Switzerland are generally thought to be the so-called 'wanderblöcke' (wandered blocks), which are erratic blocks without any preserved accompanying glacial deposits, found in the northwestern Jura Mountains (Hantke 1978, 1979). These blocks are mostly derived from southwestern Black Forest source areas. The southern margin of their distribution, a line from Grellingen (via Fehren–Himmelsried–Hölstein–Tenniken) to Olten, may represent, according to Hantke (1978), the southern limit of an Early Pleistocene Black Forest glaciation. Equally it might represent fluvial transport of blocks of Early Pleistocene or Pliocene age.

The so-called 'Deckenschotter' (cover gravels) of Switzerland are possibly older than originally assumed (Schlüchter 1988). In northeastern Switzerland they actually represent relatively high-altitude 'covers' of gravel that had been subsequently eroded in cuesta-like landforms. These 'Deckenschotter' are traditionally subdivided into 'lower' and 'upper' gravel units, and since the investigations of Frei (1912), have been assigned to the Günz and Mindel stages of the classic Alpine Quaternary stratigraphy. They contain indicators of glaciofluvial, fluvial and glacigenic deposition. The original covers have been largely destroyed by subsequent erosion and weathering. The ice advances to which these meltwater deposits originally belonged must have reached beyond the Würmian maximum. On the Belchen Plateau the 'Deckenschotter' is overlain by a sequence of redeposited loess with intercalated gravels. The weathering profile, altogether 37 m thick, contains a series of palaeosols. Schlüchter (1988) estimates that a period of more than 1.5 million years was required for the development of this soil sequence. Thus the Swiss 'Deckenschotter' would date from the very Early Pleistocene, if not the Pliocene. It cannot be determined, however, to what extent the material was already weathered prior to soil formation (Fig. 165).

It is postulated that the widespread deposition of the Swiss 'Deckenschotter' and the later partial refilling of the overdeepened valleys and troughs in the Swiss Alpine foreland represents not only a significant change of morphogenic conditions but also a time gap. The incision of the deep valleys postdates the aggradation of the 'Deckenschotter'. This may have been the result of tectonic uplift in northeastern Switzerland. The intensity of incision differs regionally. It was particularly extensive in the central and western parts of central Switzerland.

A second step in the landscape evolution was the renewed change from erosion to

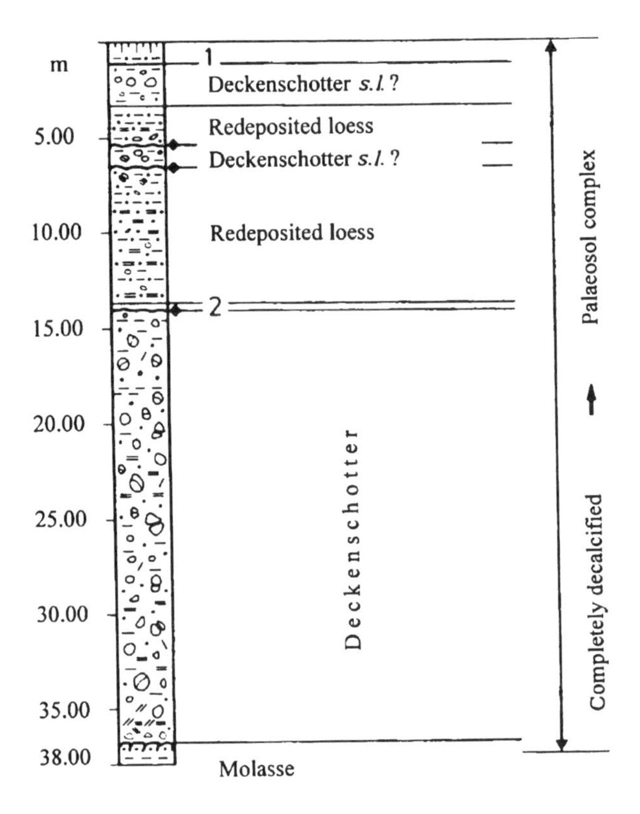

Figure 165 Schematic lithostratigraphic diagram of the S 3 borehole on the Belchen Plateau, Switzerland (from Schlüchter 1988)

accumulation in the main valleys of the Alpine foreland. Schlüchter (1981a, 1989a) assessed the corresponding sequence of strata at Thalgut and other places as an indication that this 'Middle Pleistocene morphotectonic revolution' had possibly affected the entire northern Alpine foreland, although with regional differences. The 'Middle Pleistocene morphotectonic revolution' has been identified in southwest Germany (Ellwanger *et al.* 1995), but no such extreme 'revolution' has been found so far in the Bavarian or in the Austrian sectors.

One of the longest Pleistocene stratigraphical records in the Alpine region was discovered at Thalgut (Schlüchter 1986, 1989b) (Fig. 166 and Table 7). At a depth of 60 m below the floor of a large gravel pit, a rhythmically stratified, fossil-rich interglacial sediment was cored, the Jaberg Lacustrine Clay, the base of which could not be reached (Schlüchter 1989b). According to Welten (1982a, b, 1988) this warm-stage deposit contains abundant *Fagus* and *Pterocarya*, which have never been found in any Alpine interglacials correlated with the Holsteinian or Eemian. Therefore Schlüchter (1989a, b) interprets this interglacial as older than the Mindel/Riss Interglacial of Meikirch. Its precise stratigraphical position, however, is still unknown.

In the south German Alpine foreland, early observations indicated the involvement of more

FIGURE 166 Thalgut, Quaternary stratigraphic key section for the Swiss Mittelland region, with core drilling equipment. Delta of the classic Riss Stage, overlain by gravels and till of the Würmian ice advance and underlain by different facies of ice-marginal deposits of an older glaciation (Photograph: Schlüchter 1983)

than the four glaciations postulated by Penck & Brückner (1901/09) (Table 8). Eberl (1930) identified ancient, highly elevated gravels in the Iller–Lech region that he regarded as deposits of a 'Donau Ice Age', dating from a time before the Günz Glaciation of Penck & Brückner. Graul (1949) introduced the term 'Deckschotter' (covering gravels) for these gravels in order to make a distinction from the 'Deckenschotter' (cover gravels) of the Günz and Mindel stages *sensu* Penck & Brückner (1901/09). Schaefer (1953) confirmed the existence of this 'Donau Ice Age' and later added another, 'Biber Ice Age' (represented by gravels at Staufenberg and in the Aindlingern terrace flight, east of the Lech River) that predated the 'Donau Ice Age'. These additional cold stages have been generally accepted (cf. for example by Brunnacker 1986). It is

TABLE 7 Lithostratigraphy and climatostratigraphic interpretation of the Thalgut section (after Schlüchter 1989b)

Depth / section	Major hiatuses and palaeosols	Climatostratigraphy	Lithostratigraphy: lithostratigraphic units		Lithostratigraphy: lithogenetic description
620 m			Rotachewald diamicton		lodgement till
exposed in gravel pit		Interstadial	Upper Münsingen Gravel	upper unit	meltwater gravel
				lower unit	weathered gravel
		Interglacial	Thalgut lake clays		fossiliferous lacustrine clays
			Kirchdorf delta gravels		delta foresets
		Interglacial			
			Thalgut varves		varved lacustrine clays
			'mud till'		waterlain till
			Gerzensee block till		
576 m					
borehole information					delta foresets
		Interglacial	Jaberg Lacustrine Clay		fossiliferous lake clays (with *Pterocarya*)
462 m					

Table 8 Stratigraphical subdivision of the Quaternary deposits in the Alps and Alpine foreland. Both possible Quaternary/Tertiary boundaries are marked by arrows. Ages of palaeomagnetic reversals and correlation with deep-sea oxygen-isotope stages according to Shackleton *et al.* (1990). Stratigraphy after Ellwanger *et al.* (1995), Doppler, Ellwanger and Schlüchter (personal communication)

Magneto-stratigraphy	Age	Oxygen-Isotope Stages	Chronostrat.	Bavaria	Switzerland + Baden-Württemberg
Brunhes Epoch			Late Pleist.	Würmian Cold Stage Riss/Würm Warm Stage	Würmian Cold Stage Krumbach Warm Stage (Eemian)
			Middle Pleistocene	Riss Complex	Riss/Würm Complex: Doppelwall Riss Cold Stage
				Mindel/Riss Warm Stage	Samerberg Warm Stages (Holsteinian) Older Riss Cold Stages
				Mindel Complex	
				Günz/Mindel Warm Stage	Morphotectonic Revolution / most extensive glaciation
	780 000	19		Günz Complex	
Matuyama Epoch			Early Pleistocene		'Deckenschotter' Complex: Haslach-Mindel Complex
Jaramillo Event	900 000	27		Uhlenberg Warm Stage	Uhlenberg Warm Stage (flora)
	1 070 000	31			
					Günz Complex
				Donau Complex	Morphotectonic Revolution
			←		
Olduvai Event	1 770 000	64			
					'Deckschotter' Complex: Uhlenberg Warm Stage (fauna) Donau Complex
	1 950 000	71			
				Biber Complex	
					Biber Complex
Gauss Epoch	2 600 000	104	← Tertiary		

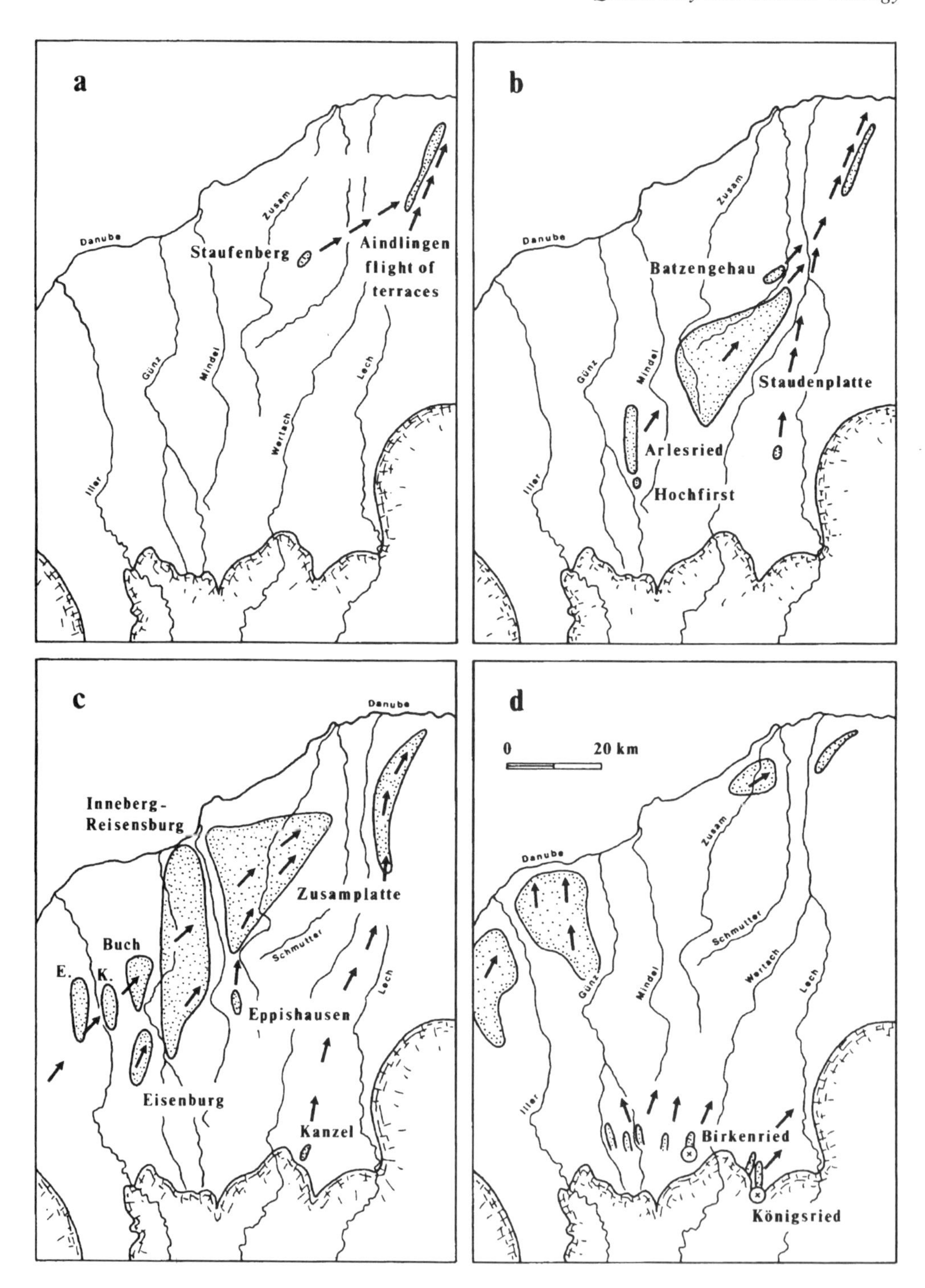
a
Danube
Staufenberg
Aindlingen flight of terraces
Zusam
Günz
Mindel
Wertach
Lech
Iller
b
Danube
Zusam
Batzengehau
Günz
Mindel
Staudenplatte
Arlesried
Hochfirst
Iller
c
Danube
Inneberg-Reisensburg
Zusamplatte
Buch
E.
K.
Schmutter
Eppishausen
Lech
Eisenburg
Kanzel
d
0
20 km
Zusam
Danube
Schmutter
Günz
Mindel
Wertach
Lech
Iller
Birkenried
Königsried

implied that these gravels are of glaciofluvial origin, though the correlative till sheets and end moraines have not yet been identified (cf. Fig. 167). Thus, three major cycles of deposition have been identified in the Alps:

- Riss/Würmian Complex
- Hiatus
- 'Deckenschotter' Complex, representing the Günz and Mindel Complexes
- Hiatus
- 'Deckschotter' Complex, representing the Biber and Donau Complexes

Each hiatus represents a period of fundamental change which is largely controlled by large-scale tectonic movement and resulting changes of catchment areas and drainage directions. Those periods are referred to as 'morphotectonic revolutions' (Schlüchter 1989a, Ellwanger *et al.* 1995).

The Early Pleistocene gravel sequences of the Iller–Lech region have been studied in much detail. According to Habbe & Rögner (1989), they should be subdivided as follows:

- The oldest gravels in the Iller–Lech region are those on Staufenberg and in the highest level of the Aindlingern flight of terraces (Fig. 167a). They were deposited by a predecessor of the present Iller River that drained towards the northeast. These gravels are assigned to the **Older Biber Cold Stage**.
- The gravels of the Staudenplatte, together with those on Hochfirst (cf. Rögner 1986) near Arlesried, in the Batzengehau, and on Stoffersberg (west of Landsberg) are assigned to the **Younger Biber Cold Stage** (Fig. 167b).
- The gravels of the Zusamplatte (= 'Untere Deckschotter'), as well as those at Eppishausen, Kellmünz, Buch, Eisenburg and Inneberg–Reisensburg are assigned to the **Older Donau Cold Stage** (Fig. 167c).
- Another gravel spread, the 'Ältere Zwischenterrassenschotter' were found to be younger than the Zusamplatte gravels. They were deposited by a northward-flowing river. The term 'Zwischenterrassenschotter' was introduced by Graul (1943) for correlative deposits in the lower Lech region (Aindlinger flight of terraces). It refers to deposits that are situated between the (Rainer) 'Hochterrasse' (high terrace) and the 'Deckschotter' mentioned above, i.e. they belong chronostratigraphically between the Donau Stage 'Deckschotter' and the 'Ältere Deckenschotter' of the Günz Stage. Within the 'Zwischenterrassenschotter', three stratigraphical levels can be distinguished, of which the lower (younger) ones are assigned to the Günz. The 'Ältere Zwischenterrassenschotter' are assigned to the **Younger Donau Cold Stage** by Habbe & Rögner (1989). Rögner (1979, 1980) believes he has found a connection between gravels of the Younger Donau Cold Stage and corresponding till units in the Wertach–Lech region. At two places, near Birkenried–Irsee and at Königsried–Stocken, these gravels are apparently intercalated with glacigenic deposits (Fig. 167d).

FIGURE 167 (opposite) a–d Pre-Günz deposits in the Iller–Lech region. (a) Older Biber Cold Stage ('Obere Deckschotter'), (b) Younger Biber Cold Stage ('Mittlere Deckschotter'), (c) Older Donau Cold Stage ('Untere Deckschotter'), (d) Younger Donau Cold Stage ('Zwischenterrassenschotter'). E. = Erolzheimer Schotter; K. = Kellmünzer Schotter. Ice margin of the Würm maximum indicated on all maps; interfingering of till/gravel indicated by crosses on Fig. 140d (after Habbe & Rögner 1989; supplemented by Doppler, personal communication)

Whilst the relative succession of these beds is not disputed, the chronological classification of the various early glaciations remains highly problematical.

Whereas Woldstedt (1958) found that the four-fold stratigraphical subdivision of Penck & Brückner was still perfectly valid, this concept is to be seriously questioned today. Apparently the old stratigraphical units partly comprise deposits of more than one cold and warm stage. Habbe (1989) concludes that at the present state of knowledge, only a rough stratigraphical subdivision of the pre-Mindel deposits can be proposed, if at all. But even the Mindel Stage deposits and the extent of the Riss Glaciation are being questioned (see below). The revision of the Alpine Quaternary stratigraphy has begun, but it is far from being completed.

12.5 UHLENBERG INTERGLACIAL

In the Alpine Quaternary subdivision, biostratigraphical investigations have played a less important role than in northern Europe because very few interglacial deposits have been preserved. Therefore an isolated occurrence of Early Pleistocene 'schieferkohle' on the Uhlenberg near Dinkelscherben (west of Augsburg) is of key importance. Unfortunately it only represents part of an interglacial (or interstadial). A sequence of three pollen zones has been distinguished here: an *Alnus–Pinus* zone with a high percentage of *Tsuga*, an *Alnus–Picea–Pinus* zone, and a *Picea–Pinus–Betula* zone with thermophilous deciduous trees. A correlation with one of the Cromerian interglacials is ruled out because of the high frequency of Early Pleistocene taxa (Schedler 1979). A palaeomagnetic investigation revealed a magnetic reversal that might be interpreted as the Jaramillo event (Strattner & Rolf 1995). But molluscs and small mammals from the overlying flood loam seem to be much older and have been correlated with the Tegelen Complex (Ellwanger *et al.* 1994, Rähle 1995). The Uhlenberg Interglacial deposits overlie the glaciofluvial 'Untere Deckschotter' (Graul 1949, 1962; Löscher 1976), the type deposits of the Donau Cold Stage (Scheuenpflug 1979) in the Zusamplatte area. This implies that there is unequivocal evidence of Early Pleistocene Alpine glaciation in the south German Alpine foreland.

At another site in the south German Alpine foothills, near Rosshaupten, the Matuyama/Brunhes boundary has been identified in the stratum immediately overlying the 'Untere Deckschotter' (Tillmanns *et al.* 1986). This does not contradict the Uhlenberg date, since a strongly developed pseudo-gley soil between the 'Deckschotter' and their sedimentary cover may represent more than one interglacial (Bibus 1995).

12.6 GÜNZ COMPLEX

Penck & Brückner (1901/09) did not consider any deposits to be older than the Günz Cold Stage. Therefore all gravel occurrences at higher altitudes than the Mindel Stage 'Jüngere Deckenschotter' were called 'Ältere Deckenschotter' and assigned to the Günz Cold Stage. Today the 'Ältere Deckenschotter' are distinguished from the older gravel spreads of the Donau and Biber cold stages (see above). The possibilities of a further subdivision are limited. On the basis of morphostratigraphic investigations, Stepp (1981) proposed a strongly differentiated subdivision of the Böhener Feld gravels. He subdivided this gravel morphologically into six individual gravel units. His results, however, have not been verified by more recent investigations (Habbe 1986b).

In the Salzach Glacier region, the glaciers of the Günz Cold Stage advanced considerably further than those of subsequent glaciations. Thus Weinberger (1950) was able to find morphostratigraphical evidence that the Günz had been a true Alpine glaciation (with piedmont glaciation). In the western marginal region of the Salzach Glacier, Grimm *et al.* (1979) and Doppler (1980) reconfirmed these findings. Kohl (1955, 1958) also assigned corresponding moraine ridges on the Traun–Enns Platte, 60 km further east, to the Günz Cold Stage. However, a correlation with sediments elsewhere is very difficult because of the dearth of datable sediments; the Günz age of the deposits is still not certain.

Originally it had been assumed that the Günz was younger than the Matuyama/Brunhes boundary. However, on the Höchsten and near Heiligenberg (north of Lake Constance) K. Fromm (1996) recently demonstrated that tills and gravels of the Günz Glaciation are reversely magnetised. Consequently, at least part of the Günz Complex must be assigned to the Mayutama Epoch (Schreiner 1992). At another site, near Rosshaupten in the Allgäu, the Mayutama/Brunhes reversal has also been found above the 'Ältere Deckenschotter' (Tillmans *et al.* 1986, Jerz & Doppler 1990). It is not yet clear, however, whether the entire Günz Complex predates the Matuyama/Brunhes boundary and whether all deposits of the Alpine region formerly assigned to the 'Günz' really form part of the same glaciation. Accordingly, different correlation schemes are possible. Ellwanger *et al.* (1995) and Doppler (personal communication) refer to the 'Günz' as a 'Günz Complex'.

12.7 GÜNZ/HASLACH INTERGLACIAL

No interglacial deposits have been found so far that directly postdate the Günz Complex. It is only known that its deposits are separated from the next youngest stratigraphical units by a fossil soil.

12.8 HASLACH GLACIATION

The exact number of glaciations in the Alps prior to the Riss Stage is uncertain. As early as 1952, Schädel subdivided the 'Jüngere Deckenschotter' of Penck & Brückner's (1901/09) Mindel Cold Stage into two separate gravel units in the Württemberg Rottal (south of Ulm). Schreiner & Ebel (1981) confirmed this subdivision and distinguished a higher (i.e. older) Haslach Gravel and a lower Tannheim Gravel. In the proximal area, the vertical distance between the two units amounts to 10 m. Schreiner & Ebel found that the Haslach Gravels clearly contained a lower crystalline rock content than the Tannheim Gravels (3.5 *vs.* 9.7%). They therefore assigned the Haslach Gravel to a newly established Haslach Cold Stage, which is generally accepted today (e.g. Ellwanger *et al.* 1995, Jerz 1995).

12.9 HASLACH/MINDEL INTERGLACIAL

Schreiner & Ebel (1981) concluded that the interglacial site at Unterpfauzenwald described by Göttlich & Werner (1974) should not be assigned to the Mindel/Riss Interglacial, but instead be placed between the Mindel and Haslach Cold Stages. The pollen diagram from this site shows a preponderance of conifer forest with high proportions of pine and fir. In addition to beech trees and *Pterocarya, Ostrya* and *Tsuga* also occur, according to new investigations (Bludau, 1996). This is evidence of a relatively warm and humid climate. An age older than the North German Holsteinian may thus be assumed (Schreiner 1992).

12.10 MINDEL GLACIATION

The 'Jüngere Deckenschotter' of the Mindel valley and their association with moraines north of Obergünzburg caused Penck to name his second Alpine glaciation the Mindel Cold Stage. Since then, there has been no end of attempts to reinterpret beds that had originally been classified as of Mindel age. Thus Sinn (1972) assigned parts of the deposits and landforms of the Böhener Feld (= Hochfeld) near Memmingen to the Riss Cold Stage. In contrast, Eichler & Sinn (1975) assigned parts of it to the Günz Cold Stage, and Löscher (1976) even proposed dates as far back as the Donau. Schaefer (1973) attempted to subdivide the Mindel-age Grönenbacher Feld altogether into ten independent gravel terraces, each of which was thought to represent a corresponding major ice advance. After comprehensive geomorphological mapping Habbe (1986b) was not able to confirm any of these suggestions. The homogeneity of the sediments and the overall palaeogeographical situation suggest a single body of gravel that was formed during one glacial period.

The 'Jüngere Deckenschotter' of the Grönenbacher Feld can be followed via the Schwaighausen Gravels (north of Memmingen) into the Günz valley. Furthermore, widespread remains of 'Jüngere Deckenschotter' are found throughout the Mindel valley.

The end moraines of the Mindel Glaciation, south of Memmingen, are perfectly shaped and include the end moraines at Brandholz–Manneberg. No continuation of the end moraine is found to the west (Sinn 1972), but the moraines can be followed to the east as far as Obergünzburg (Penck & Brückner 1901, Eberl 1930, Stepp 1981, Habbe 1986a).

East of the Lech River, 'Jüngere Deckenschotter' are hardly found beyond the outer limit of the old moraine regions. 'Jüngere Deckenschotter' with 'sand pipes' are found in the Isar valley, for instance, south of München (Munich), in the Mangfall valley and in the range of the Inn Glacier and the Salzach Glacier. After Penck & Brückner (1901/09) and Knauer (1928, 1931) the sedimentary succession of the Isar valley was recently reassessed by Jerz (1987) and the Mangfall valley was recently mapped by Grottenthaler (1985). Further to the east, Mindel-age landforms were also identified in the course of morphological investigations (Inn–Chiemsee Glacier; Troll 1924). The maps of Eichler & Sinn (1974) and Grimm *et al.* (1979), in the western and northwestern parts of the Salzach Glacier region, match the results of investigations in the eastern part of the Salzach Glacier region by Weinberger (1950).

In Austria, end moraines of the Mindel Cold Stage had already been identified by E. Brückner (1886); they were later confirmed and mapped in detail by Weinberger (1950).

The absolute age and stratigraphic position of the Mindel Cold Stage is still unclear. Reversely magnetised fine sediments from a Mindel Stage 'Jüngerer Deckenschotter' at Allschwil near Basel (Zollinger 1991) suggest that at least part of the Mindel Cold Stage may belong to the Matuyama Epoch (Ellwanger *et al.* 1995).

12.11 MINDEL/RISS INTERGLACIAL

The interglacial between the Mindel and Riss cold stages in the Alps was traditionally referred to as the 'Great Interglacial'. This concept may result from the fact that it was actually a series of two interglacials. In the Swiss section of Meikirch, which is about 100 m deep, deposits of three cold stages and two interglacials have been encountered (Fig. 168). The sediments of the Würmian Cold Stage are underlain by deposits of the last (Riss/Würm) interglacial that show a vegetational succession typical of the Eemian in Central Europe (Welten 1982b). Beneath these

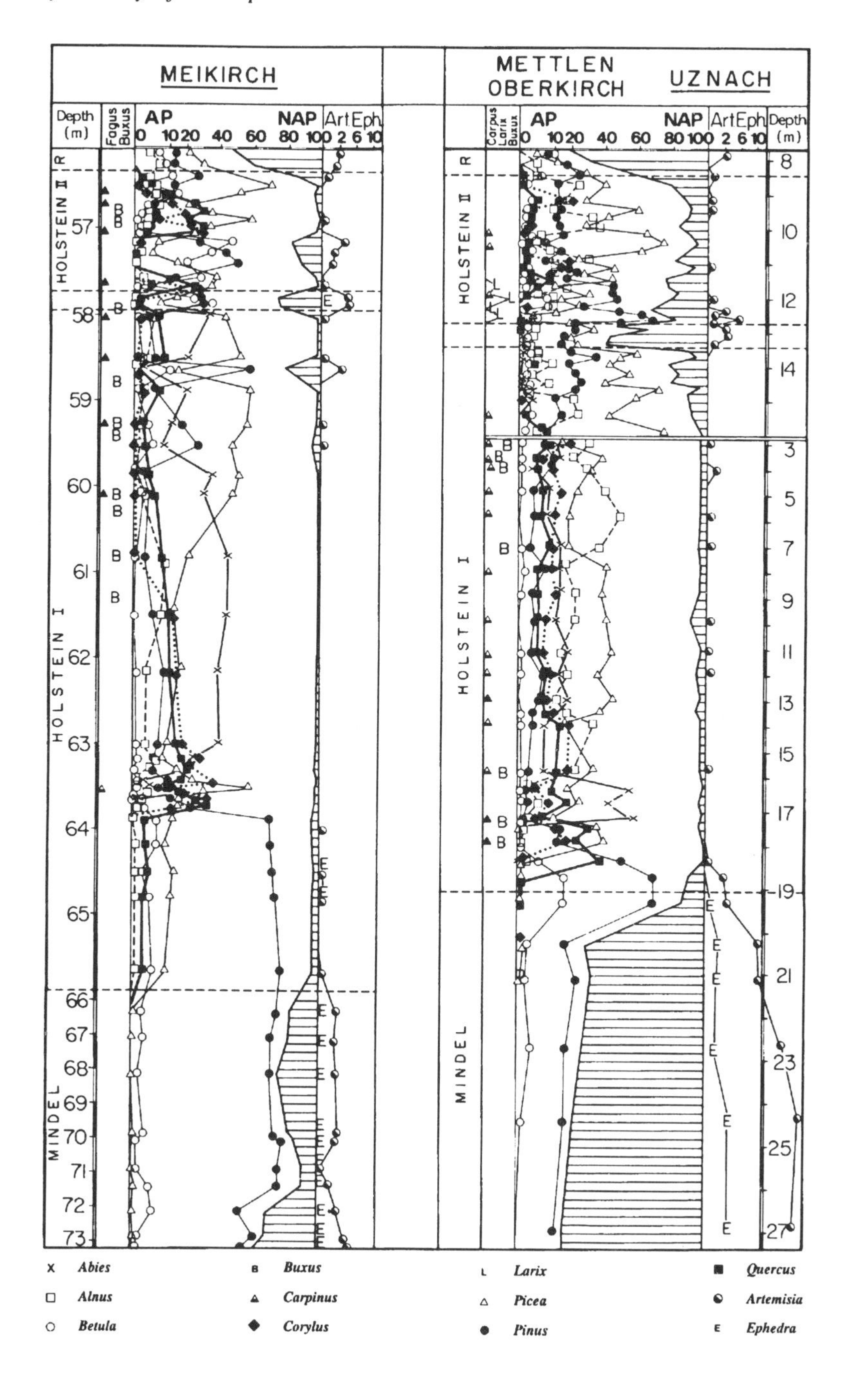

FIGURE 168 Pollen diagrams of Meikirch and Uznach with the 'double' pre-Riss integlacial (from Welten 1982b)

strata, cold-stage sediments that are subdivided by several interstadials occur. There are no glacigenic deposits (morainic materials) in this part of the sequence. The cold-stage strata are underlain by further lake sediment containing a bipartite interglacial. The latter was correlated by Welten (1982b) with the Holsteinian and Wacken interglacials of northern Germany. A similar occurrence of a 'double interglacial' was discovered by Welten near Uznach at the edge of the Linth plain. Although in the latter case the deposits of the Eemian (Riss/Würm) Interglacial are missing, there can be no doubt that the sequence predates the Eemian.

The vegetational development of this 'double interglacial' began in both Uznach and Meikirch with a mixed oak–fir–hazel phase, after which silver fir (*Abies alba*) soon reached very high frequencies. In Uznach, under the wetter and more severe climate of the Linth plain, pine and white alder were dominant, but fir persisted. In the cold period between the two interglacials a decline of pine resulted in domination of *Pinus* and/or *Betula*, with a pronounced increase in the occurrence of both herb pollen and *Artemisia*. The subsequent second interglacial is characterised by the same floral elements as the first, except for the absence of fir. The transition towards the Riss Cold Stage is characterised by increasing rates of *Artemisia* and other non-arboreal pollen (Welten 1982b).

In the Samerberg 2 borehole, in Bavaria, a 'double interglacial' has also been discovered underlying a thin Riss Stage till. This is also correlated with the North German Holsteinian and Wacken/Dömitz interglacials (Figs 169 and 170) (Grüger 1983). Here the penultimate cold period also seems to have left relatively rare traces; a Riss till of only 50 cm thickness was encountered (Jerz 1983). The question arises whether or not there were unequivocal traces of glaciation before – an assessment supported by Schlüchter (written communication), or whether the evidence has been subsequently eroded away. It cannot be excluded that Samerberg, and possibly also the Swiss sections, contain an erosional unconformity that would be difficult to identify in a borehole section.

12.12 RISS GLACIATION

In the French Alpine foreland a series of moraines exist that are known to be older than Würmian but which cannot be precisely dated. In this zone of so-called 'Riss moraines', south of Lyon, two complexes of glacigenic deposits can be distinguished that are separated by a strongly developed fossil soil. According to Mandier (1983, 1984), the latter should be considered the result of an interglacial event, which may or may not be the Mindel/Riss Interglacial.

Schlüchter (1989b) is inclined to believe that the extension of the Riss Glaciation was much smaller than has been so far assumed. Indeed, the lack of glacigenic Riss Stage deposits in the Meikirch section may support this conclusion. Cold-stage sediments of Riss age do occur, but they include no trace of a glaciation. The deposits of the Riss/Würm Interglacial, the Riss Glaciation and the Mindel/Riss Interglacial seem to form a sequence of uninterrupted lake sediments. The same applies to a similar borehole near Niederweningen, Switzerland. Schlüchter (1989b) therefore regards the penultimate cold stage in the Alps as an event of minor importance which, in terms of its ice-sheet extent, was similar to the Early Würmian ice advance. This assessment, however, is opposed to the classic Quaternary stratigraphy and requires corroboration through further research. At present, the Riss in Switzerland cannot reliably be subdivided stratigraphically because of the uncertainties that have arisen from the Meikirch and Thalgut boreholes (see above).

In southern Germany conceptions of what should be identified as the 'Riss Cold Stage' have

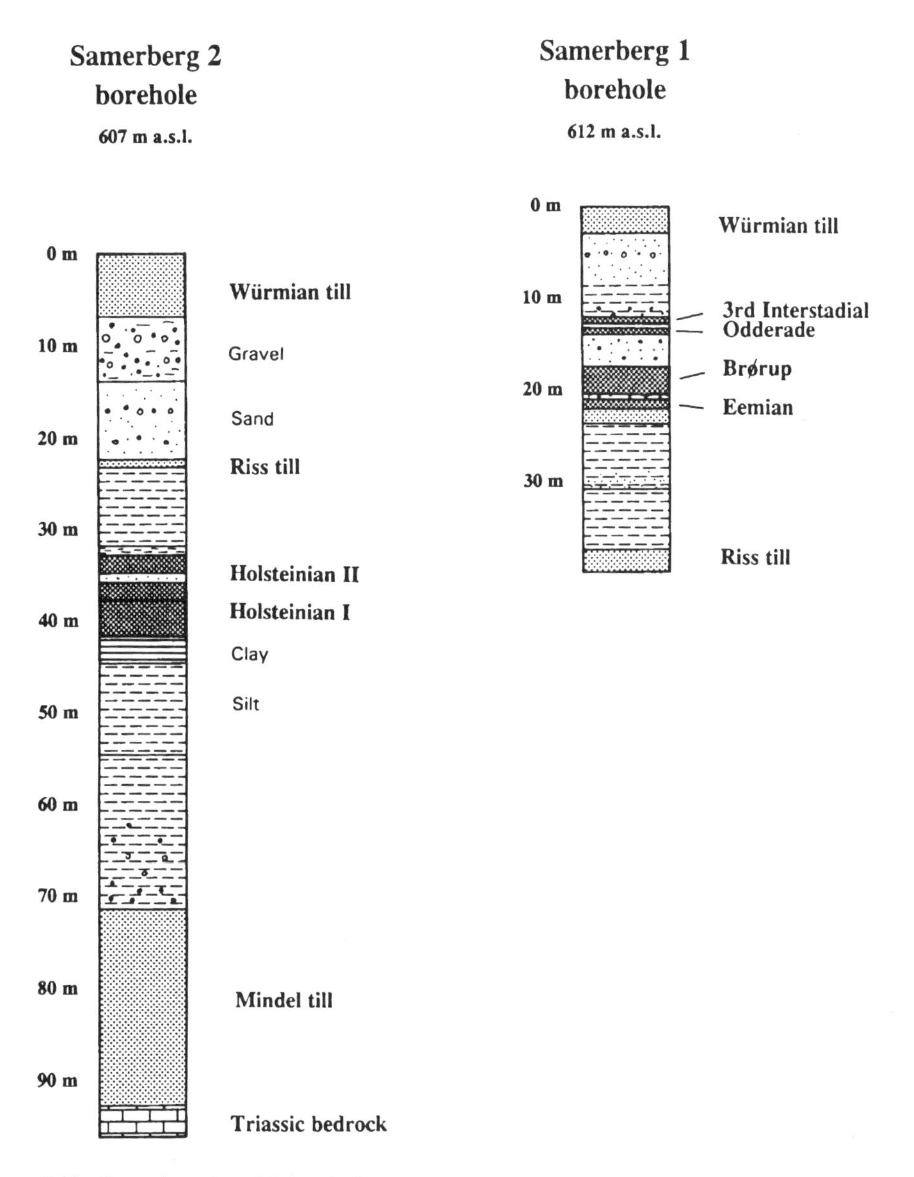

FIGURE 169 Samerberg 1 and 2 borehole logs (after Jerz *et al.* 1979, Grüger 1979 and Jerz 1983)

undergone a marked change in the course of time. When they originally established their Alpine Quaternary chronology, Penck & Brückner started from the assumption that their four cold stages represented single, uniform glaciations. Later Eberl (1930) distinguished a Riss I, II and III Glaciation, but he interpreted these as individual glacier advances within a single complex glacial stage. Schaefer (1973) assumed, on the other hand, that the original Riss Cold Stage was bipartite. He separated an independent older 'Paar Cold Stage' from the original Riss Cold Stage. In the Loisach Glacier region, in his opinion, the Riss Cold Stage glacier overrode the deposits of the 'Paar Cold Stage', and there (north of Augsburg) advanced furthest north

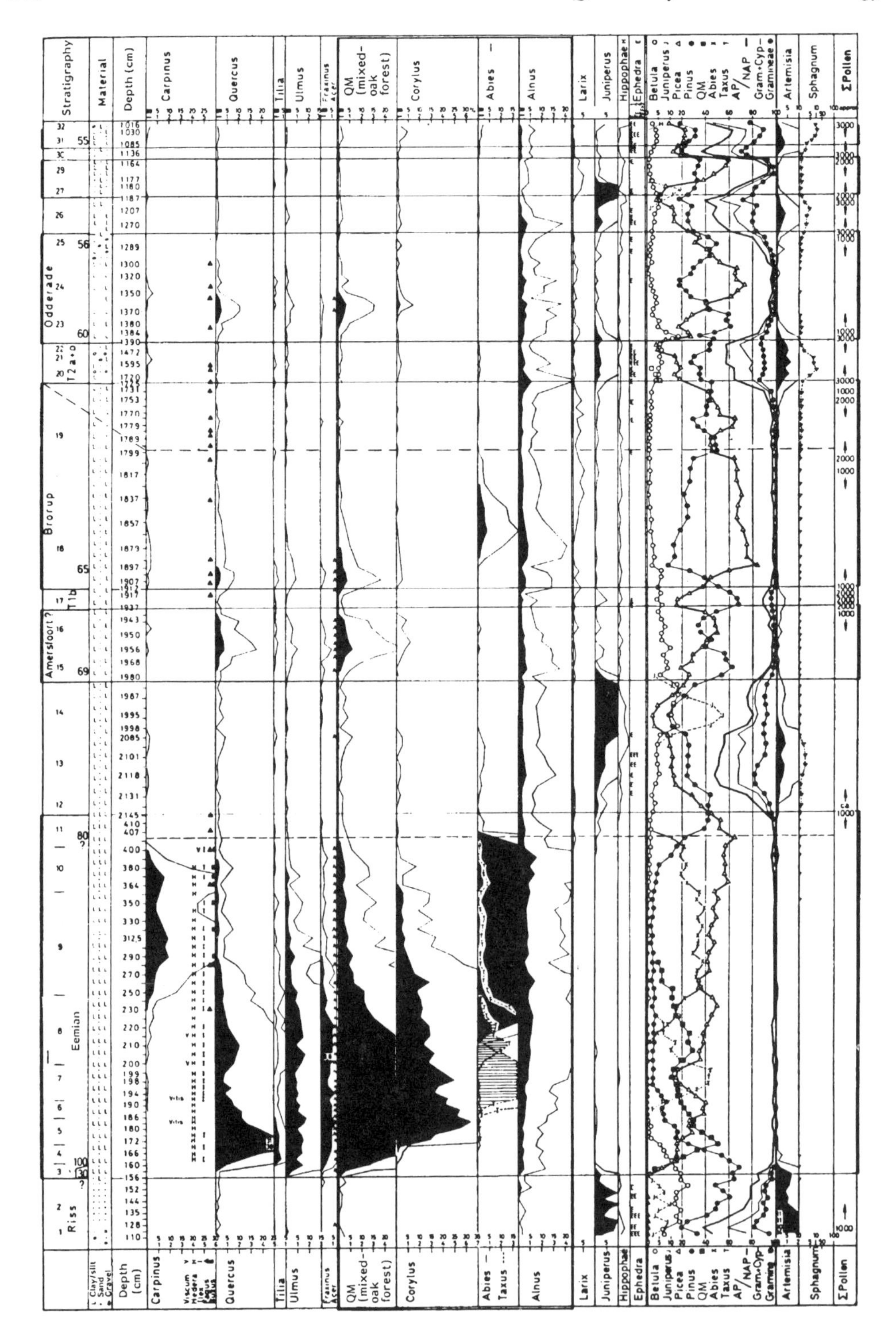
Stratigraphy
Material
Depth (cm)
Carpinus
Quercus
Tilia
Ulmus
Fraxinus
Acer
QM (mixed-oak forest)
Corylus
Abies
Taxus
Alnus
Larix
Juniperus
Hippophae
Ephedra
Betula
Picea
Pinus
QM
AP/NAP
Gram.+Cyp.
Gramineae
Artemisia
Sphagnum
ΣPollen
Odderade
Brorup
Amersfoort?
Eemian
Riss
Clay/silt
Sand
Gravel

(Schaefer 1975). Conversely, Schreiber & Müller (1991) came to the conclusion that it had not been the penultimate (Riss) glaciation that had advanced furthest into the Alpine foreland, but an older ice advance that they assigned provisionally to Schaefer's 'Paar Cold Stage'. Lack of precise dating makes it impossible currently to clarify the chronostratigraphical situation. It cannot be ruled out that the classic Riss Cold Stage really includes a number of glaciations. The separation of a 'Paar Cold Stage' in the Isar–Loisach Glacier region, however, cannot be regarded as assured, because the 'Paar' deposits might also include Mindel-age sediments.

Within the Rhine Glacier region, the Riss Glaciation can be subdivided into three ice advances. The **Older Riss** of Schreiner (1989, 1992) was first described by Schreiner & Haag (1982) as an independent advance. Its deposits are marked by a greater degree of weathering; in the Riss valley and neighbouring valleys they underlie the deposits of the Middle Riss from which they are separated by gravels. The extent of the intervening glacial retreat cannot yet be assessed. The differences in weathering give reason to believe that the break might at least have been of interstadial status. The distribution of the till of this first ice advance is restricted to former glacier tongues; accordingly Schreiner & Haag refer to the Middle Riss as the **Zungen-Riss** ('tongue Riss').

In the eastern part of the Rhine Glacier region (between Biberach and Leutkirch), the Middle Riss is confined by a double rampart of end moraines. Schreiner (1989) and Ellwanger (1990) refer to it there as **Doppelwall-Riss** ('double rampart Riss'). The end moraines are generally gravelly push moraines 10 to 30 m in height, formed at an interval of 1 to 3 km. They delineate the margin of the Middle Riss Glaciation in the eastern region of the Rhine Glacier. In contrast to this pronounced landform assemblage, the end moraine of the **Younger Riss** is incompletely developed. The moraine ridges north of Ingoldingen and south of Eberhardzell, at the eastern margin of the former Rhine Glacier, are part of it. The associated gravel fields in the Riss valley and in the Umlach valley (to north of Warthausen) have mostly succumbed to erosion (Schreiner 1992).

Riss moraines have been mapped in the Salzach Glacier region (Grimm *et al.* 1979), the Traun Glacier region (Van Husen 1977) and in the Upper Austrian Krems valley (Kohl 1971). Even in parts of the intra-Alpine area, deposits of the penultimate glaciation have been preserved. Thus, it has even been possible to identify moraines of the Riss Glaciation in the narrow valleys of the Steyr and Enns rivers (Van Husen 1968, 1975). A subdivision into individual ice advances or stadials has not been attempted in this case. A review of the state of knowledge on these matters has been presented by Kohl (1983).

The stratigraphy of the Alpine Riss Cold Stage is still far from solved – at least in major parts of the glaciated area. Neither the precise extent of the glaciers, nor the number of ice advances or their age, can yet be determined with certainty (see for instance Ellwanger *et al.* 1995). Under these conditions a direct correlation of the classic Riss or parts of it with the Saalian Glaciation of northern Europe is not yet possible.

12.13 RISS/WÜRM INTERGLACIAL

Interglacial deposits have been known from the Alpine region since the last century. Amongst the first deposits of the Alpine region that were classified as interglacial in origin were the

FIGURE 170 (opposite) Pollen diagram of the Riss/Würm Interglacial (Eemian) and Early Würmian interstadials from Samerberg (from Welten 1981)

so-called 'schieferkohlen'. These lightly carbonised formations are found in numerous places in the Alpine region. However, the first thorough examination by Baumberger *et al.* (1923) and later palynological investigations by Lüdi (1953) revealed that the 'schieferkohlen' formed under relatively cold conditions. They often form part of a sequence of deposits that began with lake sediments (clays, silts, sometimes lake marls). Eventually, as the lake silted up, this led to deposition of a series of slightly sandy clays with interstratified peat, that was later compressed mostly to thin layers of 'coal'. The Großweil and Pfefferbichl (Upper Bavaria) 'schieferkohle' deposits are palynologically correlated with the Eemian Interglacial by most researchers (e.g. Grüger & Schreiner 1993).

As early as 1953, Reich was able to demonstrate that the vegetational development of the Alpine Riss/Würm Interglacial resembled the Eemian of northern Germany. However, the 'schieferkohlen' only represent the second half of the interglacial. The first part of the interglacial interval is generally represented by the underlying lake deposits. More recent investigations of the 'schieferkohlen' of the Oberallgäu (Ebel 1983), for example, have largely reconfirmed these findings. Where interglacial floral elements are found they usually represent extremely shortened sequences from the final phase of an interglacial that render age determination by palynological means very difficult (Peschke 1983a). Thus, for instance, Peschke (1983b) used palaeobotanical evidence to correlate the 'schieferkohlen' of Herrnhausen (near Wolfratshausen, south of München) with the Mindel/Riss Interglacial, whereas geologists postulate a Riss/Würm (Eemian) age (Jerz & Ulrich 1983).

After initial investigations by Firbas (1927) in the Swiss region, Welten (1944, 1952, 1958) and his students dealt particularly with the interglacial deposits of the Alps. The sites of Meikirch (Welten 1982a, b; 1988), Thungschneit (Schlüchter 1976, Welten 1982a), Thalgut (Welten 1988), Uster (Wyssling & Wyssling 1978; Welten 1988) and Wildhaus (Welten 1988) have been found within the limits of the last glaciation. Outside this region are the Niederweningen, Sulzberg (Welten 1981) and Gondiswil (Wegmüller 1986) sites. The highest-lying interglacial occurrence found so far is in the southern Belledonne (French Alps) at an altitude of 1600 m (Hannss *et al.* 1992).

In southern Germany Beug (1972) has investigated the Eemian Interglacial at Zeifen in the vicinity of Laufen on the Salzach River, and the Eurach site on the Starnberg Lake (Beug 1979). The best-investigated pollen section of the Alpine region, comprising the last interglacial and three Early Würmian interstadials, however, is the Samerberg profile in Bavaria, southeast of Rosenheim (Grüger 1979). Here, the vegetational succession corresponds generally with that of the North German Eemian. The montane character of the site is expressed through high values of pine pollen. *Picea* appears early and soon reaches a dominance among the tree taxa. After a short pronounced *Taxus* maximum, the immigration of *Abies* and *Carpinus* follows. At the end of the interglacial all three species vanish and are replaced by *Pinus*. Samerberg is the stratotype of the Würmian Stage (Chaline & Jerz 1984). In the lower-altitude Riss/Würm Interglacial sequence at Zeifen (Jung *et al.* 1972), the montane species *Picea* and *Abies* appear in smaller numbers and *Carpinus* dominates persistently (Grüger 1989). A Riss/Würm Interglacial site comparable to the Samerberg sequence has recently been identified in the Wurzacher Becken, a tongue basin of the Riss Stage Rhine Glacier (Grüger & Schreiner 1993).

On the northern banks of Mondsee (Austria), a huge sequence of lacustrine sediments has been exposed, both overlain and underlain by till. Palynological investigations have shown that the lake sediments span the last interglacial and four Early Würmian interstadials (Klaus 1987). The interglacial vegetational succession corresponds to that at Samerberg. The climatic optimum here is characterised by a mixed-oak-forest, rich in *Ilex* and *Taxus*. This fact indicates that the average annual temperature in the Riss/Würm Interglacial was 2 to 3°C higher than during the Holocene climatic optimum. At the end of the interglacial, as in northern Germany, there was total deforestation. The Early Würmian interstadials are characterised by renewed spread of forest, in which *Pinus* and *Picea*, and to a lesser extent *Quercus* and *Taxus*, were re-established (Klaus 1987).

These interglacial sites also provide an important insight into the course of events during the Würmian Cold Stage. Samerberg and the Mondsee site alike lie far inside the maximum limits of the Würmian Glaciation. The continuous sequence of strata demonstrates that the Würm glaciers did not advance into the Alpine foreland earlier than 25,000 years ago.

In the Gail valley near Nieselach (Austria), interglacial deposits have also been preserved. Here, a sequence of strata 10 m thick, including laminated clay with sand layers, is exposed, overlain by gravel that is rich in organic material. This bed is in turn overlain by organic sediment 3 m thick (Fritz 1971, Van Husen & Draxler 1980). Above an erosional disconformity follows coarse gravel that must be assigned to the last glaciation.

Whereas the lacustrine sediments contain only scarce evidence of a cold-stage vegetation, the organic bed was deposited under interglacial conditions. Palynological investigations indicate the growth of forest dominated by *Fagus, Picea* and *Abies* (Fritz 1971). U/Th dating of the organic bed yielded an age of 113,000 ±9000 years, that indicates deposition took place towards the end of the last interglacial (Van Husen 1989). The presence of *Fagus*, however, suggests correlation earlier in the Riss/Würm Interglacial.

Frenzel (1983) believes that the alleged Riss/Würm sites actually represent two different interglacials. He makes a distinction between a Zeifen-type interglacial (central part of the interglacial characterised by a preponderance of *Carpinus* in a *Quercus–Fraxinus–Ulmus* forest, with abundant *Corylus*) and a Pfefferbichl-type interglacial (central part of the interglacial interval characterised by fir and pine forests). Whereas the Zeifen-type corresponds to the North German Eemian, the Pfefferbichl-type is thought to be part of an older interglacial event. However, this concept is not shared by other researchers (Grüger 1979, Menke & Tynni 1984).

The reconstruction of the palaeogeography of the last interglacial is made possible through a series of palaeosols which are found widespread under younger overlying strata. In Switzerland they include the fossil soils at Bümberg (Schlüchter 1976, 1982), Jaberg (Schlüchter 1982), Güntighausen/Thurtal (Schlüchter 1982) and Hurifluh (Schlüchter 1976). The Schorn section (northeast of Starnberg) is an example from the south German Alpine foothills (Fig. 171). Beneath a cover of decalcified loess loam with a Holocene argillic brown earth at the top, a Riss till with another interglacial argillic brown earth within its upper part is found. Pedological investigations have shown that the intensity of soil formation during the Riss/Würm Interglacial was similar to Holocene pedogenesis (Jerz 1982). The loess soils of the Riss/Würm Interglacial (e.g. the Stillfried A profile) are discussed in chapter 15.

Older palaeosols occur in various places in the Alpine foreland. In many cases (as in the Schorn section) their age remains uncertain. Drilling in the gravel spreads between Memmingen and Mindelheim locally identified two or more fossil soils. Deep weathering sinks (sand pipes) suggest the intensity of the weathering process. In the Iller and Salzach regions the sinks in the 'Jüngere Deckenschotter' reach to an average depth of 5 to 6 m. It must be remembered, however, that such weathering forms may have been continuously modified through a sequence of interglacials, and that part of the sections may have been subsequently

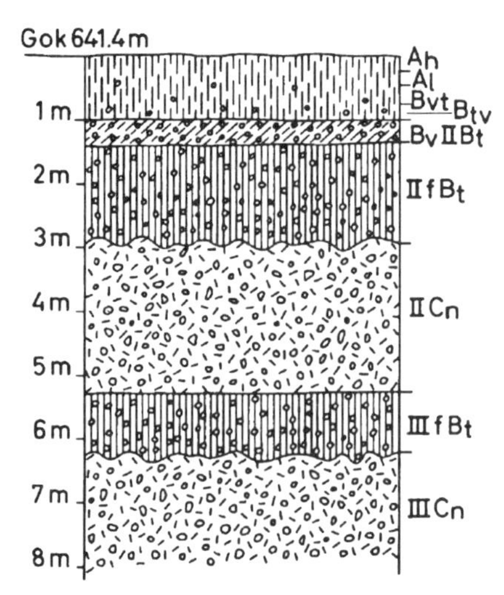

FIGURE 171 Fossil soils at Schorn (northeast of Starnberg, Bavaria); Gok = ground surface. To the right: soil horizons; the horizons marked with a II belong to the first fossil soil, the horizons with a III to the second fossil soil (after Jerz 1982)

eroded. In the München 'Deckenschotter' the sinks reach only half as deep because the Riss and Würmian gravels overlie the older gravels (Jerz 1982). In the sediment overlying the Early Pleistocene deposits of the Alpine foreland, multiple fossil soils have been locally preserved. Thus beneath Würmian Stage deposits in the Rosshaupten brickworks, the Riss/Würm Interglacial soil and two additional interglacial soils have been preserved. In the lower part of the sequence, the Matuyama/Brunhes magnetic reversal has been identified, dating the underlying 'Ältere Deckenschotter' to the pre-780,000 BP period (see chapter 12.5). The Würmian loess includes three moist soils and the Lohner Boden soil (Tillmanns *et al.* 1986, Jerz & Doppler 1990). This section has been described in detail by Leger (1988).

12.14 WÜRMIAN COLD STAGE

The subdivision of the last cold stage, the Würmian, has long been disputed in the Alpine region. Initially a single undivided Würmian Glaciation was envisaged, subdivided only by a series of recessional end moraines (Troll 1924). Knauer (1928) and Eberl (1930) proposed a sequence of three independent Würmian ice advances but their concept was refuted. However, the Alps, like northern Europe and North America, experienced an Early Würmian ice advance, the extent of which is still a matter of debate. The Subcommission of European Quaternary Stratigraphy (1983) has decided that the Würmian Stage should be subdivided into the Lower, Middle and Upper Würmian substages (Chaline & Jerz 1984) (Table 9).

In the northern French Alps, the first ice advance of the Würmian is thought to have reached the furthest; a maximum ice advance at about 70,000 BP has been advocated by Hannss (1982) and Monjuvent & Nicoud (1988). Mandier (1983, 1984) also attributes the outermost Würmian end moraines in the valleys of Rhône and Ain to an Early Würmian ice advance. This interpretation is not only incompatible with the situation in Switzerland and the northern Alpine foreland, but also with that in the southern French Alps, for which Jorda (1988) postulates a Würmian glaciation maximum after 30,000 BP.

TABLE 9 Subdivision of the Weichselian and Würmian Cold Stages in northern Germany and in the northern Alpine foreland (after Schreiner 1992)

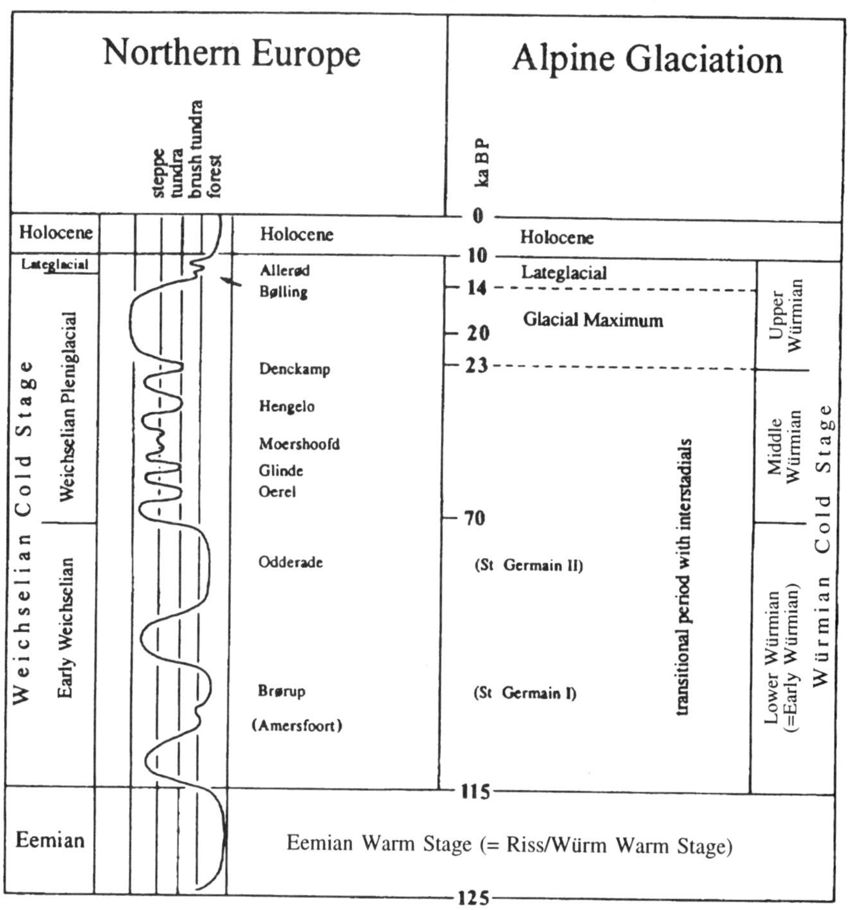

The stratigraphical framework in the northern part of the French Alpine glaciated area is still unclear. For instance, the argillic brown earth on the supposedly Early Würmian Saint Laurent Terrace (Mandier 1984) might possibly be of last interglacial age and thus would indicate that the moraines in question would be Riss (or older) (Billard & Orombelli 1986). If this were correct, the moraines at Grenay–Satolas and Lagneux, east of Lyon and Rives, to the west of Grenoble, would represent the outermost Würmian end moraine of the Rhône and Isère glaciers. They have been equated with end moraines in northern Italy assigned to the Late Würmian maximum. Ice-contact formations near Pontida (Bergamo, Lombardy) have yielded dates equivalent to the maximum stage of the Late Würmian Lecco Piedmont Glacier, i.e. shortly before 17,700 ±360 BP (Orombelli 1974; Alessio *et al.* 1979). If there was an Early Würmian ice advance in the region, it must have been of relatively minor extent.

In many parts of Switzerland where the outermost marginal positions of the last glaciation are strongly developed, a system of more distal end moraines had been mapped by previous workers and thought to be related to an earlier, more extensive Early Würmian ice advance.

The latter might correspond to what Troll (1924) called his 'advanced end moraines' in south Germany. However, strong soil development and periglacial alteration of the landforms indicate that they are older forms. Schlüchter (1989a) therefore assumes that the ice advance in question cannot be assigned to the last cold stage but must be attributed to a markedly older glaciation that might even predate the Mindel/Riss Interglacial.

In Meikirch the Riss/Würm Interglacial deposits are overlain only by deposits of one ice advance. The same is true for the Gossau site near Zurich. However, stratigraphical investigations at Thalgut, Jaberg, Thungschneit and Cossonay (Arn 1984) imply that the last cold stage in Switzerland is represented by two major ice advances, during both of which the glaciers extended into the Alpine foreland. However, the older advance did not reach as far as the younger one. The sections of Gossau and at Walensee indicate that, from 60,000 to 28,000 BP, between the two ice advances an ice-free period occurred (Schindler *et al.* 1985, Schlüchter *et al.* 1987).

As opposed to the results from France and Switzerland, the Samerberg (Bavaria) and Mondsee (Austria) profiles show an uninterrupted sequence of lake sediments between unequivocal deposits of the Riss/Würm Interglacial (Eemian) and the Late Würmian till (approximately 20,000 BP). Van Husen (1977) also only found the traces of just one Würmian ice advance in the Traun valley. Here, there is no evidence for a possible advance during the Early Würmian. The main ice advance of the last glaciation accordingly only occurred after 27,000 BP, as in northern Europe. In Vorarlberg, for example, a mammoth tusk from deposits overridden by the Würm ice gives a maximum age of the ice advance of 23,900 ±400 BP (De Graaff 1992). Traub & Jerz (1976) obtained an age of 21,650 ±250 BP for snails in loess from the edge of the Salzach Glacier; the corresponding beds had been deposited directly before the accumulation of the highest Lower Terrace gravels.

Since the investigations by Woillard (1975) in the Vosges Mountains and of the Les Echets site near Lyon (De Beaulieu & Reille 1984a, b), many details of the climatic development at the transition from the Riss/Würm Interglacial to the Würmian Cold Stage have been revealed in the circum-Alpine area. The vegetational evolution during the Würmian is comparable with that of the Weichselian in northern Central Europe. The first synthesis based on detailed investigations of various sites was offered by Welten (1981). By the end of the Riss/Würm Interglacial, the vegetation retreated markedly, resulting finally in total deforestation. During the Alpine equivalents of the Brørup and Odderade interstadials forest colonised the region once more. The interstadial forests were characterised by *Pinus* and *Picea*. In the Samerberg pollen diagram, the 'double-interstadial' character of the Brørup, with the short intervening deforestation phase, is well developed. The situation is similar to that in the Netherlands, where this treeless interval initially led to a subdivision into two independent interstadials (Amersfoort and Brørup) (see chapter 11.16). In contrast to the development in northern Germany, the equivalent of the Oerel Interstadial is also characterised by reforestation, so that three interstadials with reforestation overlie the Riss/Würm Interglacial in the Samerberg 1 borehole (Grüger 1989).

Within the loess deposits of the Alpine foreland, fossil wet-ground soils spanning the period from the Early Würmian to the Late Würmian maximum, are found in many places. The weakly developed soil profiles typically consist of a narrow decalcified layer showing gley characteristics. In most cases two or three of these gley zones are found, often containing ice-wedge casts and cryoturbations (Jerz 1976, 1982). Further north, towards the Danube, generally decalcified 'brown tundra soils' developed under the influence of a rather dry climate (Brunnacker 1957).

Within the area of the last glaciation, lacustrine sediments have been preserved in some places that could be assigned to the Lower and Middle Würmian Stage by means of radiocarbon dating. Among these are the Schabs (Fliri 1978) and Albeins sites in southern Tyrol (Fliri 1989), as well as Podlanig in the Gail valley (Fritz 1977). In many places fossiliferous deposits are preserved within these beds dating from about 35,000 BP (Schladming and Hohentauern sites, Draxler & Van Husen 1978; Freibach site, Felber & Van Husen 1976; Nieselach, Fritz 1971, 1975). At this time, a pine forest, rich in *Fagus* and *Picea*, spread into the Carinthian area. This vegetation closely resembled the local vegetation today, except that the forest limit was slightly lower. In the Ramesch cavern, in the northeastern limestone Alps, many cave bear bones and associated Levalloisian technique artefacts have been recovered (Kohl 1989). In spite of certain discrepancies in the radiocarbon dates, this period of occupation may well be correlated with the formation of the Stillfried B loess soil (Van Husen 1981; cf. chapter 15).

The most prominent occurrence of pre-Late Würmian lacustrine deposits are the varved clays of Baumkirchen, 12 km east of Innsbruck (Fliri *et al.* 1970, 1971, 1972, Fliri 1973), which seem to postdate the Middle Würmian interstadial. During their deposition the climate was already rather cool; in the vicinity of the lake shrub tundra was dominant. The sequence of varved clays, over a hundred metres thick, is overlain by ca. 70 m of gravels deposited during the ice advance that again are covered by a thick till. The varved clays were deposited in a shallow lake, as indicated by trace fossils of fish on the bedding planes. The sedimentation rate was constantly high, reaching an average of 5 cm/year (Bortenschlager & Bortenschlager 1978). Radiocarbon dates on wood (*Pinus*, *Alnus* and *Hippophae*) from the varved clays have yielded ages of between 31,600 ±1300 and 26,800 ±1300 years BP (Fliri 1973).

An important element for the determination of the cold-stage climatic conditions is the reconstruction of the snow-line altitude (Gross *et al.* 1978). Penck & Brückner (1909) had already estimated that during the Würmian Cold Stage the snow line in the Alps was 1200 m lower than today. This value is almost identical to recent calculations (Patzelt 1975: 1250 m). Haeberli (1982) assumes that during the Würmian maximum the average annual temperatures were at least 15°C below today's values. Precipitation is estimated as having been less than half the present rate.

As opposed to the Scandinavian glaciation area, where the thickness of the ice can only be determined at very few points because of the almost total absence of nunataks, the maximum height of the Alpine ice streams can be reconstructed precisely by mapping the upper limits of erratics (cf. Von Klebelsberg 1935, Jäckli 1962, 1970, Weinhardt 1973, Van Husen 1987).

The advance of the Late Würmian glaciers was a single major ice advance. However, during its retreat numerous minor readvance phases left a series of recessional end moraines. Thorough geomorphological investigations of this sequence have been undertaken in various parts of the Alpine area. The landform inventory ('glacial series') of the last glaciation has been investigated in great detail in the region of the western Rhine Glacier (Schreiner 1974). Well-developed 'glacial series' are found at many places – for instance near Winterstettenstadt at the northern limit of the former Rhine Glacier and at the northern limit of the former Iller Glacier, south of Memmingen (Graul 1973). From the Iller Glacier region new mapping of the ice-marginal positions and a reconstruction of successive Würmian Stage drainage patterns have been provided by Habbe (1986a and b) and Ellwanger (1980, 1988). Early assessments of the young moraines of the Isar–Loisach Glacier by Rothpletz (1917) and Knauer (1929, 1931) were supplemented by Grottenthaler (1980) and Feldmann (1990, 1992). A classical interpretation of the Inn Glacier region was offered by Troll (1924).

During the Würmian Glaciation maximum, transfluences developed at various points of the ice-stream network. In the Inn valley, for instance, around Landeck, five major tributary glaciers grew together to form one major valley glacier, advancing eastward down the Inn valley. At the same time, a major tributary glacier from the south had already blocked the Inn valley around Innsbruck. Congestion resulted in a rise of the ice surface until finally the threshold to the north was crossed, and ice from the Inn valley flowed over the Fernpass and the Seefelder Senke due north into the Isar and Loisach drainage systems (Van Husen 1985b). Such events are reflected in the supply of gravel and major erratics. In the foreland of the Isar–Loisach Glacier, Alpine crystalline rocks constitute up to 35% of the total, in the 20 to 31.5 mm fraction. In contrast, in the eastern marginal region of the glacier lobe, the Tölz Glacier, the contents range only between 0 and 1%, since no direct ice supply from the Inn valley developed there (Dreesbach 1985).

The marginal region of the Salzach Glacier was reinvestigated by Grimm *et al.* (1979) and J.H. Ziegler (1983). As in other areas, no traces of an Early Würmian ice advance were found here. In the foreland near Salzburg and Tittmoning, two enormously overdeepened basins were carved 200 m deep into the subsurface and then completely filled with sediments. The Salzach Glacier left a great number of end-moraine ridges, some of which dammed the drainage and formed a series of lakes. The moraine dams were breached successively by the meltwaters during the Oldest Dryas. During formation of the late- and postglacial drainage network, four to six terraces accumulated, the ages of which have not been determined (J.H. Ziegler 1983).

Numerous recent investigations of the events during the Würmian Cold Stage are available from Austria. In the Traun valley Van Husen (1977) undertook a detailed investigation of the Würmian deposits and reconstructed the extent and landform assemblages of the individual glacier fluctuations. The Würmian Stage ice-marginal positions and glacial deposits in the upper Enns valley were also mapped by Van Husen (1968) who showed that the connection between the Enns Glacier and the Traun Glacier ceased immediately after the glacial maximum. In the lower Enns valley, Van Husen (1971) found a tripartite sequence of Würmian gravel terraces, probably equivalent to the similar set of Danube terraces near Linz (Kohl 1968). Comparable investigations are also available from parts of the Mur and Drau Glacier areas (Van Husen 1976, 1980).

Van Husen (1981) demonstrated that both north and south of the main ranges of the Alps, the Late Würmian glacial maximum was followed by a long stillstand during which the ice margin stagnated within a few hundred metres of the outermost moraine. During this halt a morphologically more distinct end moraine accumulated than during the Würmian maximum, and the main gravel body of the Lower Terrace also accumulated from this moraine.

The Upper Würmian Substage of the subdivision of the INQUA Subcommission for European Quaternary Stratigraphy (SEQS) is subdivided into the so-called 'High-Glacial' and 'Lateglacial' phases (Chaline & Jerz 1984). The Lateglacial comprises the period of recession from the glacial maximum until the end of the Younger Dryas Chron. The Lateglacial in the Alps began with a period of rapid melting of the glaciers, resulting in widespread downwasting of inactive ice and accretion of various ice-contact landforms, that apparently do not represent any major stillstands (Weinberger 1955, Ebers *et al.* 1966, Van Husen 1981). However, later in the Lateglacial, five glacier halts can be recognised (Van Husen 1981).

The basic subdivision of the Lateglacial ice retreat was originally proposed by Penck & Brückner (1909) who distinguished three 'retreat stages' of the Würmian glaciers: Bühl, Gschnitz and Daun; Bühl represents the last readvance of the still-intact cold-stage ice-stream

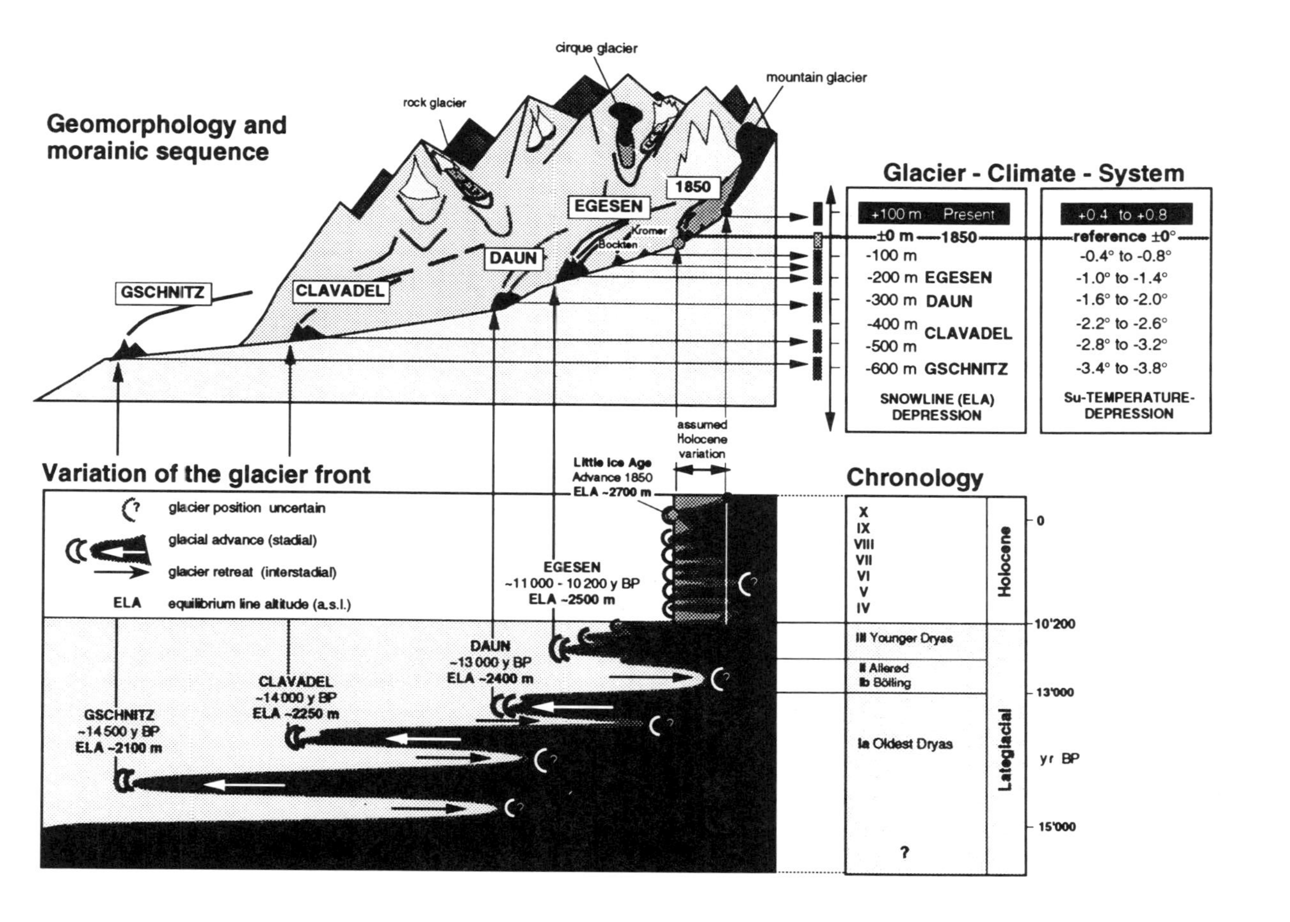

FIGURE 172 Lateglacial and Holocene ice retreat in the eastern Swiss Alps (after Maisch 1981, 1992)

network; Gschnitz and Daun are inner-Alpine local readvances. Later additional stages were added, such as the Steinach stage that predates the Gschnitz stage (Von Klebelsberg 1950, Von Senarclens-Grancy 1958) and the Egesen stage, as defined by Heuberger (1966).

Steinach cannot be distinguished from Gschnitz by a distinct difference in snow-line altitude but rather through morphological indicators similar to the distinction between the Daun and Egesen events. In both cases the older moraines (Steinach, Daun) have been overprinted by strong periglacial activity and erosion during the more recent ice advances (Gschnitz, Egesen). A survey of the respective stages, their type localities and their definitions has been offered by Mayr & Heuberger (1968) and Patzelt (1972). Heuberger (1966) mapped the central Alpine Late Glacial stages. Figure 172 also includes further oscillations that have been distinguished in the Tyrolian central Alps (Gross *et al.* 1978, Kerschner 1982) and in Graubünden (Switzerland) (Maisch 1981, Vuagneux 1983) since that time. Ice retreat from the inner Würmian end moraine probably began as early as 17,000 BP. The lateglacial Daun readvance generally took place during the Older Dryas Chron at the latest (before 11,000 BP, Heuberger, written communication), whilst Patzelt & Bortenschlager (1978) demonstrated that the Egesen stage represents the glacial advance during the **Younger Dryas** Chron.

The climatic transition from the Late Würmian to Holocene is still far from being well understood. In addition to dating problems, the various environmental indicators (glaciers, frost features, vegetation, insects, molluscs, stable isotopes) have different response times and react to more than one controlling factor (e.g. summer temperatures, mean annual temperatures, precipitation, soil formation) (Ammann *et al.* 1994). In the Alps the first reforestation occurred as early as during the Bølling Chron as clearly indicated in the central Inn valley (Bortenschlager 1984) and the Swiss Mittelland (K.F. Kaiser 1989). In northern Germany this did not happen before the Allerød. The first postglacial Alpine forest was characterised by the spread of *Betula* and *Pinus*. In northern Germany and Britain, *Pinus* was unable to grow during the Younger Dryas climatic deterioration. However, in the Alpine area, pollen analysis indicates that the cooling was less severe, at least at altitudes below 600 m. Pine apparently continued to colonise southern Germany and Switzerland until the Holocene (Ammann *et al.* 1994).

12.15 HOLOCENE

Whereas Penck & Brückner (1909) understood 'post-glacial' to be the period from the Würmian glacial maximum to the 'present time', nowadays the postglacial is equated with the Holocene. The time interval between the Würmian glacial maximum and the end of the Egesen readvance is referred to as lateglacial (see above). The late- and postglacial fluctuations of the Alpine glaciers have been thoroughly investigated in various places (Patzelt 1972, 1973a, b, Patzelt & Bortenschlager 1978). Amongst the more recent publications are the investigations of Fraedrich (1979) in the Ferwall Mountains (Tyrol/Vorarlberg), Habbe & Walz (1983) in the Val Viola (Sondrio, Italy), Porter & Orombelli (1982) in the Val d'Aosta, Hannss (1982) in the French Central Alps, Maisch (1992) in Mittelbünden and H.-N. Müller *et al.* (1983) in the western Swiss Alps. By using radiocarbon dating it has frequently been possible to determine the ages of the individual readvances. A survey of the glacial evolution during the postglacial in the Alps and other high mountain regions is provided by F. Röthlisberger (1986) and Furrer (1991).

Postglacial ice advances are calculated for 8500 BP, 7500 BP, 4500 to 5000 BP and 3000 to

3500 BP, whereby the old concept of an intensive postglacial warm period lasting many thousands of years has been definitely refuted. Although the glaciers of the Egesen stage reached far beyond today's limits, the postglacial readvances all remained within the range of the 'neo'-historic glacier fluctuations. In the most recent past, particularly remarkable ice advances occurred during the so-called 'Little Ice Age' of the 16th to 19th centuries (Kinzl 1929, 1932, Patzelt 1973a, b, Maisch 1992).

Furrer (1954) was the first to point out the great palaeogeographical and palaeoclimatic indicator value of fossil soils buried by landslides, talus slopes and moraines. Soils indicate long intervals of geomorphological stability, whereas the overlying strata represent morphodynamic activity (Furrer *et al.* 1971, Gamper 1985). A systematic assessment of these phenomena has provided the foundation of reconstructions of the postglacial history of the Alpine region.

Periglacial activity during the Holocene in the Swiss Alps can roughly be subdivided into two major phases:

1. In the early Holocene hardly any solifluction occurred.
2. Towards the end of the Atlantic period, at about 5000 years BP simultaneous glacial advances and mass movements occurred. Radiocarbon-dated fossil soils demonstrate that, since that time, enormous sediment redeposition has taken place (Gamper 1985). A major glacial readvance combined with strong periglacial activity in the Central Alps has been dated to the Middle Atlantic period, as early as 6600 to 6000 BP (Patzelt 1973a).

In the interior Alps (Inn valley), human activity can be identified palynologically far into the Mesolithic period by means of culture-indicator plants. The first indications are found as early as the 9th and 10th millennia BP, i.e. in the Boreal period. In the mid-8th millennium (early Atlantic period) unambiguous indicators of settlement and ancient agriculture (Wahlmüller 1985) are found. For the interval starting at 5000 BP, an increase in human activities has been seen (Burga 1979) that might have contributed to intensified sediment movement. In the eastern Alps, Patzelt (1987) verified several phases of increased sediment redeposition and aggradation in the river valleys during the Holocene in the middle Preboreal, the late Atlantic and around 3500 BP for example. Detailed studies of the terrace deposits, by means of dendrochronology (Becker 1982) or, in the younger parts, by using pottery sherds and other artefacts, may help to further subdivide the Holocene of the Alpine region.

13

Quaternary History of North America

During the great ice ages North America was partly covered by extensive ice sheets in a way very similar to the coverage of Europe (Fig. 173). However, climatic change during the Quaternary has affected both continents very differently. Whereas in Europe E–W-trending mountain ranges, especially the Alps, formed an effective barrier for re-immigration of higher fauna and flora, the different topography of North America with its N–S-trending mountains has led to a much more gradual shift in vegetation. As a consequence, re-immigration was much easier and may have followed much the same pattern at the end of each glaciation.

In North America, the Pleistocene glaciations developed two largely independent ice sheets: the Laurentide Ice Sheet that covered the Canadian Shield and adjoining areas, and the Cordilleran Ice Sheet in the mountains of northwestern North America. Because of the large size of the North American continent, it is impossible to cover all areas in equal detail in this short review. Therefore, only selected regions are discussed below, with emphasis on the southern margins of both ice sheets.

Whereas in Europe the Weichselian glaciers were surrounded by an almost vegetation-free belt more than 100 km wide, the last cold-stage (Wisconsinan) North American ice sheets advanced into a wooded landscape in the Midwest from Illinois to Ohio. Tree stumps below till at numerous sites provide evidence of overridden forests. The same applies to older glaciations in Nebraska and Kansas. Consequently, the question arose whether the Wisconsinan ice sheet was ever surrounded by a significant periglacial zone in which permafrost conditions prevailed (e.g. Potzger 1951). This question has to a great extent been answered. In the western Plains of Nebraska, extensive ice-wedge casts have been found up to 50 km south of the Wisconsinan glacial limit (Wayne 1991). In Wisconsin, a broad belt of permafrost features lies outside the glacial limit of the last ice sheet (Black 1965), and the same is true of Illinois, where patterned ground can be traced about 40 km south of the Wisconsinan ice maximum (Johnson 1990). Wayne (1967) has reported periglacial features from the narrow periglacial zone in Indiana. In southwestern Ontario, Morgan (1972, 1982) has recorded periglacial features formed during the late Wisconsinan ice retreat. However, in the Appalachian Mountains, a wide belt of at least discontinuous permafrost seems to have reached 300 km south of the ice margin. Towards the Atlantic coast, under the influence of a more maritime climate, the permafrost belt was extremely narrow. One reason why the North American periglacial zone was narrower than its European counterpart is that it occurred about 10° latitude further south (Fig. 174) (Péwé 1983a, b).

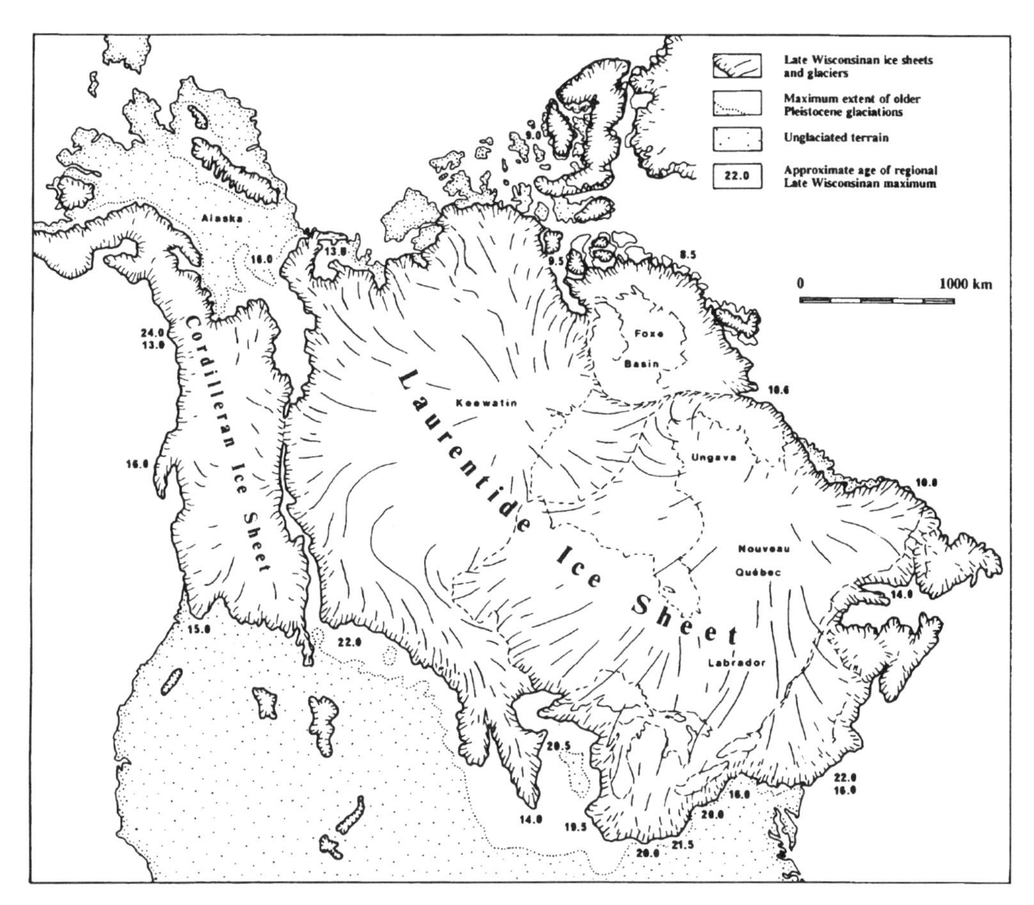

FIGURE 173 Extent of the Wisconsinan Glaciation in North America; extent of older glaciations indicated (after map distributed by the Canadian Geological Survey)

13.1 LATE TERTIARY

Cainozoic glaciation in North America began in the Tertiary. In Alaska traces of Miocene and Pliocene glaciation have been found on the Seward Peninsula, in the Wrangell Mountains and in the Upper Cook Inlet. The most striking record of Miocene glaciation has been preserved in the lower parts of the Yakataga Formation, covering an area of 30,000 km^2 on the Gulf of Alaska coast (Hamilton 1994). The Miocene glacial sedimentary sequence that was first described in some detail by Plafker & Addicott (1976), is mostly glaciomarine in origin. It comprises gravity flows of sand and gravel and lenses of diamicton, and includes channel-fill sequences up to 500 m wide and 70 m deep (Eyles 1987), similar in appearance to the buried tunnel channels of the European Pleistocene. Dating of molluscs has yielded ages of 15–16 million years but these ages are believed to be too high. In contrast, planktonic foraminifera have given ages of 5–6 million years. Correlation with neighbouring deep-sea drilling sites suggests an age for the initial tidewater glaciation between 5.0 and 6.7 million years (Lagoe *et al.* 1993).

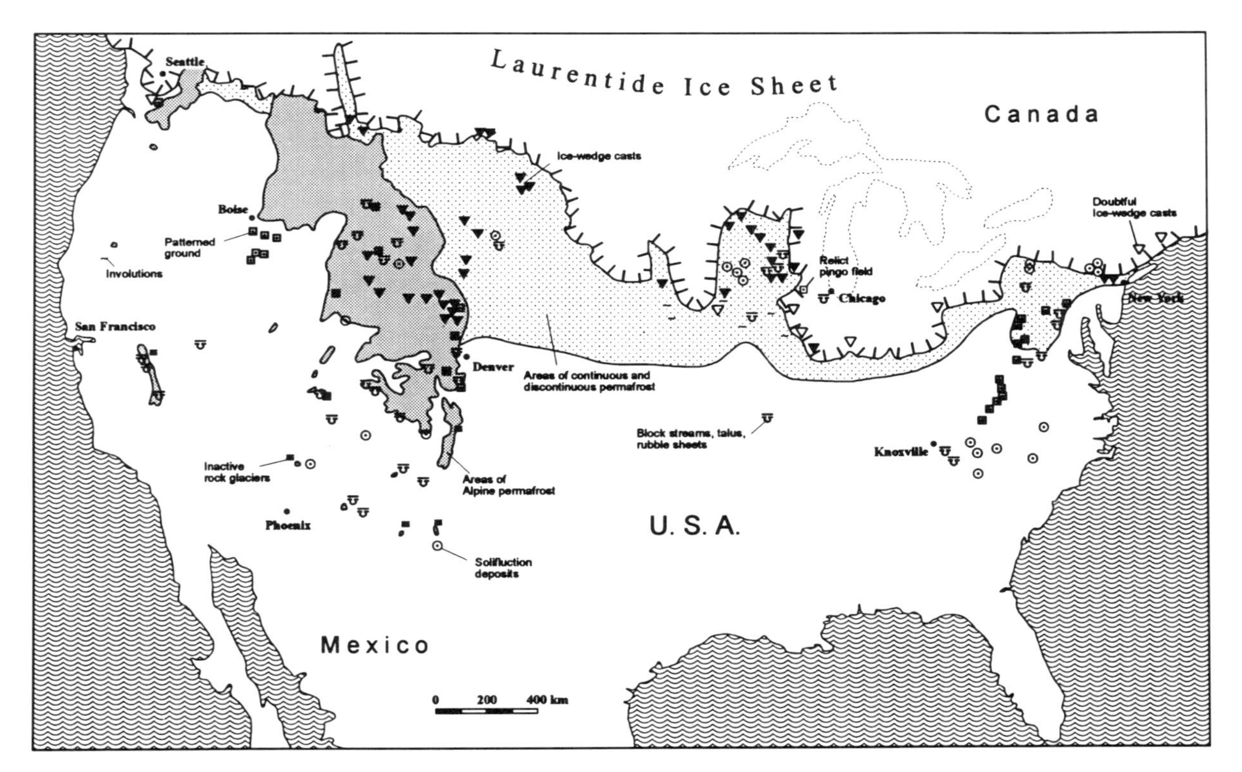

FIGURE 174 Periglacial features in the contiguous United States outside the Wisconsinan glacial limit (from Péwé 1983a)

13.2 EARLY PLEISTOCENE

The classical subdivision of the North American Quaternary stratigraphy dates from the late 19th century. Chamberlin (1894) distinguished three major glaciations: the Kansan (oldest), the East-Iowan (middle) and the East-Wisconsin (youngest). The East-Wisconsin was soon shortened simply to Wisconsin by Chamberlin (1895). One year later Chamberlin (1896) identified the Illinois glaciation. He was followed by Leverett (1898a, b, c) who introduced the terms Yarmouth, Sangamon and Peoria for the intervening interglacials. The Aftonian, the term for the pre-Kansan interglacial, had already been introduced earlier by Chamberlin (1895). The original three cold stages were supplemented by Shimek (1909) with the Nebraskan as the oldest North American glaciation.

Later it was discovered that the Iowan did not represent a separate cold stage. The term is redundant; some of the deposits originally correlated with this period are Illinoian or older, some are Early Wisconsinan in age. From this the following stratigraphical sequence evolved:

Wisconsinan Glaciation
Sangamonian Interglacial
Illinoian Glaciation
Yarmouth Interglacial
Kansan Glaciation
Aftonian Interglacial
Nebraskan Glaciation

This subdivision was correlated with the four-fold subdivision of the European Alpine Quaternary stratigraphy and has remained valid well into recent decades (cf. Flint 1971).

In the course of time, however, evidence emerged indicating that the traditional scheme might be incomplete. Using K/Ar dating of volcanic tuff from the Sierra Nevada, Dalrymple (1972) demonstrated that one of the local tills, the Sherwin Till, had an age of over 738,000 years. This age was later confirmed by fission-track dating (Izett & Naeser 1976).

These dates were first greeted with scepticism, because glacial deposits of this age were unknown from elsewhere. In Kansas and Nebraska, the age determination of the oldest tills that had been found up to that time was regarded as reliable. The relative chronology and regional correlations were largely based on what was thought to be an ash unique marker horizon, the so-called 'Pearlette Ash'. The age of this ash, which derived from volcanism in the Yellowstone region, was assumed to be 'Yarmouth to Kansan', i.e. Holsteinian to Elsterian in the European stratigraphy. Accordingly, the Kansan and Nebraskan tills were assumed to be considerably younger than the Matuyama/Brunhes magnetic reversal. It was found out only later that the so-called 'Pearlette Ash' really represented three different ash falls, the ages of which were older than previously supposed. They were dated to about 600 ka (Pearlette O), 1.2 Ma (Pearlette S) and 2.0 Ma BP (Pearlette B) by K/Ar and $^{40}Ar/^{39}Ar$ dating of sanidines and fission-track dating of zircons (Wilcox & Naeser 1992).

At this point the whole underlying concept of correlations within North American Quaternary stratigraphy collapsed. According to present knowledge, the United States may have experienced ten or eleven major pre-Illinoian glaciations. Not all of them are reliably dated, so that changes in correlation between the various sites must be anticipated. In one instance, a till was found underlying the Pearlette B Ash (pre-Illinoian K). This till would be about age-equivalent with the Praetiglian of the Netherlands. In Nebraska and Iowa, four

Magneto-stratigraphy	Age	Oxygen-Isotope Stages	Chronostrat.	The Netherlands	Great Britain	North America: Stratigraphy	North America: Tephra layers
Brunhes Epoch			Late Pleist.	Weichselian Cold Stage Eemian Warm Stage	Devensian Cold Stage Ipswichian Warm Stage	Wisconsinan Cold Stage Sangamonian Warm Stage	
			Middle Pleistocene	Saalian Cold Stage	Wolstonian Cold Stage	Illinoian Cold Stage	
						pre-Illinoian Cold Stage A	
				Holsteinian Warm Stage	Hoxnian Warm Stage		
				Elsterian Cold Stage	Anglian Cold Stage	pre-Illinoian Cold Stage B	
				Cromerian IV (Noordbergum)	Cromerian (West Runton) ?		
				Cromerian C Cold Stage		pre-Illinoian Cold Stage C	
				Cromerian III (Rosmalen)			Pearlette O Tephra
						pre-Illinoian Cold Stage D	
				Cromerian B Cold Stage			
				Cromerian II (Westerhoven)	Cromerian (West Runton) ?	pre-Illinoian Cold Stage E	Bishop Ash
	780 000	19		Cromerian A Cold Stage			
Matuyama Epoch			Early Pleistocene	Cromerian I (Waardenburg)	Hiatus		
						pre-Illinoian Cold Stage F	
				Dorst Cold Stage			
Jaramillo Event	900 000	27		Leerdam Warm Stage Linge Cold Stage			
	1 070 000	31		Bavel Warm Stage			
				Menapian Cold Stage		pre-Illinoian Cold Stage G	
							Pearlette S Tephra
				Waalian C Warm Stage			
				Waalian B Cold Stage			
				Waalian A Warm Stage			
			←	Eburonian Cold Stage		pre-Illinoian Cold Stage H	
						pre-Illinoian Cold Stage I	
Olduvai Event	1 770 000	64			Beestonian Cold Stage		
				Tiglian C5-6 Warm Stage	Pastonian Warm Stage		
				Tiglian C4e (Beerse Member)	Baventian Cold Stage	pre-Illinoian Cold Stage J	
	1 950 000	71		Tiglian C1-3 Warm Stage	Bramertonian, Antian		
							Pearlette B Tephra
				Tiglian B Cold Stage	Thurnian Cold Stage		
				Tiglian A Warm Stage	Ludhamian Warm Stage		
				Praetiglian Cold Stage	pre-Ludhamian Cold Stage	pre-Illinoian Cold Stage K	
	2 600 000	104	←				
Gauss Epoch			Tertiary				

glaciations were proved between the Pearlette S and Pearlette O ashes, two of which are reversely magnetised (Hallberg 1986). At least one till underlies reversely magnetised deposits in Illinois and Indiana (Johnson 1986) and in west-central Wisconsin a till unit with reversed polarity has also been found (R.W. Baker *et al.* 1983, Matsch & Schneider 1986). At least three tills have been identified as older than the Matuyama/Brunhes boundary in Yellowstone Park. In addition to the palaeomagnetism, their ages were determined by K/Ar dating of ash layers and lavas (Richmond 1986a). In the Californian Sierra Nevada, two glaciations have been dated by K/Ar of tephra layers as older than 738,000 BP (Fullerton 1986b). In the Puget Lowland two and in the Olympic Mountains, Washington, three glaciations were found to be older than the Matuyama/Brunhes boundary. Erosion has uncovered deposits of three glaciations, the Orting, Stuck and Salmon Springs Glaciation, which are overlain by the 1-million-year-old Lake Tapps Tephra (dated by palaeomagnetism and fission track; Easterbrook, 1986, 1992).

As a consequence, the terms 'Kansan' and 'Nebraskan' had to be abandoned because they clearly had been used for deposits of very different ages (cf. Table 10). They are preliminarily replaced by the term 'pre-Illinoian' (Richmond & Fullerton 1986) although in some areas more complete local stratigraphies are available (cf. Šibrava *et al.* 1986).

Particularly in areas close to the glaciation centres, where erosion prevailed, reconstruction of the glacial history is hampered by the lack of sediments. However, under favourable conditions the sedimentary record of adjoining sea areas can provide hints about the glacial history. For instance, a sediment core from south of the Grand Banks, off Labrador, provided a record of the last 0.9 million years. Detailed lithological, mineralogical and palynological investigations allowed distinction of material from different source areas. The record begins in Oxygen Isotope Stage 28, when the Grand Banks were still part of a forested coastal plain. Limited glaciation may have occurred in Stages 18 and 16, as indicated by detritus from Appalachian and Gulf of St Lawrence source areas. In Stages 14 and 12 large-scale erosion of Tertiary regolith and Cretaceous clays from the land areas took place, culminating in the formation of the Laurentian Channel towards the end of Stage 12. Extensive ice-sheet influence is also recorded in Stages 10, 8 and 6, with minor influence in Stages 4 and 2 (Piper *et al.* 1994).

13.3 PRE-ILLINOIAN WARM STAGES

Milder summers and warmer winters than today are indicated by a number of faunas identified in deposits older than 1.6 million years in southwestern Kansas and northern Nebraska (Wayne & Aber 1991). However, this period also includes deposition of the first regional till sheet, the so-called Elk Creek Till. The faunas therefore most likely represent warm phases of the Early Pleistocene.

Early and Middle Pleistocene vegetation has attracted little attention in North America. This partially results from a lack of adequate sites and often poor pollen preservation. Smiley *et al.* (1991) list only 27 sites in the United States from which floral investigations of pre-Wisconsinan Quaternary and Late Tertiary sediments have been published. Most of these cover the Sangamonian Interglacial. Dating, however, is often problematic.

TABLE 10 (opposite) Stratigraphical subdivision of the Quaternary deposits in the Netherlands, Britain and North America. Both possible Quaternary/Tertiary boundaries are marked by arrows. Ages of palaeomagnetic reversals and correlation with deep-sea oxygen-isotope stages according to Shackleton *et al.* (1990). Stratigraphy after Richmond & Fullerton (1986) and Gibbard *et al.* (1991)

In southeastern Indiana an organic deposit at Hadley Farm has been studied by Kapp & Gooding (1964). The lower part of the sequence is dominated by *Ostrya* and *Carpinus*, with *Quercus, Pinus* and *Corylus* also being present. It is overlain by a pollen assemblage with high contents of *Fagus, Carya, Ulmus* and grasses and herbs, representing a shift from closed to open forest. The climate at that time was warmer than at present, as indicated by the presence of *Liquidambar* and *Planera aquatica*. Both taxa are exotic to the area today although in other respects the pollen assemblage is similar to Sangamonian-age floras from Indiana and Illinois. The deposits were originally interpreted as 'Yarmouth Interglacial' in age, but palaeomagnetic investigations and amino-acid determinations have shown that the site is very old, predating the Matuyama/Brunhes boundary (Johnson, personal communication). The climate during this interglacial was warmer and drier than in the Holocene (Smiley *et al.* 1991).

A buried lagoonal deposit recovered in southern Delaware has been amino-acid dated to 500 ka to 1 Ma (Nickman & Demarest 1982) and contains three pollen assemblages. The lowest assemblage seems to represent the end of an interglacial with *Tsuga, Pinus, Fagus, Liquidambar* and *Quercus*. It is followed by pollen of a marsh community deposited during a period of lowered sea level and includes *Alnus, Cephalanthus, Vitis, Myrica, Viburnum* and Cyperaceae. The upper part of the sequence again contains tree pollen, but with fewer mesic indicators than the lower part of the profile (Smiley *et al.* 1991). At present this site represents an isolated occurrence of Early Pleistocene temperate deposits. Correlation with the European chronology or with any deep-sea isotope stage is not possible.

Since the Tertiary had been a period of worldwide regression, North America had been connected with Asia via the Bering land bridge probably throughout the Tertiary (Hopkins 1967). In the Quaternary the land bridge was also open during glacial maxima, when eustatic sea level was lowered. Nevertheless, North American Quaternary faunae differ markedly from their European equivalents. They do not contain rhinoceros, hippopotamus or hyaena proper, hunting hyaena (*Chasmaporthetes*) being the only exception. In addition to other forms which also occurred in Europe, South American mammals like marsupials (opossums) and ground sloths (*Megalonyx, Eremotherium, Nothrotheriops* and *Glossotherium*) immigrated into North America during the Pleistocene. One, *Megalonyx jeffersonii*, spread as far north as Alaska (Nilsson 1983).

The mammal faunas of North America have been subdivided into three major 'mammal ages': the Blancan, the Irvingtonian and the Rancholabrean. The **Blancan mammal fauna**, named after the type site Mount Blanco in northern Texas, still contains mainly Tertiary elements. It can be roughly correlated with the European Villafranchian. The **Irvingtonian mammal fauna**, named after a type site near San Francisco, California, was originally correlated with the Nebraskan to Illinoian Stages. In contrast to the Blancan fauna, the Irvingtonian contains true elephants (e.g. *Mammuthus meridionalis*). The **Rancholabrean mammal fauna**, named after the Rancho La Brea tar pit in Hancock Park, Los Angeles, California, is Late Pleistocene in age and seems to span the Illinoian to Wisconsinan period (Nilsson 1983). It is further discussed in chapter 13.8.3.

13.4 PRE-ILLINOIAN GLACIAL STAGES

The pre-Illinoian glaciations reached furthest south in the Central Plains of Nebraska, Iowa, Kansas and Missouri. Correlation of the pre-Illinoian tills is based on radiometric age determination of the Pearlette B, S and O ashes and the Bishop Ash. In western Iowa at least seven major pre-Illinoian till units have so far been distinguished. Clast lithology and heavy-mineral composition allows a subdivision into three till groups (A, B and C), and mineralogical criteria and physical stratigraphy are used for further subdivisions. The till units are also separated by some well-developed palaeosols. However, both dating and interregional correlation are still poorly developed (Hallberg 1986). In Kansas, all diamictons and interbedded stratified deposits are now referred to as the **Independence Formation**. The two till members within this formation are correlated with the A2 and A3 tills of Hallberg's classification. They have normal polarity and are between 0.7 and 0.6 Ma old (Aber 1991).

Ice advances into Kansas caused considerable changes to the drainage system. A major eastward-trending river was blocked and filled with sediment, and west of the ice margin a large ice-dammed lake formed (Glacial Lake Atchison). Overflow and catastrophic drainage to the south caused rapid incision of what is now the lower Big Blue River. In later ice advances this passage and the Kansas River were locally blocked by ice tongues. Temporary drainage diversions occurred, but after deglaciation the Blue/Kansas River drainage system re-established itself (Fig. 175) (Aber 1991).

13.5 PRE-ILLINOIAN/ILLINOIAN INTERGLACIAL

Originally the last interglacial before the Illinoian had been defined by the Yarmouth Palaeosol, identified in a well section near Yarmouth, Des Moines County, Iowa (Leverett 1898b). Like the Sangamon Palaeosol it consists in many places of what is referred to as 'accretion gley', i.e. a soil sediment of redeposited silt and clay with an admixture of organic material which can be up to 3 m thick (Frye & Willman 1975). However, recent investigations have shown that the Yarmouth Soil represents different times in different regions. For example, in parts of Indiana and Illinois it was formed in the period correlated with deep-sea Oxygen Isotope Stages 7–11, whereas in other regions it started to form in Stage 13 or 15 (Hallberg 1986). The geosols are thus often composites, reflecting soil formation during more than one stage.

In major parts of the American Midwest, where Illinoian deposits are absent, a combined Yarmouth–Sangamon Palaeosol is found. Using micromorphological, textural and mineralogical investigations, it is possible to identify the different processes that have contributed to the formation of this geosol complex. Woida & Thompson (1993) investigated a Yarmouth–Sangamon Palaeosol about 5 m thick near Earlham, Iowa, and were able to identify a sequence of erosional and accretional phases, repeatedly interrupted by soil formation.

At a site in Saunders County, eastern Nebraska, a peat from the pre-Illinoian/Illinoian Interglacial contains relatively little tree pollen (Fredlund, unpublished). *Quercus* at frequencies up to 9% is the most abundant type, followed by *Pinus* (4%). The other taxa include *Juniperus, Ostrya, Carpinus, Populus, Castanea, Fraxinus* cf. *nigra, Myrica, Sambucus, Juglans, Alnus, Salix, Betula, Rhus, Cornus* and *Shepherdia*. Amongst the non-arboreal pollen, Cyperaceae are most common. The assemblage suggests a moist environment. The absence of major quantities of wind-blown pollen and the dominance of non-arboreal pollen suggest that the area supported open grassland vegetation (Smiley *et al.* 1991).

In Manitoba a palaeosol has been identified in sections along the Nelson River that is probably of pre-Illinoian/Illinoian age. Microscopic investigation of charcoal remains within this Sundance Palaeosol has shown the presence of largely non-arboreal taxa which suggest a tundra environment. The thickness of the palaeosol, however, is twice that of Holocene soils on a comparable substrate. The presence of a palaeosol in this stratigraphical position instead of

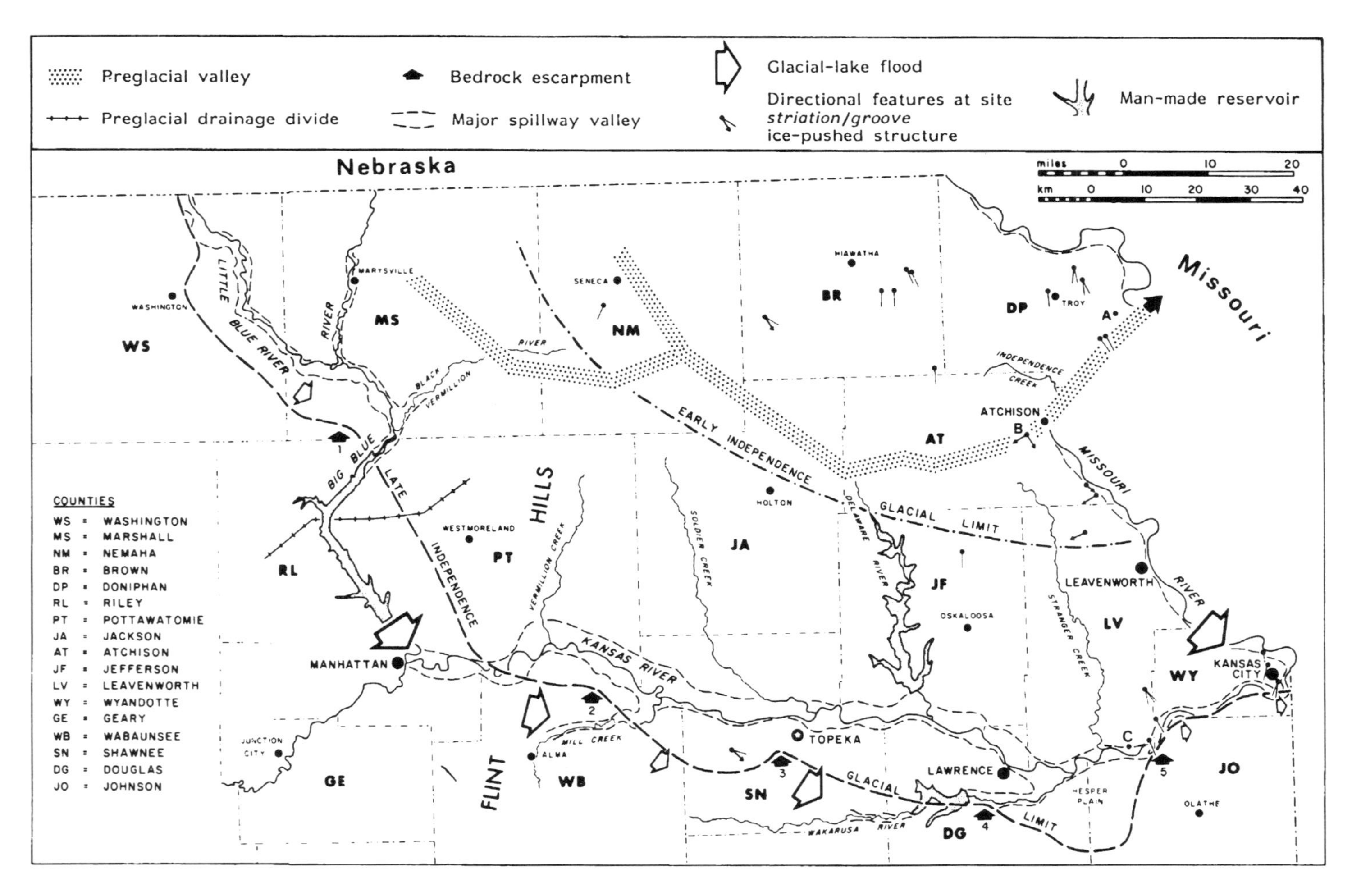

FIGURE 175 Extent of pre-Illinoian ice sheets in Kansas and ancient drainage systems (from Aber 1991)

lake deposits requires free drainage to the north and an ice-free Hudson Bay. The soil is therefore thought to represent an interglacial rather than an interstadial (Dredge *et al.* 1990).

On the whole, much less is known of the pre-Illinoian/Illinoian Warm Stage than of the second-from-last interglacial in Europe (see chapters 11 and 12). In the Alps, as well as in northern Europe, the two glaciations are separated by at least two fully developed warm stages. Nevertheless, nothing comparable has yet been identified in North America.

13.6 ILLINOIAN GLACIATION

Although the occurrence of Illinoian tills has been reported from many states, thorough investigations have mainly been carried out in Illinois, where the Illinoian glaciers advanced about 200 km beyond the limits of the subsequent Wisconsinan Glaciation. Willman & Frye (1970) provide a comprehensive review over the ice-marginal positions and till stratigraphy. The Illinoian sedimentary sequence comprises the Glasford Formation, named after Glasford in Peoria County. It comprises tills and outwash deposits of gravel, sand and silt. The Illinoian Stage was originally subdivided into three substages, the Liman, Monican and Jubileean, each of which was represented by a till unit. The substages were separated by ice-free periods, when minor soils were able to develop.

Further investigations have revealed that the Illinoian stratigraphy is more complicated. Lineback (1979) distinguished eight till members within the Illinoian sequence, and four intercalated fossil soils. He concluded that the Illinoian might represent a 'superstage' and include the deposits of several cold and warm stages. Despite this, palaeomagnetic investigations suggested that they were at least all normally magnetised.

More recent investigations have shown that two major glaciations have to be attributed to the Illinoian Stage, the deposits of which are separated by the Pike Palaeosol. The latter was originally described as a 'moderate to strongly developed palaeosol' (W.H. Allen & Ward 1979), which might suggest a significant climatic amelioration, but correlation with other buried soils suggests that it represents an interstadial (Johnson 1986). The tills of the older glaciation phase are the Kellerville and Smithboro Till Members of the Glasford Formation. The multiple tills of the younger Illinoian glaciation are subdivided into members of the Glasford and Winnebago Formations. The Winnebago Formation was originally thought to be of Early or Middle Wisconsinan age (Richmond & Fullerton 1986). However, all Illinoian deposits can almost certainly be correlated with Oxygen Isotope Stage 6; the Pike Soil probably does not represent an interglacial. Indeed, amino-acid analyses suggest that all deposits attributed to the Illinoian are of about the same age (H. Johnson, personal communication). Detailed investigations of Illinoian till stratigraphy have also been conducted in Ohio and Indiana (Fullerton 1986a, B.B. Miller *et al.* 1992, Szabo 1992). The results largely confirm the interpretations from Illinois.

In Canada, closer to the centre of the large glaciations, pre-Wisconsinan deposits have also been recorded in numerous places. The York Till (Terasmae, 1960) of the Toronto area is definitely of pre-Sangamonian age, because it underlies the last interglacial Don Beds (Terasmae 1960, Karrow 1990, Eyles & Williams 1992). In the Hudson Bay Lowland, four till beds were found underlying marine deposits of the interglacial Missinaibi Formation (Shilts 1984a). However, it is unclear whether these represent different glaciations or just local oscillations of one ice sheet. Other pre-Sangamonian tills are reported from various places in Ontario, but reliable dates are lacking (Barnett 1992). There are pre-Wisconsinan tills on Banks

Island in the Arctic, some of them magnetically reversed (e.g. the Banks Glaciation of Vincent *et al.* 1984). Even in Manitoba, at the centre of the last glaciation, the Illinoian Amery Till has been identified (Dredge & Nielsen 1985).

In the extra-glacial area during the Illinoian, the Loveland Loess was deposited. Although it has been described from sites in Iowa, Illinois, Indiana, Ohio and Arkansas, it is generally much less extensive than the overlying Wisconsinan loesses. The only fossil soil described from this unit is the Sangamonian soil developed in its upper part (Norton *et al.* 1988).

Grüger (1972a, b) and J.E. King & Saunders (1986) have studied Late Illinoian pollen sequences from two sites, at Hopwood Farm and in the Pittsburg Basin. In both cases the Late Illinoian flora is dominated by spruce (*Picea* up to 50% of total pollen) and pine (*Pinus* about 50%). This may indicate that, as in the Wisconsinan, the Illinoian forest extended close to the ice margin.

13.7 SANGAMONIAN INTERGLACIAL

The Sangamonian Stage was named after a fossil soil that was first found in hand-dug wells in Sangamon County, Illinois. The name was first used by Leverett (1898a), who defined it as 'the weathered zone between the Iowan loess (early Wisconsinan) and the Illinoian till sheet'. However, this definition did not suffice to answer all questions, and in the course of time several controversies have arisen over the Sangamon Soil. The state of this debate has been summarised by Follmer (1978) as follows.

One problem which was discussed for decades was the origin of the so-called **'gumbotil'**. The poorly drained, gleyed material on pre-Wisconsinan glacial deposits in the northern Mississippi basin is a grey, leached, deoxidised loam, showing strong signs of weathering. It was originally referred to as 'gumbo' by Leverett and earlier workers. Kay (1916) introduced the term 'gumbotil' for this material, which he considered to be largely the product of chemical weathering. Frye *et al.* (1960) pointed out that the descriptive term 'gumbotil' included two genetically distinct materials, an *in situ* soil and what they referred to as 'accretion gley', i.e. a soil sediment. They suggested the term 'gumbotil' should be restricted to *in situ* soils. However, in practice it seemed that most sections originally described as 'gumbotil' were not formed *in situ*. Frye & Willman (1963) therefore concluded that the 'gumbotil' actually is an accretion gley. This did not solve all the problems, but since then the term 'gumbotil' has become redundant (Follmer 1978).

One of the principles of soil stratigraphy is that profiles must be traced laterally to determine their spatial variation (Yaalon 1971). This has not been done for the Sangamon Soil, because it was defined much earlier, and from that results the second problem, i.e. precisely where the Sangamonian Interglacial Stage starts and ends. Leverett's Sangamon Soil is time-transgressive. It contains several merged soils of Wisconsinan, Sangamonian and possibly Late Illinoian age that may only be differentiated if careful investigations of the pedological parameters and of the geomorphological circumstances are undertaken at each site (Follmer 1978, 1982).

The Sangamonian of North America, the last interglacial before the Holocene, thus is defined, at least by some workers, differently from the Eemian of Europe. This must be considered wherever stratigraphical comparisons are attempted. Whilst the Eemian is restricted to Oxygen Isotope Stage 5e, the Sangamonian, according to the Geological Survey of Canada and the Illinois Geological Survey, represents the entire Stage 5. This means that it spans the period from about 130,000 to 75,000 BP and includes the Brørup and Odderade Interstadials of Europe (e.g. Fulton, 1984, 1989; St Onge, 1987; Curry & Follmer, 1992). Others, for instance the US Geological Survey and most participants of IGCP projects, follow the European concept (e.g. Richmond & Fullerton 1986, Barnett 1992). The latter position is supported by the fact that the Laurentide Ice Sheet began to grow during Oxygen Isotope Stage 5d, and the Appalachian ice cap during Oxygen Isotope Stage 5b. Moreover, on Baffin Island the maximal ice advance of the last glaciation occurred as early as Stage 5d (G.H. Miller *et al.* 1992).

In the Great Lakes region, the Sangamon Geosol is characterised by a strongly developed B horizon that shows clear signs of clay translocation. In oxidised profiles this soil is often reddish (7.5YR Munsell hues) or brownish (10YR hues) in colour (Curry & Follmer 1992). With its continuous clay cutans it resembles the typical Eemian soils of Europe. Although this soil has been well preserved in many localities outside the glacial limit, its remains are far more rarely found within the Wisconsinan glacial area. For instance, unequivocal Sangamonian soils have not so far been reported from northern Ohio (Szabo 1992), although an accretion gley at Garfield Heights (near Cleveland) and the Sidney Soil at Sidney, Ohio, are probably of Sangamonian age (Dreimanis 1992, B.B. Miller *et al.* 1992).

At Hopwood Farm and in the Pittsburg Basin, the Sangamonian sequences were interpreted as representing drier and warmer conditions than during any part of the Holocene, mainly because of the occurrence of the warm-climate tortoise *Geochelone* at Hopwood Farm (King & Saunders 1986). However, E.C. Grimm (1989) points out that the non-arboreal pollen sequences are similar to younger Holocene assemblages in northern Illinois and Iowa, and the predominance of Chenopodiaceae–Amaranthaceae during the Sangamonian may reflect special ecological conditions (unstable water levels, desiccated lacustrine sediments) rather than a different climate (Curry & Forester 1991).

A Sangamonian buried swamp in Washington, DC, contains a last interglacial pollen assemblage with *Quercus* and *Carya*, suggesting climatic conditions similar to those in the Holocene (A.S. Knox 1962). The same is true for the Sangamonian sequence from Long Island (Donner 1964). In the Atlantic Provinces of Canada, 30 pre-Wisconsinan interglacial sites have been identified (Mott & Grant 1985), seven of which are clearly correlated with the Sangamonian *sensu stricto* (Stage 5e). One of these (Woody Cove, Newfoundland) spans the whole interglacial (Brookes *et al.* 1982). The climatic optimum was reached early in the stage, as compared with the Holocene. During all of Oxygen Isotope Stage 5 there seem to have been non-glacial conditions in this region. The climatic optimum of the Sangamonian was significantly warmer, particularly during the summers, and more continental (Mott 1990).

In the central areas of the glaciation, the ground was still isostatically depressed when the warm stage started. Therefore in the Hudson Bay Lowland the last interglacial, like the Holocene, began with a high sea-level stand. Clays that are assumed to be of marine origin can be traced to an altitude of 120 m a.s.l (the Holocene marine deposits in the same region reach altitudes of 120–150 m a.s.l.). After the regression, resulting from isostatic rebound, climatic conditions changed from cool to warmer than today and finally back to cooler (Dredge *et al.* 1990).

One of the key sections for the study of the Sangamonian is the Don Valley Brickyards in central Toronto in Canada. The last interglacial Don Beds (now referred to as the Don Formation) were first described by Coleman (1894). Besides the deciduous trees which now grow in the area, they contain a number of plant taxa that occur at present much further south. They include *Chamaecyparis thyoides, Gleditsia, Fraxinus quadrangulata, Maclura pomifera,*

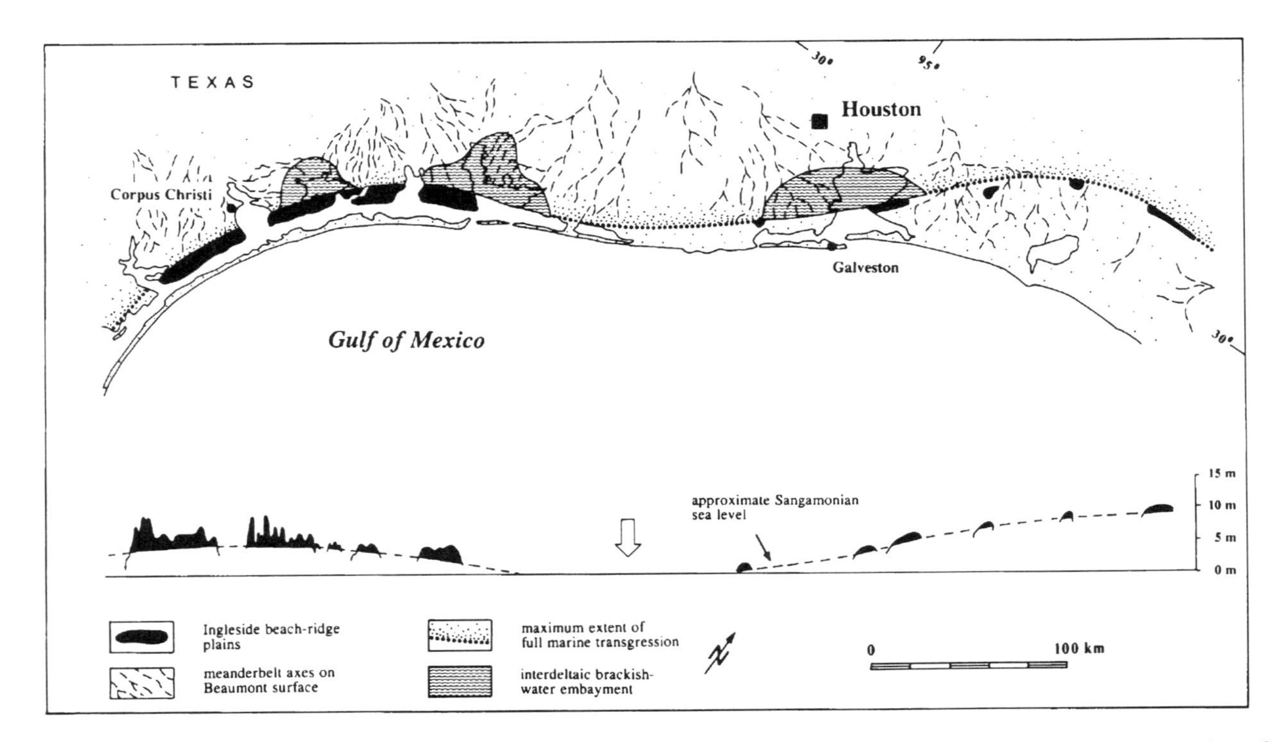

FIGURE 176 Evidence for a Late Pleistocene high sea-level stand along the Texas coastal plain and subsequent warping of the strandline. The position of the last interglacial Ingleside barrier is indicated (from Winker 1991)

Quercus stellata, Quercus muhlenbergii and *Robinia pseudoacacia*. However, McAndrews (in preparation) now disputes the identification of *Chamaecyparis thyoides*. The pollen assemblage suggests that the average climate during the Sangamon was about 3°C warmer than today (Karrow 1990).

Vertebrate remains of last interglacial age have been recovered from 20 sites in Canada, from Nova Scotia to the Old Crow River in Yukon. The faunae include a number of species which are now extinct, like *Megalonyx* sp. (ground sloth), *Castoroides ohioensis* (giant beaver), *Mammut americanum* (mastodon), *Mammuthus primigenius* (mammoth), horses, *Camelops hesternus* (western camel) and *Bison latifrons* (giant bison) (Harington 1990).

Last interglacial high sea-level stands have been recorded on both the Pacific and Atlantic coasts of North America and Hudson Bay. Their relative position with regard to present sea level depends on the regional tectonic and glacio-isostatic history of the region. In San Francisco Bay, for instance, the upper limit of the last interglacial marine Yerba Buena Mud has been identified at a depth of 25 m below present sea level (Sloan 1992). In contrast, on the Gulf Coast of Texas, the last interglacial shoreline is up to 9 m above the present (Fig. 176). This shoreline, the so-called Ingleside Barrier, can be traced from the Mexican border along the Gulf Coast into Florida (Winker 1991). Grant (1980) and Stea *et al.* (1992) identify the Sangamonian sea level in maritime Canada as 4–6 m above present sea level where it is represented by a well-developed wave-cut bench overlain by sands and gravels.

13.8 WISCONSINAN GLACIATION

13.8.1 Early Wisconsinan

Since the Sangamonian Stage of North America is regarded as representing either deep-sea Oxygen Isotope Stage 5e only, or all of Stage 5, the length of the following stage, the Wisconsinan, also varies accordingly. If the Sangominan is regarded as including all of Stage 5, the Wisconsinan is shorter than the European Weichselian and does not start before about 80,000 BP.

The formal Wisconsinan stratigraphy of the Laurentide Ice Sheet with its subdivision into Early, Middle and Late Wisconsinan substages, developed in the Eastern Great Lakes – St Lawrence Lowland – Ohio River region during the late 1950s and early 1960s. With further subdivisions into stadials and interstadials, it was proposed formally at the 24th International Geological Congress in 1972 by Dreimanis & Karrow (1972). After discussions at various field conferences of IGCP Project 24, the substage divisions (Early, Middle, Late Wisconsinan) became accepted for the entire glaciated region of North America and later also for Greenland. The main stratotype region for the stadials and interstadials is the Lake Erie basin, whilst for the lithostratigraphic units it is the entire region from the northeastern end of Lake Huron to the St Lawrence Lowlands.

In the area of the Lake Michigan and the Green Bay lobes, the Illinois Geological Survey developed a parallel stratigraphical scheme that was originally based largely on loesses and fossil soils. It subdivides the Wisconsinan into the Altonian, Farmdalian and Woodfordian substages (Willman & Frye 1970). This stratigraphical scheme is generally adopted throughout the Midwest region, especially in the loess areas.

After the Sangamonian *sensu stricto*, drier climate resulted in the relatively dry midwestern region (Pittsburg Basin, central Illinois) in a change from deciduous forest to prairie conditions

(Grüger 1972a). In more humid areas, such as Scarborough, Ontario, open woodland existed close to the ice margin (Terasmae 1960, Dreimanis *et al.* 1989).

In Toronto, the Don Beds are overlain by (but separated by an hiatus of unknown length from) the Early Wisconsinan Scarborough Formation. The deposits are dominated by *Pinus* and *Picea*, but they also contain a rich mixture of temperate deciduous taxa, which include *Castanesa, Carya, Quercus, Fagus, Ilex, Tilia, Tsuga canadensis, Ulmus* and *Shepherdia*. It is not clear if this sequence represents one of the Early Weichselian interstadials, i.e. Brørup or Odderade, although it seems likely. The floral assemblage suggests a mean annual temperature 6°C lower than at present (Terasmae 1960).

After formation of the Sangamon Geosol, loess deposition and redeposition was interrupted repeatedly by a number of minor soil-forming periods. The soils are characterised by organic-rich A horizons and weakly developed B horizons. At Athens Quarry, Illinois, where these soils were studied in a good exposure in a sequence overlying the Sangamon Geosol (Fig. 177), they were defined by Curry & Follmer (1992) as the Indian Point Geosol Complex. The formation of the Chapin Geosol, in a basal mixed zone of the loess on top of the Sangamon Soil, also falls into this period.

Throughout major regions of the mid-continental United States, the Sangamon Geosol is overlain by the Roxana Silt, a loess deposit. Accelerator mass-spectrometer radiocarbon ages of samples taken from the loess at various points along the upper Mississippi valley have yielded ages between 27,000 and 55,000 years BP (Leigh & Knox 1993, Leigh 1994).

At the top of the upper, redeposited part of the Roxana Silt, which is referred to as the Robein Silt, the weak Farmdale Geosol is developed. It consists of a distinctive AO horizon with abundant wood fragments, a black A horizon and a Bg horizon. These horizons are convoluted in places, perhaps resulting from cryoturbation (Fig. 177) (Follmer 1983).

The Laurentide Ice Sheet began to form in northeastern Labrador probably as early as Oxygen Isotope Stage 5d (Vincent & Prest 1987). Simultaneously major ice advances occurred in Greenland (Funder *et al.* 1991, 1994, Israelson *et al.* 1994). However, it was not before the Early Wisconsinan Substage *sensu stricto* that a major ice sheet formed in continental North America. There has been much discussion concerning whether or not Early Wisconsinan glaciers advanced into the United States. The Whitewater and Fairhaven tills in southwestern Ohio and southeastern Indiana were formerly regarded as remnants of such glaciations (e.g. Gooding 1963, Fullerton 1986a). However, recent investigations have revealed that the interstadial deposits separating the strata in question from undoubtedly classical Wisconsinan till may be considerably older than assumed. It is now concluded that what was originally regarded as the Middle Wisconsinan 'Sidney weathering interval' (Forsyth 1965) may in fact date from the Sangamonian, and that the two tills underlying it are Illinoian (the Fairhaven Till) and pre-Illinoian (the Whitewater Till) in age (B.B. Miller *et al.* 1992).

In New England along the coast as well as inland, remnants of a till sheet older than the Late Wisconsinan glaciation have been preserved in numerous places. For instance, the lower till at Sankaty Head in Nantucket and the Montauk Till Member on Long Island were deposited during a glaciation that was at least as extensive as the last one. However, these tills are probably pre-Wisconsinan in age. At Sankaty Head the till is overlain by marine deposits, from which a detrital coral yielded a uranium–thorium age of 133,000 ±7000 years. The fauna in these deposits indicates that they were deposited in sea water warmer than present. The underlying glacigenic strata therefore probably date from the Illinoian or are even older (Oldale & Colman 1992).

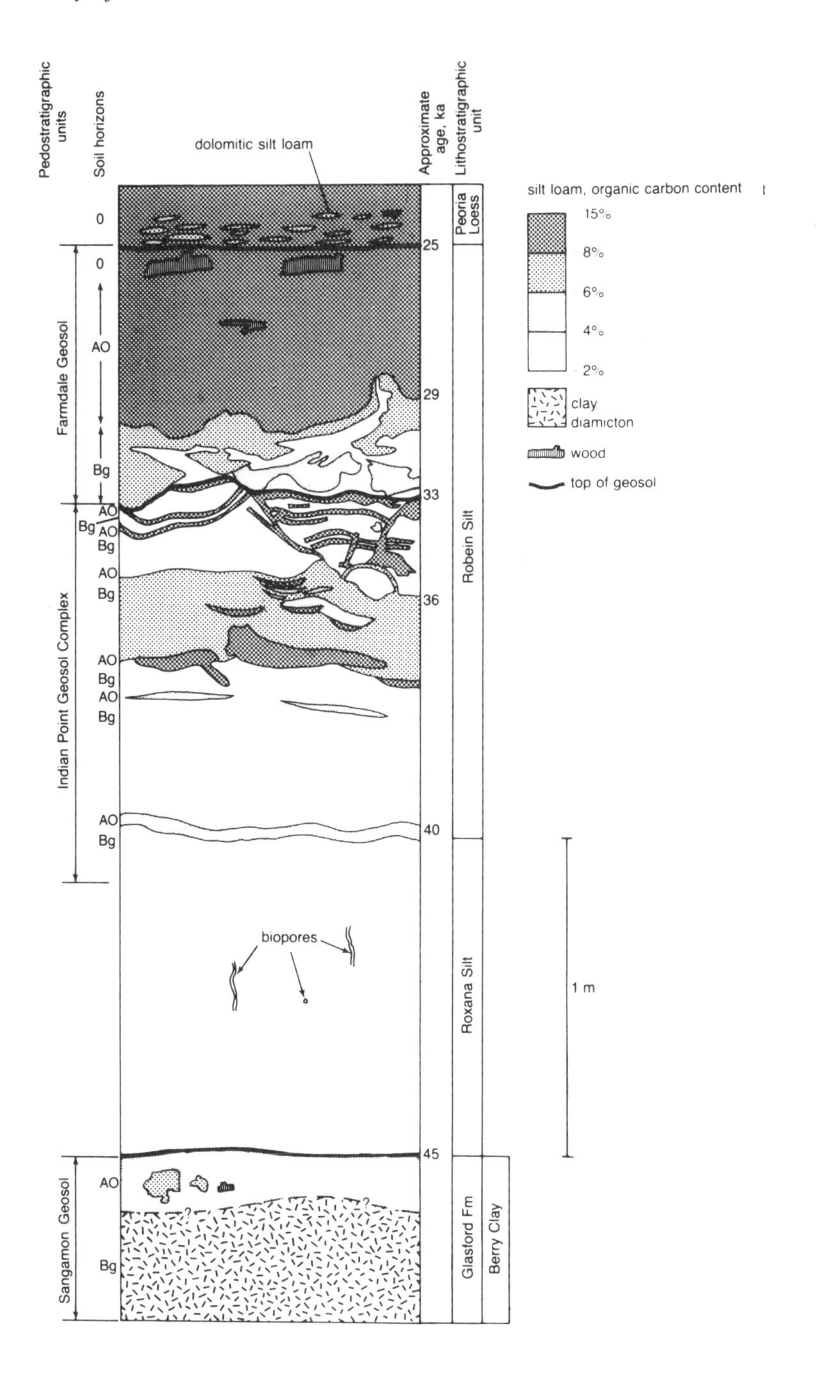

FIGURE 177 Representative profile of the loess/palaeosol sequence from Athens Quarry, Illinois. The tops of the Sangamon and Farmdale geosols and Indian Point geosol complex are emphasised with heavy lines (from Curry & Follmer 1992)

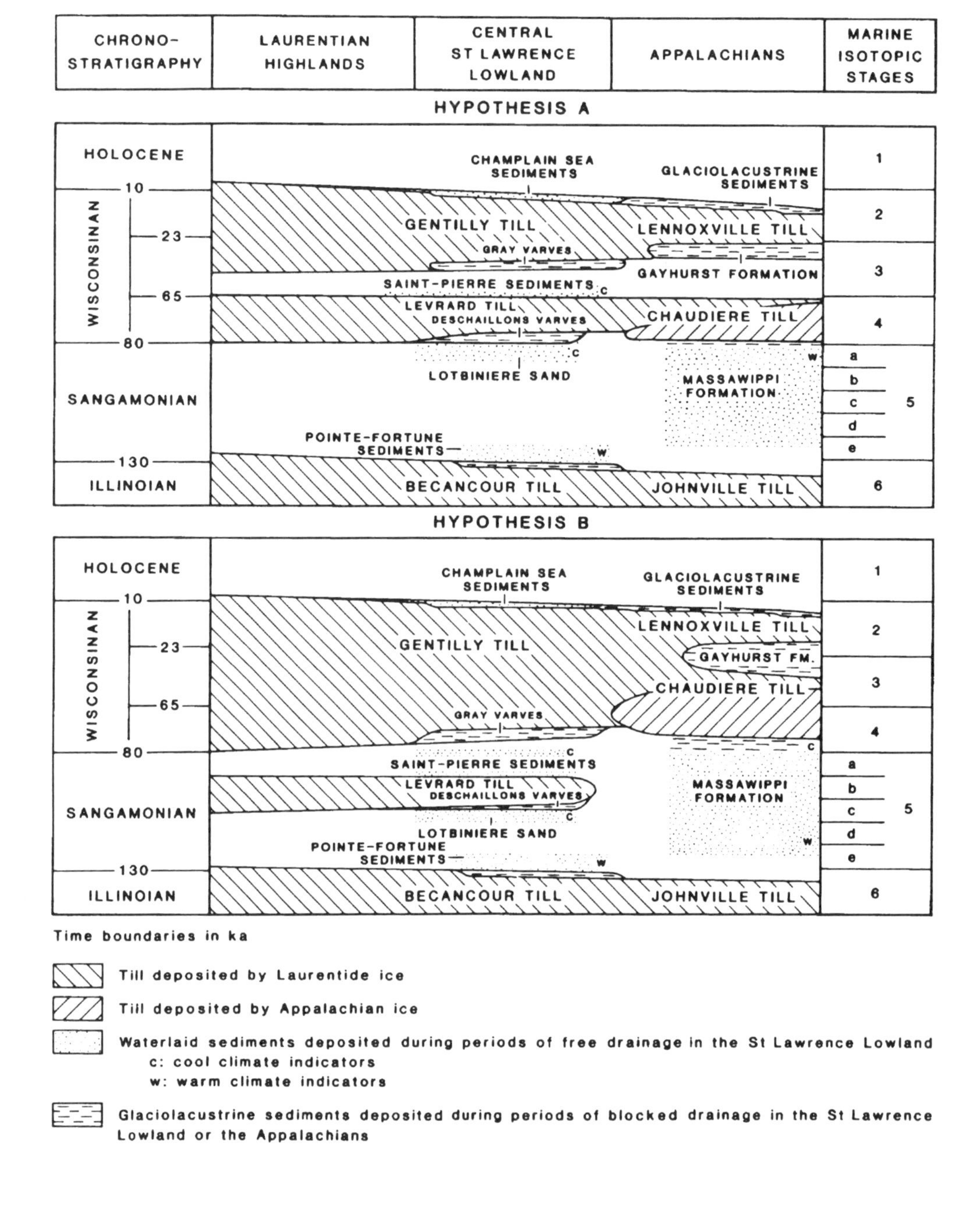

FIGURE 178 Two hypothetical time–space correlations between the Late Pleistocene depositional records of the St Lawrence Lowland and the Appalachians of southern Quebec (from Lamothe *et al.* 1992)

This interpretation fits with the results of investigations in Maine, where Weddle (1992) showed that the area had undergone only one Late Wisconsinan glaciation during the last cold stage. This event is represented by a sequence of ice-marginal deposits, involving several till units and changing ice-movement directions. On the other hand, in maritime Canada four tills from different ice centres have been found, overlying the Sangamonian Interglacial beds. It is regarded as unlikely that they are all Late Wisconsinan in age, and the same might apply to adjacent Maine as well (Stea *et al.* 1992 and Stea, personal communication).

However, it seems that Early Wisconsinan glaciation in North America was largely limited to Canada. Until now the ages assigned to the individual events are still relatively uncertain. For instance, Appalachian and Laurentide ice are thought to have coalesced as early as the Stage 5/4 boundary (i.e. about 80,000 BP). The advancing ice then blocked drainage through the St Lawrence River and dammed glacial Lake Deschaillons, which was over 100 km long and 25 km wide. Varved clays were deposited in this lake. Lamothe *et al.* (1992), presented two alternative correlation schemes for Early Wisconsinan events and deposits in this region (Fig. 178). Nova Scotia seems also to have supported a local ice cap from Early Wisconsinan times (Stea *et al.* 1992) and early Middle Wisconsinan tills are present around Hudson Bay. In other parts of Canada, including the Prairies, several till sheets postdating interglacial deposits have been recognised, but again, the dating is uncertain.

In the eastern Great Lakes region, the Early Wisconsinan Substage begins with deposition of the glaciolacustrine deltaic deposits of the Scarborough Formation (Kelly & Martini 1986). The pollen content indicates climatic development from cool-temperate to subarctic conditions in the upper parts (Terasmae 1960). As the Scarborough Formation was deposited in a lake at least 45 m higher than the present Lake Ontario, the present outlet through the St Lawrence River must have been blocked. Thus the Scarborough Formation marks the first advance of an ice sheet after the Sangamonian Interglacial into the Great Lakes region (Dreimanis 1992, Hicock & Dreimanis 1992).

13.8.2 Middle Wisconsinan

The Middle Wisconsinan includes a lengthy interstadial complex from about 65,000 to 25,000 years BP, from which no major ice advances have been recorded (Dredge & Thorleifson 1987). The age of tills, which were formerly attributed to Middle Wisconsinan ice advances, such as the Titusville Till, have been questioned by more recent research (see review in Clark & Lea 1992). At Titusville, in northwestern Pennsylvania, interstadial deposits have been radiocarbon dated to about 40,000 to 35,000 BP. Pollen and macrofossil analyses indicate that the area was colonised by spruce forest (Berti 1975). Pollen spectra in northern Illinois during the Middle Wisconsinan, for the period between ca. 47,000 and 24,000 BP, are still dominated by *Pinus* and *Picea*, with some *Betula* and *Salix*, which indicates forest or open woodland (Heusser & King 1988). In the Toronto area, the Thorncliffe Formation was laid down during the Middle Wisconsinan Substage. This has been radiocarbon and thermoluminescence-dated to between >50,000 and 28,000 BP (Hicock & Dreimanis 1992).

13.8.3 Late Wisconsinan

A vertebrate **fauna** from the Late Wisconsinan has been preserved at the Rancho La Brea site in Los Angeles. Seeping oil from underlying deposits has led to the formation of tar pools in which animals became trapped and died. In excavations, mammal bones of more than 4000 individuals have been recovered, including 39 mammal and 69 avian taxa. The beds, originally correlated with the Sangamonian, have been shown, using radiocarbon dates, to be much younger. Faunae of similar composition have been found in numerous places throughout North America. They are characterised by the immigration of new Palaearctic species like giant bison (*Bison latifrons*, now extinct) and the modern bison (*Bison bison*). In Alaska and adjoining parts of Canada the Eurasian steppe bison (*Bison priscus*) also occurred. The Eurasian woolly mammoth (*Mammuthus primigenius*) also immigrated via the Bering land bridge and spread into Canada and the northeastern United States. Other Palaearctic immigrants include caribou (*Rangifer tarandus*), elk or moose (*Alces alces*), musk ox (*Ovibos moschatus*) and forms related to the European cave lion (*Panthera leo atrox*) and brown bear (*Ursus arctos*). In addition, the broad-fronted elk (*Alces latifrons*), Asiatic yak (*Bos grunniens*) and Saiga antelope (*Saiga tatarica*) appeared in Alaska (Nilsson 1983). As in Siberia, well-preserved carcasses have been recovered from permanently frozen ground in Alaska (Péwé 1975).

In the Midwest, southeast of the Great Lakes, the Late Wisconsinan begins with deposition of the Peoria Loess. This loess was deposited under the influence of the advancing Late Wisconsinan ice sheet. The loess units are time-transgressive, aeolian deposition having begun about 25,000 BP in northern Illinois and about 23,000 BP further to the south (Curry & Follmer 1992). The primary loess sources were the valleys of the Mississippi and Missouri rivers. Here, vegetation changed from steppe to periglacial conditions. At Wedron, for instance, macrofossils of arctic and subarctic plants were recovered from proglacial lacustrine sediment, including *Dryas integrifolia* (arctic avens), *Vaccinium uliginosum* (arctic blueberry), *Selaginella selaginoides* and *Betula glandulosa* (Garry *et al.* 1990). Loess sedimentation at Wedron ended at 22,000 BP when the site was overridden by the ice sheet (Curry & Follmer 1992).

In the west, extensive sand dunes formed during the Quaternary. The Nebraska Sandhills Region with its 50,000 km^2 of stabilised dunes represents the largest such region in the western hemisphere. Landforms and internal structure suggest that the sand was supplied by northwesterly winds. Radiocarbon and thermoluminescence dating of buried soils in these dunes have shown that dune fields were still active during the Holocene, between 7000 and 3000 years BP. The penultimate pre-Holocene dune movement period ended about 13,000 BP. Stratigraphical investigations have shown that there were at least two earlier dune-movement phases (Forman & Maat 1990). Wind-blown sand is also found further east, in the loess region. The Parkland Sand of Illinois, for instance, is accumulated in sand-sheets and dunes, some of which are over 30 m high (Willman & Frye 1970). In most places the Parkland Sand is found at the present land surface, but locally it is also overlain by Late Wisconsinan loess.

Little is known about the build-up phase of the Late Wisconsinan Laurentide Ice Sheet. On the basis of the sea-level record in Atlantic Canada, Quinlan & Beaumont (1982) have suggested that a major ice dome must have formed during an early glaciation phase in the central Labrador highlands. This coincides with the observation of Klassen (1983a, b) that striae directions indicate a pre-Late Wisconsinan ice-dispersal centre between the Churchill and St Lawrence rivers. From this centre ice seems to have flowed radially in all directions. In the southwest the Harricana Interlobate Moraine is interpreted as having formed at the contact between the Labrador and Hudson ice (Vincent 1989).

When and how the Keewatin Ice Centre, west of Hudson Bay, developed is not quite clear. However, its existence during the late phase of the Wisconsinan Glaciation is well documented by large-scale geomorphological features, which have mainly been mapped from aerial

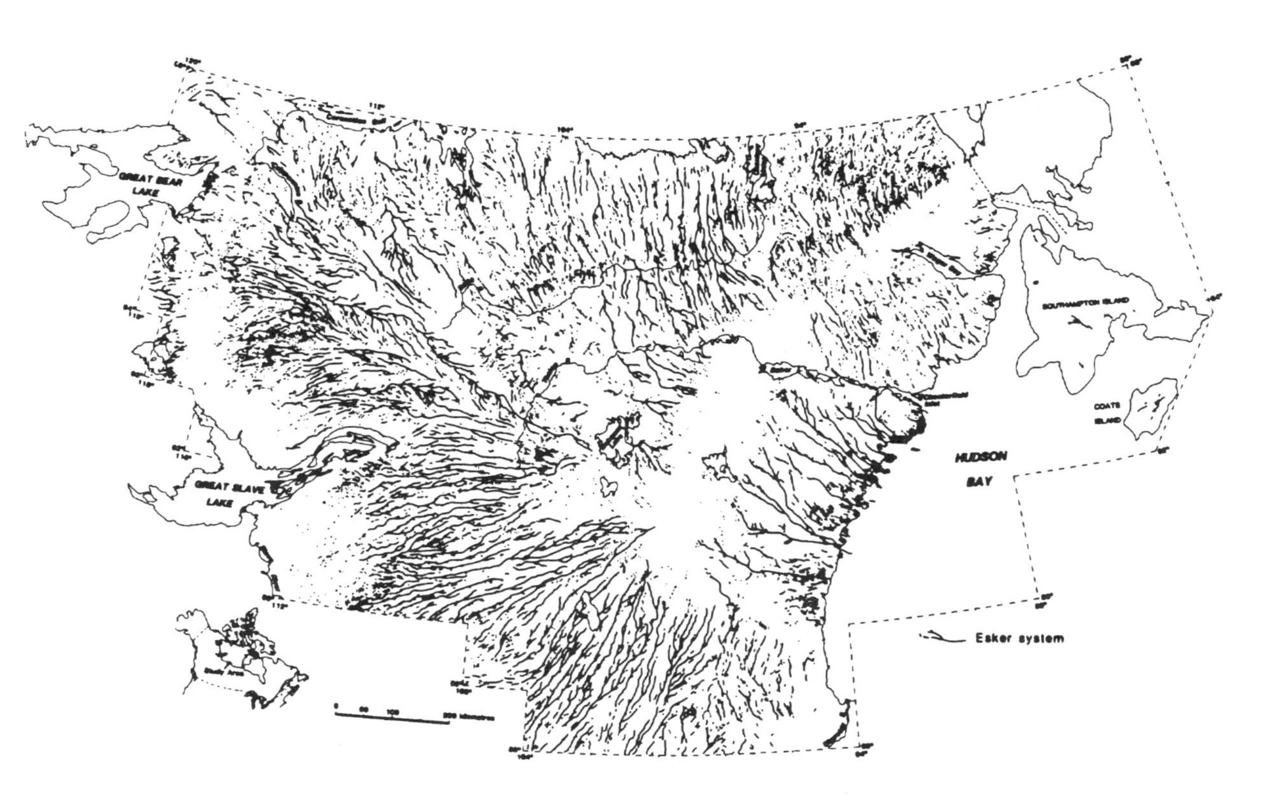

FIGURE 179 Distribution of eskers on the northwestern Canadian Shield (from Shilts *et al.* 1987)

photographs (Shilts *et al.* 1987). Eskers spread radially from a clearly marked dispersal centre (Fig. 179). Their formation has been explained by the absence of a deformable bed in the Canadian Shield region (Walder & Fowler 1994). Their preservation, however, is a result of the gradual decay of a dynamically inactive ice mass.

In most parts of North America, the Late Wisconsinan ice sheet was the most extensive of the last glacial stage. It had several flow centres, so that extent and timing of glaciation and deglaciation were different across the continent. Along the southern ice margin from Montana to Connecticut, an early glacial maximum was reached around 22,000–20,000 BP. In places, for instance in Washington, in Iowa and in Pennsylvania, a later ice advance between 16,500 and 14,000 BP reached furthest south. However, further to the north, the maximal ice-sheet advances became increasingly younger. In Newfoundland the Wisconsinan maximum did not occur before 10,000 BP, i.e. the Younger Dryas, and on Baffin Island the glaciers reached their northeasternmost positions around 8500 BP (Prest 1983).

In the western part of North America, the Late Wisconsinan Laurentide Ice Sheet was the most extensive Pleistocene ice sheet. It was less extensive than earlier glaciations in the south and in the western Arctic. In the eastern Arctic, Holocene ice extended further than any preceding glaciation.

In eastern North America the ice sheet extended on to the continental shelf, covering major parts of the Gulf of Maine. Off Nova Scotia submerged end moraines are found and drowned drumlins occur in Boston Harbour. Sea level was lowered, but the ice sheet also caused isostatic depression of the glaciated area. In Massachusetts, the Late Wisconsinan ice sheet formed a glaciomarine ice-contact delta in the Merrimack valley during early deglaciation (about 14,250 BP). This delta lies 33 m above present sea level (G.B. Edwards 1988). A submerged barrier spit/lagoon complex about 10 km off Cape Ann indicates a relatively low sea-level stand of about −43 m at about 12,000 BP. It formed when eustatic sea level had begun to rise but when the area was still strongly isostatically depressed (Oldale *et al.* 1993).

The Atlantic Provinces of Canada have a complicated glacial history. In many places along the coasts of **Nova Scotia** multiple Wisconsinan tills are exposed (Fig. 180). It is not quite clear when extensive glaciation of this region began. The earliest Wisconsinan ice flows crossed the peninsula in an eastward and southeastward direction. The ice came from an Appalachian source, the so-called Gaspereau Ice Centre, and from the Laurentide Ice Sheet (Fig. 181a). In the following glaciation phase most of Nova Scotia was under the influence of a centre north of the mainland, in the Prince Edward Island area, the so-called Escuminac Ice Centre. The ice sheet crossed the peninsula from north to south and advanced to the outer shelf (Fig. 181b). Subsequently, a local ice cap developed on Nova Scotia, dispersing ice from the centre of the peninsula in all directions. A local ice rise also possibly developed on the shelf south of Nova Scotia (Fig. 181c). During the last glaciation phase the local ice-dispersal centre retreated to northeastern Nova Scotia (Fig. 181d) (Stea *et al.* 1995).

The southern margin of the Laurentide Ice Sheet was largely subdivided into major and minor lobes. This is in contrast to the situation in Europe where the Weichselian ice sheet at its maximum possessed a rather straight margin. In the west, in Montana, the shape of the St Mary, Milk River, Shelby, Havre and Missouri Valley Lobes was largely controlled by the pre-glacial topography (Fullerton & Colton 1986). North and South Dakota were covered by the extensive James Lobe, and further to the east the Des Moines Lobe advanced via Minnesota into Iowa (Hallberg & Kemnis 1986). Between the Des Moines Lobe and the Green Bay Lobe of Wisconsin a complex pattern of minor sublobes developed in the surroundings of the ENE–

FIGURE 180 Three tills in a drumlin at Smith Cove, Nova Scotia. The tills are Hartlen Till (Early Wisconsinan, Appalachian or Laurentide ice), Lawrencetown Till (Late Wisconsinan, ice centre on the shelf north of Nova Scotia) and Beaver River Till (Late Wisconsinan, ice flow from Scotian ice divide) (Photograph: R. Stea)

WSW-trending Superior Lobe (Matsch & Schneider 1986). In the east the Green Bay Lobe, Lake Michigan Lobe and Huron Lobe followed the forms of the lake basins. Temporarily the Lake Michigan Lobe and Huron Lobe were separated by the Saginaw Sublobe (Eschman & Mickelson 1986). Further to the east the Ontario–Erie Lobe abutted a number of sublobes. In New England alone did the southern ice margin of the Laurentide Ice Sheet not have a lobate form (Mickelson *et al.* 1983).

The Wisconsinan Glaciation shaped major parts of the North American landscape, especially that of the Great Lakes region. Belts of end moraines can be traced around the southern margins of the lakes (Fig. 182). Some of these end moraines differ greatly from the push moraines of northwestern Europe. Internally some consist almost entirely of till (Fig. 183) (Wickham *et al.* 1988). For this reason, Mickelson *et al.* (1983) consequently define an end moraine as 'a ridge composed predominantly or entirely of till and formed at the ice margin during the last episode of till deposition.' These ridges often show a very subdued topography. Similar end moraines are found, for instance, in Estonia at the Pandivere ice-marginal position. However, many of the published sections are based entirely on borehole information, where glaciotectonic deformation can be easily overlooked. Dreimanis (personal communication) has observed glaciotectonic deformations in fresh road cuttings through end moraines in southwestern Ontario, and large-scale glaciotectonic deformations are, for instance, exposed at Ludington on the eastern shore of Lake Michigan (Fig. 184).

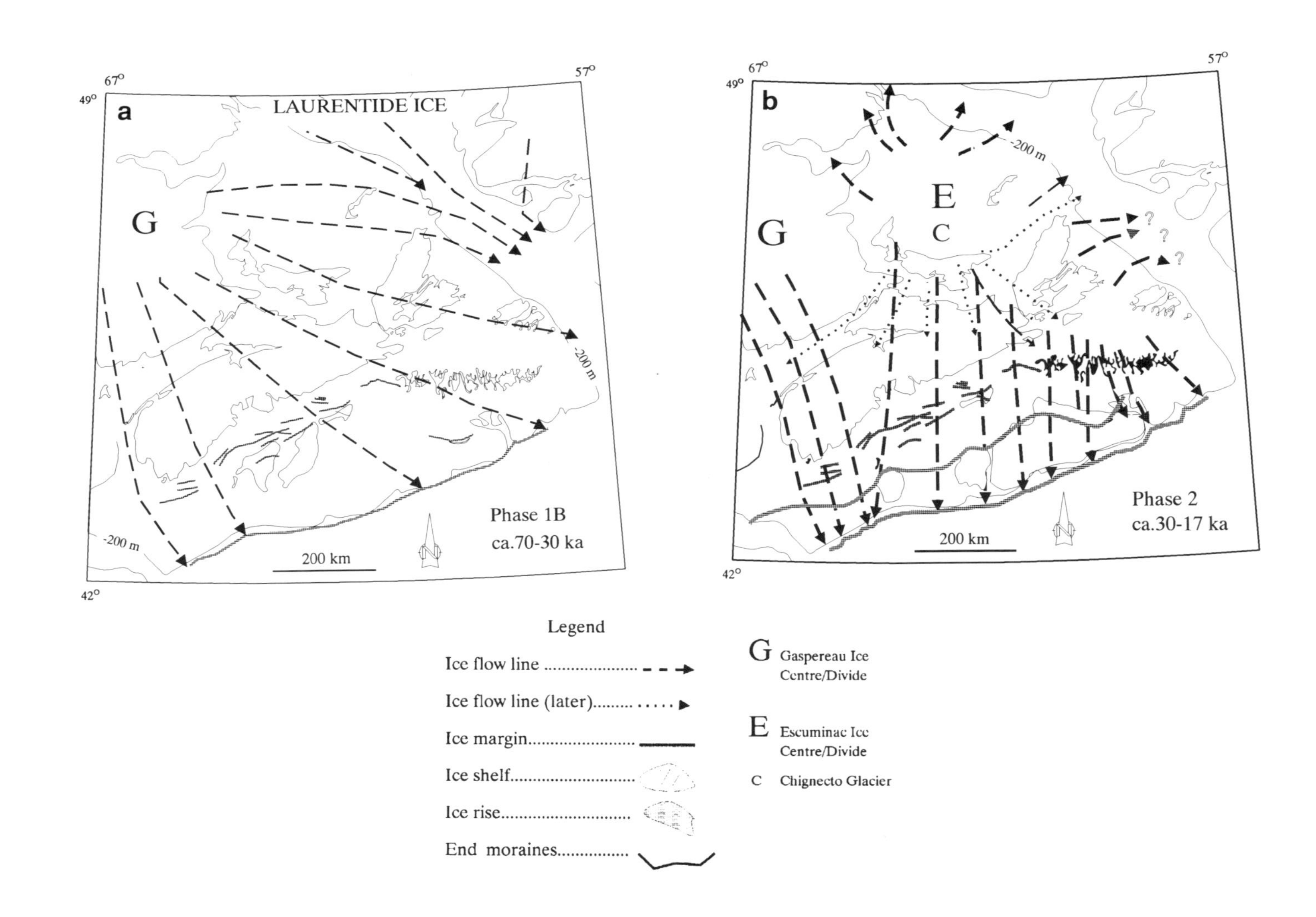
a
LAURENTIDE ICE
G
-200 m
-200 m
200 km
Phase 1B
ca.70-30 ka
67°
57°
49°
42°
b
G
E
C
-200 m
?
?
?
200 km
Phase 2
ca.30-17 ka
67°
57°
49°
42°
Legend
Ice flow line
Ice flow line (later)
Ice margin
Ice shelf
Ice rise
End moraines
G Gaspereau Ice Centre/Divide
E Escuminac Ice Centre/Divide
C Chignecto Glacier

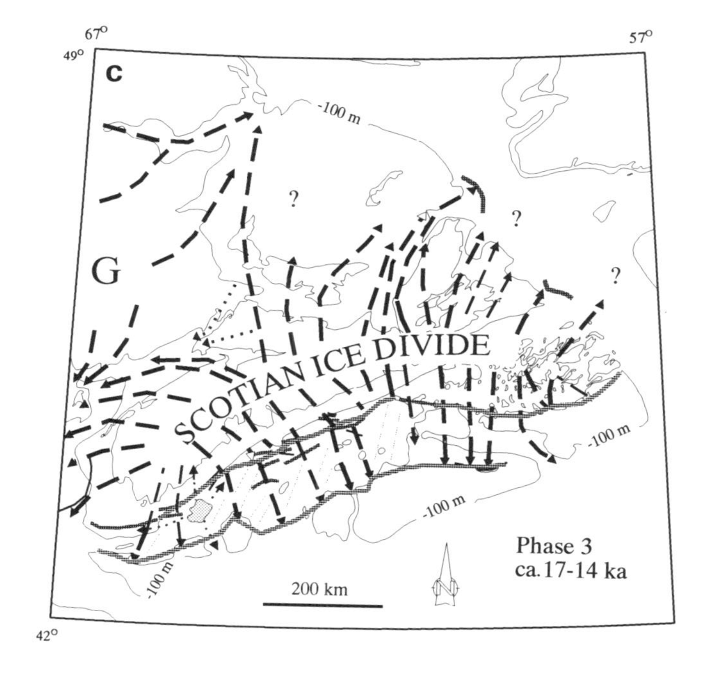

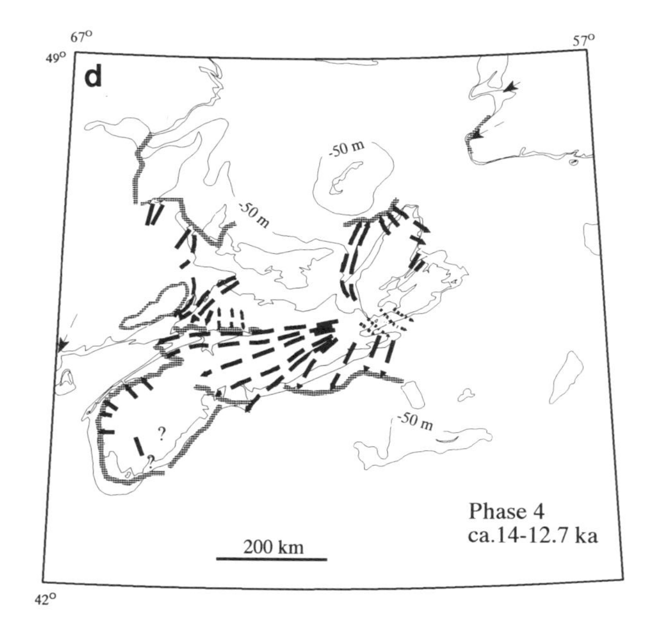

FIGURE 181 Evolution (advance and retreat) of ice divides and domes over Maritime Canada during the Wisconsinan (70–10 ka BP). (a) Advance of Phase 1B ice (70–30 ka BP). (b) Ice flow from the Escuminac Ice Centre on the shelf north of mainland Nova Scotia. (c) Advance and retreat of Phase 3 ice (Scotian Ice Divide). A short-lived ice shelf pinned on the outer banks may have formed during initial deglaciation from a larger ice mass centred on the Magdalen Plateau (Phase 2). The landward ice margin is marked by the Scotian Shelf End Moraine Complex. (d) Ice Flow Phase 4. (From Stea *et al.* 1995)

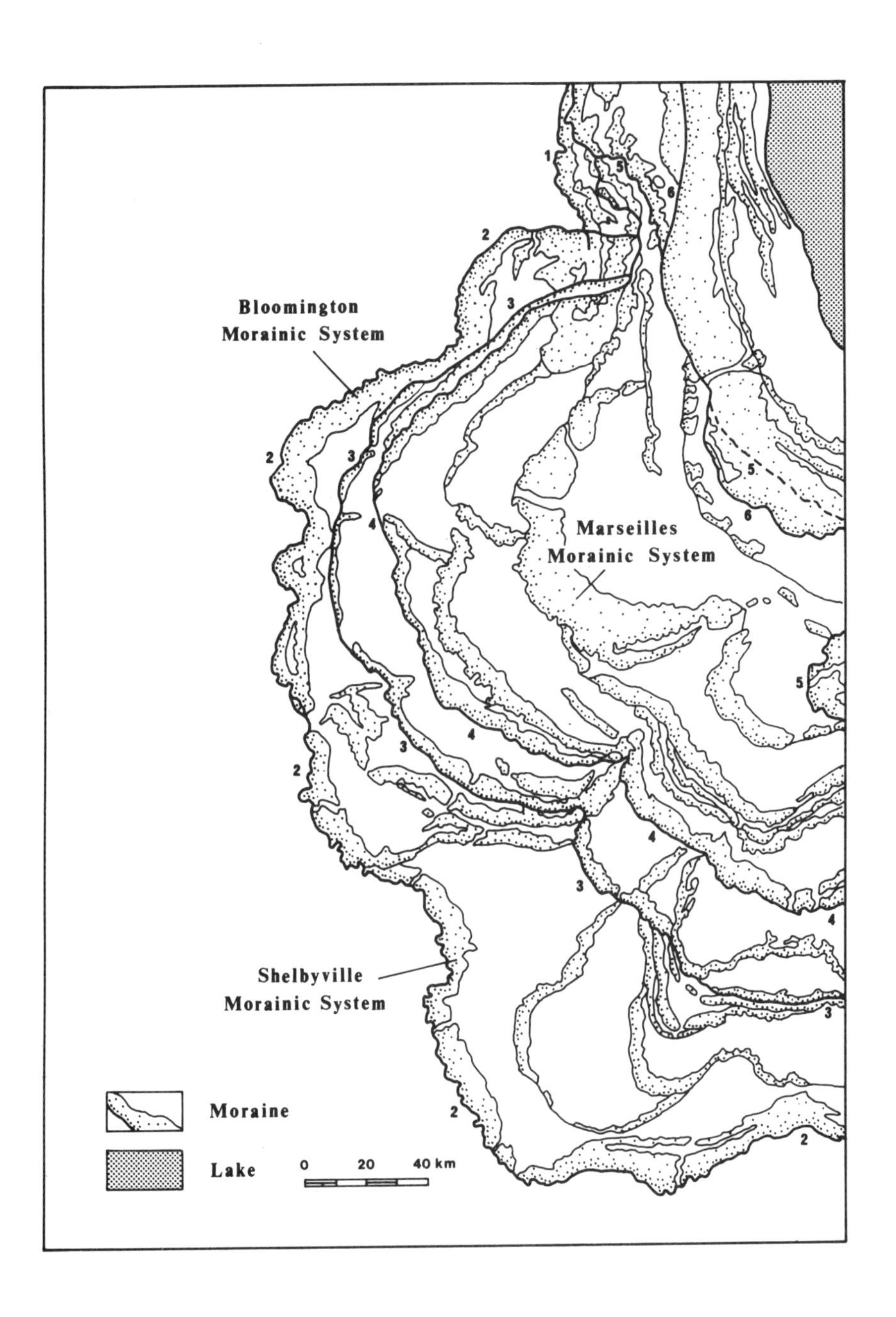

FIGURE 182 End moraines of the Late Wisconsinan Glaciation in Illinois; the major advances and readvances are marked by numerals; (1) Marengo, (2) Shelby, (3) Putnam, (4) Livingston, (5) Woodstock, (6) Crown Point (from Hansel & Johnson 1992; after Willman & Frye 1970)

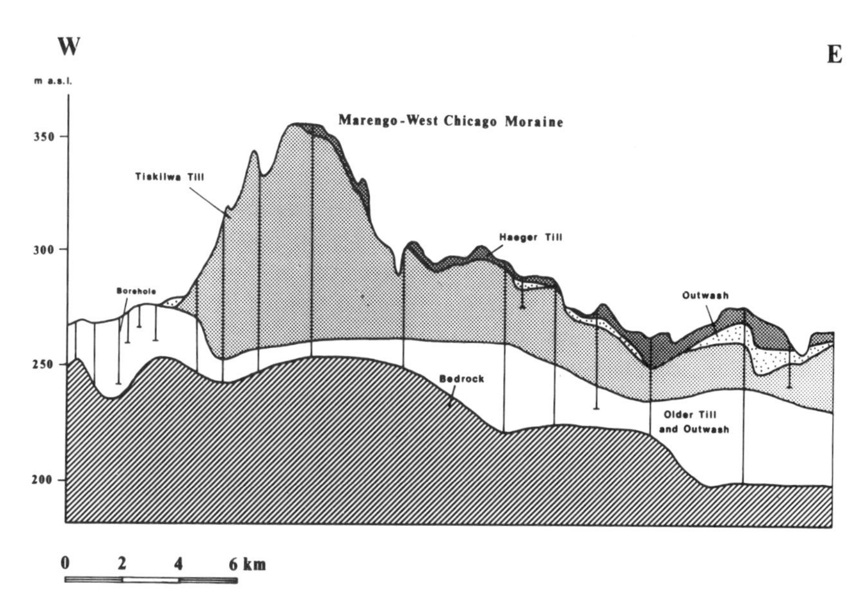

FIGURE 183 Cross section through the Marengo–West Chicago Moraine; the end moraine is composed completely of till (from Wickham *et al.* 1988)

FIGURE 184 Ice-pushed moraine with till diapirs at Ludington, Michigan (Photograph: Ehlers 1993)

The end moraines, as in Europe, also provided the first means by which the Wisconsinan Glaciation could be subdivided. They formed the basis of the morphostratigraphical schemes of Leverett (1929) and Leighton (1960). Later it was found that some of those landforms are inherited features, and that parts of many moraines in the Great Lakes region do not mark halts of the last ice sheet (Mickelson *et al.* 1983). Morphostratigraphy was replaced by a chronostratigraphical approach by Frye & Willman (1960), later visualised in time–distance diagrams (Frye *et al.* 1965). They subdivided the Late Wisconsinan into two principal phases, the Woodfordian and the Valderan. Intensive geomorphological and stratigraphical investigations in the 1970s, however, did show that the scheme required revision (Evenson 1973, Evenson & Mickelson 1974, Mickelson & Evenson 1975).

In the Great Lakes area, the Late Wisconsinan glaciation began around 25,000 BP. The Michigan Lobe area may serve as an example for the sequence of events (Fig. 185). Here the first ice advance is referred to as the Marengo Phase. In most places the deposits of the Marengo Phase were later overridden; only in northernmost Illinois does its end moraine form the outermost Wisconsinan maximum. The moraine consists almost entirely of the till of this ice advance, the Tiskilwa Till, which here averages a thickness of 90 m (Wickham *et al.* 1988). Eight glacial phases can be distinguished in the Michigan Lobe area, the first two and the latest of which are well-dated with the radiocarbon method. The second advance, called the Shelby Phase, represents the glacial maximum. The Michigan Lobe at that time advanced far south into Illinois and deposited the Shelbyville Morainic System, dated at about 20,000 BP. The subsequent Putnam, Livingston, Woodstock and Crown Point Phases were all relatively short-lived events. The ice margin finally retreated from Illinois at about 14,000 BP. In the following Mackinaw Interstadial the Great Lakes came into existence; the ice front retreated to a position

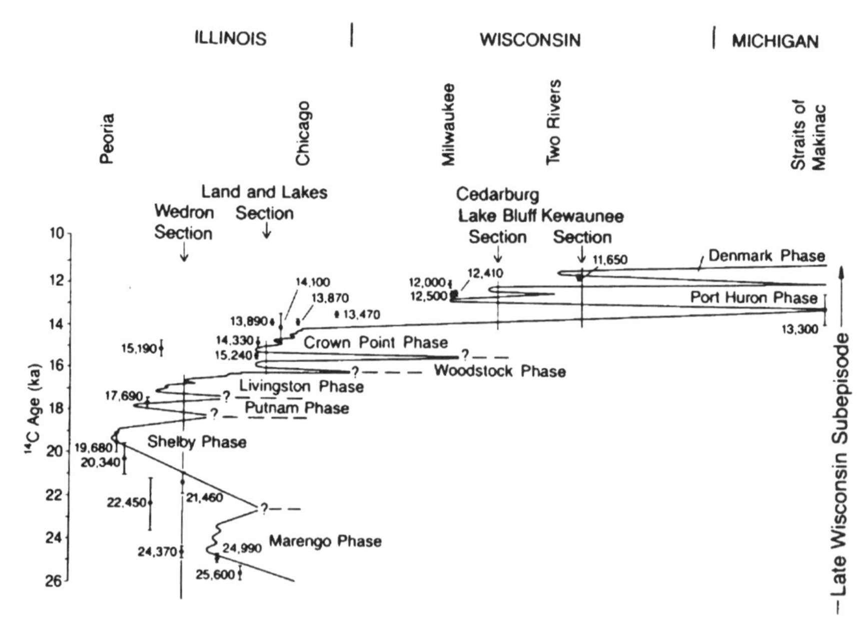

FIGURE 185 Time–distance diagram showing glacial phases during the Late Wisconsinan in the Lake Michigan Lobe area (from Hansel & Johnson 1992)

in the Michigan Lake basin. In the Port Huron Phase the ice readvanced as far south as Milwaukee. This readvance is dated to about 12,500 BP (13,000 BP in Ontario; Karrow, personal communication). Subsequently the ice melted back to the Straits of Mackinac.

A key section for this event is the Two Creeks forest bed exposed in the Lake Michigan shore bluffs in Maniwotoc County, Wisconsin. At the type section lacustrine silts and sands and the forest bed overlie and underlie Wisconsinan till. The section, first described by Thwaites & Bertrand (1957), has been radiocarbon-dated to ca. 12,050–11,750 BP (Kaiser 1994). Other sites of the Two Creeks Interstadial include the Cheboygan bryophyte bed (Larson *et al.* 1994).

During the following readvance, the ice margin reached to the Two Rivers Moraine in Wisconsin and the Manistee Moraine in Michigan. The end of this readvance marks the end of the Late Wisconsinan in the Michigan Lobe area (Hansel & Johnson 1992). Similar sequences but with different lithostratigraphical, chronostratigraphical or event-stratigraphical names are found in the other lobal areas of the Great Lakes region. The last ice advance in the Great Lakes region, the Marquette Advance (10,000 BP), did not reach Lake Michigan. It only affected the Lake Superior basin.

The major ice advances and retreat phases of the Late Wisconsinan occurred within a very short period of time. This implies that ice movement occurred in a series of extremely fast phases, at rates comparable to the flow of the recent ice streams of Antarctica. Reconstructions of ice thickness in the lobate areas suggest that the ice sheet must have been very thin (W.H. Mathews 1974, P.U. Clark 1992). Indeed, Clark concludes that very low driving stresses existed for the ice lobes, enabling them to advance by sliding and/or subglacial sediment deformation at rapid rates.

In Ohio the ice advances in the marginal zone of the Miami sublobe have been investigated using detailed lithostratigraphical analyses. After an undated first ice advance, the Late Wisconsinan glaciation reached its maximum extent at the Hartwell Moraine about 19,600 BP (Lowell *et al.* 1990b). The ice sheet overrode trees and deposited a thick till sheet. A minor recession of the ice margin followed (more than 30 km), but vegetation did not re-enter the region. A third ice advance remained about 20 km behind the maximum moraines and deposited a stone-rich lodgement till. A fourth ice advance in the northern part of the study area may have been of only local importance. The ice advances are in phase with the advances of the Lake Erie Lobe (Fullerton 1986a) and the Lake Michigan Lobe (Hansel & Johnson 1992), suggesting climatic control of the events (Lowell & Stuckenrath 1991, Ekberg *et al.* 1993).

In New England the retreat of the Laurentide Ice Sheet was accompanied by a marine transgression into coastal Massachusetts, New Hampshire and both coastal and central Maine. Rapid ice-marginal recession in the Gulf of Maine region was probably caused by the development of a calving bay, whilst the margin of the grounded ice sheet in southern New England retreated much more slowly. After 13,000 BP the remaining ice cap in New England was separated from the Laurentide Ice Sheet by rapid ice retreat along the St Lawrence valley and subsequent transgression of the Goldthwait and Champlain Sea. Final disintegration of the stagnant ice mass resulted in accumulation of ice-contact stratified deposits and a lack of end moraines in northern central New England (Mickelson *et al.* 1983).

Another region with widespread ice-decay features in the marginal area of the Laurentide Ice Sheet is the area of the James and Des Moines lobes in parts of Minnesota, South and North Dakota. Here the landscape of the lowland areas is characterised by a thin, hummocky cover of supraglacial till, whilst the till cover is thicker and the landforms are higher in the uplands (Mickelson *et al.* 1983).

Whilst the outer margins of the Laurentide Ice Sheet are well established in most areas, details of ice movement are still far from solved. Landforms in many cases only reflect the last ice movement, but the distribution of erratics suggests that earlier glacial phases saw a different pattern of ice movement. In the central areas of the Laurentide Ice Sheet the interpretation of 'dark erratics' in tills around Hudson Bay has caused considerable controversy. The 'dark erratics' are greywackes with yellowish 'eyes'. They occur together with jaspers and have been interpreted as originating from Proterozoic rocks that underlie southern Hudson Bay and outcrop on its eastern side, on the Belcher Islands. These erratics were first described by Bell (1872) and soon after the Belcher Islands were first mentioned as a probable source area (Bell 1887). The erratics are found in tills in northern Manitoba, central Saskatchewan and northwestern Ontario (Prest & Nielsen 1987) and have been interpreted as evidence of ice flow from Labrador across Hudson Bay (Shilts *et al.* 1979). However, it is not clear if the source area is really restricted to eastern Hudson Bay. The landforms, striae and ice-retreat pattern all seem to suggest that an ice-dispersal centre existed over Hudson Bay with ice spreading radially in all directions (Dredge & Cowan 1989).

In the mountains of western North America, the **Cordilleran Ice Sheet** formed through coalescence of large piedmont glaciers in the Rocky Mountains and Cascade Ranges in Canada and flowed south into the Unites States. It extended south to about 47°30′.

Based on his investigations in the Wind River Mountains, Wyoming, Richmond (1965) originally subdivided the Wisconsinan Stage in the western States into an Early Wisconsinan Bull Lake Glaciation, a Middle Wisconsinan 'Interglaciation' and a Late Wisconsinan Pinedale Glaciation. The Bull Lake and Pinedale glaciations were defined from the 'Bull Lake Till' and 'Pinedale Till' of Blackwelder (1915), and also the interpretation of the intercalated fossil soil as 'interglacial' dates from the same publication. The development of modern dating techniques, however, has caused a major revision of the original interpretation. The outer Bull Lake end moraine at Pinedale, on the west side of the Wind River Range, has been dated by uranium-trend analysis to 160 ±50 ka (Rosholt *et al.* 1985). This means that it is of late Illinoian age. It clearly predates the Sangamonian Interglacial, during which period the interglacial fossil soil had formed. The same applies to other tills correlated with the Bull Lake Glaciation, such as for instance the Horse Butte Till of the Yellowstone region (Pierce *et al.* 1976). Therefore today the Bull Lake tills and end moraines of the Rocky Mountains are correlated with the late Illinoian Stage glaciation (Richmond 1986a, b, c). The Pinedale Till and end moraines are the equivalents of the Late Wisconsinan glacigenic deposits of the Laurentide Ice Sheet.

In the Canadian Cordillera the Late Wisconsinan Glaciation is referred to as the Fraser Glaciation, and also locally as the McConnell, Macauley and Kluane Glaciation (Armstrong *et al.* 1965, Clague 1986). It was preceded by a prolonged ice-free period, the so-called Olympia Nonglacial Interval (Boutellier Nonglacial Interval in Yukon). During the period that began prior to 60,000 BP, glaciers were restricted to the high mountains. It is not clear whether the period also includes the Sangamonian Interglacial. Early Wisconsinan ice advances in British Columbia, if they existed, seem to have been rather limited. Deposits of older glaciations have been found both in British Columbia and in Yukon, but their age is uncertain (Clague 1989).

In the Cordillera the older glaciations did not advance significantly further south than those of the Wisconsinan. The extent and shape of the Cordilleran Ice Sheet was largely controlled by topography. Major ice lobes developed, in the west the Juan de Fuca Lobe and the Puget Lobe, constrained by the Cascade Range and the Olympic Mountains, and east of the Cascade Range the large Okanogan Lobe, and several smaller lobes occupied N–S-trending valleys.

In western Washington, the Puget Lowland was covered by the Late Wisconsinan Cordilleran ice (Fig. 186). The deposits are well dated. The last major glaciation, the Late Wisconsinan Fraser Glaciation, can be subdivided into three advances, the oldest of which occurred about 21,500 BP. After a short period of ice melt it readvanced to its maximal position which is radiocarbon-dated to between 15,000 and 14,000 BP in Washington (Easterbrook 1986, 1992).

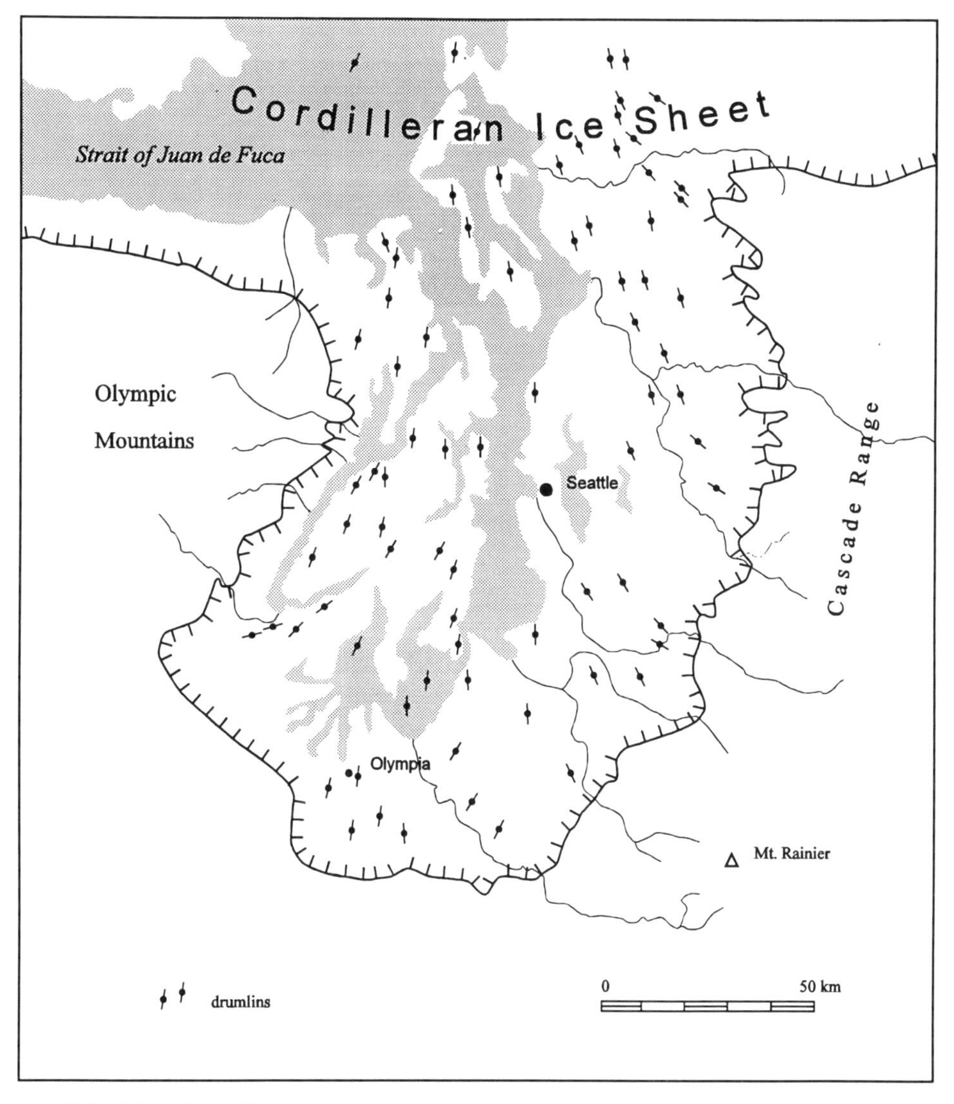

FIGURE 186 Map of the Wisconsinan Cordilleran Ice Sheet in the Puget Lowland, Washington (from Easterbrook 1986)

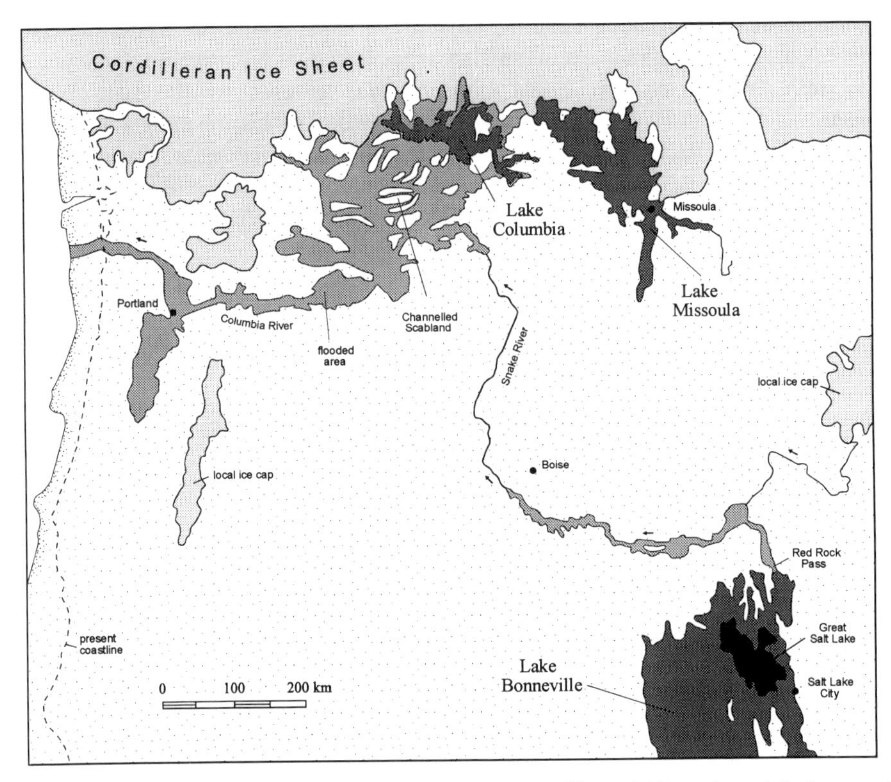

FIGURE 187 Northwestern USA with location of Lakes Bonneville and Missoula and drainage paths of the catastrophic floods during the Late Wisconsinan (from O'Connor 1993)

The Juan de Fuca and Puget Lobes underwent sudden decay between 14,500 and 13,000 BP. Because its tills are overlain by glaciomarine deposits of the so-called Emerson Drift, it has been suggested that ice retreat occurred in the form of a calving ice front (Domack 1983). However, the radiocarbon dates indicate that glaciomarine deposition was not time-transgressive, and at least locally the glaciomarine sediments overlie stagnant ice deposits. Easterbrook (1992) therefore suggests that thinning of the rapidly melting ice sheet allowed sea water to enter the lowlands and caused the remaining ice to float. This episode was followed by a last glacial readvance at about 11,300 BP

The southern margin of the Cordilleran Ice Sheet occurred in Washington, Idaho and Montana. The easternmost lobe of the Cordilleran Ice Sheet, the Flathead Lobe, reached just south of Lake Flathead, Montana, where it coalesced with Alpine glaciers from the Rocky Mountains (Richmond 1986b). Alpine mountain glaciers reached further south. In Washington, in the northern part of the Columbia Basin, the ice sheet blocked drainage of the Columbia River, leading to the formation of the large ice-dammed lakes Columbia and Missoula (Fig. 187). During the Late Wisconsinan Substage at least six flood episodes occurred between 15,000 and 12,000 BP, each comprising one or more outbursts. Similar events had also occurred repeatedly during earlier glaciations. They sculpted the so-called 'Channelled Scabland', an area 150 × 200 km in size (G.A. Smith 1993).

Further to the south, in the Great Basin, numerous pluvial lakes formed at periods during the

Quaternary. In the Late Wisconsinan about 120 lakes existed in the Great Basin, the largest of which were Lakes Lahontan and Bonneville. Lake Lahontan covered a maximum area of 21,000 km^2, which is almost the size of the present Lake Erie. Lake Bonneville, originally described by Gilbert (1890), reached a maximum size of 51,700 km^2, slightly smaller than the modern Lake Michigan. The former reached its last maximum about 17,000–15,000 BP. The lake sediments represent major parts of the Quaternary, but dating and correlation of the individual units is difficult. However, it is clear that during the Late Illinoian the lake levels had risen almost as high as during the Late Wisconsinan; shore deposits of the respective stratigraphical unit, the 'Alpine Formation', have been identified in various places (Currey 1990, Morrison 1991).

In the Late Wisconsinan, Lake Bonneville rose to the level of the Red Rock Pass, allowing overflow to the Snake River. Catastrophic drainage down the Portneuf and Snake rivers followed, with a maximum discharge of about 935,000 m^3 per second (Morrison 1991). The hydraulic conditions of the flood have been reconstructed recently from a detailed survey of the geomorphological and sedimentary features (O'Connor 1993).

It has always been a matter of debate where and how the Cordilleran and the Laurentide ice sheets met. Reconstructions of the Late Wisconsinan glaciations of North America have shown ice-free corridors reaching hundreds of kilometres north and south (Fulton 1984, Dyke & Prest 1987, Bobrowsky & Rutter 1992). However, at the western edge of the Laurentide Ice Sheet in Alberta, Canada, this concept has been challenged. It has recently been postulated that in this region the Late Wisconsinan ice sheet advanced much further than its predecessors. Liverman *et al.* (1988) found that in the Grande Prairie region of western Alberta the only till from an eastern source was that from the Late Wisconsinan glaciation. According to Young *et al.* (1994), the only Quaternary deposit underlying the Late Wisconsinan till sheet, the Saskatchewan Gravels and Sands, were deposited between 21,300 and 42,910 BP. If these dates are correct, then the Wisconsinan Laurentide Ice Sheet may have been the only one to have coalesced with the Cordilleran Ice Sheet and that confluence extended very far to the south. This in turn would mean that the ice-free corridor postulated for this area (e.g. Rutter 1984, Dyke & Prest 1987) did not exist during the last glaciation.

Since the main drainage divide in North America is near the US/Canadian border, the Laurentide ice was forced to flow uphill on its way south. This was even more so, because during the Late Wisconsinan the central parts of the ice sheet were already isostatically depressed, and most of the areas north of the Great Lakes were at or below sea level. The extent of maximum isostatic depression is not known. In the west, the present land surface of the formerly glaciated area slopes from more than 1000 m a.s.l., near the Rocky Mountains in Montana and Alberta, to below sea level in Hudson Bay. From directional indicators it is known that in Alberta the ice flowed westward, straight uphill (Mickelson *et al.* 1983).

The Wisconsinan ice sheet extensively remoulded major parts of the North American landscape. Although, in many places, the sedimentary cover is very thin, often of the order of only a few metres, the ice sheets have left behind a variety of characteristic landforms. Apart from the ice-decay features many of these landforms are streamlined. The role that subglacial meltwater played in their formation has recently been discussed. Wright *et al.* (1964) were the first to record the existence of **tunnel valleys** in North America from the Superior Lobe region. The features they described have since been discussed repeatedly (e.g. Wright 1973, Mooers 1989, Patterson 1994). The shapes and outlines of these incisions are very similar to their European counterparts. Although the exact mechanism of their formation is still a matter of debate, there is little doubt that they were formed by subglacial meltwater erosion.

Similar features have been found elsewhere in North America. When the Wisconsinan Cordilleran Ice Sheet in the Puget Lowland of Washington last advanced about 15,000 BP, it deposited a vast outwash deposit known as the 'great Lowland fill' during its advance. This sandur deposit is about 100 m thick but is dissected by a series of deep channels, incised up to 400 m into the underlying substratum. Since many troughs are mantled by basal till, formation by postglacial erosion can be excluded. The channels are therefore interpreted as subglacial meltwater erosion features (Booth & Hallet 1993, Booth 1994), like the tunnel valleys of Minnesota (Wright 1973, Mooers 1989) and Wisconsin (Nelson & Mickelson 1977, Attig *et al.* 1989).

Not only channels but also positive landforms, like drumlins, have been suggested as forming by subglacial meltwater by Shaw (1983, 1989a) and co-workers (e.g. Shaw & Kvill 1984, Shaw & Gilbert 1990, Rains *et al.* 1993). For instance, in southern Ontario and northern New York State a tract of land has been identified in which the orientation of tunnel channels and streamlined features like drumlins differs strikingly from that of adjoining areas. However, it is not clear whether these features were actually shaped by flowing water or by fast-flowing ice in the vicinity of tunnel channels. The subglacial meltwater flood hypothesis of drumlin formation has been refuted by other workers (e.g. Muller & Pair 1992). Boyce & Eyles (1991) suggest that the Peterborough drumlin field, north of Lake Ontario, was incised by streams of deforming till cutting into pre-existing sediments. The widespread occurrence of rhythmically laminated silty clays speaks against the idea that they originated as fluvial infill of subglacial cavities, as was suggested by Sharpe (1987).

A major sector of the Laurentide Ice Sheet drained through Hudson Strait towards the North Atlantic. Extensive field investigations have shown that Hudson Strait was occupied by a large ice stream, with a catchment area extending far beyond Hudson Bay, including major parts of Keewatin. Erratics from these areas have been found on the islands in western Hudson Strait (Laymon 1992). The 'Heinrich Layers' of ice-rafted detritus in the North Atlantic (see chapter 7.2) have been correlated with major meltwater outbursts from the Laurentide Ice Sheet (Andrews & Tedesco 1992, Andrews *et al.* 1994). According to P.U. Clark (1994) the ice sheet had become unstable, because under the thick ice cover the glacier sole became warmer and resulted in a deformable bed that triggered the outbursts.

13.8.4 The Great Lakes

Quaternary geological investigations in the Great Lakes region began in the mid-19th century. Agassiz (1850) visited Lake Superior and described striae, glacially sculpted terrain and raised shorelines. He was followed by Lawson (1893) who was the first to measure shoreline altitudes along the Lake Superior coast. Later, F.B. Taylor (1895, 1897) noticed that the ancient shorelines were tilted. In this century, Leverett & Taylor (1915) presented a first review of the Late Quaternary history of the Great Lakes. In 1958 Hough published a new *Geology of the Great Lakes*. The most recent comprehensive reviews have been by Karrow (1984), Karrow & Calkin (1985), Karrow (1989) and Teller & Kehew (1994).

The age of the Great Lakes is unknown. It is highly likely that the lake basins have existed through most of the Quaternary and that they were successively excavated by the Pleistocene ice sheets. Lithological differences have been influential in the shaping of the individual lakes. For instance, an escarpment of resistant Silurian Niagaran dolomite determines the arcuate shape of the western and northern shores of Lake Michigan, and the subdivision of Lake Huron into the main lake and the northeastern Georgian Bay, and parts of the Michigan, Huron and Erie lake basins have developed in easily eroded Devonian shales (Trenhaile 1990). However, the formation of the basins reflects the interaction of the lobate outermost margin of the ice sheet with the underlying bedrock. In Europe, much smaller basins were formed, like the Saalian tongue basins of the Netherlands (35 km long, 130 m deep) or the Elsterian 'Elbtalwanne' in eastern Germany (over 100 km long, over 160 m deep), but there the ice advanced over thick, unconsolidated Tertiary rocks. The basins were largely refilled at the end of each cold stage, and only in rare cases (like the Lübeck basin) were they re-excavated during subsequent glaciations. Normally, each new glaciation formed its own marginal features.

The lake basins are deeper than indicated by recent bottom contours. In the over 500-km-long Lake Superior basin off the Minnesota shore, the sediment fill reaches maxima of about 500 m. This great thickness has encouraged research into a stratigraphical subdivision of the sedimentary sequence, but to no avail. All deposits seem to be of Wisconsinan age, probably even Late Wisconsinan (Farrand & Drexler 1985). The bottom contours of eastern Lake Superior show a system of N–S-trending, interconnected channels (Farrand & Drexler 1985) that closely resemble the tunnel channels of the north European glaciations. Apparently they continue across the upper peninsula of Michigan as the Au Train–Whitefish channel and related drainage paths west of Munising, though at a much higher altitude. The latter also served as proglacial outwash channels during the Marquette Readvance about 10,000 years BP (Farrand & Drexler 1985; see below).

Because the Great Lakes are sufficiently large water bodies, the tilt of former shorelines gives an indication of the rates of isostatic uplift. For example, the elevations of Lake Washburn shorelines over a distance of 200 km between Duluth and the eastern end of this lake on the Keweenaw Peninsula (at about the middle of the south shore of Lake Superior) differ by as much as 70 m. Moreover, isostatic rebound still seems to be continuing (J.A. Clark *et al.* 1994); the northeastern part of the Superior basin is uplifted by 27 cm per century, and the southwestern part is subsiding at 21 cm per century relative to the head of St Mary's River (Farrand & Drexler 1985).

The Great Lake basins were not only shaped by the Pleistocene ice sheets; in turn they also controlled ice movement. The Ontario basin was the first to be entered by the advancing Late Wisconsinan ice. Chamberlin (1883, 1888) had already pointed out that during further advance the Lake Erie basin had deflected ice movement to the west, until the Erie Lobe coalesced with the southward-flowing Huron Lobe. During deglaciation, the Superior basin was the last part of the Great Lakes to become completely ice-free at about 9.5 ka BP (Barnett 1985).

The history of the Great Lakes after the Late Wisconsinan glacial maximum is a chronicle of changing outlets caused by the presence or absence of ice barriers, by differential isostatic uplift and by downcutting of the outlets. Drainage of ice-dammed lakes is often a catastrophic event. Consequently, outburst floods made a significant contribution to the drainage history of the southern margin of the Laurentide Ice Sheet, as has been emphasised recently by Teller (1985, 1987, 1990), Kehew & Lord (1986), Kehew (1993) and Kehew & Teller (1994). Sediment cores and high-resolution seismic investigations have contributed largely to the understanding of lake histories (e.g. Foster & Colman 1991, Colman *et al.* 1994b).

The different stages of lake development are referred to as **phases**, in contrast to the glacial stratigraphical stadials and interstadials. Only the principal phases of lake development can be outlined below. When the Wisconsinan ice sheet had reached its maximum position south of the present Great Lakes at about 18,000 BP, the ice margin was well south beyond the Great Lakes–Gulf of Mexico drainage divide, and no lakes existed (Fig. 188a). The oldest Late Wisconsinan lake in the Lake Michigan basin, **Lake Milwaukee**, formed at the end of the Woodstock Phase (A.F. Schneider 1983, A.F. Schneider & Need 1985). It is thought to have persisted from about 15,500 to 14,500 BP. At the same time, during the Erie Interstadial parts of Lake Erie became ice-free, and **Lake Leverett** formed (Mörner & Dreimanis 1973, Dreimanis 1977). Drainage from Lake Leverett was probably eastward via the Mohawk River into the Atlantic Ocean (Fig. 188b). The deposits of these early lakes were overridden by the readvancing Laurentide Ice Sheet. In the Michigan basin, the Michigan Lobe subsequently advanced to the Valparaiso ice marginal position (Fig. 188c). Only by about 14,100 BP did parts of the lake basins again become ice-free.

The first lake to form was at the southwestern margin of the Erie basin at the Ohio/Indiana boundary. This **Lake Maumee** at first drained via the Wabash River in Indiana towards the Mississippi (Fig. 188d). By about 14,000 BP the southern end of the Lake Michigan basin also became ice-free again and **Lake Chicago** formed at a high level, the so-called **Glenwood Level**. Later, when major parts of Michigan had become ice-free, drainage across Michigan through the Imlay Channel became possible (Fig. 188e). Still later, drainage shifted further north to a now buried valley. All of these outlets were directed towards the west (Barnett 1985, Eschman & Karrow 1985).

When the margin of the Huron Lobe retreated, newly formed **Lake Saginaw** drained via the Grand River into Lake Chicago, which drained in turn through an outlet at Chicago towards the Mississippi (Hansel *et al.* 1985).

In contrast to the results of earlier investigations, about 13,400 BP, during the Mackinaw Interstadial, the ice did not retreat as far north as the Straits of Mackinac (Larson *et al.* 1994). This means that eastward drainage remained blocked, so that no 'Intra-Glenwood Low Phase' could have existed in the Lake Michigan basin. However, the ice may have retreated from the Saginaw Peninsula, in the southern part of the Huron basin, far enough to allow realignment of the drainage of **Lake Arkona** towards the northwest, via Saginaw Bay into Lake Michigan (Eschman & Karrow 1985) (Fig. 188f). During this time the eastern lakes experienced low lake levels. Lake Huron drained via the isostatically depressed Fenelon Falls outlet directly into Lake Ontario, and in the Lake Erie basin the extreme low level of Lake Ypsilanti formed (Fig. 188g).

During the subsequent Port Huron glacial readvance, about 13,000 BP eastward drainage through the Mohawk River was blocked. Consequently, **Lake Whittlesey** in the Erie basin also drained via the Ubly Channel into Lake Saginaw and from here into the Michigan basin (Fig. 188h). This drainage event may have occurred in the form of a catastrophic outburst, eroding the deeply incised Grand Valley (Kehew 1993). During the following retreat of the ice margin, Lakes Saginaw and Whittlesey combined to form **Lake Warren**, which still drained westward through the Grand River (Fig. 188i).

The Two Creeks Interstadial, about 12,050–11,750 BP, saw the ice margin retreat sufficiently far north to allow drainage from the Michigan basin to the northeast, probably via the Straits of Mackinac (**Early Lake Algonquin**, Fig. 188j). Because the northern part of the basin at that time was still strongly depressed isostatically, this caused the level of Lake Michigan to drop considerably below its present level. This developmental stage is called the **Two Creeks Low Phase** (Hansel *et al.* 1985) or referred to as the **Kirkfield Phase of Lake Algonquin** (C.F.M. Lewis *et al.* 1994). During this phase, the latter drained eastward, via the

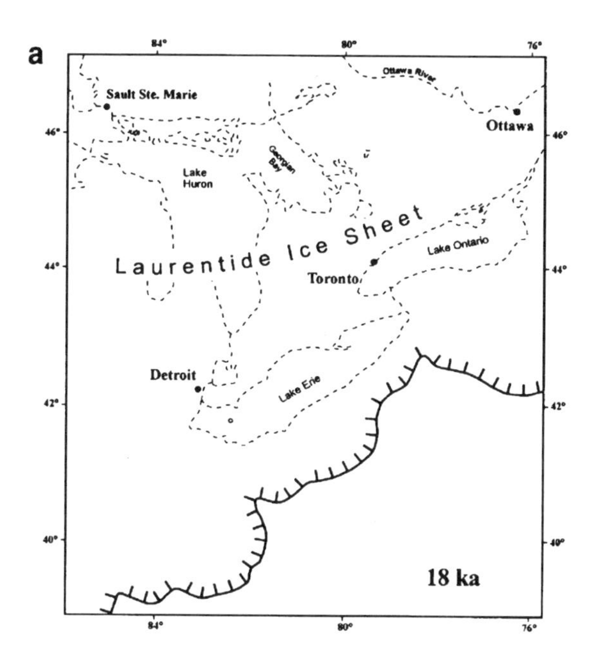

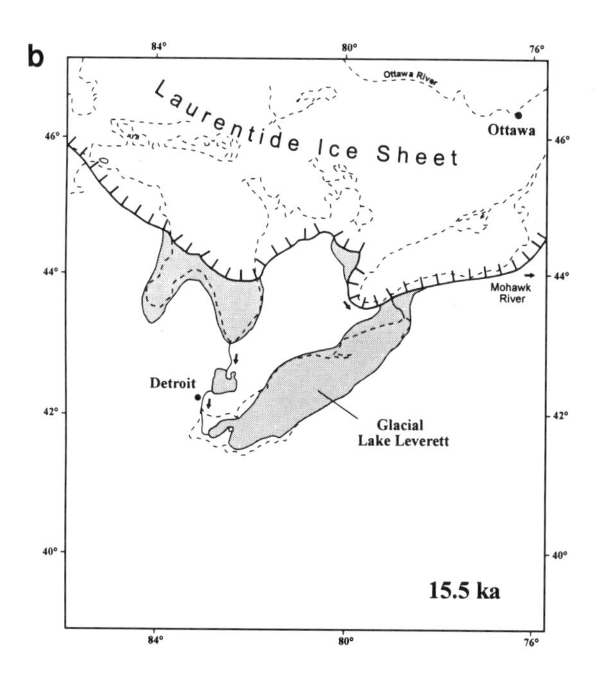

FIGURE 188 Palaeogeographic maps showing the lateglacial development of the Great Lakes: (a) Nissouri Stade (about 18 ka BP); (b) Lake Erie Interstade (about 15.5 ka BP)

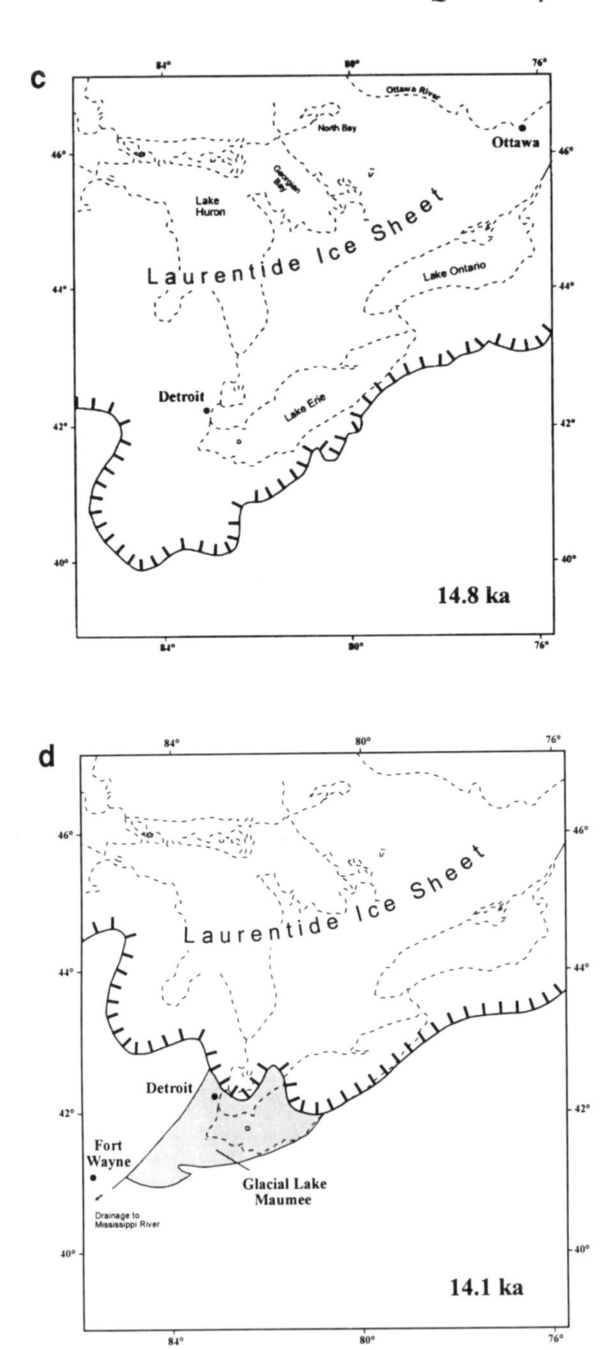

FIGURE 188 (*continued*) Palaeogeographic maps showing the lateglacial development of the Great Lakes: (c) Port Bruce Stade (about 14.8 ka BP); (d) highest Lake Maumee (about 14.1 ka BP)

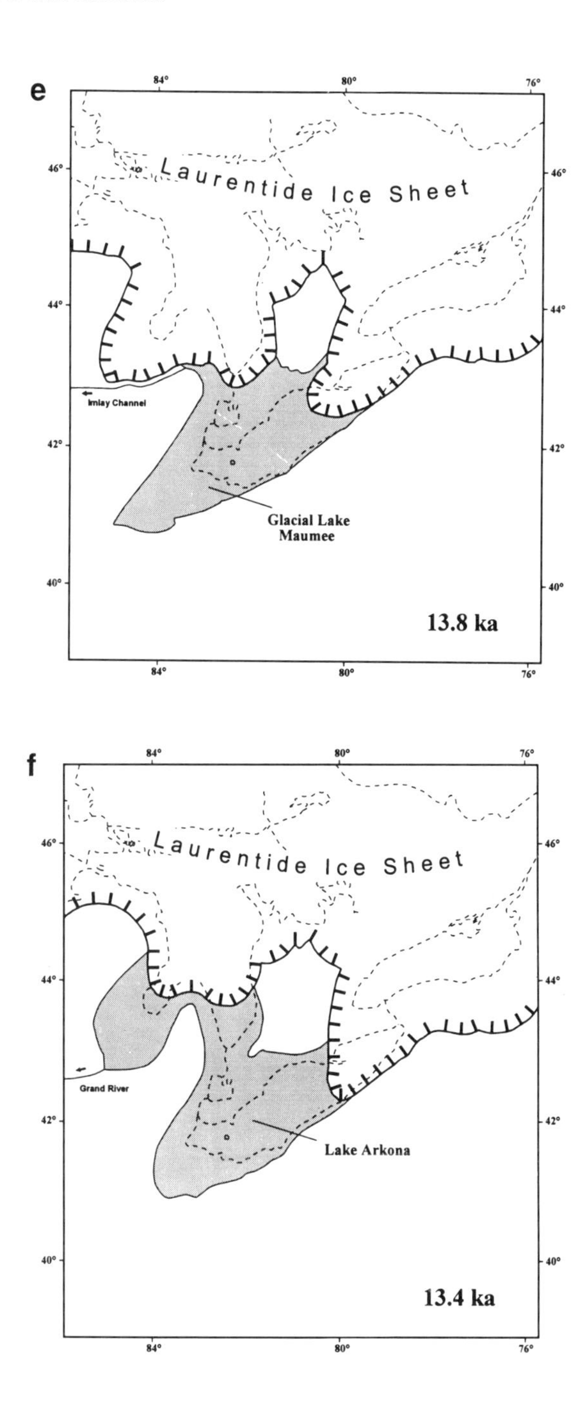

FIGURE 188 (*continued*) Palaeogeographic maps showing the lateglacial development of the Great Lakes: (e) final phase of Lake Maumee (about 13.8 ka BP); (f) Early Lake Arkona (about 13.4 ka BP)

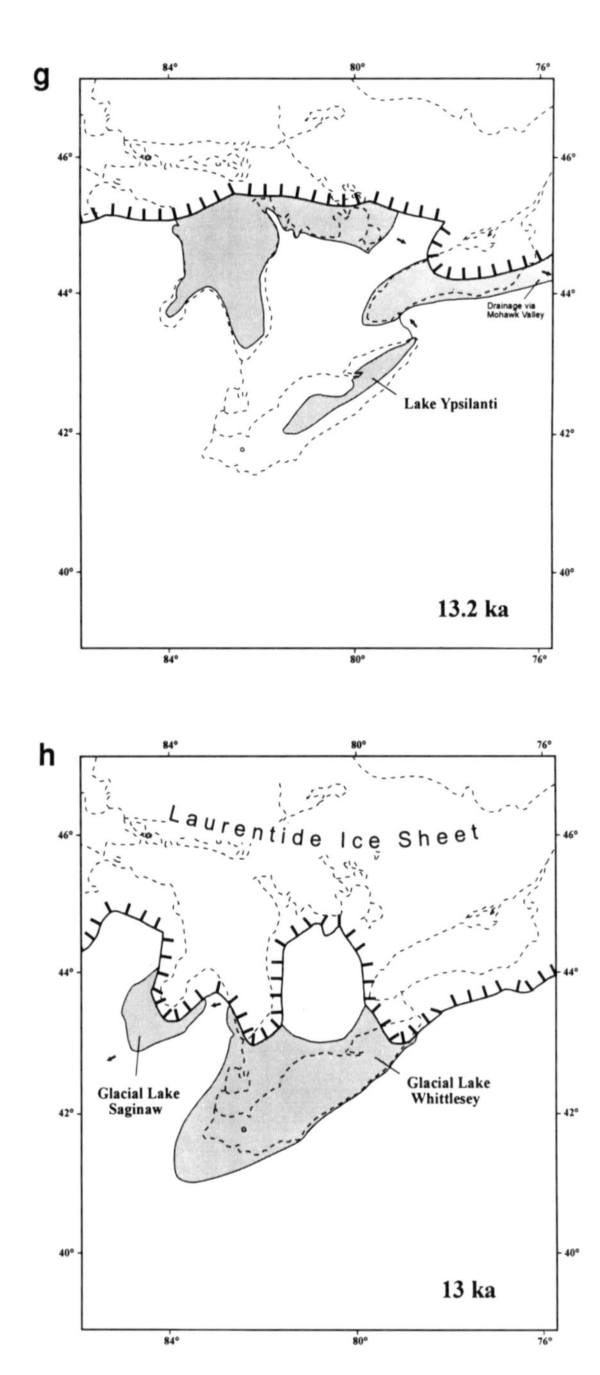

FIGURE 188 (*continued*) Palaeogeographic maps showing the lateglacial development of the Great Lakes: (g) Mackinaw Interstade (about 13.2 ka BP); (h) Glacial Lakes Whittlesey and Saginaw (about 13 ka BP)

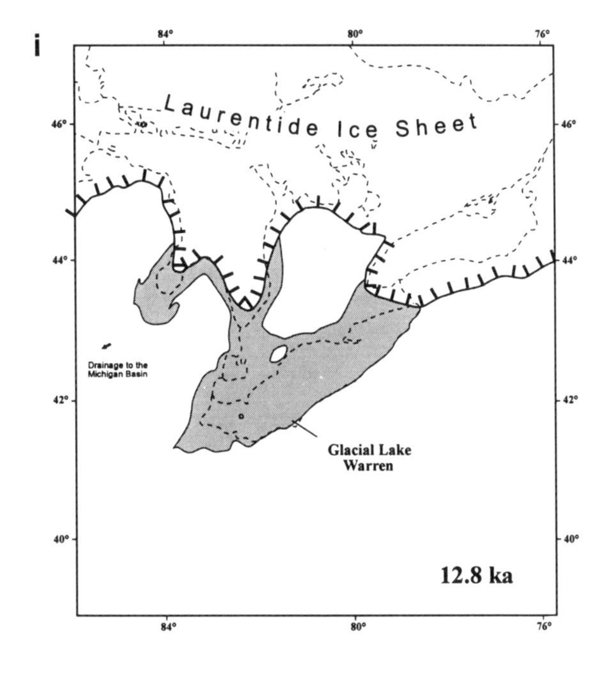

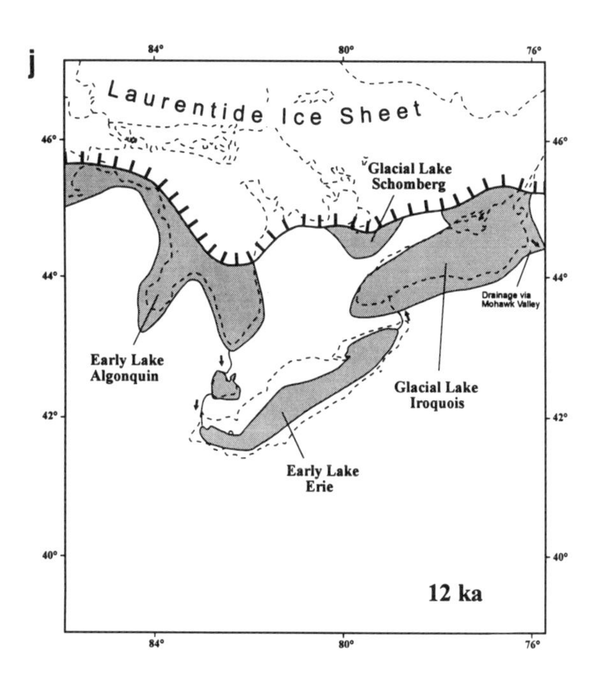

FIGURE 188 (*continued*) Palaeogeographic maps showing the lateglacial development of the Great Lakes: (i) highest Lake Warren (about 12.8 ka BP); (j) Early Lake Algonquin (about 12 ka BP)

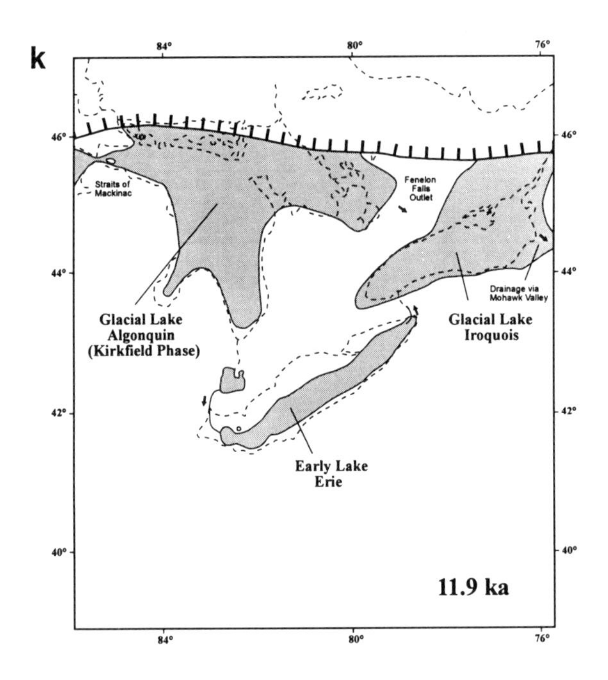

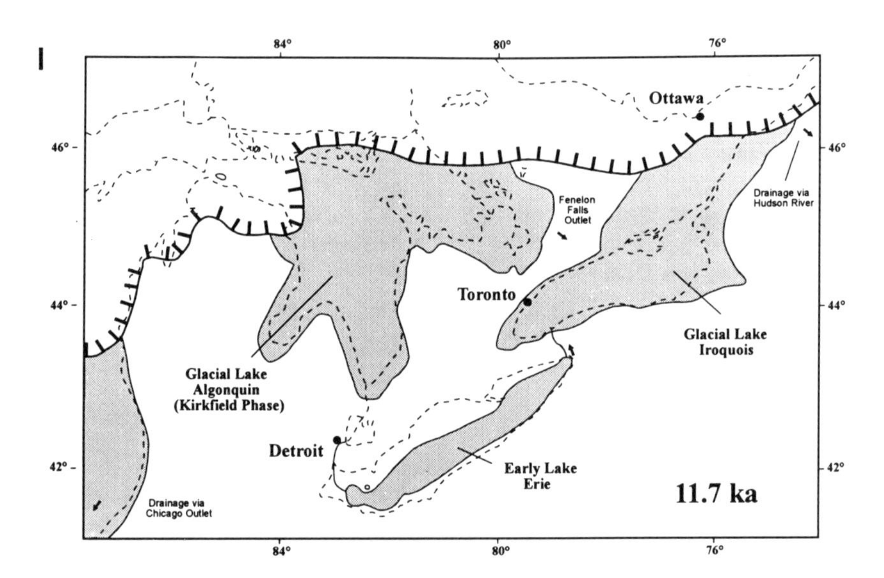

FIGURE 188 (*continued*) Palaeogeographic maps showing the lateglacial development of the Great Lakes: (k) early Kirkfield Phase of Lake Algonquin (about 11.9 ka BP); (l) Kirkfield Phase of Lake Algonquin (about 11.7 ka BP)

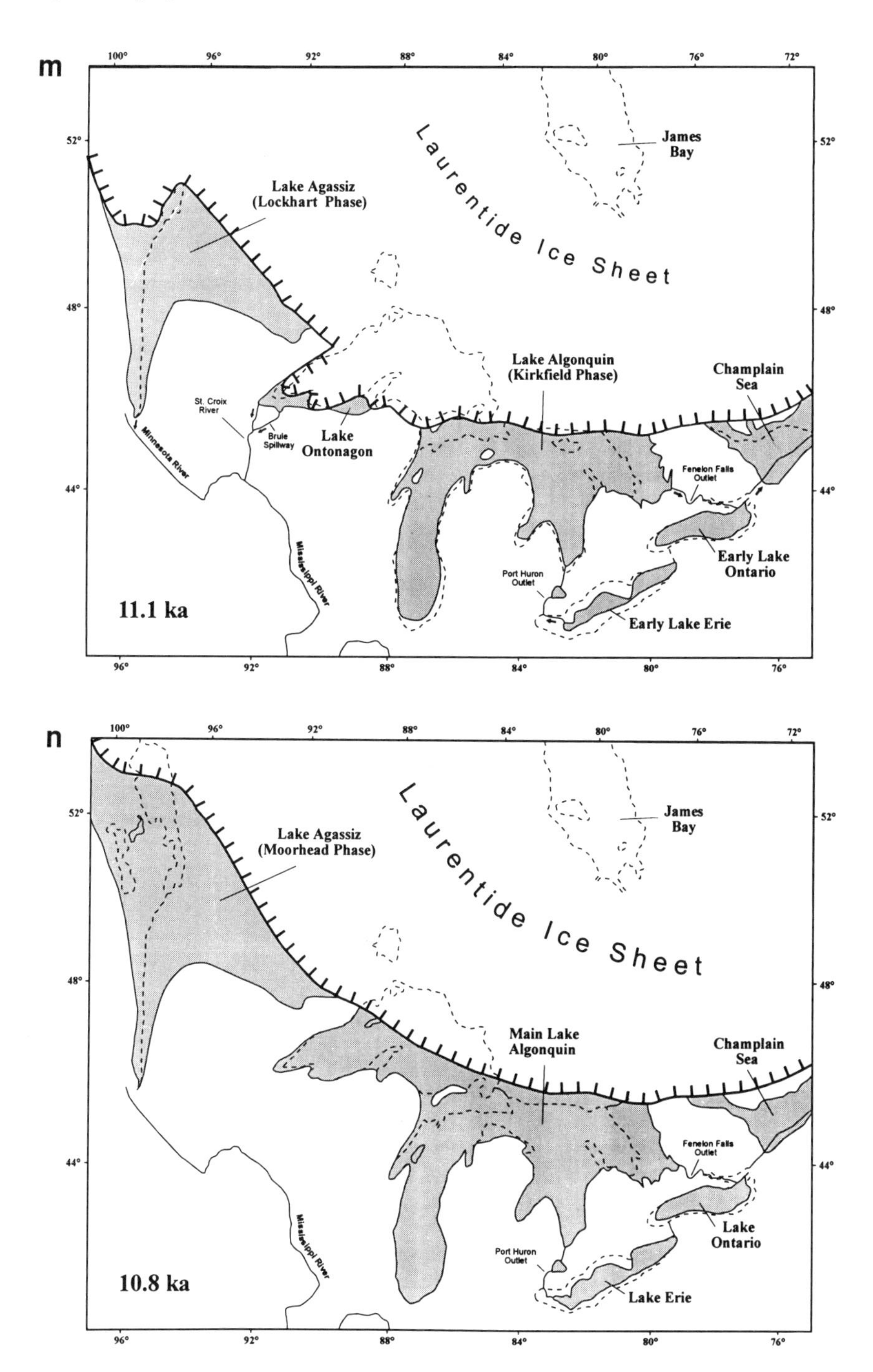

FIGURE 188 (*continued*) Palaeogeographic maps showing the lateglacial development of the Great Lakes: (m) late Kirkfield Phase of Lake Algonquin (about 11.1 ka BP); (n) Main Late Algonquin Phase (about 10.8 ka BP)

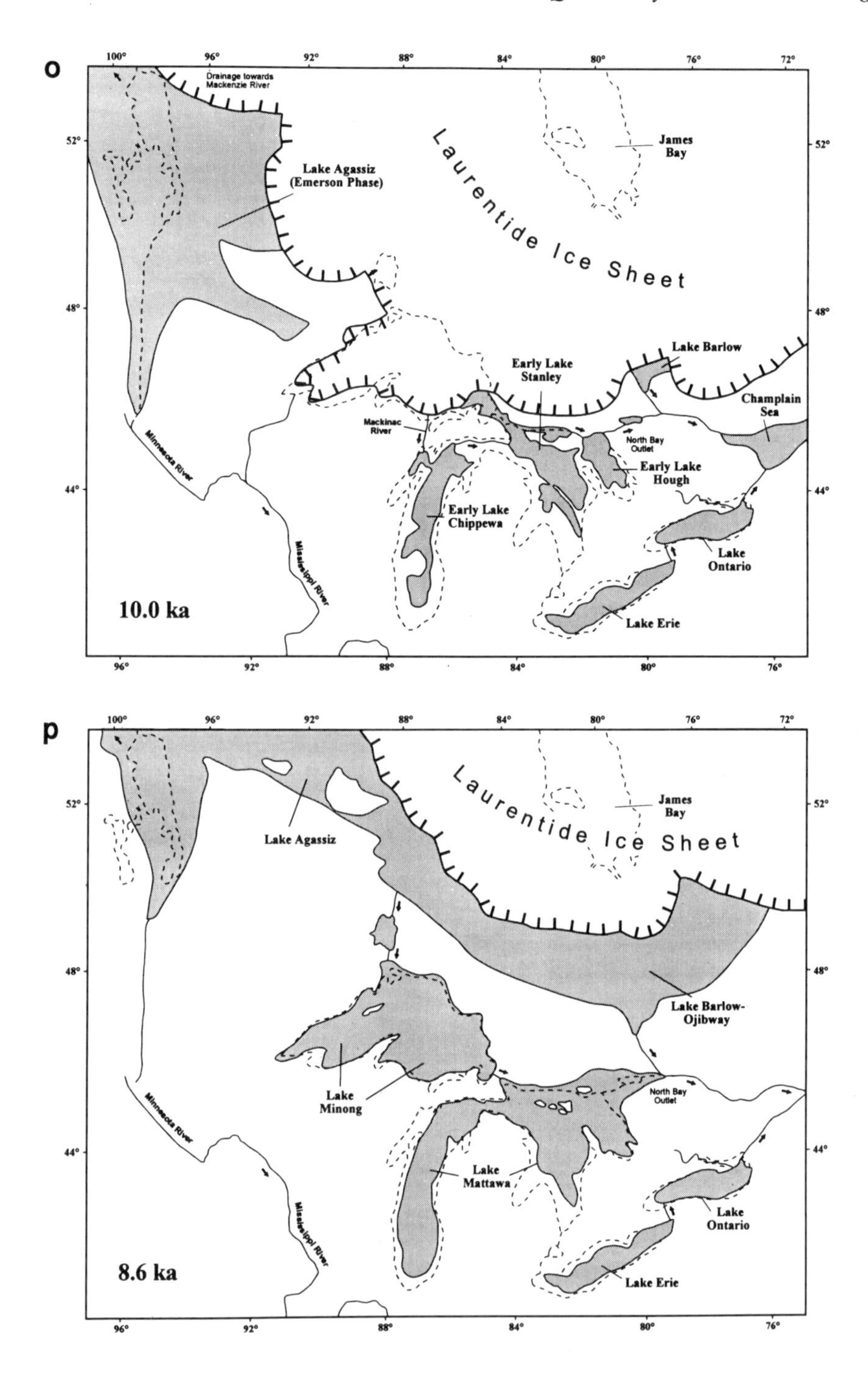

FIGURE 188 (*continued*) Palaeogeographic maps showing the lateglacial development of the Great Lakes: (o) Ottawa–Marquette Phase (about 10.0 ka BP); (p) Lake Mattawa High Phase (about 8.6 ka BP)

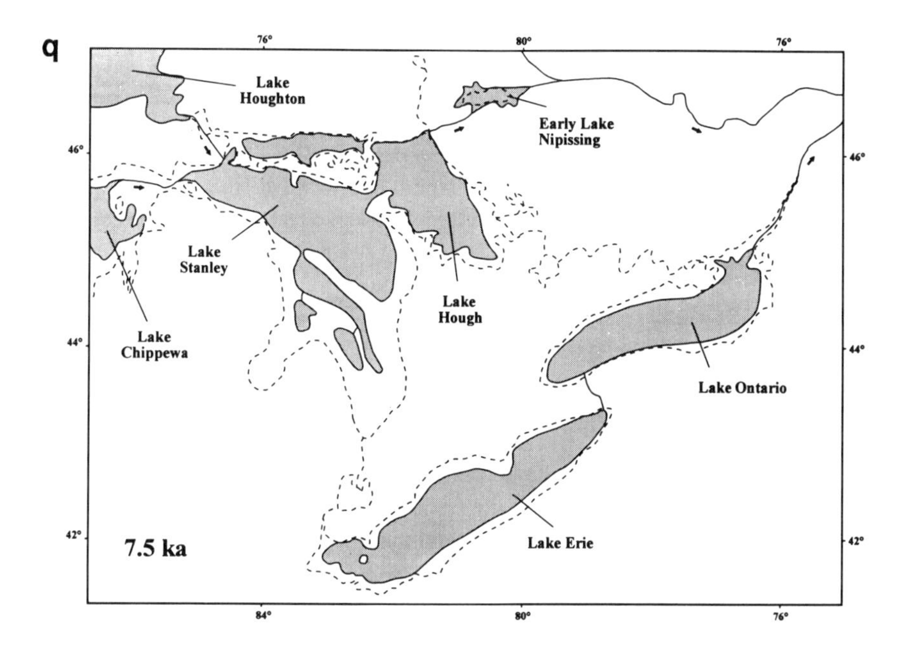

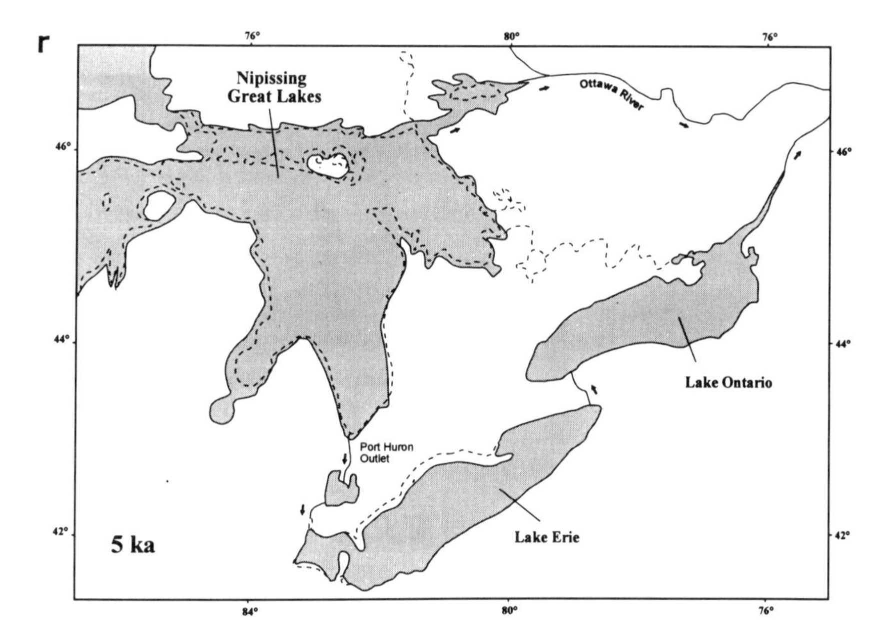

FIGURE 188 (*continued*) Palaeogeographic maps showing the lateglacial development of the Great Lakes: (q) post-Mattawa Low Phase (about 7.5 ka BP); (r) Nipissing Great Lakes (about 5 ka BP) (after Lewis *et al.* 1994)

Fenelon Falls Outlet into **Lake Iroquois** in the Ontario basin. Proglacial Lake Iroquois had developed when the Ontario Lobe had melted back far enough to re-open an outlet into the Mohawk valley at Rome in New York State (Muller & Prest 1985) (Fig. 188k).

During the subsequent Greatlakean ice advance, drainage through the Straits of Mackinac and the Indian River lowland was once again blocked, and the level of Lake Michigan rose higher than present. At this time the lake drained southward to the Mississippi River. This **Calumet Phase** seems to have been relatively short-lived, because it is not represented by any prominent shoreline features. It has been dated to approximately 11,800-11,200 BP (Hansel *et al.* 1985). This ice readvance seems not to have had much impact on the eastern lakes, in which the Kirkfield Phase of Lake Algonquin continued (C.F.M. Lewis *et al.* 1994) (Fig. 188l).

By this time the Laurentide Ice Sheet was also rapidly melting at the Atlantic coast. 'Draw-down', related to the calving front of the ice stream in the St Lawrence valley, resulted in the first Late Wisconsinan marine transgression in North America in the form of the **Champlain Sea**. The timing and extent of the transgression maximum has been a matter of debate (Gadd 1988). Shoreline features of this sea have been mapped along the margins of the St Lawrence Lowland. The maximum transgression of the Champlain Sea may have occurred around 11,700 BP, and the levels of Lake Ontario and the Champlain Sea seem to have been the same for a short period. However, the apparently strong outflow from the Ontario basin prevented the marine influence from transgressing into the lake (Pair *et al.* 1988).

Shortly before 11,000 BP the ice margin had probably retreated from northern Michigan. When the Straits of Mackinac became ice-free once more, the Kirkfield Phase Lake Algonquin expanded into the Michigan basin. Then Lake Michigan and Lake Huron joined to form **Main Lake Algonquin** (Hansel *et al.* 1985). The lake level of this phase is still a matter of debate. Eschman & Karrow (1985) assumed that the level was high, but Larsen (1987) thinks that in the southern Lake Michigan basin it was well below modern lake level. The drainage of Lake Algonquin turned eastward, first through the Fenelon Falls (Kirkfield) Outlet towards Lake Ontario and later through the Port Huron Outlet into Lake Erie (Finamore 1985, C.F.M. Lewis & Anderson 1992). At this time the Lake Superior basin also became ice-free for the first time (C.F.M. Lewis *et al.* 1994) (Fig. 188n).

Around the margins of the waning ice sheet ice-dammed lakes also formed in other areas. The largest of these was **Lake Agassiz**. It covered a total area of 950,000 km^2 in North Dakota, Minnesota, Saskatchewan, Manitoba and Ontario, although not all at the same time. Recession of the ice margin led to an extension of the lake towards the northeast, whilst isostatic uplift simultaneously caused a regression along its southwestern shores. The lake began to form around 11,700 BP when the ice margin had retreated beyond the water divide at the head of the Minnesota River (Fig. 188m). First the lake drained via the Minnesota River into the Mississippi. Later, drainage changed into Lake Algonquin, where its influx is recorded by a change in oxygen-isotope composition. The drainage of Lake Agassiz into the Gulf of Mexico via the Mississippi River was shifted temporarily to the east about 11,000 BP when the ice front retreated in western Ontario. When the ice readvanced, the outlet shifted northwest to the Mackenzie River and then again to the south (Colman *et al.* 1994a).

It has been postulated that drainage of Lake Agassiz into the North Atlantic prevented the production of North Atlantic Deep Water and triggered the Younger Dryas cooling (Broecker *et al.* 1988). However, recent investigations suggest that the Younger Dryas predates the shift of drainage direction by about 500 years (Rodrigues & Vilks 1994). For a short period (from ca. 9900 to 9500 BP), Lake Agassiz also temporarily drained to the northwest, via Clearwater and

FIGURE 189 Evidence of Younger Dryas readvance on Nova Scotia: Ice-pushed, folded Allerød peat overlain by till at Collins Pond (Photograph: R. Stea)

the Lower Athabaska River (Fisher & Smith 1994). Finally Lake Agassiz drained completely at 8000 BP with the opening of the Hudson Bay (Teller 1985).

The last major ice advance in the Great Lakes region is the **Marquette Readvance** (Farrand & Drexler 1985) that covered almost the entire Lake Superior basin. Since it is dated to approximately 10,000 BP, it can be correlated with the Younger Dryas in Europe. In Ontario, the Marks and MacKenzie moraines, and also possibly the Nipigon moraines, are correlated with this ice advance (Dredge & Cowan 1989, Barnett 1992) (Fig. 188o). On Nova Scotia, for instance, a Younger Dryas readvance of the local ice cap is recorded by a glacigenic diamicton overlying organic deposits (Stea & Mott 1989) (Fig. 189).

In pollen diagrams from interior North America, the Younger Dryas cooling is often relatively poorly represented. Whilst in the coastal regions of New England and Canada it is clearly identifiable (Anderson & MacPherson 1994, Cwynar *et al.* 1994, Mott 1994, Peteet *et al.* 1994), it is poorly reflected in pollen diagrams from Québec (Richard 1994). Lowe *et al.* (1994) conclude that poorer representation there may result from proximity to the decaying Laurentide Ice Sheet and lower temperatures being still present before the Younger Dryas began. Evidence for a Younger Dryas-like cooling has also been reported from the Pacific coast of Canada (Mathewes *et al.* 1993). However, its effects there have not been as marked as on the Atlantic coast. For a long time it seemed as if the Younger Dryas cooling had only affected the circum-North Atlantic region, but Alley *et al.* (1993) have shown that the Younger Dryas was a global cooling event.

13.9 HOLOCENE

In contrast to the situation in Europe, at the beginning of the Holocene major parts of North America were still covered by the Laurentide Ice Sheet.

Morphologically and with regard to its position in the centre of the glaciated area, **Hudson Bay** is the North American equivalent of the Baltic Sea. Its origin has been a matter of debate. Traditionally it was regarded as a very ancient feature. Palaeozoic sediments are known to occur beneath its floor, as well as remnants of a supposed Tertiary drainage network. In contrast, W.A. White (1972) challenged this view and suggested that Hudson Bay and Foxe Basin were features excavated by the Pleistocene ice sheets which, in the process, had removed about 1000 m of bedrock. This, however, cannot be reconciled with the evidence. Sugden (1976) pointed out that all points supportive of the antiquity of the feature still remained valid. The oldest tills in North America already contain clasts from the Canadian Shield, so that Precambrian rocks must have been exposed at that time. The Shield in general seems to have experienced net glacial erosion of only a few tens of metres. Greater erosion seems to have occurred near the outer zones of the ice sheet than close to the centre (Dyke *et al.* 1989). This is in common with northern Europe, where Tertiary weathered bedrock was preserved at the surface, for example, of Finland (cf. chapter 11.9).

Traditionally, it had been assumed that Hudson Bay had been covered by the Laurentide Ice Sheet throughout the Wisconsinan (Denton & Hughes 1981). However, marine evidence suggests that no related ice shelf existed across Baffin Bay and the northern Labrador Sea (Aksu 1985). Additionally, the Missinaibi Beds of the last interglacial in the Hudson Bay Lowlands are overlain by a till sequence, the individual units of which are separated by thin and discontinuous sand seams. Dredge & Cowan (1989) interpret these seams as minor subglacial meltwater features, whereas Andrews *et al.* (1983) and Shilts (1984a) think they indicate that the bay might have been ice-free periodically during the last cold stage. In the latter case, the core of the Laurentide Ice Sheet would have been less stable than had been thought but would be susceptible to rapid 'draw-down' if the fast-flowing ice stream through Hudson Strait could drain into the open sea (Dyke *et al.* 1989).

Instability of the Laurentide Ice Sheet is also reflected in an early Holocene abrupt ice-stream advance which may have occurred at about 9.9–9.6 ka BP at the mouth of Hudson Strait. This Cold Cove Advance resulted in increased iceberg release from the calving ice front over 200 km long in about 500 m deep open water. It may have been the cause of a brief cooling period noted in several high-resolution climate records of the North Atlantic for the period immediately following the Younger Dryas (Kaufman *et al.* 1993).

The last advances of active ice in the Hudson Bay area are the so-called **Cochrane Readvances**, named after the Cochrane district south of James Bay, discussed by Antevs (1925, 1928). This author considered them to be equivalents of the European Younger Dryas end moraines. However, later it became clear that they were much younger. Review of the geological and geomorphological evidence has shown that they represent minor surges (50–75 km) of an unstable ice margin into ice-dammed lakes that fringed the entire southern margin of the ice sheet. Varve counts indicate that the individual surges lasted only about 25 years. The lake sediments have been dated to about 8300 BP. It is thought that the lakes led to thinning of the ice sheet, flattening of the glacier profile and thus contributed to the rapid deglaciation of Hudson Bay (Dredge & Cowan 1989).

At this phase Hudson Bay ice was calving rapidly on its northern side into Hudson Strait and

Boothia Strait (west of Southampton Island), and the southern margin, which was bordered by lakes, also disintegrated rapidly. Eventually, buoyant forces of the invading sea lifted the entire remaining ice sheet, causing catastrophic drainage of the southern ice-dammed lakes. This drainage is recorded in 'drainage horizons', consisting of rounded pebbles of varved clay, overlain by pebbly sands (Dredge & Cowan 1989).

At this stage of deglaciation, about 8000 BP, the **Tyrrell Sea** invaded the isostatically depressed Hudson Bay lowlands and flooded a vast area, reaching as much as 300 km beyond the present coastline. At about the same time, the Baltic Sea in Europe finally connected to the ocean in its Litorina Phase. The highest shoreline of Hudson Bay is found at 180 m a.s.l. at the southern end at James Bay and in an area north of the Nelson River which was formerly covered by the Keewatin ice. Isostatic rebound has subsequently led to gradual regression to the present sea level, forming flights of elevated beach ridges. This process continues; the remaining uplift has been estimated at about 150–300 m (Andrews 1970, Walcott 1970).

West of Hudson Bay the Keewatin ice was soon reduced to a remnant centre northwest of Hudson Bay. The presence of this ice is recorded by eskers, ribbed moraine and crevasse fillings as well as by the fact that where it was present the Late Wisconsinan till, even in low areas, is not covered by marine deposits (Dredge & Nixon 1992). The larger, eastern ice sheet remnant on the Quebec/Labrador plateau did not completely melt away before 6500 BP (H.E. Wright 1984), i.e. during the hypsithermal period when northern Europe already had experienced its Holocene climatic optimum. In the North American mountains, glacier retreat was interrupted by major readvances, one of which occurred about 8400 BP in the North Cascade Mountains in Washington State (Beget 1981).

Meanwhile, in the Great Lakes area, deglaciation of successively lower outlets had resulted in a series of lakes all below present-day lake level. Water level in Lake Michigan fell to ca. 61 m below present level, which is referred to as the **Chippewa Low Phase** (Buckley 1974). The low-level **Lake Stanley** formed in the Huron basin (Hough 1955), and in the Georgian Bay low-level **Lake Hough** developed (Lewis 1969). This phase started around 10,000 BP when the North Bay outlet became ice-free (Fig. 188o) and lasted over 4000 years. Since then, the further development of the Great Lakes has been entirely controlled by isostatic rebound and outlet erosion (Figs 188p and q).

Continued uplift gradually raised the northern outlet of the lakes. In this process, Lakes Superior, Huron and Michigan coalesced to form the **Nipissing Great Lakes** (Fig. 188r). At about 5500 BP, the transgression passed the present lake level. Finally, uplift raised the lakes to the level of the abandoned southern outlets, leading to the formation of a well-developed terrace about 7 m above present lake level. Both the outlet via Chicago and the St Clair outlet via Detroit were reactivated, whilst the northeastern outlet still existed. For a brief period, all three outlets were active simultaneously, until further uplift closed the northeastern pass. By about 5000 BP the rapidly downcutting St Clair outlet captured the entire drainage system and the upper Great Lakes had a single outlet from then on (Trenhaile 1990). At this stage, the so-called **Lake Algoma** formed a beach at 182 m. Finally, further incision of the St Clair outlet resulted in a fall in lake level to the present 177 m (Eschman & Karrow 1985). At about 2200 BP differential isostatic uplift again separated Lake Superior from lakes Michigan and Huron. Since then the Superior basin has drained via the St Mary's River to the other two lakes (Farrand & Drexler 1985).

As in Europe, the Late Holocene of North America is also characterised by renewed glacier growth (Burke & Birkeland 1984), although because of inconsistent dates, a coherent chronology of glacial readvances is not yet possible. However, there is general agreement concerning the ice advances during the Little Ice Age. For example, several glacier oscillations have been determined by tree-ring dating in Alaska. Wiles & Calkin (1994) identified three major glacier advances, one about 3600 BP, one at AD 600 and one during the Little Ice Age. In the Canadian Rockies, Luckman (1993) has established a tree-ring chronology for the last 900 years that indicates several periods of cold climate. The last two periods from 1690–1705 and 1810–1825 resulted in ice advances ca. 1700–1725 and 1825–1875. The earliest Little Ice Age ice advance apparently occurred between AD 1150 and 1350.

The climatic history in North America in the Holocene, as in Europe, has been largely reconstructed from vegetational and faunal records. A period of favourable conditions occurred relatively soon after deglaciation. This climatic optimum, or **hypsithermal**, as in Europe, occurred in different regions at different times (Pielou 1991). Isopoll maps have been used in attempts to reconstruct the changing vegetational conditions on a continental scale. Overall, the climatic development has been found to have been rather complex, and there are considerable regional variations.

In eastern North America, general warming after the ice sheet retreat reached a maximum at about 6000 BP. At that time mean July temperature in Chicago was about 2°C higher than at present, and annual precipitation was about 200 mm greater than today. Subsequently, a slight cooling occurred. In the northeastern United States after 9000 BP, summer temperatures fell but moisture increased, leading to the replacement of *Pinus* by *Betula* and *Fagus* (Webb III *et al.* 1993).

In the western plains, prairie began to replace the earlier pioneer forest as early as 11,000 BP, whilst further east a brief period of deciduous forest growth preceded the final dominance of prairie, which started about 9500 BP (Wright 1984). Antevs (1948, 1952) proposed that the Holocene climatic development of the interior West of the United States could be subdivided into three phases:

Anathermal	cool + moist	9000–7000 BP
Altithermal	warm + dry	7000–4500 BP
Medithermal	cooler + moister	4500–present

This model is now invalid because the periods of maximum warmth and moisture are time-transgressive. The minimum effective moisture, for instance, occurred in the early Holocene in the northwest and in the Sierra Nevada, in the middle Holocene in the Great Basin and on the Colorado Plateau, and during the late Holocene in the southern deserts (Thompson *et al.* 1993).

14

Quaternary History of the Rivers

Since the early days of Quaternary research, the river terraces were considered to be the product of climatic fluctuations (Penck & Brückner 1901/09). Every cold phase was thought to correspond to one gravel aggradation and every warm phase to an incision. On the basis of this concept, Soergel (1924) established what he regarded as a 'complete subdivision of the Ice Age'. Thorough investigations of the terrace sequence of the Thuringian Ilm River led him to distinguish ten distinct cold phases separated by intervals of a warmer climate. An eleventh cold phase, the Pomeranian Phase of the Weichselian, was said not to be represented in the terrace sequence. After the climatic curve of the last 650,000 years had been computed on the basis of orbital variations and published the same year (Koeppen & Wegener 1924), Soergel believed he had found a complete conformity between his sequence of terraces and the climatic curve (Soergel 1925). However, serious doubts arose because the number of aggradational phases appeared too high. Indeed Woldstedt (1929) observed that there was no reason at that point to abandon the traditional three-fold subdivision of the glacial period as then established in northern Germany.

Later discoveries were to show that Soergel was wrong, but in a way different from Woldstedt's ideas. Today it is known that there have been 25 cold phases within the Quaternary since the end of the Olduvai Event (cf. Shackleton *et al.* 1990). During each of these, terrace aggradation would have been possible. The distinction and chronological classification of the individual terrace gravels, however, is far more difficult than Soergel had anticipated in his time. Today the well-investigated Middle and Lower Rhine sequence includes 12 to 15 Pleistocene terraces (Bibus 1980, Schirmer 1990); on most of the other European rivers fewer terraces have been so far identified (Table 11). The main Central European rivers are shown on Figure 190.

In North America, the palaeoclimatic interpretation of Quaternary river terraces is met with scepticism. Flint (1957) and Schumm (1965) have pointed out that influences other than those driven by climate can play a decisive role in terrace formation, and Morrison (1968) concluded that terraces are unreliable indicators of glacial–interglacial cycles. Similar conclusions are drawn by Flint (1976). However, because fluvial deposits often provide the only key to Quaternary history in the non-glaciated parts of North America, their study has continued, though with great caution. Fluvial successions hold vast amounts of evidence of palaeo-hydrological, climatic and chronological importance and therefore their study is promising, as long as oversimplistic interpretations are avoided. Examples of studies are given in Baker (1983) and Morrison (1991). One major example, the Mississippi River, will be discussed below.

TABLE 11 Stratigraphic subdivision of the Quaternary river deposits of the Lower Rhine, Thames and Mississippi. Both possible Quaternary/Tertiary boundaries are marked by arrows. Ages of palaeomagnetic reversals and correlation with deep-sea oxygen-isotope stages according to Shackleton *et al.* (1990). Stratigraphy after Gibbard (1985), Schirmer (1990) and Autin *et al.* (1991)

Magneto-stratigraphy	Age	Oxygen-Isotope Stages	Chronostrat.	Lower Rhine	Thames	Mississippi
Brunhes Epoch			Late Pleist.	Lower Terraces NT1-3 Eemian Warm Stage	Shepperton Gravel Kempton Park Gravel Ipswichian Warm Stage	Prairie Complex Sangamonian Warm Stage
Brunhes Epoch			Middle Pleistocene	Middle Terrace MT4 Middle Terrace MT3 Krefeld Warm Stage Middle Terrace MT2 Frimmersdorf Warm Stage Middle Terrace MT1 Main Terrace HT4 Ville Warm Stage Main Terrace HT3	Boyn Hill Gravel Lynch Hill Gravel Taplow Gravel brackish water deposits at Swanscombe Black Park Gravel Winter Hill Gravel Gerrards Cross Gravel Kesgrave Formation	Intermediate Complex
	780 000	19				
Matuyama Epoch			Early Pleistocene	Main Terrace HT2	Kesgrave Formation	Upland Complex
Jaramillo Event	900 000	27				
	1 070 000	31				
Matuyama Epoch			Early Pleistocene	Main Terrace HT1 Frechen Warm Stage III Gravel d Frechen Warm Stage II Gravel c Frechen Warm Stage I Gravel b2		Upland Complex
			←			
Olduvai Event	1 770 000	64				
	1 950 000	71				
Matuyama Epoch			Early Pleistocene	Fortuna Warm Stage Gravel b1		Upland Complex
	2 600 000	104	←			
Gauss Epoch			Tertiary			

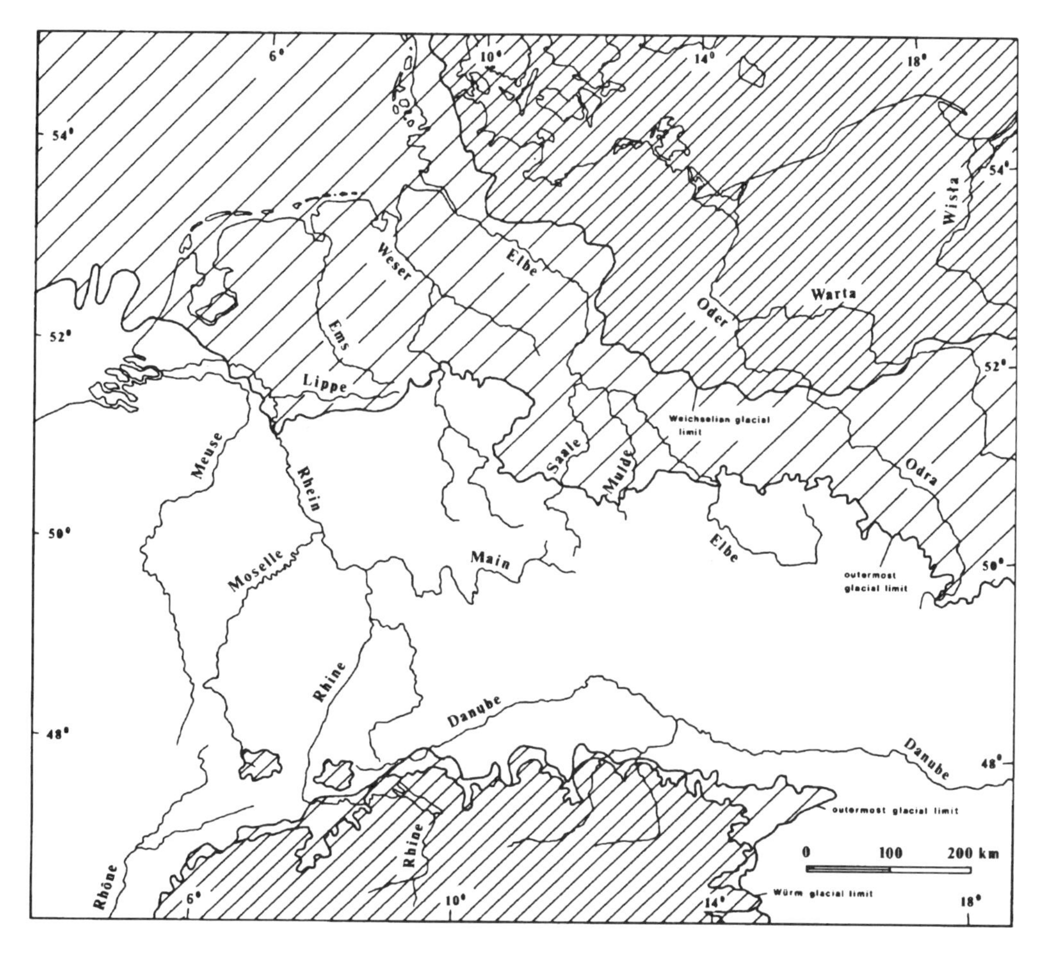

FIGURE 190 Map of the periglacial area between the Nordic and Alpine glaciations and location of the most important rivers

Each river has its own characteristics. Therefore in this chapter it is only possible to summarise a few selected river histories. The Danube is presented because it represents the largest drainage system of the Alpine glaciation, the Rhine as a river that drained both Alpine and North European glaciated areas, the Thames as the principal drainage system of the British Ice Sheet, the Siberian rivers because of their drainage reversals, and the Mississippi because it is the largest North American glacial drainage system, taking water both from the Cordilleran and the Laurentide ice sheets.

14.1 RIVER DANUBE (DONAU)

The present drainage of the Danube system is directed southeastwards towards the Black Sea. In the Late Tertiary, during deposition of the upper freshwater molasse in what is now the south German Alpine foothills region, drainage towards the west still prevailed (Füchtbauer 1967). Only in the Pliocene, with the uplift of the Alps and their foothills, was this drainage direction reversed, and the rivers began to flow eastwards. The Danube was formed at this stage (Lemcke *et al.* 1953). The incision by the newly formed Danube resulted in large-scale erosion of the upper freshwater molasse sediments (Mackenbach 1984, Tillmanns 1984).

In the Pliocene and Early Pleistocene the Danube catchment extended much further north than it does today; the River Main was a tributary at this time. Therefore the ancient Danube was supplied not only with Alpine gravel but also with quantities of ancient Palaeozoic rocks such as lydites from the northeastern Bavarian basement complex (Fig. 191). Whether the Aare River in the southwest (Hantke 1979, Villinger 1986) and the Alpine Rhine (Alpenrhein) (Villinger 1986, 1989) also drained into the Danube or not is not completely settled. The river history of the Danube can be only partially reconstructed, for only minor fragments of the older

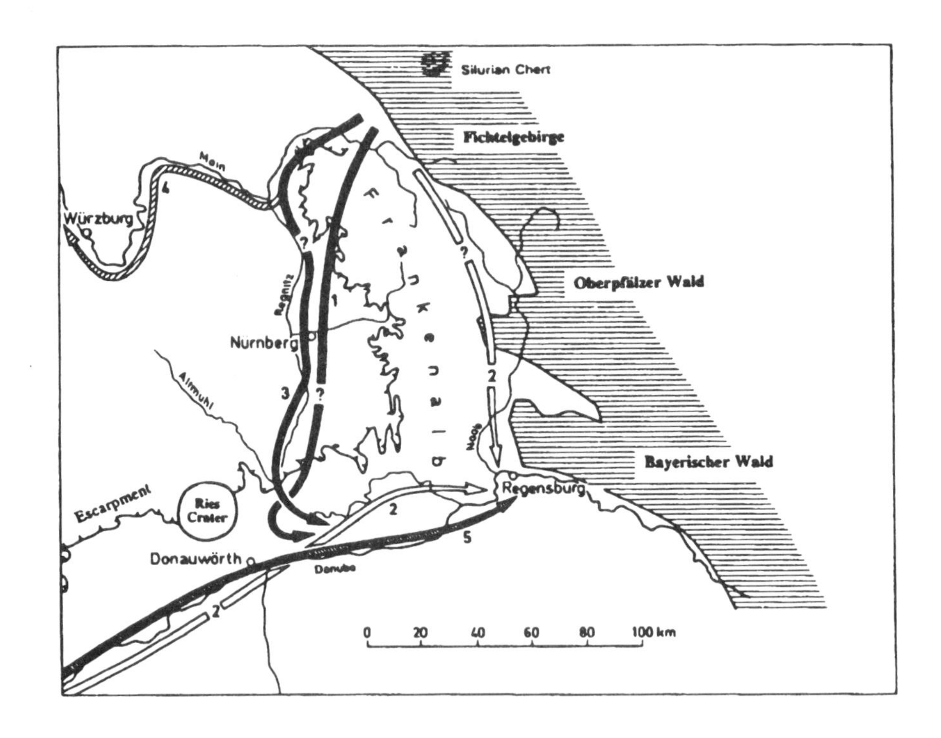

FIGURE 191 Fluvial development of the Danube River in Bavaria; 1 = high-altitude gravels of Pliocene proto-Main River; 2 = high-altitude gravels of Pliocene proto-Naab and Danube Rivers; 3 = valley terrace gravels of Early Pleistocene Main River; 4 = Main River since Early Pleistocene; 5 = Danube River since Riss Cold Stage (after Tillmanns 1980 and personal communication)

terraces have been preserved. The most comprehensive subdivision of the Danube terrace deposits has been established in the Vienna area, where it is possible to distinguish seven or eight cold-stage gravel aggradations. The direct correlation with corresponding Alpine glaciations, however, is still problematic (Fink & Majdan 1954, Fink 1960, Fink & Piffl 1975).

On the basis of the occurrences of 'non-Alpine' gravel from Swabian Alp (Schwäbische Alb) and Black Forest (Schwarzwald) source areas, an Early Pleistocene course of the Danube could be reconstructed that runs in part over 15 km south of the present Danube course. Scheuenpflug (1970) first discovered relics of this river course in the central Zusamplatte area and it has subsequently been attributed to the Alpine Biber Cold Stage (Jerz & Doppler 1990). In 1976 Scheuenpflug detected a second, 'Pliocene Danube', even further south. At the same time, Löscher (1976) presented a map of the Early Pleistocene river based on numerous geological sections. The gravel of this early Danube contains a high proportion of White Jura clasts. Alpine clasts are generally absent from the coarse fractions but do occur in the finer fractions (Scheuenpflug 1970). The gravels directly overlie the molasse and are, in turn, overlain by Alpine gravels (Löscher 1976). The existence of this ancient Danube course has been questioned on various grounds. For example, Schaefer (1980) pointed out that the gradient of the supposed river was inconsistent with the relief, whilst Aktas (1987) explained the occurrence of the White Jura clasts by reworking of the so-called 'Brockenhorizont' (clast horizon) of the molasse. The 'Brockenhorizont' consists of the ejecta from the Nördlinger Ries crater that formed by the meteorite impact at approximately 14.7 million years BP. Since the paper by Reuter (1925) on the dispersion of big ejected blocks (the so-called 'Reuterschen Blöcke') their distribution area has been known to include the Zusamplatte. The bedding directions measured from numerous exposures show a northeasterly palaeodrainage direction (Aktas 1987). However, the measurements do not contradict Scheuenpflug's interpretation. Although the question is not yet settled, most workers tend to accept Scheuenpflug's concept (e.g. Schreiner 1992, Jerz 1993).

The oldest Danube deposits along the present course of the river are found in the Kelheim–Regensburg region. The clast assemblage of these Höhenhofer Gravels (Tillmanns 1977) and related high-altitude gravels records an input from the Moldanubican and Saxothuringican source areas (lydites, amongst others) as well as from the Alpine region (radiolarites). The base of the gravels occurs between 45 to 120 m above the modern valley floor. According to Tillmanns (1977, 1984) their stratigraphic position can be established by correlation via the Altmühl–Danube valley to the Swabian Danube and further to the glaciofluvial gravels of the Iller–Lech Platte region. The base of the Early Pleistocene high-altitude gravels, correlated with the Biber Stage, is 70 m above the Danube valley floor. The latter gravels predate deposition of the 'Untere Deckschotter' (see chapter 12.5), the base of which lies at about 45 m above the valley floor. The 'Untere Deckschotter' were deposited before the Matuyama/Brunhes palaeomagnetic boundary, and most likely even before the Jaramillo Event (Tillmanns *et al.* 1986, Münzing & Aktas 1987, Ellwanger *et al.* 1995).

According to Villinger (1989), at around the end of the Donau Cold Stage, the Alpine Rhine abandoned its earlier course, which was directed towards Ehingen and the Danube, and found access to the Upper Rhine valley (Fig. 192). During the course of the subsequent glaciations, the glaciers advanced increasingly further north along this former river course, eroding the Schussen–Federsee basin, which was more than 200 m deep and has since been refilled. Thus the Danube became a type of northern Alpine ice-marginal stream, collecting meltwater from a large part of the Alpine glaciated area. By the Riss Glaciation, the Rhine Glacier reached its

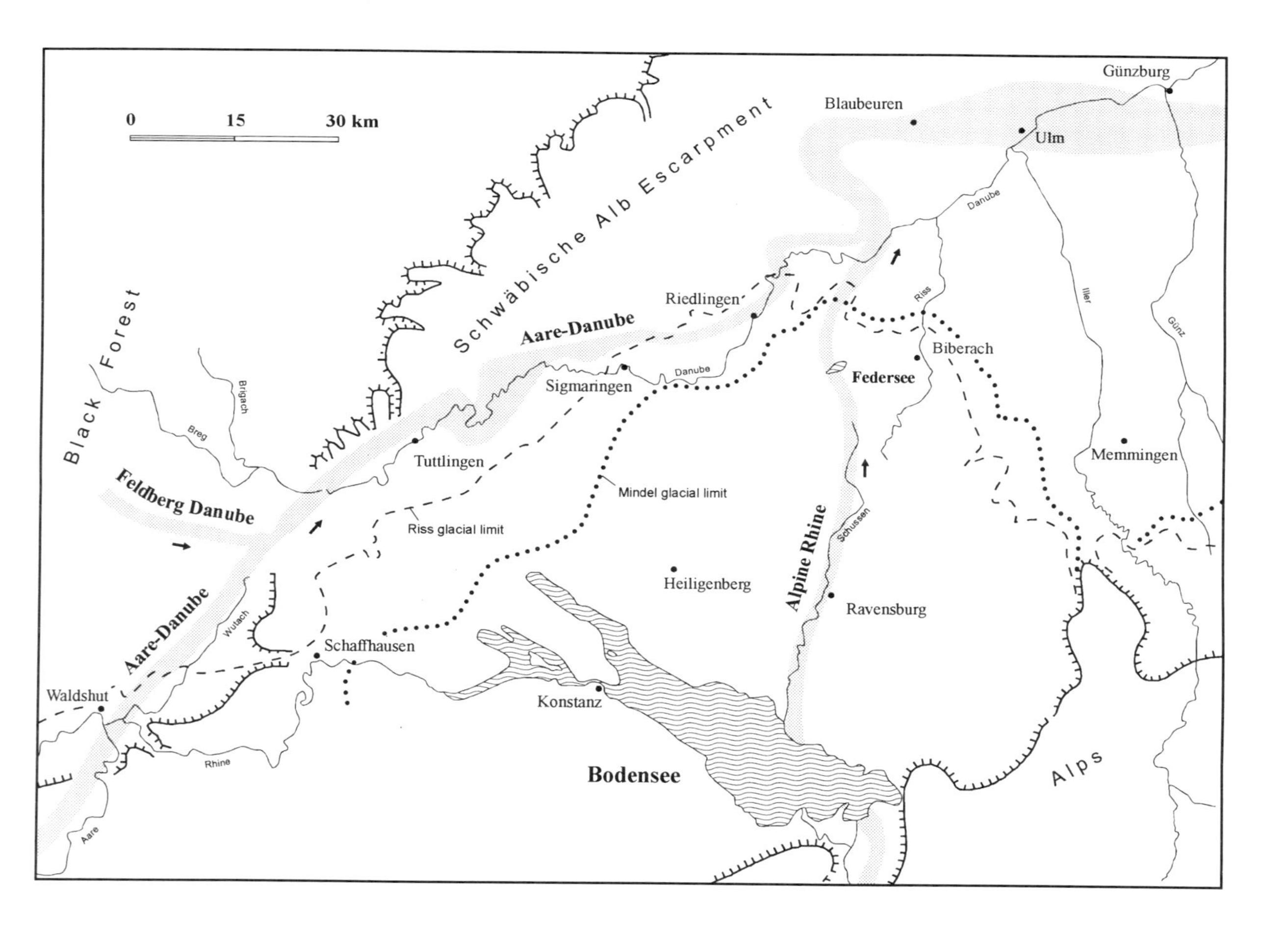

FIGURE 192 Courses of the Aare–Danube and Alpine Rhine rivers in the Upper Mio-/Pliocene and maximum extents of Mindel and Riss glaciations in the Rhine Glacier area (after Villinger 1986 and personal communication)

greatest extent and blocked the Danube in the Sigmaringen–Riedlingen area, forcing it to change course. Shortly before the maximum of the Middle Riss Glaciation it eventually occupied its current valley along the southern edge of the Schwäbische Alb ridge (Schädel & Werner 1965, Villinger 1985, 1986).

The narrow valley of the Danube between Stepperg and Neuburg was also only occupied by the river since the Riss Glaciation. In the last century, Von Gümbel (1889) was able to prove that the Danube originally flowed about 30 km further north, through what is now the Wellheim dry valley and the lower Altmühl valley. The diversion of the Danube took place in two stages: capture by the Schutter stream caused the Danube course to shift first into the Schutter valley (drainage via Rennertshofen and Hütting). Only after a second capture by the 'Neuburg River' occurred at the end of the Riss Glaciation was it diverted into its present course (Tillmanns 1977).

The geological structure of the more recent Danube terrace deposits is relatively well-known today. Because of a need to plan groundwater exploration and water resources, the Danube valley between Ulm and Passau in Germany has been thoroughly explored geologically and geophysically (Homilius *et al.* 1983). Similar detailed investigations have been carried out in Austria by Piffl (1971) and Fink & Piffl (1975). However, apprehensions concerning the stratigraphical interpretation of the deposits and the morphogenetical interpretation of the underlying mechanisms controlling the morphodynamics of the river still differ greatly between workers.

Recently Schellmann (1990) distinguished three Lower Terraces, that are all 'loess-free', in his morphological investigation of the Danube between Regensburg and Passau, and the Isar between Landshut and its confluence with the Danube. These three lower terraces are similar to those found by Schirmer (1983b) in the Main and Regnitz valleys. Schellmann assigned the highest Lower Terrace to the Würmian glacial maximum and concluded that the Danube river bed morphodynamics are directly controlled by the climatic history of the Quaternary. Whilst his Lower Terrace (NT1) was formed during the Würmian maximum, he attributes the next younger Lower Terrace (NT2) to the Oldest Dryas, and the (NT3) to the transition period between Older Dryas, Allerød and Younger Dryas. The subsequent younger terraces H1 to H7 are dated by archaeology and by radiocarbon analysis (Schellmann 1990).

Direct linkeage of the Danube Würmian Stage terraces with climatic events has been questioned by others. Whereas Schirmer (1983b) correlated the oldest of the three Würmian terraces on the upper Main and on the Regnitz River (the Reundorfer Terrasse) with the Alpine Würmian glacial maximum, Buch (1988), on the basis of investigations in the Regensburg–Straubing region, concluded that the aggradation of these terrace gravels had occurred before the Würmian maximum. During mapping of the Passau sheet, Bauberger & Unger (1984) found 'genuine, highly carbonate-bearing loess' resting on the gravels of the Lower Terrace. Moreover, on the Landau sheet Unger (1983) had also found the two levels of the Lower Terrace gravels buried by a Late Würmian loess. Loess up to 4 m thick is also found on the Lower Terrace of the Iller River, for instance (Fellheimer Feld and Rothtal Niederterrasse; Brunnacker 1953). Here the loess overlies Late Würmian outwash gravels, which were deposited before the Iller River changed its course during the glacial maximum (Doppler, written communication; cf. Habbe 1986a, b). Overlying loess spreads on the older Lower Terrace levels are also found, for example, in the northern foreland of the Harz Mountains (Ricken 1983), on the River Enz (northwest of Stuttgart; Bibus 1989) and in the Saale–Mulde area (Hiller *et al.* 1991).

According to Buch (1988) the entire Lower Terrace gravel unit should be considered as a continuous accretion from the base to the top. If the oldest Lower Terrace is older than the Würmian maximum, then this applies to the entire gravel body. Buch's dating is based on his investigations of the loess cover. This 4-m-thick aeolian deposit, which is only subdivided by weakly developed fossil soil, contains a full-glacial molluscan fauna at its base. Since no equivalent of the prominent Middle Würmian 'Lohner Boden' palaeosol (Stillfried B) has been found, Buch (1988) concluded that aggradation of the underlying terrace gravel had begun in the Early Würmian and continued until the Würmian glacial maximum.

Based mainly on the grounds of their divergent interpretation of the loess-covered oldest 'Lower Terrace', Buch and Schellmann also reach a different interpretation of the course of events. Buch (1988) concludes that lateglacial and Holocene terrace formation on the Danube was not only determined by the development of the Alpine glaciation but mostly controlled by developments within the Danube system itself. The alternation of wide basins and narrow gorges in the river course results in the operation of a specific dynamics for each respective river section. This view has also been proposed in a similar way by Fink (1977) and Kohl (1978) for the Austrian section of the Danube.

14.2 RHINE (RHEIN) AND TRIBUTARIES

To an even greater degree than the Danube, the River Rhine can be subdivided into very diverse sections the development of which has been affected by utterly different controlling parameters. The Rhine has carried meltwaters from the Alpine glaciations as well as those from the northern European ice sheets. Both glaciations forced diversions of the river course. During the Donau Glaciation, the Alpenrhein (Alpine Rhine) joined the Upper Rhine; previously it had drained into the Danube. During the Alpine glaciations, the Lake Constance (Bodensee) basin was carved out, consequently preventing any further fluvial transport of Alpine gravel from the upstream area beyond this sink. On the other hand, for a short period the Lower Rhine in the German/Dutch border region was diverted by the North European Saalian Glaciation (see below).

The development of the Rhine has been comprehensively described by many researchers. The history of the Middle Rhine has been investigated in detail by Bibus (1980) and Hoselmann (1994). Comprehensive summaries of the whole fluvial history have been recently presented by Boenigk (1990), Schirmer (1990) and Hantke (1993). The following paragraphs are based on these syntheses.

The oldest discernible river system draining from south of the Rheinisches Schiefergebirge (Rhenisch Slate Mountains) towards the present Lower Rhine and Meuse dates from the Late Eocene/Early Oligocene. This drainage system deposited the Vallendar Gravels, remnants of which are found in high positions in the Slate Mountains. At this time the river originated in the Vosges and the Saar region. Fluvial drainage along the tectonic axis of the modern Rhine from south to north can be demonstrated from sediments in the Lower Rhine Embayment no earlier than the Middle Miocene. At that time the sea gradually retreated from North Germany, giving way to deposition of brown coal up to 135 m thick. Simultaneously, fluvial channels up to 10 m deep were carved out in the Lower Rhine region, with a sediment filling with a heavy-mineral assemblage indicating derivation from middle Upper Rhine source areas. The headwaters of the Rhine at this time were in the area around the Kaiserstuhl Mountain (Boenigk 1987); the regions further south were drained through the Belfort Gap into the Rhône catchment.

During the Upper Miocene and Pliocene the first unambiguously identifiable Rhine sediments were deposited along the Middle Rhine, the so-called Kieseloolite Gravels (E. Kaiser 1903) (Fig. 193). On the basis of their content of fossiliferous and petrographically distinct 'Kieseloolites' from the Muschelkalk and the Jurassic of Lorraine, it must be assumed that the main drainage of the Rhine at that time went via the Moselle, possibly with a tributary existing in what is the modern Rhine valley (Boenigk 1981). Uplift in the area of the Rhenish Slate Mountains during the accretion of these gravels led to the formation of a first flight of terraces with three aggradational levels. In the Lower Rhine Embayment, however, subsistence continued causing stacked terrace deposits to accumulate. Boenigk (1982) distinguishes two to three Kieseloolite accumulations.

The catchment area of the Rhine expanded to the south until it reached the Alpine foothills in the latest Pliocene. This seems to have been the consequence of uplift of the Swiss Jura Mountains and the Molasse basin. This widening of the catchment area is reflected in the lithological composition of the Upper Rhine gravels (Bartz 1976). It also caused a marked change in the heavy-mineral assemblage of the Rhine sediments. Whereas older deposits were characterised by stable minerals, this change is shown by a significant increase in unstable heavy minerals (titanite, epidote and green hornblende) (Boenigk 1982). At that time the Rhine became the pre-eminent stream and the Moselle a tributary.

Development in the individual sections of the Rhine system differed very distinctly. Terrace development in the uplifted Middle Rhine area differed from that in the subsiding Lower Rhine region. Whereas a rough stratigraphical correspondence can be established between the Middle Rhine (Rhenish Slate Mountains; Bingen–Bonn) and Lower Rhine regions (between Bonn and the Rhine estuary), correlation with the Upper Rhine causes great difficulties. The fluvial history of the Upper Rhine (Basel–Bingen) was controlled by tectonic activity in the Upper Rhine Graben and Mainz Basin. The alternating deposition of coarse gravels and fine-grained sediments in this section of the river partly results from tectonic movements. On the basis of borehole evidence, several gravel units can be distinguished, the youngest of which mostly accumulated during the last cold stage (Würmian/Weichselian). However, in the lower part of this gravel a rich interglacial fauna has been found that is assigned to the Eemian Interglacial (Von Koenigswald 1988) (Fig. 194). The top of the last interglacial deposits lies at a depth of about 25 m below the valley floor between Karlsruhe and Worms (Löscher 1988). Gravel bodies more than 100 m thick in the eastern part of the graben largely reflect cold-stage deposition and must be attributed to the Alpine glaciations. Here they are underlain by sandy and silty sediments of Early Pleistocene age. Thus far, no detailed stratigraphy has been established for this area. As a rule, along the Upper Rhine subsidence has prevailed since the Tertiary. There are some parts like the most northerly part of the Mainz basin (Kandler 1970, Semmel 1969b), or parts of the Upper Rhine Graben (Bartz 1976, Boenigk 1987, Schreiner 1981), that were not involved in the subsidence, thus an incomplete flight of terraces has developed in this area.

North of the Slate Mountains, the rivers Rhine and Meuse today run almost parallel, 30–50 km apart. During the Early Pleistocene, uplift of the Slate Mountains and subsidence of the Lower Rhine Embayment continued. Initially (until the Tiglian Stage), north of the Slate Mountains the influence of the Rhine dominated; the corresponding deposits of the River Meuse being restricted to the westernmost part of the Lower Rhine Embayment (Fig. 195). Afterwards, however, tectonic changes caused the influence of the River Meuse to be extended far eastwards, so that eventually its flint-rich gravel can be traced in the Mönchengladbach–Grevenbroich area (Boenigk 1978). Apparently by the Bavel Interglacial, the Rhine had again

FIGURE 193 Pliocene 'Kieseloolithschotter' of a pre-Rhine river (Karmelenberg, Lonninger Höhe site) (Photograph: Ehlers 1990)

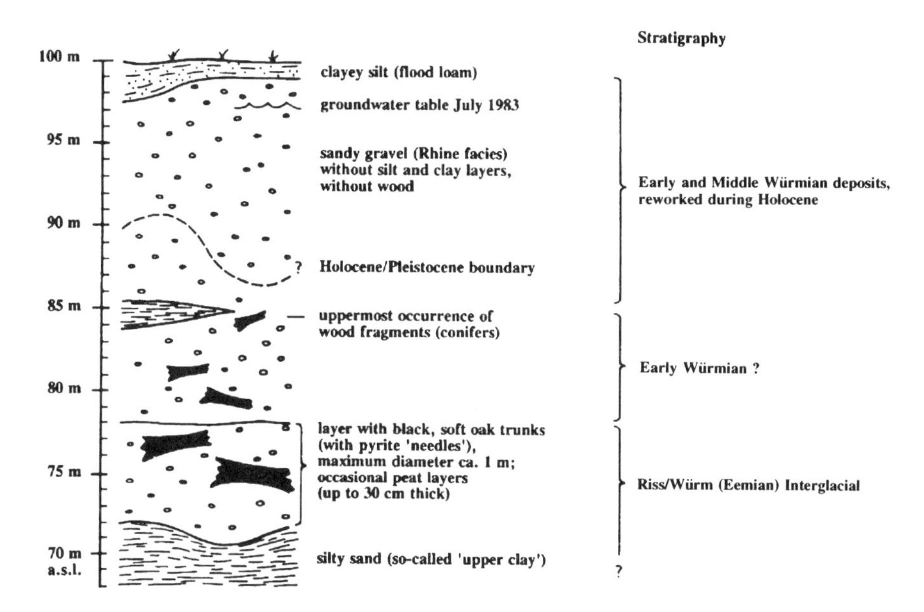

FIGURE 194 Philipp gravel pit in the floodplain of the Rhine, ca. 1 km south of Huttenheim. Gravels of the last cold stage are underlain by gravelly deposits containing numerous oak trunks and mammal remains of the last interglacial (from Löscher 1988)

expelled the River Meuse from the central part of the Lower Rhine Embayment and deposited an enormous gravel body, the so-called 'Hauptterrassenfolge' (Main Terrace sequence). This consists of four individual terraces. Simultaneously a flight of terraces formed in the Rhenish Slate Mountains, spanning a difference in altitude of 100 to 150 m. An earlier, largely morphological distinction into six 'Hauptterrassen' (Main Terraces, t_{R1}–t_{R6}) (Bibus & Semmel 1977) is problematic in a tectonically active area. New detailed mapping and petrographical analyses allowed Hoselmann (1994) to identify four 'Hauptterrassen' complexes, each characterised by a distinct petrographical composition and each sloping unidirectionally downvalley. Direct correlation with the sedimentary accumulations that typify the Lower Rhine is not yet possible. Overall the flight of terraces is less well preserved than the fluvial sedimentary sequence. The sediment body of the 'Hauptterrassen' is characterised by the dominance of epidote in heavy-mineral assemblages. The younger 'Hauptterrasse' of the Lower Rhine Embayment (HT2) can be traced morphologically into the Netherlands, where it merges into the 'Weert Zone' of the Sterksel Formation of the Dutch lithostratigraphic subdivision (Zagwijn 1985, Boenigk 1990). This heavy-mineral zone is of 'Cromerian Complex' age.

According to Bibus (1980) the so-called 'Mittelterrassen' (Middle Terraces) accumulated in the period from 'Cromerian Glacial C' until the end of the Saalian Stage. On the basis of morphology, four terraces are distinguished. According to Bibus (written communication), they represent the interval from the fifth-but-last cold stage until the end of the Saalian. By contrast Hoselmann (1994) considers the 'Mittelterrassen' might well be considerably younger. Petrographically, the 'Mittelterrassen' can be easily distinguished from the 'Hauptterrassen' deposits, because with the onset of volcanism in the Eifel they contain a very distinct mineral assemblage (Boenigk 1990). The stratigraphical subdivision of the 'Mittelterrassen' is based on

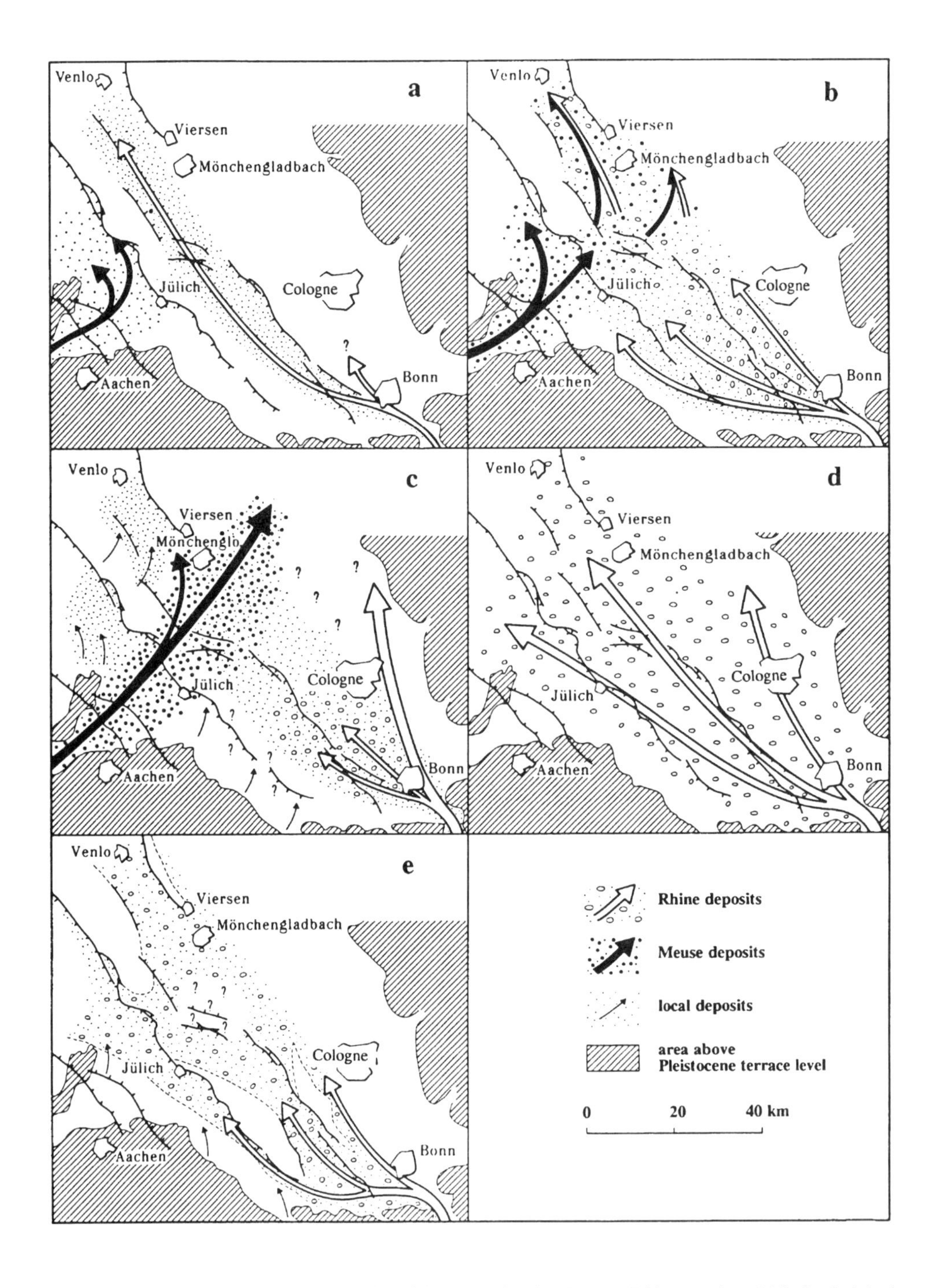

FIGURE 195 Sedimentation in the Early Pleistocene in the Lower Rhine region ('Niederrheinische Bucht') (from Boenigk 1978)

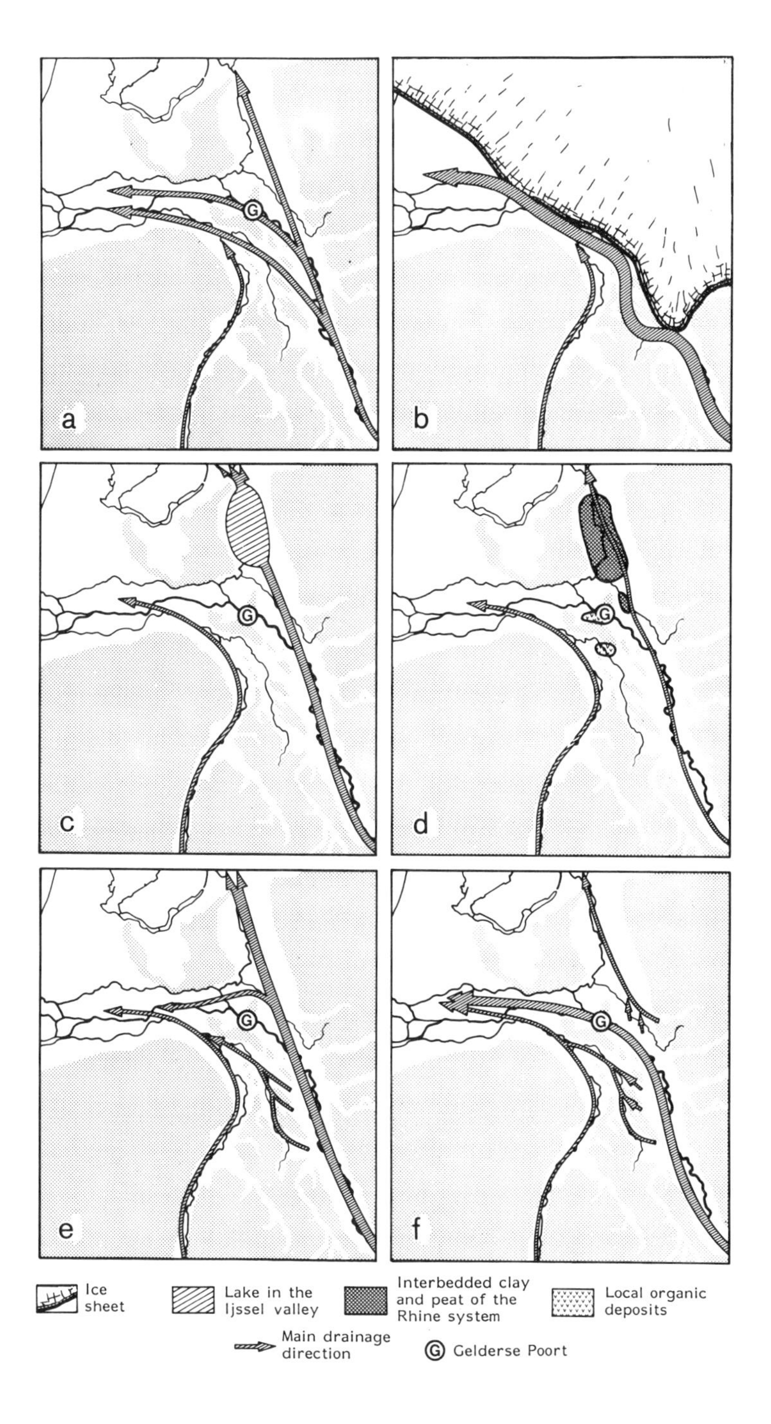

FIGURE 196 Palaeogeographic development of the Rhine and Meuse rivers since the Middle Saalian (from Van de Meene & Zagwijn 1978)

their overlying loess cover and its contained palaeosols. Moreover, the lower 'Mittelterrasse' (t_{R9}) makes part of the Saalian and the middle 'Mittelterrasse' (t_{R8}) of the fourth-but-last cold stage (Bibus 1980).

The Nordic Ice Sheet advanced twice or possibly even three times as far as the Lower Rhine. Unfortunately, the stratigraphical connection between the glacigenic deposits and the Rhine terraces is not yet entirely clear (Schirmer 1990), and the interpretation of the sedimentary sequence is, in part, disputed. It seems certain that two Saalian ice advances reached the Lower Rhine; both occurring during the Older Saalian Substage of the northwest German classification (Klostermann 1992).

During the Older Saalian ice advance, the Lower Rhine was diverted (Fig. 196). Whilst the glaciers advanced to Düsseldorf, the Rhine drained south of the Reichswald through the Nierstal valley. After the melting of the ice sheet, the Rhine changed its course to the north into the tongue basin along the Ijssel valley and curved in a northwesterly direction near Zwolle in the Netherlands. Initially a lake formed in the overdeepened tongue basin about 50 km long (Jelgersma & Breeuwer 1975). This lake was completely infilled with sediments before the end of the Saalian Stage; yet the Rhine retained its course during the Eemian and the Early Weichselian Substage. Only when this outlet was almost filled with sediments did a gradual shift of the drainage line towards the west begin. At first the Rhine made its way north of Montferland through the Liemers towards the west. Shortly afterwards the Rhine finally adopted its present course through the Gelderse Poort. This ultimate diversion only took place after the Early Weichselian Brørup Interstadial, as demonstrated by finds of peat in the Gelderse Poort area. Both the Nierstal valley and particularly the Ijssel valley continued to take part of the Rhine drainage during major floods in the Late Weichselian. The modern upper Ijssel valley, i.e. the section between Westervoort and Deventer, was formed only between 1000 BC and AD 0 (Van de Meene & Zagwijn 1978).

During the Weichselian Stage, additional terrace bodies accumulated along the Middle and Lower Rhine. The term 'Niederterrasse' (Lower Terrace) for the youngest set of Rhine terraces was introduced by E. Kaiser (1903). Later W. Ahrens (1927) discovered that the Lower Terrace was bipartite; he distinguished an Older Lower Terrace that lacked the Laacher See tephra and a Younger Lower Terrace that contained the tephra. Since the eruption of the Laacher See volcano can be dated to the Allerød, there can be no doubt that the Younger Lower Terrace must have been formed during the Younger Dryas. Near Düsseldorf and in the Neuwied Basin, a three-fold subdivision of the Lower Terrace could be established. The additional terrace occurs intermediate between the other two. It was laid down before the Bølling, according to Schirmer (1990).

14.3 THAMES

Like the Danube and the Rhine, the River Thames has been directly affected by Pleistocene glaciation and forced to change its course (Figs 197 and 198). The history of the Middle and Lower Thames regions has been summarised by Gibbard (1985, 1994) and synthesised by Bridgland (1994). Overviews of the whole development of the river system have been presented by Gibbard (1988) and Gibbard & Allen (1994), and it is on these that the following summary is largely based.

In the late Pliocene and earliest Pleistocene, the London Basin was beneath the sea, having probably been submerged during the Miocene. Fossiliferous sands with a lithology and fauna

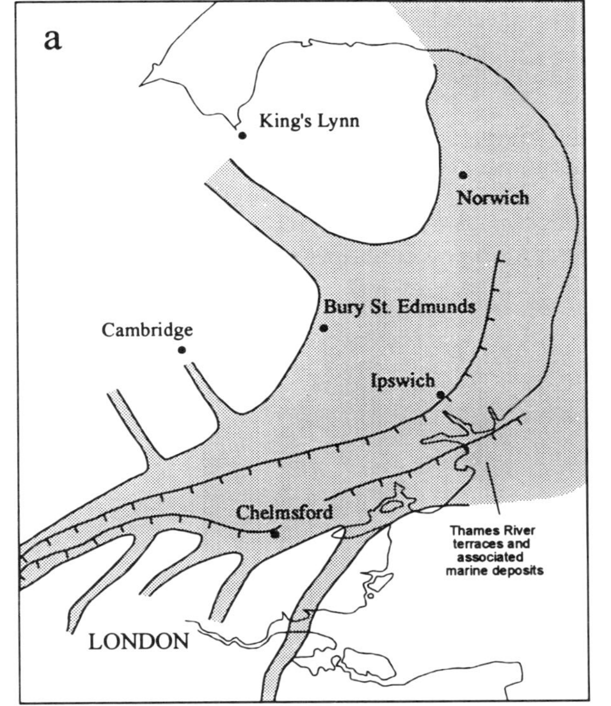

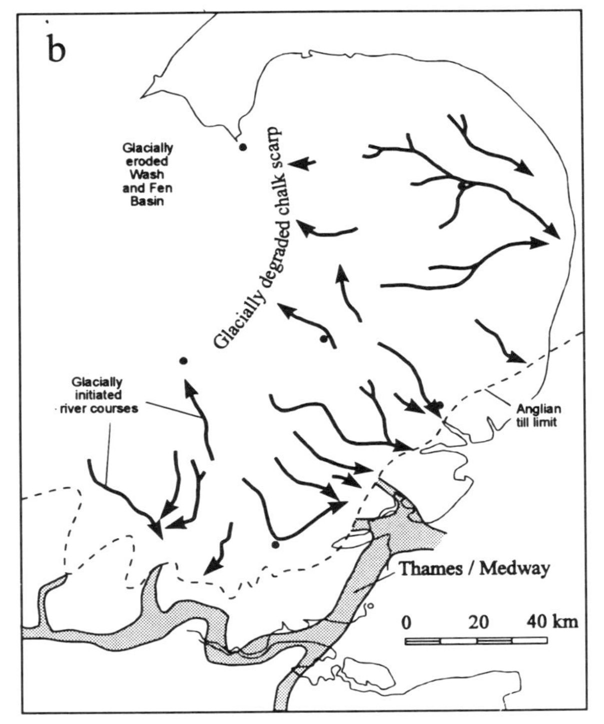

FIGURE 197 The landscape of eastern England (a) at the time of the Barham Soil and (b) after the Anglian glaciation (after Rose *et al.* 1985a, Ehlers and Gibbard 1991 and Gibbard, personal communication)

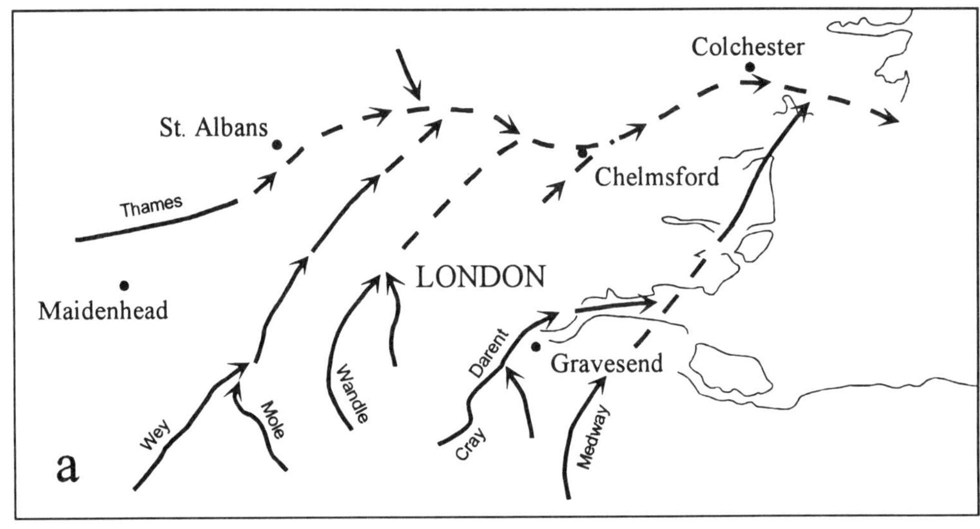

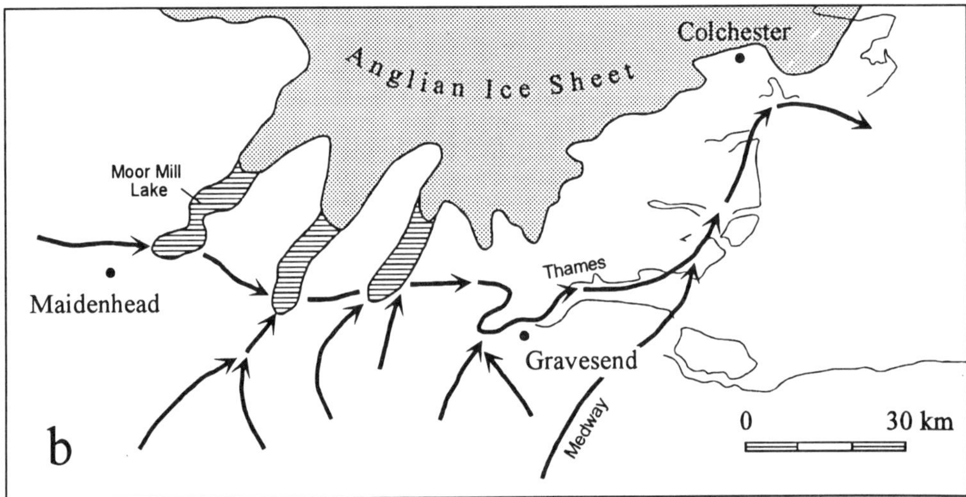

FIGURE 198 The Thames drainage system (a) before the Anglian glaciation and (b) during the maximum Anglian glaciation (from Ehlers & Gibbard 1991 and Gibbard, personal communication)

similar to that of the Red Crag Formation of Suffolk occur today at heights of up to 180 m on both the north and south sides of the basin (Chatwin 1927, Dines & Chatwin 1930). Contours of the sub-Red Crag surface indicate a relative tilting of ca. 180 m between the western London Basin and the Suffolk coast localities. Assuming a broadly uniform age (pre-Ludhamian = ?Praetiglian (Zalasiewicz *et al.* 1988)), either the eastern part of the basin must have subsided (West 1972; Moffat & Catt 1986), or the western part been uplifted in the Early Pleistocene. Although the courses of the river systems of this period are not known, it is thought that a southeastward-aligned Thames precusor and possibly other streams were entering the basin from surrounding regions (Gibbard, personal communication).

The subsequent marine regression was followed by the development of an early Thames drainage system. Remnants of this are the so-called Pebble Gravel Formation deposits, which occur at isolated localities 120–130 m a.s.l. on the South Hertfordshire Plateau, north of London. The gravels are composed predominantly of flint, much of which is reworked from Tertiary rocks. However, in some areas they contain a characteristic chert derived from the Lower Cretaceous rocks of the Weald. This indicates the existence of south-bank tributaries (Wooldridge & Linton 1955). The Pebble Gravels, the most prominent member of which was termed Nettlebed Gravel by Gibbard (1985), fall in height towards the northeast. Their precise age cannot at present be determined.

After a period of incision, a marked change in gravel composition is recorded in the next lowest unit, the Stoke Row Gravel (Gibbard 1985). This is dominated by a drastic increase in the quartz and quartzite clast frequency; this material apparently derived from Triassic conglomerates in the English West Midlands. The Baylham Common Gravel of Allen (1984) forms the downstream continuation of this unit. In southern East Anglia it forms the uppermost member of the Kesgrave Formation (or Group: Whiteman & Rose 1992). The marked lithological change may result from capture in the headwater region, although it is possible that it reflects input of meltwater from multiple glaciations in the North Welsh Snowdon and Berwyn districts (Whiteman 1983, Bowen *et al.* 1986). This is because the Kesgrave gravels contain ice-rafted blocks and smaller clasts of volcanic rocks derived from North Wales (Hey 1980, Green *et al.* 1980). In northern East Anglia, at Beeston Regis, the first influx of quartz-rich gravel was found in sediments of pre-Pastonian age (= Tiglian Substage C4: Gibbard *et al.* 1991) by Hey (1980, 1991).

After deposition of the Stoke Row Gravel, several additional units accumulated, each separated by an incision phase. The next youngest member; the Westland Green Gravel (Hey 1965, 1980) can be traced throughout the area and contains evidence of deposition by the river in a braided mode under a periglacial climate. The same sedimentary structures and lithologies occur in the younger members of these quartz- and quartzite-rich deposits. On the basis of increased flint content in younger members the Kesgrave unit has been subdivided into two subunits: the Sudbury (older) and Colchester (younger) formations by Whiteman (1992). Temperate fossil-bearing sediments of 'Cromerian Complex' age are intercalated in the younger members. The biostratigraphical evidence indicates that the Kesgrave Formation spans the period from the pre-Pastonian (Tiglian C4c) to the early Anglian (Elsterian) inclusive. The quartz-rich gravels are also found beneath the southern North Sea, where they form part of the Yarmouth Roads Formation (Balson & Cameron 1985, Bowen *et al.* 1986). At this time, therefore, the Thames was contributing to the greatly expanded deltas in the southern North Sea basin. A characteristic palaeosol, the Valley Farm Soil, formed on the terrace surfaces; it provides an important marker horizon (Rose *et al.* 1976, 1985b, Rose & Allen 1977, Kemp 1985, Rose 1994).

The advance of the continental ice sheet in the Anglian (Elsterian) Stage buried the Kesgrave Formation in all but the southernmost part of East Anglia. The ice sheet overrode the contemporary valley of the Thames in Hertfordshire and also advanced into the southern tributary valleys. This dammed the rivers and the resulting lakes progressively overspilled until the water reached the unglaciated Medway valley, via which the river drained into its earlier course in easternmost Essex (Gibbard 1977, 1979, 1985, Bridgland 1980, 1983, 1988). In the Anglian the supply of quartz-rich material ceased; the late Anglian Black Park Gravel and all subsequent units being poor in quartz in the Middle and Lower Thames (Gibbard 1985, 1994). The same is true of the contemporaneous gravels in the Upper Thames (Briggs & Gilbertson 1973, 1980).

The Black Park Gravel can be traced to southeast Essex where it passes into a delta probably formed where the river entered an ice-dammed lake in the southern North Sea basin (Gibbard 1994, 1995). All subsequent Thames gravel and sand deposits do not extend to the north beneath the sea but to the south and are aligned through the Dover Strait (Gibbard 1988, 1995, Bridgland & d'Olier 1995). This passage seems to have been formed initially in the Anglian/ Elsterian as a spillway for the ice-lake and has been subsequently repeatedly modified by both marine and fluvial activity.

In the Hoxnian (Holsteinian) Interglacial, the eustatic sea-level rise caused inundation of the Lower Thames. During the course of the transgression estuarine sands, silts and clays were deposited on freshwater organic sediments at Clacton and in substantial channel fillings in southeast Essex (West 1972, Bridgland 1980, 1983, 1988, Roe 1994). Upstream the brackish water extended to Swanscombe, east of London, where fossil- and Palaeolithic artefact-bearing fine-channel and overbank sediments are found. Further upstream the Hoxnian is poorly represented. The river seems to have established a meandering pattern at this time.

During the Wolstonian (Saalian) Stage a series of gravel and sand units were deposited in the Thames valley under periglacial conditions. In the Middle and Lower Thames three major units accumulated: the Boyn Hill, Lynch Hill and Taplow Gravels. In the Upper Thames they can be correlated with the Hanborough Terrace gravels, equivalent of the Boyn Hill and Lynch Hill members. Particularly in the Middle Thames region, the Boyn Hill and Lynch Hill gravels often contain considerable quantities of late Middle Acheulian artefacts (Wymer 1968, 1994, Gibbard 1985). The Upper Thames' Wolvercote Terrace gravels, the equivalent of the Taplow Gravels, contain non-local flint, thought to be derived from till at Moreton-in-Marsh (Bishop 1958, Briggs & Gilbertson 1980). Since the till was deposited by the Wolstonian glaciation of the Midlands, the gravels must either be contemporary with or post date the glacial event.

Recently there has been considerable discussion regarding the number of interglacial events represented in the Thames system, particularly in the Lower Thames area, east of London. Here there are numerous deposits, many of which include evidence of estuarine conditions at a higher level than present. Based particularly on the interpretation of amino-acid racemisation measurements, Bridgland (1988, 1994) proposed that the deposits, many of which were previously correlated with the Ipswichian (Eemian) Interglacial (e.g. by West 1972), represent additional post-Hoxnian and pre-Ipswichian events. By contrast Gibbard (1994) concluded that the sequences are indeed of Ipswichian age and represent the dissected remnants of the interglacial estuary. He correlates the sites including Trafalgar Square, Peckham, Ilford, West Thurrock and Grays and supports Hollin's (1977) view that the contemporary water level rose to ca. 10 m a.s.l. Resolution of these contrasting views must await further evidence, especially the validity and reliability of amino-acid geochronology.

The beginning of the Devensian (Weichselian) was marked by the eustatic sea-level regression that resulted in incision into the Ipswichian estuarine sediments. Again, gravel and sand were deposited under a periglacial climate. Intensified solution of Chalk bedrock during cold stages often associated with scour has considerably modified the gravel and sand accumulations leading to collapse or local thickening of sequences (Gibbard 1985). Fluvial deposits present represent three aggradational phases, in the Early, Middle and Late Devensian. The Early Devensian Summertown–Radley Terrace gravel and downstream equivalents are not well represented throughout the valley. By contrast, the Middle Devensian gravels and sands are well developed. In the Upper Thames they occur below the modern floodplain, but emerge downstream and are found above the valley floor. In the London area, organic deposits within the Kempton Park Gravel contain a herb-dominated flora indicating full-glacial to interstadial environments (ca. 45–30 ka). Further downcutting and aggradation of valley-bottom Shepperton Gravel and equivalents occurred during the Late Devensian. In the Middle and Upper Thames regions organic deposits intercalated in these gravels have been dated to ca. 15–10 ka, although slightly older dates have been obtained from the tributary Lea valley in northeast London. They also contain fossils of cold-climate, treeless shrub- and herb-dominated vegetation. The dates imply a period of downcutting and/or non-deposition in the Thames system that lasted from ca. 30 to 15 ka. During this interval, clayey silt, the so-called 'brickearth' (Langley Silt Complex), accumulated in the Thames Valley. This sediment, largely of local origin, contains a loess component and has been dated by thermoluminescence to a period about 17 ka BP (Gibbard *et al.* 1987).

In the Flandrian (Holocene), organic sediment, tufa, clay, silt and sand were deposited on the gravels and sands underlying the modern floodplain. Former depressions were infilled, and the river adopted a single meandering/anastomosing course. Since the Neolithic period, land clearance has resulted in accumulation of overbank silt and clay by flooding.

14.4 THE SIBERIAN RIVERS

The major glaciations of the Pleistocene also affected large parts of western Siberia. During their maximum extent the ice sheets blocked all drainage towards the north and considerably enlarged catchment areas, directing part of the meltwater discharge towards the Mediterranean via the Black Sea. A reconstruction has been attempted by Arkhipov *et al.* (1995); however, all reconstructions of the palaeodrainage in Siberia are hampered by the fact that the events are not well-dated.

During the maximum extent of the Pleistocene ice sheets, the drainage systems of both eastern Europe and western Siberia were altered dramatically. It is assumed that in Siberia as in Europe during the **Older Saalian** glacial maximum (Drenthe Substage), as well as during the Elsterian, the ice sheet was most extensive (Fig. 199). In western Siberia the Ob, Irtysh, Yenisei and Tunguska rivers were dammed, forming large proglacial lakes that drained to the south all the way to the Black Sea and the Mediterranean.

In western Siberia the **Tazovian (Warthe)** ice sheet reached far south and blocked the northward drainage of the Ob and Yenisei rivers (Arkhipov *et al.* 1986). Overflow to the Aral Sea is certain, but whether the waters reached the Mediterranean via the Caspian and Black Sea is not confirmed. Therefore the drainage pattern shown on the map (Fig. 200) is a minimal version. The maximal version would more closely approach the picture for the Pleistocene glacial maximum. However, major ice-dammed lakes in western Siberia would have been larger than during the Drenthe and they may have suffered higher losses by evaporation.

In the west Siberian lowlands east of the Ural Mountains, the extent of the **Weichselian** (Valdaian) glaciation is still unclear. In Russian publications different versions are given (Grosswald 1980, Arkhipov 1984, Arkhipov *et al.* 1986). Recent studies near the Arctic coast (Astakhov 1992) suggest that in the Late Weichselian the Kara Sea Ice Sheet, which originated on the Arctic shelf, sent a single ice stream southward onto the mainland, but it was not confluent with local glaciers from the Putorama upland to the east, so that the north-flowing rivers were not blocked. During the Early Weichselian, however, the Kara Sea Ice Sheet did extend well on to the mainland, forming arcuate glaciotectonic ridges and burying marine sediments and peat of the last interglacial (Eemian). In fact some of the Early Weichselian glacial ice still exists

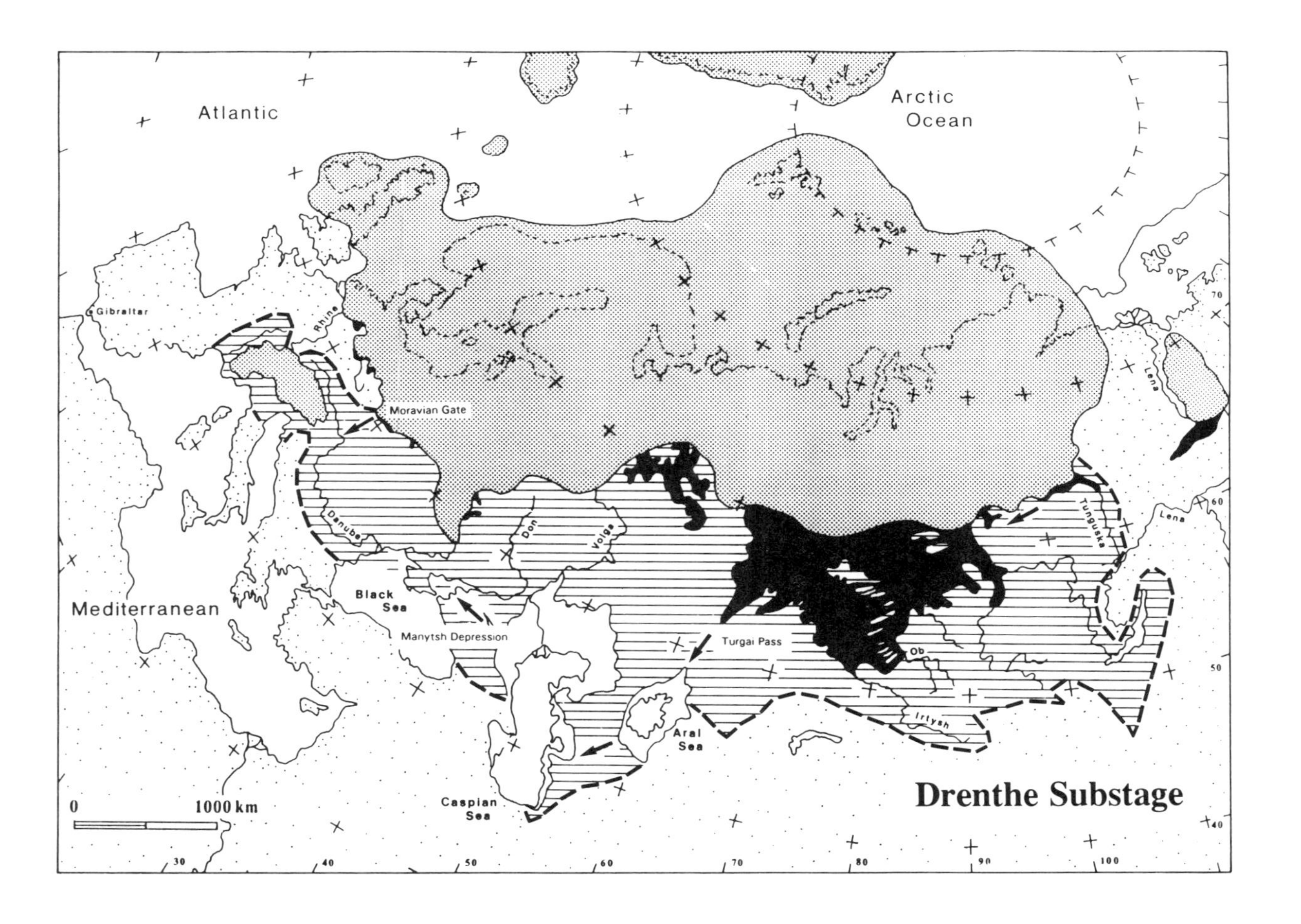

FIGURE 199 Ice sheets and drainage conditions during the Drenthe Substage of the Saalian. Ice sheets = grey, ice-dammed lakes = black, drainage towards Mediterranean = horizontal lines (after Grosswald 1980, Arkhipov 1984, Arkhipov *et al.* 1986, Astakhov 1992, Ehlers 1994)

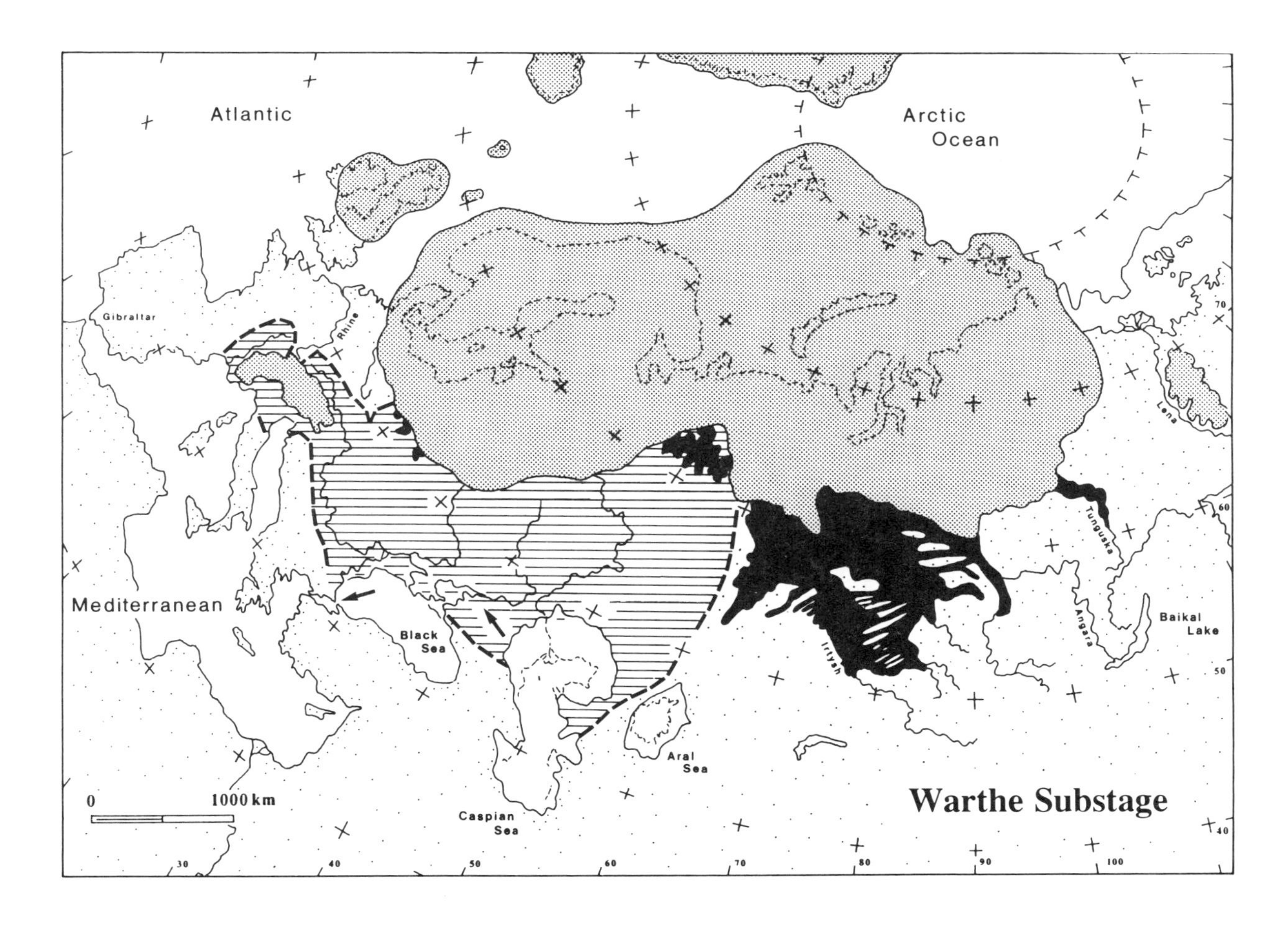

FIGURE 200 Ice sheets and drainage conditions during the Warthe Substage of the Saalian. Shading as Fig. 199 (from Grosswald 1980, Arkhipov 1984, Arkhipov *et al.* 1986, Astakhov 1992, Ehlers 1994)

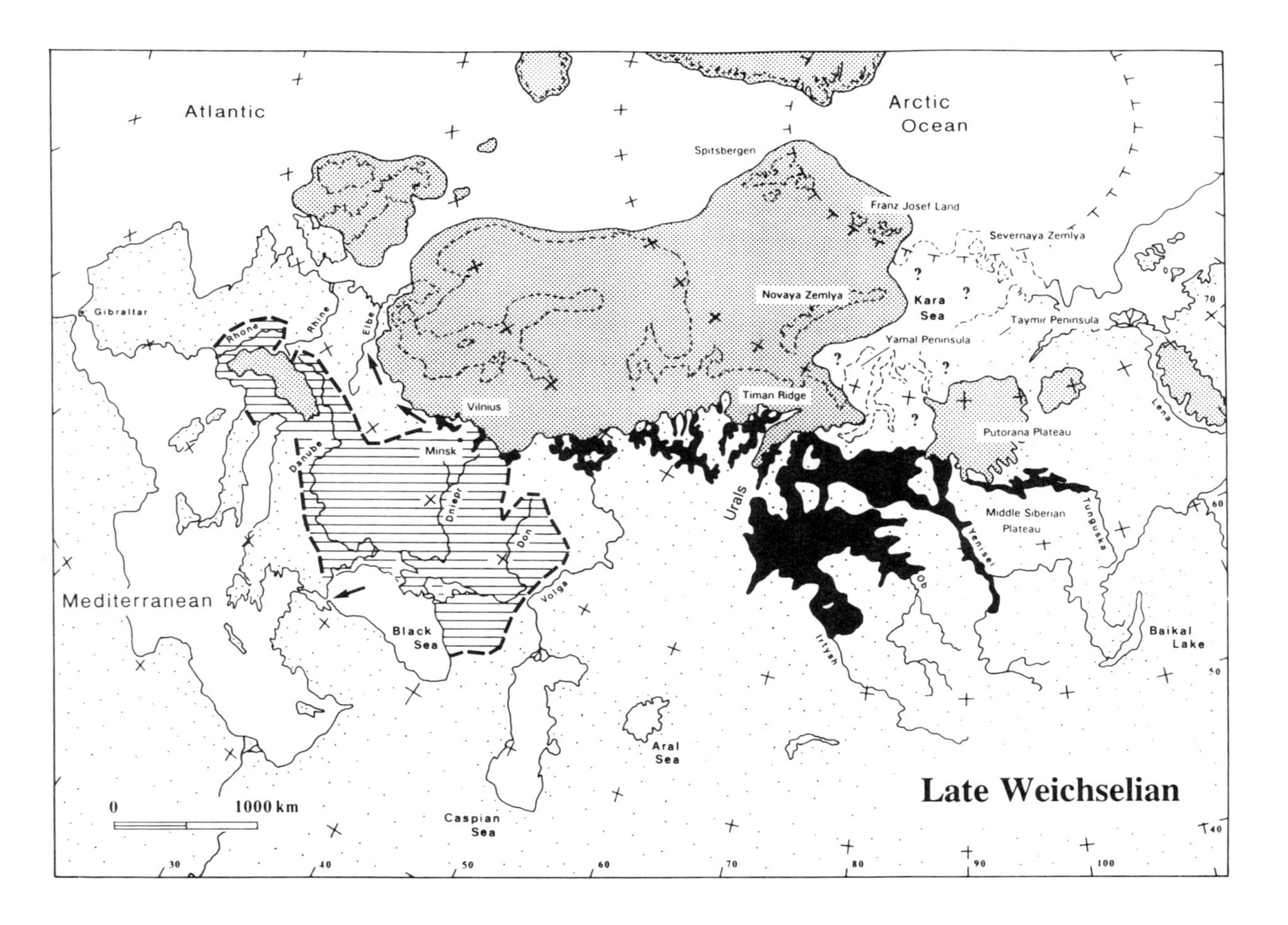

FIGURE 201 Ice sheets and drainage conditions during the Late Weichselian. Shading as Fig. 199 (after Grosswald 1980, Arkhipov 1984, Arkhipov *et al.* 1986, Astakhov 1992, Ehlers 1994)

beneath loess, radiocarbon-dated to the Late Weichselian, and is slowly wasting away in the permafrost climate that has prevailed ever since this ice advance. Because of the persistence of dead ice and permafrost the landscape has the fresh appearance that suggests a Late Weichselian age (Grosswald, 1980), but the stratigraphy and dating indicate otherwise.

Thus in the Early Weichselian the northward drainage of the Ob, Irtysh and possibly the Yenisei rivers were dammed (Fig. 201), forming proglacial lakes in the West Siberian lowlands, which are surrounded by mountains and uplands on the west, south and east (Ural Mountains, Kazakh upland, Altay, Sayan and Yenisei mountains, and Middle Siberian Plateau). The only possible outlet for such lakes is to the southwest through the Turgai Pass. The lake sediments in the lowlands consist of silt and clay with some interbedded sand. Much of the sediment is redeposited loess, and Astakhov (1992) believes that a major portion of the sediment is loess rather than lacustrine deposits. However, shoreline features first described by Volkov & Volkova (1975, 1979) and Volkov *et al.* (1978) indicate the existence of large-scale ice-dammed lakes. Terrace deposits at 70–80 m a.s.l. have been radiocarbon-dated as 22 to 12.3 ka BP. At this level no outflow would occur today through the Turgai Pass, which is 126 m a.s.l. This would pose a drainage problem if the lake level was 80 m and a northern outlet was blocked by an ice sheet (unless the lake level was kept lower by evaporation). It was pointed out by Astakhov (1992), however, that the Turgai Pass has been filled to its modern level by colluvium and other post-overflow sediments, and that the buried spillway beneath is at only 40 m a.s.l. Thus the pass could have served as the outlet for any lake with an elevation above 40 m. Features indicative of higher shorelines at 105–110 and 127–130 m were identified by Volkov and dated by thermoluminescence as 70 ±11 ka BP, with the lake sediments filling palaeovalleys of Eemian age (dated by TL as 130 ±31 ka BP). Such older lakes certainly must have drained southwards.

The level of the Caspian Sea in the Early Weichselian was initially 76 m above the present, high enough to overflow via the Maytsh Depression to the Black Sea, according to Chepalyga (1984), who attributes the high level to low regional temperatures and extensive permafrost, which in turn reduced evaporation. After erosion of the outlet the level was lowered, and it subsequently became a closed basin with moderate salinity, according to the fossil molluscan fauna (Chepalyga 1984). On the other hand, the high levels might also be attributed to influx of meltwater from the dammed lakes of the West Siberian lowlands by way of the Turgai Pass. In fact, oxygen-isotope values of −12 to −14.5, compared to about 0 today, imply the addition of glacial meltwaters (Chepalyga 1984).

Despite the temporary overflow of the Caspian Sea to the Black Sea, the latter did not overflow to the Mediterranean. The eustatic lowering of the Mediterranean sea level prevented the inflow of that water via the Bosporus to the Black Sea, which became isolated like the Caspian Sea today (Chepalyga 1984).

In summary, according to these reconstructions, in the Early Weichselian the ice-dammed lakes in the West Siberian lowlands overflowed via the Turgai Pass to the Aral and Caspian seas and temporarily to the Black Sea. However, the lack of any influx of Scandinavian meltwater at this time kept the Black Sea at a relatively low level and isolated it from the Mediterranean. In the Late Weichselian, on the other hand, the Siberian rivers were not dammed, and Scandinavian meltwater from the Russian plain, via the Volga River, was insufficient to fill the Caspian to overflow level. In this case the Black Sea received no water from the Caspian, and overflow to the Mediterranean depended entirely on the influx of meltwater from further west in Europe via the Dniepr and Danube rivers.

A comparison of the maps shows that the Scandinavian and West Siberian ice sheets caused major enlargements of the catchment area of the Mediterranean Sea at different times. The drainage system changes are not well-dated, however, and the various receiving lake basins involved apparently reacted differently to the climatic changes (for the Weichselian see Chepalyga 1984). During the Saalian Drenthe Substage the Mediterranean catchment reached its maximum size and may have even included Lake Baikal. However, during the subsequent Warthe Substage, a smaller area may have drained towards the Mediterranean via the Volga River and the Caspian Sea. Finally, during the Late Weichselian only the eastern part of the Scandinavian Ice Sheet provided meltwater that reached the Black Sea and the Mediterranean.

14.5 MISSISSIPPI

In North America the impact of Pleistocene glaciations on the fluvial system differed greatly from that in Europe. The North American main water divide between the southward-directed Mississippi River system and the northward and northeastward-flowing rivers runs in a west–east direction across the continent. Therefore almost from the start the growing ice sheets were moving up a gently rising surface. The original drainage was blocked and large ice-dammed lakes formed. Finally, when the ice sheet advanced towards the original water divide, the ponded water overspilled and found a new outlet towards the south. Thus for a certain period of time the catchment area of the Mississippi also included a major proportion of the Laurentide Ice Sheet. The large water masses enabled the rivers to cut through the sills which originally separated the two drainage basins and to reverse the original gradient. Thus new parts were added to the southerly directed drainage system. Since there was no reversal in drainage direction when the ice finally melted down, the upper reaches of these rivers became permanent additions to the Mississippi system.

This phenomenon has been known since the last century, and from morphological investigations it is clear that, for instance, the middle and upper parts of the Missouri River owe their present course to the interaction with continental ice sheets. As Flint (1955) pointed out, in South Dakota the river today flows at right angles to the regional slope. Its valley is narrow and trench-like, whilst its tributaries from the west are much better developed. It can be seen that these valleys originally continued beyond the present Missouri; the abandoned segments, however, are now filled and blocked with Quaternary sediments. At Kansas City the Missouri River valley widens considerably and turns east with a rather abrupt bend. It is assumed that at this place the river had reached the outermost edge of the ice sheet and continues in a valley of preglacial origin, which was the original drainage path of the Kansas River (Heim & Howe 1963).

The second major addition to the Mississippi drainage system occurred in the northeast, in Pennsylvania, Ohio and Indiana. These areas were originally partly drained northeastwards towards the Erie basin and partly by the River Teays towards the northwest. A pre-Illinoian ice sheet blocked drainage and in an ice-dammed lake rhythmically laminated clay and clayey silt were deposited, the 'Mintford Silt' of Stout & Schaaf (1931). This event predates the Matuyama/Brunhes boundary (Jacobsen *et al.* 1987). Under the ice cover major portions of the Teays river system were buried by thick Quaternary deposits, locally over 120 m, and after repeated blocking by ice sheets the drainage system was added to the Ohio River catchment area. The course of this early so-called 'Deep Stage' Ohio River was aligned via Hamilton and Venice. Its valley was about 30 m deeper than the present Ohio valley. The 'Deep Stage' drainage system is today buried beneath Illinoian and Wisconsinan deposits. During the Illinoian glacial maximum the ice margin reached as far south as Cincinnati. It blocked the ancient drainage system. An ice-dammed lake formed which finally spilled over directly westward from Cincinnati to form the present Ohio River valley via Anderson Ferry and North Bend (Durrell 1977).

In Iowa and Illinois, the Mississippi River was forced by the ice sheets to change its course. Various abandoned valley segments have been detected, some of which are filled with younger deposits. From the stratigraphical relationships and the bedrock valley-floor elevations it has been possible to reconstruct the sequence of events which led to these changes (Leverett 1921, 1942).

Like the Rhine delta in Europe, the Mississippi delta is an area of subsidence. Quaternary deposits here are over 1300 m thick. Marine deposits alternate with fluvial strata, which were deposited during cold-stage low stands of the sea (Akers & Holck 1957). Early investigators distinguished five major Quaternary stratigraphical cycles in the Mississippi Delta region. Each of them starts with a series of coarse-grained deposits, followed by finer sediments. The sequences were correlated with the then known glacial–interglacial cycles (Fisk & McFarlan 1955, Doering 1956). With the progress of dating techniques, however, it became obvious that the early reconstruction was far from complete.

Further upstream, a sequence of terraces can be distinguished above the Mississippi floodplain (Figs 202 and 203). Fisk (1944) assumed that a narrow Holocene floodplain was flanked by four pairs of terraces. The idea was that during glacial cold stages downcutting occurred and during the warm stages the terraces accumulated. Also underlying this scheme was the concept of a narrowing valley, the oldest one being about seven times wider than the modern one. This concept is no longer valid; the present opinion is that the terrace remnants are not pairs but left on one valley side only, with lateral erosion on the other. The Mississippi valley is now possibly wider than ever before (Autin *et al.* 1991).

Under the influence of the old, four-fold subdivision of the North American Pleistocene, Woldstedt (1965) concluded that the Mississippi terraces could be correlated as follows:

Prairie Terrace	Wisconsinan
Montgomery Terrace	Illinoian
Bentley Terrace	Kansan
Williana Terrace	Nebraskan
Lafayette Gravels	preglacial Early Pleistocene

This interpretation is now out of date. The Lafayette Gravel of the Ohio and Tennessee valley region is part of what is now called the 'Upland Complex', a spread of gravelly deposits reaching from Illinois to southwestern Alabama. It continues in a westward widening belt across central Louisiana from Sicily Island into Texas. The deposits are erosional remnants of a former continuous blanket. They cap hilltops and interfluve areas. Therefore it is clear that they were deposited before formation of the recent Mississippi valley. Because the recent drainage system was well established in the early Pleistocene (Saucier 1987), it is clear that the upland deposits must be older than that. The Williana and Bentley terraces of Fisk (1938) and Krinitzsky (1949) are also part of this Upland Complex. Boulders of chert, quartzite, sandstone and petrified wood have been found in these gravels in Louisiana, eastern Arkansas and western Mississippi, some of which weigh hundreds of kilograms. They are interpreted as ice-rafted (Autin *et al.* 1991).

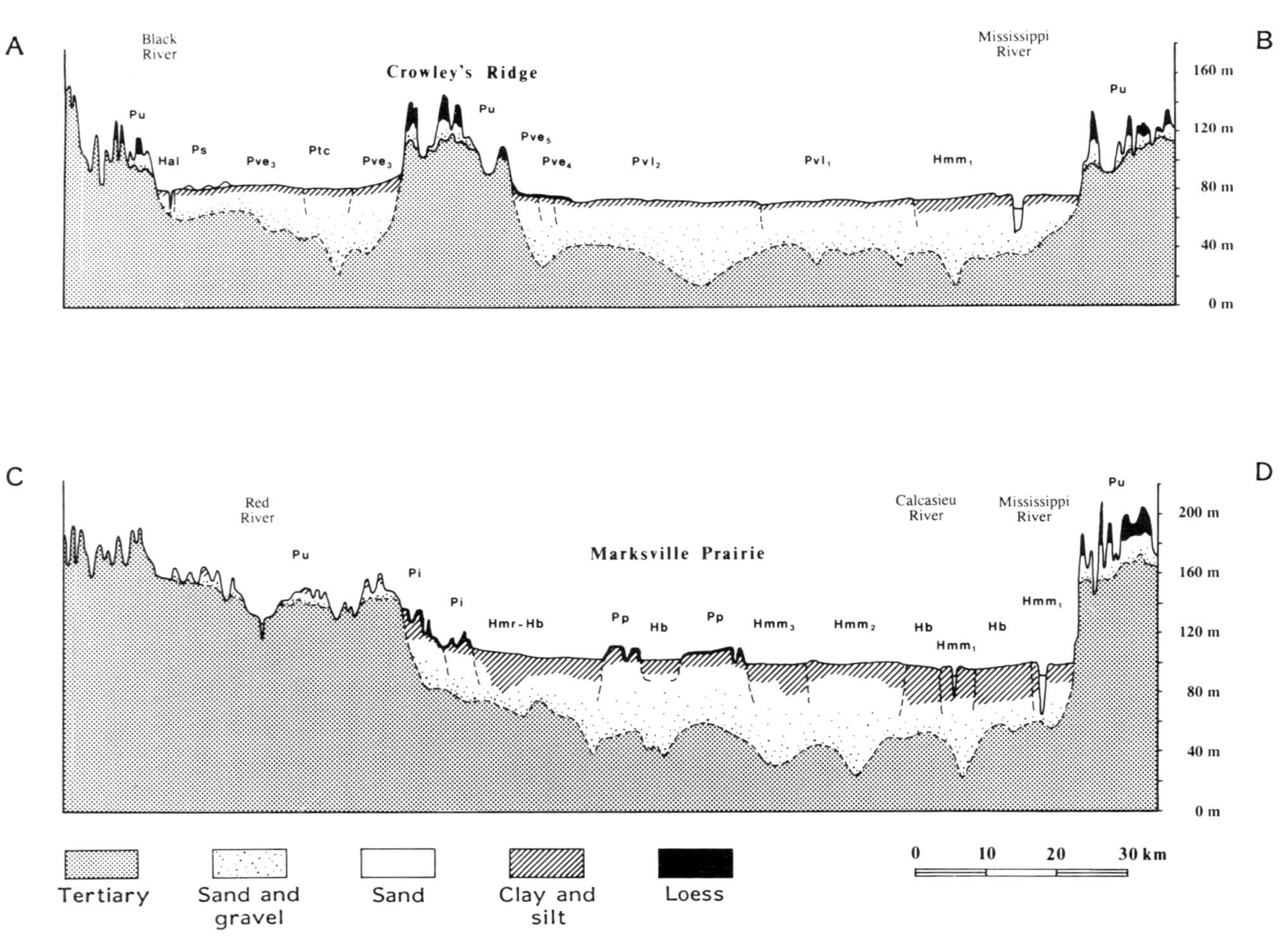

Figure 202 Cross sections through the Mississippi valley; for locations see Fig. 203 (from Autin *et al.* 1991)

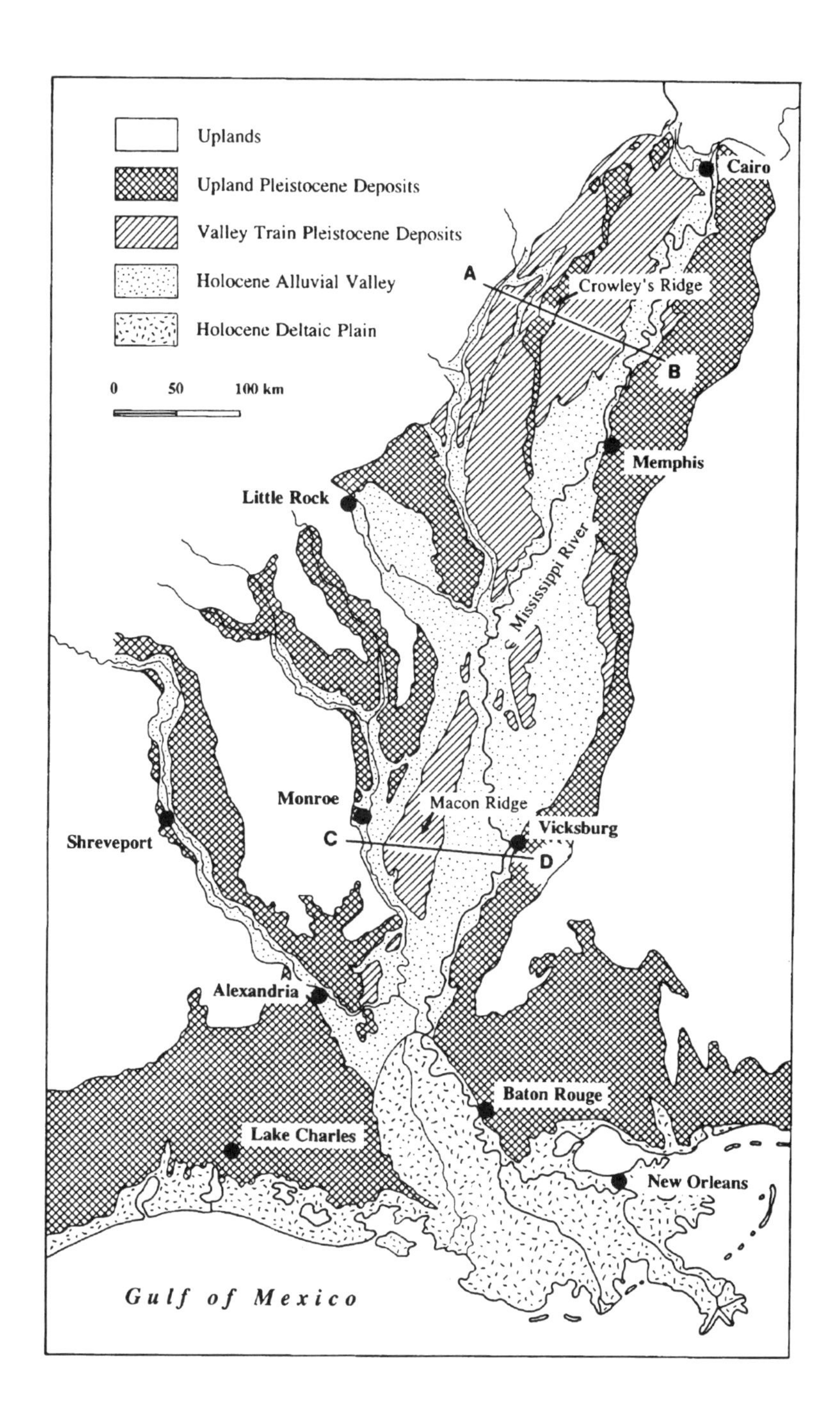

FIGURE 203 Pleistocene terraces of the lower Mississippi valley; the position of the profiles (Fig. 202) is indicated (from Autin *et al.* 1991)

The next youngest sequence of deposits, the 'Intermediate Complex' (Fig. 202), comprises the Montgomery Terrace of Fisk (1938) and a few other terrace remnants. Though very limited in extent, it probably represents most of the Pleistocene. In west-central Louisiana the deposits form a sequence of interbedded sand, silt and clay 100 to 300 m thick. In southeastern Louisiana the 'Intermediate Complex' sediments fine upward. The deposits are capped by a well-developed thick geosol and overlain by loess (Autin *et al.* 1991).

The 'Prairie Complex' comprises the Prairie Terrace of Fisk (1938). In the lower Mississippi valley this terrace is covered by a veneer of loess (Fig. 202). Fisk (1940) correlated it with the Beaumont Formation of Texas and the Pamlico and Pensacola terraces of Florida, incorporating in it the Ingleside Barrier of Price (1933). The ancient transgressive coastline has been identified along the Gulf Coast of Mississippi (Otvos 1982) and Texas (Fig. 176). However, it is now believed that the morphostratigraphical units of the 'Prairie Complex' result from a series of events. The Ingleside Barrier and their equivalents, the Biloxi and Gulfport Formations in Mississippi and Alabama (Otvos 1982, 1992) are regarded as Sangamonian in age. They formed when the local sea level was 6–7 m above the present. This accretional phase was followed by slight valley incision during the subsequent regression and renewed deposition during the Farmdale Interstadial, the deposits of which are also included in the 'Prairie Complex'. A clear subdivision is needed into pre-Wisconsinan and Wisconsinan subunits (Autin *et al.* 1991).

The Wisconsinan Glaciation led to deposition of a valley train of glaciofluvial sediments. The sand and gravel reaches a thickness of 30–60 m down to the base of the Quaternary deposits. However, not all of this is neccessarily Late Wisconsinan in age; the lower part of this sequence may include outwash of older glaciations. In the outwash areas several terrace levels can be distinguished. Most of these have been correlated with various phases of the Wisconsinan Glaciation. Only from the Natchez region have older outwash terraces been reported. This so-called Natchez Formation (Chamberlin & Salisbury 1906, Leighton & Willman 1949) contains crystalline rocks derived from the Laurentide Ice Sheets, but it is covered by several loesses, so that an Early or Middle Pleistocene age has been postulated. The terrace surfaces exhibit traces of an anastomosing network of abandoned braided stream channels (Autin *et al.* 1991). In the lower Mississippi valley (south of Cairo) two valley trains have been distinguished, the older one being either Early Wisconsinan (Saucier 1978) or, more likely, pre-Wisconsinan in age (Rutledge *et al.* 1985).

The effects of meltwater drainage through the Mississippi have been felt in the Gulf of Mexico. In a borehole in the northeastern Gulf (Ocean Drilling Programme Leg 100, hole 625B) negative $\delta^{18}O$ values in planktonic foraminifera have been recorded since approximately 2.30 million years BP. The observed anomalies would require a freshwater source about five to seven times the volume of the largest historical floods. The oldest anomaly has been dated to between 2.30 and 2.26 Ma and may be correlated with a till from Iowa which has been dated to 2.4 Ma (Boellstorff 1978).

Southward meltwater drainage occurs when drainage through the St Lawrence River is blocked by ice. This requires an extent of an ice sheet as far south as approximately 40°N (Andrews 1987, Teller 1987). Because the global $\delta^{18}O$ curves seem to indicate that the early Pleistocene and Pliocene were periods of smaller ice volume by about 50% (Mix 1987), this means that in order to reach such southerly positions the early ice sheets must have been thinner than the Late Pleistocene ones.

The influence of meltwater is also felt at the end of the last glacial stage. The negative $\delta^{18}O$

values end abruptly between 11 and 10 ka, when eastward drainage through Hudson and St Lawrence rivers replaced the southward-directed flow of meltwater (Joyce *et al.* 1993).

14.6 FLUVIAL DEVELOPMENT DURING THE HOLOCENE

Knowledge about Holocene fluvial dynamics has increased considerably during recent decades. Amongst the more recent studies are the works of Brakenridge (1981, 1987), J.C. Knox (1983), Schirmer (1983a, d), Schreiber (1985), Buch (1988, 1989), Schellmann (1990) and Hiller *et al.* (1991). The postglacial fluvial deposits are by no means an undifferentiated accumulation; in some river systems they can be subdivided into a sequence of terraces. Schirmer (1983b) distinguished three different types of fluvial deposits, according to their morphological expression:

1. *Terrace stairways* consist of a sequence of terraces, the surfaces of which are sculpted by fluvial action and in which the surface of each younger terrace is situated at a lower level than the former.
2. *Terrace rows* or *fill terraces* are a sequence of terraces the surfaces of which are roughly at the same altitude.
3. *Stacked terrace deposits* are formed by several aggradation cycles, accumulated one on top of the other. The younger units cover the older ones, so that only the youngest aggradation is visible as a terrace on the land surface.

Often such sequences can only be reconstructed from boreholes. In some cases it is possible to distinguish the individual sedimentary units based on grain-size distribution. Within each accumulation normally a fining-upward tendency is registered.

Schirmer (1981) distinguished two types of terrace deposits, referred to as V and L types. The V type is accumulated by a braided, vertically accumulating (aggrading) river. Accumulation starts with a coarse basal layer (gravel lag), which grades upwards into sand-rich gravels with an upward-increasing sand content. In contrast, the L terrace type is deposited by a meandering, laterally accumulating river on the lee side of oxbows. The river changes its course laterally. In Central Europe this type of terrace deposition begins with an extreme matrix-poor gravel. Schirmer (1983b) refers to it as 'skeleton gravel'. The sand content strongly increases upwards, and the sediments grade into pure sands. The development of Central European river terraces from the last glacial maximum until now has been summarised by Schirmer (1991) (Fig. 204). North American examples are discussed in detail by Brakenridge (1987).

Fluvial aggradations show sedimentary sequences that are repeated regularly. They are composed of river-bed sediment, floodplain channel sediment, floodplain sediment and soil. This rhythm has been termed **fluvial series** by Schirmer (1979) in accordance with Troll's (1977) 'fluvioglacial series'. The floodplain channels of a uniform terrace aggradation normally form regular bundles or groups striking in the same direction. In L terraces they mainly reflect meander displacement. Where the floodplain channels of L terraces of different age abut, they mostly strike in different directions, so that they normally form a morphological discordance. The outermost floodplain channel of a terrace which bounds against the older terrace is called 'floodplain edge channel' (Schirmer 1983b) (Fig. 205). The cover of the youngest terraces, as a rule, is formed by alluvium (flood loam), a fine-grained sediment that accumulates by vertical accretion following high floods in distal reaches of the floodplains. Redeposited loess and inwashed soil or regolith play a dominant role in flood loam accumulation.

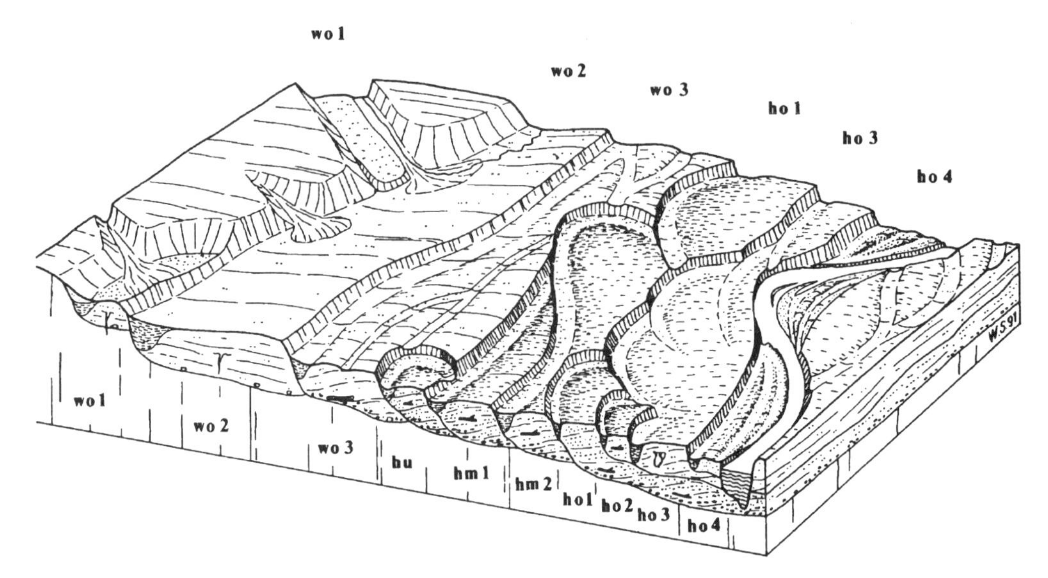

FIGURE 204 Schematic sequence and stratigraphy of the valley floors of Central European rivers; wo = Upper Würmian, hu = lower Holocene; hm = Middle Holocene, ho = Upper Holocene (after Schirmer 1991)

Where more than one Holocene terrace occurs, they can often be differentiated on the basis of their soils. Soil-forming intensity depends on the time available for pedogenesis. In Central Europe, argillic brown earths are restricted to sediments that accumulated before the Atlantic period. On the younger terraces they are substituted by brown earths and A/C soils, on the youngest terraces by poorly-developed A/C soils. Regionally different soil-ecological conditions, such as different carbonate contents of the parent material, must be taken into consideration (Schirmer 1991).

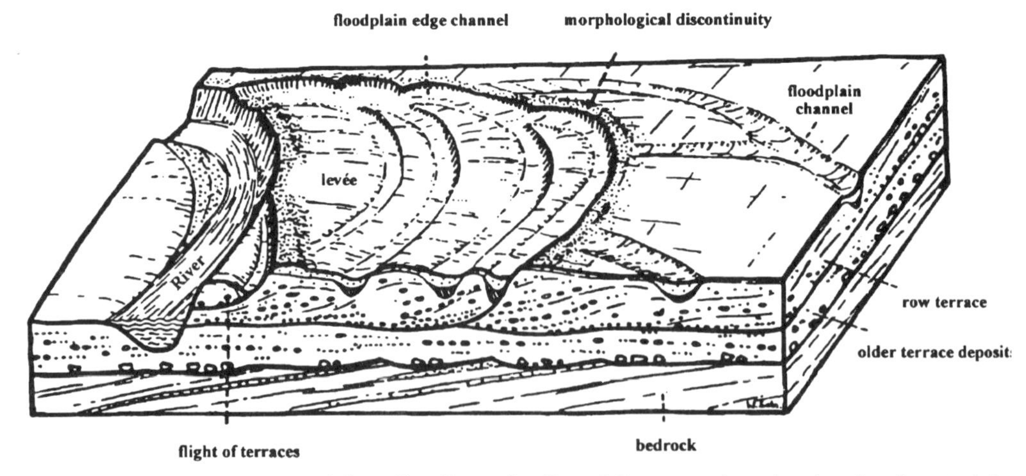

FIGURE 205 Block diagram of the valley floor of a Central European river showing the characteristic landforms and geological units (from Schirmer 1983b)

In many cases other methods should be used for differentiating the floodplain sediments. Dendrochronological investigations of wood remains allowed the construction of a continuous tree-ring chronology for western Germany covering the last 9000 years (Becker *et al.* 1989). The ages of older, 'floating' chronologies can be roughly estimated with the aid of radiocarbon dating. Dendrochronology offers the advantage that large numbers of wood samples, often complete tree trunks, can be investigated in a short period of time, which allows a far better determination of terrace sediment ages than radiometric dating of individual wood fragments, some of which may be redeposited older material.

A second method that is well-suited for dating Central European Holocene fluvial terraces is archaeology. Archaeological investigations are especially useful in cases like the River Main sediments, where tree trunks are not preserved in the youngest terrace sediments. The youngest wood samples there date back to about AD 870. In many terrace sediments, however, fragments of pottery have been found, which can be used to provide a date. Because of the advanced state of modern archaeology in the Frankonian region this method is very helpful. Medieval and early modern ceramics can be well-determined (Schirmer 1983c). In other regions the ceramics can at least be utilised for relative dating purposes.

Pollen analysis under favourable conditions also allows the dating of floodplain channel sediments. Best results are achieved in former depressions, where mud and peat could accumulate. In contrast, in terrestrial sediments the pollen is mostly destroyed by oxidation. When evaluating pollen diagrams, however, it must be remembered that the fill of the floodplain channels does not necessarily represent the maximum age of that landform element, but that it may have been re-mobilised by later high floods. The reliability of the results increases with the growing amount of data available.

The question of whether Holocene terrace formation of Central European rivers was controlled by extra-regional, climatic events has been hotly debated. In Europe, Becker (1983) and Havlicek (1983) regard the early Holocene conditions rather as a continuum, whereas Starkel (1983) and Schirmer (1983b), because of the distinct terrace bodies in their areas, favour clearly discernible periods of climate-controlled terrace formation. Extra-regional similarities and possible correlations with glacier fluctuations suggest an environmentally controlled process. The extent of human influence on sediment redistribution is not quite clear. It is well known that in Europe, since the Neolithic period, humans have actively interfered in the shaping of valleys and rivers (Schirmer 1993). Agriculture and forest clearing have greatly increased runoff and led to increased input of fine sediments to river systems. The results, however, seem to differ from one river to another (Hiller *et al.* 1991). A review of the Central European valley development during the last 25,000 years is given in Figure 206. However, it should be kept in mind that this scheme applies to an area comparable in size to little more than one of the US federal states.

In **North America** the Holocene development of the rivers has been the subject of numerous studies; J.C. Knox (1983) has summarised the state of research. River development here has been largely climate-controlled. Rapid warming and drier conditions between 10,000 and 8000 BP favoured fluvial deposition in most regions. In the following period rivers in the southwestern United States experienced major incision, which was less intensive in the Prairies and non-existent in the eastern woodlands. This period was followed by renewed alluviation. At about 6000 BP a period of lateral channel migration and fluvial incision affected most areas except the dry southwest. The reason for this change may have been a change in atmospheric circulation towards a more meridional pattern, causing higher precipitation and more frequent

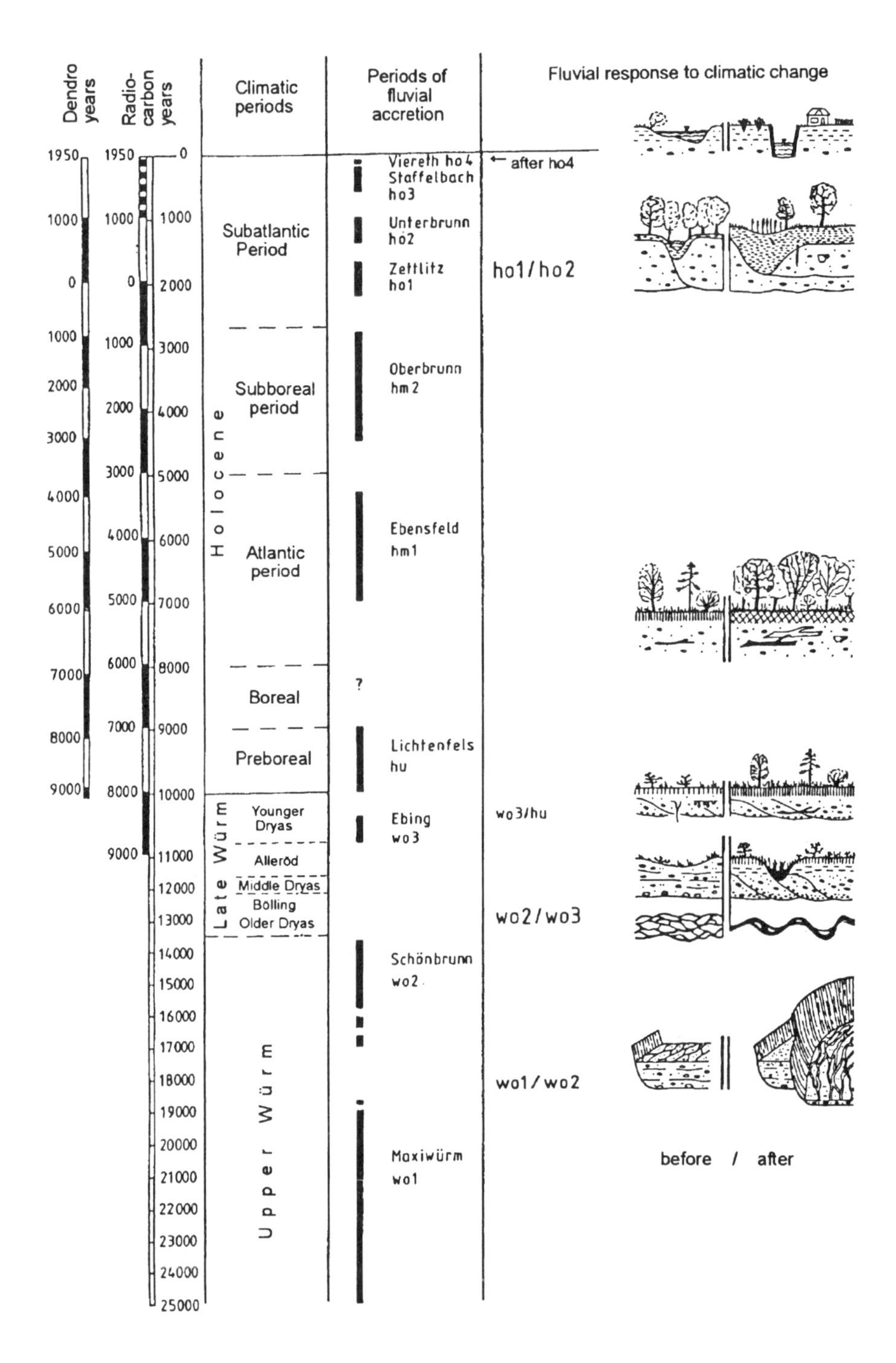

FIGURE 206 Shifts in the valley development in Central Europe within the last 25,000 years; dendrochronological years according to Suess (1980) and Becker *et al.* (1991) (from Schirmer 1993)

floods. After a period of relative stability starting about 4500–4000 BP, a new phase of channel migration and incision followed in most regions between 3000 and 1800 BP. From then on fluvial activity remained modest until about 800 BP when minor alluviation began, ending in many areas in the late 19th century because of increasing human interference.

Much work has concentrated on the Mississippi valley. Transformation from a braided to a meandering river was shown to have been a gradual change. It started near the coast. At Baton Rouge the transformation occurred about 12,000 BP (Krinitzsky & Smith 1969), and in the vicinity of Cairo 3000 years later at about 9000 BP (Saucier 1981). Fisk (1944) had worked out a detailed chronology of the meander belts and of sequences of channel positions within those belts. However, Saucier (1974) showed that major parts of Fisk's original concept had to be revised. Today it is known that the Mississippi and Ohio rivers never flowed in separate meander belts south of Cairo, that the abandonment of one meander belt and formation of another is a slow process, and that two or more meander belts were used simultaneously for centuries (Saucier 1985, Autin *et al.* 1991).

Major changes of the river course from one side of the Mississippi valley to the other directly influenced behaviour of the tributaries by shortening or lengthening their courses. The resulting stream adjustment in some cases led to the formation of new terraces, as could be demonstrated in the Tunica Hills north of Baton Rouge (Delcourt & Delcourt 1977, Alford *et al.* 1983). The development of the Mississippi River and its tributaries therefore are not easy to compare and extra-regional correlations are difficult (Autin *et al.* 1991).

The lower courses of many modern rivers are characterised by a different fluvial regime. Anastomosing rivers are not restricted to sandurs (see chapter 4.2.4) but occur also, for instance, in the lower reaches of the Mississippi (Fisk 1944) and the Rhine (De Gans 1991). Under the influence of a rapid postglacial sea-level rise, an anastomosing fluvial system with a rapidly rising base level developed. Törnqvist (1993a, b, 1994) has analysed the sedimentological processes in the Holocene Rhine Delta in detail. He has demonstrated that the anastomosing fluvial system was caused by rapid vertical accretion (>1.5 mm/yr). After the rate of sea-level rise decelerated about 4300 years BP the avulsion frequency also decreased considerably.

15

Loess Stratigraphy

The palaeoenvironmental information stored in the loess sedimentary record approaches that of the deep-sea sediments. The longest loess sequences have been identified in **China** and during recent decades the Chinese sequence has undergone detailed investigation. It is now clear that it spans the entire Pleistocene, since loess deposition began around 2.5 million years ago in China (Ding *et al.* 1992). The source areas of the Chinese loess are the great Asian deserts, including the Gobi. Transport distances seem to have been greater than in Europe or North America, indicating stronger winds or less vegetation that might have trapped the silt (Pye 1987).

The loess sequence is underlain by the Red Clay Formation, which is found in various places beneath the Loess Plateau. At least its upper part is interpreted as Pliocene loess, strongly altered by pedogenesis (Bronger & Heinkele 1989). It varies in thickness between less than 5 m and over 100 m. The upper part of the sequence represents a strongly developed palaeosol complex being characterised by rubification, carbonate leaching and reprecipitation, and clay translocation. This soil is interpreted as having formed under subtropical, relatively humid conditions (Ding *et al.* 1992).

The loess overlies the Red Clay with a sharply defined contact, indicating a rapid and dramatic change of climate. From that time onwards loess deposition has remained continuous, but with reduced rates during the interglacials, as indicated by grain-size distribution and thickness of layers (Ding *et al.* 1992, 1994). Pedological investigations of the loess sequence suggest that the loess was deposited during the glacial stages under an arid climate with sparse vegetation and a high aeolian dust influx, whereas the intercalated soils developed under relatively humid conditions (Liu *et al.* 1986, Bronger & Heinkele 1989).

Until today research has been generally concentrated on a few standard sections, including the classic Luochuan section (Heller & Liu 1982, Liu *et al.* 1986), the section at Xifeng (Liu *et al.* 1987) and a 330-m-thick sequence at Lanzhou (Burbank & Li 1985). The last mentioned provides a detailed record of the last 1.3 million years. The high accumulation rate has been explained by the proximity to the northern source areas and favourable depositional conditions in the Lanzhou Basin.

Most recently, research has focused on a section 5 km north of the city of Baoji. This section has been more strongly affected by pedogenic processes, 37 major soils having been identified (Rutter *et al.* 1991). Ding *et al.* (1994) have investigated the grain-size distribution of the 160-m-thick Baoji sequence and the upper part of the underlying Pliocene Red Clay. It has been found that the grain-size distribution clearly reflects changes in aeolian transport conditions. The source areas of the loess are the inland deserts of northwestern China, closely related to the winter monsoon winds from Siberia (Liu *et al.* 1986, 1989). Ding *et al.* (1994) have tuned the Baoji sequence to orbital forcing, using the calculations of Berger & Loutre (1991). The magnetic reversals provide a valuable means of age control, allowing correlation with the oxygen-isotope curves of DSDP Site 607 (Raymo *et al.* 1989) and ODP Site 677 (Shackleton *et al.* 1990).

Less is known about the thick loess sequences in the **Central Asian** republics of Tajikistan and Uzbekistan. Here Dodonov (1979) distinguished 20 pedocomplexes, of which nine were younger than the Matuyama/Brunhes boundary. The oldest pedocomplex is thought to be 1.79 million years old. The distribution and geomorphological relationships of the loess in Uzbekistan and Tajikistan suggest that they were derived from local rivers, partially fed by glacial meltwater. Using magnetic susceptibility, it is now possible to correlate the younger part of the Central Asian loess sequence (the last 800,000 years) with that in China and with the deep-sea oxygen-isotope curve (Shackleton *et al.* 1995). Detailed palaeopedological investigations of the 70-m-thick Chashmanigar loess sequence have shown that in the Tajik loess sequence at least 30 palaeosols can be identified, 21 of which are in the Matuyama palaeomagnetic epoch (Bronger *et al.* 1995). It seems that the Central Asian loess sequence may allow palaeoclimatic reconstruction at least as detailed as the Chinese loess sequence.

In **North America** four or five loess units can characteristically be distinguished (Ruhe 1976, 1983, Leigh & Knox 1994). For this reason loess deposition in America has until recently been regarded as only a result of relatively young glaciations, although the deposits are known to reach considerable thickness locally. The Palouse Loess on the Columbia Plateau, for example, is over 75 m thick. Richmond *et al.* (1965) regarded this loess sequence as containing three pre-Wisconsinan loess units, separated by distinct palaeosols. Detailed investigations of sections have since revealed 21 palaeosols, the lower half of the sequence being reversely magnetised, i.e. older than 780,000 years. The section investigated is 26 m deep, but is underlain by at least a further 10 m of loess. It is clear that the sequence spans a minimum of 1 million years (Busacca 1991). However, detailed investigations, using the intercalated tephra layers as marker horizons, have only recently begun (Busacca *et al.* 1992).

Until 1975, Péwé still assumed the entire Alaskan loess sequence was no older than Illinoian. However, near Fairbanks a fossiliferous loess profile over 30 m thick has been found that spans the last 3 million years. Several intercalated tephra layers have been dated using the fission-track technique, and palaeomagnetic investigations revealed the major magnetostratigraphical units reaching back to the Mammoth Subchron of the Gauss Normal Epoch (Westgate *et al.* 1990).

In central North America loess underlies most of the Great Plains down to the southern boundaries of Kansas and Missouri. In the east, loess is still widespread in Pennsylvania. Further east it becomes scarce, probably as a result of the more humid climate. Loess also occurs in the Mississippi valley in a belt about 150 km wide, reaching south to the delta. Most of this loess is of Wisconsinan age.

The youngest loess unit in mid-continental North America is the Late Wisconsinan **Peoria Loess**. The term 'Peorian' dates from Leverett (1898c). Kay & Leighton (1933) restricted its use to the loess deposits outside the Shelbyville Moraines in Illinois. In the Mississippi River valley, loess deposits can be followed in a band 25–30 km wide from western Kentucky to south of Baton Rouge. The greatest loess thickness occurs east of the river. The average thickness of the Peoria Loess there is about 15 m, but a maximum of about 27 m occurs in the Natchez–Vicksburg area. The first review of the loess deposits of the Mississippi valley was provided by Wascher *et al.* (1948). Miller (1991) has summarised the more recent state of knowledge. He distinguished five loess sheets. However, apart from the Peoria Loess neither the age of the loess sheets nor their distribution is known in any detail (Autin *et al.* 1991).

The **Roxana Loess**, originally called 'Roxana Silt' (Frye & Willman 1960) is Early Wisconsinan in age. In Illinois it reaches a maximum thickness of slightly over 5 m along the Mississippi valley (McKay 1979). Equivalents of the Roxana Silt of Illinois are also found in the northern part of the Mississippi valley region. The southernmost occurrences are on the southern end of Crowley's Ridge (Miller *et al.* 1985). The unit contains buried Inceptisols and is capped by a weakly-developed geosol correlated with the Wisconsinan Farmdale Soil of Willman & Frye (1970). In the Driftless Area of Wisconsin and northern Illinois, Leigh & Knox (1993) have dated the formation of the Roxana Loess to between 27,000 and 55,000 BP. It has been suggested in the past that the Roxana Loess was not directly related to glaciation (Norton *et al.* 1988, Winters *et al.* 1988). However, thickness trend, mineralogy and geochemistry strongly suggest a proglacial origin (Leigh 1994).

Illinoian loess has been identified in the United States in many places. In the Midwest it is referred to as the **Loveland Loess** (Shimek 1909, 1910) after its type locality in Loveland, Iowa (Daniels & Handy 1959). Its distribution reaches from northern Kansas and south-central Nebraska in the west to the Ohio valley in the east. In Illinois it has been described by Willman & Frye (1970). The Loveland Loess is widely distributed in Iowa and Illinois; it also occurs in the Driftless Area of Wisconsin (Leigh & Knox 1994). The Loveland Loess underlies the Sangamon Soil, and thermoluminescence dates have confirmed its Illinoian age (Forman *et al.* 1992).

The 'Sicily Island Loess' of the Mississippi valley is an equivalent of the Loveland Loess further to the north. It is widely distributed on surfaces older than the Prairie Complex. B.J. Miller *et al.* (1985) identified it as pre-Peoria in age. The geosol at its top is a well-developed Alfisol, slightly more strongly developed than the modern surface soils on the Peoria loess. The soil represents the last interglacial (McKay & Follmer 1985, Rutledge *et al.* 1985). The age determination of the Loveland Loess is also supported by the results of amino-acid dating (McCoy *et al.* 1990).

Some workers prefer a broader definition for the Loveland Loess. On the Osage Plains it is said to contain one or more palaeosols older than the Sangamonian (Madole *et al.* 1991). In addition, south of the Platte River in Nebraska the Loveland Loess contains two strongly developed palaeosols (Morrison 1987, Wayne & Aber 1991) which suggest that it represents aeolian deposits of at least three glaciations.

The next older unit in the Mississippi valley, the **Crowley's Ridge Loess**, has been identified at many sites on Crowley's Ridge and as far south as Natchez, Mississippi. McKay & Follmer (1985) correlate the soil at its top with a palaeosol underlying the Illinoian till. This geosol is more strongly developed than other geosols and must have formed during a long and warm interglacial (B.J. Miller 1991).

Older loesses are known from only very few places. In Wisconsin, Leigh & Knox (1994) identified a pre-Loveland loess which they termed the **Wyalusing Formation**. Like the Crowley's Ridge Loess, it postdates the Matuyama/Brunhes boundary, so that it must be Middle Pleistocene in age, but its exact age is not known. The oldest loess of the Mississippi valley, informally termed **Marianna Loess**, has been described only from Crowley's Ridge in

Arkansas (Miller *et al.* 1985). It contains a strongly developed palaeosol (Alfisol) suggesting formation during a major Middle Pleistocene interglacial. In Nebraska, a number of sites with multiple palaeosols in loess sequences have been found. Most of the several tens of metres thick sequence overlies the Lava Creek B Ash, which has been dated to 620,000 BP. However, in places this ash layer also overlies older loess (Morrison 1987) (Fig. 207).

Since the formation of the most recent loess units in North America can be correlated with the existence of extensive ice sheets, the preservation of so few older loesses in the mid-continental USA and the Mississippi valley seems to indicate glaciation that reached as far south as the Upper Mississippi valley occurred only very rarely throughout the Quaternary. Because all known loess units postdate the Matuyama/Brunhes boundary, this seems to suggest that large continental ice sheets did not develop prior to the Middle Pleistocene. However, early Pleistocene tills are known from Iowa and Nebraska (Hallberg 1986). Thus it appears more likely that intensive erosion, probably under a climate more humid than present, has caused removal of most of the older tills and loesses (Leigh & Knox 1994).

In **Eastern Europe** a thick mantle of loess covers the plains of the Ukraine and adjoining parts of Russia. The oldest parts of the sequence date from the end of the Pliocene. A relatively reliable chronology from the Matuyama/Brunhes boundary has been developed, based on fossil fauna and palaeomagnetic dating. The youngest part of the sequence is radiocarbon-dated. In contrast to Central Europe, a strongly developed palaeosol, the Bryansk Soil, has been identified within the Weichselian sequence and dated to 26,000–32,000 BP (Velichko *et al.* 1984, Velichko 1990).

In the Central Danube Basin of Hungary in **Central Eastern Europe** a thick sedimentary sequence was deposited, that spans the entire Neogene and Quaternary. The Pannonian Basin, as it is called, was submerged beneath the Pannonian Sea in the Miocene, and evolved into the Pannonian freshwater lake which dried out at the end of the Pliocene. These lacustrine sediments thus are overlain by fluvial and aeolian Quaternary deposits. Regionally up to 50–100-m-thick loess–palaeosol–sand sequences occur in this sequence, containing 20–30 palaeosols. However, the sequence is not quite complete, and only the younger part has so far been adequately dated (Rónai 1985, Pécsi & Schweitzer 1991). Several long profiles both from deep exposures (such as the Matra opencast brown-coal mine) and boreholes have been analysed using grain size, mineral content and palaeomagnetism. The Plio-Pleistocene subaerial deposits are characterised by red soils developed in clays, which are overlain by the loess sequence. The loess sequence has been shown to contain up to 12 palaeosols (Pécsi & Schweitzer 1992).

West of the Pannonian Basin loess sequences are generally not very thick. However, because much research has been concentrated on this relatively small area, some detailed regional chronologies have evolved. It was recognised relatively early that many loess sequences contain buried soils. Penck & Brückner (1909) termed these layers ‘leimenzonen’ and regarded them as forest soils of the Pleistocene warm stages. Götzinger (1938) pointed out that each ‘leimenzone’ need not necessarily represent an interglacial soil. Instead he distinguished a ‘Kremser Leimenzone’, ‘Göttweiger Leimenzone’ and ‘Paudorfer Leimenzone’, which he correlated with the Mindel/Riss Interglacial, the Riss/Würm Interglacial and a Würmian interstadial. Further investigations in the 1950s (Brandtner 1954, Fink 1954) resulted in a revision of the profiles and in the introduction of new names. The Göttweig soil profile was no longer used because the sequence was thought to be incomplete, and the stratigraphical position of the soil uncertain. The following standard sequence arose:

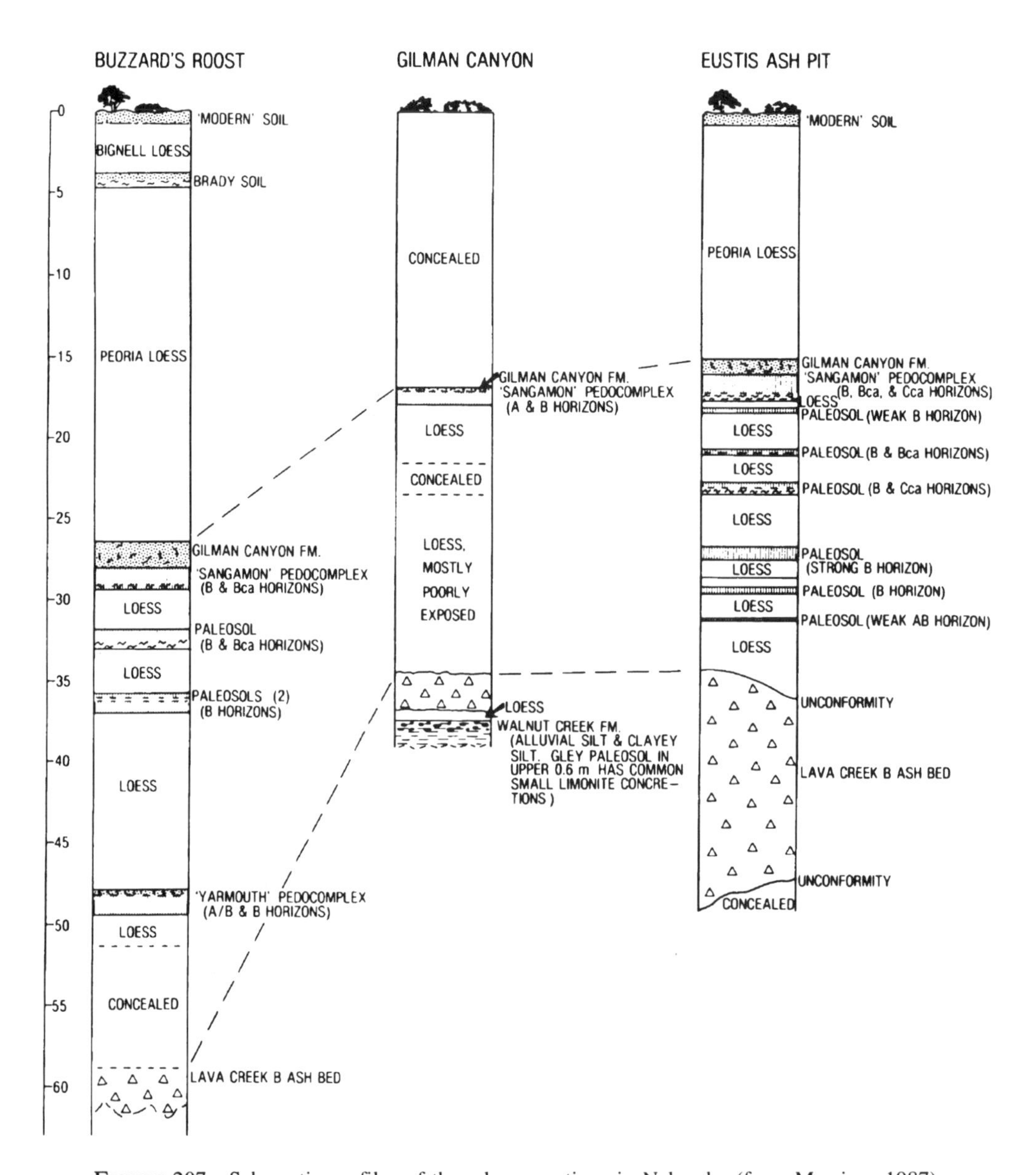

FIGURE 207 Schematic profiles of three loess sections in Nebraska (from Morrison 1987)

1. Modern soil, often a Chernozem
2. Loess
3. **Stillfried B**
4. Loess
5. **Stillfried Complex** (Stillfried A after Fink 1954), an argillic brown earth, overlain by three strongly developed humus zones, separated from each other by thin layers of loess.
6. Loess

7. **Krems Soil** (after Brandtner 1954), a layer 1–2 m thick of a strong red-brown or brick-red soil

In the early 1950s Brandtner (1954) still considered that the Krems Soil was that from the last interglacial. However, thorough investigations by Fink & Kukla (1977) showed that the Krems loess profile is far more complex than originally thought. Today 16 fossil soils are distinguished. The soil complex that originally had been interpreted as representing the last interglacial was found to represent three different Early Pleistocene warm stages. Palaeomagnetism and the small mammal faunas both indicate that these soils are over one million years old (Kohl 1986, Fig. 208). The Stillfried Complex, according to present knowledge, is composed of the B_t horizon of an argillic brown earth of the last interglacial and the soils of three Early Würmian interstadials. The interstadial Stillfried B soil has been dated to 28,000 BP (Kohl 1986).

In Germany, the Weichselian loess sequence is subdivided by a number of palaeosols. As in North America, the Early Weichselian loess can be clearly distinguished from that deposited in the Late Weichselian. The Early Weichselian loess sequence contains a bleached horizon and two or three humic zones, which are referred to as **Mosbach Humic Zones** in the Rhine–Main region (Schönhals *et al.* 1964). They can be correlated with the Early Weichselian interstadials (Brørup and Odderade). The Middle Weichselian begins with a solifluction deposit altered by pedogenetic processes. This is referred to in Hessen as the **Niedereschbacher Zone** (Semmel 1968) and is characterised by cryoclastic clay formation. In Rheinhessen it contains a molluscan fauna, which resembles those of the Early Weichselian interstadials (Remy 1969).

Within the Weichselian loess sequence, the **Lohner Boden** is the most prominent fossil soil (Schönhals *et al.* 1964). It is the equivalent of the Austrian Stillfried B soil and the North American Farmdale soil. In spite of strong solifluction and redeposition it is widespread. Whilst this fossil soil is easily recognised from Austria to southern Niedersachsen, its identification in the Lower Rhine region has been difficult. Only after identification and dating of the Eltville Tephra by Rohdenburg & Semmel (1971) were earlier misinterpretations corrected (see below). The Lohner Boden is also found in Belgium and the Netherlands. Within the overlying Late Weichselian (Late Würmian) loess, Semmel (1968) identified a sequence of up to four weakly developed moist soils, which he termed 'Erbenheim Soils'. After pioneering work by Semmel (1967, 1968), Bibus (1974) also established the soil stratigraphical subdivision of the second from last glacial loess in the Rhine–Main region (Fig. 209).

Within Germany older, pre-Saalian loesses have been found in a number of places, for instance in the sections at Frimmersdorf, Brühl and Erkelenz (Brunnacker *et al.* 1982), Heitersheim in southern Baden (Bronger 1966, Bibus & Pasda 1991), Bötzingen and Buggingen (Bleich *et al.* 1982). In Switzerland, in the Allschwil brickyard at Basel, Bibus (1990) identified the fossil soils of five warm stages. This find is of major importance, because the loess sequence here overlies the 'Jüngere Deckenschotter'. From Bad Soden in the Taunus Mountains Semmel (1967) described a loess profile in which eight fossil B_t horizons were exposed. Each of these clay-enrichment horizons represents one warm stage. The age correlation of the older parts of the profile has so far only been partially possible. With the help of palaeomagnetic investigations, Semmel & Fromm (1976) demonstrated that the Matuyama/Brunhes boundary lies between the fifth and sixth soil, and the Jaramillo Event between the seventh and eighth fossil B_t horizon.

In incomplete sequences such as those in Central Europe, age determination of the older

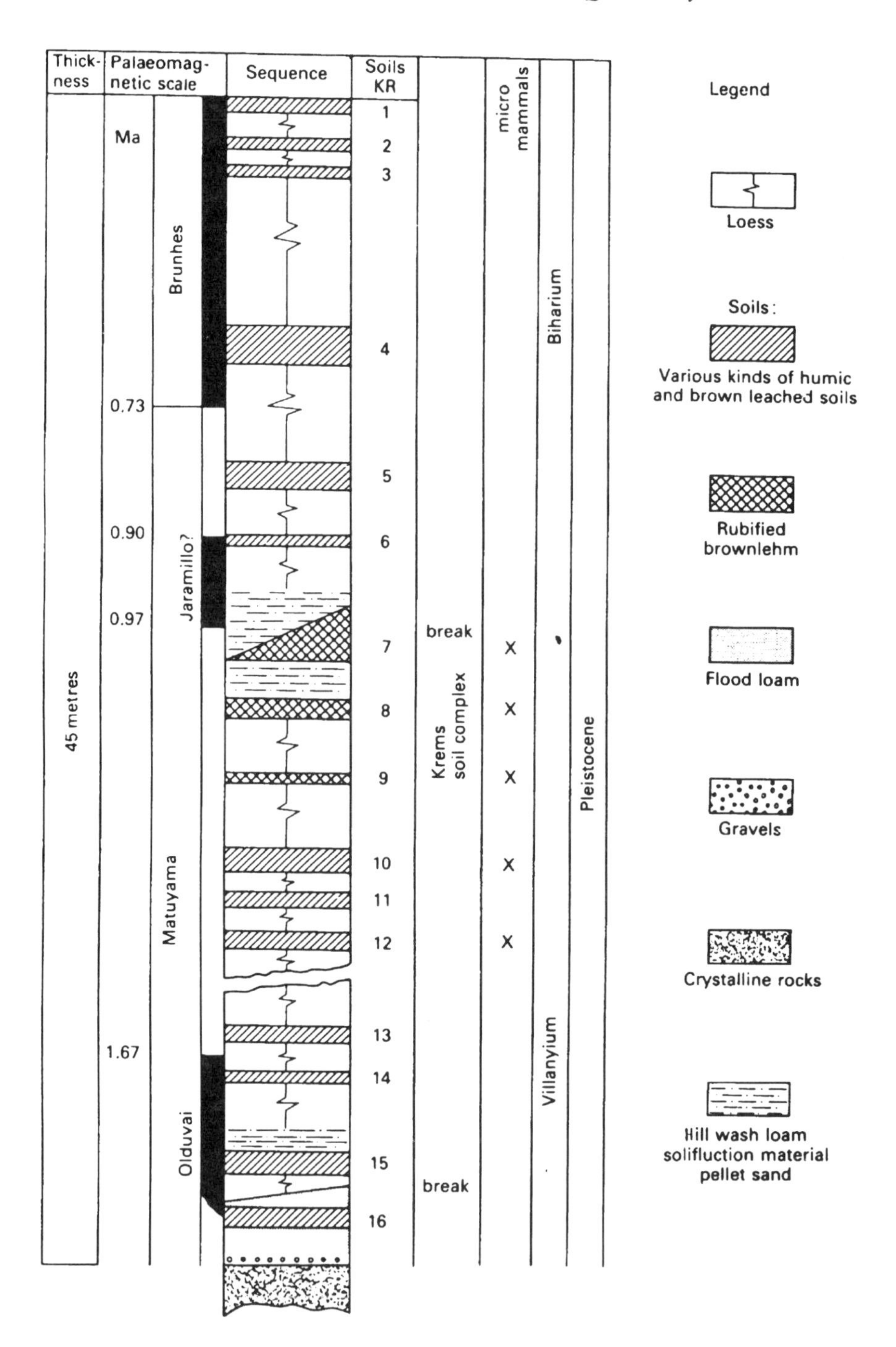

FIGURE 208 The loess section at Krems (Austria). The 'Kremser Boden', which originally was regarded as the soil of the last interglacial, actually comprises three palaeosols which are Early Pleistocene in age (from Kohl 1986)

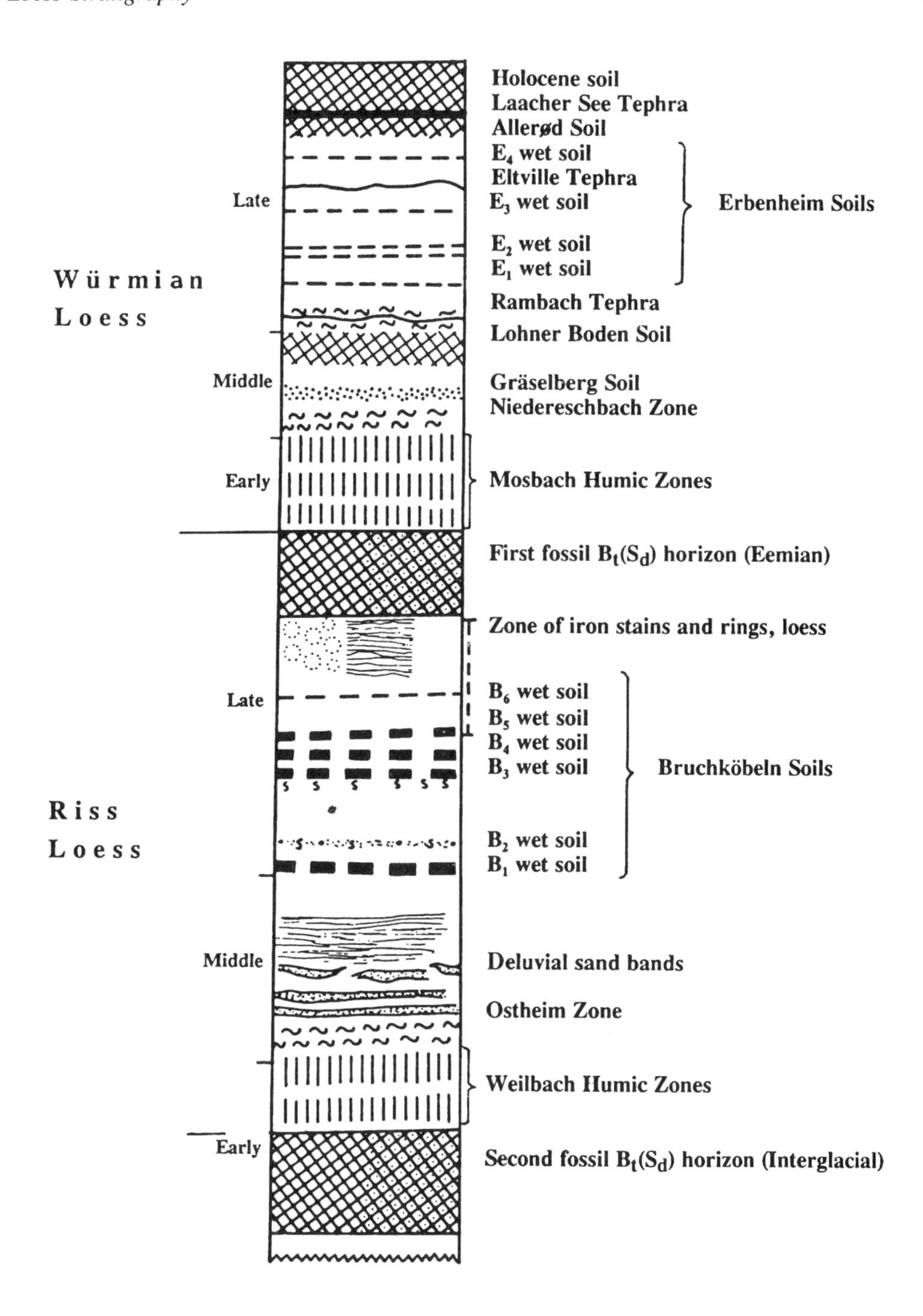

FIGURE 209 Stratigraphic subdivision of the Würmian and Riss loess sequences in Germany (from Bibus 1974; Würmian loess after Semmel 1967)

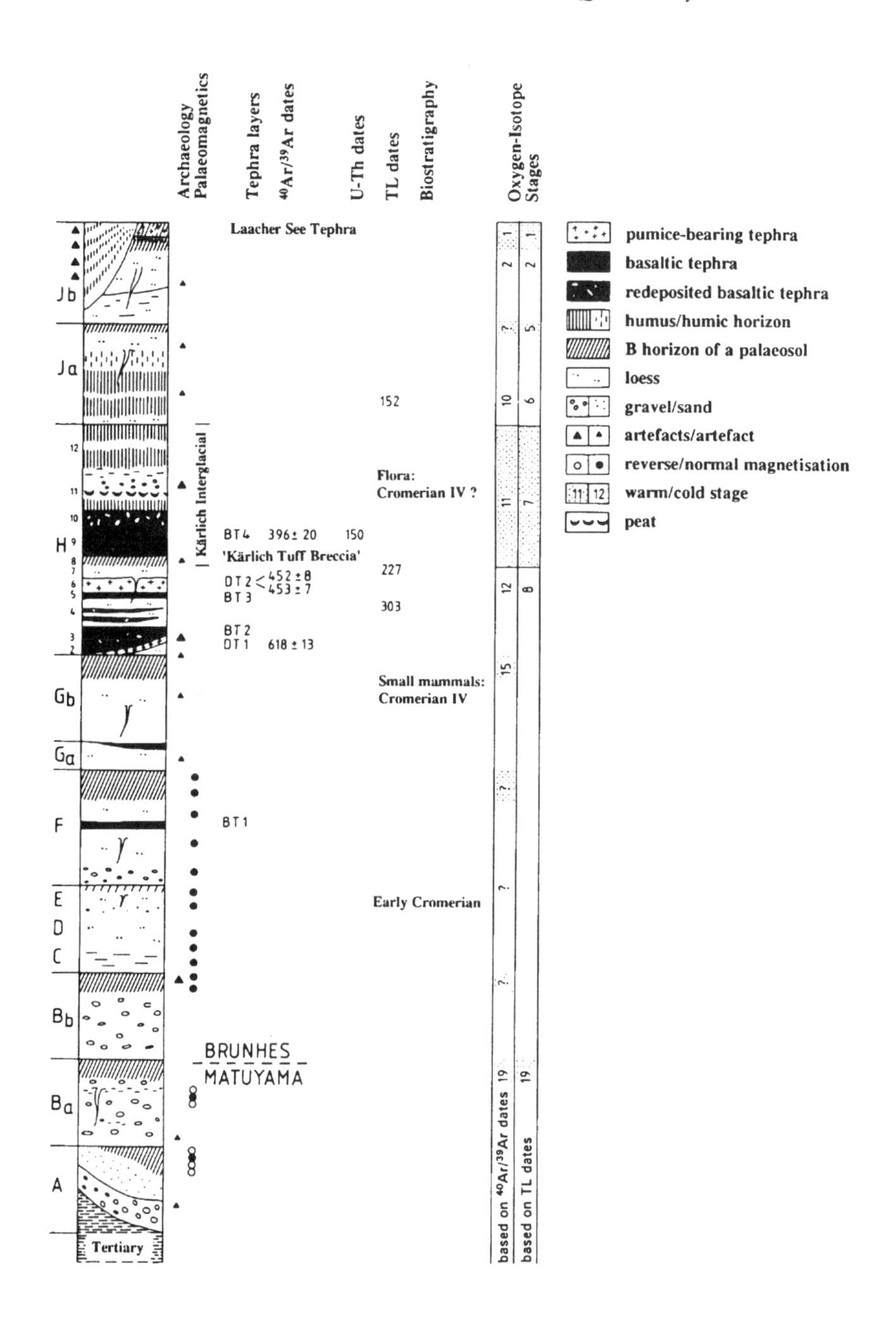

FIGURE 210 Schematic profile of the Kärlich clay pit with the most important dates regarding stratigraphy (after Brunnacker in Bosinski *et al.* 1980, Brunnacker *et al.* 1980, Bosinski *et al.* 1986, Zöller *et al.* 1988, Van den Bogaard *et al.* 1989, flora after Bittmann 1990, small mammals after Van Kolfschoten 1990). Major parts of the loess sequence are now dated to the Cromerian Complex (after Schirmer 1990)

loess soils is often problematic. The Kärlich section in the Neuwied Basin (Middle Rhine region) (Fig. 210) may serve as an example of the problems. The profile in the Kärlich clay pit was first described by Mordziol (1913). He interpreted the 20-m-thick loess sequence as being mainly Riss loess overlain by Würmian loess which contained one fossil soil. He correlated this soil with that at Paudorf (Stillfried B) in the Austrian loess stratigraphy (Mückenhausen 1959). Moreover, a layer of volcanic tephra clasts 1–2 m thick was found. J. Frechen (1959) developed a detailed stratigraphy of the volcanic strata within the exposure which he correlated with two separate eruptions. The Kärlich Tephra was regarded as an indicator horizon for the Würmian glacial maximum (Schönhals 1959).

A first revision of this age estimate was due when Brunnacker (1968) reinvestigated the section. He was able to show that the loess above the Kärlich Tephra contained a sequence of humus zones and humic solifluction deposits that resembled the Early Würmian humus zones of the Czech loess stratigraphy (Kukla & Ložek 1961). Thus the fossil soil under the Kärlich Tephra, which until then had been regarded as a Würmian interstadial soil (Stillfried B) was then reinterpreted as the Riss/Würm Interglacial soil (Eemian). Consequently, the Kärlich Tephra could not be identical with the 'Kärlich Tephra' of the Hessian loess profiles, the position of which was clearly above the Stillfried B soil (Schönhals *et al.* 1964). The 'Kärlich Tephra' of Hessen was thus renamed the Eltville Tephra.

In the 1980s, doubt was cast on the last interglacial age of the Kärlich Interglacial, i.e. the fossil soil underlying the Kärlich 'Brockentuff' (Tuff Breccia). A palynological investigation by Urban (1983) suggested that it was a Middle Pleistocene interglacial of unknown age, showing no close resemblance to either the Eemian or the Holsteinian or the Wacken Interglacial. Zöller *et al.* (1988) dated the loess immediately under the 'Brockentuff' and Kärlich Interglacial soil using the thermoluminescence method to 227,000 years BP. However, dating of the 'Brockentuff' with the $^{40}Ar/^{39}Ar$ laser method yielded an age of 396,000 ±20,000 years BP (Van den Bogaard *et al.* 1989). This equates with the results of new palaeobotanical investigations that would suggest correlation of the Kärlich Interglacial with the 'Cromerian Interglacial IV' (Bittmann 1990).

Thus within a period of 30 years, the Kärlich Interglacial has moved from an intra-Weichselian attribution to the Cromerian Complex Stage. However, it must be remembered that the $^{40}Ar/^{39}Ar$ laser dates are no absolute ages. M. Frechen (1991) points out that they represent maximum ages, and that they are based on individual measurements.

In comparison to parts of the neighbouring Continent, aeolian deposits such as loess and wind-blown sand are relatively rare in **Britain**. Pre-Devensian sediments are very fragmentary. The oldest currently recognised are the Barham Coversand and equivalents found in Essex and Suffolk, beneath Anglian till (Rose *et al.* 1985a). Wolstonian age loess is known from Northfleet (Kerney & Sieveking 1977), Bobbitshole (West 1957) and Hitchin (Gibbard 1974; Gibbard, Catt & Wintle, unpublished). More continuous spreads have been recognised from the Devensian. These occur over much of southern and eastern England, where they form part of a spread of so-called 'brickearth' sediments (Catt 1977). Thermoluminescence dating of these spreads indicate deposition during the Late Devensian (e.g. Wintle 1981, Parks & Rendell 1992). Wind-blown sand present in Lincolnshire and central East Anglia is also of Late Devensian age (Catt 1991b, Straw 1991) although it has been locally remobilised during the Holocene. The same applies to the Shirdley Hill Sand of southwest Lancashire (Wilson *et al.* 1981). Similar deposits of loess and sand loess are recorded from southwestern England (Catt & Staines 1982, Scourse 1986) as well as further north (Lee 1979).

SECTION IV

PERSPECTIVE

16

Overview

Since the time when Penck & Brückner (1901/09) published their classic account of the Alpine Ice Ages, their division of the Pleistocene into four cold stages with intervening warm stages has become a model that has been applied globally for numerous local stratigraphies. Some followers even went as far as classifying deposits in northern Germany or other regions with the Alpine Quaternary stratigraphical nomenclature. A uniformity of names, however, only makes sense where deposits can be reliably demonstrated to be of the same age. Penck & Brückner's system has been understood as a chronostratigraphical subdivision, but actually it is based upon morphostratigraphical correlations that are not easily transferred to different regions. Regardless of these shortcomings, Penck & Brückner's old Ice Age subdivision is still of great philosophical value; in the Alps its terminology still underlies most of the recent works, such as in the extensive Swiss Ice Age geology summary by Hantke (1992).

Although being an internationally recognised Quaternary stratigraphical framework, the original stratigraphic subdivision into Günz, Mindel, Riss and Würm no longer corresponds with the results of the deep-sea stratigraphy and other recent discoveries of international Quaternary research. Even in the Alpine region, the results of recent research can only be fitted into this old system with great difficulty (cf. chapter 12). These problems became evident within the framework of International Geological Correlation Project 'Quaternary Glaciations in the Northern Hemisphere' (IGCP Project 24). As a result, one of the conclusions in the final report was: 'In the Alpine region the classic nomenclature of Penck & Brückner must be abandoned.' (Šibrava 1986).

Deep-sea research has succeeded in establishing an almost complete Quaternary stratigraphy that records, at least in broad detail, the course of palaeoclimatical changes. Comparable long sequences in the continents are only provided by the long lake-deposit sequences, such as that at Funza in Colombia (Hooghiemstra 1984) or the loess sections in China (Ding *et al.* 1992) and Central Asia (Dodonov 1979). In the meantime, the time scale of the deep-sea stratigraphy has been improved so that it must be considered reliable for the last 1.5 million years. Thus the oceanic sediments provide a Quaternary time scale to which the continental stratigraphies might eventually be fixed.

The extent to which the continental glacial history can be correlated with the deep-sea chronology is still uncertain today. It has been demonstrated that the two major glaciations of the Weichselian Cold Stage are depicted in the oxygen-isotope curve (Stages 2 and 4). On the other hand, it is also certain that the subdivision of the Weichselian maximum into various ice

Magneto-stratigraphy	Age	Oxygen-Isotope Stages	Chronostrat.	The Netherlands	North Germany	Bavaria	Switzerland + Baden-Württemberg
Brunhes Epoch			Late Pleist.	Weichselian Cold Stage Eemian Warm Stage	Weichselian Cold Stage Eemian Warm Stage	Würmian Cold Stage Riss/Würm Warm Stage	Würmian Cold Stage Krumbach Warm Stage (Eemian)
			Middle Pleistocene	Saalian Cold Stage	Saalian Cold Stage s.s. Wacken Warm Stage Fuhne Cold Stage	Riss Complex	Riss/Würm Complex: Doppelwall Riss Cold Stage
				Holsteinian Warm Stage	Holsteinian Warm Stage	Mindel/Riss Warm Stage	Samerberg Warm Stages (Holsteinian)
				Elsterian Cold Stage	Elsterian Cold Stage		Older Riss Cold Stages
				Cromerian IV (Noordbergum) Cromerian C Cold Stage		Mindel Complex	
				Cromerian III (Rosmalen)			Morphotectonic Revolution / most extensive glaciation
				Cromerian B Cold Stage Cromerian II (Westerhoven)		Günz/Mindel Warm Stage	
	780 000	19		Cromerian A Cold Stage		Günz Complex	
Matuyama Epoch			Early Pleistocene	Cromerian I (Waardenburg)			'Deckenschotter' Complex: Haslach-Mindel Complex
Jaramillo Event	900 000	27		Dorst Cold Stage Leerdam Warm Stage Linge Cold Stage	Pinneberg Warm Stage Elmshorn Cold Stage	Uhlenberg Warm Stage	Uhlenberg Warm Stage (flora)
	1 070 000	31		Bavel Warm Stage	Lieth Warm Stage		
				Menapian Cold Stage	Pinnau Cold Stage		Günz Complex
				Waalian C Warm Stage Waalian B Cold Stage Waalian A Warm Stage	Tornesch Warm Stage (TC) Cryomer (TB) Tornesch Warm Stage (TA)	Donau Complex	Morphotectonic Revolution
			←	Eburonian Cold Stage	Lieth Cold Stage		
Olduvai Event	1 770 000	64		Tiglian C5-6 Warm Stage Tiglian C4e (Beerse Member)			'Deckschotter' Complex: Uhlenberg Warm Stage (fauna) Donau Complex
	1 950 000	71		Tiglian C1-3 Warm Stage	Ellerhoop Warm Stage		
				Tiglian B Cold Stage Tiglian A Warm Stage	Krückau Cold Stage Nordende Warm Stage Ekholt Cold Stage Meinweg Warm Stage	Biber Complex	
				Praetiglian Cold Stage	Praetiglian Cold Stage		Biber Complex
Gauss Epoch	2 600 000	104	← Tertiary				

advances (such as Brandenburg Phase, Pomeranian Phase etc. in Germany, or the Nissouri, Port Bruce and Port Huron in the Eastern Great Lakes region of North America) is not depicted. Even if these advance and retreat phases represent events that manifest themselves worldwide in perceptible changes of ice volume, identification would be improbable because of their short duration and the restricted resolution of the oxygen-isotope curve, caused by low sedimentation rates and mixing by bioturbation. Likewise, even the far more detailed ice-core records from Greenland and Antarctica do not reflect the minor ice-sheet oscillations. However, unlike in the deep-sea cores, the Younger Dryas event is clearly visible in the ice-core records (Reeh *et al.* 1991) (Fig. 127). In contrast, the ice cores do show oscillations which are difficult to interpret. These include the Dansgaard–Oeschger cycles, strong alterations in oxygen-isotope composition in the cold-stage part of the sequence (Dansgaard *et al.* 1984, Oeschger *et al.* 1984). The cycles have a duration of 500 to 2000 years. It is assumed that they may result from changes in the intensity or direction of the North Atlantic current, associated with changes in deep water formation (Paterson 1994).

Whereas the climatic changes of the Late Pleistocene are relatively well known, the events before the Eemian Interglacial can only be reconstructed in broad outline. Thus correlation of the terrestrial stratigraphy with the oxygen-isotope curve is impeded. For instance, it remains unclear what status the ice-free phase between the Drenthe and Warthe glaciations in northern Germany and Poland has within the oxygen-isotope stratigraphy or where the Fuhne Cold Stage and the Wacken Interglacial should be placed. Through palaeomagnetic analyses, however, some fixed points have been established (Table 12).

The negatively magnetised 'Cromerian Interglacial I' probably corresponds to Oxygen Isotope Stage 21. Consequently, 'Cromerian Interglacials II to IV' might be equivalent to Stages 19, 17 and 15. Yet assuming that traces of extensive glaciation in 'Cromerian Glacial A' should be clearly reflected in the oxygen-isotope sequence, it should probably be correlated with Stage 16. This would move the 'Cromerian Interglacials II to IV' to Stages 15, 13 and 11. In this case the strongly developed Stage 10 would correspond to the Elsterian Cold Stage, Stage 9 to the Holsteinian, Stage 7 to the Wacken/Dömnitz and Stage 6 to the Saalian Cold Stage. If this is correct, no continental equivalents would currently be known for Oxygen Isotope Stages 17 and 19. It would be conceivable that weakly developed Stage 19 was not an interglacial but only of interstadial rank. Such a classification, however, is merely speculative for the time being. The Holsteinian may still be Stage 11 (with the Elsterian assigned to the pronounced Oxygen Isotope Stage 12), or Wacken and Holsteinian may be represented by the 'double' Oxygen Isotope Stage 7. For the time span from the Matuyama/Brunhes boundary to the present time, the Chinese loess sequence includes seven interglacial soils, of which soils 2 and 5 are pedocomplexes of two and three separate soils respectively, each likely to indicate an interglacial interval (cf. An Zhisheng *et al.* 1990, Rutter *et al.* 1991) (Fig. 211). The soils S/2-I and S/2-II may be easily correlated with the double peak of Oxygen Isotope Stage 7, the three-fold soil S/5 might fall within the span of Stages 13 and 15.

TABLE 12 (opposite) Stratigraphical subdivision of the Quaternary deposits in northern Germany and the Alps. Both possible Quaternary/Tertiary boundaries are marked by arrows. Ages of palaeomagnetic reversals and correlation with deep-sea oxygen-isotope stages according to Shackleton *et al.* (1990). Stratigraphy after Menke (1975), Gibbard *et al.* (1991), Ellwanger *et al.* (1995), Doppler, Ellwanger & Schlüchter (personal communication)

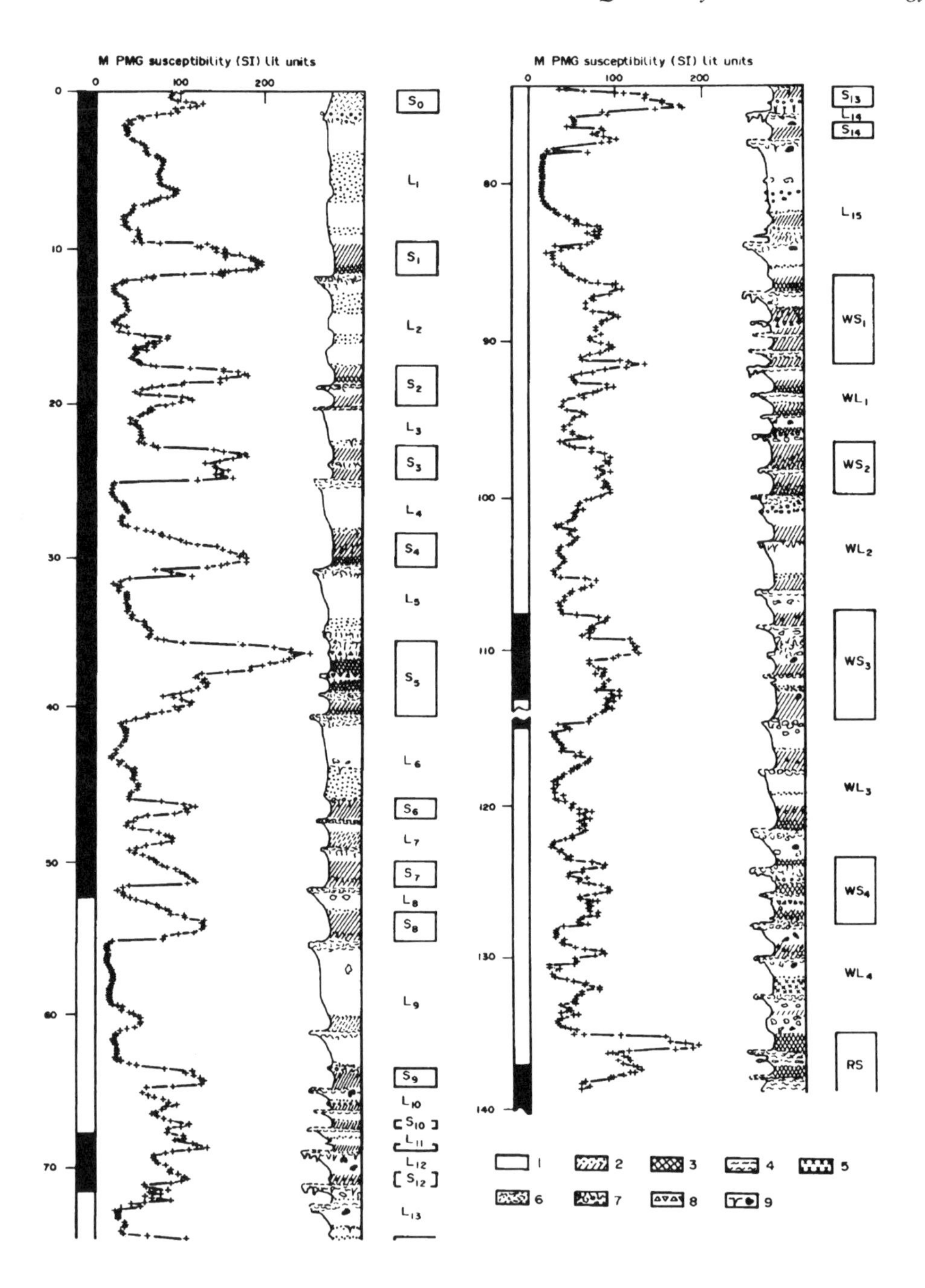

FIGURE 211 Palaeomagnetic susceptibility of the Luochuan loess/palaeosol sequences (China). 1 = loess; 2 = bioturbated soil; 3 = B_t horizons; 4 = clay; 5 = pseudogley; 6 = soft carbonate precipitates; 7 = loesskindl and carbonate banks; 8 = corroded concretions; 9 = wedges and crotowinas (from An Zhisheng *et al.* 1990)

The discrepancies are even greater between the continental and the deep-sea chronologies in the older part of the geological time scale. There is no obvious correlation for the Alpine Donau and Biber Cold Stages. Whereas the Dutch Pleistocene stratigraphy currently includes six interglacials for the period before the Jaramillo Event, 25 interglacial fossil soils have been recognised for the same period in the Chinese loess sequences. Neither of these successions offers sufficient events to match the 40 'interglacial periods' in the deep-sea oxygen-isotope record between the base of the Jaramillo Event and the Gauss/Matuyama boundary. However, it must be emphasised that the oxygen isotope curve does not directly indicate temperature oscillations, so that some of the peaks need not represent warm intervals in the continental record.

Recent discoveries from North America and from the Alpine region demonstrate that the number of cold stages including extensive glaciations were many more than previously assumed. However, in most regions very few traces of these early glaciations have been preserved. This is particularly the case in Europe.

The first indications of Early Pleistocene glaciations come from **North America** (see chapter 13). The earliest glaciations in the northern part of the continent are indicated by ice-rafted debris in the glaciomarine Yakataga Formation of Alaska. Our present knowledge suggests that North America may have experienced 10 or 11 major glaciations prior to the Illinoian. Tephra layers within the Quaternary sequence allow radiometric dating. One till, for example, underlies the Pearlette B Ash. This 'pre-Illinoian Stage K' till would be about the age-equivalent of the Praetiglian of the Netherlands. In Nebraska and Iowa four glaciations have been proven between Pearlette S and Pearlette O ash beds, two of which are in the reversely magnetised area (Hallberg 1986). In the Puget Lowland, in the Olympic Mountains and in Yellowstone Park at least three tills have been recognised as being older than the Matuyama/Brunhes boundary. In addition, in Illinois, Indiana and Wisconsin at least one till underlies reversely magnetised deposits (Richmond & Fullerton 1986).

At about the same time as the North American Early Pleistocene glaciations were discovered, Mercer *et al.* (1975) and Mercer (1976) showed by dating of volcanic deposits that there had also been early glaciation in **South America**. Patagonia was repeatedly glaciated between 3.0 and 1.8 million years ago. The glaciation of the Southern Andes reached its largest extent at about 1.0 million years ago, i.e. approximately the same time as the Menapian Cold Stage of northern Europe (cf. Rabassa & Clapperton 1990).

There is evidence that the **Greenland** Ice Sheet formed about 8 million years ago (Srivastava & Arthur 1987). Investigations of deep-sea cores from the North Atlantic have shown that massive ice-rafted detritus deposition set in at about 2.6 to 2.65 million years ago (Raymo *et al.* 1989). The glaciomarine deposits of Lodin Elv and Kap København (Greenland) are dated to this period (Penney 1993). A corresponding ^{18}O enrichment in sediment samples from the North Atlantic (around 2.6 million years old) also coincides with the estimated age for the oldest tillites in **Iceland** (Eiríksson & Geirsdóttir 1991).

In Iceland the deposits of 18 glaciations have been distinguished that can be dated by intercalated volcanic deposits. Rough estimates have also been made of the extent of the individual glaciations. From about two million years ago ice sheets developed to a size approximately corresponding to that of the local Weichselian glaciation. Eight of these glaciations occurred prior to the Matuyama/Brunhes boundary and four afterwards. The younger part of this sequence partly alternates with marine interglacial deposits, best exposed in the cliffs of the Tjörnes Peninsula in northern Iceland (Einarsson & Albertsson 1989). Near

Húsafell, west of Langjökull, tillites of the Gauss and Matuyama epochs outcrop between young lavas, and represent the oldest part of the Icelandic glacial sequence (Sæmundsson & Noll 1974). Recent investigations of tillites in the Borgarfjörður and Hvalfjörður region (north of Reykjavik) have also identified the Gauss/Matuyama boundary within the glacial sequence (Geirsdóttir 1990).

Recent investigations in the central **North Sea** (Fladen Ground) have also produced evidence of old glaciations (Sejrup *et al.* 1987). In a core drilling 200.6 m deep, a till was found at a depth of 146.55 to 158.0 m, within the Aberdeen Ground Formation that is reversely magnetised. This implies an age of greater than 780,000 years. In neighbouring drillings, Stoker & Bent (1985) detected till in the normally magnetised part of the same formation which they correlated to the 'Cromerian Glacial A'.

From the continental sector of northern Europe there is little concrete evidence of glaciation prior to the Elsterian (Fig. 212). In **Denmark**, near Harreskov (Jutland) Cromerian interglacial lacustrine sediments, found in the mid-1960s, are underlain by meltwater sands 5 m thick.

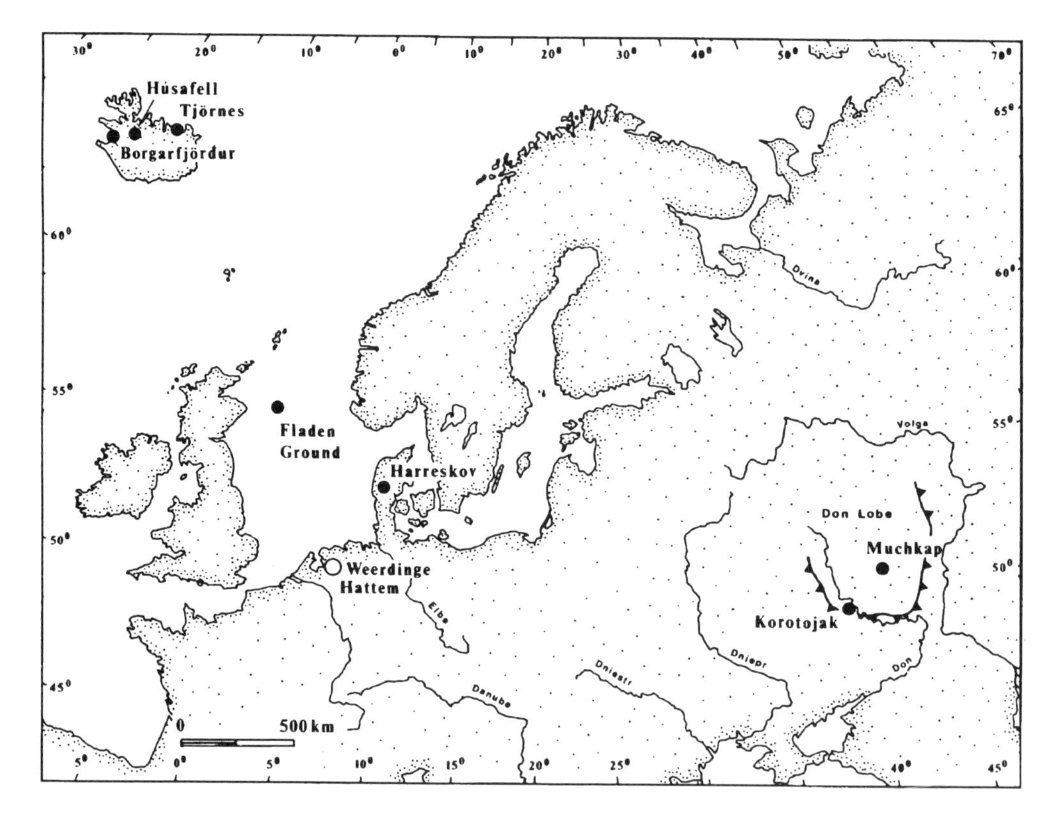

FIGURE 212 Evidence of pre-Elsterian glaciations in Europe. The occurrences are not of the same age. Tjörnes: 4 glaciations in the Brunhes, 8 in the Matuyama Epoch; Borgarfjörður and Hvalfjörður: 5 or 6 glaciations in the Gauss and Matuyama epochs. Húsafell: 7 tillites in the Gauss, 3 in the Matuyama Epoch. Fladen Ground: till from the Matuyama Epoch; Harreskov: till of unknown age underlying Cromerian deposits; Hattem: gravelly sands with Nordic clasts from the Matuyama Epoch; Weerdinge: gravelly sands with Scandinavian erratics from Cromerian Glacial Stage C; Don Lobe (Muchkap, Korotojak): Cromerian tills and possibly earlier glaciation indicated by Scandinavian erratics in fluvial deposits

These in turn overlie a till-like deposit about 5 m thick (S.Th. Andersen 1967). Almost the same sequence was encountered at Ølgod, as described by Jessen & Milthers (1928). However, at neither of these sites have the supposed glacial deposits been studied in detail.

In **Russia** the Quaternary glaciations have advanced in two vast ice lobes far towards the south. Originally, they were assumed to represent advances of the Dniepr Stage, the Russian equivalent of the Saalian (cf. Woldstedt 1958). Meanwhile clear evidence suggests that the eastern or Don Lobe is considerably older. In the 1970s, it became evident that the Don Lobe tills were overlain by deposits containing the Holsteinian Singil mammal fauna. The glaciation was thus correlated with the Elsterian Cold Stage. In the meantime, however, at sites near Muchkap and Korotojak, it has been demonstrated that the Don Till is intercalated with deposits containing the Cromerian Tiraspol mammal fauna. This indicates that a very extensive Cromerian glaciation (younger than the Matuyama/Brunhes boundary) occurred in Russia. The existence of another, even older glaciation has been discussed, but the evidence is disputed (Velichko & Faustova 1986). The existence of similarly early glaciations has been claimed in **Poland** (e.g. by Rzechowski 1986). However, the age estimates are largely based on thermoluminescence dating.

Corresponding deposits have not yet been demonstrated in **northern Germany**. In the karst hollows on salt domes that served as sediment traps, several pre-Elsterian interglacial sequences have been found but no tills are present. Within the minerogenic sediments Scandinavian indicator clasts were also absent (Menke 1975, H. Müller 1992). Only the so-called 'Hattem Beds' of the Dutch/German border region provide strong indirect evidence of glaciation (Fig. 213). These beds are gravel-rich units containing strongly weathered Nordic material (Lüttig & Maarleveld 1961). The strata, which are a member of the Dutch Enschede Formation, are assigned chronostratigraphically to the Menapian Stage. The Hattem Beds contain Scandinavian rocks as well as material from the Weser catchment area and the Thuringian Forest in central Germany. It is possible that this material was transported by drift ice for part of its way. Amongst the erratics are crystalline rocks, 'Backsteinkalk' clasts and Silurian chert. Flint is lacking, however.

A second bed with Scandinavian indicator erratics is found in the Netherlands in the so-called 'mixed zone' at the base of the Urk Formation ('Weerdinge Member'; Zandstra 1971). Ruegg & Zandstra (1977) are of the opinion that an ice sheet must have been in close proximity during deposition of the Weerdinge Member which is correlated with the 'Cromerian Glacial C' (cf. Zagwijn 1985).

Not only is current knowledge of the older glaciations still sketchy, but many questions still remain unanswered about the more recent glaciations. Even the precise extent of the last glaciation has not yet been unambiguously determined for most of the formerly glaciated area in Eurasia. It is certain that in the Novaya Zemlya region, an independent glaciation centre formed, from where glacial debris was transported far into the Russian plain. Moreover, the Putorana Mountains in the northwestern part of the central Siberian uplands were glaciated during the Weichselian (Valdaian) Stage. Whether this glaciation became confluent with the North European Ice Sheet in the Late Weichselian remains doubtful. In the late-1970s a series of latitude-parallel end-moraine ridges were discovered in the West Siberian lowland. It seems that the Ob and Irtysh rivers were temporarily impounded to form an ice-dammed lake over 1000 km long and over 200 km wide (Grosswald 1980). Also in the Yenisei region a glacial lake several hundred kilometres long temporarily existed. The age of these events, however, remains equivocal (see also chapter 14.4). For this reason Russian publications concerning the

FIGURE 213 Hattem Beds exposed in the Itterbeck–Uelsen thrust moraine, Niedersachsen (Photograph: Ehlers 1991)

Weichselian Glaciation present both a maximum and a minimum version of ice-sheet extent side by side (Arkhipov 1984, Arkhipov *et al.* 1986). Recent investigations suggest that the maximum version represents an older glaciation (Biryukov *et al.* 1988, Astakhov 1992). Sediment cores from the Arctic sea floor show no signs of major ice-rafting after the Eemian. The Barents Sea Ice Sheet has been discussed in chapter 7.3.2. There, as well as northwest of Franz Josef Land and Severnaya Zemlya, the shelf margin formed a natural border for the maximum extent of any terrestrial glaciation.

In East Siberia a Late Weichselian ice cap proper formed only in the Werchojansk Mountains. Arkhipov *et al.* (1986) assume that free drainage of the Lena River occurred during the Weichselian Stage, whereas Isayeva (1984) considers that the Lena must have been temporarily impounded. As a consequence of the worldwide marine regression during the Weichselian Glaciation, large portions of the epicontinental seas such as the North Sea and the Laptev Sea became dry land. The Black Sea also suffered a pronounced regression and temporarily became an inland sea. The Asov Sea also dried out. On the other hand, the Caspian Sea greatly expanded – partly because of reduced evaporation, and partly as a consequence of the glacially enlarged catchment area of the Volga River.

This brief overview demonstrates that some fundamental questions of Quaternary stratigraphy still remain unsolved. The climatic history of the Ice Age will only be elucidated when a sufficient basis of well-established regional investigations is available. As long as this remains unavailable for major parts of the world, our reconstruction attempts of the extent and effect of the Quaternary climatic oscillations will inevitably still contain fundamental errors.

17

Outlook

Quaternary research does not deal with a random period of the earth's history but with the period in which we live. The Holocene (sometimes still referred to as the 'Recent' by geologists) is an integrated part of the Quaternary. For this reason Paul Woldstedt, when he founded the new yearbook of the German Quaternary Research Association (DEUQUA) after the war, called it *Eiszeitalter und Gegenwart* (Ice Age and Recent). In his introductory remarks in the first volume he expressly suggests that one of the main aims of Quaternary research must be 'to contribute to our understanding of the recent (modern) period and of (the consequences of) our presence in it' (Woldstedt 1951: 15).

The extreme climatic oscillations of the Quaternary are well recorded worldwide because of their relatively young age. Causes, courses and consequences of these oscillations may provide indications about possible future climatic trends. Whilst the Quaternary stratigraphical framework has been greatly improved during the last few decades, knowledge of the impact of climatic changes is still at an early stage. Projects such as the 'Climate Mapping, Analysis and Prediction' project (CLIMAP) and the 'Co-operative Holocene Mapping Project' (COHMAP) are important steps towards that aim (Wright *et al.* 1993).

Whilst major ice sheets have a response time of possibly thousands of years, mountain glaciers react rapidly to climatic changes. The retreat of the Alpine glaciers in the second half of this century reflects the reality of the effects of gradual warming. Less visible, but of great ecological and economical significance, is the simultaneous reduction of permafrost (Nelson *et al.* 1993). In a recent unpublished study of the Inter-governmental Panel on Climate Change (IPCC), a global reduction of permafrost areas by 16% is anticipated by the year 2050. This is based on the assumption that gradual climatic change will occur over this period.

Over the last few years it has been known that, apart from the major warm and cold stages of the Quaternary, a number of brief but rather intense climatic oscillations have affected Quaternary climates. In a benchmark paper, Heinrich (1988) demonstrated that within the Weichselian sediments of the North Atlantic, a number of layers with a relatively high content of coarse clastic sediment were present, which must have been deposited by drift ice; the so-called 'Heinrich layers'. The youngest of the related so-called 'Heinrich events' are correlated with major meltwater outbursts from the North American Laurentide Ice Sheet (Andrews *et al.* 1994). These are attributed by P.U. Clark (1994) to the North American Ice Sheet becoming unstable, in the context of warming from the base and the presence of a deformable bed.

The Holocene climate has so far remained extraordinarily stable. Oscillations like the 'Little Ice Age' were characterised by a drop in temperature of only a few degrees. Recent investigations of Greenland ice cores, however, suggest that the climate of the preceding warm stage was possibly much less stable. The ice samples from the Eemian seem to indicate that incursions of cold-stage climate in the middle of the interglacial occurred repeatedly, and that the transition took only a few decades (Anklin *et al.* 1993, Dansgaard *et al.* 1993). However, indications of such dramatic changes have not yet been found in any of the organic deposits from the last warm stage in Europe. The pollen diagrams from the Eemian give no hints of any sudden, intensive climatic deteriorations. Further investigations are needed before the results of the ice-core investigations can be properly assessed.

Climatic changes have a strong influence on global sea level. During the last interglacial, the Eemian, the climatic optimum seems to have been slightly warmer than the Holocene optimum. Moreover, the CO_2 content of the atmosphere during the early Eemian seems to have been markedly higher than during the early Holocene (Barnola *et al.* 1987). If and to what degree sea level has reacted to these differences is only poorly known. It is assumed that during the Eemian sea level was slightly higher than during the Holocene. However, the Holocene example also shows that the sea-level maximum is by no means contemporaneous with the climatic optimum. Because of strong anthropogenic alteration of the atmospheric composition during the last 100 years questions about the impact of those changes on climate and sea level have focused current investigations. A considerable contribution to answering these questions must come from Quaternary research.

References

Aario, R. 1977a. Classification and terminology of morainic landforms in Finland. *Boreas* 6: 87–100.

Aario, R. 1977b. Associations of flutings, drumlins, hummocks and transverse ridges. *GeoJournal* 1: 65–72.

Aartolahti, T. 1974. Ring ridge hummocky moraines in northern Finland. *Fennia* 134: 22 pp.

Aartolahti, T. 1995. Glacial morphology in Finland. In: Ehlers, J., Kozarski, S. & Gibbard, P.L. (eds): *Glacial Deposits in North-East Europe:* 37–50. Rotterdam: Balkema.

Abele, G. 1994. Large rockslides: their causes and movement on internal sliding planes. *Mountain Research and Development* 14: 315–320.

Abelmann, A. 1985. Palökologische und ökostratigraphische Untersuchungen von Diatomeenassoziationen an holozänen Sedimenten der zentralen Ostsee. *Berichte – Reports, Geologisch-Paläontologisches Institut der Universität Kiel* 9: 200 pp.

Abelson, P. H. 1954. Amino acids in fossils. *Science* 119: 576.

Abelson, P. H. 1955. Paleobiochemistry. *Carnegie Institute, Washington, Yearbook* 54: 107–109.

Aber, J. S. 1989. Spectrum of constructional glaciotectonic landforms. In: Goldthwait, R.P. & Matsch, C.L. (eds): *Genetic Classification of Glacigenic Deposits*: 281–292. Rotterdam: Balkema.

Aber, J. S. 1991. The glaciation of northeastern Kansas. *Boreas* 20: 297–314.

Aber, J. S. & Bluemle, J. P. 1991. *Great Plains Glaciotectonics*. North Dakota Geological Survey, Miscellaneous Map 31.

Aber, J. S., Croot, D. G. & Fenton, M. M. 1989. *Glaciotectonic Landforms and Structures.* Dordrecht: Kluwer. 201 pp.

Āboltiņš, O. P. 1989. *Gliatsiostruktura i lednikovyi morfogenez.* Riga: Zinatne. 284 pp.

Āboltiņš, O. P. & Dreimanis, A. 1995. Glacigenic deposits in Latvia. In: Ehlers, J., Kozarski, S. & Gibbard, P. L. (eds): *Glacial Deposits in North-East Europe*: 115–124. Rotterdam: Balkema.

Āboltiņš, O. P. & Markots, A. 1995. Skujene plateau-like hills area. In: *INQUA Excursion C 3, Guide.*

Adam, K. D. 1951. Der Waldelefant von Lehringen – eine Jagdbeute des diluvialen Menschen. *Quartär* 5: 79–92.

Adam, K. D. 1954. Die mittelpleistozänen Faunen von Steinheim an der Murr (Württemberg). *Quarternaria* 1: 131–144.

Agassiz, L. 1840. *Études sur les glaciers.* Neuchâtel: Jent & Gaßmann. 346 pp. + Atlas.

Agassiz, L. 1850. *Lake Superior: Its Physical Character, Vegetation and Animals Compared to Those of Other and Similar Regions*. Boston: Gould, Kendall & Lincoln. 428 pp.

Agrell, H. 1976. The highest coastline in south-eastern Sweden. *Boreas* 5: 143–154.

Ahlbrandt, T. S., Swineheart, J. B. & Maroney, D. G. 1983. The dynamic Holocene dunefields of the Great Plains and Rocky Mountain basins, U.S.A. In: Brookfield, M. E. & Ahlbrandt, T. S. (eds): *Eolian sediments and processes*: 379–406. Amsterdam: Elsevier.

Ahlmann, H. W. 1935. Contribution to the physics of glaciers. *Geographical Journal* 86: 97–113.

Åhman, R. 1976. The structure and morphology of minerogenic palsas in northern Norway. *Biuletyn Peryglacjalny* 26: 25–31.

Ahrens, H. & Lotsch, W. 1976. Zum Problem des Pliozäns in Brandenburg. *Jahrbuch für Geologie* 7/8: 277–323.

Ahrens, W. 1927. Das Alter des großen mittelrheinischen Bimssteinausbruchs und sein Verhältnis zu den jüngsten Rheinterrassen. *Geologische Rundschau* 18: 45–59.

Aikens, C. M. 1984. Environmental Archaeology in the Western United States. In: H. E. Wright (ed.): *Late Quaternary Environments of the United States, Vol. 2, The Holocene*: 239–251. London: Longman.

Aitken, M. J. 1985. *Thermoluminescence dating*. London: Academic Press. 359 pp.

Aitken, M. J. 1994. Optical dating: A non-specialist review. *Quaternary Science Reviews* 13: 503–508.

Akers, W. H. & Holck, A. J. J. 1957. Pleistocene beds near the edge of the continental shelf, southeastern Louisiana. *Geological Society of America Bulletin* 68: 983–992.

Aksu, A. 1985. Climatic and oceanographic changes of the past 400,000 years: evidence from deep-sea cores on Baffin Bay and Davis Strait. In: Andrews, J. T. (ed.): *Quaternary Environments: Eastern Canadian Arctic, Baffin Bay, and West Greenland*: 181–209. London: Allen & Unwin.

Aktas, A. 1987. Altquartäre Schotter der Zusam-Platte, Bayerisch Schwaben. *Geologisches Institut der Universiät zu Köln, Sonderveröffentlichungen* 62: 100 pp.

Aldrich, L. T. & Nier, A. L. 1948. Argon-40 in potassium minerals. *Physical Reviews* 74: 876–877.

Alessio, M., Allegri, L., Bella, F., Belluomini, G., Calderoni, G., Cortesi, C., Improta, S., Manfra, L. & Orombelli, G. 1979. I depositi lacustri di Rovagnate, di Pontida e di Pianico in Lombardia: datazione con il ^{14}C. *Geografia Fisica e Dinamica Quaternaria* 1: 131–137.

Alford, J. J., Kolb, C. R. & Holmes, J. C. 1983. Terrace stratigraphy in the Tunica Hills of Louisiana. *Quaternary Research* 19: 55–63.

Allen, P. (ed.) 1984. *Field guide (revised edition, October 1984) to the Gipping and Waveney Valleys. May, 1982.* Cambridge: Quaternary Research Association. 116 pp.

Allen, P. 1991. Deformation structures in British Pleistocene sediments. In: Ehlers, J., Gibbard, P. & Rose, J. (eds): *Glacial Deposits in Great Britain and Ireland*: 455–469. Rotterdam: Balkema.

Allen, W. H., Jr. & Ward, R. A. 1979. The Brussels Formation – A stratigraphic re-appraisal. In: Mahaney, W. C. (ed.): *Quaternary Soils*: 167–185. Norwich: Geo Abstracts.

Alley, R. B. 1989. Water-pressure coupling of sliding and bed deformation: II. Velocity–depth profiles. *Journal of Glaciology* 35: 119–129.

Alley, R. B. 1991. Deforming-bed origin for southern Laurentide till sheets? *Journal of Glaciology* 37: 67–76.

Alley, R. B., Blankenship, D. D., Bentley, C. R. & Rooney, S. T. 1986. Deformation of till beneath ice stream B, West Antarctica. *Nature* 322: 57–59.

Alley, R. B., Blankenship, D. D., Rooney, S. T. & Bentley, C. R. 1987. Continuous till deformation beneath ice sheets. In: Waddington, E. D. & Walder, J. S. (eds): *The physical basis of ice sheet modelling. Proceedings of the Vancouver Symposium, August 1987. IAHS Publication* 170: 81–91.

Alley, R. B., Bond, J., Chappelaz, C., Clapperton, A., Del Genio, L., Keigwin, L. & Peteet, D. 1993. Global Younger Dryas? *EOS*: 586–588.

Ammann, B. & Lotter, A. F. 1989. Late-Glacial radiocarbon- and palynostratigraphy on the Swiss Plateau. *Boreas* 18: 109–126.

Ammann, B., Lotter, A. F., Eicher, U., Gaillard, M.-J., Wohlfarth, B., Haeberli, W., Lister, G., Maisch, M., Niessen, F. & Schlüchter, Ch. 1994. The Würmian Late-glacial in lowland Switzerland. *Journal of Quaternary Science* 9: 119–125.

Ampferer, O. 1939. Über einige Formen der Bergzerreißung. *Sitzungsberichte der Akademie der Wissenschaften, Wien, Mathematisch-Naturwissenschaftliche Klasse, Abteilung* I, 148: 1–14.

An Zhisheng, Liu Tungsheng, Porter, S. C., Kukla, G., Wu Xihao & Hua Yingming 1990. The long-term paleomonsoon variation record by the loess–paleosol sequence in Central China. *Quaternary International* 7/8: 91–95.

Andersen, B. G. 1979. The deglaciation of Norway 15,000–10,000 B.P. *Boreas* 8: 79–87.

Andersen, B. G. 1981. Late Weichselian Ice Sheets in Eurasia and Greenland. In: Denton, G. H. & Hughes, T. (eds): *The Last Great Ice Sheets*: 3–35. New York: Wiley.

Andersen, B. G. & Borns, H. W. Jr. 1994. *The Ice Age World*. Oslo: Scandinavian University Press. 208 pp.

Andersen, B. G. & Mangerud, J. 1989. The Last Interglacial–Glacial cycle in Fennoscandia. *Quaternary International* 3/4: 21–29.

Andersen, B. G. & Nesje, A. 1992. Quantification of Late Cenozoic glacial erosion in a fjord landscape. *Sveriges Geologiska Undersökning* Ca 81: 15–20.
Andersen, B. G., Lundqvist, J. & Saarnisto, M. 1995. The Younger Dryas margin of the Scandinavian Ice Sheet – An introduction. *Quaternary International* 28: 145–146.
Andersen, S. Th. 1957. New investigations of interglacial fresh-water deposits in Jutland. A preliminary report. *Eiszeitalter und Gegenwart* 8: 181–186.
Andersen, S. Th. 1961. Vegetation and its environment in Denmark in the Early Weichselian Glacial (Last Glacial). *Danmarks Geologiske Undersøgelser, Række* 2, 75: 1–175.
Andersen, S. Th. 1965. Interglacialer og interstadialer i Danmarks kvartær. *Meddelelser fra Dansk Geologisk Forening* 15: 486–506.
Andersen, S. Th. 1967. Istider og mellemistider. In: Nørrevang & Meyer (eds): *Danmarks Natur*, Vol. 1: 199–250. Copenhagen: Politiken.
Anderson, T. W. & MacPherson, J. B. 1994. Wisconsinan Late-glacial environmental change in Newfoundland: a regional review. *Journal of Quaternary Science* 9: 171–178.
Andersson, I. G. 1906. Solifluction, a component of subaerial denudation. *Journal of Geology* 14: 91–112.
Andrews, J. T. 1970. Differential crustal recovery and glacial chronology (6,700–0 BP), west Baffin Island, N.W.T., Canada. *Arctic and Alpine Research* 2: 115–134.
Andrews, J. T. 1987. The Late Wisconsin glaciation and deglaciation of the Laurentide Ice Sheet. In: Ruddiman, W. F. & Wright, H. E. Jr. (eds): *North American and Adjacent Oceans During the Last Deglaciation. Geological Society of America. The Geology of North America* K-3: 13–37.
Andrews, J. T. 1989. Quaternary geology of the northeastern Canadian Shield. In: Fulton, R. J. (ed.): *Quaternary Geology of Canada and Greenland. Geological Society of America, The Geology of North America* K-1: 276–317.
Andrews, J. T. & Tedesco, K. 1992. Detrital carbonate-rich sediments, northwestern Labrador Sea: Implications for ice-sheet dynamics and iceberg rafting (Heinrich) events in the North Atlantic. *Geology* 20: 1087–1090.
Andrews, J. T., Shilts, W. W. & Miller, G. H. 1983. Multiple deglaciations of the Hudson Bay Lowlands, Canada, since deposition of the Missinaibi (last-interglacial?) Formation. *Quaternary Research* 19: 18–37.
Andrews, J. T., Erlenkeuser, H., Tedesco, K., Aksu, A. E. & Jull, A. J. T. 1994. Late Quaternary (Stage 2 and 3) Meltwater and Heinrich Events, Northwest Labrador Sea. *Quaternary Research* 41: 26–34.
Anklin, M., Barnola, J. M., Beer, J., Blunier, T., Chappellaz, J., Clausen, H. B., Dahl-Jensen, D., Dansgaard, W., De Angelis, M., Delmas, R. P., Duval, P., Fratta, M., Fuchs, A., Fuhrer, K., Gundestrup, N., Hammer, C., Iversen, P., Johnsen, S., Jouzel, J., Kipfstuhl, J., Jegrand, M., Lorius, C., Maggi, V., Miller, H., Moore, J. C., Oeschger, H., Orombelli, G., Peel, D. A., Raisbeck, G., Raynaud, D., Schött-Hvidberg, C., Schwander, J., Shoji, H., Souchez, R., Stauffer, B., Steffensen, J. P., Stievenard, M., Sveinbjörnsdottir, A., Thorsteinsson, T. & Wolff, E. W. 1993. Climate instability during the last interglacial period recorded in the GRIP ice core. *Nature* 364: 203–218.
Antevs, E. 1925. Retreat of the last Ice-sheet in eastern Canada. *Geological Survey of Canada, Memoir* 146: 142 pp.
Antevs, E. 1928. The last glaciation. *American Geographical Society, Research Series* 17: 292 pp.
Antevs, E. 1948. The Great Basin, with emphasis on glacial and post-glacial times: Climatic changes and Pre-White man. *Bulletin of the University of Utah* 38: 168–191.
Antevs, E. 1952. Cenozoic climates of the Great Basin. *Geologische Rundschau* 40: 94–108.
Arkhipov, S. A. 1984. Late Pleistocene Glaciation of Western Siberia. In: A. A. Velichko (ed.): *Late Quaternary Environments of the Soviet Union*: 13–19. London: Longman.
Arkhipov, S. A., Bespaly, V. G., Faustova, M. A., Glushkova, O. Yu, Isaeva, L. L. & Velichko, A. A. 1986. Ice-sheet reconstructions. *Quaternary Science Reviews* 5: 475–483.
Arkhipov, S. A., Ehlers, J., Johnson, R. G. & Wright, H. E., Jr 1995. Glacial drainage towards the Mediterranean during the Middle and Late Pleistocene. *Boreas* 24: 196–206.
Armstrong, J. E., Crandell, D. R., Easterbrook, D. J. & Noble, J. B. 1965. Late Pleistocene stratigraphy and chronology in southwestern British Columbia and northwestern Washington. *Geological Society of America Bulletin* 76: 321–330.
Arn, R. 1984. Contribution à l'étude stratigraphique du Pléistocène de la région lémanique. Dissertation, Lausanne. 307 pp.

Arnold, J. R. 1956. Beryllium-10 produced by cosmic rays. *Science* 124: 584–585.
Aschrafi, A. 1981. Schwermineral-Untersuchungen in den Terrassensedimenten der Leine und ihrer Nebenflüsse Rhume und Innerste. Dissertation, Universität Hannover. 76 pp.
Aseev, A. A. 1968. Dynamik und geomorphologische Wirkung der europäischen Eisschilde. *Petermanns Geographische Mitteilungen* 112: 112–115.
Ashley, G. M. 1975. Rhythmic sedimentation in Glacial Lake Hitchcock, Massachusetts–Connecticut. In: Jopling, A. V. & McDonald, B. C. (eds): *Glaciofluvial and glaciolacustrine sedimentation. Society of Economic Paleontologists and Mineralogists, Special Publication* 23: 304–320.
Astakhov, V. 1992. The last glaciation in West Siberia. *Sveriges Geologiska Undersökning* Ca 81: 21–30.
Attig, J. W., Mickelson, D. M. & Clayton, L. 1989. Late Wisconsin landform distribution and glacier-bed conditions in Wisconsin. *Sedimentary Geology* 62: 399–405.
Augustowski, B. 1965. Pattern and development of ice marginal streamways of the Kashubian coast. *Geographia Polonica* 6: 35–42.
Autin, W. J., Burns, S. F., Miller, B. J., Saucier, R. T. & Snead, J. I. 1991. Quaternary geology of the Lower Mississippi Valley. In: Morrison, R. B. (ed.): *Quaternary Nonglacial Geology: Conterminous U.S. Geological Society of America. The Geology of North America* K-2: 547–582.
Averdieck, F.-R. 1967. Die Vegetationsentwicklung des Eem-Interglazials und der Frühwürm-Interstadiale von Odderade/Schleswig-Holstein. *Fundamenta* B 2: 101–125.
Azzaroli, A., de Guili, C., Ficcarelli, G. & Torre, D. 1988. Late Pliocene to early mid-Pleistocene mammals in Eurasia: Faunal succession and dispersal events. *Palaeogeography, Palaeoclimatology, Palaeoecology* 66: 77–100.
Backman, J., Shackleton, N. J. & Tauxe, L. 1983. Quantitative nannofossil correlation to open deep sea sections from Plio-Pleistocene boundary at Vrica Italy. *Nature* 304: 156–158.
Bajc, A. F. 1986. Molluscan paleoecology and Superior basin water levels, Marathon, Ontario. M.Sc. thesis, University of Waterloo. 271 pp.
Baker, R. W., Diehl, J. F., Simpson, T. W., Zelazny, L. W. & Beske-Diehl, S. 1983. Pre Wisconsinan glacial stratigraphy, chronology, and paleomagnetics of west-central Wisconsin. *Geological Society of America Bulletin* 94: 1442–1449.
Baker, V. R. 1983. Late-Pleistocene fluvial systems. In: Wright, H. E. (ed.): *Late-Quaternary Environments of the United States, Volume 1: The Late Pleistocene*: 115–129. Minneapolis: University of Minnesota Press.
Ballantyne, C. K. 1984. The Late Devensian periglaciation of upland Scotland. *Quaternary Science Reviews* 3: 311–343.
Ballantyne, C. K. 1987. The present-day periglaciation of upland Britain. In: Boardman, J. (ed.): *Periglacial processes and landforms in Britain and Ireland*: 113–126. Cambridge: Cambridge University Press.
Ballantyne, C. K. & Harris, Ch. 1994. *The Periglaciation of Great Britain*. Cambridge: Cambridge University Press. 330 pp.
Balson, P. S. & Cameron, T. D. J. 1985. Quaternary mapping offshore East Anglia. *Modern Geology* 9: 221–239.
Banham, P. H. 1975. Glacitectonic structures: a general discussion with particular reference to the contorted drift of Norfolk. In: Wright, A. E. & Moseley, F. (eds): *Ice Ages: Ancient and Modern*: 69–94. Liverpool: Seel House Press.
Bard, E., Arnold, M., Mangerud, J., Paterne, M., Labeyrie, L., Duprat, J., Mélières, M.-A., Sønstegaard, E. & Duplessy, J.-C. 1994. The North Atlantic atmosphere – sea surface ^{14}C gradient during the Younger Dryas climatic event. *Earth and Planetary Science Letters* 126: 275–287.
Barnes, J. W., Lang, E. J. & Potratz, K. A. 1956. Ratio of ionium to uranium in coral limestone. *Science* 124: 175–176.
Barnes, P. W., Reimnitz, E. & Fox, D. 1982. Ice rafting of fine-grained sediment, a sorting and transport mechanism, Beaufort Sea, Alaska. *Journal of Sedimentary Petrology* 52: 493–502.
Barnett, P. J. 1985. Glacial retreat and lake levels, North Central Lake Erie Basin, Ontario. In: Karrow, P. F. & Calkin, P. E. (eds): *Quaternary Evolution of the Great Lakes. Geological Association of Canada Special Paper* 30: 185–194.
Barnett, P. J. 1992. Quaternary geology of Ontario. In: *Geology of Ontario, Ontario Geological Survey, Special Volume* 4, Part 2: 1011–1088. Ontario: Ministry of Northern Development and Mines.

Barnola, J. M., Raynaud, D., Korotkevich, Y. S. & Lorius, C. 1987. Vostok ice core provides 160,000-year record of atmospheric CO_2. *Nature* 329: 408–414.

Barnosky, A. D. 1986. 'Big Game' Extinction Caused by Late Pleistocene Climatic Change: Irish Elk (*Megaloceros giganteus*) in Ireland. *Quaternary Research* 25: 128–135.

Barrie, J. V. 1980. Iceberg–sea bed interaction (northern Labrador Sea). *Annals of Glaciology* 1: 71–76.

Barsch, D. 1969. Studien und Messungen an Blockgletschern in Macun, Unterengadin. *Zeitschrift für Geomorphologie, Supplement-Band* 8: 11–30. *Also*: Studies and measurements on rock glaciers at Macun in the Lower Engadine. In: Evans, D. J. A. (ed.) 1994: *Cold Climate Landforms*: 457–473. Chichester: Wiley.

Barsch, D. 1992. Permafrost creep and rock glaciers. *Permafrost and Periglacial Processes* 3: 175–188.

Barsch, D. & Treter, U. 1976. Zur Verbreitung von Periglaziärphänomenen in Rondane/Norwegen. *Geografiska Annaler* 58 A: 83–93.

Bartz, J. 1976. Quartär und Jungtertiär im Raum Rastatt. *Jahreshefte des Geologischen Landesamts Baden-Württemberg* 18: 121–178.

Battarbee, R. W. 1973. Preliminary studies of Lough Neagh sediments, II. Diatom analysis from the uppermost sediment. In: Birks, H. J. B. & West, R. G. (eds): *Quaternary Plant Ecology*: 279–288. Oxford: Blackwell.

Bauberger, W. & Unger, H. J. 1984. *Geologische Karte von Bayern 1:25 000, Erläuterungen zum Blatt Nr. 7446 Passau.* München: Bayerisches Geologisches Landesamt. 175 pp.

Baumann, K.-H., Lackschewitz, K. S., Mangerud, J., Spielhagen, R. F., Wolf-Welling, T. C. W., Henrich, R. & Kassens, H. 1995. Reflection of Scandinavian Ice Sheet fluctuations in Norwegian Sea sediments during the past 150,000 years. *Quaternary Research* 43: 185–197.

Baumann, W. & Mania, D. 1983. Die paläolithischen Neufunde von Markkleeberg bei Leipzig. *Veröffentlichungen des Landesmuseums für Vorgeschichte Dresden* 16: 280 pp.

Baumberger, E., Gerber, E., Jeannet, A. & Weber, J. 1923. Die diluvialen Schieferkohlen der Schweiz. *Beiträge zur Geologie der Schweiz, Geotechnische Serie 8.*

Becker, B. 1982. Dendrochronologie und Paläoökologie subfossiler Baumstämme aus Flußablagerungen. Ein Beitrag zur nacheiszeitlichen Auentwicklung im südlichen Mitteleuropa. *Mitteilungen der Kommission für Quartärforschung der Österreichischen Akademie der Wissenschaften* 5: 120 pp.

Becker, B. 1983. Postglaziale Auwaldentwicklung im mittleren und oberen Maintal anhand dendrochronologischer Untersuchungen subfossiler Baumstammablagerungen. *Geologisches Jahrbuch* A 71: 45–59.

Becker, B. & Kromer, B. 1986. Extension of the Holocene dendrochronology by the Preboreal pine series, 8800 to 10,100 BP. *Radiocarbon* 28: 961–967.

Becker, B., Jäger, K.-D., Kaufmann, D. & Litt, T. 1989. Dendrochronologische Datierungen von Eichenhölzern aus den frühbronzezeitlichen Hügelgräbern bei Helmdorf und Leubingen (Aunjetitzer Kultur) und an bronzezeitlichen Flußeichen bei Merseburg. *Jahresschriften für Mitteldeutsche Vorgeschichte* 72: 299–312.

Becker, B., Kromer, B. & Trimborn, P. 1991. A stable-isotope tree-ring timescale of the Late Glacial/ Holocene boundary. *Nature* 353: 647–649.

Beer, J., Shen, C., Heller, F., Liu, T., Bonani, G., Dittrich, B., Suter, M. & Kubik, P. W. 1993. ^{10}Be and magnetic susceptibility in Chinese loess. *Geophysical Research Letters* 20: 57–60.

Beget, J. 1981. Early Holocene glacier advance in the North Cascade Range, Washington. *Geology* 9: 409–413.

Beget, J. 1987. Low profile of the northwest Laurentide ice sheet. *Arctic and Alpine Research* 19: 81–88.

Behre, K.-E. 1962. Pollen- und diatomeenanalytische Untersuchungen an letztinterglazialen Kieselgurlagern der Lüneburger Heide (Schwindebek und Grevenhof im oberen Luhetal). *Flora* 152: 326–370.

Behre, K.-E. 1988. The rôle of man in European vegetation history. In: Huntley, B. & Webb, T. III (eds): *Vegetation History*: 633–672. Dordrecht: Kluwer.

Behre, K.-E. 1989. Biostratigraphy of the Last Glacial Period in Europe. *Quaternary Science Reviews* 8: 25–44.

Behre, K.-E. & Lade, U. 1986. Eine Folge von Eem und 4 Weichsel-Interstadialen in Oerel/Niedersachsen und ihr Vegetationsablauf. *Eiszeitalter und Gegenwart* 36: 11–36.

Bell, R. 1872. Report on the country between Lake Superior and the Albany River. *Report of Progress for 1871–72, Geological Survey of Canada*: 101–114.

Bell, R. 1887. Report on an exploration of portions of the Attawapiskat and Albany Rivers. *Report of Progress for 1886, Geological Survey of Canada*: 1G–39G.
Benda, L. (ed.) 1995. *Das Quartär Deutschlands*. Berlin: Gebrüder Borntraeger. 408 pp.
Ber, A. 1987. Glaciotectonic deformation of glacial landforms and deposits in the Suwalki Lakeland (NE Poland). In: Van der Meer, J. J. M. (ed.): *Tills and Glaciotectonics*: 135–143. Rotterdam: Balkema.
Ber, A. 1990. Stratigraphy of the Quaternary of the Suwalki Lakeland and its substrate based on recent data. *Kwartalnik Geologiczny* 38: 463–478.
Berendt, G. 1879. Gletschertheorie oder Drifttheorie in Norddeutschland. *Zeitschrift der Deutschen Geologischen Gesellschaft* 31: 1–20.
Berger, A. L. 1977. Support for the astronomical theory of climate change. *Nature* 269: 44–45.
Berger, A. L. & Loutre, M. F. 1991. Insolation values for the climate of the last 10 million years. *Quaternary Science Reviews* 10: 297–317.
Berger, A. L., Imbrie, J., Hays, J., Kukla, G. & Saltzman, B. (eds) 1984. *Milankovitch and Climate*. Dordrecht: Reidel.
Berger, G. W., Pillans, B. J. & Palmer, A. S. 1992. Dating loess up to 800 ka by thermoluminescence. *Geology* 20: 403–406.
Bergersen, O. F. & Garnes, K. 1983. Glacial deposits in the culmination zone of the Scandinavian ice sheet. In: Ehlers, J. (ed.): *Glacial Deposits in North-West Europe*: 29–40. Rotterdam: Balkema.
Berglund, B. E. 1964. The Post-Glacial shore displacement in eastern Blekinge, southeastern Sweden. *Sveriges Geologiska Undersökning* C 559: 47 pp.
Berglund, B. E. & Lagerlund, E. 1981. Eemian and Weichselian stratigraphy in South Sweden. *Boreas* 10: 323–362.
Berglund, B. E., Björck, S., Lemdahl, G., Bergsten, H., Nordberg, K. & Kolstrup, E. 1994. Late Weichselian environmental change in southern Sweden and Denmark. *Journal of Quaternary Science* 9: 127–132.
Bergström, R. 1992. The Fennoscandian End Moraine Zone in eastern Sweden. *Sveriges Geologiska Undersökning* Ca 81: 43–50.
Bernhardi, A. 1832. Wie kamen die aus dem Norden stammenden Felsbruchstücke und Geschiebe, welche man in Norddeutschland und den benachbarten Ländern findet, an ihre gegenwärtigen Fundorte? *Jahrbuch für Mineralogie* 3: 257–267.
Berthelsen, A. 1975. Weichselian ice advances and drift successions in Denmark. *Bulletin of the Geological Institutions of the University of Uppsala, New Series* 5: 21–29.
Berthelsen, A. 1978. The methodology of kineto-stratigraphy as applied to glacial geology. *Bulletin of the Geological Society of Denmark* 27: 25–38.
Berti, A. A. 1975. Pollen and seed analysis of the Titusville Section (Mid-Wisconsinan), Titusville, Pennsylvania. *Canadian Journal of Earth Sciences* 12: 1675–1684.
Beskow, G. 1935. Praktiska och kvartärgeologiska resultat av grusinventeringen i Norrbottens län. *Geologiska Föreningens i Stockholm Förhandlingar* 57: 120–123.
Beug, H. J. 1972. Das Riß/Würm-Interglazial von Zeifen, Landkreis Laufen a.d. Salzach. *Bayerische Akademie der Wissenschaften, Mathematisch-Naturwissenschaftliche Klasse N.F.* 151: 46–75.
Beug, H. J. 1979. Vegetationsgeschichtlich-pollenanalytische Untersuchungen am Riß/Würm-Interglazial von Eurach am Starnberger See/Obb. *Geologica Bavarica* 80: 91–106.
Bibus, E. 1974. Abtragungs- und Bodenbildungsphasen im Rißlöß. *Eiszeitalter und Gegenwart* 25: 166–182.
Bibus, E. 1980. *Zur Relief-, Boden und Sedimententwicklung am unteren Mittelrhein. Frankfurter Geowissenschaftliche Arbeiten* D 1: 295 pp.
Bibus, E. 1989. Zur Gliederung, Ausbildung und stratigraphischen Stellung von Enzterrassen in Großbaustellen bei Vaihingen an der Enz. *Jahreshefte des Geologischen Landesamts Baden-Württemberg* 31: 7–22.
Bibus, E. 1990. Das Mindestalter des 'jüngeren Deckenschotters' bei Basel aufgrund seiner Deckschichten in der Ziegelei Allschwil. *Jahreshefte des Geologischen Landesamts Baden-Württemberg* 32: 223–234.
Bibus, E. 1995. Äolische Deckschichten, Paläoböden und Mindestalter der Terrassen in der Iller-Lech-Platte. *Geologica Bavarica* 99: 135–164.
Bibus, E. & Pasda, C. 1991. Zur feinstratigraphischen Gliederung und Einstufung eines Artefaktfundes im Lößprofil Heitersheim (Südbaden). *Quartär* 41/42: 195–202.

Bibus, E. & Semmel, A. 1977. Über die Auswirkung quartärer Tektonik auf die altpleistozänen Mittelrhein-Terrassen. *Catena* 4: 385–408.

Bijlsma, S. 1981. Fluvial sedimentation from the Fennoscandian area into the north-west European basin during the late Cenozoic. *Geologie en Mijnbouw* 60: 337–345.

Bik, M. J. J. 1969. The origin and age of the prairie mounds of southern Alberta, Canada. *Biuletyn Peryglacjalny* 19: 85–130.

Billard, A. 1985. Quaternary chronologies around the Alps. In: Mahaney, W. C. (ed.): *Correlation of Quaternary Chronologies*: 177–189. Norwich: GeoBooks.

Billard, A. & Orombelli, G. 1986. Quaternary Glaciations in the French and Italian Piedmonts of the Alps. *Quaternary Science Reviews* 5: 407–411.

Billard, A., Bucha, V., Horacek, J. & Orombelli, G. 1983. Preliminary paleomagnetic investigations on Pleistocene sequences in Lombardy, Northern Italy. *Rivista italiana di paleontologia i stratigrafia* 88: 295–317.

Bini, A., Cita, M. B. & Gaetani, M. 1978. Southern Alpine lakes – Hypothesis of an erosional origin related to the Messinian entrenchment. *Marine Geology* 27: 271–288.

Birks, H. H., Paus, A., Svendsen, J. I., Alm, T., Mangerud, J. & Landvik, J. Y. 1994. Late Weichselian environmental change in Norway, including Svalbard. *Journal of Quaternary Science* 9: 133–145.

Birks, H. J. B. & Birks, H. H. 1980. *Quaternary Paleoecology*. London: Edward Arnold. 289 pp.

Biryukov, V. Y., Faustova, M. A., Kaplin, P. A., Pavlidis, Y. A., Romanova, E. A. & Velichko, A. A. 1988. The palaeogeography of arctic shelf and coastal zone of Eurasia at the time of the last glaciation (18,000 yr B.P.). *Palaeogeography, Palaeoclimatology, Palaeoecology* 68: 117–125.

Bishop, W. W. 1958. The Pleistocene geology and geomorphology of three gaps in the Middle Jurassic escarpment. *Philosophical Transactions of the Royal Society of London* B 241: 255–306.

Bittmann, F. 1990. Neue biostratigraphische Korrelierung des Kärlicher Interglazials (Neuwieder Becken/ Mittelrhein). In: Schirmer, W. (ed.): Rheingeschichte zwischen Mosel und Maas. *Deuqua-Führer* 1: 67–70.

Björck, S. & Digerfeldt, G. 1984. Climatic changes at Pleistocene/Holocene boundary in the Middle Swedish end moraine zone, mainly inferred from stratigraphic indications. In: Mörner, N.-A. & Karlén, W. (eds): *Climatic Changes on a Yearly to Millennial Basis*: 37–56. Dordrecht: Reidel.

Björck, S. & Digerfeldt, G. 1989. Lake Mullsjön – a key site for understanding the final stage of the Baltic Ice Lake east of Mt. Billingen. *Boreas* 18: 209–219.

Björnsson, H. 1988. Hydrology of Ice Caps in Volcanic Regions. *Vísindafélag íslendinga Rit* XLV: 139 pp. + volume of maps.

Björnsson, H. 1992. Jökulhlaups in Iceland: prediction, characteristics and simulation. *Annals of Glaciology* 16: 95–106.

Black, R. F. 1965. Ice wedge casts of Wisconsin. *Wisconsin Academy of Sciences, Arts and Letters* 54: 187–222.

Black, R. F. 1976. Periglacial features indicative of permafrost. *Quaternary Research* 6: 3–26.

Blackwelder, E. 1915. Post-Cretaceous history of the mountains of central western Wyoming. *Journal of Geology* 23: 97–117, 193–217, 307–340.

Bleich, K., Hädrich, F., Hummel, P., Müller, S., Ortlam, D. & Werner, J. 1982. Paläoböden in Baden-Württemberg. *Geologisches Jahrbuch* F 14: 63–100.

Bloom, A. L., Broecker, W. S., Chappell, M. A., Matthews, R. K. & Mesolella, K. J. 1974. Quaternary sea level fluctuations on a tectonic coast: New $^{230}Th/^{234}U$ dates from the Huon Peninsula, New Guinea. *Quaternary Research* 4: 185–205.

Bludau, W. 1996. Pollenanalytische Untersuchungen interglazialer Sedimente des Profils Unterpfauzenwald (östliches Rheingletschergebiet). *Abhandlungen aus dem Geologischen Landesamt Baden-Württemberg* 15 (in press).

Bluemle, J. P. & Clayton, L. 1984. Large-scale glacial thrusting and related processes in North Dakota. *Boreas* 13: 279–299.

Blytt, A. 1876. *Immigration of the Norwegian Flora*. Christiania: Alb. Cammermeyer. 89 pp.

Bobrowsky, P. & Rutter, N. W. 1992. The Quaternary geologic history of the Canadian Rocky Mountains. *Géographie physique et Quaternaire* 46: 5–50.

Bock, W., Menke, B., Strehl, E. & Ziemus, H. 1985. Neuere Funde des Weichselspätglazials in Schleswig-Holstein. *Eiszeitalter und Gegenwart* 35: 161–180.

Boellstorff, J. 1978. North American Pleistocene stages reconsidered in light of probable Pliocene–Pleistocene continental glaciation. *Science* 202: 305–307.

Boenigk, W. 1970. Zur Kenntnis des Altquartärs bei Brüggen (westlicher Niederrhein). *Geologisches Institut der Universität Köln, Sonderveröffentlichungen* 17: 138 pp.

Boenigk, W. 1978. Gliederung der altquartären Ablagerungen in der Niederrheinischen Bucht. *Fortschritte in der Geologie von Rheinland und Westfalen* 28: 135–212.

Boenigk, W. 1981. Die Gliederung der tertiären Braunkohlendeckschichten in der Ville (Niederrheinische Bucht). *Fortschritte in der Geologie im Rheinland und Westfalen* 29: 193–263.

Boenigk, W. 1982. Der Einfluß des Rheingraben-Systems auf die Flußgeschichte des Rheins. *Zeitschrift für Geomorphologie, Supplement-Band* 42: 167–175.

Boenigk, W. 1983. *Schwermineralanalyse.* Stuttgart: Enke. 158 pp.

Boenigk, W. 1987. Petrographische Untersuchungen jungtertiärer und quartärer Sedimente am linken Oberrhein. *Jahresberichte und Mitteilungen des Oberrheinischen Geologischen Vereines N.F.* 69: 357–394.

Boenigk, W. 1990. Die pleistozänen Rheinterrassen und deren Bedeutung für die Gliederung des Eiszeitalters in Mitteleuropa. In: Liedtke, H. (ed.): *Eiszeitforschung*: 130–140. Darmstadt: Wissenschaftliche Buchgesellschaft.

Boenigk, W., Von der Brelie, G., Brunnacker, K., Koci, A., Schlickum, W. R. & Strauch, F. 1974. Zur Plio-Pleistozän-Grenze im Bereich der Ville (Niederrheinische Bucht). *Newsletters in Stratigraphy* 3: 219–241.

Bohncke, S., Vandenberghe, J. & Huizer, A. S. 1993. Periglacial environments during the Weichselian Late Glacial in the Maas Valley, The Netherlands. *Geologie en Mijnbouw* 72: 193–210.

Booth, D. B. 1994. Glaciofluvial infilling and scour of the Puget Lowland, Washington, during ice-sheet glaciation. *Geology* 22: 695–698.

Booth, D. B. & Hallet, B. 1993. Channel networks carved by subglacial water – Observations and reconstruction in the eastern Puget Lowland of Washington. *Geological Society of America Bulletin* 105: 671–683.

Bortenschlager, S. 1984. Beiträge zur Vegetationsgeschichte Tirols I. Inneres Ötztal und unteres Inntal. *Berichte des Naturwissenschaftlich-Medizinischen Vereins in Innsbruck* 71: 19–56.

Bortenschlager, I. & Bortenschlager, S. 1978. Pollenanalytische Untersuchung am Bänderton von Baumkirchen (Inntal, Tirol). *Zeitschrift für Gletscherkunde und Glazialgeologie* 14: 95–103.

Böse, M. 1979. Die geomorphologische Entwicklung im westlichen Berlin nach neueren stratigraphischen Untersuchungen. *Berliner Geographische Abhandlungen* 28: 46 pp.

Böse, M. 1990. Methodisch-stratigraphische Studien und paläomorphologische Untersuchungen zum Pleistozän südlich der Ostsee. *Berliner Geographische Abhandlungen* 51: 114 pp.

Böse, M. 1991. A palaeoclimatic interpretation of frost-wedge casts and aeolian sand deposits in the lowlands between Rhine and Vistula in the Upper Pleniglacial and Late Glacial. *Zeitschrift für Geomorphologie N.F. Supplement-Band* 90: 15–28.

Bosinski, G. 1974. Paläolithikum und Mesolithikum. In: Woldstedt, P. & Duphorn, K. (eds): *Norddeutschland und angrenzende Gebiete im Eiszeitalter*: 432–461. Stuttgart: Koehler.

Bosinski, G. 1985. *Der Neandertaler und seine Zeit. Kunst und Altertum am Rhein* 118: 74 pp. + plates.

Bosinski, G., Brunnacker, K., Lanser, K. P., Stephan, S., Urban, B. & Würges, K. 1980. Altpaläolithische Funde von Kärlich, Kreis Mayen-Koblenz (Neuwieder Becken). *Archäologisches Korrespondenzblatt* 10: 295–314.

Bosinski, G., Kröger, K., Schäfer, J. & Turner, E. 1986. Altsteinzeitliche Siedlungsplätze auf den Osteifelvulkanen. *Jahrbuch des Römisch-Germanischen Zentralmuseums* 33: 97–130.

Boulton, G. S. 1968. Flow tills and related deposits on some Vestspitsbergen glaciers. *Journal of Glaciology* 7: 391–412.

Boulton, G. S. 1970. On the deposition of subglacial and meltout tills at the margins of certain Svalbard glaciers. *Journal of Glaciology* 9: 231–245.

Boulton, G. S. 1971. Till genesis and fabric in Svalbard, Spitsbergen. In: R. P. Goldthwait (ed.): *Till – a Symposium*: 41–72. Columbus: Ohio State University Press.

Boulton, G. S. 1982. Subglacial processes and the development of glacial bedforms. In: Davidson-Arnott, R., Nickling, W. & Fahrey, B. D. (eds): *Research in glacial, glacio-fluvial and glacio-lacustrine systems. Proceedings of the 6th Guelph Symposium on Geomorphology, 1980*: 1–31. Norwich:

GeoBooks.

Boulton, G. S. 1986. A paradigm shift in glaciology? *Nature* 322: 18.

Boulton, G. S. 1987. A theory of drumlin formation by subglacial sediment deformation. In: Menzies, J. & Rose, J. (eds): *Drumlin Symposium*: 25–80. Rotterdam: Balkema.

Boulton, G. S. 1990. Sedimentary and sea level changes during glacial cycles and their control on glacimarine facies architecture. In: Dowdeswell, J. A. & Scourse, J. D. (eds): *Glacimarine Environments: Processes and Sediments. Geological Society Special Publication* No. 53: 15–52.

Boulton, G. S. & Clark, C. D. 1990. The Laurentide ice sheet through the last glacial cycle: the topology of drift lineations as a key to the dynamic behaviour of former ice sheets. *Transactions of the Royal Society of Edinburgh: Earth Sciences*, 81: 327–347.

Boulton, G. S. & Hindmarsh, R. C. A. 1987. Sediment deformation beneath glaciers: rheology and geological consequences. *Journal of Geophysical Research* 92: 9059–9082.

Boulton, G. S. & Jones, A. S. 1979. Stability of temperate ice caps and ice sheets resting on beds of deformable sediment. *Journal of Glaciology* 24: 29–43.

Boulton, G. S., Smith, G. D., Jones, A. S. & Newsome, J. 1985. Glacial geology and glaciology of the last mid-latitude ice-sheets. *Journal of the Geological Society of London* 142: 447–474.

Bourdier, F. 1963. *Le bassin du Rhône au Quaternaire*. Editions du Centre National de la Recherche Scientifique, Paris. 2 volumes, 363 pp.

Bowen, D. Q. 1978. *Quaternary geology*. Oxford: Pergamon Press. 221 pp.

Bowen, D. Q. 1991. Time and space in the glacial sediment systems of the British Isles. In: Ehlers, J., Gibbard, P. & Rose, J. (eds): *Glacial Deposits in Great Britain and Ireland*: 3–11. Rotterdam: Balkema.

Bowen, D. Q., Rose, J., McCabe, A. M. & Sutherland, D. G. 1986. Correlation of Quaternary glaciations in England, Ireland, Scotland and Wales. *Quaternary Science Reviews* 5: 299–340.

Boyce, J. I. & Eyles, N. 1991. Drumlins carved by deforming till streams below the Laurentide ice sheet. *Geology* 19: 187–790.

Boyd, D., Scott, D. B. & Douma, M. 1988. Glacial tunnel valleys and Quaternary history of the outer Scotian Shelf. *Nature* 333: 61–64.

Bradley, R. S. 1985. *Quaternary Paleoclimatology – Methods of Paleoclimatic Reconstruction*. Boston: Allen & Unwin. 472 pp.

Brakenridge, C. G. 1981. Late Quaternary floodplain sedimentation along the Pomme de Terre River, Southern Missouri. *Quaternary Research* 15: 62–76.

Brakenridge, C. G. 1987. Fluvial systems in the Appalachians. In: Graf, W. L. (ed.): *Geomorphic Systems of North America. The Geological Society of America, Centennial Special Volume* 2: 37–46.

Brandtner, F. 1954. Jungpleistozäner Löß und fossile Böden in Niederösterreich. *Eiszeitalter und Gegenwart* 4/5: 49–82.

Bremer, H. 1989. On the geomorphology of the South German Scarplands. *Catena Supplement* 15: 45–67.

Brenchley, P. J., Marshall, J. D., Carden, G. A. F., Robertson, D. B. R., Long, D. G. F., Meidla, T., Hints, L. & Anderson, T. F. 1994. Bathymetric and isotopic evidence for a short-lived Late Ordovician glaciation in a greenhouse period. *Geology* 22: 295–298.

Bridge, D. McC. & Hopson, P. M. 1985. Fine gravel, heavy mineral and grain-size analyses of Mid-Pleistocene glacial deposits in the lower Waveney valley, East Anglia. *Modern Geology* 9: 129–144.

Bridgland, D. R. 1980. A reappraisal of Pleistocene stratigraphy in north Kent and eastern Essex, and new evidence concerning the former courses of the Thames and Medway. *Quaternary Newsletter* 32: 15–24.

Bridgland, D. R. 1983. The Quaternary fluvial deposits of north Kent and east Essex. Ph.D. thesis (CNAA). City of London Polytechnic: 660 pp.

Bridgland, D. R. 1986. Clast lithological analysis. *Quaternary Research Association, Technical Guide* 3: 207 pp.

Bridgland, D. R. 1988. The Pleistocene fluvial stratigraphy and palaeogeography of Essex. *Proceedings of the Geologists' Association* 99: 291–314.

Bridgland, D. R. (ed.) 1994. *Quaternary of the Thames*. London: Chapman & Hall. 441 pp.

Bridgland, D. R. & d'Olier, B. 1995. The Pleistocene evolution of the Thames and Rhine drainage systems in the southern North Sea Basin. In: Preece, R. C. (ed.): *Island Britain: a Quaternary perspective. Geological Society of London, Special Publication* 96: 27–45.

Bridgland, D. R., Gibbard, P. L. & Preece, R. C. 1990. The geology and significance of the interglacial sediments at Little Oakley, Essex. *Philosophical Transactions of the Royal Society of London* B 328: 307–339.

Briggs, D. J. & Gilbertson, D. D. 1973. The age of the Hanborough Terrace of the River Evenlode, Oxfordshire. *Proceedings of the Geologists' Association* 84: 155–173.

Briggs, D. J. & Gilbertson, D. D. 1980. Quaternary processes and environments in the Upper Thames basin. *Transactions of the Institute of British Geographers* 5: 53–65.

Brinkmann, R. 1933. Über Kreuzschichtung im deutschen Buntsandsteinbecken. *Nachrichten der Akademie der Wissenschaften Göttingen* 24, *Mathematisch-Physikalische Klasse* IV: 32.

Bristow, C. R. & Cox, F. C. 1973. The Gipping Till: A reappraisal of East Anglian glacial stratigraphy. *Journal of the Geological Society of London* 129: 1–37.

Brodzikowski, K. & van Loon, A. J. 1987. A systematic classification of glacial and periglacial environments, facies and deposits. *Earth Science Reviews* 24: 297–381.

Broecker, W. S. 1965. Isotope geochemistry and the Pleistocene climatic record. In: Wright, H. E. Jr. & Frey, D. G. (eds): *The Quaternary of the United States*: 737–753. Princeton: Princeton University Press.

Broecker, W. S. & Denton, G. H. 1989. The role of ocean–atmosphere reorganisations in glacial cycles. *Geochimia Cosmochimia Acta* 53: 2465–2501.

Broecker, W. S. & Denton, G. H. 1990. The role of ocean–atmosphere reorganisations in glacial cycles. *Quaternary Science Reviews* 9: 305–343.

Broecker, W. S. & Thurber, D. L. 1965. Uranium series dating of corals and oolites from Bahaman and Floridan Key Limestones. *Science* 149: 58–60.

Broecker, W. S., Turekian, K. K. & Heezen, B. C. 1958. The relation of deep sea sedimentation rates to variations in climate. *American Journal of Science* 256: 503–517.

Broecker, W. S., Thurber, D. L., Goddard, J., Ku, T., Matthews, R. K. & Mesolella, K. J. 1968. Milankovitch hypothesis supported by precise dating of coral reefs and deep-sea sediments. *Science* 159: 1–4.

Broecker, W. S., Andree, M., Wolfli, W., Oeschger, H., Bonani, G., Kennett, J. & Peteet, D. 1988. The chronology of the last deglaciation: Implications to the cause of the Younger Dryas event. *Paleoceanography* 3: 1–19.

Bronger, A. 1966. Lösse, ihre Verbraunungszonen und fossilen Böden – ein Beitrag zur Stratigraphie des oberen Pleistozäns in Südbaden. *Schriften des Geographischen Instituts der Universität Kiel* XXIV (2): 113 pp.

Bronger, A. & Heinkele, T. H. 1989. Micromorphology and genesis of palaeosols in the Luochuan loess section, China: Pedostratigraphic and environmental implications. *Geoderma* 45: 123–143.

Bronger, A., Winter, R., Derevjanko, O. & Aldag, S. 1995. Loess–palaeosol sequences in Tadjikistan as a palaeoclimatic record of the Quaternary in Central Asia. *Quaternary Proceedings* 4: 69–81.

Brookes, I. A. 1970. New evidence for an independent Wisconsin-age ice cap over Newfoundland. *Canadian Journal of Earth Sciences* 7: 1374–1382.

Brookes, I. A., McAndrews, J. H. & Von Bitter, P. 1982. Quaternary interglacial and associated deposits in southwest Newfoundland. *Canadian Journal of Earth Sciences* 19: 410–423.

Brooks, L. D. 1979. Another hypothesis about iceberg draft. *Proceedings of the 5th International Conference on Port & Ocean Engineering under Arctic conditions*, Technical University Trondheim: 241–252.

Broster, B. E. & Dreimanis, A. 1981. Deposition of multiple lodgement tills by competing glacial flows in a common ice sheet: Cranbrook, British Columbia. *Arctic and Alpine Research* 13: 197–204.

Brückner, E. 1886. Die Vergletscherung des Salzachgebietes. *Pencks Geographische Abhandlungen* I (1): 183 pp. Wien.

Brückner, H. 1980. Marine Terrassen in Süditalien. Eine quartärmorphologische Studie über das Küstentiefland von Metapont. *Düsseldorfer Geographische Schriften* 14: 235 pp.

Brückner, H. 1986. Stratigraphy, evolution and age of Quaternary marine terraces in Morocco and Spain. *Zeitschrift für Geomorphologie N.F., Supplement-Band* 62: 83–101.

Brückner, H. 1988. Indicators for formerly higher sea levels along the east coast of India and on the Andaman Islands. *Hamburger Geographische Studien* 44: 47–72.

Brückner, H. & Halfar, A. 1994. Evolution and age of shorelines along Woodfiord, northern Spitsbergen. *Zeitschrift für Geomorphologie N.F., Supplement-Band* 97: 75–91.

Brugam, R. B. 1980. Postglacial diatom stratigraphy of Kirchner Marsh, Minnesota. *Quaternary Research* 13: 133–146.
Brunhes, B. 1906. Recherches par la direction d'aimantation des roches volcaniques. *Journal de Physique Théorique et Appliquée* 4 (5): 705–724.
Brunnacker, K. 1953. Die bodenkundlichen Verhältnisse der würmeiszeitlichen Schotterfluren im Illergebiet. *Geologica Bavarica* 18: 113–130.
Brunnacker, K. 1957. Die Geschichte der Böden im jüngeren Pleistozän in Bayern. *Geologica Bavarica* 34: 95 pp.
Brunnacker, K. 1968. Das Quartärprofil von Kärlich/Neuwieder Becken. *Geologie en Mijnbouw* 47: 206–208.
Brunnacker, K. 1986. Quaternary Stratigraphy in the Lower Rhine Area and Northern Alpine Foothills. *Quaternary Science Reviews* 5: 371–379.
Brunnacker, K., Boenigk, W., Koči, A. & Tillmanns, W. 1976. Die Matuyama/Brunhes-Grenze am Rhein und an der Donau. *Neues Jahrbuch für Geologie und Paläontologie, Abhandlungen* 151: 358–378.
Brunnacker, K., Würges, K. & Urban, B. 1980. Kärlich (Terrassenschotter mit Lößdeckschichten). In: *Tagung der Deutschen Quartärvereinigung, Aachen 1980, Exkursion 1: Mittel- und Niederrhein*: 79–86.
Brunnacker, K., Butzke, H., Dahm, H.-D., Dahm-Arens, H., Dubber, H.-J., Erkwoh, F.-D., Mertens, H., Mückenhausen, E., Paas, W., Schalich, J., Skupin, K., Will, K.-H., Wirth, W. & Von Zeschwitz, E. 1982. Paläoböden in Nordrhein-Westfalen. *Geologisches Jahrbuch* F 14: 165–253.
Brunnberg, L. & Possnert, G. 1992. Radiocarbon dating of the Goti-Finiglacial boundary of the Swedish Time Scale. *Boreas* 21: 89–96.
Brunotte, E. 1978. Zur quartären Formung von Schichtkämmen und Fußflächen im Bereich des Markoldendorfer Beckens und seiner Umrahmung (Leine–Weser–Bergland). *Göttinger Geographische Abhandlungen* 72: 142 pp.
Bruns, J. 1989. Stress indicators adjacent to buried channels of Elsterian age in North Germany. *Journal of Quaternary Science* 4: 267–272.
Bryant, I. D., Holyoak, D. T. & Moseley, K. A. 1983. Late Pleistocene deposits at Brimpton, Berkshire, England. *Proceedings of the Geologists' Association* 94: 321–343.
Buch, M. W. 1988. Spätpleistozäne und holozäne fluviale Geomorphodynamik im Donautal zwischen Regensburg und Straubing. *Regensburger Geographische Schriften* 21: 197 pp. + Appendix.
Buch, M. W. 1989. Late Pleistocene and Holocene development of the Danube Valley east of Regensburg. *Catena Supplement* 15: 279–287.
Buckland, P. C. & Dugmore, A. 1991. 'If this is a refugium, why are my feet so bloody cold?' – The origins of the Icelandic biota in the light of recent research. In: Maizels, J. K. & Caseldine, C. (eds): *Environmental change in Iceland: past and present*: 107–125. Dordrecht: Kluwer.
Buckland, W. 1823. *Reliquiae Diluvianae.* London: Murray. 303 pp.
Buckley, S. B. 1974. Study of Post-Pleistocene ostracod distribution in the soft sediments of southern Lake Michigan. Ph.D. thesis, University of Illinois, Urbana, Illinois. 189 pp.
Büdel, J. 1948. Die klimageomorphologischen Zonen der Polarländer. *Erdkunde* 2: 22–53.
Büdel, J. 1951. Die Klimazonen des Eiszeitalters. *Eiszeitalter und Gegenwart* 1: 16–26.
Büdel, J. 1959. Periodische und episodische Solifluktion im Rahmen der klimatischen Solifluktionstypen. *Erdkunde* 13: 297–314.
Büdel, J. 1969. Der Eisrindeneffekt als Motor der Tiefenerosion in der exzessiven Talbildungszone. *Würzburger Geographische Arbeiten* 25: 41 pp.
Buol, S. W., Hole, F. D. & McCracken, R. J. 1989. *Soil genesis and classification*, 3rd edition. Ames, Iowa: University Press.
Burbank, D. W. & Cheng, K. J. 1991. Relative dating of Quaternary moraines, Rongbuk Valley, Mount Everest, Tibet: Implications for an ice sheet on the Tibetan Plateau. *Quaternary Research* 36: 1–18.
Burbank, D. W. & Li, J.L. 1985. Age and palaeoclimatic significance of the loess of Lanzhou, North China. *Nature* 316: 429–431.
Burga, C. 1979. Postglaziale Klimaschwankungen in Pollendiagrammen der Schweiz. *Vierteljahresschrift der Naturforschenden Gesellschaft in Zürich* 124: 265–283.
Burger, A. W. 1986. Sedimentpetrographie am Morsum Kliff, Sylt (Norddeutschland). *Mededelingen van het Werkgroep voor Tertiaire en Kwartaire Geologie* 23: 99–109.

Burke, R. M. & Birkeland, P. W. 1984. Holocene glaciation in the mountain ranges of the western United States. In: Wright, H. E. (ed.) *Late Quaternary Environments of the United States, Vol. 2: The Holocene*: 3–11. London: Longman.
Busacca, A. J. 1991. Loess deposits and soils of the Palouse and vicinity. In: Morrison, R. B. (ed.): *Quaternary Nonglacial Geology: Conterminous U.S. The Geological Society of America. The Geology of North America* K-2: 216–228.
Busacca, A. J., Nelstead, K. T., McDonald, E. V. & Purser, M. D. 1992. Correlation of distal tephra layers in loess in the channeled Scabland and Palouse of Washington State. *Quaternary Research* 37: 281–303.
Cailleux, A. 1942. Les actions éoliennes périglaciaires en Europe. *Societé Géologique de France* 21 (Mémoir 46): 176 pp.
Cailleux, A. 1952. Morphoskopische Analyse der Geschiebe und Sandkörner und ihre Bedeutung für die Paläoklimatologie. *Geologische Rundschau* 40: 11–19.
Cameron, T. D. J., Stoker, M. S. & Long, D. 1987. The history of Quaternary sedimentation in the UK sector of the North Sea Basin. *Journal of the Geological Society of London* 144: 43–58.
Canada Soil Survey Committee (eds) 1978. The Canadian system of soil classification. *Canada Department of Agriculture, Publication* 1646: 164 pp.
Carlson, A. B., Raastad, H. & Sollid, J. L. 1979. Inlandsisens avsmelting i sørøstlige Jotunheimen og tilgrensende områder. *Norsk Geografisk Tidsskrift* 33: 173–186.
Castel, I. I. Y. 1991. Late Holocene eolian drift sands in Drenthe (The Netherlands). *Nederlandse Geografische Studies* 133: 156 pp.
Castel, I., Koster, E. & Slotboom, R. 1989. Morphogenetic aspects and age of Late Holocene eolian drift sands in Northwest Europe. *Zeitschrift für Geomorphologie N.F.* 33: 1–26.
Cato, I. 1987. On the definitive connection of the Swedish geochronological time scale with the present. *Sveriges Geologiska Undersökning* Ca 68: 55 pp.
Catt, J. A. 1977. Loess and cover sands. In: Shotton, F. W. (ed.): *British Quaternary Studies, Recent Advances*: 221–229. Oxford: Clarendon Press.
Catt, J. A. 1987. Effects of the Devensian cold stage on soil characteristics and distribution in eastern England. In: Boardman, J. (ed.): *Periglacial processes and landforms in Britain and Ireland*: 145–152. Cambridge: Cambridge University Press.
Catt, J. A. 1988a. *Quaternary Geology for Scientists and Engineers*. Chichester: Ellis Horwood. 340 pp.
Catt, J. A. 1988b. Soils of the Plio-Pleistocene: do they distinguish types of interglacial? *Philosophical Transactions of the Royal Society of London* B 318: 539–557.
Catt, J. A. 1990. Paleopedology Manual. *Quaternary International* 6: 95 pp.
Catt, J. A. 1991a. Late Devensian glacial deposits and glaciations in eastern England and the adjoining offshore region. In: Ehlers, J., Gibbard, P. & Rose, J. (eds): *Glacial Deposits in Great Britain and Ireland*: 61–68. Rotterdam: Balkema.
Catt, J. A. 1991b. The Quaternary history and glacial deposits of East Yorkshire. In: Ehlers, J., Gibbard, P. & Rose, J. (eds): *Glacial Deposits in Great Britain and Ireland*: 185–191. Rotterdam: Balkema.
Catt, J. A. 1992. Quaternary environments and their impact on British soils and agriculture. *Quaternary Proceedings* 2: 17–24.
Catt, J. A. & Penny, L. F. 1966. The Pleistocene deposits of Holderness, East Yorkshire. *Proceedings of the Yorkshire Geological Society* 35: 375–420.
Catt, J. A. & Staines, S. J. 1982. Loess in Cornwall. *Proceedings of the Ussher Society* 5: 368–376.
Cepek, A. G. 1962. Zur Grundmoränenstratigraphie in Brandenburg. *Berichte der Geologischen Gesellschaft der DDR* 7: 275–278.
Cepek, A. G. 1965. Stratigraphie der quartären Ablagerungen des Norddeutschen Tieflandes. In: Gellert, J. F. (ed.): *Die Weichsel-Eiszeit im Gebiet der Deutschen Demokratischen Republik*: 45–65.
Cepek, A. G. 1967. Stand und Probleme der Quartärstratigraphie im Nordteil der DDR. *Berichte der Deutschen Gesellschaft für Geologische Wissenschaften, A, Geologie/Paläontologie* 12: 375–404.
Cepek, A. G. 1972. Zum Stand der Stratigraphie der Weichsel-Kaltzeit in der DDR. *Wissenschaftliche Zeitschrift der Ernst-Moritz-Arndt-Universität Greifswald, Mathematisch-Naturwissenschaftliche Reihe* 21: 11–21.
Chaline, J. & Jerz, H. 1984. Arbeitsergebnisse der Subkommission für Europäische Quartärstratigraphie, Stratotypen des Würm-Glazials. *Eiszeitalter und Gegenwart* 35: 185–206.

Chalmers, R. 1895. Report on the surface geology of eastern New Brunswick, northwestern Nova Scotia and a portion of Prince Edward Island. *Geological Survey of Canada, Annual Report 1894*, Vol. 1 no. 7, pt. m: 144 pp.

Chamberlin, T. C. 1877. Geology of eastern Wisconsin. In: *Geology of Wisconsin, survey of 1873–1877*, Vol. 2: 97–246. Madison: Commissioners of Public Printing.

Chamberlin, T. C. 1883. Preliminary paper of the terminal moraine of the second glacial epoch: *United States Geological Survey, Third Annual Report, 1881–82*: 291–402.

Chamberlin, T. C. 1888. The rock-scorings of the great ice invasions. *7th Annual Report of the US Geological Survey 1885–86*: 147–248.

Chamberlin, T. C. 1891. Classification of the Pleistocene glacial formations. *Compte rendu du V. Congrès Geologique International, Washington.*

Chamberlin, T. C. 1894. Glacial phenomena of North America. In: Geikie, J.: *The Great Ice Age*, 3rd edition: 724–775. London: Stanford.

Chamberlin, T. C. 1895. The classification of American glacial deposits. *Journal of Geology* 3: 270–277.

Chamberlin, T. C. 1896. Nomenclature of glacial formations. *Journal of Geology* 4: 872–876.

Chamberlin, T. C. 1897. Supplementary hypothesis respecting the origin of the loess of the Mississippi Valley. *Journal of Geology* 5: 795–802.

Chamberlin, T. C. 1899. An attempt to frame a working hypothesis of the cause of glacial periods on an atmospheric basis. *Journal of Geology* 7: 545–584, 667–685, 751–787.

Chamberlin, T. C. & Salisbury, R. D. 1906. *The Pleistocene or Glacial Period. Geology*, Vol. 3. London.

Chappell, J. & Shackleton, N. J. 1986. Oxygen isotopes and sea level. *Nature* 324: 137–140.

Charlesworth, J. K. 1957. *The Quaternary era, with special reference to its glaciations*, 2 volumes. London: Edward Arnold. 1700 pp.

Chatwin, C. P. 1927. Fossils from the iron sands of Netley Heath (Surrey). *Memoirs of the Geological Survey, Summary of Progress 1926*: 154–157.

Cheney, R. E., Douglas, B. C., Sandwell, D. T., Marsh, J. G., Martin, T. V. & MacCarthy, J. J. 1984. Applications of satellite altimetry to oceanography and geophysics. *Marine Geophysical Research* 7: 17–32.

Chepalyga, A. L. 1984. Inland sea basins. In: Velichko, A. A. (ed.): *Late Quaternary Environments of the Soviet Union*: 229–247. London: Longman.

Cheremisinava, E. 1970. On diatom flora of Mindel-Riss interglacial deposits of the south-eastern part of Kaliningrad region (Russian with English summary). In: Danilans, I. (ed.): *Problems of Quaternary Geology* V: 49–63. Riga: Academy of Sciences of the Latvian SSR.

Christiansen, E. A. 1971a. Tills in Southern Saskatchewan, Canada. In: Goldthwait, R. P. (ed.): *Till – a Symposium*: 167–183. Columbus: Ohio State University Press.

Christiansen, E. A. 1971b. *Geology and groundwater resources of the Melville Area (62K,L) Saskatchewan.* Saskatchewan Research Council, Geology Division, Map 12.

Church, M. 1972. Baffin Island sandurs: a study of arctic fluvial processes. *Geological Survey of Canada Bulletin* 216: 208 pp.

Clague, J. J. 1986. The Quaternary stratigraphic record of British Columbia – evidence for episodic sedimentation and erosion controlled by glaciation. *Canadian Journal of Earth Sciences* 23: 885–894.

Clague, J. J. 1989. Quaternary geology of the Canadian Cordillera. In: Fulton, R. J. (ed.): *Quaternary Geology of Canada and Greenland. Geological Society of America. The Geology of North America* K-1: 17–95.

Clark, G. M. 1968. Sorted patterned ground; New Appalachian localities south of the glacial border. *Science* 161: 355–356.

Clark, J. A., Hendriks, M., Timmermans, T. J., Struck, C. & Hilverda, K. J. 1994. Glacial isostatic deformation of the Great Lakes region. *Geological Society of America Bulletin* 106: 19–31.

Clark, P. U. 1992. Surface form of the Laurentide Ice Sheet and its implications to ice-sheet dynamics. *Geological Society of America Bulletin* 104: 595–605.

Clark, P. U. 1994. Unstable behaviour of the Laurentide Ice Sheet over deforming sediment and its implications for climatic change. *Quaternary Research* 41: 19–25.

Clark, P. U. & Hansel, A. K. 1989. Clast ploughing, lodgement and glacier sliding over a soft glacier bed. *Boreas* 18: 201–207.

Clark, P. U. & Lea, P. D. (eds) 1992. The Last Interglacial–Glacial Transition in North America. *Geological Society of America Special Paper* 270: 317 pp.
Clark, P. U. & Walder, J. H. 1994. Subglacial drainage, eskers, and deforming beds beneath the Laurentide and Eurasian ice sheets. *Geological Society of America Bulletin* 106: 304–314.
Clarke, G. K. C. 1987. Fast glacier flow: ice streams, surging and tidewater glaciers. *Journal of Geophysical Research* 92: 8835–8841.
Clayton, L., Teller, J. T. & Attig, J. W. 1985. Surging of the southwestern part of the Laurentide Ice Sheet. *Boreas* 14: 235–241.
Close, M. H. 1867. Notes on the general glaciation of Ireland. *Journal of the Royal Geological Society of Ireland* 1: 207–242.
Coleman, A. P. 1894. Interglacial fossils from the Don Valley. *American Geologist* 13: 85–95.
Collins, E. I., Lichvar, R. W. & Evert, E. F. 1984. Description of the only known fen-palsa in the contiguous United States. *Arctic and Alpine Research* 16: 255–258.
Colman, S. M., Pierce, K. L. & Birkeland, P. W. 1987. Suggested terminology for Quaternary dating methods. *Quaternary Research* 28: 314–319.
Colman, S. M., Clark, J. A., Clayton, L., Hansel, A. K. & Larsen, C. R. 1994a. Deglaciation, lake levels, and meltwater discharge in the Lake Michigan basin. *Quaternary Science Reviews* 13: 879–890.
Colman, S. M., Forester, R. M., Reynolds, R. L., Sweetkind, D. S., King, J. W., Gangemi, P., Jones, G. A., Keigwin, L. D. & Foster, D. S. 1994b. Lake-level history of Lake Michigan for the past 12,000 Years: The record from deep lacustrine sediments. *Journal of Great Lakes Research* 20: 73–92.
Conard, N. J. 1990. The Paleolithic finds from Tönchesberg. In: Schirmer, W. (ed.): *Rheingeschichte zwischen Mosel und Maas. Deuqua-Führer* 1: 46–49.
Conrad, T. 1839. Notes on American geology. *American Journal of Science* 35: 237–251.
Coope, G. R. 1959. A Late Pleistocene insect fauna from Chelford, Cheshire. *Proceedings of the Royal Society of London* B 151: 70–86.
Coope, G. R. 1977. Quaternary Coleoptera as aids in the interpretation of environmental history. In: Shotton, F. W. (ed.): *British Quaternary studies, recent advances*: 55–68. Oxford: Clarendon Press.
Coope, G. R. & Angus, R. B. 1975. An ecological study of a temperate interlude in the middle of the last glaciation, based on fossil Coleoptera from Isleworth, Middlesex. *Journal of Animal Ecology* 44: 365–391.
Coope, G. R., Shotton, F. W. & Strachan, I. 1961. A Late Pleistocene fauna and flora from Upton Warren, Worcestershire. *Philosophical Transactions of the Royal Society of London* B 244: 379–417.
Cornelius, M. 1984. Sedimentpetrographische und geochemische Untersuchungen saalezeitlicher Moränen in Hamburg-Tinsdal. *Mitteilungen aus dem Geologisch-Paläontologischen Institut der Universität Hamburg* 57: 33–56.
Cotta, B. 1848. *Briefe über Alexander von Humboldt's Kosmos. Ein Commentar zu diesem Werke für gebildete Laien. Erster Theil.* Leipzig: Weigel. 356 pp.
Cotta, B. 1867. *Die Geologie der Gegenwart*, 2nd edition. Leipzig: J. J. Weber. 484 pp.
Cox, F. C. 1981. The 'Gipping Till' revisited. In: Neale, J. & Flenley, J. (eds): *The Quaternary in Britain*: 32–42. Oxford: Pergamon.
Coxon, P. 1978. The first record of fossil naled in Britain. *Quaternary Newsletter* 24: 9–11.
Coxon, P. 1988. Remnant periglacial features on the summit of Truskmore, Counties Sligo and Leitrim, Ireland. *Zeitschrift für Geomorphologie N.F.* 71: 81–91.
Coxon, P. & Browne, P. 1991. Glacial deposits and landforms of central and western Ireland. In: Ehlers, J., Gibbard, P. & Rose, J. (eds): *Glacial Deposits in Great Britain and Ireland*: 355–365. Rotterdam: Balkema.
Coxon, P. & Flegg, A. 1985. A Middle Pleistocene interglacial deposit from Ballyline, Co. Kilkenny. *Proceedings of the Royal Irish Academy* 85 B (9): 107–120.
Coxon, P. & O'Callaghan, P. 1987. The distribution and age of pingo remnants in Ireland. In: Boardman, J. (ed.): *Periglacial processes and landforms in Britain and Ireland*: 195–202. Cambridge: Cambridge University Press.
Creer, K. M., Gross, D. L. & Lineback, J. A. 1976. Origin of regional geomagnetic variations recorded by Wisconsinan and Holocene sediments from Lake Michigan, U.S.A., and Lake Windermere, England. *Geological Society of America Bulletin* 87: 531–540.

Croll, J. 1864. On the physical cause of the change of climate during geological epochs. *Philosophical Magazine* 28: 121–137.

Croot, D. G. 1988. Morphological, structural and mechanical analysis of neoglacial ice-pushed ridges in Iceland. In: Croot, D.G. (ed.): *Glaciotectonics: Forms and Processes*: 33–47. Rotterdam: Balkema.

Currey, D. R. 1990. Quaternary palaeolakes in the evolution of semidesert basins, with special emphasis on Lake Bonneville and the Great Basin, U.S.A. *Palaeogeography, Palaeoclimatology, Palaeoecology* 76: 189–214.

Curry, B. B. & Follmer, L. R. 1992. The last interglacial–glacial transition in Illinois: 123–25 ka. In: Clark, P. U. & Lea, P. D. (eds): *The Last Interglacial–Glacial Transition in North America. Geological Society of America Special Paper* 270: 71–88.

Curry, B. B. & Forester, R. M. 1991. Paleoenvironments and lithology of the Hopwood Farm site, Montgomery County, Illinois. *Geological Society of America, Abstracts with Programs* 18 (6): 557.

Cwynar, L. C., Levesque, A. J., Mayle, F. E. & Walker, I. 1994. Wisconsinan Late-glacial environmental change in New Brunswick: a regional synthesis. *Journal of Quaternary Science* 9: 161–164.

Czarnetzki. A. 1983. Zur Entwicklung des Menschen in Südwestdeutschland. In: Müller-Beck (ed.): *Urgeschichte in Baden Württemberg*: 217–240. Stuttgart: Theiss.

Czerwonka, J. A. & Krzyszkowski, D. 1992. Pleistocene stratigraphy of the central part of Silesian Lowland, Southwestern Poland. *Bulletin of the Polish Academy of Sciences, Earth Sciences* 40: 203–233.

Czudek, T. 1995. Cryoplanation terraces – a brief review and some remarks. *Geografiska Annaler* 77 A: 95–105.

Dahl, R. 1965. Plastically sculptured detail forms on rock surfaces in northern Nordland, Norway. *Geografiska Annaler* 47 A: 83–140.

Dalrymple, G. B. 1972. Potassium–argon dating of geomagnetic reversals and North American glaciations. In: Bishop, W. W. & Miller, J. A. (eds): *Calibration of Hominoid Evolution, Recent Advances in Isotopic and Other Dating Methods Applicable to the Origin of Man*: 107–134. Toronto: University of Toronto Press.

Daly, R. A. 1940. *Strength and Structure of the Earth.* New York: Prentice Hall. 434 pp.

Dana, J. D. 1849. Geology. *United States Exploring Expedition, during the years 1838, 1839, 1840, 1841, 1842 under the command of Charles Wilkes U.S.N., Vol. X.* Philadelphia: C. Sherman. 756 pp.

Dana, J. D. 1863. *Manual of geology*. Philadelphia: Theodore Bliss & Co. 798 pp.

Dana, J. D. 1895. *Manual of geology*, 4th edition. New York: American Book Company. 1087 pp.

Daniels, F., Boyd, C. A. & Saunders, D. F. 1953. Thermoluminescence as a research tool. *Science* 117: 343–349.

Daniels, R. B. & Handy, R. L. 1959. Suggested new type section for the Loveland Loess in western Iowa. *Journal of Geology* 67: 114–119.

Danilāns, I. Ya. 1970. Some results of studying the composition of Latvian till deposits for the clarification of lithological distinctions of tills of different age (Russian with English summary). *Problems of Quaternary Geology* V: 7–36.

Danilāns, I. Ya. 1973. *Chetvertichnie otlozheniia Latvii.* Riga: Zinatne. 314 pp.

Dansgaard, W., Johnsen, S. J., Möller, J. & Langway Jr., C. C., 1969. One thousand centuries of climatic record from Camp Century on the Greenland ice sheet. *Science* 166: 377–381.

Dansgaard, W., Johnsen, S. J., Clausen, H. B. &. Langway Jr., C. C. 1971. Climatic record revealed by the Camp Century ice core. In: Turekian, K. (ed.): *The Late Cenozoic Glacial Ages*: 37–56. New Haven: Yale University Press.

Dansgaard, W., Clausen, H. B., Gundestrup, N., Hammer, C. U., Johnsen, S. F., Kristinsdottir, P. M. & Reeh, N. 1982. A new Greenland deep ice core. *Science* 218: 1273–1277.

Dansgaard, W., Johnsen, S. J., Clausen, H. B., Dahl-Jensen, D., Gundestrup, N., Hammer, C. U. & Oeschger, H. 1984. North Atlantic climatic oscillations revealed by deep Greenland ice cores. In: Hansen, J. E. & Takahashi, T. (eds): *Climate Processes and Climate Sensitivity. American Geophysical Union, Geophysical Monographs* 29: 288–298.

Dansgaard, W., Johnsen, S. J., Clausen, H. B., Dahl-Jensen, D., Gundestrup, N. S., Hammer, C. U., Hvidberg, C. S., Steffensen, J. P., Sveinbjörnsdottir, A. E., Jouzel, J. & Bond, G. 1993. Evidence for general instability of past climate from a 250-kyr ice-core record. *Nature* 364: 218–220.

Dapples, E. C. & Rominger, J. F. 1945. Orientation analysis of fine gravel sediments: a report of progress. *Journal of Geology* 53: 246–261.
Dardis, G. F., McCabe, A. M. & Mitchell, W. J. 1984. Characteristics and origins of lee-side stratification sequences in Late Pleistocene drumlins, Northern Ireland. *Earth Surface Processes and Landforms* 9: 409–424.
Dare-Edwards, A. J. 1984. Aeolian clay deposits of southeastern Australia: parna or loessic clay? *Transactions of the Institute of British Geographers N.S.* 9: 337–344.
David, P. P. 1977. *Sand dune occurrences of Canada.* Canada Department of Indian and Northern Affairs, National Parks Branch, Contract 74–230: 183 pp.
Dawson, A. G. 1977. A fossil lobate rock glacier in Jura. *Scottish Journal of Geology* 13: 37–42.
Dawson, A. G. 1992. *Ice Age Earth. Late Quaternary Geology and Climate.* London: Routledge. 293 pp.
Dawson, A. G., Matthews, J. A. & Shakesby, R. A. 1987. Rock platform erosion on periglacial shores: a modern analogue for Pleistocene rock platforms in Britain. In: Boardman, J. (ed.): *Periglacial processes and landforms in Britain and Ireland*: 173–182. Cambridge: Cambridge University Press.
De Beaulieu, J. L. & Reille, M. 1984a. A long Upper Pleistocene pollen record from Les Echets, near Lyon, France. *Boreas* 13: 111–132.
De Beaulieu, J. L. & Reille, M. 1984b. The pollen sequence of Les Echets (France): a new element for the chronology of the Upper Pleistocene. *Géographie physique et Quaternaire* XXXVIII: 3–9.
De Charpentier, J. 1834. Sur la cause probable du transport des blocs erratiques de la Suisse. *Annales des mines, 3e série*, t. 8: 219–236. Paris 1835.
De Charpentier, J. 1841. *Essai sur les glaciers et sur le terrain erratique du bassin du Rhône.* Lausanne: Marc Ducloux. 363 pp.
De Charpentier, J. 1842. Sur l'application de l'hypothése de M. Venetz aux phénomènes erratiques du Nord. *Bibliothèque universelle de Genève* XXXIX: 327–346.
De Gans, W. 1991. Kwartairgeologie van West-Nederland. *Grondboor en Hamer* 45: 103–114.
De Gans, W., De Groot, Th. & Zwaan, H. 1986. The Amsterdam basin, a case study of a glacial basin in The Netherlands. In: Van der Meer, J. J. M. (ed.): *Tills and Glaciotectonics*: 205–216. Rotterdam: Balkema.
De Geer, G. 1884. Autoreport. *Geologiska Föreningens i Stockholm Förhandlingar* 7: 3.
De Geer, G. 1888/90. Om Skandinaviens nivåförändringar under Kvartärperioden. *Geologiska Föreningens i Stockholm Förhandlingar* 10: 366–379, 12: 61–110.
De Geer, G. 1896. *Om Skandinaviens geografiska utveckling efter istiden.* Stockholm: P. A. Norstedt & Söner. 160 pp.
De Geer, G. 1897. Om rullstenåsarnes bildningssätt. *Geologiska Föreningens i Stockholm Förhandlingar* 19: 366–386
De Geer, G. 1912. A geochronology of the last 12,000 years. *XIth International Geological Congress, Stockholm 1910, Comptes Rendus* 1: 241–258.
De Geer, G. 1940. Geochronologia Suecica Principles. *Kungliga Svenska Vetenskabs-Akademiens Handlingar* Series 3, 18 (6): 367 pp.
De Graaff, L. W. S. 1992. Zur Morpho- und Chronostratigraphie des Oberen Würm in Vorarlberg. *Jahrbuch der Geologischen Bundesanstalt* 135: 809–824.
De Jong, J. 1988. Climatic variability during the past three million years, as indicated by vegetational evolution in northwest Europe and with emphasis on data from The Netherlands. *Philosophical Transactions of the Royal Society of London* B 318: 603–617.
De Jong, J. & Maarleveld, G. C. 1983. The glacial history of the Netherlands. In: Ehlers, J. (ed.): *Glacial Deposits in North-West Europe*: 353–356. Rotterdam: Balkema.
De Jong, M. G. G., Rappol, M. & Rupke, J. 1982. Sedimentology and geomorphology of drumlins in western Allgäu, South Germany. *Boreas* 11: 37–45.
De Lamothe, L. 1899. Note sur les anciennes plages et terraces du bassin de l'Isser (Département d'Alger) et de quelques autres bassins de la côte algérienne. *Bulletin de la Société Géologique de France (3. sér)* 27: 257–303.
De Ridder, N. A. & Wiggers, A. J. 1956. De korrelgrootte-verdeling van de keileem en het proglaciale zand. *Geologie en Mijnbouw* 35: 287–311.
De Saussure, H. B. 1779–96. *Voyages dans les Alpes* (4 volumes). Neuchâtel.
Delcourt, P. A. & Delcourt, H. R. 1977. The Tunica Hills, Louisiana–Mississippi; Late glacial locality for spruce and deciduous forest species. *Quaternary Research* 7: 218–237.

Delorme, L. D. 1968. Pleistocene freshwater Ostracoda from Yukon, Canada. *Canadian Journal of Zoology* 46: 859–876.
Delorme, L. D. 1990. Freshwater Ostracodes. In: Warner, B. G. (ed.): *Methods in Quaternary Ecology. Geoscience Canada Reprint Series* 5: 93–100.
Demek, J. 1968. Cryoplanation terraces in Yakutia. *Biuletyn Peryglacjalny* 17: 91–116.
Demek, J. 1969. Cryogene processes and the development of cryoplanation terraces. *Biuletyn Peryglacjalny* 18: 115–125.
Demorest, M. 1938. Ice flowage as revealed by glacial striae. *Journal of Geology* 46: 700–725.
Denton, G. H. & Hughes, T. J. (eds) 1981. *The Last Great Ice Sheets.* New York: Wiley. 484 pp.
Depéret, C. 1918. Essai de coordination chronologique générale des temps quaternaires. *Paris, Acad. Sci. C.R.* 167: 418–422.
Derbyshire, E. 1983. Origin and characteristics of some Chinese loess at two locations in China. In: Brookfield, M. E. & Ahlbrandt, T. S. (eds): *Eolian Sediments and Processes*: 69–90. Amsterdam: Elsevier.
Derbyshire, E., Shi, Y., Li, J., Zheng, B., Li, S. & Wang, J. 1991. Quaternary glaciation of Tibet: The geological evidence. *Quaternary Science Reviews* 10: 485–510.
Derry, D. R. 1933. Heavy minerals of the Pleistocene beds of the Don Valley, Toronto, Ontario. *Journal of Sedimentary Petrology* 3: 113–118.
Devoy, R. J. N. 1983. Late Quaternary shorelines in Ireland: an assessment of their implications for isostatic land movement and relative sea-level changes. In: Smith, D. E. & Dawson, A. G. (eds): *Shorelines and Isostasy*: 227–254. London: Academic Press.
Devoy, R. J. N. (ed.) 1987. *Sea Surface Studies.* London: Croom Helm. 649 pp.
Deynoux, M., Miller, J. H. G., Domack, E. W., Eyles, N., Fairchild, I. J. & Young, G. M. (eds) 1994. *Earth's Glacial Record. World and Regional Geology* 5. Cambridge: Cambridge University Press. 266 pp.
Dietrich, J., Zorn, R. & Nielsen, H. 1979. Iceberg investigation along the west coast of Greenland. *Proceedings of the 5th International Conference on Port & Ocean Engineering under Arctic conditions, Technical University Trondheim*: 241–252.
Dijkmans, J. W. A. 1990. Aspects of geomorphology and thermoluminescence dating of cold-climate eolian sands. Thesis, University of Utrecht: 250 pp.
Dines, H. G. & Chatwin, C. P. 1930. Pliocene sandstone from Rothamstead (Hertfordshire). *Memoirs of the Geological Survey, Summary of Progress 1929*: 1–7.
Ding, Z., Rutter, N. W., Jingtai, H. & Liu, T. 1992. A coupled environmental system formed at about 2.5 Ma in East Asia. *Palaeogeography, Palaeoclimatology, Palaeoecology* 94: 223–242.
Ding, Z., Yu, Z., Rutter, N. W. & Liu, T. 1994. Towards an orbital time scale for Chinese loess deposits. *Quaternary Science Reviews* 13: 39–70.
Dionne, J.-C. 1979. Ice action in the lacustrine environment. A review with particular reference to Subarctic Quebec, Canada. *Earth Science Reviews* 15: 185–212.
Dionne, J.-C. 1981. A boulder-strewn tidal flat, north shore of the Gulf of St. Lawrence, Québec. *Géographie physique et Quaternaire* XXXV: 261–267.
Dionne, J.-C. 1992. Ice-push features. *The Canadian Geographer* 36: 86–91.
Dionne, J.-C. 1993. Sediment load of shore ice and ice rafting potential, Upper St. Lawrence Estuary, Québec, Canada. *Journal of Coastal Research* 9: 628–646.
Dionne, J.-C. 1994. Les erratiques lointains de l'embouchure du Sagunenay, Québec. *Géographie physique et Quaternaire* 48: 179–194.
Dobson, P. 1826. Remarks on bowlders. *American Journal of Science* 10: 217–218.
Dodonov, A. E. 1979. Stratigraphy of the Upper Pliocene – Quaternary deposits of Tajikistan (Soviet Central Asia). *Acta Geologica Academiae Scientiarum Hungariae* 22: 63–73.
Doeglas, D. J. 1949. Loess; an eolian product. *Journal of Sedimentary Petrology* 19: 112–117.
Doering, J. A. 1956. Review of Quarternary surface formations of the Gulf Coast region. *American Association of Petroleum Geologists Bulletin* 40: 1816–1862.
Dokuchaev, V. V. 1883. *Russian chernoziom* (in Russian). St Petersburg.
Domack, E. W. 1983. Facies of late Pleistocene glacial sediments on Whidbey Island, Washington. In: Molnia, B. F. (ed.): *Glacial-marine sedimentation*: 535–570. New York: Plenum Press.
Domack, E. W. & Lawson, D. E. 1985. Pebble fabric in an ice-rafted diamicton. *Journal of Geology* 93: 577–591.

Donner, J. 1964. Pleistocene geology of eastern Long Island, New York. *American Journal of Science* 262: 355–376.
Donner, J. 1978. The dating of the levels of the Baltic Ice Lake and the Salpausselkä moraines in South Finland. *Societas Scientiarum Fennica, Commentationes Physico-Mathematicae* 48: 11–38.
Donner, J. 1980. The determination and dating of synchronous Late Quaternary shorelines in Fennoscandia. In: N.-A. Mörner (ed.): *Earth Rheology, Isostasy and Eustasy*: 285-293. Chichester: Wiley.
Donner, J. 1982. Fluctuations in water level of the Baltic Ice Lake. *Annales Academiae Scientiarum Fennicae* A III, 134: 13–26.
Donner, J. 1983. The identification of Eemian interglacial and Weichselian interstadial deposits in Finland. *Annales Academiæ Scientiarum Fennicæ A* III, 136: 38 pp.
Donner, J. 1988. The Eemian site of Norinkylä compared with other interglacial and interstadial sites in Ostrobothnia, western Finland. *Annales Academiæ Scientiarum Fennicæ A* III, 149: 31 pp.
Donner, J. 1995. The Quaternary History of Scandinavia. *World and Regional Geology* 7. Cambridge: Cambridge University Press. 198 pp.
Donner, J. & Gardemeister, R. 1971. Redeposited Eemian marine clay in Somero, south-west Finland. *Geological Society of Finland, Bulletin* 43: 73–88.
Donner, J., Korpela, K. & Tynni, R. 1986. Veiksel-jääkauden alajaotus Suomessa. (The subdivision of the Weichselian stage in Finland). *Terra* 98: 240–247.
Doppler, G. 1980. Das Quartär im Raum Trostberg an der Alz im Vergleich mit dem nordwestlichen Altmoränengebiet des Salzachvorlandgletschers (Südostbayern). Dissertation, München. 198 pp.
Doss, B. 1896. Ueber das Vorkommen von Drumlins in Livland. *Zeitschrift der Deutschen Geologischen Gesellschaft* XLVIII: 1–13.
Dowdeswell, J. A. & Sharp, M. J. 1986. Characterisation of pebble fabrics in modern terrestrial glacigenic sediments. *Sedimentology* 33: 699–710.
Dowdeswell, J. A., Hambrey, M. J. & Wu, R. T. 1985. A comparison of clast fabric and shape in late Precambrian and modern glacigenic sediments. *Journal of Sedimentary Petrology* 55: 691–704.
Drake, R. E. & Curtis, G. H. 1987. K–Ar geochronology of the Laetoli fossil localities. In: Leakey, M. D. & Harris, J. M. (eds): *Laetoli: A Pliocene site in northern Tanzania*: 48–51. Oxford: Clarendon Press.
Draxler, I. & Van Husen, D. 1978. Zur Einstufung innerwürmzeitlicher Sedimente von Ramsau/ Schladming und Hohentauern (Steiermark). *Zeitschrift für Gletscherkunde und Glazialgeologie* 14: 105–114.
Draxler, I. & Van Husen, D. 1989. Ein ^{14}C-datiertes Profil in der Niederterrasse bei Neurath (Stainz, Stmk.). *Zeitschrift für Gletscherkunde und Glazialgeologie* 25: 123–130.
Dredge, L. A. 1982. Relict ice scour marks and late phases of Lake Agassiz in northernmost Manitoba. *Canadian Journal of Earth Sciences* 19: 1079–1087.
Dredge, L. A. & Cowan, W. R. 1989. Quaternary geology of the southwestern Canadian Shield. In: Fulton, R. J. (ed.): *Quaternary Geology of Canada and Greenland. Geological Society of America. The Geology of North America* K-1: 214–249.
Dredge, L. A. & Nielsen, E. 1985. Glacial and interglacial deposits in the Hudson Bay Lowlands: A summary of sites in Manitoba. *Current Research, Part A, Geological Survey of Canada, Paper* 85-1A: 247–257.
Dredge, L. A. & Nixon, F. M. 1992. Glacial and environmental geology of northeastern Manitoba. *Geological Survey of Canada Memoir* 432: 80 pp.
Dredge, L. A. & Thorleifson, L. H. 1987. The Middle Wisconsinan history of the Laurentide Ice Sheet. *Géographie physique et Quaternaire* XLI: 215–235.
Dredge, L. A., Morgan, A. V. & Nielsen, E. 1990. Sangamon and pre-Sangamon interglaciations in the Hudson Bay Lowlands of Manitoba. *Géographie physique et Quaternaire* 44: 319–336.
Dreesbach, R. 1985. Sedimentpetrographische Untersuchungen zur Stratigraphie des Würmglazials im Bereich des Isar-Loisachgletschers. Dissertation, München. 176 pp.
Dreimanis, A. 1939. Eine neue Methode der quantitativen Geschiebeforschung. *Zeitschrift für Geschiebeforschung und Flachlandsgeologie* 15: 17–36.
Dreimanis, A. 1947. An Improved Petrological Method (1939) for the Investigation of Boulder Clays. *Contributions of Baltic University* 46: 8 pp.

Dreimanis, A. 1953. Studies of friction cracks along shores of Cirrus Lake and Kasakokwog Lake, Ontario. *American Journal of Science* 251: 769–783.
Dreimanis, A. 1959. Rapid macroscopic fabric studies in drill-cores and hand specimens of till and tillite. *Journal of Sedimentary Petrology* 29: 459–463.
Dreimanis, A. 1976. Tills, their origin and properties. In: Legget, R. F. (ed.): *Glacial Till. The Royal Society of Canada Special Publication* 12: 11–49.
Dreimanis, A. 1977. Correlation of Wisconsin glacial events between the eastern Great Lakes and the St. Lawrence Lowlands. *Géographie physique et Quaternaire* XXXI: 37–51.
Dreimanis, A. 1982. Middle Wisconsinan Substage in its type region, the Eastern Great Lakes, and Ohio River basin, North America. *Quaternary Studies in Poland* 3 (2): 21–28.
Dreimanis, A. 1989. Tills: Their genetic terminology and classification. In: Goldthwait, R. P. & Matsch, C. L. (eds): *Genetic Classification of Glacigenic Deposits*: 17–83. Rotterdam: Balkema.
Dreimanis, A. 1990, Formation, deposition and identification of subglacial and supraglacial till. In: Kujansuu, R. & Saarnisto, M. (eds): *Glacial Indicator Tracing*: 35–59. Rotterdam: Balkema.
Dreimanis, A. 1992. Early Wisconsinan in the north-central part of the Lake Erie basin: A new interpretation. In: Clark, P. U. & Lea, P. D. (eds): *The Last Interglacial–Glacial Transition in North America. Geological Society of America Special Paper* 270: 109–118.
Dreimanis, A. & Karrow, P. F. 1972. Glacial history of the Great Lakes – St. Lawrence Region, the classification of the Wisconsin(an) Stage, and its correlatives. *Proceedings of the 24th International Geological Congress, Montreal, Section* 12: 5–15.
Dreimanis, A. & Vagners, J. J. 1971. Bimodal distribution of rock and mineral fragments in basal tills. In: Goldthwait, R. P. (ed.): *Till: A Symposium*: 237–250. Columbus: Ohio State University Press.
Dreimanis, A. & Zelčs, V. 1995 Pleistocene stratigraphy of Latvia. In: Ehlers, J., Kozarski, S. & Gibbard, P. L. (eds): *Glacial Deposits in North-East Europe*: 105–113. Rotterdam: Balkema.
Dreimanis, A., Reavely, G. H., Cook, R. J. B., Knox, K. S. & Moretti, F. J. 1957. Heavy mineral studies of till in Ontario and adjacent areas. *Journal of Sedimentary Petrology* 27: 148–161.
Dreimanis, A., Liivrand, E. & Raukas, A. 1989. Glacially redeposited pollen in tills of southern Ontario, Canada. *Canadian Journal of Earth Sciences* 26: 1667–1676.
Drewry, D. 1986. *Glacial Geologic Processes*. London: Edward Arnold. 276 pp.
Drewry, D. J. 1991. The response of the Antarctic ice sheet to climatic change. In: Harris, C. & Stonehouse, B. (eds): *Antarctica and global climate change*: 90–106. Cambridge: Scott Polar Research Institute & Belhaven Press.
Driever, B. M. W. 1988. *Calcareous Nannofossil Biostratigraphy and Palaeoenvironmental Interpretation of the Mediterranean Pliocene. Utrecht Micropaleontological Bulletins.*
Duphorn, K., Kögler, F. C. & Stay, B. 1981. Late-glacial varved clay in the southern Baltic Sea. *Quaternary Studies in Poland* 3: 29–36.
Duphorn, K., Stay, B. & Stiller, D. 1979. Deglaciation of the Bara Basin near Malmö, Sweden. *Boreas* 8: 141–144.
Durrell, R. H. 1977. A recycled landscape. *Quarterly of the Cincinnati Museum of Natural History* 14 (2): 9 pp.
Dyke, A. S. 1993. Landscapes of cold-centred Late Wisconsinan ice caps, Arctic Canada. *Progress in Physical Geography* 17: 223–247.
Dyke, A. S. & Prest, V. K. 1987. Late Wisconsinan and Holocene retreat of the Laurentide Ice Sheet. *Géographie physique et Quaternaire* 41: 237–263.
Dyke, A. S., Vincent, J.-S., Andrews, J. T., Dredge, L. A. & Cowan, W. R. 1989. The Laurentide Ice Sheet and an introduction to the Quaternary geology of the Canadian Shield. In: Fulton, R. J. (ed.): *Quaternary Geology of Canada and Greenland. Geological Society of America, The Geology of North America* K-1: 178–188.
Dylik, J. 1967. Solifluxion, congelifluxion and related slope processes. *Geografiska Annaler* 49 A: 167–177.
Dzierżek, J., Lindner, L. & Nitychoruk, J. 1986. Late Quaternary deglaciation of the Eastern Polish Tatra Mts. *Bulletin of the Polish Academy of Sciences, Earth Sciences* 34: 395–407.
Easterbrook, D. J. 1986. Stratigraphy and chronology of Quaternary deposits of the Puget Lowland and Olympic Mountains of Washington and the Cascade Mountains of Washington and Oregon. *Quaternary Science Reviews* 5: 135–159.

Easterbrook, D. J. 1992. Advance and retreat of Cordilleran Ice Sheets in Washington, U.S.A. *Géographie physique et Quaternaire* 46: 51–68.
Ebel, R. 1983. Die Lagerungsverhältnisse der Schieferkohlen zwischen der Ostrach und der Iller bei Sonthofen im Oberallgäu. *Geologica Bavarica* 84: 123–146.
Eberl, B. 1930. *Die Eiszeitenfolge im nördlichen Alpenvorlande. – Ihr Ablauf, ihre Chronologie auf Grund der Aufnahmen des Lech- und Illergletschers.* Augsburg: Benno Filser. 427 pp.
Ebers, E. 1925. Die bisherigen Ergebnisse der Drumlinforschung. Eine Monographie des Drumlins. *Neues Jahrbuch für Mineralogie, Abhandlungen* 53, Abteilung B: 153–270.
Ebers, E. 1937. Zur Entstehung der Drumlins als Stromlinienkörper. Zehn weitere Jahre Drumlinforschung (1926–1936). *Neues Jahrbuch für Mineralogie, Beilage-Band* 78, Abt. B: 200–240.
Ebers, E., Weinberger, L. & Del Negro, W. 1966. Der pleistozäne Salzachvorlandgletscher. *Veröffentlichungen der Gesellschaft für Bayerische Landeskunde* 19–22: 216 pp.
Edelman, C. H. 1933. Petrologische provincies in het Nederlandsche Kwartair. Thesis Amsterdam. Amsterdam: Centen Uitg. Mij. 104 pp.
Edelman, C. H., Florschütz, F. & Jeswiet, J. 1936. Über spätpleistozäne und frühholozäne kryoturbate Ablagerungen in den östlichen Niederlanden. *Verhandelingen van het Geologisch-Mijnbouwkundig Genootschap voor Nederland en Koloniën, Geologische Serie* 11: 301–360.
Edwards, G. B., Jr. 1988. Late Quaternary geology of Northeastern Massachusetts and the Merrimack Embayment, Western Gulf of Maine. M.S. thesis, Boston University, Boston.
Edwards, M. B. 1979. Late Precambrian loessites from North Norway and Svalbard. *Journal of Sedimentary Petrology* 49: 85–91.
Edwards, R. L., Chen, J. H. & Wasserburg, G. J. 1986/87. ^{238}U- ^{234}U- ^{230}Th- ^{232}Th systematics and the precise measurement of time over the past 500,000 years. *Earth and Planetary Science Letters* 81: 175–192.
Edwards, R. L., Chen, J. H., Ku, T. L. & Wasserburg, G. J. 1987. Precise timing of the last interglacial period from mass spectrometric determination of thorium-230 in corals. *Science* 236: 1547–1553.
Ehlers, J. 1978. Die quartäre Morphogenese der Harburger Berge und ihrer Umgebung. *Mitteilungen der Geographischen Gesellschaft in Hamburg* 68: 181 pp.
Ehlers, J. (ed.) 1983a. *Glacial Deposits in North-West Europe*. Rotterdam: Balkema. 470 pp.
Ehlers, J. 1983b. Different till types in North Germany. In: Evenson, E. B., Schlüchter, Ch. & Rabassa, J. (eds): *Tills and Related Deposits*: 61–80. Rotterdam: Balkema.
Ehlers, J. 1988. *The Morphodynamics of the Wadden Sea*. Rotterdam: Balkema. 397 pp.
Ehlers, J. 1990a. Untersuchungen zur Morphodynamik der Vereisungen Norddeutschlands unter Berücksichtigung benachbarter Gebiete. *Bremer Beiträge zur Geographie und Raumplanung* 19: 166 pp.
Ehlers, J. 1990b. Reconstructing the dynamics of the North-West European Pleistocene Ice Sheets. *Quaternary Science Reviews* 9: 71–83.
Ehlers, J. 1992. Origin and distribution of red tills in North Germany. *Sveriges Geologiska Undersökning* Ca 81: 97–105.
Ehlers, J. 1994. *Allgemeine und historische Quartärgeologie*. Stuttgart: Enke. 358 pp.
Ehlers, J. & Gibbard, P. L. 1991. Anglian glacial deposits in Britain and the adjoining offshore regions. In: Ehlers, J., Gibbard, P. & Rose, J. (eds): *Glacial Deposits in Great Britain and Ireland*: 17–24. Rotterdam: Balkema.
Ehlers, J. & Grube, F. 1983. Meltwater deposits in north-west Germany. In: Ehlers, J. (ed.): *Glacial Deposits in North-West Europe*: 249–256. Rotterdam: Balkema.
Ehlers, J. & Iwanoff, A. 1983. Geophysical well-logging and its stratigraphical interpretation. In: Ehlers, J. (ed.): *Glacial Deposits in North-West Europe*: 263–265. Rotterdam: Balkema.
Ehlers, J. & Linke, G. 1989. The origin of deep buried channels of Elsterian age in Northwest Germany. *Journal of Quaternary Science* 4: 255–265.
Ehlers, J. & Stephan, H.-J. 1979. Forms at the base of till strata as indicators of ice movement. *Journal of Glaciology* 22: 345–355.
Ehlers, J. & Stephan, H.-J. 1983. Till fabric and ice movement. In: Ehlers, J. (ed.): *Glacial Deposits in North-West Europe*: 267–274. Rotterdam: Balkema.
Ehlers, J. & Wingfield, R. 1991. The extension of the Late Weichselian/Late Devensian ice sheets in the North Sea Basin. *Journal of Quaternary Science* 6: 313–326.

Ehlers, J., Meyer, K.-D. & Stephan, H.-J. 1984. Pre-Weichselian glaciations of North-West Europe. *Quaternary Science Reviews* 3: 1–40.
Ehlers, J., Gibbard, P. & Whiteman, C. A. 1991. The glacial deposits of northwestern Norfolk. In: Ehlers, J., Gibbard, P. & Rose, J. (eds): *Glacial Deposits in Great Britain and Ireland*: 223–232. Rotterdam: Balkema.
Ehrenberg, C. G. 1854. *Microgeologie*. Leipzig: Leopold Voss. 374 pp.
Ehrmann, W. U. 1994. Die känozoische Vereisungsgeschichte der Antarktis. *Berichte zur Polarforschung* 137: 152 pp.
Eichler, H. & Sinn, P. 1974. Zur Gliederung der Altmoränen im westlichen Salzachgletschergebiet. *Zeitschrift für Geomorphologie N.F.* 18: 132–158.
Eichler, H. & Sinn, P. 1975. Zur Definition des Begriffes 'Mindel' im schwäbischen Alpenvorland. *Neues Jahrbuch für Geologie und Paläontologie, Monatshefte*: 705–718.
Einarsson, T. & Albertsson, K. J. 1989. The glacial history of Iceland during the past three million years. *Philosophical Transactions of the Royal Society of London* B 318: 637–644.
Einarsson, T., Hopkins, D. M. & Doell, R. R. 1967. The stratigraphy of Tjörnes, northern Iceland, and the history of the Bering Land Bridge. In: Hopkins, D. M. (ed.): *The Bering Land Bridge*: 312–325. California: Stanford University Press.
Eiríksson, J. & Geirsdóttir, Á. 1991. A record of Pliocene and Pleistocene glaciations and climatic changes in the North Atlantic based on variations in volcanic and sedimentary facies. *Marine Geology* 101: 147–159.
Eissmann, L. 1967. Glaziäre Destruktionszonen (Rinnen, Becken) im Altmoränengebiet des Norddeutschen Tieflandes. *Geologie* 16: 804–833.
Eissmann, L. 1974. Die Begründung der Inlandeistheorie für Norddeutschland durch den Schweizer Adolph von Morlot im Jahre 1844. *Abhandlungen und Berichte des Naturkundlichen Museums 'Mauritianum' Altenburg* 8: 289–318.
Eissmann, L. 1975. Das Quartär der Leipziger Tieflandsbucht und angrenzender Gebiete um Saale und Elbe. Modell einer Landschaftsentwicklung am Rand der europäischen Kontinentalvereisung. *Schriftenreihe für Geologische Wissenschaften* 2: 263 pp. + volume of illustrations.
Eissmann, L. 1978. Mollisoldiapirismus. *Zeitschrift für Angewandte Geologie* 24: 130–138.
Eissmann, L. 1981. Periglaziäre Prozesse und Permafroststrukturen aus sechs Kaltzeiten des Quartärs. Ein Beitrag zur Periglazialgeologie aus der Sicht des Saale-Elbe-Gebietes. *Altenburger Naturwissenschaftliche Forschungen* 1: 171 pp.
Eissmann, L. 1982. Zum Ablauf der Elstereiszeit in der Leipziger Tieflandsbucht unter besonderer Berücksichtigung geschiebeanalytischer Befunde. *Zeitschrift für Geologische Wissenschaften* 10: 771–781.
Eissmann, L. 1986. Quartärgeologie und Geschiebeforschung im Leipziger Land mit einigen Schlußfolgerungen zu Stratigraphie und Vereisungsablauf im Norddeutschen Tiefland. *Altenburger Naturwissenschaftliche Forschungen* 3: 105–133.
Eissmann, L. 1987. Lagerungsstörungen im Lockergebirge. Exogene und endogene Tektonik im Lockergebirge des nördlichen Mitteleuropa. Geophysik und Geologie. *Geophysikalische Veröffentlichungen der Karl-Marx-Universität Leipzig* III (4): 7–77.
Eissmann, L. 1990. Das mitteleuropäische Umfeld der Eemvorkommen des Saale-Elbe-Gebietes und Schlußfolgerungen zur Stratigraphie des jüngeren Quartärs. *Altenburger Naturwissenschaftliche Forschungen* 5: 11–48.
Eissmann, L. 1995. The pre-Elsterian Quaternary deposits of central Germany. In: Ehlers, J., Kozarski, S. & Gibbard, P. L. (eds): *Quaternary deposits in North-East Europe*: 423–437. Rotterdam: Balkema.
Eissmann, L. & Litt, Th. 1995. Late Pleistocene deposits in central Germany. In: Ehlers, J., Kozarski, S. & Gibbard, P. L. (eds): *Quaternary deposits in North-East Europe*: 465–472. Rotterdam: Balkema.
Eissmann, L. & Müller, A. 1979. Leitlinien der Quartärentwicklung im Norddeutschen Tiefland. *Zeitschrift für Geologische Wissenschaften* 7: 451–462.
Eissmann, L., Litt, Th. & Wansa, St. 1995. Elsterian and Saalian deposits in their type area in central Germany. In: Ehlers, J., Kozarski, S. & Gibbard, P. L. (eds): *Quaternary deposits in North-East Europe*: 439–464. Rotterdam: Balkema.
Ekberg, M. P., Lowell, T. V. & Stuckenrath, R. 1993. Late Wisconsin glacial advance and retreat patterns in southwestern Ohio, USA. *Boreas* 22: 189–204.

Ekman, I. & Iljin, V. 1995. Deglaciation, the Younger Dryas end moraines and their correlation in Russian Karelia and adjacent areas. In: Ehlers, J., Kozarski, S. & Gibbard, P. L. (eds): *Glacial deposits in North-East Europe*: 195–209. Rotterdam: Balkema.
Elias, S. A. 1994. *Quaternary Insects and Their Environments*. Washington: Smithsonian Institution Press. 284 pp.
Ellwanger, D. 1980. Rückzugsphasen des würmzeitlichen Illergletschers. *Arbeiten aus dem Institut für Geologie und Paläontologie der Universität Stuttgart N.F.* 76: 93–126.
Ellwanger, D. 1988. Würmeiszeitliche Rinnen und Schotter bei Leutkirch/Memmingen. *Jahreshefte des Geologischen Landesamts Baden-Württemberg* 30: 207–229.
Ellwanger, D. 1990. Zur Riß-Stratigraphie im Andelsbach-Gebiet (Baden-Württemberg). *Jahreshefte des Geologischen Landesamts Baden-Württemberg* 32: 235–245.
Ellwanger, D., Fejfar, O. & Von Koenigswald, W. 1994. Die biostratigraphische Aussage der Arvicolidenfauna vom Uhlenberg bei Dinkelscherben und ihre morpho- und lithostratigraphische Konsequenz. *Münchner Geowissenschaftliche Abhandlungen* (A) 26: 173–191.
Ellwanger, D., Bibus, E., Bludau, W., Kösel, M. & Merkt, J. 1995. Baden-Württemberg. In: Benda, L. (ed.): *Das Quartär Deutschlands*: 255–295. Berlin: Gebrüder Borntraeger.
Elson, J. A. 1961. The geology of tills. In: Penner, E. & Butler, J. (eds): *Proccedings of the 14th Canadian Soil Mechanics Conference, National Research Council of Canada, Associate Committee on Soil and Snow Mechanics. Technical Memorandum* 69: 5–36.
Elverhøi, A., Lønne, Ø. & Seland, R. 1983. Glaciomarine sedimentation in a modern fjord environment, Spitsbergen. *Polar Research (new series)* 1: 127–149.
Elverhøi, A., Fjeldskaar, W., Solheim, A., Nyland-Berg, M. & Russwurm, L. 1993. The Barents Sea ice sheet – a model of its growth and decay during the last ice maximum. *Quaternary Science Reviews* 12: 863–873.
Elverhøi, A, Svendsen, J. I., Solheim, A., Andersen, E. S., Milliman, J., Mangerud, J. & Hooke, R. LeB. 1995. Late Quaternary sediment yield from the High Arctic Svalbard area. *Journal of Geology* 103: 1–17.
Elzenga, W., Schwan, J., Baumfalk, Y. A., Vandenberghe, J. & Krook, L. 1987. Grain surface characteristics of periglacial aeolian and fluvial sands. *Geologie en Mijnbouw* 65: 273–286.
Embleton, C. & King, C. A. M. 1975. *Periglacial geomorphology*. London: Edward Arnold. 203 pp.
Emery, K. O. 1963. Organic transportation of marine sediments. In: Hill, M. N. (ed.): *The Sea*: 776–793. New York: Wiley Interscience.
Emery, K. O. & Uchupi, E. 1984. *The Geology of the Atlantic Ocean.* New York: Springer. 1050 pp. + maps.
Emery, K. O., Merrill, A. S. & Druffel, E. R. M. 1988. Changed Late Quaternary marine environments on Atlantic continental shelf and upper slope. *Quaternary Research* 30: 251–269.
Emiliani, C. 1955. Pleistocene paleotemperatures. *Journal of Geology* 63: 539–578.
Enquist, F. 1916. Der Einfluß des Windes auf die Verteilung der Gletscher. *Bulletin of the Geological Institutions of the University of Uppsala* 14: 1–108.
Eppensteiner, W., Van Husen, D. & Krzemien, R. 1973. Beobachtungen an pleistozänen Driftblöcken des Marchfeldes. *Verhandlungen der Geologischen Bundesanstalt* 1993 (2): 331–336.
Erd, K. 1970. Pollenanalytical classification of the Middle Pleistocene in the German Democratic Republic. *Palaeogeography, Palaeoclimatology, Palaeoecology* 8: 129–145.
Erd, K. 1973. Vegetationsentwicklung und Biostratigraphie der Dömnitz-Warmzeit (Fuhne/Saale 1) im Profil von Pritzwalk/Prignitz. *Abhandlungen aus dem Zentralen Geologischen Institut* 18: 9–48.
Erd, K. 1978. Pollenstratigraphie im Gebiet der skandinavischen Vereisungen. *Schriftenreihe für Geologische Wissenschaften* 9: 99–119.
Erd, K. 1987. Die Uecker-Warmzeit von Röpersdorf bei Prenzlau als neuer Interglazialtyp im Saale-Komplex der DDR. *Zeitschrift für Geologische Wissenschaften* 15: 297–313.
Erd, K., Palme, H. & Präger, F. 1987. Holsteininterglaziale Ablagerungen von Rossendorf bei Dresden. *Zeitschrift für Geologische Wissenschaften* 15: 281–295.
Eriksson, B. 1993. The Eemian pollen stratigraphy and vegetational history of Ostrobothnia, Finland. *Geological Survey of Finland Bulletin* 372: 36 pp.
Eriksson, B., Grönlund, T. & Kujansuu, R. 1980. Interglasiaalikerrostuma Evijärvellä, Pohjanmaalla (An interglacial deposit at Evijärvi in the Pohjanmaa region, Finland). *Geologi* 32: 65–71.

Eriksson, K. 1983. Till investigations and mineral prospecting. In: Ehlers, J. (ed.): *Glacial Deposits in North-West Europe*: 107–113. Rotterdam: Balkema.
Eschman, D. F. & Karrow, P. F. 1985. Huron Basin Glacial Lakes: A review. In: Karrow, P. F. & Calkin, P. E. (eds): *Quaternary Evolution of the Great Lakes. Geological Association of Canada Special Paper* 30: 79–93.
Eschman, D. F. & Mickelson, D. M. 1986. Correlation of glacial deposits in the Huron, Lake Michigan and Green Bay Lobes in Michigan and Wisconsin. *Quaternary Science Reviews* 5: 53–57.
Esmark, J. 1824. Bidrag till vor jordklodes historie. *Magazin for Naturvidenskaberne* III (1): 29–54.
Evans, D. J. A. (ed.) 1994: *Cold Climate Landforms*. Chichester: Wiley 526 pp.
Evenson, E. B. 1973. Late Pleistocene shorelines and stratigraphic relations in the Lake Michigan Basin. *Geological Society of America Bulletin* 84: 2281–2298.
Evenson, E. B. & Clinch, J. M. 1987. Debris transport mechanisms at active alpine glacier margins: Alaskan case studies. *Geological Survey of Finland, Special Paper* 2: 111–136.
Evenson, E. B. & Mickelson, D. M. 1974. A reevaluation of the lobation and red till stratigraphy and nomenclature in part of eastern Wisconsin. In: Knox, J. C. & Mickelson, D. M. (eds): *Late Quaternary Environments of Wisconsin*: 102–117. Madison: Wisconsin Geological and Natural History Survey.
Evenson, E. B., Schlüchter. Ch. & Rabassa, J. (eds) 1983. *Tills and Related Deposits*. Rotterdam: Balkema. 454 pp.
Evzerov, V. & Koshechkin, B. 1981. Glaciation flow stages in the western part of the Kola Peninsula. In: Gorbunov, G. I. (ed.): *Glacial Deposits and Glacial History in Eastern Fennoscandia*: 38–47. Apatity: Kola Branch of the USSR Academy of Sciences.
Eybergen, F. A. 1987. Glacier snout dynamics and contemporary push moraine formation at the Turtmannglacier, Wallis, Switzerland. In: Van der Meer, J. J. M. (ed.): *Tills and Glaciotectonics*: 217–231. Rotterdam: Balkema.
Eyles, C. H. 1987. Glacially influenced submarine-channel sedimentation in the Yakataga Formation, Middleton Island, Alaska. *Journal of Sedimentary Petrology* 57: 1004–1017.
Eyles, C. H. & Eyles, N. 1984. Glaciomarine sediments of the Isle of Man as a key to late Pleistocene stratigraphic investigations in the Irish Sea Basin. *Geology* 12: 359–264.
Eyles, N. 1993. Earth's glacial record and its tectonic setting. *Earth-Science Reviews* 35: 1–248.
Eyles, N. & Clark, B. N. 1988. Storm-influenced deltas and ice scouring in a late Pleistocene glacial lake. *Geological Society of America Bulletin* 100: 793–809.
Eyles, N. & McCabe, A. M. 1989. The Late Devensian (22,000 BP) Irish Sea Basin: The sedimentary record of a collapsed ice sheet margin. *Quaternary Science Reviews* 8: 304–351.
Eyles, N. & McCabe, A. M. 1991. Glaciomarine deposits of the Irish Sea Basin: The role of glacio-isostatic disequilibrium. In: Ehlers, J., Gibbard, P. L. & Rose, J. (eds): *Glacial Deposits in Great Britain and Ireland*: 311–331. Rotterdam: Balkema.
Eyles, N. & Williams, N. E. 1992. The sedimentary and biological record of the last interglacial–glacial transition at Toronto, Canada. In: Clark, P. U. & Lea, P. D. (eds): *The Last Interglacial–Glacial Transition in North America. Geological Society of America Special Paper* 270: 119–137.
Eyles, N., Eyles, C. H. & Miall, A. D. 1983. Lithofacies types and vertical profile model; an alternative approach to the description and environmental interpretation of glacial diamict and diamictite sequences. *Sedimentology* 30: 393–410.
Eyles, N., McCabe, A. M. & Bowen, D. Q. 1994. The stratigraphic and sedimentological significance of Late Devensian ice sheet surging in Holderness, Yorkshire, U.K. *Quaternary Science Reviews* 13: 727–759.
Farrand, W. R. & Drexler, Ch. W. 1985. Late Wisconsinan and Holocene history of the Lake Superior Basin. In: Karrow, P. F. & Calkin, P. E. (eds): *Quaternary Evolution of the Great Lakes. Geological Association of Canada Special Paper* 30: 17–32.
Faustova, M. A. 1995. Glacial stratigraphy of the Late Pleistocene in the northwestern part of the Russian Plain. In: Ehlers, J., Kozarski, S. & Gibbard, P. L. (eds): *Quaternary deposits in North-East Europe*: 179–182. Rotterdam: Balkema.
Feeser, V. 1988. On the mechanics of glaciotectonic contortion of clays. In: Croot, D. G. (ed.): *Glaciotectonics: Forms and Processes*: 63–76. Rotterdam: Balkema.
Felber, H. & Van Husen, D. 1976. Eine innerwürmzeitliche Seeablagerung im Freibachtal (Kärnten). *Zeitschrift für Gletscherkunde und Glazialgeologie* 11: 195–201.

Felber, M., Frei, W. & Heitzmann, P. 1991. Seismic evidence of pre-Pliocene valley formation near Novazzano (Ticino, Switzerland). *Eclogae Geologicae Helvetiae* 84: 753–761.

Feldmann, L. 1990. Jungquartäre Gletscher- und Flußgeschichte im Bereich der Münchener Schotterebene. Dissertation, Universität Düsseldorf. 355 pp.

Feldmann, L. 1992. Ehemalige Ammerseestände im Hoch- und Spätglazial des Würm. *Eiszeitalter und Gegenwart* 42: 52–61.

Felix-Henningsen, P. 1979. Merkmale, Genese und Stratigraphie fossiler und reliktischer Bodenbildungen in saalezeitlichen Geschiebelehmen Schleswig-Holsteins und Süd-Dänemarks. Dissertation, Universität Kiel. 219 pp.

Felix-Henningsen, P. 1983. Palaeosols and their stratigraphical interpretation. In: Ehlers, J. (ed.): *Glacial Deposits in North-West Europe*: 289–295. Rotterdam: Balkema.

Felix-Henningsen, P. 1990. Die mesozoisch-tertiäre Verwitterungsdecke (MTV) im Rheinischen Schiefergebirge. Aufbau, Genese und quartäre Überprägung. *Relief, Boden, Paläoklima* 6: 192 pp.

Felix-Henningsen, P. & Stephan, H.-J. 1982. Stratigraphie und Genese fossiler Böden im Jungmoränengebiet südlich von Kiel. *Eiszeitalter und Gegenwart* 32: 163–175.

Felix-Henningsen, P. & Urban, B. 1982. Paleoclimatic interpretation of a thick intra-Saalian paleosol, the 'Bleached Loam' on the Drenthe Moraines of Northern Germany. *Catena* 9: 1–8.

Fenton, M. M. 1987. Deformation terrain on the northern Great Plains. In: Graf, W. L. (ed.): *Geomorphic Systems of North America. The Geological Society of America, Centennial Special Volume* 2: 176–182.

Ferrians, O. J., Jr. 1994. Permafrost in Alaska. In: Plafker, G. & Berg, H. C. (eds): *The Geology of Alaska. Geological Society of America. The Geology of North America* G-1: 845–854.

Feustel, R. 1989. Der Homo sapiens und das Jungpaläolithikum. In: Herrmann, J. (ed.): *Archäologie in der Deutschen Demokratischen Republik – Denkmale und Funde*: 41–47. Stuttgart: Theiss.

Feyling-Hanssen, R. W. 1954. Late Pleistocene Foraminifera from the Oslofjord Area, Southeast Norway. *Norsk Geologisk Tidsskrift* 33: 109–152.

Finamore, P. F. 1985. Glacial Lake Algonquin and the Fenelon Falls Outlet. In: Karrow, P. F. & Calkin, P. E. (eds): *Quaternary Evolution of the Great Lakes. Geological Association of Canada Special Paper* 30: 125–132.

Finckh, P. 1978. Are southern Alpine lakes former Messinian canyons? Geophysical evidence for preglacial erosion in the southern lakes. *Marine Geology* 27: 289–302.

Fink, J. 1954. Die fossilen Böden im österreichischen Löß. *Quartär* 6: 85–108.

Fink, J. 1960. Leitlinien einer österreichischen Quartärstratigraphie. *Mitteilungen der Geologischen Gesellschaft Wien* 53: 249–266.

Fink, J. 1973. Zur Morphogenese des Wiener Raumes. *Zeitschrift für Geomorphologie, N.F., Supplement-Band* 17: 91–117.

Fink, J. 1977. Jüngste Schotterakkumulation im Österreichischen Donauabschnitt. *Erdwissenschaftliche Forschungen* XIII: 190–211.

Fink, J. & Kukla, G. J. 1977. Pleistocene climates in Central Europe: at least 17 interglacials after the Olduvai Event. *Quaternary Research* 7: 363–371.

Fink, J. & Majdan, H. 1954. Zur Gliederung der pleistozänen Terrassen des Wiener Raumes. *Jahrbuch der Geologischen Bundesanstalt* 97: 211–249.

Fink, J. & Piffl, L. 1975. The Danube from Krems to Vienna. *The Royal Society of New Zealand Bulletin* 13: 127–132.

Firbas, F. 1927. Beiträge zur Kenntnis der Schieferkohlen des Inntals und der interglazialen Waldgeschichte der Ostalpen. *Zeitschrift für Gletscherkunde* 15: 261–277.

Firbas, F. 1949. *Spät- und nacheiszeitliche Waldgeschichte Mitteleuropas nördlich der Alpen* (2 volumes). Jena: Gustav Fischer. 480 + 256 pp.

Fisher, T. G. & Smith, D. G. 1994. Glacial Lake Agassiz: Its northwest maximum extent and outlet in Saskatchewan (Emerson Phase). *Quaternary Science Reviews* 13: 845–858.

Fisk, H. N. 1938. Geology of Grant and LaSalle parishes. *Louisiana Geological Survey Geological Bulletin* 10: 246 pp.

Fisk, H. N. 1940. Geology of Avoyelles and Rapides parishes. *Louisiana Geological Survey Geological Bulletin* 18: 240 pp.

Fisk, H. N. 1944. *Geological investigations of the alluvial valley of the Lower Mississippi River.* Vicksburg, Mississippi: US Army Corps of Engineers, Mississippi River Commission. 78 pp.

Fisk, H. N. & McFarlan, E., Jr. 1955. Late Quaternary deltaic deposits of the Mississippi River. *Geological Society of America Special Paper* 62: 279–302.
Flemal, R. C. 1976. Pingos and pingo scars: Their characteristics, distribution and utility in reconstructing former permafrost environments. *Quaternary Research* 6: 37–53.
Fletcher, K. & Wehmiller, J. F. (eds) 1992. Quaternary coasts of the United States: Marine and lacustrine systems. *SEPM Special Publication* 48: 450 pp.
Flint, R. F. 1937. Pleistocene drift border of eastern Washington. *Bulletin of the Geological Society of America* 48: 203–231.
Flint, R. F. 1943. Growth of the North American ice sheet during the Wisconsin age. *Bulletin of the Geological Society of America* 54: 325–362.
Flint, R. F. 1947. *Glacial geology and the Pleistocene Epoch.* New York: Wiley. 589 pp.
Flint, R.F. 1955. *Pleistocene geology of eastern South Dakota. US Geological Survey Professional Paper* 262: 173 pp.
Flint, R. F. 1957. *Glacial and Pleistocene geology*. New York: Wiley. 553 pp.
Flint, R. F. 1965. Introduction: Historical Perspectives. In: Wright, H. E., Jr. & Frey, D. G. (eds): *The Quaternary of the United States*: 3–11. Princeton: Princeton University Press.
Flint, R. F. 1971. *Glacial and Quaternary geology*. New York: Wiley. 892 pp.
Flint, R. F. 1976. Physical evidence of Quaternary climatic change. *Quaternary Research* 6: 519–528.
Flint, R. F., Sanders, J. E. & Rodgers, J. 1960. Diamictite: A substitute term for symmictite. *Geological Society of America Bulletin* 71: 1809–1810.
Fliri, F. 1973. Beiträge zur Geschichte der alpinen Würmvereisung: Forschungen am Bänderton von Baumkirchen (Inntal, Nordtirol). *Zeitschrift für Geomorphologie N.F., Supplement-Band* 16: 1–14.
Fliri, F. 1978. Die Stellung des Bändertonvorkommens von Schabs (Südtirol) in der alpinen Würm-Chronologie. *Zeitschrift für Gletscherkunde und Glazialgeologie* 14: 115–118.
Fliri, F. 1989. Eine erste Bestimmung des Beginnes der Haupt-Würmvereisung im Zentralraum der Alpen (Albeins bei Brixen). *Der Schlern*: 62–65.
Fliri, F., Bortenschlager, S., Felber, H., Heissel, W., Hilscher, H. & Resch, W. 1970. Der Bänderton von Baumkirchen (Inntal, Tirol). Eine neue Schlüsselstelle zur Kenntnis der Würm-Vereisung der Alpen. *Zeitschrift für Gletscherkunde und Glazialgeologie* VI: 5–35.
Fliri, F., Hilscher, H. & Markgraf, V. 1971. Weitere Untersuchungen zur Chronologie der alpinen Vereisung (Bändertone von Baumkirchen, Inntal, Nordtirol). *Zeitschrift für Gletscherkunde und Glazialgeologie* VII: 5–24.
Fliri, F., Felber, H. & Hilscher, H. 1972. Weitere Ergebnisse der Forschung am Bänderton von Baumkirchen (Inntal, Norditrol). *Zeitschrift für Gletscherkunde und Glazialgeologie* VIII: 203–213.
Follmer, L. R. 1978. The Sangamon Soil in its type area – A review. In: Mahaney, W. C. (ed.): *Quaternary Soils*: 125–165. Toronto: York University.
Follmer, L. R. 1982. The geomorphology of the Sangamon surface: its spatial and temporal attributes. In: Thorn, C. E. (ed.): *Space and Time in Geomorphology. The 'Binghampton' Symposia in Geomorphology, International Series* 12: 117–146. London: Allen & Unwin.
Follmer, L. R. 1983. Sangamonian and Wisconsinan pedogenesis in the midwestern United States. In: Porter, S. C. (ed.): *Late Quaternary environments of the United States, Volume 1, The late Pleistocene*: 138–144. Minneapolis: University of Minnesota Press.
Forbes, E. 1846. On the connexion between the distribution of the existing fauna and flora of the British Isles, and the geological changes which have affected their area, especially during the epoch of the northern drift. *Great Britain Geological Survey, Memoir* 1: 336–432.
Forman, S. L. & Maat, P. 1990. Stratigraphic evidence for late Quaternary dune activity near Hudson on the Piedmont of northern Colorado. *Geology* 18: 745–748.
Forman, S. L. & Nachette, M. N. 1991. Thermoluminescence dating. In: Morrison, R. B. (ed.): *Quaternary Nonglacial Geology: Conterminous U.S. Geological Society of America. The Geology of North America* K-2: 61–65.
Forman, S. L., Bettis III, E. A., Kemnis, T. J. & Miller, B. B. 1992. Chronologic evidence for multiple periods of loess deposition during the Late Pleistocene in the Missouri and Mississippi River Valley, United States: Implications for the activity of the Laurentide Ice Sheet. *Palaeogeography, Palaeoclimatology, Palaeoecology* 93: 71–83.
Forsström, L. & Eronen, M. 1985. Flandrian and Eemian shore levels in Finland and adjacent areas – a

discussion. *Eiszeitalter und Gegenwart* 35: 135–145.
Forsström, L., Eronen, M. & Grönlund, T. 1987. On marine phases and shore levels of the Eemian Interglacial and Weichselian interstadials on the coast of Ostrobothnia, Finland. In: Perttunen, M. (ed.): *Fennoscandian land uplift. Proceedings of a symposium at Tvärminne, April 10–11, 1986, arranged by the Finnish National Committee for Quaternary Research. Geological Survey of Finland, Special Paper* 2: 37–42.
Forsström, L., Aalto, M., Eronen, M. & Grönlund, T. 1988. Stratigraphic evidence for Eemian crustal movements and relative sea-level changes in Eastern Fennoscandia. *Palaeogeography, Palaeoclimatology, Palaeoecology* 68: 317–335.
Forsyth, J. L. 1965. Age of the buried soil in the Sidney, Ohio, area. *American Journal of Science* 263: 251–297.
Foster, D. S. & Colman, S. M. 1991. Preliminary interpretation of the high-resolution seismic stratigraphy beneath Lake Michigan. *US Geological Survey, Open-File Report* 91–21: 42 pp.
Fraedrich, R. 1979. Spät- und postglaziale Gletscherschwankungen in der Ferwallgruppe (Tirol/Vorarlberg). *Düsseldorfer Geographische Schriften* 12: 161 pp.
Francis, E. A. 1975. Glacial sediments: a selected review. In: Wright, A. E. & Moseley, F. (eds): *Ice Ages: Ancient and Modern. Geological Journal Special Issue* 6: 43–68.
Frank, H. 1979. Glazial übertiefte Täler im Bereich des Isar-Loisach-Gletschers. Neue Erkenntnisse über Aufbau und Mächtigkeit des Quartärs in den alpinen Tälern, im Gebiet des 'Murnauer Schotters' und im 'Tölzer Lobus' (erste Mitteilung). *Eiszeitalter und Gegenwart* 29: 77–99.
Franz, H.-J. & Weisse, R. 1965. Das Brandenburger Stadium. In: Gellert, J. F. (ed.): *Die Weichsel-Eiszeit im Gebiet der Deutschen Demokratischen Republik* 69–81. Berlin: Akademie-Verlag.
Frechen, J. 1959. Die basaltischen 'Ausbläser' von Kärlich (Neuwieder Becken) und die Verbreitung ihrer Tuffe. *Fortschritte in der Geologie von Rheinland und Westfalen* 4: 301–312.
Frechen, M. 1991. Interpretation physikalischer Datierungen in der Quartärforschung. *Geologisches Institut der Universität Köln, Sonderveröffentlichungen* 82: 99–112.
Fredén, C. 1979. The Quaternary history of the Baltic. The western part. In: Gudelis, V. & Königsson, L.-K. (eds): *The Quaternary History of the Baltic. Acta Universitatis Upsaliensis, Symposia Universitatis Upsaliensis Annum Quingentesimum Celebrantis* 1: 59–74.
Frei, R. 1912. Monographie des Schweizerischen Deckenschotters. *Beiträge zur Geologischen Karte der Schweiz, N.F.* 37.
French, H. M. 1972. Asymmetrical slope development in the Chiltern Hills. *Biuletyn Peryglacjalny* 21: 51–73.
French, H. M. 1976. *The periglacial environment.* London: Longman. 309 pp.
French, H. M. 1986. Periglacial involutions and mass displacement structures, Banks Island, Canada. *Geografiska Annaler* 68 A: 167–174.
French, H. M. & Harry, D. G. 1988. Nature and origin of ground ice, Sandhills Moraine, southwest Banks Island, Western Canadian Arctic. *Journal of Quaternary Science* 3: 19–30.
Frenzel, B. 1967. *Die Klimaschwankungen des Eiszeitalters.* Braunschweig: Vieweg. 291 pp.
Frenzel, B. 1983. Die Vegetationsgeschichte Süddeutschlands im Eiszeitalter. In: Müller-Beck (ed.): *Urgeschichte in Baden Württemberg*: 90–166. Stuttgart: Theiss.
Fritz, A. 1971. Das Interglazial von Nieselach, Kärnten. *Carinthia II, Sonderheft* 28 (Festschrift Kahler): 317–330.
Fritz, A. 1975. Beitrag zur würmglazialen Vegetation Kärntens. *Carinthia* II, 197–222.
Fritz, A. 1977. Pollenanalytische Untersuchung der lignitführenden Sedimente von Podlaning im unteren Lesachtal (Kärnten). *Carinthia* II, 167: 189–215.
Fromm, E. 1938. Geochronologisch datierte Pollendiagramme und Diatomeenanalysen aus Ångermanland. *Geologiska Föreningens i Stockholm Förhandlingar* 60: 365–381.
Fromm, K. 1996. Paläomagnetik an altpleistozänen Sedimenten bei Heiligenberg (Oberschwaben). *Abhandlungen des Geologischen Landesamts Baden-Württemberg* 15. (In press).
Frye, J. C. & Willman, H. B. 1960. Classification of the Wisconsinan Stage in the Lake Michigan glacial lobe. *Illinois Geological Survey Circular* 285: 16 pp.
Frye, J. C. & Willman, H. B. 1963. Loess stratigraphy, Wisconsinan classification and accretion gleys in central-western Illinois. (Midwestern Section) Friends of the Pleistocene, 14th Annual Meeting. *Illinois Geological Survey Guidebook* 5: 37 pp.

Frye, J. C. & Willman, H. B. 1975. Quaternary System. In: Willman, H. B., Atherton, E., Buschbach, T. C., Collinson, C., Frye, J. C., Hopkins, M. E., Lineback, J. A. & Simon, J. A. (eds): *Handbook of Illinois Stratigraphy. Illinois State Geological Survey Bulletin* 95: 211–239.
Frye, J. C., Shaffer, P. R., Willman, H. B. & Ekblaw, G. E. 1960. Accretion-gley and the gumbotil dilemma. *American Journal of Science* 258: 185–190.
Frye, J. C., Willman, H. B. & Black, R. F. 1965. Outline of glacial geology of Illinois and Wisconsin. In: Wright, H. E., Jr. & Frey, D. G. (eds): *The Quaternary of the United States*: 43–61. Princeton: Princeton University Press.
Füchtbauer, H. 1967. Die Sandsteine in der Molasse nördlich der Alpen. *Geologische Rundschau* 56: 266–300.
Fuhrmann, R. 1976. Die stratigraphische Stellung der Lösse in Mittel- und Westsachsen. *Zeitschrift für Geologische Wissenschaften* 4: 1241–1270.
Fuhrmann, R. 1990. Die Molluskenfauna des Interglazials von Gröbern. *Altenburger Naturwissenschaftliche Forschungen* 5: 148–167.
Fuhrmann, R. & Pietrzeniuk, E. 1990a. Die Ostrakodenfauna des Interglazials von Gröbern (Kreis Gräfenhainichen). *Altenburger Naturwissenschaftliche Forschungen* 5: 168–193.
Fuhrmann, R. & Pietrzeniuk, E. 1990b. Die Ostrakodenfauna des Interglazials von Grabschütz (Kreis Delitzsch). *Altenburger Naturwissenschaftliche Forschungen* 5: 202–227.
Fuller, M. D. 1962. A magnetic fabric in till. *Geological Magazine* 49: 233–237.
Fullerton, D. S. 1986a. Stratigraphy and correlation of glacial deposits from Indiana to New York and New Jersey. *Quaternary Science Reviews* 5: 23–52.
Fullerton, D. S. 1986b. Chronology and correlation of glacial deposits in the Sierra Nevada, California. *Quaternary Science Reviews* 5: 161–169.
Fullerton, D. S. & Colton, R. B. 1986. Stratigraphy and correlation of the glacial deposits on the Montana Plains. *Quaternary Science Reviews* 5: 69–82.
Fulton, R. J. (ed.) 1984. Quaternary stratigraphy of Canada – A Canadian contribution to IGCP Project 24. *Geological Survey of Canada Paper* 84–10: 210 pp.
Fulton, R. J. (ed.) 1989. *Quaternary Geology of Canada and Greenland. Geological Society of America. The Geology of North America* K-1: 839 pp.
Fulton, R. J. 1989. Foreword. In: Fulton, R. J. (ed.): *Quaternary Geology of Canada and Greenland. Geological Society of America. The Geology of North America* K-1: 1–11.
Funder, S., Hjort, Ch. & Kelly, M. 1991. Isotope Stage 5 (130–74 ka) in Greenland, a review. *Quaternary International* 10–12: 107–122.
Funder, S., Hjort, Ch. & Landvik, J. Y. 1994. The last glacial cycles in East Greenland, an overview. *Boreas* 23: 283–293.
Funnell, B. M. & West, R. G. 1962. The early Pleistocene of Easton Bavents, Suffolk. *Quarterly Journal of the Geological Society of London* 118: 125–141.
Furrer, G. 1954. Solifluktionsformen im Schweizerischen Nationalpark. *Ergebnisse der wissenschaftlichen Untersuchungen im Schweizerischen Nationalpark* IV (29): 203–276.
Furrer, G. 1991. 25 000 Jahre Gletschergeschichte dargestellt an einigen Beispielen aus den Schweizer Alpen. *Neujahrsblatt herausgegeben von der Naturforschenden Gesellschaft in Zürich* 193: 4–52.
Furrer, G. & Bachmann, F. 1972. Solifluktionsdecken im schweizerischen Hochgebirge als Spiegel der postglazialen Landschaftsentwicklung. *Zeitschrift für Geomorphologie N.F., Supplement-Band* 13: 163–172.
Furrer, G., Bachmann, F. & Fitze, P. 1971. Erdströme als Formelemente von Solifluktionsdecken im Raum Munt Chavagl/Munt Buffalora (Schweiz. Nationalpark). *Ergebnisse der wissenschaftlichen Untersuchungen im Schweizerischen Nationalpark* XI (65): 189–269.
Furrer, G., Leuzinger, H. & Ammann, K. 1975. Klimaschwankungen während des alpinen Postglazials im Spiegel fossiler Böden. *Vierteljahrsschrift der Naturforschenden Gesellschaft in Zürich* 120: 15–31.
Gadd, N. R. (ed.) 1988. The Late Quaternary development of the Champlain Sea Basin. *Geological Association of Canada Special Paper* 35: 312 pp.
Galloway, J. P. & Carter, L. D. 1993. Late Holocene longitudinal and parabolic dunes in northern Alaska: preliminary interpretations of age and palaeoclimatic significance. In: Dusel-Bacon, C. & Till, A. B. (eds): *Geologic studies in Alaska by the U.S. Geological Survey, 1992. U.S. Geological Survey Bulletin* 2068: 3–11.

Galon, R. 1965. Some new problems concerning subglacial channels. *Geographia Polonica* 6: 19–28.
Galon, R., Lankauf, K. & Noryskiewicz, B. 1983. Zur Entstehung der subglaziären Rinnen im nordischen Vereisungsgebiet an einem Beispiel aus der Tuchola-Heide. *Petermanns Geographische Mitteilungen, Ergänzungsheft* 282: 176–183.
Gamper, M. 1985. Morphochronologische Untersuchungen an Solifluktionszungen, Moränen und Schwemmkegeln in den Schweizer Alpen. Eine Gliederung mit Hilfe der 14C-Altersbestimmung fossiler Böden. *Physische Geographie* 17: 115 pp.
García Ambrosiani, K. 1990. Pleistocene stratigraphy in central and northern Sweden – a reinvestigation of some classical sites. *Stockholm University, Department of Quaternary Research, Report* 16: 15 pp.
Gareis, J. 1978. Die Toteisfluren des Bayerischen Alpenvorlandes als Zeugnis für die Art des spätwürmzeitlichen Eisschwundes. *Würzburger Geographische Arbeiten* 46: 101 pp.
Garleff, K. 1968. Geomorphologische Untersuchungen an geschlossenen Hohlformen ('Kaven') des Niedersächsischen Tieflandes. *Göttinger Geographische Abhandlungen* 44: 142 pp.
Garleff, K., Brunotte, E. & Stingl, H. 1988. Fußflächen im zentralen Teil der Hessischen Senke. *Berliner Geographische Abhandlungen* 47: 63–76.
Garrison, R. E. 1981. Diagenesis of oceanic carbonate sediments: A review of the DSDP perspective. In: Warme, J. E., Douglas, R. G. & Winterer, E. L. (eds): *The Deep Sea Drilling Project: A Decade of Progress. Society of Economic Paleontologists and Mineralogists, Special Publication* 32: 181–207.
Garry, C. E., Schwert, D. P., Baker, R. G., Kemnis, T. J., Horton, D. G. & Sullivan, A. E. 1990. Plant and insect remains from the Wisconsinan interstadial/stadial transition at Wedron, north-central Illinois. *Quaternary Research* 33: 387–399.
Gasser, U. & Nabholz, W. 1969. Zur Sedimentologie der Sandfraktion im Pleistozän des schweizerischen Mittellandes. *Eclogae Geologicae Helvetiae* 62: 467–516.
Gassert, D. 1975. Stausee- und Rinnenbildung an den südlichsten Eisrandlagen in Norddeutschland. *Würzburger Geographische Arbeiten* 43: 55–65.
Gauthier, R. C. 1980. Decomposed granite, Big Bald Mountain area, New Brunswick. *Current Research, Part B, Geological Survey of Canada, Paper* 80-1B: 277–282.
Geikie, A. 1863. On the phenomena of the glacial drift of Scotland. *Transactions of the Geological Society of Glasgow* 1: 190 pp.
Geikie, J. 1874. *The Great Ice Age and its relationship to the antiquity of man*. London: Isbister. 575 pp.
Geikie, J. 1894. *The Great Ice Age and its relationship to the antiquity of man*, 3rd edition. London: Stanford. 850 pp.
Geirsdóttir, Á. 1990. Diamictites of late Pliocene age in western Iceland. *Jökull* 40: 3–25.
Gellert, J. F. (ed.) 1965. *Die Weichselvereisung auf dem Territorium der Deutschen Demokratischen Republik.* Berlin: Akademie-Verlag. 261 pp.
Geyh, M. A. & Schleicher, H. 1990. *Absolute Age Determination. Physical and Chemical Dating Methods and Their Application.* Berlin: Springer. 503 pp.
Gibbard, P. L. 1974. Pleistocene stratigraphy and vegetational history of Hertfordshire, 2 volumes: 286 + 129 pp. Unpublished Ph.D. thesis, University of Cambridge.
Gibbard, P. L. 1977. Pleistocene history of the Vale of St Albans. *Philosophical Transactions of the Royal Society of London* B 280: 445–483.
Gibbard, P. L. 1979. Middle Pleistocene drainage in the Thames valley. *Geological Magazine* 116: 35–44.
Gibbard, P. L. 1985. *The Pleistocene History of the Middle Thames Valley*. Cambridge: Cambridge University Press. 155 pp.
Gibbard, P. L. 1986. Flint gravels in the Quaternary of southeast England. In: Sieveking, G. de G. & Hart, M. B. (eds): *The Scientific Study of Flint and Chert*: 141–149.
Gibbard, P. L. 1988. The history of the great northwest European rivers during the past three million years. *Philosophical Transactions of the Royal Society of London* B 318: 559–602.
Gibbard, P. L. 1991. The Wolstonian Stage in East Anglia. In: Lewis, S. G., Whiteman, C. A. & Bridgland, D. R. (eds): *Central East Anglia & the Fen Basin Field Guide*: 7–13. Cambridge: Quaternary Research Association.
Gibbard, P. L. 1994. *Pleistocene History of the Lower Thames Valley*. Cambridge: Cambridge University Press. 229 pp.
Gibbard, P. L. 1995. The formation of the Strait of Dover. In: Preece, R. C. (ed.): *Island Britain: a Quaternary perspective. Geological Society of London, Special Publication* 96: 15–26.

Gibbard, P. L. & Allen, L. G. 1994. Drainage history of south and east England during the Pleistocene. *Terra Nova* 6: 444–452.
Gibbard, P. L. & Stuart, A. J. 1974. Trace fossils from proglacial lake sediments. *Boreas* 3: 69–74.
Gibbard, P. L. & Turner, C. 1988. In defence of the Wolstonian stage. *Quaternary Newsletter* 54: 9–14.
Gibbard, P. L. & Turner, C. 1990. Cold stage type sections: some thoughts on a difficult problem. *Quaternaire* 1: 33–40.
Gibbard, P. L., Wintle, A. G. & Catt, J. A. 1987. Age and origin of clayey silt 'brickearth' in West London, England. *Journal of Quarternary Science* 2: 3–9.
Gibbard, P. L., West, R. G., Zagwijn, W. H., Balson, P. S., Burger, A. W., Funnell, B. M., Jeffery, D. H., de Jong, J., van Kolfschoten, T., Lister, A. M., Meijer, T., Norton, P. E. P., Preece, R. C., Rose, J., Stuart, A. J., Whiteman, C. A. & Zalasiewicz, J. A. 1991. Early and Middle Pleistocene correlations in the southern North Sea Basin. *Quaternary Science Reviews* 10: 23–52.
Gibbard, P. L., West, R. G., Andrew, R. & Pettit, M. 1992. The margin of a Middle Pleistocene ice advance at Tottenhill, Norfolk, England. *Geological Magazine* 129: 59–76.
Gignoux, M. 1913. Les formations marines pliocènes et quaternaires de l'Italie du Sud et de la Sicilie. *Annales de l'Université de Lyon, N.S.* I. *Sc. Méd.* 36: 693 pp.
Gilbert, G. K. 1871. On certain glacial and post-glacial phenomena of the Maumee Valley. *American Journal of Science* 1: 339–345.
Gilbert, G. K. 1890. Lake Bonneville. *US Geological Survey Monograph* 1: 438 pp.
Gilbert, R. 1990. Rafting in glacimarine environments. In: Dowdeswell, J. A. & Scourse, J. D. (eds): *Glacimarine Environments: Processes and Sediments. Geological Society Special Publication* No. 53: 105–120.
Girard, H. 1855. *Die norddeutsche Ebene, insbesondere zwischen Elbe und Weichsel, geologisch dargestellt.* Berlin: Reimer. 265 pp.
Glapa, H. 1971. Warthezeitliche Eisrandlagen im Gebiet der Letzlinger Heide. *Geologie* 20: 1087–1110.
Glückert, G. 1986. The First Salpausselkä at Lohja, southern Finland. *Bulletin of the Geological Society of Finland* 58 (1): 45–55.
Glückert, G. 1995. The Salpausselkä End Moraines in southwestern Finland. In: Ehlers, J., Kozarski, S. & Gibbard, P. L. (eds): *Glacial Deposits in North-East Europe*: 51–55. Rotterdam: Balkema.
Gold, L. W. & Lachenbruch, A. H. 1973. Thermal conditions in permafrost – A review of North American literature. *North American Contribution, Permafrost Second International Conference (Yakutsk, USSR, 13–28 July 1973)*: 3–23.
Goldthwait, R. P. 1951. Development of end moraines in east-central Baffin Island. *Journal of Geology* 59: 567–577.
Goldthwait, R. P. 1959. Scenes in Ohio during the last ice age. *The Ohio Journal of Science* 59: 193–216.
Goldthwait, R. P. & Matsch, C. L. (eds) 1989. *Genetic Classification of Glacigenic Deposits.* Rotterdam: Balkema. 294 pp.
Gooding, A. M. 1963. Illinoian and Wisconsinan glaciations in the Whitewater basin, southeastern Indiana, and adjacent areas. *Journal of Geology* 71: 665–682.
Gordon, J. E. 1993. Beinn Alligin. In: Gordon, J. E. & Sutherland, D. G. (eds): *Quaternary of Scotland. Geological Conservation Review Series* 6: 118–122. London: Chapman & Hall.
Gordon, J. E. & Sutherland, D. G. (eds) 1993. Quaternary of Scotland. *Geological Conservation Review Series* 6. London: Chapman & Hall. 695 pp.
Goslar, T., Kuc, T., Ralska-Jasiewiczkowa, M., Rozanski, K., Arnold, M., Bard, E. & Van Geel, B. 1993. High resolution lacustrine record of the Late Glacial/Holocene transition in Central Europe. *Quaternary Science Reviews* 12: 287–294.
Göttlich, K. H. & Werner, J. 1974. Vorrißzeitliche Interglazialvorkommen in der Altmoräne des östlichen Rheingletschers. *Geologisches Jahrbuch* A 18: 49–79.
Gottsche, C. 1897a. Die tiefsten Glacialablagerungen der Gegend von Hamburg. Vorläufige Mittheilung. *Mittheilungen der Geographischen Gesellschaft in Hamburg* XIII: 131–140.
Gottsche, C. 1897b. Die Endmoränen und das marine Diluvium Schleswig-Holstein's, im Auftrage der Geographischen Gesellschaft in Hamburg untersucht. Theil I: Die Endmoränen. *Mittheilungen der Geographischen Gesellschaft in Hamburg* XIII: 57 pp.
Gottsche, C. 1898. Die Endmoränen und das marine Diluvium Schleswig-Holstein's, im Auftrage der

Geographischen Gesellschaft in Hamburg untersucht. Theil II: Das marine Diluvium. *Mittheilungen der Geographischen Gesellschaft in Hamburg* XIV: 74 pp.
Götzinger, G. 1938. Das Quartär im österreichischen Alpenvorland. *Verhandlungen der III. Internationalen Quartär-Konferenz, Wien, September 1936*: 51–56.
Goudie, A. S. 1977. *Environmental Change*. Oxford: Clarendon Press. 244 pp.
Goudie, A. S. 1993. *Environmental Change: Contemporary Problems in Geography*. Oxford: Clarendon Press. 329 pp.
Goudie, A. S., Cooke, R. U. & Doornkamp, J. C. 1979. The formation of silt from quartz dune sand by salt processes in deserts. *Journal of Arid Environments* 2: 105–112.
Goździk, J. S. 1973. Geneza i pozycja stratygraficzna struktur peryglacjalnych w środkowej Polsce (Summary: Origin and stratigraphic position of periglacial structures in Middle Poland). *Acta Geographica Lodziensia* 31: 1–119.
Goździk, J. S. 1995. Periglacial impact on some features of glacial deposits in central Poland. In: Ehlers, J., Kozarski, S. & Gibbard, P. L. (eds): *Quaternary deposits in North-East Europe*: 319–327. Rotterdam: Balkema.
Graf, W. L. (ed.) 1987. Geomorphic Systems of North America. *The Geological Society of America, Centennial Special Volume* 2: 643 pp.
Grahle, H.-O. 1936. Die Ablagerungen der Holstein-See (Mar. Interglaz. I), ihre Verbreitung, Fossilführung und Schichtenfolge in Schleswig-Holstein. *Abhandlungen der Preußischen Geologischen Landesanstalt N.F.* 172: 1–110.
Grahmann, R. 1925. Diluvium und Pliozän in Nordwestsachsen. *Abhandlungen der mathematisch-physikalischen Klasse der Sächsischen Akademie der Wissenschaften, Leipzig* 39 (4): 82 pp.
Grahmann, R. 1932. Der Löß in Europa. *Mitteilungen der Gesellschaft für Erdkunde Leipzig* 51: 5–24.
Grahmann, R. 1955. The Lower Palaeolithic site of Markkleeberg and other comparable localities near Leipzig. *Transactions of the American Philosophical Society Philadelphia, N.S.* 45/46: 509–687.
Grant, D. R. 1980. Quaternary sea-level change in Atlantic Canada as an indication of crustal delevelling. In: Mörner, N.-A. (ed.): *Earth Rheology, Isostasy and Eustasy*: 201–214. London: Wiley.
Grant, D. R. 1989. Quaternary geology of the Atlantic Appalachian region of Canada. In: Fulton, R. J. (ed.): *Quaternary Geology of Canada and Greenland. Geological Society of America. The Geology of North America* K-1: 393–440.
Graul, H. 1943. Zur Morphologie der Ingolstädter Ausräumungslandschaft. Die Entwicklung des unteren Lechlaufes und des Donaumoosbeckens. *Forschungen zur Deutschen Landeskunde* 43: 114 pp.
Graul, H. 1949. Zur Gliederung des Altdiluviums zwischen Wertach-Lech und Flossach-Mindel. *Berichte der Naturforschenden Gesellschaft Augsburg* 2: 3–31.
Graul, H. 1962. Eine Revision der pleistozänen Stratigraphie des schwäbischen Alpenvorlandes. *Petermanns Geographische Mitteilungen* 106: 253–271.
Graul, H. 1973. Der Stand der Quartärforschung im S der BRD in lithostratigraphischer, pedologischer und geomorphologischer Hinsicht. *Heidelberger Geographische Arbeiten* 38: 251–265.
Gravenor, C. P. 1955. The origin and significance of prairie mounds. *American Journal of Science* 253: 475–481.
Gravenor, C. P. & Kupsch, W. O. 1959. Ice-disintegration features in western Canada. *Journal of Geology* 67: 48–64.
Gravenor, C. P. & Wong, T. 1987. Magnetic and pebble fabrics and origin of the Sunnybrook Till, Scarborough, Ontario, Canada. *Canadian Journal of Earth Sciences* 24: 2038–2046.
Gray, J. M. 1991. Glaciofluvial landforms. In: Ehlers, J., Gibbard, P. L. & Rose, J. (eds): *Glacial Deposits in Great Britain and Ireland*: 443–454. Rotterdam: Balkema.
Gray, J. M. & Coxon, P. 1991. The Loch Lomond Stadial glaciation in Britain and Ireland. In: Ehlers, J., Gibbard, P. & Rose, J. (eds): *Glacial Deposits in Great Britain and Ireland*: 89–105. Rotterdam: Balkema.
Green, C. P., Hey, R. W. & McGregor, D. F. M. 1980. Volcanic pebbles in Pleistocene gravels of the Thames in Buckinghamshire and Hertfordshire. *Geological Magazine* 117: 59–64.
Grimm, E. C. 1989. Palynological and plant-macrofossil studies of the Tonica thermokarst, La Salle County, Illinois. *Geological Society of America Abstracts with Programs* 21 (4): 13.
Grimm, W.-D., Bläsig, H., Doppler, G., Fakhrai, M., Goroncek, K., Hintermaier, G., Just, J., Kiechle, W., Lobinger, W. H., Ludewig, H., Muzavor, S., Pakzad, M., Schwarz, U. & Sidiropoulos, T. 1979.

Quartärgeologische Untersuchungen im Nordwestteil des Salzach-Vorlandgletschers (Oberbayern). In: Schlüchter, Ch. (ed.): *Moraines and Varves – Origin, Genesis, Classification*: 101–119. Rotterdam: Balkema.

Gripp, K. 1924. Über die äußerste Grenze der letzten Vereisung in Nordwest-Deutschland. *Mitteilungen der Geographischen Gesellschaft in Hamburg* 36: 159–245.

Gripp, K. 1929. Glaciologische und geologische Ergebnisse der Hamburgischen Spitzbergen-Expedition 1927. *Abhandlungen aus dem Gebiete der Naturwissenschaften, herausgegeben vom Naturwissenschaftlichen Verein in Hamburg* XXII (3/4): 145–249.

Gripp, K. 1938. Endmoränen. *Comptes Rendus Cong. Int. Geogr. Amsterdam, T.II, Sect. IIa Geographie Physique*: 215–228, Amsterdam. *Also*: End moraines. In: Evans, D. J. A. (ed.) 1994: *Cold Climate Landforms*: 255–267. Chichester: Wiley.

Gripp, K. 1943. Die Rengeweihe von Stellmoor, Ahrensburger Stufe. In: Rust, A. (ed.): *Die alt- und mittelsteinzeitlichen Funde von Stellmoor*: 106–122. Neumünster: Wachholtz.

Gripp, K. 1964. *Erdgeschichte von Schleswig-Holstein*. Neumünster: Wachholtz. 411 pp.

Gripp, K. 1974. Untermoräne – Grundmoräne – Grundmoränenlandschaft. *Eiszeitalter und Gegenwart* 25: 5–9.

Grönlund, T. 1991. The diatom stratigraphy of the Eemian Baltic Sea on the basis of sediment discovieries in Ostrobothnia, Finland. *Geological Survey of Finland, Report of Investigation* 102: 26 pp.

Grootes, P. M., Stuiver, M., White, J. W. C., Johnsen, S. & Jouzel, J. 1993. Comparison of oxygen isotope records from the GISP2 and GRIP Greenland ice cores. *Nature* 366: 552–554.

Gross, G., Kerschner, H. & Patzelt, G. 1978. Methodische Untersuchungen über die Schneegrenze in alpinen Gletschergebieten. *Zeitschrift für Gletscherkunde und Glazialgeologie* XII: 223–251.

Grosswald, M. G. 1980. Late Weichselian ice sheet of northern Eurasia. *Quaternary Research* 13: 1–32.

Grottenthaler, W. 1980. *Geologische Karte von Bayern 1:25 000, Erläuterungen zum Blatt Nr. 7833 Fürstenfeldbruck*. München: Bayerisches Geologisches Landesamt. 82 pp.

Grottenthaler, W. 1985. *Geologische Karte von Bayern 1:25 000, Erläuterungen zum Blatt Nr. 8036 Otterfing und zum Blatt Nr. 8136 Holzkirchen*. München: Bayerisches Geologisches Landesamt. 189 pp.

Grottenthaler, W. 1989. Lithofazielle Untersuchungen von Moränen und Schottern in der Typusregion des Würm. In: Rose, J. & Schlüchter, Ch. (eds): *Quaternary Type Sections: Imagination or Reality?*: 101–112. Rotterdam: Balkema.

Grube, F. 1967. Die Gliederung der Saale–(Riß–)Kaltzeit im Hamburger Raum. *Fundamenta* B 2: 168–195.

Grube, F. 1979. Zur Morphogenese und Sedimentation im quartären Vereisungsgebiet Norddeutschlands. *Verhandlungen des Naturwissenschaftlichen Vereins in Hamburg (NF)* 23: 69–80. *Also*: On morphogenesis and sedimentation in the Quaternary glaciation area of Northwest Germany. In: Evans, D. J. A. (ed.) 1994: *Cold Climate Landforms*: 313–321. Chichester: Wiley.

Grube, F. 1981. The Subdivision of the Saalian in the Hamburg Region. *Mededelingen Rijks Geologische Dienst* 34 (4): 15–22.

Grube, F. 1983. Tunnel valleys. In: Ehlers, J. (ed.): *Glacial deposits in North-West Europe*: 257–258. Rotterdam: Balkema.

Grüger, E. 1968. Vegetationsgeschichtliche Untersuchungen an cromerzeitlichen Ablagerungen im nördlichen Randgebiet der deutschen Mittelgebirge. *Eiszeitalter und Gegenwart* 18: 204–235.

Grüger, E. 1972a. Late Quaternary vegetation development in south-central Illinois. *Quaternary Research* 2: 217–231.

Grüger, E. 1972b. Pollen and seed studies of Wisconsin vegetation in Illinois. *Geological Association of America Bulletin* 83: 2715–2734.

Grüger, E. 1979. Spätriß, Riß/Würm und Frühwürm am Samerberg in Oberbayern – ein vegetationsgeschichtlicher Beitrag zur Gliederung des Jungpleistozäns. *Geologica Bavarica* 80: 5–64.

Grüger, E. 1983. Untersuchungen zur Gliederung und Vegetationsgeschichte des Mittelpleistozäns am Samerberg in Oberbayern. *Geologica Bavarica* 84: 21–40.

Grüger, E. 1989. Palynostratigraphy of the last interglacial/glacial cycle in Germany. *Quaternary International* 3/4: 69–70.

Grüger, E. & Schreiner, A. 1993. Riß/Würm- und würmzeitliche Ablagerungen im Wurzacher Becken (Rheingletschergebiet). *Neues Jahrbuch für Geologie und Paläontologie, Abhandlungen* 189: 81–117.

Gry, H. 1974. Ledeblokkenes kornstörrelses forhold og transportmåde. *Dansk Geologisk Forening, Årsskrift for 1973*: 140–151.

Gubler, E. 1976. Beitrag des Landesnivellements zur Bestimmung vertikaler Krustenbewegungen in der Gotthard-Region. *Schweizer Mineralogisch-Petrographische Mitteilungen* 56: 675–678.

Gudelis, V. (ed.) 1971. *Crystalline indicator boulders in the East Baltic area* (in Russian). Vilnius: Mintis. 95 pp. + plates.

Gudmundsson, G. H. 1994. An order-of-magnitude estimate of the current uplift-rates in Switzerland caused by the Würm Alpine deglaciation. *Eclogae Geologicae Helvetiae* 87: 545–557.

Guérin, C. 1980. Les Rhinoceros (Mammalia perissodactyla) du Miocène Terminal au Pleistocène supérieur en Europe occidentale. Comparaison avec les espèces actuelles. *Documents du laboratoire de Géologie de Lyon* 79: 1185 pp. Lyon.

Guilbault, J.-P. 1993. Quaternary foraminiferal stratigraphy in sediments of the eastern Champlain Sea basin, Québec. *Géographie physique et Quaternaire* 47: 43–68.

Guiot, J., De Beaulieu, J. L., Cheddadi, R., David, F., Ponel, P. & Reille, M. 1993. The climate in Western Europe during the last Glacial/Interglacial cycle derived from pollen and insect remains. *Palaeogeography, Palaeoclimatology, Palaeoecology* 103: 73–93.

Gupta, S. K. & Sharma, P. 1992. On the nature of the ice cap on the Tibetan Platau during the late Quaternary. *Palaeogeography, Palaeoclimatology, Palaeoecology* 97: 339–343.

Gustavson, T. C. 1975. Sedimentation and physical limnology in proglacial Malaspina Lake, S.E. Alaska. In: Jopling, A. V. & McDonald, B. C. (eds): *Glaciofluvial and glaciolacustrine sedimentation. Society of Economic Paleontologists and Mineralogists, Special Publication* 23: 249–263.

Gwyn, Q. H. J. & Dreimanis, A. 1979. Heavy mineral assemblages in tills and their use in distinguishing glacial lobes in the Great Lakes region. *Canadian Journal of Earth Sciences* 16: 2219–2235.

Haase, G., Lieberoth, I. & Ruske, R. 1970. Sedimente und Paläoböden im Lößgebiet. In: Richter, H., Haase, G., Lieberoth, I. & Ruske, R. (eds): *Periglazial – Löß – Paläolithikum im Jungpleistozän der Deutschen Demokratischen Republik. Petermanns Geographische Mitteilungen, Ergänzungsheft* 274: 99–212.

Habbe, K. A. 1969. Die würmzeitliche Vergletscherung des Gardasee-Gebietes. *Freiburger Geographische Arbeiten* 3: 254 pp.

Habbe, K. A. 1986a. Bemerkungen zum Altpleistozän des Illergletscher-Gebietes. *Eiszeitalter und Gegenwart* 36: 121–134.

Habbe, K. A. 1986b. Zur geomorphologischen Kartierung von Blatt Grönenbach (I) – Probleme, Beobachtungen, Schlußfolgerungen. *Erlanger Geographische Arbeiten* 47: 119 pp.

Habbe, K. A. 1988. Zur Genese der Drumlins im süddeutschen Alpenvorland – Bildungsräume, Bildungszeiten, Bildungsbedingungen. *Zeitschrift für Geomorphologie, Supplement-Band* 70: 33–50.

Habbe, K. A. 1989. Die pleistozänen Vergletscherungen des süddeutschen Alpenvorlandes. *Mitteilungen der Geographischen Gesellschaft in München* 74: 27–51.

Habbe, K. A. & Rögner, K. 1989. Second International Conference on Geomorphology in Frankfurt, Field Trip C 10: Bavarian Alpine Foreland between Rivers Iller and Lech. *Geoöko-Forum* 1: 181–222.

Habbe, K. A. & Walz, H.-G. 1983. Beobachtungen zu Gletschervorstößen des älteren Holozäns in Val Viola (Oberes Addatal, Prov. Sondrio/Italien). In: Schroeder-Lanz, H. (ed.): *Late- and Postglacial Oscillations of Glaciers: Glacial and Periglacial Forms*: 1–13. Rotterdam: Balkema.

Haeberli, W. 1975a. Untersuchungen zur Verbreitung von Permafrost zwischen Flüelapaß und Piz Grialetsch (Graubünden). *Mitteilungen der Versuchsanstalt für Wasserbau, Hydrologie und Glaziologie der Eidgenössischen Technischen Hochschule Zürich* 17: 221 pp.

Haeberli, W. 1975b. Eistemperaturen in den Alpen. *Zeitschrift für Gletscherkunde und Glazialgeologie* XI (2): 203–220.

Haeberli, W. 1979. Holocene push-moraines in Alpine permafrost. *Geografiska Annaler* 61 A: 43–48.

Haeberli, W. 1982. Klimarekonstruktionen mit Gletscher-Permafrost-Beziehungen. *Materialien zur Physiogeographie* 4: 9–17.

Haeberli, W. 1985. Creep of Mountain Permafrost: Internal Structure and Flow of Alpine Rock Glaciers. *Mitteilungen der Versuchsanstalt für Wasserbau, Hydrologie und Glaziologie der Eidgenössischen Technischen Hochschule Zürich* 77: 142 pp.

Haeberli, W., King, L. & Flotron, A. 1979. Surface movement and lichen-cover studies at the active rock glacier near the Grubengletscher, Wallis, Swiss Alps. *Arctic and Alpine Research* 11: 421–441.

Haflidason, H., Aarseth, I., Haugen, J.-E., Sejrup, H. P., Lövlie, R. & Reither, E. 1991. Quaternary stratigraphy of the Draugen area, Mid-Norwegian Shelf. *Marine Geology* 101: 125–146.

Hahn, J. 1983. Eiszeitliche Jäger zwischen 35 000 und 15 000 Jahren vor heute. In: Müller-Beck (ed.): *Urgeschichte in Baden Württemberg*: 273–330. Stuttgart: Theiss.

Hajdas, I., Ivy, S. D., Beer, J., Bonani, G., Imboden, D., Lotter, A. F., Sturm, M. & Suter, M. 1993. AMS radiocarbon dating and varve chronology of lake Soppensee: 6000 to 12,000 ^{14}C years BP. *Climate Dynamics* 9: 107–116.

Haldorsen, S. & Shaw, J. 1982. The problem of recognizing melt-out till. *Boreas* 11: 261–277.

Haldorsen, S., Jörgensen, P., Rappol, M. & Riezebos, P. A. 1989. Composition and source of the clay-sized fraction of Saalian till in The Netherlands. *Boreas* 18: 89–97.

Hall, A. M. 1980. Late Pleistocene deposits at Wing, Rutland. *Philosophical Transactions of the Royal Society of London* B 289: 135–164.

Hall, A. M. 1991. Pre-Quaternary landscape evolution in the Scottish Highlands. *Transactions of the Royal Society of Edinburgh, Earth Sciences* 82: 1–26.

Hall, A. M. & Bent, A. J. A. 1990. The limits of the last British ice sheet in northern Scotland and the adjacent shelf. *Quaternary Newsletter* 61: 2–12.

Hall, C. M., Walter, R. C., Westgate, J. A. & York, D. 1984. Geochronology, stratigraphy and geochemistry of Cindrey Tuff in the Pliocene hominid-bearing sediments of the Middle Awash, Ethiopia. *Nature* 308: 26–31.

Hallberg, G. R. 1986. Pre-Wisconsin glacial stratigraphy of the Central Plains Region in Iowa, Nebraska, Kansas and Missouri. *Quaternary Science Reviews* 5: 11–15.

Hallberg, G. R. & Kemnis, T. J. 1986. Stratigraphy and correlation of the glacial deposits of the Des Moines and James Lobes and adjacent areas in North Dakota, South Dakota, Minnesota, and Iowa. *Quaternary Science Reviews* 5: 65–68.

Hallik, R. 1960. Die Vegetationsentwicklung der Holstein-Warmzeit in Nordwestdeutschland und die Altersstellung der Kieselgurlager der südlichen Lüneburger Heide. *Zeitschrift der Deutschen Geologischen Gesellschaft* 112: 326–333.

Hallik, R. 1975. Moortypen Nordeuropas, unter besonderer Berücksichtigung der Verhältnisse in Schweden. *Abhandlungen und Verhandlungen des Naturwissenschaftlichen Vereins in Hamburg NF* 18/19: 33–41.

Hambrey, M. 1994. *Glacial Environments*. London: University College London Press. 296 pp.

Hambrey, M. & Alean, J. 1992. *Glaciers*. Cambridge: Cambridge University Press. 208 pp.

Hambrey, M. J. & Harland, W. B. (eds) 1981. *Earth's Pre-Pleistocene Glacial Record*. Cambridge: Cambridge University Press. 1004 pp.

Hamelin, L.-E. 1961. Périglaciaire du Canada: idées nouvelles et perspective globales. *Cahiers de Géographie de Québec* 13: 205–216.

Hamilton, T. D. 1994. Late Cenozoic glaciation of Alaska. In: Plafker, G. & Berg, H. C. (eds): *The Geology of Alaska. Geological Society of America. The Geology of North America*, G-1: 813–844.

Hannss, Ch. 1982. Spätpleistozäne bis postglaziale Talverschüttungs- und Vergletscherungsphasen im Bereich des Sillon Alpin der französischen Nordalpen. *Mitteilungen der Kommission für Quartärforschung der Österreichischen Akademie der Wissenschaften* 4: 213 pp.

Hannss, Ch., Wegmüller, S. & Biju-Duval, J. 1992. Les dépôts interglaciaires de l'Arselle (chaîne de Belledonne, Alpes Françaises). *Revue de Géographie Alpine* LXXX: 7–20.

Hansel, A. K. & Johnson, W. H. 1992. Fluctuations of the Lake Michigan lobe during the late Wisconsin subepisode. *Sveriges Geologiska Undersökning* Ca 81: 133–144.

Hansel, A. K., Mickelson, D. M., Schneider, A. F. & Larsen, C. L. 1985. Late Wisconsinan and Holocene History of the Lake Michigan Basin. In: Karrow, P. F. & Calkin, P. E. (eds): *Quaternary Evolution of the Great Lakes. Geological Association of Canada Special Paper* 30: 39–53.

Hansen, J. M. 1979. Palynology of some Danish glacial sediments. *Bulletin of the Geological Society of Denmark* 28: 131–134.

Hanson, K. L., Lettis, W. R., Wesling, J. R., Kelson, K. I. & Metzger, L. 1992. Quaternary marine terraces, south-central coastal California: Implications for crustal deformation and coastal evolution. In: Fletcher, K. & Wehmiller, J. F. (eds): *Quaternary Coasts of the United States: Marine and Lacustrine Systems. SEPM Special Publication* 48: 324–332.

Hantke, R. 1978. *Eiszeitalter. Die jüngste Erdgeschichte der Schweiz und ihrer Nachbargebiete*, Band 1. Thun: Ott. 468 pp.
Hantke, R. 1979. Die Geschichte des Alpen-Rheintales in Eiszeit und Nacheiszeit. *Jahresberichte und Mitteilungen des Oberrheinischen Geologischen Vereins, N.F.* 61: 279–295.
Hantke, R. 1980. *Eiszeitalter. Die jüngste Erdgeschichte der Schweiz und ihrer Nachbargebiete*, Band 2. Thun: Ott. 703 pp.
Hantke, R. 1983. *Eiszeitalter. Die jüngste Erdgeschichte der Schweiz und ihrer Nachbargebiete*, Band 3. Thun: Ott. 730 pp.
Hantke, R. 1992. *Eiszeitalter Die jüngste Erdgeschichte der Alpen und ihrer Nachbargebiete*. Landsberg/ Lech: Ecomed. 1908 pp.
Hantke, R. 1993. *Flußgeschichte Mitteleuropas. Skizzen zu einer Erd-, Vegetations- und Klimageschichte der letzten 40 Millionen Jahre*. Stuttgart: Enke. 460 pp.
Harington, C. R. 1990. Vertebrates of the last interglaciation in Canada: a review, with new data. *Géographie physique et Quaternaire* 44: 375–387.
Harland, W. B. 1981. Chronology of Earth's glacial and tectonic record. *Journal of the Geological Society of London* 138: 197–203.
Harland, W. B., Armstrong, R. L., Cox, A. V., Craig, L. E., Smith, A. G. & Smith, D. G. 1990. *A geologic time scale 1989*. Cambridge: Cambridge University Press. 263 pp.
Harris, Ch. & Donnelly, R. 1991. The glacial deposits of South Wales. In: Ehlers, J., Gibbard, P. & Rose, J. (eds): *Glacial Deposits in Great Britain and Ireland*: 279–290. Rotterdam: Balkema.
Hart, J. & Boulton, G. S. 1991. The glacial drifts of northeastern Norfolk. In: Ehlers, J., Gibbard, P. & Rose, J. (eds): *Glacial Deposits in Great Britain and Ireland*: 233–243. Rotterdam: Balkema.
Harting, P. 1874. De bodem van het Eemdal. *Verslag Koninklijke Akademie van Wetenschappen, Afdeling* N, II, Deel VIII: 282–290.
Hartz, N. & Milthers, V. 1901. Det senglaciale Ler i Allerød Teglværksgrav. *Dansk Geologisk Forening, Meddelelser* 1: 31–60.
Havlicek, P. 1983. Late Pleistocene and Holocene Fluvial Deposits of the Morava River (Czechoslovakia). *Geologisches Jahrbuch* A 71: 209–217.
Hays, J. D., Imbrie, J. & Shackleton, N. J. 1976. Variations in the earth's orbit: pacemaker of the ice ages. *Science* 194: 1121–1132.
Hedberg, H. D. (ed.) 1976. *International Stratigraphic Guide*. New York: Wiley. 200 pp.
Heerdt, S. 1965. Zur Stratigraphie des Jungpleistozäns im mittleren N-Mecklenburg. *Geologie* 14: 589–609.
Heim, A. 1870. Die Schliff-Flächen an den Porphyr-Bergen von Hohburg. *Neues Jahrbuch für Mineralogie und Geologie* (1870): 608–610.
Heim, A. 1874. Über die Schliffe an den Porphyr-Bergen von Hohburg. *Neues Jahrbuch für Mineralogie und Geologie* (1874): 953–959.
Heim, A. 1885. Handbuch der Gletscherkunde. Bibliothek *Geographischer Handbücher* XVI. Stuttgart: Engelhorn. 560 pp.
Heim, A. 1919. *Geologie der Schweiz; Band 1: Molasseland und Juragebirge*. Leipzig: Tauchnitz.
Heim, G. E. & Howe, W. B. 1963. Pleistocene drainage and depositional history in northwestern Missouri. *Kansas Academy of Sciences* 66: 378–392.
Heine, K. 1970. Fluß- und Talgeschichte im Raum Marburg. *Bonner Geographische Abhandlungen* 42: 195 pp.
Heinrich, H. 1988. Origin and consequences of cyclic ice rafting in the Northeast Atlantic Ocean during the past 130,000 years. *Quaternary Research* 29: 142–152.
Heinrich, W.-D. 1981. Zur stratigraphischen Stellung der Wirbeltierfaunen aus den Travertinfundstätten von Weimar-Ehringsdorf und Taubach in Thüringen. *Zeitschrift für Geologische Wissenschaften* 9: 1031–1055.
Heinrich, W.-D. 1982. Zur Evolution und Biostratigraphie von *Arvicola* (Rodentia, Mammalia) im Pleistozän Europas. *Zeitschrift für Geologische Wissenschaften* 10: 683–735.
Heinrich, W.-D. 1987. Neue Ergebnisse zur Evolution und Biostratigraphie von *Arvicola* (Rodentia, Mammalia) im Quartär Europas. *Zeitschrift für Geologische Wissenschaften* 15: 389–406.
Heller, F. & Liu, T. 1982. Magnetostratigraphic dating of loess deposits in China. *Nature* 300: 431–433.
Heller, F., Xiuming, L., Liu, T. & Tongchun, X. 1991. Magnetic susceptibility of loess in China. *Earth and Planetary Science Letters* 103: 301–310.

Hemleben, C. & Spindler, M. 1983. Recent advances in research on living planktonic foraminifera. In: Meulenkamp, J. E. (ed.): *Reconstruction of Marine Paleoenvironments, Utrecht Micropalaeontology Bulletin* 30: 141–170.

Hennig, G. J. & Grün, R. 1983. ESR dating in Quaternary geology. *Quaternary Science Reviews* 2: 157–238.

Henrich, R., Wagner, T., Goldschmidt, P. & Michels, K. 1995. Depositional regimes in the Norwegian–Greenland Sea: the last two glacial to interglacial transitions. *Geologische Rundschau* 84: 28–48.

Hesemann, J. 1939. Diluvialstratigraphische Geschiebeuntersuchungen zwischen Elbe und Rhein. *Abhandlungen des Naturwissenschaftlichen Vereins zu Bremen* XXXI: 247–285.

Hesemann, J. 1975. *Kristalline Geschiebe der nordischen Vereisungen*. Krefeld: Geologisches Landesamt Nordrhein-Westfalen. 267 pp.

Hess, H. 1904. *Die Gletscher*. Braunschweig: Vieweg. 426 pp.

Heuberger, H. 1966. Gletschergeschichtliche Untersuchungen in den Zentralalpen zwischen Sellrain und Ötztal. *Wissenschaftliche Alpenvereinshefte* 20: 126 pp.

Heuberger, H. 1968. Die Alpengletscher im Spät- und Postglazial. *Eiszeitalter und Gegenwart* 19: 270–275.

Heuberger, H. 1974. Alpine Quaternary Glaciation. In: Ives, J. D. & Barry, R. G. (eds): *Arctic and Alpine Environments*: 319–338. London: Methuen.

Heusser, L. E. & King, J. E. 1988. North America, with special emphasis on the development of the Pacific coastal forest and prairie/forest boundary prior to the last glacial maximum. In: Huntley, B. & Webb, T. III (eds): *Vegetational history*: 193–236. Dordrecht: Kluwer.

Heusser, L. E. & Shackleton, N. J. 1979. Direct marine–continental correlation: 150,000-year oxygen isotope–pollen record from the North Pacific. *Science* 204: 837–839.

Hey, R. W. 1965. Highly quartzose pebble gravels in the London Basin. *Proceedings of the Geologists' Association* 76: 403–420.

Hey, R. W. 1978. Horizontal Quaternary shorelines of the Mediterranean. *Quaternary Research* 10: 197–203.

Hey, R. W. 1980. Equivalents of the Westland Green Gravels in Essex and East Anglia. *Proceedings of the Geologists' Association* 91: 279–290.

Hey, R. W. 1991. Pre-Anglian glacial deposits and glaciations in Britain. In: Ehlers, J., Gibbard, P. L. & Rose, J. (eds): *Glacial Deposits in Great Britain and Ireland*: 13–16. Rotterdam: Balkema.

Hicock, S. R., Dreimanis, A. & Broster, B. E. 1981. Submarine flow tills at Victoria, British Columbia. *Canadian Journal of Earth Sciences* 18: 71–80.

Hicock, S. R. & Dreimanis, A. 1992. Sunnybrook Drift in the Toronto Area, Canada: Reinvestigation and reinterpretation. In: Clark, P. U. & Lea, P. D. (eds): *The Last Interglacial–Glacial Transition in North America. Geological Society of America Special Paper* 270: 139–161.

Hillaire-Marcel, C., Occhietti, S. & Vincent, J.-S. 1981. Sakami moraine, Quebec: a 500-km-long moraine without climatic control. *Geology* 9: 210–214.

Hiller, A., Litt, Th. & Eissmann, L. 1991. Zur Entwicklung der jungquartären Tieflandstäler im Saale–Elbe–Raum unter besonderer Berücksichtigung von ^{14}C-Daten. *Eiszeitalter und Gegenwart* 41: 26–46.

Hinsch, W. 1979. Rinnen an der Basis des glaziären Pleistozäns in Schleswig-Holstein. *Eiszeitalter und Gegenwart* 29: 173–178.

Hinsch, W. 1993. Marine Molluskenfaunen in Typusprofilen des Elster-Saale-Interglazials und des Elster-Spätglazials. *Geologisches Jahrbuch* A 138: 9–34.

Hirvas, H. 1991. Pleistocene stratigraphy of Finnish Lapland. *Geological Survey of Finland, Bulletin* 354: 123 pp.

Hirvas, H. 1995. Glacial stratigraphy of northern Finland. In: Ehlers, J., Kozarski, S. & Gibbard, P. L. (eds): *Glacial Deposits in North-East Europe*: 29–36. Rotterdam: Balkema.

Hirvas, H. & Nenonen, K. 1987. The till stratigraphy of Finland. *Geological Survey of Finland, Special Paper* 3: 49–63.

Hitchcock, E. 1841. First anniversary address before the Association of American Geologists, and their second annual meeting in Philadelphia, April 5, 1841. *American Journal of Science* 41: 232–275.

Hoare, P. G. 1991. Pre-Midlandian glacial deposits in Ireland. In: Ehlers, J., Gibbard, P. & Rose, J. (eds): *Glacial Deposits in Great Britain and Ireland*: 37–45. Rotterdam: Balkema.

Höfle, H.-Ch. 1980. Klassifikation von Grundmoränen in Niedersachsen. *Verhandlungen des Naturwissenschaftlichen Vereins in Hamburg (NF)* 23: 81–91.

Höfle, H.-Ch. 1983a. Periglacial phenomena. In: Ehlers, J. (ed.): *Glacial Deposits in North-West Europe*: 297–298. Rotterdam: Balkema.

Höfle, H.-Ch. 1983b. Strukturmessungen und Geschiebeanalysen an eiszeitlichen Ablagerungen auf der Osterholz-Scharmbecker Geest. *Abhandlungen des Naturwissenschaftlichen Vereins zu Bremen* 40: 39–53.

Höfle, H.-Ch. 1991. Über die innere Struktur und die stratigraphische Stellung mehrerer Endmoränenwälle im Bereich der Nordheide bis östlich Lüneburg. *Geologisches Jahrbuch* A 126: 151–169.

Höfle, H.-Ch. & Lade, U. 1983. The stratigraphic position of the Lamstedter Moraine within the Younger Drenthe substage (Middle Saalian). In: Ehlers, J. (ed.): *Glacial Deposits in North-West Europe*: 343–346. Rotterdam: Balkema.

Högbom, A. G. 1912. Summary of lecture. *Bulletin of the Geological Institutions of the University of Uppsala* 11: 302.

Hoinkes, H. 1970. Methoden und Möglichkeiten von Massenhaushaltsstudien auf Gletschern. Ergebnisse der Meßreihe Hintereisferner (Ötztaler Alpen) 1953–1968. *Zeitschrift für Gletscherkunde und Glazialgeologie* VI: 37–90.

Hollin, J. T. 1977. Thames interglacial sites, Ipswichian sea levels and Antarctic ice surges. *Boreas* 6: 33–52.

Hollingworth, S. E. 1934. Some solifluction phenomena in the northern part of the English Lake District. *Proceedings of the Geologists' Association* 2: 167–188.

Holmes, Ch. D. 1941. Till fabric. *Geological Society of America Bulletin* 52: 1299–1354.

Holmes, Ch. D. 1944. Origin of loess – a criticism. *American Journal of Science* 242: 442–446.

Holmes, Ch. D. 1947. Kames. *American Journal of Science* 245: 240–249.

Hölting, B. 1958. Die Entwässerung des würmzeitlichen Eisrandes in Mittelholstein. *Meyniana* 7: 61–98.

Homci, H. 1974. Jungpleistozäne Tunneltäler im Nordosten von Hamburg (Rahlstedt-Meiendorf). *Mitteilungen aus dem Geologisch-Paläontologischen Institut der Universität Hamburg* 43: 99–126.

Homilius, J., Weinig, H., Brost, E. & Bader, K. 1983. Geologische und geophysikalische Untersuchungen im Donauquartär zwischen Ulm und Passau. *Geologisches Jahrbuch* E 25: 73 pp.

Hooghiemstra, H. 1984. Vegetational and climatic history of the High Plain of Bogotá, Colombia: a continuous record of the last 3.5 million years. Thesis, University of Amsterdam. *Dissertationes Botanicae* 79: 368 pp.

Hooke, R. LeB. 1984. On the role of mechanical energy in maintaining subglacial water conduits at atmospheric pressure. *Journal of Glaciology* 30: 180–187.

Hopkins, D. M. (ed.) 1967. *The Bering land bridge*. Stanford: Stanford University Press. 495 pp.

Hopkins, O. B. 1923. Some structural features of the plains area of Alberta caused by Pleistocene glaciation. *Geological Society of America Bulletin* 34: 419–430.

Hoppe, G. 1952. Hummocky moraine regions, with special reference to the interior of Norrbotten. *Geografiska Annaler* 34: 1–72.

Hoppe, G. 1959. Glacial morphology and inland ice recession in northern Sweden. *Geografiska Annaler* 41 A: 193–212.

Hopson, P. M. & Bridge, D. McM. 1987. Middle Pleistocene stratigraphy in the lower Waveney valley, East Anglia. *Proceedings of the Geologists' Association* 98: 171–186.

Horne, D. J., Lord, A. R., Robinson, J. E. & Whittaker, J. E. 1990. Ostracods as climatic indicators in interglacial deposits or, On some new and little-known British Quaternary Ostracoda. *Courier Forschungs-Institut Senckenberg* 123: 129–140.

Horton, A. 1989. Quinton. In: Keen, D. H. (ed.): *West Midlands Field Guide*: 69–76. Cambridge: Quaternary Research Association.

Hoselmann, Ch. 1994. Stratigraphie des Hauptterrassen-Bereichs am Unteren Mittelrhein. *Geologisches Institut der Universität zu Köln, Sonderveröffentlichungen* 96: 235 pp.

Hough, J. L. 1955. Lake Chippewa, a low stage of Lake Michigan indicated by bottom sediments. *Geological Society of America Bulletin* 66: 957–968.

Hough, J. L. 1958. *Geology of the Great Lakes*. Urbana: University of Illinois Press. 313 pp.

Houmark-Nielsen, M. 1983. Glacial stratigraphy and morphology of the northern Bælthav region. In: Ehlers, J. (ed.): *Glacial Deposits in North-West Europe*: 211–217. Rotterdam: Balkema.

Houmark-Nielsen, M. 1987. Pleistocene stratigraphy and glacial history of the central part of Denmark. *Bulletin of the Geological Society of Denmark* 36: 1–189.

Houmark-Nielsen, M. 1988. Glaciotectonic unconformities in Pleistocene stratigraphy as evidence for the behaviour of former Scandinavian icesheets. In: Croot, D. G. (ed.): *Glaciotectonics: Forms and Processes*: 91–99. Rotterdam: Balkema.

Houmark-Nielsen, M. 1989. The last interglacial–glacial cycle in Denmark. *Quaternary International* 3/4: 31–39.

Houmark-Nielsen, M. 1994. Late Pleistocene stratigraphy, glaciation chronology and Middle Weichselian environmental history from Klintholm, Møn, Denmark. *Bulletin of the Geological Society of Denmark* 41: 181–202.

Houmark-Nielsen, M. & Berthelsen, A. 1981. Kineto-stratigraphic evaluation and presentation of glacial-stratigraphic data, with examples from northern Samsø, Denmark. *Boreas* 10: 411–422.

Houmark-Nielsen, M. & Sjørring, S. 1991. *Om Istiden i Danmark*. København: Geologisk Centralinstitut. 45 pp.

Hövermann, J. 1956. Beiträge zum Problem der saale-eiszeitlichen Eisrandlagen in der Lüneburger Heide. *Abhandlungen der Braunschweigischen Wissenschaftlichen Gesellschaft* VIII: 36–54.

Howard, A. D., Fairbridge, R. W. & Quinn, J. H. 1968. Terraces – fluvial. In: Fairbridge, R. W. (ed.): *The Encyclopedia of Geomorphology*: 1117–1123. New York: Reinhold Book Corporation.

Hu, Q., Smith, P. E., Evensen, N. M. & York, D. 1994. Lasing the Holocene: Extending the ^{40}Ar–^{39}Ar laser probe method into the ^{14}C age range. *Earth and Planetary Science Letters* 123: 331–336.

Hucke, K. & Voigt, E. 1967. *Einführung in die Geschiebeforschung (Sedimentärgeschiebe)*. Oldenzaal: Nederlandse Geologische Vereniging. 132 pp.

Hughes, O. L. 1965. Surficial geology of part of the Cochrane District, Ontario, Canada. In: Wright, H. E., Jr. & Frey, D. G. (eds): *International Studies on the Quaternary. Geological Society of America Special Paper* 84: 535–565.

Hughes, T. 1973. Is the West Antarctic ice sheet disintegrating? *Journal of Geophysical Research* 78: 7884–7910.

Hughes, T. 1981. Topographic criteria for reconstructing former ice sheets. In: Denton, G. H. & Hughes, T. (eds): *The Last Great Ice Sheets*: 231–235. New York: Wiley.

Hughes, T. 1987. Ice dynamics and deglaciation models when ice sheets collapsed. In: Ruddiman, W. F. & Wright, H. E., Jr. (eds): *North American and Adjacent Oceans During the Last Deglaciation. Geological Society of America. The Geology of North America* K-3: 183–220.

Hughes, T. 1992. Abrupt climatic change related to unstable ice-sheet dynamics: toward a new paradigm. *Palaeogeography, Palaeoclimatology, Palaeoecology* 97: 203–234.

Huntley, D. J., Godfrey-Smith, D. I. & Thewalt, M. L. W. 1985. Optical dating of sediments. *Nature* 313: 105–107.

Hüser, M. 1982. Die Feldspatgehalte quartärzeitlicher Sande Niedersachsens. *Mitteilungen aus dem Geologischen Institut der Universität Hannover* 22: 81 pp.

Hustedt, F. 1953. Die Systematik der Diatomeen in ihren Beziehungen zur Geologie und Ökologie nebst einer Revision des Halobiensystems. *Svensk Botanisk Tidsskrift* 47: 509–519.

Hütt, G., Jungner, H., Kujansuu, R & Saarnisto, M. 1993. OSL and TL dating of buried podsols and overlying sands in Ostrobothnia, western Finland. *Journal of Quaternary Science* 8: 125–132.

Hutter, K. 1983. *Theoretical glaciology*. Rotterdam: Reidel. 510 pp.

Hutton, J. 1795. *Theory of the Earth*, vol. 2. Edinburgh: Creech. 567 pp.

Hyvärinen, H. & Eronen, M. 1979. The Quaternary history of the Baltic. The northern part. In: Gudelis, V. & Königsson, L.-K. (eds): *The Quaternary History of the Baltic. Acta Universitatis Upsaliensis, Symposia Universitatis Upsaliensis Annum Quingentesimum Celebrantis* 1: 7–27.

Ignatius, H., Axberg, St., Niemistö, L. & Winterhalter, B. 1981. Quaternary geology of the Baltic Sea. In: Voipio, A. (ed.): *The Baltic Sea*: 54–121. Amsterdam: Elsevier.

Illies, H. 1952. Die eiszeitliche Fluß- und Formengeschichte des Unterelbe-Gebietes. *Geologisches Jahrbuch* 66: 525–558.

Imbrie, J. & Imbrie, K. P. 1986. *Ice Ages – solving the mystery*. Cambridge: Harvard University Press. 224 pp.

Imbrie, J., Hays, J. D., Martinson, D. G., MacIntyre, A., Mix, A. C., Morley, J. J., Pisias, N. G., Prell, W. L. & Shackleton, N. J., 1984. The orbital theory of Pleistocene climate: support from a revised chronology of the marine δ^{18}O record. In: Berger, A., Imbrie, J., Hays, J., Kukla, G. & Saltzman, B. (eds): *Milankovitch and Climate*: 269–305. Dordrecht: Reidel.

Isayeva, L. L. 1984. Late Pleistocene Glaciation of North-Central Siberia. In: Velichko, A. A. (ed.): *Late Quaternary Environments of the Soviet Union*: 21–30. London: Longman.
Israelson, C., Funder, S. & Kelly, M. 1994. The Aucellaelv stade at Aucellaelv, the first Weichselian glacier advance in Scoresby Sund, East Greenland. *Boreas* 23: 424–431.
Iversen, J. 1954. The Late-Glacial Flora of Denmark and its relation to climate and soil. *Danmarks Geologiske Undersøgelse*, II. Række, 80: 87–119.
Ives, J. D. 1962. Indications of recent extensive glacierization in north central Baffin Island, N.W.T. *Journal of Glaciology* 4: 197–205.
Ives, J. D. 1974. Biological refugia and the nunatak hypothesis. In: Ives, J. D. & Barry, R. G. (eds): *Arctic and Alpine Environments*: 605–636. London: Methuen.
Ives, J. D. & Andrews, J. T. 1963. Studies in the physical geography of north central Baffin Island. *Geographical Bulletin* 19: 5–48.
Ives, J. D., Andrews, J. T. & Barry, R. G. 1975. Growth and decay of the Laurentide Ice Sheet and comparisons with Fenno-Scandinavia. *Naturwissenschaften* 62: 118–125.
Izett, G. A. & Naeser, C. W. 1976. Age of the Bishop Tuff of eastern California as determined by the fission-track method. *Geology* 4: 587–590.
Jäckli, H. 1962. Die Vergletscherung der Schweiz im Würmmaximum. *Eclogae Geologicae Helvetiae* 55: 285–294.
Jäckli, H. 1965. Pleistocene glaciation of the Swiss Alps and signs of postglacial differential uplift. In: Wright, H. E., Jr. & Frey, D. G. (eds): *International studies on the Quaternary. INQUA USA 1965. Geological Survey of America, Special Papers* 84: 153–157.
Jäckli, H. 1970. *Die Schweiz zur letzten Eiszeit. Karte 1:550 000, Atlas Schweiz.*
Jackson, L. E. & Clague, J. J. 1991. The Cordilleran Ice Sheet: One hundred and fifty years of exploration and discovery. *Géographie physique et Quaternaire* 45: 269–280.
Jacobson, R. B., Elston, D. P. & Heaton, J. W. 1987. Stratigraphy and magnetic polarity of the High Terrace Remnants in the Upper Ohio and Monongahela Rivers in West Virginia, Pennsylvania and Ohio. *Quaternary Research* 29: 216–232.
Jäger, K.-D. & Heinrich, W.-D. 1982. The Travertin at Weimar-Ehringsdorf – an interglacial site of Saalian age? In: Sibrava, V. (ed.): *Quaternary Glaciations in the Northern Hemisphere, IGCP Project 73/1/24*, Report No. 7: 98–113.
Jahn, A. 1975. *Problems of the periglacial zone (Zagadnienia strefy peryglacjalnej).* Warszawa: Panstwowe Wydawnictwo Naukowe. 223 pp.
Jamieson, T. F. 1865. On the history of the last geological changes in Scotland. *Quarterly Journal of the Geological Society of London* 21: 161–203.
Jamieson, T. F. 1874. On the last stage of the Glacial Period in north Britain. *Quarterly Journal of the Geological Society of London* 30: 317–338.
Janczyk-Kopikowa, Z. 1975. Flory interglacjalu mazowieckiego w Ferdynandowie. *Biuletyn Instytutu Geologicznego* 290: 5–96.
Jánossy, D. 1969. Stratigraphische Auswertung der europäischen mittelpleistozänen Wirbeltierfauna. *Berichte der Deutschen Gesellschaft für Geologische Wissenschaften* A 14 (1): 367–438.
Jansen, E. & Erlenkeuser, H. 1985. Ocean circulation in the Norwegian Sea during the last deglaciation: isotopic evidence. *Palaeogeography, Palaeoclimatology, Palaeoecology* 49: 189–206.
Jelgersma, S. & Breeuwer, J. B. 1975. Toelichting bij de kaart Glaciale verschijnselen gedurende het Saalien, 1:600 000. In: Zagwijn, W. H. & Van Staalduinen, C. J. (eds): *Toelichting bij geologische overzichtskaarten van Nederland*: 93–103. Haarlem: Rijks Geologische Dienst.
Jerz, H. 1976. *Bodenkarte von Bayern 1:25 000, Erläuterungen zu Blatt Nr. 7927 Amendingen.* München: Bayerisches Geologisches Landesamt. 78 pp.
Jerz, H. 1979. Das Wolfratshausener Becken – seine glaziale Anlage und Übertiefung. *Eiszeitalter und Gegenwart* 29: 63–69.
Jerz, H. 1981. Quartär. In: *Erläuterungen zur Geologischen Karte von Bayern 1:500 000, 3. Auflage*: 134–151. München: Bayerisches Geologisches Landesamt.
Jerz, H. 1982. Paläoböden in Südbayern (Alpenvorland und Alpen). *Geologisches Jahrbuch* F 14: 27–43.
Jerz, H. 1983. Die Bohrung Samerberg 2 östlich Nußdorf am Inn. *Geologica Bavarica* 84: 5–16.
Jerz, H. 1987. *Geologische Karte von Bayern 1:25 000, Erläuterungen zum Blatt Nr. 8034 Starnberg-Süd.* München: Bayerisches Geologisches Landesamt. 173 pp.

Jerz, H. 1993. *Das Eiszeitalter in Bayern – Erdgeschichte, Gesteine, Wasser, Boden. Geologie von Bayern, Bd. 2*. Stuttgart: Schweizerbart. 243 pp.
Jerz, H. 1995. Bayern. In: Benda, L. (ed.): *Das Quartär Deutschlands*: 296–326. Berlin: Gebrüder Borntraeger.
Jerz, H. & Doppler, G. 1990. *Paläoböden in Bayerisch Schwaben. 9. Tagung des Arbeitskreises 'Paläoböden' der Deutschen Bodenkundlichen Gesellschaft vom 24.5. bis 26.5. in Günzburg. Programm und Exkursionsführer*: 30 pp. München: Bayerisches Geologisches Landesamt.
Jerz, H. & Linke, G. 1987. Arbeitsergebnisse der Subkommission für Europäische Quartärstratigraphie: Typusregion des Holstein-Interglazials (Berichte der SEQS 8). *Eiszeitalter und Gegenwart* 37: 145–148.
Jerz, H. & Ulrich, R. 1983. Das Schieferkohlevorkommen von Herrnhausen südlich von Wolfratshausen (Obb.). *Geologica Bavarica* 84: 101–106.
Jerz, H., Bader, K. & Pröbstl, M. 1979. Zum Interglazialvorkommen von Samerberg bei Nußdorf am Inn. *Geologica Bavarica* 80: 65–71.
Jessen, A. 1931. Lønstrup Klint. *Danmarks Geologiske Undersøgelse*, 2. Række 49: 142 pp. + Atlas.
Jessen, K. & Milthers, V. 1928. Stratigraphical and paleontological studies of interglacial fresh-water deposits in Jutland and Northwest Germany. *Danmarks Geologiske Undersøgelse*, 2. Række 48: 379 pp. + Atlas.
Johnsen, S. J., Clausen, H. B., Dansgaard, W., Iversen, P., Jouzel, J., Stauffer, B. & Steffensen, J. P. 1992. Irregular glacial interstadials recorded in a new Greenland ice core. *Nature* 359: 311–313.
Johnsen, S. J., Clausen, H. B., Dansgaard, W., Gundestrup, N. S., Hammer, C. U. & Tauber, H. 1995. The Eem stable isotope record along the GRIP Ice Core and its interpretation. *Quaternary Research* 43: 117–124.
Johnson, H., Richards, P. C., Long, D. & Graham, C. C. 1993. *The geology of the northern North Sea*. London: HMSO.
Johnson, W. H. 1986. Stratigraphy and correlation of the glacial deposits of the Lake Michigan Lobe prior to 14 ka BP. *Quaternary Science Reviews* 5: 17–22.
Johnson, W. H. 1990. Ice-wedge casts and relict patterned ground in Central Illinois and their environmental significance. *Quaternary Research* 33: 51–72.
Johnstrup, F. 1874. *Om hævningsfænomenerne i Møens Klint*. København: Schultz. 45 pp.
Jones, R. L. & Keen, D. H. 1993. *Pleistocene Environments in the British Isles*. London: Chapman & Hall. 346 pp.
Joon, B., Laban, C. & Van Der Meer, J. J. M. 1990. The Saalian glaciation in the Dutch part of the North Sea. *Geologie en Mijnbouw* 69: 151–158.
Jorda, M. 1988. Modalités paléoclimatiques et chronologiques de la déglaciation würmienne dans les Alpes Françaises du Sud. *Bulletin de l'Association française pour l'étude du Quaternaire* 1988 (2/3): 111–122.
Jørgensen, G. & Rasmussen, J. 1986. Glacial striae, roches moutonnées and ice movements in the Faeroe Islands. *Danmarks Geologiske Undersøgelse* C 7: 114 pp.
Josenhans, H. W. & Zevenhuizen, J. 1990. Dynamics of the Laurentide Ice Sheet in Hudson Bay, Canada. *Marine Geology* 92: 1–26.
Jouzel, J., Lorius, C., Petit, J. R., Barkov, N. I., Kotlyakov, V. M. & Petrov, V. M. 1987. Vostok ice core. A continuous isotopic temperature record over the last climatic cycle (160,000 years). *Nature* 329: 403–408.
Joyce, J. E., Tjalsma, L. R. C. & Pritzman, J. M. 1993. North American glacial meltwater history for the past 2.3 m.y.: Oxygen isotope evidence from the Gulf of Mexico. *Geology* 21: 483–486.
Jung, W., Beug, H.-J. & Dehm, R. 1972. Das Riß/Würm-Interglazial von Zeifen, Landkreis Laufen a.d. Salzach. *Bayerische Akademie der Wissenschaften, Mathematisch-Naturwissenschaftliche Klasse, Abhandlungen, Neue Folge* 151: 1–131.
Kahlke, H. D. 1975. Zur chronologischen Stellung der Travertine von Weimar-Ehringsdorf. In: *III. Internationales Paläontologisches Kolloquium 1968. Das Pleistozän von Weimar-Ehringsdorf, Teil 2. Abhandlungen des Zentralen Geologischen Instituts* 23: 591–596.
Kahlke, H. D. 1981. *Das Eiszeitalter*. Leipzig: Urania. 192 pp.
Kahlke, R.-D. 1994. Die Entstehungs-, Entwicklungs- und Verbreitungsgeschichte des oberpleistozänen *Mammuthus–Coelodonta*-Faunenkomplexes in Eurasien (Großsäuger). *Abhandlungen der Senckenbergischen Naturforschenden Gesellschaft* 546: 164 pp.

Kaiser, E. 1903. Die Ausbildung des Rhein-Tales zwischen Neuwieder Becken und Bonn-Cölner Bucht. *Verhandlungen des 14. Deutschen Geographentages Cöln*: 206–215. Berlin.

Kaiser, K. 1958. Wirkungen des pleistozänen Bodenfrostes in den Sedimenten der Niederrheinischen Bucht. *Eiszeitalter und Gegenwart* 9: 110–129.

Kaiser, K. F. 1989. Late glacial reforestation in the Swiss Mittelland, as illustrated by the Dättnau Valley. In: Rose, J. & Schlüchter, Ch. (eds): *Quaternary Type Sections: Imagination or Reality?*: 161–178. Rotterdam: Balkema.

Kaiser, K. F. 1994.Two Creeks Interstade dated through dendrochronology and AMS. *Quaternary Research* 42: 288–298.

Kaltwang, J. 1992. Die pleistozäne Vereisungsgrenze im südlichen Niedersachsen und im östlichen Westfalen. *Mitteilungen aus dem Geologischen Institut der Universität Hannover* 33: 161 pp.

Kamb, B. 1987. Glacier surge mechanism based on linked-cavity configuration of the basal water conduit system. *Journal of Geophysical Research* 92 B9: 9083–9100.

Kamb, B., Raymond, C. F., Harrison, W. D., Engelhardt, H., Echelmeyer, K. A., Humphrey, N., Brugman, M. M. & Pfeffer, T. 1985. Glacier surge mechanism: 1982–1983 surge of Variegated Glacier, Alaska. *Science* 227: 469–479.

Kandler, O. 1970. Untersuchungen zur quartären Entwicklung des Rheintales zwischen Mainz/Wiesbaden und Bingen/Rüdesheim. *Mainzer Geographische Studien* 3: 92 pp.

Kapp, R. O. & Gooding, A. M. 1964. Pleistocene vegetational studies in the Whitewater Basin, southeastern Indiana. *Journal of Geology* 72: 307–326.

Karlén, W. 1988. Scandinavian glacial and climatic fluctuations during the Holocene. *Quaternary Science Reviews* 7: 199–209.

Karrow, P. F. 1967. Pleistocene geology of the Scarborough area. *Ontario Department of Mines, Geological Report* 46: 108 pp.

Karrow, P. F. 1976. The texture, mineralogy and petrography of North American tills. In: Legget, R. F. (ed.): *Glacial Till: An Inter-disciplinary Study. Royal Society of Canada Special Publication* 12: 83–98.

Karrow, P. F. 1984. Quaternary stratigraphy and history, Great Lakes – St. Lawrence region. In: Fulton, R. J. (ed.): *Quaternary stratigraphy of Canada – A Canadian contribution to IGCP Project 24. Geological Survey of Canada Paper* 84-10: 137–153.

Karrow, P. F. 1989. Quaternary geology of the Great Lakes subregion. In: Fulton, R. J. (ed.): *Quaternary Geology of Canada and Greenland. Geological Society of America. The Geology of North America* K-1: 326–350.

Karrow, P. F. 1990. Interglacial beds at Toronto, Ontario. *Géographie physique et Quaternaire* 44: 289–297.

Karrow, P. F. & Calkin, P. E. (eds) 1985. Quaternary Evolution of the Great Lakes. *Geological Association of Canada Special Paper* 30: 258 pp.

Karte, J. 1979. Räumliche Abgrenzung und regionale Differenzierung des Periglaziärs. *Bochumer Geographische Arbeiten* 35: 211 pp.

Karte, J. 1987. Pleistocene periglacial conditions and geomorphology in north central Europe. In: Boardman, J. (ed.) *Periglacial processes and landforms in Britain and Ireland*: 67–75. Cambridge: Cambridge University Press.

Karukäpp, R., Raukas, A. & Hyvärinen, H. 1992. The deglaciation of the territory. In: Raukas, A. & Hyvärinen, H. (eds): *Geology of the Gulf of Finland* (in Russian): 112–136. Tallinn.

Kasprzak, L. 1985. A model of push moraine development in the marginal zone of the Leszno Phase, west central Poland. *Quaternary Studies in Poland* 7: 23–54.

Kasprzak, L. & Kozarski, S. 1989. Ice-lobe contact sedimentary scarps in marginal zones of the major Vistulian ice-sheet positions, west central Poland. *Quaestiones Geographicae, Special Issue* 2: 69–81.

Kasse, C. 1988. Early-Pleistocene tidal and fluviatile environments in the southern Netherlands and northern Belgium. Thesis, Free University Press, Amsterdam. 190 pp.

Katzenberger, O. 1989. Experimente zu Grundlagen der ESR-Datierung von Molluskenschalen. *Geologisches Institut der Universität zu Köln, Sonderveröffentlichungen* 72: 71 pp.

Kaufman, D. S., Miller, G. H., Stravers, J. A. & Andrews, J. T. 1993. Abrupt early Holocene (9.9–9.6 ka) ice-stream advance at the mouth of Hudson Strait, Arctic Canada. *Geology* 21: 1063–1066.

Kay, G. F. 1916. Gumbotil, a new term in Pleistocene geology. *Science, new series*, 44: 637–638.

Kay, G. F. & Leighton, M. M. 1933. Eldoran Epoch of the Pleistocene Period. *Geological Society of America Bulletin* 44: 669–674.
Keen, D. H. 1982. Late Pleistocene land mollusca in the Channel Islands. *Journal of Chonchology* 31: 57–61.
Kehew, A. E. 1993. Glacial-lake outburst erosion of the Grand Valley, Michigan, and its impacts on glacial lakes in the Lake Michigan Basin. *Quaternary Research* 39: 36–44.
Kehew, A. E. & Lord, M. L. 1986. Origin of large-scale erosional features of glacial-lake spillways in the northern Great Plains. *Geological Society of America Bulletin* 97: 162–177.
Kehew, A. E. & Teller, J. T. 1994. History of Late Glacial runoff along the southwestern margin of the Laurentide Ice Sheet. *Quaternary Science Reviews* 13: 859–877.
Keigwin, L. D. 1978. Pliocene closing of the Isthmus of Panama, based on biostratigraphic evidence from nearby Pacific Ocean and Caribbean Sea cores. *Geology* 6: 630–634.
Keigwin, L. D. 1982. Isotopic paleoceanography of the Caribbean and east Pacific: role of Panama uplift in late Neogene time. *Science* 217: 350–353.
Keilhack, K. 1896. Die Geikiessche Gliederung der nordeuropäischen Glazialablagerungen. *Jahrbuch der Königlich Preußischen Geologischen Landesanstalt für 1895*: 111–124.
Keilhack, K. 1898a. Glaciale Hydrographie. In: Berendt, G., Keilhack, K., Schröder, H. & Wahnschaffe, F. (eds): *Neuere Forschungsergebnisse auf dem Gebiete der Glacialgeologie in Norddeutschland erläutert an einigen Beispielen. Jahrbuch der Königlich Preußischen Geologischen Landesanstalt und Bergakademie zu Berlin für das Jahr 1897*, XVIII: 113–129.
Keilhack, K. 1898b. Die Stillstandslagen des letzten Inlandeises und die hydrographische Entwicklung des pommerschen Küstengebietes. *Jahrbuch der Königlich Preußischen Geologischen Landesanstalt* 19: 90–152.
Kelly, R. I. & Martini, I. P. 1986. Pleistocene glaciolacustrine deltaic deposits of the Scarborough Formation, Ontario, Canada. *Sedimentary Geology* 47: 27–52.
Kelts, K. R. 1978. Geological and sedimentological evolution of Lake Zurich and Lake Zug. Dissertation, ETH Zürich. 256 pp.
Kemp, R. A. 1985. The Valley Farm Soil in southern East Anglia. In: Boardman, J. (ed.): *Soils and Quaternary Landscape Evolution*: 179–196. Chichester: Wiley.
Kemp, R. A. 1987. The interpretation and environmental significance of a buried Middle Pleistocene soil near Ipswich Airport, Suffolk, England. *Philosophical Transactions of the Royal Society of London* B 317: 365–391.
Kemp, R. A., Whiteman, C. A. & Rose, J. 1993. Palaeoenvironmental and stratigraphic significance of the Valley Farm and Barham soils in eastern England. *Quaternary Science Reviews* 12: 833–848.
Kennedy, G. L., Wehmiller, J. F. & Rockwell, T. K. 1992. Paleoecology and paleozoogeography of Late Pleistocene Marine-Terrace faunas of southwestern Santa Barbara County, California. In: Fletcher, K. & Wehmiller, J. F. (eds): *Quaternary Coasts of the United States: Marine and Lacustrine Systems. SEPM Special Publication* 48: 343–361.
Kerney, M. P. 1977. British Quaternary non-marine Mollusca: a brief review. In: Shotton, F. W. (ed.): *British Quaternary studies, recent advances*: 31–42. Oxford: Clarendon Press.
Kerney, M. P. & Sieveking, G. de G. 1977. Northfleet. In: Shephard-Thorn, I.R & Wymer, J.J. (eds): *Southeast England and the Thames Valley. X INQUA Congress, Guidebook for Excursion* A 5: 44–47. Norwich: Geo Abstracts.
Kerschner, H. 1982. Outlines of the climate during the Egesen Advance (Younger Dryas, 11 000–10 000 BP) in the Central Alps of the western Tyrol, Austria. *Zeitschrift für Gletscherkunde und Glazialgeologie* 16: 229–240.
Kessler, P. 1925. *Das eiszeitliche Klima und seine geologischen Wirkungen im nicht vereisten Gebiet.* Stuttgart: Schweizerbart. 210 pp.
King, J. E. & Saunders, J. J. 1986. *Geochelone* in Illinois and the Illinoian–Sangamonian vegetation of the type region. *Quaternary Research* 25: 89–99.
King, L. & Åkerman, J. 1993. Mountain permafrost in Europe. *Permafrost, Proceedings of the Sixth International Conference* 2: 1022–1027.
King, L. H., MacLean, B. & Drapeau, G. 1972. The Scotian Shelf submarine end-moraine complex. In: *International Geological Congress, 24th session, Section 8: Marine Geology and Geophysics*: 237–249.

Kinzl, H. 1929. Beiträge zur Geschichte der Gletscherschwankungen in den Ostalpen. *Zeitschrift für Gletscherkunde* 17: 66–121.

Kinzl, H. 1932. Die größten nacheiszeitlichen Gletschervorstöße in den Schweizer Alpen und in der Mont-Blanc-Gruppe. *Zeitschrift für Gletscherkunde* 20: 269–397.

Klassen, R. A. 1983a. A preliminary report on drift prospecting studies in Labrador. *Current Research, Part A, Geological Survey of Canada, Paper* 83-1A: 353–355.

Klassen, R. A. 1983b. A preliminary report on drift prospecting studies in Labrador, Part II. *Current Research, Part A, Geological Survey of Canada, Paper* 84-1A: 247–254.

Klassen, R. W., Delorme, L. D. & Mott, R. J. 1967. Geology and paleontology of Pleistocene deposits in southwestern Manitoba. *Canadian Journal of Earth Sciences* 4: 433–447.

Klaus, W. 1987. Das Mondsee-Profil: R/W-Interglazial und vier Würm-Interstadiale in einer geschlossenen Schichtfolge. *Mitteilungen der Kommission für Quartärforschung der österreichischen Akademie der Wissenschaften* 7: 3–18.

Kleman, J. 1990. On the use of glacial striae for reconstruction of paleo-ice sheet flow patterns – With application to the Scandinavian ice sheet. *Geografiska Annaler* 72 A: 217–236.

Klostermann, J. 1985. Versuch einer Neugliederung des späten Elster- und des Saale-Glazials der Niederrheinischen Bucht. *Geologisches Jahrbuch A* 83: 46 pp.

Klostermann, J. 1992. *Das Quartär der Niederrheinischen Bucht. Ablagerungen der letzten Eiszeit am Niederrhein*. Krefeld: Geologisches Landesamt Nordrhein-Westfalen. 200 pp.

Klostermann, J., Rehagen, H.-W. & Wefels, U. 1988. Hinweise auf eine saalezeitliche Warmzeit am Niederrhein. *Eiszeitalter und Gegenwart* 38: 115–127.

Kluiving, S. J. 1994. Glaciotectonics in the Itterbeck–Uelsen push moraines, Germany. *Journal of Quaternary Science* 9: 235–244.

Kluiving, S. J., Rappol, M. & Van der Wateren, F. M. 1991. Till stratigraphy and ice movements in eastern Overijssel, The Netherlands. *Boreas* 20: 193–205.

Knauer, J. 1928. Glazialgeologische Ergebnisse aus dem Isargletschergebiet. *Zeitschrift der Deutschen Geologischen Gesellschaft* 80: 294–303.

Knauer, J. 1929. *Erläuterungen zur Geognostischen Karte von Bayern 1:100 000, Blatt München West (Nr. XXVII), Teilblatt Landsberg*: 47 pp.

Knauer, J. 1931. *Erläuterungen zur Geognostischen Karte von Bayern 1:100 000, Blatt München West (Nr. XXVII), Teilblatt München-Starnberg*: 48 pp.

Knoth, W. 1964. Zur Kenntnis der pleistozänen Mittelterrassen der Saale und Mulde nördlich von Halle. *Geologie* 13: 598–616.

Knox, A. S. 1962. Pollen from the Pleistocene terrace deposits of Washington, D.C. *Pollen and Spores* 4: 356–358.

Knox, J. C. 1983. Responses of river systems to Holocene climates. In: Wright, H. E. (ed.): *Late-Quaternary Environments of the United States, Volume 2: The Holocene*: 26–41. Minneapolis: University of Minnesota Press.

Knudsen, K. L. 1980. Foraminiferal faunas in marine Holsteinian Interglacial deposits of Hamburg–Hummelsbüttel. *Mitteilungen aus dem Geologisch-Paläontologischen Institut der Universität Hamburg* 49: 193–214.

Knudsen, K. L. 1985. Correlation of Saalian, Eemian and Weichselian foraminiferal zones in North Jutland. *Bulletin of the Geological Society of Denmark* 33: 325–339.

Knudsen, K. L. 1987. Foraminifera in Late Elsterian–Holsteinian deposits of the Tornskov area in South Jutland, Denmark. *Danmarks Geologiske Undersøgelse* B 10: 7–31.

Knudsen, K. L. 1993. Late Elsterian–Holsteinian foraminiferal stratigraphy in borings of the Lower Elbe area, NW Germany. *Geologisches Jahrbuch* A 138: 97–119.

Knudsen, K. L. 1994. The marine Quaternary in Denmark: a review of new evidence from glacial–interglacial studies. *Bulletin of the Geological Society of Denmark* 41: 203–218.

Knudsen, K. L. & Lykke-Andersen, A.-L. 1982. Foraminifera in Late Saalian, Eemian, Early and Middle Weichselian of the Skærumhede I boring. *Bulletin of the Geological Society of Denmark* 30: 97–109.

Knudsen, K. L. & Sejrup, H. P. 1988. Amino acid geochronology of selected interglacial sites in the North Sea area. *Boreas* 17: 347–354.

Koeppen, W. & Wegener, A. 1924. *Die Klimate der geologischen Vorzeit*. Berlin: Borntraeger. 266 pp.

Kohl, H. 1955. Altmoränen und pleistozäne Schotterfluren zwischen Laudach und Krems. *Jahrbuch des Oberösterreichischen Musealvereins* 100: 321–344.
Kohl, H. 1958. Unbekannte Altmoränen in der südwestlichen Traun-Enns-Platte. *Mitteilungen der Geographischen Gesellschaft Wien* 100: 131–143.
Kohl, H. 1968. Beiträge über Aufbau und Alter der Donausohle bei Linz. *Naturkundliches Jahrbuch der Stadt Linz*: 7–60.
Kohl, H. 1971. Das Quartärprofil von Kremsmünster in Oberösterreich. *Geographische Jahresberichte aus Österreich* 33: 82–88.
Kohl, H. 1974. Die Entwicklung des quartären Flußnetzes im Bereich der Traun-Enns-Platte/ Oberösterreich. *Heidelberger Geographische Arbeiten* 40: 31–44.
Kohl, H. 1978. Zur Jungpleistozän- und Holozänstratigraphie in den oberösterreichischen Donauebenen. In: Nagl, H. (ed.): *Beiträge zur Quartär- und Landschaftsforschung*: 269–290. Wien: Hirt.
Kohl, H. 1983. Beiträge zur Quartärstratigraphie aus dem oberösterreichischen Raum. *Innsbrucker Geographische Studien* 8: 13–33.
Kohl, H. 1986. Pleistocene glaciations in Austria. *Quaternary Science Reviews* 5: 421–427.
Kohl, H. 1989. Zur Frage der Korrelation unterschiedlicher Sedimentfolgen am nördlichen Alpenrand sowie in den Ostalpen und deren Vorland im Jungpleistozän. In: Rose, J. & Schlüchter, Ch. (eds): *Quaternary Type Sections: Imagination or Reality?*: 71–78. Rotterdam: Balkema.
Kolp, O. 1982. Entwicklung und Chronologie des Vor- und Neudarßes. *Petermanns Geographische Mitteilungen* 126: 85–94.
Kolp, O. 1986. Entwicklungsphasen des Ancylus-Sees. *Petermanns Geographische Mitteilungen* 130: 79–94.
Kolstrup, E. 1980. Climate and stratigraphy in Northwestern Europe between 30 000 BP and 13 000 BP with special reference to The Netherlands. *Mededelingen Rijks Geologische Dienst* 32: 181–253.
Kolstrup, E. & Houmark-Nielsen, M. 1991. Weichselian palaeoenvironments at Kobbelgård, Møn. *Boreas* 20: 169–182.
Kolstrup, E. & Jørgensen, J. B. 1982. Older and Younger Coversand in southern Jutland (Denmark). *Bulletin of the Geological Society of Denmark* 30: 71–77.
Kondratiene, O. P., 1981. Main regularities of development of vegetation in the southern Baltic Region during interglacials (in Russian). In: *Pleistocene glaciations on the East European Plain*: 126–132. Moscow: Nauka.
Kondratjeva, K. A., Khrutzky, S. F. & Romanovsky, N. N. 1993. Changes in the extent of permafrost during the Late Quaternary Period in the territory of the former Soviet Union. *Permafrost and Periglacial Processes* 4: 113–119.
Korn, J. 1927. *Die wichtigsten Leitgeschiebe der nordischen kristallinen Gesteine im norddeutschen Flachlande*. Preußische Geologische Landesanstalt: 64 pp.
Kosack, B. & Lange, W. 1985. Das Eem-Vorkommen von Offenbüttel/Schnittlohe und die Ausbreitung des Eem-Meeres zwischen Nord- und Ostsee. *Geologisches Jahrbuch* A 86: 3–17.
Koster, E. A. 1978. De stuifzanden van de Veluwe; een fysisch-geografische studie. Thesis, University of Amsterdam. 195 pp.
Koster, E. A. 1982. Terminology and lithostratigraphic division of (surficial) sandy eolian deposits in The Netherlands: an evaluation. *Geologie en Mijnbouw* 61: 121–129.
Koster, E. A. 1988. Ancient and modern cold-climate aeolian sand deposition: a review. *Journal of Quaternary Science* 3: 69–83.
Koster, E. A. 1995. Progress in cold-climate aeolian research. *Quaestiones Geographicae* (in press).
Koster, E. A. & Dijkmans, J. W. A. 1988. Niveo-aeolian deposits and denivation forms, with special reference to the Great Kobuk Sand Dunes, northwestern Alaska. *Earth Surface Processes and Landforms* 13: 153–170.
Koster, E. A., Castel, I. Y. & Nap, R. L. 1993. Genesis and sedimentary structures of late Holocene aeolian drift sands in northwest Europe. In: Pye, K. (ed.): *The Dynamics and Environmental Context of Aeolian Sedimentary Systems. Geological Society Special Publication* 72: 247–267.
Kozarski, S. 1967. The origin of subglacial channels in the North Polish and North German Plain. *Bulletin de la Societé des Amis des Sciences et Lettres de Poznan*, B 20: 21–36.
Kozarski, S. 1975. Oriented kettle holes in outwash plains. *Quaestiones Geographicae* 2: 99–112.
Kozarski, S. 1980. An outline of Vistulian stratigraphy and chronology of the Great Poland Lowland.

Quaternary Studies in Poland 2: 21–35.
Kozarski, S. 1983. The Holocene generation of paleomeanders in the Warta River Valley, Great Polish Lowlands. *Geologisches Jahrbuch* A 71: 109–118.
Kozarski, S. (ed.) 1991. Late Vistulian (=Weichselian) and Holocene aeolian phenomena in Central and Northern Europe. *Zeitschrift für Geomorphologie, Supplement-Band* 90: 207 pp.
Kozarski, S. 1992. Lithostratigraphy of Upper Plenivistulian deposits in the Great Poland Lowland within the area of the last glaciation. *Sveriges Geologiska Undersökning* Ca 81: 157–162.
Kozarski, S. 1993. Late Plenivistulian deglaciation and the expansion of the periglacial zone in NW Poland. *Geologie en Mijnbouw* 72: 143–157.
Kozarski, S. 1995. The periglacial impact on the deglaciated area of northern Poland after 20 kyr BP. *Biuletyn Peryglacjalny* 34 (in press).
Krigström, A. 1962. Geomorphological studies of sandur plains and their braided rivers in Iceland. *Geografiska Annaler* 44: 328–346.
Krinitzsky, E. L. 1949. Geological investigations of gravel deposits in the Lower Mississippi Valley and adjacent uplands. *US Army Engineer Waterways Experiment Station Technical Memorandum* 3–272: 58 pp.
Krinitzsky, E. L. & Smith, F. L. 1969. Geology of backswamp deposits in the Atchafalaya Basin, Louisiana. *US Army Engineer Waterways Experiment Station Technical Report* S-69-8: 58 pp.
Krinsley, D. H. & Donahue, J. 1968. Environmental interpretation of sand grain surface textures by electron microscopy. *Geological Society of America Bulletin* 79: 743–748.
Krinsley, D. H. & Doornkamp, J. C. 1973. *Atlas of quartz sand surface textures.* Cambridge: Cambridge University Press. 91 pp.
Kröger, K., Bittmann, F., Van den Bogaard, P. & Turner, E. 1988. Der Fundplatz Kärlich-Seeufer. Neue Untersuchungen zum Altpaläolithikum im Rheinland. *Jahrbuch des Römisch-Germanischen Zentralmuseums Mainz* 35: 111–135.
Krüger, J. 1970. Till fabric in relation to direction of ice movement – A study from the Fakse Banke, Denmark. *Geografisk Tidsskrift* 69: 133–170.
Krüger, J. 1983. Glacial morphology and deposits in Denmark. In: Ehlers, J. (ed.): *Glacial Deposits in North-West Europe*: 181–191. Rotterdam: Balkema.
Krüger, J. 1987. Træk af et glaciallandskabs udvikling ved nordranden af Myrdalsjökull, Island. *Dansk Geologisk Forening, Årsskrift for 1986*: 49–65.
Krüger, J. & Thomsen, H. H. 1981. Till fabric i et recent bundmorænelandskab, Island. *Dansk Geologisk Forening, Årsskrift for 1980*: 19–28.
Krüger, J. & Thomsen, H. H. 1984. Morphology, stratigraphy, and genesis of small drumlins in front of the glacier Myrdalsjökull, south Iceland. *Journal of Glaciology* 30: 94–105.
Krupinski, K. M. 1986. Sediments of the Eemian Interglacial at Komorów near Pruszków (Mazovian Lowland). *Bulletin of the Polish Academy of Sciences, Earth Sciences* 34: 363–467.
Krygowski, B. 1969. New data to glacial till classification. *Zeszyty Naukowe Unywersytetu Im. Adama Mickiewicza w Poznaniu, Geografia* 8: 84–93.
Krygowski, B. & Krygowski, T. M. 1965. Mechanical method of estimating of abrasion grade of sand grains (Mechanical graniformametry). *Journal of Sedimentary Petrology* 35: 496–499.
Krzyszkowski, D. 1991a. Saalian sediments in the Bełchatów outcrop, central Poland. *Boreas* 20: 29–46.
Krzyszkowski, D. (ed.) 1991b. The polyinterglacial Czyzów Formation in the Kleszczów Graben (Central Poland). *Folia Quaternaria* 61–62: 257 pp.
Krzyszkowski, D. 1993. Pleistocene glaciolacustrine sedimentation in a tectonically active zone, Kleszczów Graben, central Poland. *Sedimentology* 40: 623–644.
Krzyszkowski, D. 1994. Forms at the base of till units indicating deposition by lodgement and melt-out, with examples from the Wartanian tills near Bełchatów, central Poland. *Sedimentary Geology* 91: 229–238.
Kubiena, W. L. 1931. Mikropedologische Studien. *Archiv für Pflanzenbau* A 5: 613–648.
Kubiena, W. L. 1938. *Micropedology*. Ames, Iowa.
Kuhle, M. 1989. Die Inlandvereisung Tibets als Basis einer in der Globalstrahlungsgeometrie fußenden, reliefspezifischen Eiszeittheorie. *Petermanns Geographische Mitteilungen* 113: 265–285.
Kujansuu, R. 1990. Glacial flow indicators in air photographs. In: R. Kujansuu & M. Saarnisto (eds): *Glacial Indicator Tracing*: 71–86. Rotterdam: Balkema.

Kukla, G. J. & Ložek, V. 1961. Loesses and related deposits. *Instytut Geologiczny, Prace* XXXIV: 11–28.
Kukla, G. J., Heller, F., Ming, L. M., Chun, X. T. & An, Z. 1988. Pleistocene climates in China dated by magnetic susceptibility. *Geology* 16: 811–814.
Kupetz, M., Schubert, G., Seifert, A. & Wolf, L. 1989. Quartärbasis, pleistozäne Rinnen und Beispiele glazitektonischer Lagerungsstörungen im Niederlausitzer Braunkohlengebiet. *Geoprofil* 1: 2–17.
Kupsch, W. O. 1962. Ice-thrust ridges in Western Canada. *Journal of Geology* 70: 582–594.
Kuster, H. & Meyer, K.-D. 1979. Glaziäre Rinnen im mittleren und nordöstlichen Niedersachsen. *Eiszeitalter und Gegenwart* 29: 135–156.
Kvasov, D. D. 1978. The Late-Quaternary history of large lakes and inland seas of Eastern Europe. *Annales Academiæ Scientiarum Fennicæ, Series A, III. Geologica – Geographica* 127: 71 pp.
Lade, U. & Hagedorn, H. 1982. Sedimente und Relief einer eiszeitlichen Hohlform bei Krempel (Elbe–Weser–Dreieck). *Eiszeitalter und Gegenwart* 32: 93–108.
Lagerbäck, R. 1988a. The Veiki moraines in northern Sweden – widespread evidence of an Early Weichselian deglaciation. *Boreas* 17: 469–486.
Lagerbäck, R. 1988b. Periglacial phenomena in the wooded areas of Northern Sweden – relicts from the Tärendö Interstadial. *Boreas* 17: 487–499.
Lagerbäck, R. & Robertsson, A.-M. 1988. Kettle holes – stratigraphical archives for Weichselian geology and palaeoenvironment in northernmost Sweden. *Boreas* 17: 439–468.
Lagerlund, E. 1980. Litostratigrafisk indelning av Västskånes Pleistocen och en ny glaciationsmodell för Weichsel. *University of Lund, Department of Quaternary Geology, Thesis* 5: 106 pp.
Lagerlund, E. & Houmark-Nielsen, M. 1993. Timing and pattern of the last deglaciation in the Kattegat region, southwest Scandinavia. *Boreas* 22: 337–347.
Lagoe, M. B., Eyles, C. N., Eyles, N. & Hale, Ch. 1993. Timing of late Cenozoic tidewater glaciation in the far North Pacific. *Geological Society of America Bulletin* 105: 1542–1560.
Lamb, H. H. & Woodroffe, A. 1970. Atmospheric circulation during the last ice age. *Quaternary Research* 1: 29–58.
Lambeck, K. 1993a. Glacial rebound of the British Isles. I: Preliminary model results. *Geophysical Journal International* 115: 941–959.
Lambeck, K. 1993b. Glacial rebound of the British Isles. II: A high-resolution, high-precision model. *Geophysical Journal International* 115: 960–990.
Lambert, A. M. & Hsü, K. J. 1979. Varve-like sediments of the Walensee, Switzerland. In: Schlüchter, Ch. (ed.): *Moraines and Varves – Origin, Genesis, Classification*: 287–294. Rotterdam: Balkema.
Lamothe, M., Parent, M. & Shilts, W. W. 1992. Sangamonian and early Wisconsinan events in the St. Lawrence Lowland and Appalachians of southern Quebec, Canada. In: Clark, P. U. & Lea, P. D. (eds): *The Last Interglacial–Glacial Transition in North America. Geological Society of America Special Paper* 270: 171–184.
Lang, G. 1994. *Quartäre Vegetationsgeschichte Europas – Methoden und Ergebnisse.* Jena: Gustav Fischer. 462 pp.
Lankauf, K. R. 1982. Budowa geologiczna rynny (podwójnej) strzyzynskiej wraz z jej najblizszym otoczeniem w Borach Tucholskich oraz charakterystyka sedymentologiczna osadów profilu Zamrzenica (The geological structure of the Strzyzyny (double) Channel together with its close surroundings in Bory Tucholskie (Tuchola Forests) and a sedimentological description of the deposits of the profile Zamrzenica). *Acta Universitatis Nicolai Copernici, Geografia* XVII: 10–26.
LaRocque, J. A. A. 1966. Pleistocene mollusca of Ohio, Part 1. *Ohio Division of the Geological Survey Bulletin* 62: 111 pp.
LaRocque, J. A. A. 1967. Pleistocene mollusca of Ohio, Part 2. *Ohio Division of the Geological Survey Bulletin* 62: 113–356.
LaRocque, J. A. A. 1968. Pleistocene mollusca of Ohio, Part 3. *Ohio Division of the Geological Survey Bulletin* 62: 357–553.
LaRocque, J. A. A. 1970. Pleistocene mollusca of Ohio, Part 4. *Ohio Division of the Geological Survey Bulletin* 62: 555–800.
Larsen, C. E. 1987. Geological history of glacial Lake Algonquin and the Upper Great Lakes. *United States Geological Survey Bulletin* 1801: 36 pp.
Larsen, E., Sejrup, H. P., Johnsen, S. F. & Knudsen, K. L. 1995. Do Greenland ice cores reflect NW European interglacial climate variations? *Quaternary Research* 43: 125–132.

Larson, G. J., Lowell, T. V. & Ostrom, N. E. 1994. Evidence for the Two Creeks interstade in the Lake Huron basin. *Canadian Journal of Earth Sciences* 31: 793–797.

Lauritzen, S.-E. 1995. High-resolution paleotemperature proxy record for the last interglaciation based on Norwegian speleothems. *Quaternary Research* 43: 133–146.

Lawson, A. C. 1893 Sketch of the Coastal Topography of the North Side of Lake Superior with Special Reference to the Abandoned Strands of Lake Warren. *Geological and Natural History Survey of Minnesota, 20th Annual Report, 1891*: 181–289.

Lawson, D. E. 1979. Sedimentological analysis of the western terminus region of the Matanuska Glacier, Alaska. *CRREL Report* 79-9: 112 pp.

Lawson, D. E. 1989. Glacigenic resedimentation: Classification concepts and application to mass-movement processes and deposits. In: Goldthwait, R. P. & Matsch, C. L. (eds): *Genetic Classification of Glacigenic Deposits*: 147–169. Rotterdam: Balkema.

Lawson, D. E. 1993. Glaciohydrologic and glaciohydraulic effects on runoff and sediment yield in glacierized basins. *CRREL Monograph* 93-2: 108 pp.

Laymon, C. A. 1992. Glacial geology of western Hudson Strait, Canada, with reference to Laurentide Ice Sheet dynamics. *Geological Society of America Bulletin* 104: 1169–1177.

Lazarenko, A. A. 1984. The loess of Central Asia. In: Velichko, A. A. (ed.): *Late Quaternary Environments of the Soviet Union*: 125–131. London: Longman.

Lee, M. P. 1979. Loess from the Pleistocene of the Wirral Peninsula, Merseyside. *Proceedings of the Geologists' Association* 90: 21–26.

Leffingwell, E. de K. 1915. Ground-ice wedges; the dominant form of ground-ice on the north coast of Alaska. *Journal of Geology* 23: 635–654.

Leger, M. 1988. Géomorphologie de la vallée subalpine du Danube entre Sigmaringen et Passau. Dissertation, Paris. 3 volumes: 621 pp.

Leigh, D. S. 1994. Roxana silt of the Upper Mississippi Valley: Lithology, source, and palaeoenvironment. *Geological Society of America Bulletin* 106: 430–442.

Leigh, D. S. & Knox, J. C. 1993. AMS radiocarbon age of the Upper Mississippi Valley Roxana Silt. *Quaternary Research* 39: 282–289.

Leigh, D. S. & Knox, J. C. 1994. Loess of the Upper Mississippi Valley Driftless Area. *Quaternary Research* 42: 30–40.

Leighton, M. M. 1960. The classification of the Wisconsin glacial stage of the north-central United States. *Journal of Geology* 68: 529–552.

Leighton, M. M. & Willman, H. B. 1949. Loess formations of the Mississippi Valley. *Geological Society of America Bulletin* 60: 1904–1905.

Lembke, H., Altermann, M., Markuse, G. & Nitz, B. 1970. Die periglaziäre Fazies im Alt- und Jungmoränengebiet nördlich des Lößgürtels. In: Richter, H., Haase, G., Lieberoth, I. & Ruske, R. (eds): *Periglazial – Löß – Paläolithikum im Jungpleistozän der Deutschen Demokratischen Republik. Petermanns Geographische Mitteilungen, Ergänzungsheft* 274: 213–268.

Lemcke, K., Von Engelhard, W. & Füchtbauer, H. 1953. Geologische und sedimentpetrographische Untersuchungen im Westteil der ungefalteten Molasse des Süddeutschen Alpenvorlandes. *Beihefte zum Geologischen Jahrbuch* 11: 110 pp. + Appendix.

Lemdahl, G. 1988. Palaeoclimatic and palaeoecological studies based on subfossil insects from Late Weichselian sediments in southern Sweden. *LUNDQUA Thesis* 22: 11 pp.

Leverett, F. 1898a. The weathered zone (Sangamon) between the Iowan loess and the Illinoian till sheet. *Journal of Geology* 6: 171–181.

Leverett, F. 1898b. The weathered zone (Yarmouth) between the Illinoian and Kansan till sheets. *Journal of Geology* 6: 238–243.

Leverett, F. 1898c. The Peorian soil and weathered zone (Toronto Formation?) *Journal of Geology* 6: 244–249.

Leverett, F. 1921. Outline of Pleistocene history of Mississippi Valley. *Journal of Geology* 29: 615–626.

Leverett, F. 1929. Moraines and shorelines of the Lake Superior region. *US Geological Survey Professional Paper* 154: 72 pp.

Leverett, F. 1942. Shiftings of the Mississippi River in relation to glaciation. *Geological Society of America Bulletin* 53: 1283–1298.

Leverett, F. & Taylor, F. B. 1915. The Pleistocene of Indiana and Michigan and the history of the Great Lakes. *United States Geological Survey Monograph* 53: 529 pp.
Lewis, C. F. M. 1969. Late Quaternary history of lake levels in the Huron and Erie basins. *Proceedings of the 12th Conference on Great Lakes Research, Ann Arbor, Michigan*: 250–270.
Lewis, C. F. M. & Anderson, T. W. 1992. Stable isotope (O and C) and pollen trends in Lake Erie, evidence for a locally-induced climatic reversal of Younger Dryas age in the Great Lakes basin. *Climate Dynamics* 6: 241–250.
Lewis, C. F. M., Moore, T. C., Jr., Rea, D. K., Dettman, D. L., Smith, A. M. & Mayer, L. A. 1994. Lakes of the Huron Basin: their record of runoff from the Laurentide Ice Sheet. *Quaternary Science Reviews* 13: 891–922.
Lewis, H. C. 1894. *Papers and Notes on the Glacial Geology of Great Britain and Ireland.* London: Longmans, Green, and Co. 469 pp.
Li Shije & Shi Jafeng 1992. Glacial and lake fluctuations in the area of the west Kunlun mountains during the last 45 000 years. *Annals of Glaciology* 16: 79–84.
Libby, W. F. 1952. *Radiocarbon Dating.* Chicago: University of Chicago Press.
Lidén, R. 1913. Geokronologiska studier öfver det finiglaciala skedet i Ångermanland. *Sveriges Geologiska Undersökning* Ca 9: 39 pp.
Lidén, R. 1938. Den senkvartära strandförskjutningens förlopp och kronologi i Ångermanland. *Geologiska Föreningens i Stockholm Förhandlingar* 92: 5–20.
Lie, S. E., Stabell, B. & Mangerud, J. 1983. Diatom stratigraphy related to Late Weichselian sea-level changes in Sunnmøre, Western Norway. *Norges Geologiske Undersøkelse* 380: 203–219.
Liedtke, H. 1957. Beiträge zur geomorphologischen Entwicklung des Thorn-Eberswalder Urstromtales zwischen Oder und Havel. *Wissenschaftliche Zeitschrift der Humboldt-Universität Berlin, Mathematisch-Naturwissenschaftliche Reihe* 6: 3–49.
Liedtke, H. 1962. Glaziale Urstromtäler und Eisrandlagen am Südrande der nordischen Vereisung. *Verhandlungen des 33. Deutschen Geographentages Köln 1961*: 385–392.
Liedtke, H. 1975. Die nordischen Vereisungen in Mitteleuropa. *Forschungen zur deutschen Landeskunde* 204: 160 pp.
Liedtke, H. 1981. Die nordischen Vereisungen in Mitteleuropa, 2. Auflage. *Forschungen zur deutschen Landeskunde* 204: 307 pp.
Liivrand, E. 1991. Biostratigraphy of the Pleistocene deposits in Estonia and correlations in the Baltic region. *Stockholm University, Department of Quaternary Research, Report* 19: 114 pp.
Lilliesköld, M. 1990. Lithology and transport distance of glaciofluvial material. In: R. Kujansuu & M. Saarnisto (eds): *Glacial Indicator Tracing*: 151–164. Rotterdam: Balkema.
Lindén, A. 1975. Till petrographic studies in an Archaean bedrock area in southern central Sweden. *Striae* 1: 57 pp.
Lindner, L. 1988. Stratigraphy and extents of Pleistocene continental glaciations in Europe. *Acta Geologica Polonica* 38: 63–83.
Lindner, L. (ed.) 1992. *Czwartorzęd. Osady metody badań stratigrafia.* Warszawa: Wydawnictwo PAE. 683 pp.
Lindner, L. 1995. Till sequences and local moraines in the Holy Cross Mountains area in central Poland. In: Ehlers, J., Kozarski, S. & Gibbard, P. L. (eds): *Glacial Deposits in North-East Europe*: 329–337. Rotterdam: Balkema.
Lindner, L. & Brykczyńska, E. 1980. Organogenic deposits at Zbójno by Predbórz, western slopes of the Holy Cross Mts., and their bearing on stratigraphy of the Pleistocene of Poland. *Acta Geologica Polonica* 30: 153–163.
Lindner, L. & Marks, L. 1994. Pleistocene glaciations and interglacials in the Vistula, the Oder, and the Elbe drainage basins (Central European Lowland). *Acta Geologica Polonica* 44: 153–165.
Lindsay, J. F. 1970. Clast fabric of till and its development. *Journal of Sedimentary Petrology* 40: 629–641.
Lindström, G. 1886. Om postglaciala sänkningar af Gotland. *Geologiska Föreningens i Stockholm Förhandlingar* 8: 251–281.
Lineback, J. A. 1979. The status of the Illinoian glacial stage. In: *Wisconsinan, Sangamonian and Illinoian stratigraphy in central Illinois. Illinois State Geological Survey, Guidebook* 13: 69–78.
Liniger, H. 1966. Das Plio-Altpleistozäne Flußnetz der Nordschweiz. *Regio Basiliensis* 7: 158–177.

Linke, G. 1983. *Geologische Übersichtskarte Raum Hamburg 1:50 000, Quartärbasis, Blatt 1: Morphologie.* Hamburg: Geologisches Landesamt.
Linke, G. & Hallik, R. 1993. Die pollenanalytischen Ergebnisse der Bohrungen Hamburg–Dockenhuden (qho 4), Wedel (qho 2) und Hamburg–Billbrook. *Geologisches Jahrbuch* A 138: 169–184.
Lipps, S. 1985. Relief- und Sedimententwicklung an der Mittellahn. *Marburger Geographische Schriften* 98: 93 pp.
Litt, Th. 1990. Stratigraphie und Ökologie des eeminterglazialen Waldelefanten-Schlachtplatzes von Gröbern, Kreis Gräfenhainichen. *Veröffentlichungen des Landesmuseums für Vorgeschichte in Halle* 43: 193–208.
Litt, Th. 1994. Paläoökologie, Paläobotanik und Stratigraphie des Jungquartärs im nordmitteleuropäischen Tiefland unter besonderer Berücksichtigung des Elbe–Saale–Gebietes. *Dissertationes Botanicæ* 227: 185 pp.
Litt, Th & Turner, Ch. 1993. Arbeitsergebnisse der Subkommission für Europäische Quartärstratigraphie: Die Saalesequenz in der Typusregion. *Eiszeitalter und Gegenwart* 43: 125–128.
Liu, T. S., Shouxin, Z. & Jiamao, H. 1986. Stratigraphy and paleoenvironmental changes in the loess of central China. *Quaternary Science Reviews* 5: 489–495.
Liu, T. S., Ding, Z. L., Chen, M. Y. & An, Z. S. 1989. The global surface energy system and the geological role of wind stress. *Quaternary International* 2: 43–54.
Liu, X. M., Liu, T. S., Xu, T. C., Liu, C. & Chen, M. Y. 1987. A preliminary study on magnetostratigraphy of a loess profile in Xifeng area, Gansu province. In: Liu T. S. (ed.): *Aspects of Loess Research*: 164–174. Beijing: China Ocean Press.
Liverman, D. G. E., Catto, N. R. & Rutter, N. W. 1988. Laurentide glaciation in west-central Alberta: a single (Late Wisconsinan) event. *Canadian Journal of Earth Sciences* 26: 266–274.
Ljungner, E. 1930. Spaltentektonik und Morphologie der schwedischen Skagerrak-Küste. *Bulletin of the Geological Institutions of the University of Uppsala* 21: 478 pp.
Ljungner, E. 1949. East–west balance of the Quaternary ice caps in Patagonia and Scandinavia. *Bulletin of the Geological Institutions of the University of Uppsala* 33: 11–96.
Lliboutry, L. 1964/65. *Traité de glaciologie*. Paris: Masson. 2 volumes, 428 + 162 pp.
Lliboutry, L. 1968. General theory of subglacial cavitation and sliding of temperate glaciers. *Journal of Glaciology* 7: 21–58.
Lliboutry, L. 1971. Permeability, brine content and temperature of temperate ice. *Journal of Glaciology* 10: 15–29.
Lliboutry, L. 1983. Modifications to the theory of intra-glacial waterways for the case of subglacial ones. *Journal of Glaciology* 29: 216–226.
Lona, F. 1950. Contributi alla storia della vegetazione e del clima nella Val Padana. Analisi pollinica del giacimento villafranchiano di Leffe (Bergamo). *Atti della Società Italiana di Scienze Naturali* 89: 123–178.
Lona, F. & Bertoldi, R. 1973. La storia del Plio-Pleistocene italiano in alcune sequenze vegetazionali lacustre e marine. *Atti Academia Nazionale dei Lincei, Memorie* 11 (8): 45 pp.
Long, D. & Stoker, M. S. 1986. Channels in the North Sea: The nature of a hazard. *Advances in Underwater Technology, Ocean Science and Offshore Engineering* 6: 339–351.
Long, D., Laban, C., Streif, H., Cameron, T. D. J. & Schüttenhelm, R. T. E. 1988. The sedimentary record of climatic variation in the southern North Sea. *Philosophical Transactions of the Royal Society of London* B 318: 523–537.
Lord, A. R. & Robinson, J. E. 1978. Marine Ostracoda from the Quaternary Nar Valley Clay, West Norfolk. *Bulletin of the Geological Society of Norfolk* 30: 113–118.
Löscher, M. 1976. Die präwürmzeitlichen Schotterablagerungen in der nördlichen Iller-Lech-Platte. *Heidelberger Geographische Arbeiten* 45: 157 pp.
Löscher, M. 1988. Stratigraphische Interpretation der jungpleistozänen Sedimente in der Oberrheinebene zwischen Bruchsal und Worms. In: Von Koenigswald, W. (ed.): *Zur Paläoklimatologie des letzten Interglazials im Nordteil der Oberrheinebene*: 79–104. Stuttgart: Gustav Fischer.
Lotter, A. F. 1991. Absolute dating of the Late-Glacial period in Switzerland using annually laminated sediments. *Quaternary Research* 35: 321–330.
Louis, H. 1952. Zur Theorie der Gletschererosion in den Tälern. *Eiszeitalter und Gegenwart* 2: 12–24.
Løvlie, R. 1989. Paleomagnetic stratigraphy: A correlation method. *Quaternary International* 1: 129–149.

Lowe, J. J. & Walker, M. J. C. 1984. *Reconstructing Quaternary Environments*. London: Longman. 389 pp.
Lowe, J. J., Ammann, B., Birks, H. H., Björck, S., Coope, G. R., Cwynar, L., de Beaulieu, J.-L., Mott, R. J., Peteet, D. M. & Walker, M. J. C. 1994. Climatic changes in areas adjacent to the North Atlantic during the last glacial–interglacial transiton (14–9 ka BP): a contribution to IGCP-253. *Journal of Quaternary Science* 9: 185–198.
Lowell, T. V. & Stuckenrath, R. 1991. Advance and retreat patterns in the Miami Lobe of the Laurentide Ice Sheet. *Annals of Glaciology* 14: 172–175.
Lowell, T. V., Kite, J. S., Calkin, P. E. & Halter, E. F. 1990a. Analysis of small-scale erosional data and a sequence of Late Pleistocene flow reversal, northern New England. *Geological Society of America Bulletin* 102: 74–85.
Lowell, T. V., Savage, K. M., Brockmann, C. S. & Stuckenrath, R. 1990b. Radiocarbon analysis from Cincinnati, Ohio and their implications for glacial stratigraphic interpretations. *Quaternary Research* 34: 1–11.
Ložek, V. 1964. Quartärmollusken der Tschechoslowakei. *Rozpravy ústredního ústavu geologického* 31: 374 pp.
Ložek, V. 1965. Das Problem der Lößbildung und die Lößmollusken. *Eiszeitalter und Gegenwart* 16: 61–75.
Łoziński, M. W. 1909. Über die mechanische Verwitterung der Sandsteine im gemäßigten Klima. *Bulletin International de l'Académie des Sciences de Cracovie, Classe des Sciences Mathématiques et Naturelles* 1: 1–25. *Also*: On the mechanical weathering of sandstones in temperate climates. In: Evans, D. J. A. (ed.) 1994: *Cold Climate Landforms*: 119–134. Chichester: Wiley.
Luckman, B. H. 1993. Glacier fluctuation and tree-ring records for the last millennium in the Canadian Rockies. *Quaternary Science Reviews* 12: 441–450.
Luckman, B. H. & Crocket, K. J. 1979. Distribution and characteristics of rock glaciers in the southern part of Jasper National Park, Alberta. *Canadian Journal of Earth Sciences* 15: 540–550.
Lüdi, W. 1953. Die Pflanzenwelt des Eiszeitalters im nördlichen Vorland der Alpen. *Veröffentlichungen des Geobotanischen Instituts Rübel, Zürich* 27: 208 pp.
Lugn, A. L. 1935. The Pleistocene of Nebraska. *Nebraska Geological Survey Bulletin* 10: 223 pp.
Lugn, A. L. 1968. The origin of loesses and their relation to the Great Plains in North America. In: Schultz, C. B. & Frye, J. C. (eds): *Loess and Related Eolian Deposits of the World*: 139–182. Lincoln: University of Nebraska Press.
Lundelius, E. L., Jr., Graham, R. W., Anderson, E., Guilday, J., Holman, J. A., Steadman, D. W. & Webb, S. D. 1983. Terrestrial vertebrate faunas. In: Wright, H. E. (ed.): *Late-Quaternary Environments of the United States, Volume 1: The Late Pleistocene*: 311–353. Minneapolis: University of Minnesota Press.
Lundqvist, G. 1958. Kvartärgeologisk forskning i Sverige under ett sekel. *Sveriges Geologiska Undersökning* C 561: 57 pp.
Lundqvist, J. 1967. Submoräna sediment i Jämtlands län. *Sveriges Geologiska Undersökning* C 618: 267 pp.
Lundqvist, J. 1969. Beskrivning till jordartskarta över Jämtlands län. *Sveriges Geologiska Undersökning* Ca 45: 418 pp.
Lundqvist, J. 1971. The interglacial deposit at the Leveäniemi mine, Svappavaara, Swedish Lapland. *Sveriges Geologiska Undersökning* C 658: 163 pp.
Lundqvist, J. 1979. Morphogenetic classification of glaciofluvial deposits. *Sveriges Geologiska Undersökning* C 767: 72 pp.
Lundqvist, J. 1983a. The glacial history of Sweden. In: Ehlers, J. (ed.): *Glacial Deposits in North-West Europe*: 77–82. Rotterdam: Balkema.
Lundqvist, J. 1983b. Tills and moraines in Sweden. In: Ehlers, J. (ed.): *Glacial Deposits in North-West Europe*: 83–90. Rotterdam: Balkema.
Lundqvist, J. 1985. Deep-weathering in Sweden. *Fennia* 163: 287–292.
Lundqvist, J. 1987. Glaciodynamics of the Younger Dryas marginal zone in Scandinavia. *Geografiska Annaler* 69 A: 305–319.
Lundqvist, J. 1988. Younger Dryas – Preboreal moraines and deglaciation in southwestern Värmland, Sweden. *Boreas* 17: 301–316.
Lundqvist, J. 1989. Late glacial ice lobes and glacial landforms in Scandinavia. In: Goldthwait, R. P. & Matsch, C. L. (eds): *Genetic Classification of Glacigenic Deposits*: 217–225. Rotterdam: Balkema.

Lundqvist, J. 1990. Glacial morphology as an indicator of the direction of glacial transport. In: Kujansuu, R. & Saarnisto, M. (eds): *Glacial Indicator Tracing*: 61–70. Rotterdam: Balkema.

Lundqvist, J. 1992. Glacial stratigraphy in Sweden. In: Kauranne, K. (ed.): *Glacial stratigraphy, engineering geology and earth construction. Geological Survey of Finland, Special Paper* 15: 43–59.

Lunkka, J. P. 1994. Sedimentation and lithostratigraphy of the North Sea Drift and Lowestoft Till Formations in the coastal cliffs of northeast Norfolk, England. *Journal of Quaternary Science* 9: 209–233.

Lüttig, G. 1958. Methodische Fragen der Geschiebeforschung. *Geologisches Jahrbuch* 75: 361–418.

Lüttig, G. 1965. Interglacial and interstadial periods. *Journal of Geology* 73: 579–591.

Lüttig, G. 1991. Erratic boulder statistics as a stratigraphic aid – Examples from Schleswig-Holstein. *Newsletters in Stratigraphy* 25: 61–74.

Lüttig, G. & Maarleveld, G. C. 1961. Nordische Geschiebe in Ablagerungen prä Holstein in den Niederlanden (Komplex von Hattem). *Geologie en Mijnbouw* 40: 163–174.

Lyell, Ch. 1834. Observations on the loamy deposit called 'Loess' in the valley of the Rhine. *The Geological Society of London, Proceedings* 2 (36): 83–85.

Lyell, Ch. 1840a. On the Boulder Formation, or drift and associated freshwater deposits composing the mud-cliffs of eastern Norfolk. *The London and Edinburgh Philosophical Magazine and Journal of Science, Third Series* 16 (104): 345–380.

Lyell, Ch. 1840b. On the Cretaceous and Tertiary strata of the Danish Islands of Seeland and Möen. *Transactions of the Geological Society of London, Second Series* V: 243–257.

Lyell, Ch. 1847. On the delta and alluvial deposits of the Mississippi River, and other points of the geology of North America, observed in the years 1845, 1846. *American Journal of Science* 3: 34–39, 267–269.

Lyell, Ch. 1863. *The Geological Evidences of the Antiquity of Man with Remarks on Theories of the Origin of Species by Variation*. London: John Murray. 520 pp.

Lykke-Andersen, A.-L. 1986. On the buried Nørreå Valley – a contribution to the geology of the Nørreå-Valley, Jylland – and other buried overdeepened valleys. *Geoskrifter* 24: 211–223.

Lykke-Andersen, A.-L. 1987. A Late Saalian, Eemian and Weichselian marine sequence at Nørre Lyngby, Vendsyssel, Denmark. *Boreas* 16: 345–357.

Lykke-Andersen, A.-L. & Knudsen, K. L. 1991. Saalian, Eemian and Weichselian in the Vendsyssel–Kattegat Region, Denmark. *Striae* 34: 135–140.

Maarleveld, G. C. 1956. *Grindhoudende midden-pleistocene sedimenten. Mededelingen van de Geologische Stichting* C-VI-6: 105 pp.

Maarleveld, G. C. 1981. The sequence of ice-pushing in the Central Netherlands. *Mededelingen Rijks Geologische Dienst* 34 (1): 2–6.

Maarleveld, G. C. & Van den Toorn, J. C. 1955. Pseude-sölle in Noord-Nederland. *Tijdschrift van het Koninklijk Nederlandsch Aardrijkskundig Genootschap* LXXII: 347–360.

MacClintock, P. & Dreimanis, A. 1964. Reorientation of till fabric by overriding glacier in the St. Lawrence valley. *American Journal of Science* 262: 133–142.

MacDonald, G. M. 1990. Palynology. In: Warner, B. G. (ed.): *Methods in Quaternary Ecology*: 37–52. Geosience Canada, Reprint Series 5.

Mackay, J. R. 1973. The growth of pingos, Western Arctic Coast, Canada. *Canadian Journal of Earth Sciences* 10: 979–1004.

Mackay, J. R. 1974. Ice-wedge cracks, Garry Island, Northwest Territories. *Canadian Journal of Earth Sciences* 11: 1366–1383.

Mackay, J. R. 1978. Contemporary pingos: A discussion. *Biuletyn Peryglcjalny* 27: 133–154.

Mackay, J. R. 1979. An equilibrium model for hummocks (non-sorted circles), Garry Island, Northwest Territories. *Geological Survey of Canada, Paper* 79-1A: 165–167.

Mackay, J. R. 1980. The origin of hummocks, western Arctic coast, Canada. *Canadian Journal of Earth Sciences* 17: 996–1006.

Mackay, J. R. 1990. Some observations on the growth and deformation of epigenetic, syngenetic and anti-syngenetic ice wedges. *Permafrost and Periglacial Processes* 1: 15–29.

Mackay, J. R. & MacKay, D. K. 1976. Cryostatic pressures in nonsorted circles (mud hummocks), Inuvik, Northwest Territories. *Canadian Journal of Earth Sciences* 13: 889–897.

Mackenbach, R. 1984. Jungtertiäre Entwässerungsrichtungen zwischen Passau und Hausrück (O. Österreich). *Geologisches Institut der Universität Köln, Sonderveröffentlichungen* 55: 175 pp.

Maddy, D., Keen, D. H., Bridgland, D. R. & Green, C. P. 1991. A revised model for the Pleistocene development of the River Avon, Warwickshire. *Journal of the Geological Society of London* 148: 473–484.

Madole, R. F., Ferring, C. R., Guccione, M. J., Hall, S. A., Johnson, W. C. & Sorenson, C. J. 1991. Quaternary geology of the Osage Plains and Interior Highlands. In: Morrison, R. B. (ed.): *Quaternary Nonglacial Geology: Conterminous U.S. Geological Society of America. The Geology of North America* K-2: 503–546.

Madsen, V., Nordmann, V. & Hartz, N. 1908. *Eem-Zonerne. Studier over Cyprinaleret og andre Eem-Aflejringer i Danmark, Nord-Tyskland og Holland. Danmarks Geologiske Undersøgelse*, II.Række 17: 302 pp. + Atlas.

Mahaney, W. C., Vaikmae, R. & Vares, K. 1991. Scanning electron microscopy of quartz grains in supraglacial debris, Adishy Glacier, Caucasus Mountains, USSR. *Boreas* 20: 395–404.

Mai, D. H. 1990. Die Flora des Interglazials von Gröbern (Kreis Gräfenhainichen). *Altenburger Naturwissenschaftliche Forschungen* 5: 106–115.

Maisch, M. 1981. Glazialmorphologische und gletschergeschichtliche Untersuchungen im Gebiet zwischen Landwasser- und Albulatal (Kt. Graubünden, Schweiz). *Physische Geographie* 3: 215 pp.

Maisch, M. 1992. *Die Gletscher Graubündens – Rekonstruktion und Auswertung der Gletscher und deren Veränderungen seit dem Hochstand von 1850 im Gebiet der östlichen Schweizer Alpen (Bündnerland und angrenzende Regionen)*. Zürich: Geographisches Institut der Universität. 324 pp. + Supplement.

Makowska, A. 1979. Interglacjal eemski w dolinie dolnej Wisły (Eemian Interglacial in valley of the lower Vistula River). *Studia Geologica Polonica* 63: 1–90.

Makowska, A. 1986. Morza plejstoceńskie w Polsce – osady, wiek i paleogeografia (The Pleistocene seas in Poland – sediments, age and palaeogeography). *Prace Instytutu Geologicznego* 120: 1–74.

Mamakowa, K. 1989. Late Middle Polish Glaciation, Eemian and Early Vistulian vegetation at Imbranowice near Wroclaw and the pollen stratigraphy of this part of the Pleistocene in Poland. *Acta Palaeobotanica* 29: 11–176.

Mandier, P. 1983. Pluralité des glaciations dans la région lyonnaise et la moyenne vallée du Rhône. In: *Projet 73/1/24, Glaciations Quaternaires dans l'hémisphère nord. Programme International de Corrélation Géologique* 9: 184–204. Paris.

Mandier, P. 1984. *Le relief de la moyenne vallée du Rhône au Tertiaire et au Quaternaire.* Université de Lyon II, 3 vol.: 871 pp. Lyon.

Mangerud, J. 1970. Interglacial sediments at Fjøsanger, near Bergen, with the first Eemian pollen spectra from Norway. *Norsk Geologisk Tidsskrift* 50: 167–181.

Mangerud, J. 1982. The chronostratigraphical subdivision of the Holocene in Norden; a review. *Striae* 16: 65–70.

Mangerud, J. 1983. The glacial history of Norway. In: Ehlers, J. (ed.): *Glacial Deposits in North-West Europe*: 3–9. Rotterdam, Balkema.

Mangerud, J. 1991a. The Last Ice Age in Scandinavia. *Striae* 34: 15–30.

Mangerud, J. 1991b. The Scandinavian Ice Sheet through the last interglacial/glacial cycle. *Paläoklimaforschung* 1: 307–330.

Mangerud, J. & Svendsen, J. I. 1992. The last interglacial–glacial period on Spitsbergen, Svalbard. *Quaternary Science Reviews* 11: 633–664.

Mangerud, J., Andersen, S. T., Berglund, B. E. & Donner, J. J. 1974. Quaternary stratigraphy of Norden, a proposal for terminology and classification. *Boreas* 3: 109–128.

Mangerud, J., Sønstegaard, E., Sejrup, H. P. & Haldorsen, S. 1981. A continuous Eemian–Early Weichselian sequence containing pollen and marine fossils at Fjøsanger, western Norway. *Boreas* 10: 137–208.

Mangerud, J., Lie, S. E., Furnes, H., Kristiansen, I. L. & Lømo, L. 1984. A Younger Dryas ash bed in Western Norway and its possible correlation with tephra in cores from the Norwegian Sea and the North Atlantic. *Quaternary Research* 21: 85–104.

Mangerud, J., Bolstad, M., Elgersma, A., Helliksen, D., Landvik, J. Y., Jønne, I., Lycke, A. K., Salvigsen, O., Sandahl, T. & Svendsen, J. I. 1992. The Last Glacial Maximum on Spitsbergen, Svalbard. *Quaternary Research* 38: 1–31.

Mangerud, J., Jansen, E. & Landvik, J. Y. 1996. Late Cenozoic history of the Scandinavian and Barents Sea ice sheets. *Global and Planetary Change* 12: 11–26.
Mania, D. 1967. Pleistozäne und holozäne Ostracodengesellschaften aus dem ehemaligen Aschersiebener See. *Wissenschaftliche Zeitschrift der Universität Halle* 16: 500–550.
Mania, D. 1973. Paläoökologie, Faunenentwicklung und Stratigraphie des Eiszeitalters im mittleren Elbe–Saalegebiet auf Grund von Molluskengesellschaften. *Geologie* 21, Beiheft 78/79: 175 pp.
Mania, D. 1989a. Die ältesten Spuren des Urmenschen im eiszeitlichen Altpaläolithikum. In: Herrmann, J. (ed.): *Archäologie in der Deutschen Demokratischen Republik – Denkmale und Funde*: 24–33. Stuttgart: Theiss.
Mania, D. 1989b. Archäologische Kulturen des Mittelpaläolithikums. In: Herrmann, J. (ed.): *Archäologie in der Deutschen Demokratischen Republik – Denkmale und Funde*: 34–40. Stuttgart: Theiss.
Mania, D. 1990. Stratigraphie, Ökologie und mittelpaläolithische Jagdbefunde des Interglazials von Neumark-Nord (Geiseltal). *Veröffentlichungen des Landesmuseums für Vorgeschichte in Halle* 43: 9–130.
Mania, D. & Stechemesser, H. 1970. Jungpleistozäne Klimazyklen im Harzvorland. In: Richter, H., Haase, G., Lieberoth, I. & Ruske, R. (eds): *Periglazial–Löß–Paläolithikum im Jungpleistozän der Deutschen Demokratischen Republik. Petermanns Geographische Mitteilungen, Ergänzungsheft* 274: 39–55.
Mannerfelt, G. M. 1945. Några glacialmorfologiska formelement. *Geografiska Annaler* 27: 1–239.
Marcinek, J. 1961. Die großen Urstromtäler im Jungmoränengebiet westlich der Neiße und Oder. *Geologie* 10: 435–441.
Marcinek, J., Präger, F. & Steinmüller, A. 1970. Periglaziäre Gestaltung der Täler. In: Richter, H., Haase, G., Lieberoth, I. & Ruske, R. (eds): *Periglazial–Löß–Paläolithikum im Jungpleistozän der Deutschen Demokratischen Republik. Petermanns Geographische Mitteilungen, Ergänzungsheft* 274: 281–328.
Marks, L. 1994. Dead-ice features at the maximum extent of the last glaciation in northeastern Poland. *Zeitschrift für Geomorphologie N.F., Supplement-Band* 95: 77–83.
Marsh, J. G. & Martin, Th. V. 1982. The SEASAT Altimeter Mean Surface Model. *Journal of Geophysical Research* 87 C5: 3269–3280.
Martin, P. S. 1990. 40,000 years of extinctions on the 'planet of doom'. *Palaeogeography, Palaeoclimatology, Palaeoecology* 82: 187–201.
Mathewes, R. H., Heusser, L. E. & Patterson, R. T. 1993. Evidence for a Younger Dryas-like cooling event on the British Columbia coast. *Geology* 21: 101–104.
Mathews, W. H. 1967. Profiles of late Pleistocene glaciers in New Zealand. *New Zealand Journal of Geology and Geophysics* 10: 146–163.
Mathews, W. H. 1974. Surface profiles of the Laurentide Ice Sheet in its marginal areas. *Journal of Geology* 13: 37–43.
Mathews, W. H. & Mackay, J. R. 1960. Deformation of soils by glacier ice and the influence of pore pressures and permafrost. *Transactions of the Royal Society of Canada* LIV, III, 4: 27–36.
Matoshko, A. V. & Chugunny, Y. G. 1995. Geological activity and dynamic evolution of the Dnieper Glaciation. In: Ehlers, J., Kozarski, S. & Gibbard, P. L. (eds): *Glacial Deposits in North-East Europe*: 225–229. Rotterdam: Balkema.
Matsch, C. L. & Schneider, A. L. 1986. Stratigraphy and correlation of the glacial deposits of the glacial lobe complex in Minnesota and northwestern Wisconsin. *Quaternary Science Reviews* 5: 59–68.
Matuyama, M. 1929. On the direction of magnetization of Basalt in Japan, Tyôsen and Manchuria. *Imperial Academy of Japan Proceedings* 5: 203–205.
May, R. W. & Dreimanis, A. 1976. Compositional variability in tills. In: Legget, R. F. (ed.): *Glacial Till: An Inter-disciplinary Study. Royal Society of Canada Special Publication* 12: 99–120.
Mayhew, D. F. & Stuart, A. J. 1986. Stratigraphic and taxonomic revision of the fossil mole remains (*Rodentia: Microtinae*) from the Lower Pleistocene deposits of eastern England. *Philosophical Transactions of the Royal Society of London* B 312: 431–485.
Mayr, F. & Heuberger, H. 1968. Type areas of late glacial and post-glacial deposits in Tyrol, Eastern Alps. In: Richmond, G. M. (ed.): *Glaciation of the Alps. University of Colorado Studies, Series in Earth Sciences* 7: 143–165.
McAndrews, J. H. 1984. Pollen analysis of the 1973 ice core from Devon Island Glacier, Canada. *Quaternary Research* 22: 68–76.

McAndrews, J. H. & Boyko-Diakonow, M. 1989. Pollen analysis of varved sediment at Crawford Lake, Ontario: evidence of Indian and European farming. In: Fulton, R. J. (ed.): *Quaternary Geology of Canada and Greenland, Geological Survey of Canada, Geology of Canada* 1: 528–530.

McCabe, A. M. 1985. Glacial geomorphology. In: Edwards, K. J. & Warren, W. P. (eds): *The Quaternary History of Ireland*: 67–93. London: Academic Press.

McCabe, A. M. 1991. The distribution and stratigraphy of drumlins in Ireland. In: Ehlers, J., Gibbard, P. L. & Rose, J. (eds): *Glacial Deposits in Great Britain and Ireland*: 421–435. Rotterdam: Balkema.

McCabe, A. M. 1993. The 1992 Farrington Lecture: Drumlin bedforms and related ice-marginal depositional systems in Ireland. *Irish Geography* 26: 22–44.

McCabe, A. M. & Dardis, G. F. 1994. Glaciotectonically induced water-throughflow structures in a Late Pleistocene drumlin, Kanrawer, County Galway, western Ireland. *Sedimentary Geology* 91: 173–190.

McCabe, A. M. & Eyles, N. 1988. Sedimentology of an ice-contact glaciomarine delta, Carey Valley, Northern Ireland. *Sedimentary Geology* 59: 1–14.

McCabe, A. M., Dardis, G. F. & Hanvey, P. M. 1992. *Glacial Sedimentology in Northern and Western Ireland. Pre- and Post-Symposium Field Excursion Guide Book, Symposium on Subglacial Processes, Sediments and Landforms, University of Ulster, July 1992.* 236 pp.

McCabe, A. M., Bowen, D. Q. & Penney, D. N. 1993. Glaciomarine facies from the western sector of the last British ice sheet, Malin Beg, County Donegal, Ireland. *Quaternary Science Reviews* 12: 35–45.

McCabe, A. M., Carter, R. G. W. & Haynes, J. R. 1994. A shallow marine emergent sequence from the northwestern sector of the last British ice sheet, Portballintrae, Northern Ireland. *Marine Geology* 117: 19–34.

McCarroll, D. & Nesje, A. 1993. The vertical extent of ice sheets in Nordfjord, western Norway: measuring degree of rock surface weathering. *Boreas* 22: 255–265.

McCartney, M. C. & Mickelson, D. M. 1982. Late Woodfordian and Greatlakean history of the Green Bay Lobe, Wisconsin. *Geological Society of America Bulletin* 93: 297–302.

McCoy, W. D., Oches, E. A. & Clark, P. U. 1990. Results of aminostratigraphic investigations at Wittsburg, Crowleys Ridge, Arkansas. In: Guccione, M. J. & Rutledge, E. M. (eds): *Field guide to the Mississippi Alluvial valley, northeast Arkansas and southeast Missouri. Friends of the Pleistocene South Central Cell Field Trip Guide*: 99–101.

McCrea, W. H. 1975. Ice ages and the galaxy. *Nature* 255: 607–609.

McKay, E. D. 1979. Wisconsinan loess stratigraphy of Illinois. In: *Wisconsinan, Sangamonian, and Illinoian stratigraphy in central Illinois. Midwest Friends of the Pleistocene 26th Field Conference, May 4–6, 1979. Illinois State Geological Survey Guidebook* 13: 95–108.

McKay, E. D. III & Follmer, L. R. 1985. A correlation of Lower Mississippi Valley loesses to the glaciated Midwest. *Geological Society of America Abstracts with Programs* 17: 167.

Meier, M. F. 1984. Contribution of small glaciers to global sea level. *Science* 226: 1418–1421.

Meier, M. F. & Post, A. 1969. What are glacier surges? *Canadian Journal of Earth Sciences* 6: 807–817.

Meijer, T. 1990. Notes on Quaternary freshwater Mollusca of the Netherlands, with description of some new species. *Mededelingen van het Werkgroep voor Tertiaire en Quartaire Geologie* 26.

Mejdahl, V. & Funder, S. 1994. Luminescence dating of Late Quaternary sediments from East Greenland. *Boreas* 23: 525–535.

Menke, B. 1968. Beiträge zur Biostratigraphie des Mittelpleistozäns in Norddeutschland. *Meyniana* 18: 35–42.

Menke, B. 1975. Vegetationsgeschichte und Florenstratigraphie Nordwestdeutschlands im Pliozän und Frühquartär. Mit einem Beitrag zur Biostratigraphie des Weichsel-Frühglazials. *Geologisches Jahrbuch* A 26: 3–151.

Menke, B. 1980a. Vegetationskundlich-ökologisches Modell eines Interglazial–Glazial-Zyklus in Nordwestdeutschland. *Phytocoenologia* 7: 100–120.

Menke, B. 1980b. Wacken, Elster-Glazial, marines Holstein-Interglazial und Wacken-Warmzeit. In: Stremme, H. E. & Menke, B. (eds): *Quartär-Exkursionen in Schleswig-Holstein*: 26–35. Kiel: Geologisches Landesamt Schleswig-Holstein.

Menke, B. 1981. Vegetation, Klima und Verwitterung im Eem-Interglazial und Weichsel-Frühglazial Schleswig-Holsteins. *Verhandlungen des Naturwissenschaftlichen Vereins in Hamburg (NF)* 24 (2): 123–132.

Menke, B. 1982. On the Eemian Interglacial and the Weichselian Glacial in northwestern Germany (vegetation, stratigraphy, paleosols, sediments). *Quaternary Studies in Poland* 3: 61–68.
Menke, B. 1984. Wie stabil ist das Ökosystem 'Wald'? *Allgemeine Forst Zeitschrift* 1984 (6): 122–126.
Menke, B. 1985a. Palynologische Untersuchungen zur Transgression des Eem-Meeres im Raum Offenbüttel/Nord-Ostsee-Kanal. *Geologisches Jahrbuch* A 86: 19–26.
Menke, B. 1985b. Eem-Interglazial und 'Treene-Warmzeit' in Husum/Nordfriesland. Mit einem Beitrag von Risto Tynni, unter Mitarbeit von Holger Ziemus. *Geologisches Jahrbuch* A 86: 63–99.
Menke, B. & Ross, P.-H. 1967. Der erste Fund von Kieselgur in Schleswig-Holstein bei Brokenlande, südlich von Neumünster. (Mit einem Beitrag zur Biostratigraphie des 'Saale-Spätglazials'). *Eiszeitalter und Gegenwart* 18: 113–126.
Menke, B. & Tynni, R. 1984. Das Eeminterglazial und das Weichselfrühglazial von Rederstall/Dithmarschen und ihre Bedeutung für die mitteleuropäische Jungpleistozän-Gliederung. *Geologisches Jahrbuch* A 76: 120 pp.
Menzies, J. 1984. *Drumlins – a bibliography.* Norwich: GeoBooks.
Menzies, J. 1987. Towards a general hypothesis on the formation of drumlins. In: Menzies, J. & Rose, J. (eds): *Drumlin Symposium*: 9–24. Rotterdam: Balkema.
Menzies, J. 1989. Subglacial hydraulic conditions and their possible impact upon subglacial bed formation. *Sedimentary Geology* 62: 125–150.
Menzies, J. & Rose, J. (eds) 1987. *Drumlin Symposium.* Rotterdam: Balkema. 360 pp.
Mercer, J. H. 1976. Glacial history of southernmost South America. *Quaternary Research* 6: 125–166.
Mercer, J. H. 1978. West Antarctic ice sheet and CO_2 greenhouse effect: a threat of disaster. *Nature* 271: 321–325.
Mercer, J. H., Fleck, R. J., Mankinen, E. A. & Sander, W. 1975. Southern Patagonia: glacial events between 4 m.y. and 1 m.y. ago. *The Royal Society of New Zealand Bulletin* 13: 223–230.
Merrihue, C. & Turner, G. 1976. Potassium–argon dating by activation with fast neutrons. *Journal of Geophysical Research* 71: 2852–2857.
Meyer, H.-H. 1981. Zur klimastratigraphischen und morphogenetischen Auswertbarkeit von Flugdecksandprofilen im Norddeutschen Altmoränengebiet – erläutert an Beispielen aus der Kellenberg-Endmoräne (Landkreis Diepholz). *Bochumer Geographische Arbeiten* 40: 21–30.
Meyer, K.-D. 1965. Das Quartärprofil am Steilufer der Elbe bei Lauenburg. *Eiszeitalter und Gegenwart* 16: 47–60.
Meyer, K.-D. 1973. Zur Entstehung der abflußlosen Hohlformen auf der Neuenwalder Geest. *Jahrbuch der Männer vom Morgenstern* 53: 23–29.
Meyer, K.-D. 1983a. Saalian end moraines in Lower Saxony. In: Ehlers, J. (ed.): *Glacial Deposits in North-West Europe*: 335–342. Rotterdam: Balkema.
Meyer, K.-D. 1983b. Indicator pebbles and stone count methods. In: Ehlers, J. (ed.): *Glacial Deposits in North-West Europe*: 275–287. Rotterdam: Balkema.
Meyer, K.-D. 1983c. Zur Anlage der Urstromtäler in Niedersachsen. *Zeitschrift für Geomorphologie N.F.* 27: 147–160.
Meyer, K.-D. 1987. Ground and end moraines in Lower Saxony. In: Van der Meer, J. J. M. (ed.): *Tills and Glaciotectonics*: 197–204. Rotterdam: Balkema.
Meyer, K.-D. 1991. Zur Entstehung der westlichen Ostsee. *Geologisches Jahrbuch* A 127: 429–446.
Meyer, K.-J. 1974. Pollenanalytische Untersuchungen und Jahresschichtenzählungen an der holsteinzeitlichen Kieselgur von Hetendorf. *Geologisches Jahrbuch* A 21: 87–105.
Mickelson, D. M. & Evenson, E. B. 1975. Pre-Twocreekan age of the type Valders till, Wisconsin. *Geology* 3: 587–590.
Mickelson, D. M., Acomb, L. J. & Bentley, C. R. 1981. Possible mechanism for the rapid advance and retreat of the Lake Michigan Lobe between 13,000 and 11,000 years BP. *Annals of Glaciology* 2: 185–186.
Mickelson, D. M., Clayton, L., Fullerton, D. S. & Borns, H. W., Jr. 1983. The Late Wisconsin glacial record of the Laurentide Ice Sheet in the United States. In: Wright, H. E., Jr. (ed.): *Late-Quaternary Environments of the United States*, Vol. 1: 3–37. Minneapolis: University of Minnesota Press.
Milankovitch, M. 1941. Kanon der Erdbestrahlung und seine Anwendung auf das Eiszeitenproblem. *Königlich Serbische Akademie Belgrad, Spec. Publ.* 133: 633 pp.

Miller, B. B. & Bajc, A. F. 1990. Non-marine Molluscs. In: Warner, B. G. (ed.): *Methods in Quaternary Ecology. Geoscience Canada Reprint Series* 5: 101–112.
Miller, B. B., McCoy, W. D., Wayne, W. J. & Brockman, C. S. 1992. Ages of the Whitewater and Fairhaven tills in southwestern Ohio and southeastern Indiana. In: Clark, P. U. & Lea, P. D. (eds): *The Last Interglacial–Glacial Transition in North America. Geological Society of America Special Paper* 270: 89–98.
Miller, B. J. 1991. Pedology. In: Morrison, R. B. (ed.): *Quaternary Nonglacial Geology: Conterminous U.S. Geological Society of America, The Geology of North America* K-2: 564–573.
Miller, B. J., Lewis, G. C., Alford, J. J. & Day, W. J. 1985. Loesses in Louisiana and at Vicksburg, Mississippi. *Friends of the Pleistocene South-Central Cell Field Trip Guidebook.* 126 pp.
Miller, G. H. & Brigham-Grette, J. 1989. Amino acid geochronology: Resolution and precision in carbonate fossils. *Quaternary International* 1: 111–128.
Miller, G. H. & Mangerud, J. 1985. Aminostratigraphy of European marine interglacial deposits. *Quaternary Science Reviews* 4: 215–278.
Miller, G. H., Funder, S., de Vernal, A. & Andrews, J. T. 1992. Timing and character of the last interglacial–glacial transition in the Eastern Canadian Arctic and Northwest Greenland. In: Clark, P. U. & Lea, P. D. (eds): *The Last Interglacial–Glacial Transition in North America. Geological Society of America Special Paper* 270: 89–98.
Miller, H. 1884. On boulder-glaciation. *Royal Physical Society of Edinburgh, Proceedings* 8: 156–189.
Miller, N. G. 1973. Lateglacial plants and plant communities in northwestern New York State. *Journal of the Arnold Arboretum* 54: 123–159.
Miller, N. G. 1987. Phytogeography and paleoecology of a Late Pleistocene moss assemblage from northern Vermont. *Memoirs of the New York Botanical Garden* 45: 242–258.
Miller, N. G. 1989. Pleistocene and Holocene floras of New England as a framework for interpreting aspects of plant rarity. *Rhodora* 91: 49–69.
Miller, N. G. & Calkin, P. E. 1992. Paleoecological interprettion and age of an interstadial lake bed in western New York. *Quaternary Research* 37: 75–88.
Miller, N. G. & Thompson, G. G. 1979. Boreal and western North American plants in the Late Pleistocene of Vermont. *Journal of the Arnold Arboretum* 60: 167–218.
Miller, U. 1977. Pleistocene deposits of the Alnarp Valley, southern Sweden. *Lund University, Department of Quaternary Research, Thesis* 4: 125 pp.
Milthers, V. 1934. Die Verteilung skandinavischer Leitgeschiebe im Quartär von Westdeutschland. *Abhandlungen der Preußischen Geologischen Landesanstalt, N.F.* 156: 74 pp.
Mitchell, G. F. 1948. Two inter-glacial deposits in south-east Ireland. *Proceedings of the Royal Irish Academy* 52 B: 1–14.
Mitchell, G. F., Penny, L. F., Shotton, F. W. & West, R. G. 1973. A correlation of Quaternary deposits in the British Isles. *Geological Society of London, Special Report* 4: 99 pp.
Mix, A. C. 1987. The oxygen-isotope record of glaciation. In: Ruddiman, W. F. & Wright, H. E., Jr. (eds): *North America and adjacent oceans during the last deglaciation. Geological Society of America, Geology of North America* K-3: 111–136.
Moffat, A. J. & Catt, J. A. 1986. A re-examination of the evidence for a Plio-Pleistocene marine transgression on the Chiltern Hills. III. Deposits. *Earth Surface Processes and Landforms* 11: 233–247.
Møhl-Hansen, U. 1955. Første sikre spor af mennesker fra interglacialtid i Danmark. *Aarbøger for nordisk Oldkyndighed og Historie,* 1955: 101–126.
Mohn, H. & Nansen, F. 1893. Wissenschaftliche Ergebnisse von Dr. F. Nansens Durchquerung von Grönland 1888. *Petermanns Geographische Mitteilungen, Ergänzungsband* XXIII (105): 111 pp.
Mojski, J. E. 1982. Outline of the Pleistocene stratigraphy in Poland. *Biuletyn Instytutu Geologicznego* 343: 9–29.
Mojski, J. E. 1992. Vistulian stratigraphy and TL dates in Poland. *Sveriges Geologiska Undersökning* Ca 81: 195–200.
Mojski, J. E. 1993. *Europa w Plejstocenie – ewolucja środowiska przyrodniczego.* Warszawa: Wydawnictwo Polska Agencja Ekologiczna. 333 pp.
Mojski, J. E. 1995. Pleistocene glacial events in Poland. In: Ehlers, J., Kozarski, S. & Gibbard, P. L. (eds): *Glacial Deposits in North-East Europe*: 287–292. Rotterdam: Balkema.
Monaghan, M. C., Krishnaswami, S. & Thomas, J. H. 1983. ^{10}Be concentrations and the long-term fate of

particle-reactive nuclides in five soil profiles from California. *Earth and Planetary Science Letters* 65: 51–60.
Monjuvent, G. & Nicoud, G. 1988. Modalités et chronologie de la déglaciation Würmienne dans l'Arc alpin occidental et les massifs français: synthèse et réflexions. *Bulletin de l'Association française pour l'étude du Quaternaire* 1988 (2/3): 147–156.
Mooers, H. D. 1989. On the formation of the tunnel valleys of the Superior lobe, central Minnesota. *Quaternary Research* 32: 24–35.
Moran, S. R. 1971. Glaciotectonic structures in drift. In: Goldthwait, R. P. (ed.): *Till – a Symposium*: 127–148. Columbus: Ohio State University Press.
Moran, S. R., Clayton, L., Hooke, R. LeB., Fenton, M. M. & Andriashek, L. D. 1980. Glacier-bed landforms of the Prairie region of North America. *Journal of Glaciology* 25: 457–476.
Mordziol, C. 1913. Geologische Wanderungen durch das Diluvium und Tertiär der Umgebung von Koblenz (Neuwieder Becken). *Die Rheinlande* 5: 82 pp.
Morgan, A. V. 1971. Polygonal patterned ground of Late Weichselian age in the area north and west of Wolverhampton, England. *Geofrafiska Annaler* 53 A: 146–156.
Morgan, A. V. 1972. Late Wisconsinan ice-wedge polygons near Kitchener, Ontario, Canada. *Canadian Journal of Earth Sciences* 9: 607–617.
Morgan, A. V. 1982. Distribution and probable age of relict permafrost features in southwestern Ontario. In: French, H. M. (ed.): *The Roger J. E. Brown Memorial Volume, Proceedings 4th Canadian Permafrost Conference*: 91–100. National Research Council.
Morgan, A. V. & Morgan, A. 1990. Beetles. In: Warner, B. G. (ed.): *Methods in Quaternary Ecology. Geoscience Canada, Reprint Series* 5: 113–126.
Mörner, N.-A. 1976. Eustasy and geoid changes. *Journal of Geology* 84: 123–151.
Mörner, N.-A. 1980. The Fennoscandian Uplift: Geological data and their geodynamical implication. In: Mörner, N.-A. (ed.): *Earth Rheology, Isostasy and Eustasy*: 251–284. Chichester: Wiley.
Mörner, N.-A. & Dreimanis, A. 1973. The Erie Interstade. In: Black, R. F., Goldthwait, R. P. & Willman, H. B. (eds): *The Wisconsinan Stage. Geological Society of America Memoir* 136: 107–134.
Morrison, R. B. 1968. Means of time-stratigraphic division and long-distance correlation of Quaternary successions. In: Morrison, R. B. & Wright, H. E. (eds): *Means of Correlation of Quaternary Successions*: 1–113. Salt Lake City, University of Utah Press.
Morrison, R. B. 1987. Long-term perspective; Changing rates and types of Quaternary surficial processes; erosion, deposition, stablility cycles. In: Graf, W. L. (ed.): *Geomorphic systems in North America. The Geological Society of North America, Centennial Special Volume* 2: 163–210.
Morrison, R. B. (ed.) 1991. *Quaternary Nonglacial Geology: Conterminous U.S. The Geology of North America* K-2: 672 pp.
Mott, R. J. 1990. Sangamonian forest history and climate in Atlantic Canada. *Géographie physique et Quaternaire* 44: 257–270.
Mott, R. J. 1994. Wisconsinan Late-glacial environmental change in Nova Scotia: a regional synthesis. *Journal of Quaternary Science* 9: 155–160.
Mott, R. J. & Grant, D. R. G.. 1985. Pre-Late Wisconsinan paleoenvironments in Atlantic Canada. *Géographie physique et Quaternaire* 39: 239–254.
Mougenot, D., Boillot, G. & Rehoult, J.-P. 1983. Prograding shelfbreak types on passive continental margins: some European examples. In: Stanley, D. J. & Moore, G. T. (eds): *The Shelfbreak: Critical Interface on Continental Margins. Society of Economic Paleontologists and Mineralogists, Special Publication* 33: 61–77.
Mückenhausen, E. 1959. Die stratigraphische Gliederung des Löß-Komplexes von Kärlich im Neuwieder Becken. *Fortschritte in der Geologie von Rheinland und Westfalen* 4: 283–300.
Müller, A. 1973. Beitrag zum Quartär des Elbegebietes zwischen Riesa und Wittenberg unter besonderer Brücksichtigung der Elbtalwanne. *Zeitschrift für Geologische Wissenschaften* 1: 1105–1122.
Müller, A. 1988. Das Quartär im mittleren Elbegebiet zwischen Riesa und Dessau. Dissertation, Halle. 129 pp.
Muller, E. H. & Pair, D. L. 1992. Comment on 'Evidence for large-scale subglacial meltwater flood events in southern Ontario and northern New York State'. *Geology* 20: 90–91.
Muller, E. H. & Prest, V. K. 1985. Glacial lakes in the Ontario Basin. In: Karrow, P. F. & Calkin, P. E. (eds): *Quaternary Evolution of the Great Lakes. Geological Association of Canada Special Paper* 30:

213–229.
Müller, H. 1965. Eine pollenanalytische Neubearbeitung des Interglazialprofils von Bilshausen (Unter-Eichsfeld). *Geologisches Jahrbuch* 83: 327–352.
Müller, H. 1974a. Pollenanalytische Untersuchungen und Jahresschichtenzählungen an der eemzeitlichen Kieselgur von Bispingen/Luhe. *Geologisches Jahrbuch* A 21: 149–169.
Müller, H. 1974b. Pollenanalytische Untersuchungen und Jahresschichtenzählungen an der holsteinzeitlichen Kieselgur von Munster-Breloh. *Geologisches Jahrbuch* A 21: 107–140.
Müller, H. 1992. Climate changes during and at the end of the interglacials of the Cromerian Complex. In: Kukla, G. H. & Went, E. (eds): *Start of a Glacial*: 51–69. Berlin: Springer.
Müller, H.-N., Kerschner, H. & Küttel, M. 1983. The Val de Nendaz (Valais, Switzerland) – a type locality for the Egesen advance and the Daun advance in the Western Alps. In: Schroeder-Lanz, H. (ed.): *Late- and Postglacial Oscillations of Glaciers: Glacial and Periglacial Forms*: 73–82. Rotterdam: Balkema.
Müller, U., Rühberg, N. & Krienke, H.-D. 1995. The Pleistocene Sequence in Mecklenburg-Vorpommern. In: Kozarski, S., Ehlers, J. & Gibbard, P. L. (eds): *Glacial Deposits in North-East Europe*: 501–514. Rotterdam: Balkema.
Müller, W.-H., Huber, M., Isler, A. & Kleboth, P. 1984. *Erläuterungen zur Geologischen Karte der zentralen Nordschweiz 1:100 000 mit angrenzenden Gebieten von Baden-Württemberg*. 234 pp.
Müller-Beck, H. 1983. Sammlerinnen und Jäger von den Anfängen bis vor 35 000 Jahren. In: Müller-Beck (ed.): *Urgeschichte in Baden Württemberg*: 241–272. Stuttgart: Theiss.
Müller-Beck, H. 1995. Urgeschichte. In: Benda, L. (ed.): *Das Quartär Deutschlands*: 327–348. Berlin: Gebrüder Borntraeger.
Mullins, H. T. & Hinchey, E. J. 1989. Erosion and infill of New York Finger Lakes: implications for Laurentide ice sheet deglaciation. *Geology* 17: 622–625.
Munthe, H. 1902. Beskrivning til kartbladet Kalmar. *Sveriges Geologiska Undersökning* Ca 4: 213 pp.
Munthe, H. 1927. Studier över Ancylussjöns avlopp. *Sveriges Geologiska Undersökning* C 346: 107 pp.
Münzing, Kl. & Aktas, A. 1987. Weitere Funde molluskenführender Mergellagen im unteren Deckenschotter von Bayerisch Schwaben. *Jahresberichte und Mitteilungen des Oberrheinischen Geologischen Vereins N.F.* 69: 181–193.
Musil, R. 1975. Die Equiden aus dem Travertin von Ehringsdorf. In: *III. Internationales Paläontologisches Kolloquium 1968. Das Pleistozän von Weimar-Ehringsdorf, Teil 2. Abhandlungen des Zentralen Geologischen Instituts Berlin* 23: 265–335.
Musil, R. 1977. Die Equidenreste aus dem Travertin von Taubach. *Quartärpaläontologie* 2: 237–264.
Naeser, C. W. & Naeser, N. D. 1991. Fission-track dating. In: Morrison, R. B. (ed.): *Quaternary Nonglacial Geology: Conterminous U.S. Geological Society of America. The Geology of North America* K-2: 53–54.
Nahon, D. & Trompette, R. 1982. Origin of siltstones: Glacial grinding vs. weathering. *Sedimentology* 29: 25–35.
Nansen, F. 1922. The strandflat and isostasy. *Videnskabsselskabets i Kristiania skrifter 1921, I. Matematisk-Naturvidenskabelig Klasse* 11: 313 pp.
Nathorst, A.G. 1873. Om Skånes nivåförändringar. *Geologiska Föreningens i Stockholm Förhandlingar* 1: 181–194.
Naumann, C. F. 1848. Über die Felsenschliffe der Hohburger Porphyrberge unweit Wurzen. *Berichte und Verhandlungen der Sächsischen Gesellschaft für Wissenschaften* 1: 392–410.
Nelson, A. R. & Mickelson, D. M. 1977. Landform distribution and genesis in the Langlade and Green Bay glacial lobes, north-central Wisconsin. *Wisconsin Academy of Sciences, Arts and Letters* 65: 41–57.
Nelson, F. E., Lachenbruch, A. H., Woo, M.-K., Koster, E. A., Osterkamp, T. E., Gavrilova, M. K. & Cheng, G. 1993. Permafrost and changing climate. *Permafrost Sixth International Conference Proceedings* 2: 987–1005. Wushan: South China University of Technology Press.
Nenonen, K. 1995. Pleistocene stratigraphy of southern Finland. In: Ehlers, J., Kozarski, S. & Gibbard, P. L. (eds): *Glacial Deposits in North-East Europe*: 11–28. Rotterdam: Balkema.
Nenonen, K., Eriksson, B. & Grönlund, T. 1991. The till stratigraphy of Ostrobothnia, Western Finland, with reference to new Eemian Interglacial sites. *Striae* 34: 65–76.
Nesje, A. & Dahl, S. O. 1990. Autochthonous block fields in southern Norway: implications for the

geometry, thickness, and isostatic loading of the Late Weichselian Scandinavian ice sheet. *Journal of Quaternary Science* 5: 225–234.
Nesje, A. & Sejrup, H. P. 1988. Late Weichselian/Devensian ice sheets in the North Sea and adjacent land areas. *Boreas* 17: 371–384.
Nesje, A., Kvamme, M., Rye, N. & Lövlie, R. 1991. Holocene glacial and climate history of the Jostedalsbreen region, western Norway; evidence from lake sediments and terrestrial deposits. *Quaternary Science Reviews* 10: 87–114.
Nickman, R. J. & Demarest, J. M., III 1982. Pollen and macrofossil study of an interglacial deposit in Nova Scotia. *Géographie physique et Quaternaire* 36: 197–208.
Niemelä, J. 1971. Die quartäre Stratigraphie von Tonablagerungen und der Rückzug des Inlandeises zwischen Helsinki und Hämeenlinna in Südfinnland. *Geological Survey of Finland, Bulletin* 253: 79 pp.
Niemelä, J. & Tynni, R. 1979. Interglacial and interstadial sediments in the Pohjanmaa region, Finland. *Geological Survey of Finland, Bulletin* 302: 42 pp.
Niewiarowski, W. 1965. Kemy i formy pokrewne w Danii oraz rozmieszczenie obszarów kemowych na terenie Peribalticum w obrebie ostatniego zlodowacenia (Kames and related landforms in Denmark, and the distribution of kame landscapes in the Peribalticum within the area of the last glaciation). *Zeszyty Naukowe Uniwersystetu Mikolaja Kopernika w Toruniu, Nauki Matematyczno-Przyrodnicze, Zeszyt 11, Geografia* VI: 116 pp.
Niewiarowski, W. 1992. Geneza i ewolucja rynny zninskiej w okresie peinego i poznego Vistulianu. *Acta Universitatis Nicolai Copernici, Geografia* XXV: 3–30.
Nilsson, T. 1935. Die pollenanalytische Zonengliederung der spät- und post-glazialen Bildungen Schonens. *Geologiska Föreningens i Stockholm Förhandlingar* 57: 385–562.
Nilsson, T. 1972. *Pleistocen. Den geologiska och biologiska utvecklingen under istidsålderen.* Scandinavian University Books. 508 pp.
Nilsson, T. 1983. *The Pleistocene. Geology and Life in the Quaternary Ice Age.* Reidel: Dordrecht. 651 pp.
Nishiizumi, K., Lal, D., Klein, J., Middleton, R. & Arnold, J. R. 1986. Production of ^{10}Be and ^{26}Al by cosmic rays in terrestrial quartz in situ and implications for erosion rates. *Nature* 319: 134–136.
Norton, L. D., West, L. T. & McSweeney, K. 1988. Soil development and loess stratigraphy of the midcontinental USA. In: Eden, D. N. & Furkert, R. J. (eds): *Loess – Its distribution, geology and soils*: 145–159. Rotterdam: Balkema.
Nowaczyk, B. 1976. Eolian cover sands in central-west Poland. *Quaestiones Geographicae* 3: 57–77.
Nye, J. F. 1952. The mechanics of glacier flow. *Journal of Glaciology* 2: 82–93.
Nye, J. F. 1973. Water at the bed of the glacier. In: *Symposium on the Hydrology of Glaciers, Proceedings of the Cambridge Symposium, 7–13 September 1969. International Association of Scientific Hydrology Publication* 95: 189–194.
Nye, J. F. 1976. Water flow in glaciers: jökulhlaups, tunnels and veins. *Journal of Glaciology* 17: 181–207.
Nye, J. F. & Frank, F. C. 1973. Hydrology of the intergranular veins in a temperate glacier. In: *Symposium on the Hydrology of Glaciers (Proceedings of the Cambridge Symposium, 7–13 September 1969). International Association of Scientific Hydrology Publicaton* 95: 157–161.
O'Connor, J. E. 1993. *Hydrology, Hydraulics, and Geomorphology of the Bonneville Flood. Geological Society of America Special Paper* 274: 83 pp.
Oeschger, H., Beer, J., Siegenthaler, U., Stauffer, B., Dansgaard, W. & Langway, C. C., Jr. 1984. Late glacial climate history derived from ice cores. In: Hansen, J. E. & Takahashi, T. (eds): *Climate Processes and Climate Sensitivity. American Geophysical Union, Geophysical Monographs* 29: 299–306.
Officer, Ch. B., Newman, W. S., Sullivan, J. M. & Lynch, D. R. 1988. Glacial isostatic adjustment and mantle viscosity. *Journal of Geophysical Research* 93, B 6: 6397–6409.
Ohmert, W. 1979. Die Ostracoden der Kernbohrung Eurach 1 (Riß–Eem). *Geologica Bavarica* 80: 127–158.
Oldale, R. N. & Colman, S. M. 1992. On the age of the penultimate full glaciation of New England. In: Clark, P. U. & Lea, P. D. (eds): *The Last Interglacial–Glacial Transition in North America. Geological Society of America Special Paper* 270: 163–170.

Oldale, R. N. & O'Hara, C. J. 1984. Glaciotectonic origin of the Massachusetts coastal end moraines and fluctuating late Wisconsinan ice margin. *Geological Society of America Bulletin* 95: 61–74.

Oldale, R. N., Colman, S. M. & Jones, G. A. 1993. Radiocarbon ages from two submerged strandline features in the Western Gulf of Maine and a sea-level curve for the Northeastern Massachusetts coastal region. *Quaternary Research* 40: 38–45.

Olszewski, A. 1982. Icings and their geomorphological significance exemplified from Oscar II Land and Prins Karls Forland, Svalbard. *Acta Universitatis Nicolai Copernici, Geografia* XVI: 91–122.

Orombelli, G. 1974. Alcune date ^{14}C per il Quaternario lombardo. *Studi Trentini di Scienze Naturali* 51: 125–127.

Osterman, L. E. & Nelson, A. R. 1989. Latest Quaternary and Holocene paleoceanography of the eastern Baffin Island continental shelf, Canada: benthic foraminiferal evidence. *Canadian Journal of Earth Sciences* 26: 2236–2248.

Østrem, G, Haakensen, N. & Melander, O. 1973. *Atlas over Breer i Nord-Skandinavia.* Norges Vassdrags og Elektrisitetsvesen og Stockholm Universitet. 315 pp.

Ostry, R. C. & Deane, R. E. 1963. Microfabric analysis of till. *Geological Society of America Bulletin* 74: 165–168.

Otvos, E. G., Jr. 1982. Coastal geology of Mississippi, Alabama, and adjacent Louisiana areas. *New Orleans Geological Society Field Trip Guidebook*: 67 pp.

Otvos, E. G., Jr. 1992. Quaternary evolution of the Apalachicola coast, northeastern Gulf of Mexico. *Society for Sedimentary Geology, Special Publication* 48: 221–232.

Overbeck, F. 1975. *Botanisch-geologische Moorkunde unter besonderer Berücksichtigung der Moore Nordwestdeutschlands als Quellen zur Vegetations-, Klima- und Siedlungsgeschichte.* Neumünster: Wachholtz. 719 pp.

Paepe, R. & Baeteman, C. 1979. The Belgian coastal plain during the Quaternary. In: Oele, E., Schüttenhelm, R. T. E. & Wiggers, A. J. (eds): *The Quaternary History of the North Sea. Acta Universitatis Upsaliensis Annum Quingentesimum Celebrantis* 2: 143–146.

Pair, D., Karrow, P. F. & Clark, P. U. 1988. History of the Champlain Sea in the Central St. Lawrence Lowland, New York, and its relationship to water levels in the Lake Ontario Basin. In: Gadd, N. R. (ed.): *The Late Quaternary Development of the Champlain Sea Basin. Geological Association of Canada Special Paper* 35: 107–123.

Panzig, W.-A. 1995. The tills of NE-Rügen – lithostratigraphy, gravel composition and relative deposition directions in the southwestern Baltic Region. In: Ehlers, J., Gibbard, P. L. & Kozarski, S. (eds): *Glacial Deposits in North-East Europe*: 521–533. Rotterdam: Balkema.

Pardi, R. R. & Newman, W. S. 1987. Late Quaternary sea levels along the Atlantic coast of North America. *Journal of Coastal Research* 3: 325–330.

Parizek, R. R. 1969. Glacial ice-contact rings and ridges. In: Schumm, S. A. & Bradley, W. C. (eds): *United States Contributions to Quaternary Research. Geological Society of America Special Paper* 123: 49–102.

Parks, D. A. & Rendell, H. M. 1992. Thermoluminescence dating and geochemistry of loessic deposits in southeast England. *Journal of Quaternary Science* 7: 99–107.

Paschinger, H. 1950. Morphologische Ergebnisse einer Analyse der Höttinger Breccie bei Innsbruck. *Schlern-Schriften* 75: 86 pp.

Paschinger, H. 1957. Klimamorphologische Studien im Quartär des alpinen Inntals. *Zeitschrift für Geomorphologie N.F.* 1: 237–270.

Pasierbski, M. 1979. Remarks on the Genesis of Subglacial Channels in Northern Poland. *Eiszeitalter und Gegenwart* 29: 189–200.

Påsse, T., Robertsson, A.-M., Miller, U. & Klingberg, F. 1988. A Late Pleistocene sequence at Margreteberg, southwestern Sweden. *Boreas* 17: 141–163.

Paterson, T. T. 1940. The effects of frost action and solifluction around Baffin Bay and the Cambridge District. *Quarterly Journal of the Geological Society of London* 96: 99–130.

Paterson, W. S. B. 1969. *The Physics of Glaciers*. Oxford: Pergamon. 250 pp.

Paterson, W. S. B. 1981. *The Physics of Glaciers*, 2nd edition. Oxford: Pergamon Press. 380 pp.

Paterson, W. S. B. 1994. *The Physics of Glaciers*, 3rd edition. Oxford: Pergamon. 480 pp.

Patience, A. J. & Kroon, D. 1991. Oxygen isotope chronostratigraphy. In: Smart, P. L. & Frances, P. D. (eds): *Quaternary dating methods – a user's guide. Quaternary Research Association, Technical Guide*

4: 199–228.

Patrick, R. 1984. The history of the science of diatoms in the United States of America. In: Mann, D. G. (ed.): *Proceedings of the Seventh International Diatom Symposium*: 11–20. Königstein: Otto Koeltz.

Patterson, C. J. 1994. Tunnel-valley fans of the St. Croix moraine, east-central Minnesota, USA. In: Warren, W. P. & Croot, D. G. 1994. *Formation and Deformation of Glacial Deposits*: 69–87. Rotterdam: Balkema.

Patzelt, G. 1972. Die spätglazialen Stadien und postglazialen Schwankungen von Ostalpengletschern. *Berichte der Deutschen Botanischen Gesellschaft* 85: 47–57.

Patzelt, G. 1973a. Die neuzeitlichen Gletscherschwankungen in der Venedigergruppe (Hohe Tauern, Ostalpen). *Zeitschrift für Gletscherkunde und Glazialgeologie* 9: 5–57.

Patzelt, G. 1973b. Die postglazialen Gletscher- und Klimaschwankungen in der Venedigergruppe (Hohe Tauern, Ostalpen). *Zeitschrift für Geomorphologie N.F., Supplement-Band* 16: 25–72.

Patzelt, G. 1975. Unterinntal – Zillertal – Pinzgau – Kitzbühl (Spät- und postglaziale Landschaftsentwicklung). In: *Tirol – ein geographischer Exkursionsführer. Innsbrucker Geographische Studien* 2: 309–329.

Patzelt, G. 1987. Untersuchungen zur nacheiszeitlichen Schwemmkegel- und Talentwicklung in Tirol. 1. Teil: Das Inntal zwischen Mütz und Wattens. *Veröffentlichungen des Museums Ferdinandeum, Innsbruck* 67: 93–123.

Patzelt, G. & Bortenschlager, S. 1978. Spät- und nacheiszeitliche Gletscher- und Vegetationsentwicklung im inneren Ötztal. In: Horie, S. (ed.): *Paleolimnology of Lake Biwa and the Japanese Pleistocene* 6: 312–325.

Pazdur, A., Pazdur, M. F., Wicik, B. & Wieckowski, K. 1987. Radiocarbon chronology of annually laminated sediments from the Gosciaz Lake. *Bulletin of the Polish Academy of Sciences, Earth Sciences* 35: 139–145.

Peacock, J. D. 1983. Quaternary geology of the Inner Hebrides. *Proceedings of the Royal Society of Edinburgh* 83 B: 83–89.

Peacock, J. D. 1984. Quaternary geology of the Outer Hebrides. *Report of the British Geological Survey* 16: 1–26.

Peacock, J. D. 1991. Glacial deposits of the Hebridean region. In: Ehlers, J., Gibbard, P. & Rose, J. (eds): *Glacial Deposits in Great Britain and Ireland*: 109–119. Rotterdam: Balkema.

Peacock, J. D. 1993. Late Quaternary marine mollusca as palaeoenvironmental proxies: a compilation and assessment of basic numerical data for NE Atlantic species found in shallow water. *Quaternary Science Reviews* 12: 263–275.

Pécsi, M. 1968. Loess. In: Fairbridge, R. W. (ed.): *The Encyclopaedia of Geomorphology*. 674–678. New York: Reinhold.

Pécsi, M. & Schweitzer, F. 1991. Short- and long-term terrestrial records of the Middle Danubian Basin. In: Pécsi, M. & Schweitzer, F. (eds): *Quaternary Environment in Hungary. Studies in Geography in Hungary* 26: 9–26.

Pécsi, M. & Schweitzer, F. 1992. Long-term terrestrial records of the Middle Danubian Basin. *Quaternary International* 17: 5–14.

Penck, A. 1879. Die Geschiebeformation Norddeutschlands. *Zeitschrift der Deutschen Geologischen Gesellschaft* 31: 117–203.

Penck, A. 1882. *Die Vergletscherung der deutschen Alpen, ihre Ursachen, periodische Wiederkehr und ihr Einfluß auf die Bodengestaltung*. Leipzig: Barth. 484 pp.

Penck, A. 1894. *Morphologie der Erdoberfläche*. Stuttgart: Engelhorn. 2 volumes, 471 + 696 pp.

Penck, A. 1899. Die vierte Eiszeit im Bereich der Alpen. *Schriften der Vereinigung zur Verbreitung naturwissenschaftlicher Kenntnisse* 39: 1–20.

Penck, A. 1901. Die Übertiefung der Alpen-Thäler. *Verhandlungen des Siebenten Internationalen Geographen-Kongresses Berlin 1899*, Zweiter Theil: 232–240. Berlin: Kühl.

Penck, A. 1922a. Die Eem-Schwingung. *Verhandelingen van het Geologisch-Mijnbouwkundig Genootschap voor Nederland en Koloniën, Geologische Serie* VI: 91–105.

Penck, A. 1922b. Ablagerungen und Schichtstörungen der letzten Interglazialzeit in den nördlichen Alpen. *Sitzungsberichte der Preußischen Akademie der Wissenschaften, Physikalisch-Mathematische Klasse* 20: 214–251.

Penck, A. & Brückner, E. 1901/09. *Die Alpen im Eiszeitalter*. Leipzig: Tauchnitz. 3 volumes, 1199 pp.

Penney, D. N. 1987. Ostracoda of the Holsteinian Interglacial in Jutland, Denmark. *Danmarks Geologiske Undersøgelse* B 10: 33–67.
Penney, D. N. 1989. Microfossils (Foraminifera, Ostracoda) from an Eemian (Last Interglacial) tidal flat sequence in South-West Denmark. *Quaternary International* 3/4: 85–91.
Penney, D. N. 1993. Late Pliocene to Early Pleistocene ostracod stratigraphy and palaeoclimate of the Lodin Elv and Kap København formations, East Greenland. *Palaeogeography, Palaeoclimatology, Palaeoecology* 101: 49–66.
Perrin, R. M. S., Rose, J. & Davies, H. 1979. The distribution, variation and origins of pre-Devensian tills in eastern England. *Philosophical Transactions of the Royal Society of London* B 287: 535–570.
Persson, Ch. 1985. Beskrivning till jordartskartorna Östhammar NO. *Sveriges Geologiska Undersökning*, Ae 73: 65 pp.
Persson, Ch. 1986. Beskrivning till jordartskartorna Österlövsta SO/Grundkallen SV. *Sveriges Geologiska Undersökning*, Ae 76–77: 72 pp.
Peschke, P. 1983a. Palynologische Untersuchungen interstadialer Schieferkohlen aus dem schwäbisch-oberbayerischen Alpenvorland. *Geologica Bavarica* 84: 69–99.
Peschke, P. 1983b. Pollenanalysen der Schieferkohlen von Herrnhausen (Wolfratshausener Becken/Obb.) – ein Beitrag zum Problem interglazialer Ablagerungen in Oberbayern. *Geologica Bavarica* 84: 107–121.
Peteet, D. M., Daniels, R., Heusser, L. E., Vogel, J. S., Southon, J. R. & Nelson, D. E. 1994. Wisconsinan Late-glacial environmental change in southern New England: a regional synthesis. *Journal of Quaternary Science* 9: 151–154.
Peters, Tj. 1969. Tonmineralogie einiger Glazialablagerungen im schweizerischen Mittelland. *Eclogae Geologicae Helvetiae* 62: 517–525.
Petersen, J. 1899. Geschiebestudien. Beiträge zur Kenntniss der Bewegungsrichtungen des diluvialen Inlandeises, I. Theil. *Mittheilungen der Geographischen Gesellschaft in Hamburg* XV: 45–65.
Petersen, J. 1900. Geschiebestudien. Beiträge zur Kenntniss der Bewegungsrichtungen des diluvialen Inlandeises, II. Theil. *Mittheilungen der Geographischen Gesellschaft in Hamburg* XVI: 139–230.
Petersen, K. S. 1983. Redeposited biological material. In: Ehlers, J. (ed.): *Glacial Deposits in North-West Europe*: 203–206. Rotterdam: Balkema.
Petersen, K. S. & Konradi, P. B. 1974. Lithologiske og palæontologiske beskrivelser af profiler i kvartæret pa Sjælland. *Dansk Geologisk Forening, Årsskrift for 1973*: 47–56.
Petersen, K. S. & Kronborg, K. 1991. Late Pleistocene history of the inland glaciation in Denmark. *Paläoklimaforschung* 1: 331–342.
Pettijohn, E. J. 1975. *Sedimentary Rocks*, 3rd edition. New York: Harper & Row. 628 pp.
Péwé, T. L. 1959. Sand wedge polygons (tesselations) in the McMurdo Sound region, Antarctica. *American Journal of Science* 257: 542–552.
Péwé, T. L. 1975. Quaternary geology of Alaska. *US Geological Survey Professional Paper* 835: 145 pp.
Péwé, T. L. 1983a. The periglacial environment in North America during Wisconsin time. In: Wright, H. E. (ed.): *Late-Quaternary Environments of the United States, Volume 1: The Late Pleistocene*: 157–189. University of Minnesota Press.
Péwé, T. L. 1983b. Alpine permafrost in the contiguous United States: a review. *Arctic and Alpine Research* 15: 145–156.
Phillips, L. 1974. Vegetational history of the Ipswichian/Eemian Interglacial in Britain and Continental Europe. *New Phytologist* 73: 589–604.
Phillips, L. 1976. Pleistocene vegetational history and geology in Norfolk. *Philosophical Transactions of the Royal Society of London* B 275: 215–286.
Phleger, F. B., Parker, F. L. & Peirson, J. F. 1953. North Atlantic Foraminifera. *Reports of the Swedish Deep-Sea Expedition 1947–1948*, 7: 1–122. Göteborg: Elanders.
Picard, K. 1959. Gliederung pleistozäner Ablagerungen mit fossilen Böden bei Husum/Nordsee. *Neues Jahrbuch für Geologie und Paläontologie, Monatshefte* (6): 259–272.
Pielou, E. C. 1991. *After the Ice Age. The Return of Life to Glaciated North America.* Chicago: The University of Chicago Press. 366 pp.
Pierce, K. L., Obradovich, J. D. & Friedman, I. 1976. Obsidian hydration dating and correlation of Bull Lake and Pinedale glaciations near West Yellowstone, Montana. *Geological Society of America Bulletin* 87: 703–710.

Pietrzeniuk, E. 1991. Die Ostrakodenfauna des Eem-Interglazials von Schönfeld, Kr. Calau (Niederlausitz). *Natur und Landschaft in der Niederlausitz, Sonderheft*: 92–116.
Piffl, L. 1971. Zur Gliederung des Tullner Feldes. *Annalen des Naturhistorischen Museums Wien* 75: 293–310.
Pike, K. & Godwin, H. 1953. The Interglacial at Clacton-on-Sea, Essex. *Quarterly Journal of the Geological Society of London* 108: 261–272.
Piotrowski, J. A. 1989. Relationship between drumlin length and width as a manifestation of the subglacial processes. *Zeitschrift für Geomorphologie N.F.* 33: 429–441.
Piotrowski, J. A. 1992. Till facies and depositional environments of the upper sedimentary complex from the Stohler Cliff, Schleswig-Holstein, North Germany. *Zeitschrift für Geomorphologie N.F., Supplement-Band* 84: 37–54.
Piotrowski, J. A. 1993. Salt diapirs, pore-water traps and permafrost as key controls for glaciotectonism in the Kiel area, Northwestern Germany. In: Aber. J. S. (ed.): *Glaciotectonics and Mapping Glacial Deposits*: 86–98. Winnipeg: Hingnell.
Piotrowski, J. A. 1994a. Tunnel-valley formation in northwest Germany – geology, mechanisms of formation and subglacial bed conditions for the Bornhöved tunnel valley. *Sedimentary Geology* 89: 107–141.
Piotrowski, J. A. 1994b. Waterlain and lodgement till facies of the lower sedimentary complex from the Dänischer-Wohld Cliff, Schleswig-Holstein, North Germany. In: Warren, W. P. & Croot, D. G. (eds): *Formation and Deformation of Glacial Deposits*: 3–8. Rotterdam: Balkema.
Piotrowski, J. A. & Vahldiek, J. 1991. Elongated hills near Schönhorst, Schleswig-Holstein: Drumlins or terminal push-moraines? *Bulletin of the Geological Society of Denmark* 38: 231–242.
Piper, D. J. W. & Aksu, A. E. 1992. Architecture of stacked Quaternary deltas correlated with global oxygen isotopic curve. *Geology* 20: 415–418.
Piper, D. J. W., Mudie, P. J., Aksu, A. E. & Skene, K. I. 1994. A 1 Ma record of sediment flux south of the Grand Banks used to infer the development of glaciation in southeastern Canada. *Quaternary Science Reviews* 13: 23–37.
Pirazzoli, P. A. 1989. Present and near-future global sea-level changes. *Palaeogeography, Palaeoclimatology, Palaeoecology* 75: 241–258.
Pirazzoli, P. A. 1991. World atlas of Holocene sea-level changes. *Elsevier Oceanography Series* 58: 300 pp.
Pirazzoli, P. A., Radtke, U., Hantoro, W. S., Jouannic, C., Hoang, C. T., Causse, C. & Borel Best, M. 1993. A one million-year long sequence of marine terraces on Sumba Island, Indonesia. *Marine Geology* 109: 221–236.
Pissart, A. 1976. Les dépôts et la morphologie périglaciares de la Belgique. In: Pissart, A. (ed.): *Géomorphologie de la Belgique*: 115–135. Université de Liège.
Pissart, A. 1987. Weichselian periglacial structures and their environmental significance: Belgium, the Netherlands, and northern France. In: Boardman, J. (ed.): *Periglacial processes and landforms in Britain and Ireland*: 77–85. Cambridge: Cambridge University Press.
Plafker, G. & Addicott, W. O. 1976. Glaciomarine deposits of Miocene through Holocene age in the Yakataga Formation along the Gulf of Alaska margin, Alaska. In: Miller, T. P. (ed.): *Recent and ancient sedimentary environments in Alaska. Alaska Geological Society Symposium Proceedings*: Q1–Q23.
Plafker, G. & Berg, H. C. (eds) 1994. *The Geology of Alaska. Geological Society of America. The Geology of North America* G-1: 1055 pp.
Pointon, W. K. 1978. The Pleistocene succession at Corton, Suffolk. *Bulletin of the Geological Society of Norfolk* 30: 55–76.
Pollard, W. H. & French, H. M. 1985. The internal structure and ice crystallography of seasonal frost mounds. *Journal of Glaciology* 31: 157–162.
Porsild, A. E. 1938. Earth mounds in unglaciated arctic northwestern America. *Geographical Review* 28: 46–58.
Porter, S. C. & Denton, G. H. 1976. Chronology of neoglaciation in the North American Cordillera. *American Journal of Science* 265: 177–210.
Porter, S. C. & Orombelli, G. 1982. Late-glacial ice advances in the western Italian Alps. *Boreas* 11: 125–140.

Porter, S. C. & Orombelli, G. 1985. Glacier contraction during the middle Holocene in the western Italian Alps: Evidence and implications. *Geology* 13: 296–298.
Poser, H. 1948. Boden- und Klimaverhältnisse in Mittel- und Westeuropa während der Würm-Eiszeit. *Erdkunde* 2: 53–68. *Also*: Soil and climate relations in Central and Western Europe during the Würm Glaciation. In: Evans, D. J. A. (ed.) 1994: *Cold Climate Landforms*: 3–22. Chichester: Wiley.
Poser, H. & Müller, Th. 1951. Studien an den asymmetrischen Tälern des Niederbayerischen Hügellandes. *Nachrichten der Akademie der Wissenschaften, Göttingen, Mathematisch-Physikalische Klasse* IIb 1951 (1): 1–32.
Potzger, J. E. 1951. The fossil record near the glacial border. *Ohio Journal of Science* 51: 126–133.
Praeg, D., MacLean, B., Piper, D. J. W. & Shor, A. N. 1987. Study of iceberg scours across the continental shelf and slope off southeast Baffin Island using SeaMARC I midrange sidescan sonar. *Current Research, Part A: Geological Survey of Canada Paper* 87-1A: 847–857.
Prange, W. 1978. Der letzte weichselzeitliche Gletschervorstoß in Schleswig-Holstein – das Gefüge überfahrener Schmelzwassersande und die Entstehung der Morphologie. *Meyniana* 30: 61–75.
Prange, W. 1979. Geologie der Steilufer von Schwansen, Schleswig-Holstein. *Schriften des Naturwissenschaftlichen Vereins in Schleswig-Holstein* 49: 1–24.
Preece, R. C. 1990. The molluscan fauna of the Middle Pleistocene interglacial deposits at Little Oakley, Essex, and its environmental and stratigraphical implications. *Philosophical Transactions of the Royal Society of London* B 328: 387–407.
Prest, V. K. 1983. Canada's heritage of glacial features. *Geological Survey of Canada, Miscellaneous Report* 28: 119 pp.
Prest, V. K. 1984. The Late Wisconsinan Glacier Complex. In: Fulton, R. J. (ed.): *Quaternary Stratigraphy of Canada – A Canadian Contribution to IGCP Project 24. Geological Survey of Canada, Paper* 84-10, 21–36.
Prest, V. K. & Grant, D. R. 1969. Retreat of the last ice sheet from the Maritime Provinces – Gulf of St. Lawrence region. *Geological Survey of Canada, Paper* 69-33: 15 pp.
Prest, V. K. & Nielsen, E. 1987. The Laurentide Ice Sheet and long-distance transport. In Kujansuu, R. & Saarnisto, M. (eds): *INQUA Till Symposium, Finland 1985. Geological Survey of Finland, Special Paper* 3: 91–101.
Prest, V. K., Grant, D. R., MacNeill, R. H., Brooks, I. A., Borns, H. W., Ogden, J. G., III, Jones, J. F., Lin, C. L., Hennigar, T. W. & Parsons, M. L. 1972. *Quaternary geology, geomorphology and hydrogeology of the Atlantic Provinces: 24th International Geological Congress, Excursion Guidebook*, A61-C61: 79 pp.
Price, P. B. & Walker, R. M. 1962: A new detector for heavy particle studies. *Physical Letters* 3: 113–115.
Price, P. B. & Walker, R. M. 1963. Fossil tracks of charged particles in mica and the age of minerals. *Journal of Geophysical Research* 68: 4747–4862.
Price, R. J. 1966. Eskers near the Casement Glacier, Alaska. *Geografiska Annaler* 48 A: 111–125.
Price, W. A. 1933. Role of diastrophism in topography of Corpus Christi area, south Texas. *American Association of Petroleum Geologists, Bulletin* 17: 907–962.
Priesnitz, K. & Schunke, E. 1978. An approach to the ecology of permafrost in central Iceland. In: *Third International Conference on Permafrost, Edmonton, Proceedings* 1: 473–479.
Punkari, M. 1980. The ice holes of the Scandinavian ice sheet during the deglaciation in Finland. *Boreas* 9: 307–310.
Punkari, M. 1984. The relations between glacial dynamics and tills in the eastern part of the Baltic Shield. *Striae* 20: 49–54.
Punkari, M. 1985. Glacial geomorphology and dynamics in Soviet Karelia interpreted by means of satellite imagery. *Fennia* 163: 113–153.
Pye, K. 1984. Loess. *Progress in Physical Geography* 8: 176–217.
Pye, K. 1987. *Aeolian Dust and Dust Deposits*. London: Academic Press, 334 pp.
Pye, K. & Sperling, C. H. B. 1983. Experimental investigation of silt formation by static breakage processes: the effect of temperature, moisture and salt on quartz dune sand and granitic regolith. *Sedimentology* 30: 49–62.
Pye, K. & Tsoar, H. 1990. *Aeolian sand and sand dunes*. London: Unwin Hyman, 396 pp.
Pyritz, E. 1972. Binnendünen und Flugsandebenen im Niedersächsischen Tiefland. *Göttinger Geographische Abhandlungen* 61: 153 pp.

Quinlan, G. & Beaumont, C. 1982. The deglaciation of Atlantic Canada as reconstructed from the post-glacial relative sea-level record. *Canadian Journal of Earth Sciences* 19: 2232–2246.
Rabassa, J. & Clapperton, C. M. 1990. Quaternary Glaciations of the Southern Andes. *Quaternary Science Reviews* 9: 153–174.
Radtke, U. 1983. Genese und Altersstellung der marinen Terrassen zwischen Civitavecchia und Monte Argentario (Mittelitalien) unter besonderer Berücksichtigung der Elektronenspin-Resonanz-Altersbestimmungsmethode. *Düsseldorfer Geographische Schriften* 22: 179 pp.
Radtke, U. 1989. Marine Terrassen und Korallenriffe – das Problem der quartären Meeresspiegelschwankungen erläutert an Fallstudien aus Chile, Argentinien und Barbados. *Düsseldorfer Geographische Studien* 27: 246 pp.
Radtke, U., Grün, R. & Schwarcz, H. P. 1988. ESR dating of Pleistocene coral reef tracts of Barbados (W.I.). *Quaternary Research* 29: 197–215.
Rähle, W. 1995. Altpleistozäne Molluskenfaunen aus den Hangendschichten der Zusamplattenschotter vom Uhlenberg und Lauterbrunn (Iller–Lech-Platte, Bayerisch Schwaben). *Geologica Bavarica* 99: 103–117.
Rainio, H. 1995. Large ice-marginal formations and deglaciation in southern Finland. In: Ehlers, J., Kozarski, S. & Gibbard, P. L. (eds): *Glacial Deposits in North-East Europe*: 57–66. Rotterdam, Balkema.
Rainio, H. & Kukkonen, M. 1985. Before the Glaciofluvium. *Striae* 22: 9–15.
Rainio, H. & Saarnisto, M. (eds) 1991. Eastern Fennoscandian Younger Dryas end moraines. Field Conference North Karelia, Finland, and Karelian ASSR. *Geological Survey of Finland Guide* 32: 149 pp. Espoo: Geological Survey of Finland.
Rains, B., Shaw, J., Skoye, R., Sjogren, D. & Kvill, D. 1993. Late Wisconsin subglacial megaflood paths in Alberta. *Geology* 21: 323–326.
Ralska-Jasiewiczowa, M., Van Geel, B., Goslar, T. & Kuc, T. 1992. The record of the Late Glacial/Holocene transition in the varved sediments of lake Gosciaz, central Poland. *Sveriges Geologiska Undersökning* Ca 81: 257–268.
Rampino, M. R. & Self, S. 1982. Historic eruptions of Tambora (1815), Krakatau (1883), and Agung (1963), their stratospheric aerosols, and climatic impact. *Quaternary Research* 18: 127–143.
Rampino, M. R. & Self, S. 1992. Volcanic winter and accelerated glaciation following the Toba super-eruption. *Nature* 359: 50–52.
Rampino, M. R. & Self, S. 1993. Climate–volcanism feedback and the Toba Eruption of c.74,000 years ago. *Quaternary Research* 40: 269–280.
Rampton, V. N., Gauthier, R. T., Thibault, J. & Seaman, A. A. 1984. Quaternary geology of New Brunswick. *Geological Survey of Canada, Memoir* 416: 77 pp.
Ramsay, W. 1924. On relations between crustal movements and variations of sea-level during the late Quaternary time, especially in Fennoscandia. *Fennia* 44 (5): 39 pp.
Ramsden, J. & Westgate, J. A. 1971. Evidence for reorientation of a till fabric in the Edmonton Area, Alberta. In: Goldthwait, R. P. (ed.): *Till – A Symposium*: 335–344. Columbus: Ohio State University Press.
Rappol, M. 1983. Glacigenic properties of till – Studies in glacial sedimentology from the Allgäu Alps and The Netherlands. Thesis, Universiteit van Amsterdam. 225 pp.
Rappol, M. 1987. Saalian till in The Netherlands: A review. In: Van der Meer, J. J. M. (ed.): *Tills and Glaciotectonics*: 3–21. Rotterdam: Balkema.
Rappol, M. 1989. Glacial history of northwestern New Brunswick. *Géographie physique et Quaternaire* 43: 191–206.
Rappol, M. & Stoltenberg, H. M. P. 1985. Compositional variability of Saalian till in The Netherlands and its origin. *Boreas* 14: 33–50.
Rappol, M., Haldorsen, S., Jörgensen, P., Van der Meer, J. J. M. & Stoltenberg, H. M. P. 1989. Composition and origin of petrographically-stratified thick till in the northern Netherlands and a Saalian glaciation model for the North Sea basin. *Mededelingen van het Werkgroep voor Tertiaire en Kwartaire Geologie* 26: 31–64.
Raukas, A. 1992. Ice marginal formations of the Palivere zone in the eastern Baltic. *Sveriges Geologiska Undersökning* Ca 81: 277–284.
Raukas, A. & Tavast, E. 1994. Drumlin location as a response to bedrock topography on the southeastern slope of the Fennoscandian Shield. *Sedimentary Geology* 91: 373–382.

Raukas, A., Mickelson, D. M. & Dreimanis, A. 1978. Methods of till investigation in Europe and North America. *Journal of Sedimentary Petrology* 48: 285–294.
Raymo, M. E. 1991. Geochemical evidence supporting T. C. Chamberlin's theory of glaciation. *Geology* 19: 344–347.
Raymo, M. E. & Ruddiman, W. F. 1992. Tectonic forcing of Late Cenozoic climate. *Nature* 359: 117–122.
Raymo, M. E. & Ruddiman, W. F. 1993. Tectonic forcing of Late Cenozoic climate: Reply. *Nature* 361: 124.
Raymo, M. E., Ruddiman, W. F. & Froelich, P. N. 1988. Influence of Late Cenozoic mountain building on ocean geochemical cycles. *Geology* 16: 649–653.
Raymo, M. E., Ruddiman, W. F., Backman, J., Clement, B. M. & Martinson, D. J. 1989. Late Pliocene variation in Northern Hemisphere ice sheets and North Atlantic deep water circulation. *Paleoceanography* 2: 413–446.
Reeh, N. 1982. A plasticity theory approach to the steady-state shape of a three-dimensional ice sheet. *Journal of Glaciology* 28: 431–455.
Reeh, N., Oerter, H., Letréguilly, A., Miller, H. & Hubberten, H.-W. 1991. A new, detailed ice-age oxygen-18 record from the ice-sheet margin in central West Greenland. *Palaeogeography, Palaeoclimatology, Palaeoecology* 90: 373–383.
Reich, H. 1953. Die Vegetationsentwicklung der Interglaziale von Großweil-Ohlstadt und Pfefferbichl im Bayerischen Alpenvorland. *Flora* 140: 386–443.
Reille, M., Guiot, J. & de Beaulieu, J.-L. 1992. The Montaigu Event: an abrupt climatic change during the Early Würm in Europe. In: Kukla, G. H. & Went, E. (eds): *Start of a Glacial*: 85–95. Berlin: Springer.
Reimnitz, E., Kempema, E. W. & Barnes, P. W. 1987. Anchor ice, seabed freezing, and sediment dynamics in shallow arctic seas. *Journal of Geophysical Research* 92: 14671–14678.
Remy, H. 1969. Würmzeitliche Molluskenfaunen aus Lößserien des Rheingaues und des nördlichen Rheinhessens. *Notizblatt des Hessischen Landesamtes für Bodenforschung* 97: 98–116.
Reuter, L. 1925. Die Verbreitung jurassischer Kalkblöcke aus dem Ries im südbayerischen Diluvial-Gebiet. *Jahresberichte und Mitteilungen des Oberrheinischen Geologischen Vereines, N.F.* 14: 191–218.
Rice, R. J. & Douglas, T. 1991. Wolstonian glacial deposits and glaciation in Britain. In: Ehlers, J., Gibbard, P. & Rose, J. (eds): *Glacial Deposits in Great Britain and Ireland*: 25–35. Rotterdam: Balkema.
Richard, P. J. H. 1994. Wisconsinan Late-glacial environmental change in Québec: a regional synthesis. *Journal of Quaternary Science* 9: 165–170.
Richards, D. A. & Smart, P. L. 1991. Potassium–argon and argon–argon dating. In: Smart, P. L. & Frances, P. D. (eds): *Quaternary dating methods – a user's guide. Quaternary Research Association, Technical Guide* 4: 37–44.
Richmond, G. M. 1965. Glaciation of the Rocky Mountains. In: Wright, H. E., Jr. & Frey, D. G. (eds): *The Quaternary of the United States*: 217–230. Princeton: Princeton University Press.
Richmond, G. M. 1986a. Stratigraphy and correlation of glaciations in the Yellowstone National Park. *Quaternary Science Reviews* 5: 83–98.
Richmond, G. M. 1986b. Stratigraphy and correlation of glacial deposits of the Rocky Mountains, the Colorado Plateau and the ranges of the Great Basin. *Quaternary Science Reviews* 5: 99–127.
Richmond, G. M. 1986c. Tentative correlation of deposits of the Cordilleran Ice-Sheet in the Northern Rocky Mountains. *Quaternary Science Reviews* 5: 129–144.
Richmond, G. M. & Fullerton, D. S. 1986. Summation of Quaternary Glaciations in the United States of America. *Quaternary Science Reviews* 5: 183–196.
Richmond, G. M., Fryxell, R., Neff, G. E. & Weis, P. L. 1965. The Cordilleran Ice Sheet of the Northern Rocky Mountains, and related Quaternary history of the Columbia Plateau. In: Wright, H. E., Jr. & Frey, D. G. (eds): *The Quaternary of the United States:* 231–242. Princeton: Princeton University Press.
Richter, K. 1932. Die Bewegungsrichtung des Inlandeises, rekonstruiert aus den Kritzen und Längsachsen der Geschiebe. *Zeitschrift für Geschiebeforschung* 8: 62–66.
Richter, K. 1936a. Gefügestudien im Engebrae, Fondalsbrae und ihren Vorlandsedimenten. *Zeitschrift für Gletscherkunde* XXIV: 22–30.

Richter, K. 1936b. Ergebnisse und Aussichten der Gefügeforschung im Pommerschen Diluvium. *Geologische Rundschau* 32: 196–206.
Richter, W., Schneider, H. & Wager, R. 1951. Die saalezeitliche Stauchzone von Itterbeck-Uelsen (Grafschaft Bentheim). *Zeitschrift der Deutschen Geologischen Gesellschaft* 102: 60–75.
Ricken, W. 1983. Mittel- und jungpleistozäne Lößdecken im südwestlichen Harzvorland. Stratigraphie, Paläopedologie, fazielle Differenzierung und Konnektierung in Flußterrassen. *Catena Supplement* 3: 95–138.
Riggs, S. R., York, L. L., Wehmiller, J. F. & Snyder, S. W. 1992. Depositional patterns resulting from high-frequency Quaternary sea-level fluctuations in northeastern North Carolina. In: Fletcher, K. & Wehmiller, J. F. (eds): *Quaternary Coasts of the United States: Marine and Lacustrine Systems. SEPM Special Publication* 48: 141–153.
Ringberg, B. 1988. Late Weichselian geology of southernmost Sweden. *Boreas* 17: 243–263.
Ringberg, B. 1991. Late Weichselian clay varve chronology and glaciolacustrine environment during deglaciation in southeastern Sweden. *Sveriges Geologiska Undersökning* Ca 79: 42 pp.
Ringberg, B. 1994. *The Swedish Clay Varve Chronology.* Pact 41: 25–34. Strasbourg: Council of Europe.
Ringberg, B. & Miller, U. 1992. Lithology and stratigraphic position of Old Baltic tills in the southernmost part of Sweden. *Sveriges Geologiska Undersökning* Ca 81: 285–292.
Risberg, J., Miller, U. & Brunnberg, L. 1991. Deglaciation, Holocene shore displacement and coastal settlements in Eastern Svealand, Sweden. *Quaternary International* 9: 33–37.
Roberts, M. B., Stringer, C. B. & Parfitt, S. A. 1994. A hominid tibia from Middle Pleistocene sediments at Boxgrove, UK. *Nature* 369: 311–313.
Robertsson, A.-M. & García Ambrosiani, K. 1992. The Pleistocene in Sweden – a review of research, 1960–1990. *Sveriges Geologiska Undersökning* Ca 81: 299–306.
Robertsson, A.-M. & Rodhe, L. 1988. A Late Pleistocene sequence at Seitevare, Swedish Lapland. *Boreas* 17: 501–509.
Rodrigues, C. G. & Vilks, G. 1994. The impact of glacial lake runoff on the Goldthwait and Champlain Seas: The relationship between Glacial Lake Agassiz runoff and the Younger Dryas. *Quaternary Science Reviews* 13: 923–944.
Roe, H. M. 1994. Pleistocene buried valleys in Essex. Ph.D. thesis, University of Cambridge.
Roebroeks, W. & Van Kolfschoten, Th. 1994. The earliest occupation of Europe: a short chronology. *Antiquity* 68: 489–503.
Roebroeks, W., Conard, N. J. & Van Kolfschoten, Th. 1992. Dense forests, cold steppes and the Palaeolithic settlement of Northern Europe. *Current Anthropology* 33: 551–586.
Rögner, K. 1979. Die glaziale und fluvioglaziale Dynamik im östlichen Lechgletscher-Vorland. *Heidelberger Geographische Arbeiten* 49: 67–138.
Rögner, K. 1980. Die pleistozänen Schotter und Moränen zwischen oberem Mindel- und Wertachtal (Bayerisch Schwaben). *Eiszeitalter und Gegenwart* 30: 125–144.
Rögner, K. 1986. Genese und Stratigraphie der ältesten Schotter der südlichen Iller-Lechplatte (Bayerisch-Schwaben). *Eiszeitalter und Gegenwart* 36: 111–119.
Rohdenburg, H. & Semmel, A. 1971. Bemerkungen zur Stratigraphie des Würm-Lösses im westlichen Mitteleuropa. *Notizblätter des Hessischen Landesamtes für Bodenforschung* 99: 246–252.
Romanovsky, N. H. 1985. Distribution of recently active ice and soil wedges in the U.S.S.R. In: Church, M. & Slaymaker, S. (eds): *Field and Theory: Lectures in geocryology*: 154–165. University of British Columbia.
Romanovsky, N. N. 1978. Cryogenic (permafrost) geologic processes and phenomena (in Russian). In: Kudryavtsev, V. A. (ed.): *Permafrost Studies*: 231–282. Moscow: Moscow University Publication.
Rónai, A. 1985. Limnic and terrestrial sedimentation and the N/Q boundary in the Carpathian Basin. In: Kretzoi, M. & Pécsi, M. (eds): *Problems of the Neogene and Quaternary in the Carpathian Basin, Geological and Geomorphological Studies. Studies in Geography in Hungary* 19: 21–49.
Rose, J. 1974. Small-scale spatial variability of some sedimentary properties of lodgement till and slumped till. *Proceedings of the Geologists' Association* 85: 223–237.
Rose, J. 1985. The Dimlington Stadial/Dimlington Chronozone: a proposal for naming the main glacial episode of the Late Devensian in Britain. *Boreas* 14: 225–230.
Rose, J. 1987. The status of the Wolstonian glaciation in the British Quaternary. *Quaternary Newsletter* 53: 1–9.

Rose, J. 1988. Stratigraphic nomenclature for the British Middle Pleistocene – procedural dogma or stratigraphic common sense? *Quaternary Newsletter* 54: 15–20.
Rose, J. 1994. Major river systems of central and southern Britain during the Early and Middle Pleistocene. *Terra Nova* 6: 435–443.
Rose, J. & Allen, P. 1977. Middle Pleistocene stratigraphy in south-east Suffolk. *Quarterly Journal of the Geological Society of London* 133: 83–102.
Rose, J., Allen, P. & Hey, R. W. 1976. Middle Pleistocene stratigraphy in southern East Anglia. *Nature* 263: 492–494.
Rose, J., Allen, P., Kemp, R. A., Whiteman, C. A. & Owen, N. 1985a. The early Anglian Barham Soil in southern East Anglia. In: Boardman, J. (ed.): *Soils and Quaternary Landscape Evolution*: 197–229. Chichester: Wiley.
Rose, J., Boardman, J., Kemp, R. A. & Whiteman, C. A. 1985b. Palaeosols and the interpretation of the British Quaternary stratigraphy. In: Richards, K. S., Arnett, R. R. & Ellis, S. (eds): *Geomorphology and Soils*: 348–375. London: Allen & Unwin.
Rosholt, J. N., Buh, C. A., Shroba, R. R., Pierce, K. L. & Richmond, G. M. 1985. Uranium-trend dating and calibrations for Quaternary sediments. *U.S. Geological Survey, Open-File Report* 85–299: 48 pp.
Röthlisberger, F. 1986. *10 000 Jahre Gletschergeschichte der Erde*. Aarau: Sauerländer. 416 pp.
Röthlisberger, H. 1972. Water pressure in intra- and subglacial channels. *Journal of Glaciology* 11: 177–203.
Röthlisberger, H. & Lang, H. 1987. Glacial hydrology. In: Gurnell, A. M. & Clark, M. J. (eds): *Glacio-fluvial Sediment Transfer*: 207–284. John Wiley.
Rothpletz, A. 1917. Die Osterseen und der Isar-Vorlandgletscher. *Mitteilungen der Geographischen Gesellschaft in München* 12: 99–314.
Rotnicki, K. 1974. Slope development of Riss Glaciation end moraines during the Würm; its morphological and geological consequences. *Quaestiones Geographicae* 1: 109–139.
Rotnicki, K. 1975. Stratigraphic evidences of the survival of Riss glaciotectonic structures and forms in the marginal zone of the maximum extent of the last glaciation. *Quaestiones Geographicae* 2: 113–137.
Rousseau, D.-D. 1991. Climatic transfer function from Quaternary molluscs in European loess deposits. *Quaternary Research* 36: 195–209.
Rousseau, D.-D. & Puisségur, J.-J. 1990. A 350,000-year climatic record from the loess sequence of Achenheim, Alsace, France. *Boreas* 19: 203–216.
Rousseau, D.-D., Limondin, N., Magnin, F. & Puisségur, J.-J. 1994. Temperature oscillations over the last 10,000 years in western Europe estimated from terrestrial mollusc assemblages. *Boreas* 23: 66–73.
Różycki, S. Z. 1965. Die stratigraphische Stellung des Warthe-Stadiums in Polen. *Eiszeitalter und Gegenwart* 16: 189–201.
Ruddiman, W. F. & Kutzbach, J. E. 1990. Late Cenozoic plateau uplift and climate change. *Transactions of the Royal Society of Edinburgh: Earth Sciences* 81: 301–314.
Ruddiman, W. F. & Raymo, M. E. 1988. Northern hemisphere climate régimes during the past 3 Ma: possible tectonic connections. *Philosophical Transactions of the Royal Society of London* B 318: 411–430.
Ruddiman, W. F. & Wright, H. E., Jr. (eds) 1987. *North America and Adjacent Oceans during the Last Deglaciation. Geological Society of America. The Geology of North America* K-3: 501 pp.
Ruegg, G. H. J. 1983. Periglacial eolian evenly laminated sandy deposits in the Late Pleistocene of N.W. Europe, a facies unrecorded in modern sedimentological handbooks. In: Brookfield, M. E. & Ahlbrandt, T. S. (eds). *Eolian sediments and processes*: 455–482. Amsterdam: Elsevier.
Ruegg, G. H. J. & Zandstra, J. G. 1977. Pliozäne und pleistozäne gestauchte Ablagerungen bei Emmerschans (Drenthe, Niederlande). *Mededelingen Rijks Geologische Dienst N.S.* 28 (4): 65–99.
Rühberg, N. 1987. Die Grundmoräne des jüngsten Weichselvorstoßes im Gebiet der DDR. *Zeitschrift für Geologische Wissenschaften* 15: 759–767.
Ruhe, R. V. 1976. Stratigraphy of mid-continent loess, U.S.A. In: Mahaney, W. C. (ed.): *The Quaternary Stratigraphy of North America*: 197–211. Stroudsburg, PA: Dowden, Hutchinson, and Ross Inc.
Ruhe, R. V. 1983. Depositional environment of late Wisconsin loess in the midcontinental United States. In: Wright, H. E., Jr. (ed.): *Quaternary Environments of the United States, Vol. 1, The Late Pleistocene:* 130–136. Minneapolis: University of Minnesota Press.

Russell, R. J. 1944. Lower Mississippi Valley loess. *Geological Society of America Bulletin* 55: 1–40.
Rust, A. 1958. *Die jungpaläolithischen Zeltanlagen von Ahrensburg. Offa-Bücher* 15: 146 pp. + Appendix.
Ruszczynska-Szenajch, H. 1982. Depositional processes of Pleistocene lowland end moraines, and their possible relation to climatic conditions. *Boreas* 11: 249–260.
Rutledge, E. M., West, L. T. & Omakupt, M. 1985. Loess deposits on a Pleistocene age terrace in eastern Arkansas. *Soil Science Society of America Journal* 49: 1231–1238.
Rutter, N., Ding Z. L., Evans, M. E. & Liu, T. S. 1991. Baoji-type pedostratigraphic section, Loess Plateau, North-Central China. *Quaternary Science Reviews* 10: 1–22.
Rutter, N. W. 1984. Pleistocene history of the western Canadian ice-free corridor. In: Fulton, R. J. (ed.) 1984. *Quaternary Stratigraphy of Canada – A Canadian Contribution to IGCP Project 24. Geological Survey of Canada Paper* 84-10: 49–56.
Ruuhijärvi, R. 1960. Über die regionale Einteilung der nordfinnischen Moore. *Annales Botanici Societas Zoologicæ Botanicæ Fennicæ 'Vanamo'* 31 (1): 360 pp.
Ruuhijärvi, R. 1983. The Finnish mire types and their regional distribution. In: Gore, A. J. P. (ed.): *Mires: Swamp, Bog, Fen and Moor, B. Regional Studies*: 47–67. Amsterdam: Elsevier.
Rzechowski, J. 1986. Pleistocene till stratigraphy in Poland. *Quaternary Science Reviews* 5: 365–372.
Saarnisto, M. 1990. An outline of glacial indicator tracing. In: Kujansuu, R. & Saarnisto, M. (eds): *Glacial indicator tracing*: 1–13. Rotterdam: Balkema.
Saarnisto, M. & Salonen, V.-P. 1995. Glacial history of Finland. In: Ehlers, J., Kozarski, S. & Gibbard, P. L. (eds): *Glacial Deposits in North-East Europe*: 3–10. Rotterdam: Balkema.
Salisbury, R. D. 1888. Terminal moraines in North Germany. *American Journal of Science* 35: 401–407.
Salisbury, R. D. 1892. A preliminary paper on Drift or Pleistocene Formations of New Jersey. *Annual Report of the Geological Survey of New Jersey, 1891*: 35–108.
Salisbury, R. D. 1894. Surface geology – Report of progress, 1892. *New Jersey Geological Survey, Annual Report of the State Geologist for the year 1893*: 35–328.
Salonen, V.-P. 1986. Glacial transport distance distributions of surface boulders in Finland. *Geological Survey of Finland, Bulletin* 338: 57 pp. + Appendix.
Salonen, V.-P. 1987. Observations on boulder transport in Finland. *Geological Survey of Finland, Special Paper* 3: 103–110.
Salonen, V.-P. 1991. Glacial dispersal of Jotnian sandstone fragments in southwestern Finland. *Geological Survey of Finland, Special Paper* 12: 127–130.
Saltzman, B. & Maasch, K. A. 1990. A first-order global model of late Cenozoic climatic change. *Transactions of the Royal Society of Edinburgh: Earth Sciences* 81: 315–325.
Salvador, A. (ed.) 1993. *International Stratigraphic Guide. A Guide to Stratigraphic Classification, Terminology and Procedure*, 2nd edition. The International Union of Geological Sciences and The Geological Society of America, Inc.: 214 pp.
Saucier, R. T. 1974. Quaternary geology of the Lower Mississippi valley. *Arkansas Archaeological Survey Research Series* 6: 26 pp.
Saucier, R. T. 1978. Sand dunes and related eolian features of the Lower Mississippi River alluvial valley. *Geoscience and Man* 19: 23–40.
Saucier, R. T. 1981. Current thinking on riverine processes and geologic history as related to human settlement in the southeast. *Geoscience and Man* 22: 7–18.
Saucier, R. T. 1985. Fluvial response to Late Quaternary climatic change in the Lower Mississippi Valley. *Geological Society of America Abstracts with Programs* 17: 190.
Saucier, R. T. 1987. Geomorphological interpretation of Late Quaternary terraces in western Tennessee and their regional tectonic implications. *US Geological Survey Professional Paper* 1336 A: 19 pp.
Sauramo, M. 1918. Geochronologische Studien über die spätglaziale Zeit in Südfinnland. *Bulletin de la Commission Géologique de Finlande* 50: 44 pp.
Sauramo, M. 1923. Studies on the Quaternary varve sediments in southern Finland. *Bulletin de la Commission Géologique de Finlande* 60: 164 pp.
Sauramo, M. 1929. The Quaternary geology of Finland. *Bulletin de la Commission Géologique de Finlande* 86: 110 pp.
Sauramo, M. 1958. Die Geschichte der Ostsee. *Annales Academiae Scientiarum Fennicae* A III, 51: 522 pp.

Sæmundsson, K. & Noll, H. 1974. K/Ar ages of rocks from Húsafell, Western Iceland, and the Development of the Húsafell Central Volcano. *Jökull* 24: 40–59.
Schädel, K. 1952. Die Stratigraphie des Altdiluviums im Rheingletschergebiet. *Mitteilungen des Oberrheinischen Geologischen Vereins N.F.* 34: 1–20.
Schädel, K. & Werner, J. 1965. Untersuchungen zur Aufdeckung glazial verfüllter Täler im Donaugebiet von Sigmaringen–Riedlingen. *Jahreshefte des Geologischen Landesamts Baden-Württemberg* 7: 387–422.
Schaefer, I. 1951. Über methodische Fragen der Eiszeitforschung im Alpenvorland. *Zeitschrift der Deutschen Geologischen Gesellschaft* 102: 287–310.
Schaefer, I. 1953. Die donaueiszeitlichen Ablagerungen an Lech und Wertach. *Geologica Bavarica* 19: 13–64.
Schaefer, I. 1956. Sur la division du Quaternaire dans l'avant-pays des Alpes en Allemagne. *Actes IV Congres INQUA, Rome/Pise 1953*, Vol. 2: 910–914.
Schaefer, I. 1973. Das Grönenbacher Feld. Ein Beispiel für Wandel und Fortschritt der Eiszeitforschung seit Albrecht Penck. *Eiszeitalter und Gegenwart* 23/24: 168–200.
Schaefer, I. 1975. Die Altmoränen des diluvialen Isar-Loisachgletschers und ihr Verständnis aus der Kenntnis der Paareiszeit. *Mitteilungen der Geographischen Gesellschaft in München* 60: 115–153.
Schaefer, I. 1980. Der angebliche 'altpleistozäne Donaulauf' im schwäbischen Alpenvorland. *Jahresberichte und Mitteilungen des Oberrheinischen Geologischen Vereines N.F.* 62: 167–198.
Schafer, J. P. 1949. Some periglacial features in Central Montana. *Journal of Geology* 57: 154–174.
Schedler, J. 1979. Neue pollenanalytische Untersuchungen am Schieferkohlevorkommen des Uhlenberges bei Dinkelscherben (Schwaben). *Geologica Bavarica* 80: 165–182.
Scheffer, F. & Schachtschabel, P. (eds) 1989. *Lehrbuch der Bodenkunde,* 12. Auflage. Stuttgart: Enke. 491 pp.
Schellmann, G. 1990. Fluviale Geomorphodynamik im jüngeren Quartär des unteren Isar- und angrenzenden Donautales. *Düsseldorfer Geographische Schriften* 29: 131 pp.
Schenck, H. G. 1945. Geologic application of biometrical analysis of molluscan assemblages. *Journal of Paleontology* 19: 504–521.
Scheuenpflug, L. 1970. Weißjurablöcke und -gerölle der Alb in pleistozänen Schottern der Zusamplatte (Bayerisch Schwaben). *Geologica Bavarica* 63: 177–194.
Scheuenpflug, L. 1976. Erste Hinweise auf eine pliozäne Donau in der östlichen Iller–Lech-Platte (Bayerisch Schwaben). Vorläufige Mitteilung. *Eiszeitalter und Gegenwart* 27: 26–29.
Scheuenpflug, L. 1979. Der Uhlenberg in der östlichen Iller–Lech-Platte (Bayerisch Schwaben). *Geologica Bavarica* 80: 159–164.
Schilling, W. & Wiefel, H. 1962. Jungpleistozäne Periglazialbildungen und ihre regionale Differenzierung in einigen Teilen Thüringens und des Harzes. *Geologie* 11: 428–460.
Schindler, C. M., Fisch, W., Streiff, P., Ammann, B. & Tobolski, K. 1985. Vorbelastete Seeablagerungen und Schieferkohlen südlich des Walensees. Untersuchungen während des Baus der Nationalstr. N3. *Eclogae Geologicae Helvetiae* 78: 176–196.
Schirmer, W. 1979. Das Quartär des Regnitztales. In: *Geologische Karte von Bayern 1:25 000, Erläuterungen zu Blatt Nr. 6132 Buttenheim:* 81–89. München: Bayerisches Geologisches Landesamt.
Schirmer, W. 1981. Abflußverhalten des Mains im Jungquartär. *Geologisches Institut der Universität zu Köln, Sonderverölichungen* 41: 197–208.
Schirmer, W. (ed.) 1983a. Holozäne Talentwicklung – Methoden und Ergebnisse. *Geologisches Jahrbuch* A 71: 370 pp.
Schirmer, W. 1983b. Die Talentwicklung an Main und Regnitz seit dem Hochwürm. *Geologisches Jahrbuch* A 71: 11–43.
Schirmer, W. 1983c. Symposium 'Franken': Ergebnisse zur holozänen Talentwicklung und Ausblick. *Geologisches Jahrbuch* A 71: 355–370.
Schirmer, W. 1983d. Criteria for the differentiation of Late Quaternary river terraces. *Quaternary Studies in Poland* 4: 199–205.
Schirmer, W. (ed.) 1990. Rheingeschichte zwischen Mosel und Maas. *Deuqua-Führer* 1: 295 pp.
Schirmer, W. 1991. Breaks within the Late Quaternary river development of Middle Europe. *Aardkundige Mededelingen* 6: 115–120.
Schirmer, W. 1993. Der menschliche Eingriff in den Talhaushalt. *Kölner Jahrbuch* 26: 577–584.

Schlüchter, Ch. 1976. Geologische Untersuchungen im Quartär des Aaretals südlich von Bern (Stratigraphie, Sedimentologie, Paläontologie). *Beiträge zur Geologischen Karte der Schweiz N.F.* 148: 117 pp.

Schlüchter, Ch. 1977. *Grundmoräne* versus *Schlammoräne* – two types of lodgement till in the Alpine Foreland of Switzerland. *Boreas* 6: 181–188.

Schlüchter, Ch. 1980a. Die fazielle Gliederung der Sedimente eines Ufermoränenkomplexes – Form und Inhalt. *Verhandlungen des Naturwissenschaftlichen Vereins in Hamburg (NF)* 23: 101–117.

Schlüchter, Ch. 1980b. Bemerkungen zu einigen Grundmoränenvorkommen in den Schweizer Alpen. *Zeitschrift für Gletscherkunde und Glazialgeologie* 16 (2): 203–212.

Schlüchter, Ch. 1981a. Remarks on the Pleistocene morphogenetic evolution of the Swiss Plain. *Zeitschrift für Geomorphologie N.F., Supplement-Band* 40: 61–66.

Schlüchter, Ch. 1981b. Bemerkungen zur Korngrößenanalyse der Lockergesteine, insbesondere der Moränen, in der Schweiz. *Verhandlungen des Naturwissenschaftlichen Vereins in Hamburg (NF)* 24 (2): 155–160.

Schlüchter, Ch. 1982. Die lithostratigraphische Gliederung der Ablagerungen seit der letzten Zwischeneiszeit. *Geographica Helvetica* 37 (2): 109–115.

Schlüchter, Ch. 1986. The Quaternary Glaciations of Switzerland, with special reference to the northern Alpine Foreland. *Quaternary Science Reviews* 5: 413–419.

Schlüchter, Ch. 1988. Exkursion vom 11.Oktober 1987 der Schweizerischen Geologischen Gesellschaft im Rahmen der SNG-Jahrestagung in Luzern: Ein eiszeitgeologischer Überblick von Luzern zum Rhein – unter besonderer Berücksichtigung der Deckenschotter. *Eclogae Geologicae Helvetiae* 81: 249–258.

Schlüchter, Ch. 1989a. A non-classical summary of the Quaternary stratigraphy in the northern Alpine Foreland of Switzerland. *Bulletin de la Société neuchâteloise de Géographie* 32–33: 143–157.

Schlüchter, Ch. 1989b. The most complete Quaternary record of the Swiss Alpine Foreland. *Palaeogeography, Palaeoclimatology, Palaeoecology* 72: 141–146.

Schlüchter, Ch., Maisch, M., Suter, J., Fitze, P., Keller, W. A., Burga, C. A. & Wynistorf, E. 1987. Das Schieferkohlen-Profil von Gossau (Kanton Zürich) und seine stratigraphische Stellung innerhalb der letzten Eiszeit. *Vierteljahrsschrift der Naturforschenden Gesellschaft in Zürich* 132: 135–174.

Schneider, A. F. 1983. Wisconsinan stratigraphy and glacial sequence in southwestern Wisconsin. In: Mickelson, D. M. & Clayton, L. (eds): *Late Pleistocene History of Southeastern Wisconsin. Geoscience Wisconsin* 7: 59–85.

Schneider, A. F. & Need, E. A. 1985. Lake Milwaukee: an 'early' proglacial lake in the Lake Michigan basin. In: Karrow, P. F. & Calkin, P. E. (eds): *Quaternary Evolution of the Great Lakes. Geological Association of Canada Special Paper* 30: 55–62.

Schneider, H. 1976. Über junge Krustenbewegungen in der voralpinen Landschaft zwischen dem südlichen Rheingraben und dem Bodensee. *Mitteilungen der Naturforschenden Gesellschaft Schaffhausen* XXX: 4–99.

Schönhals, E. 1951. Über fossile Böden im nicht vereisten Gebiet. *Eiszeitalter und Gegenwart* 1: 109–130.

Schönhals, E. 1959. Der Basalttuff von Kärlich als Leithorizont des Würm-Hochglazials. *Fortschritte in der Geologie von Rheinland und Westfalen* 4: 313–322.

Schönhals, E., Rohdenburg, H. & Semmel, A. 1964. Ergebnisse neuerer Untersuchungen zur Würmlöß-Gliederung in Hessen. *Eiszeitalter und Gegenwart* 15: 199–206.

Schott, W. 1935. Die Foraminiferen in dem äquatorialen Teil des Atlantischen Ozeans. *Deutsche Atlantik-Expedition des Forschungsschiffes 'Meteor' 1925–1927, Wissenschaftliche Ergebnisse* 3: 43–134.

Schreiber, U. 1985. Das Lechtal zwischen Schongau und Rain im Hoch-, Spät- und Postglazial. *Geologisches Institut der Universität zu Köln, Sonderveröffentlichungen* 58: 192 pp.

Schreiber, U. & Müller, D. 1991. Mittel- und jungpleistozäne Ablagerungen zwischen Landsberg und Augsburg (Lech). *Sonderveröffentlichungen Geologisches Institut der Universität zu Köln* 82: 265–282.

Schreiner, A. 1974. *Erläuterungen zur Geologischen Karte des Landkreises Konstanz und Umgebung 1:50 000. 2., berichtigte Auflage.* Stuttgart: Geologisches Landesamt Baden-Württemberg. 286 pp.

Schreiner, A. 1981. Quartär. In: Groschopf, R., Kessler, G., Leiber, J., Maus, H., Ohmert, W., Schreiner, A. & Wimmenauer, W. (eds): *Erläuterungen zur Geologischen Karte 1:50 000 Freiburg i.Br. und Umgebung:* Stuttgart. 354 pp.

Schreiner, A. 1989. Zur Stratigraphie der Rißeiszeit im östlichen Rheingletschergebiet (Baden-Württemberg). *Jahreshefte des Geologischen Landesamts Baden-Württemberg* 31: 183–196.

Schreiner, A. 1992. *Einführung in die Quartärgeologie.* Stuttgart: Schweizerbart. 257 pp.

Schreiner, A. & Ebel, R. 1981. Quartärgeologische Untersuchungen in der Umgebung von Interglazialvorkommen im östlichen Rheingletschergebiet (Baden-Württemberg). *Geologisches Jahrbuch A* 59: 3–64.

Schreiner, A. & Haag, Th. 1982. Zur Gliederung der Rißeiszeit im östlichen Rheingletschergebiet (Baden-Württemberg). *Eiszeitalter und Gegenwart* 32: 137–161.

Schumacher, R. 1981. Untersuchungen zur Entwicklung des Gewässernetzes seit dem Würmmaximum im Bereich des Isar-Loisach-Vorlandgletschers. Dissertation, Universität München. 204 pp.

Schumm, S. A. 1965. Quaternary paleohydrology. In: Wright, H. E., Jr. & Frey, D. G. (eds): *The Quaternary of the United States*: 783–794. Princeton: Princeton University Press.

Schunke, E. 1975. Die Periglazialerscheinungen Islands in Abhängigkeit von Klima und Substrat. *Abhandlungen der Göttinger Akademie der Wissenschaften, Mathematisch-Physikalische Klasse*, 3. Folge 30: 273 pp.

Schwan, J. 1988. Sedimentology of coversands in northwestern Europe. Thesis, Vrije Universiteit te Amsterdam: 137 pp.

Schwarcz, H. P. 1994. Current challenges to ESR dating. *Quaternary Science Reviews* 13: 601–605.

Schwarcz, H. P., Grün, R., Latham, A. G., Mania, D. & Brunnacker, K. 1988. The Bilzingsleben Archaeological Site. New dating evidence. *Archaeometry* 30: 5–17.

Schwarzbach, M. 1974. *Das Klima der Vorzeit,* 3.Auflage. Enke: Stuttgart. 380 pp.

Schytt, V., Hoppe, G., Blake, W., Jr. & Grosswald, M. G. 1968. The extent of the Wurm glaciation in the European Arctic. *International Association of Scientific Hydrology, Publication* 79: 207–216.

Scourse, J. (ed.) 1986. *The Isles of Scilly. Field Guide.* Coventry: Quaternary Research Association. 151 pp.

Scourse, J. 1991. Glacial deposits of the Isles of Scilly. In: Ehlers, J., Gibbard, P. & Rose, J. (eds): *Glacial Deposits in Great Britain and Ireland:* 291–300. Rotterdam: Balkema.

Sefström, N. G. 1836. Ueber die Spuren einer sehr großen urweltlichen Fluth. *Poggendorfs Annalen der Physik und Chemie* 38: 614–618.

Sefström, N. G. 1838. Untersuchung über die auf den Felsen Skandinaviens in bestimmter Richtung vorhandenen Furchen und deren wahrscheinliche Entstehung. *Poggendorfs Annalen der Physik und Chemie* 43: 533–567.

Seibold, E. & Berger, W. H. 1993. *The Sea Floor. An Introduction to Marine Geology*, 2nd edition. Berlin: Springer. 356 pp.

Seidenkrantz, M.-S. 1993. Benthic foraminiferal and stable isotope evidence for a 'Younger Dryas-style' cold spell at the Saalian–Eemian transition, Denmark. *Palaeogeography, Palaeoclimatology, Palaeoecology* 102: 103–120.

Seifert, G. 1952. Gletscherschrammen auf Fehmarn (Schleswig-Holstein). *Die Naturwissenschaften* 39: 551.

Seifert, G. 1954. Das mikroskopische Korngefüge des Geschiebemergels als Abbild der Eisbewegung, zugleich Geschichte des Eisabbaus in Fehmarn, Ostwagrien und dem Dänischen Wohld. *Meyniana* 2: 124–190.

Seiler, K.-P. 1979. Glazial übertiefte Talabschnitte in den Bayerischen Alpen. Ergebnisse glazialgeologischer, hydrogeologischer und geophysikalischer Untersuchungen. *Eiszeitalter und Gegenwart* 29: 35–48.

Sejrup, H. P., Aarseth, I., Ellingsen, K. L., Reither, E., Jansen, E., Løvlie, R., Bent, A., Brigham-Grette, J., Larsen, E. & Stoker, M. 1987. Quaternary stratigraphy of the Fladen area, central North Sea: a multidisciplinary study. *Journal of Quaternary Science* 2: 35–58.

Sejrup, H. P. 1987. Molluscan and foraminiferal biostratigraphy of an Eemian–Early Welchselian section at Karmøy, southwestern Norway. *Boreas* 16: 27–42.

Sejrup, H. P., Haflidason, H., Aarseth, I., King, E., Forsberg, C. F., Long, D. & Rokoengen, K. 1994. Late Weichselian glaciation history of the northern North Sea. *Boreas* 23: 1–13.

Sejrup, H. P., Aarseth, I., Haflidason, H., Løvlie, R., Bratten, Å, Tjøstheim, G., Forsberg, C. F. & Ellingsen, K. L. 1995. Quaternary of the Norwegian Channel; paleoceanography and glaciation history. *Norsk Geologisk Tidsskrift* 75: 65–87.

Selle, W. 1962. Geologische und vegetationskundliche Untersuchungen an einigen wichtigen Vorkommen des letzten Interglazials in Nordwestdeutschland. *Geologisches Jahrbuch* 79: 295–352.
Semmel, A. 1964. Junge Schuttdecken in hessischen Mittelgebirgen. *Notizblatt des Hessischen Landesamtes für Bodenforschung* 92: 275–285.
Semmel, A. 1967. Über Prä-Würm-Lösse in Hessen. *Notizblatt des Hessischen Landesamtes für Bodenforschung* 95: 239–241.
Semmel, A. 1968. Studien über den Verlauf jungpleistozäner Formung in Hessen. *Frankfurter Geographische Hefte* 45: 133 pp.
Semmel, A. 1969a. Verwitterungs- und Abtragungserscheinungen in rezenten Periglazialgebieten (Lappland und Spitzbergen). *Würzburger Geographische Arbeiten* 26: 82 pp.
Semmel, A. 1969b. Quartär. In: Kümmerle, E. & Semmel, A.: *Geologische Karte von Hessen 1:25 000, Erläuterungen zu Blatt 5916 Hochheim a.M.*, 3.Auflage: 51–99. Wiesbaden: Hessisches Landesamt für Bodenforschung.
Semmel, A. 1972. Untersuchungen zur jungpleistozänen Talentwicklung in deutschen Mittelgebirgen. *Zeitschrift für Geomorphologie N.F., Supplement-Band* 14: 105–112.
Semmel, A. 1990. Periglaziale Formen und Sedimente. In: Liedtke, H. (ed.): *Eiszeitforschung*: 250–260. Darmstadt: Wissenschaftliche Buchgemeinschaft.
Semmel, A. & Fromm, K. 1976. Ergebnisse paläomagnetischer Untersuchungen an quartären Sedimenten des Rhein–Main–Gebiets. *Eiszeitalter und Gegenwart* 27: 18–25.
Senftl, E. & Exner, Ch. 1973. Rezente Hebung der Hohen Tauern und geologische Interpretation. *Verhandlungen der Geologischen Bundesanstalt* 1973: 209–234.
Seppälä, M. 1971. Evolution of eolian relief of the Kaamasjoki–Kiellajoki river basin in Finnish Lapland. *Fennia* 104: 88 pp.
Seppälä, M. 1976. Seasonal thawing of a palsa at Enontekiö, Finnish-Lapland in 1974. *Biuletyn Peryglacjalny* 26: 17–24.
Seppälä, M. 1988, Palsas and related forms. In: Clark, M. J. (ed.): *Advances in Periglacial Geomorphology*: 247–277. Wiley.
Sernander, R. 1894. Studier öfver den Gotländska vegetationens utvecklingshistoria. Dissertation, Uppsala. 112 pp.
Shackleton, N. J. 1967. Oxygen isotope analyses and Pleistocene temperatures re-assessed. *Nature* 215: 15–17.
Shackleton, N. J. 1977. Oxygen isotope stratigraphy of the Middle Pleistocene. In: Shotton, F. W. (ed.): *British Quaternary Studies, Recent Advances*: 1–16. Oxford: Clarendon Press.
Shackleton, N. J. 1987. Oxygen isotopes, ice volume and sea level. *Quaternary Science Reviews* 6: 183–190.
Shackleton, N. J. 1989. The Plio-Pleistocene ocean: Stable isotope history. In: Rose, J. & Schlüchter, Ch. (eds): *Quaternary Type Sections: Imagination or Reality?*: 11–24. Rotterdam: Balkema.
Shackleton, N. J. & Opdyke, N. D. 1973. Oxygen isotope and palaeomagnetic stratigraphy of equatorial Pacific core V28-238: oxygen isotope temperatures and ice volumes on a 10^5 and 10^6 year scale. *Quaternary Research* 3: 39–55.
Shackleton, N. J. & Opdyke, N. D. 1976. Oxygen isotope and palaeomagnetic stratigraphy of equatorial Pacific core V28-239, late Pliocene to latest Pleistocene. In: Cline, R. M. & Hays, R. D. (eds): *Investigation of Late Quaternary Paleoceanography and Paleoclimatology. Memoirs of the Geological Society of America* 145: 449–464.
Shackleton, N. J., Berger, A. & Peltier, W. A. 1990. An alternative astronomical calibration of the lower Pleistocene timescale based on ODP Site 677. *Transactions of the Royal Society of Edinburgh: Earth Sciences* 81: 251–261.
Shackleton, N. J., An, Z., Dodonov, A. E., Gavin, J., Kukla, G. J., Ranov, V. A. & Zhou, L. P. 1995. Accumulation rate of loess in Tadjikistan and China: Relationship with global ice volume cycles. *Quaternary Proceedings* 4: 1–6.
Sharp, M., Campbell Gemmell, J. & Tison, J.-L. 1989. Structure and stability of the former subglacial drainage system of the Glacier de Tsanfleuron, Switzerland. *Earth Surface Processes and Landforms* 14: 119–134.
Sharp, R. P. 1988. *Living Ice. Understanding Glaciers and Glaciation.* Cambridge: Cambridge University Press. 225 pp.

Sharpe, D. R. 1987. The stratified nature of drumlins from Victoria Island and southern Ontario. In: Menzies, J. & Rose, J. (eds): *Drumlin Symposium:* 195–213. Rotterdam: Balkema.
Sharpe, D. R. & Barnett, P. J. 1985. Significance of sedimentological studies in the Wisconsinan stratigraphy of southern Ontario. *Géographie physique et Quaternaire* 39: 255–273.
Shaw, J. 1972. Sedimentation in the ice-contact environment, with examples from Shropshire, England. *Sedimentology* 18: 23–62.
Shaw, J. 1977. Sedimentation in an Alpine lake during deglaciation. Okanagan Valley, British Columbia, Canada. *Geografiska Annaler* 59 A: 221–240.
Shaw, J. 1983. Drumlin formation related to inverted erosion marks. *Journal of Glaciology* 29: 461–479.
Shaw, J. 1989a. Drumlins, subglacial meltwater floods, and ocean responses. *Geology* 17: 853–856.
Shaw, J. 1989b. Sublimation till. In: Goldthwait, R. P. & Matsch, C. L. (eds): *Genetic Classification of Glacigenic Deposits*: 141–142. Rotterdam: Balkema.
Shaw, J. & Gilbert, R. 1990. Evidence for large-scale subglacial meltwater flood events in southern Ontario and northern New York State. *Geology* 18: 1169–1172.
Shaw, J. & Kvill, D. 1984. A glaciofluvial origin for drumlins of the Livingston Lake area, Saskatchewan. *Canadian Journal of Earth Sciences* 21: 1442–1459.
Shen, C., Beer, J., Liu, T., Oeschger, H., Bonani, G., Suter, M. & Wölfli, W. 1992. ^{10}Be in Chinese loess. *Earth and Planetary Science Letters* 109: 169–177.
Shennan, I. 1987. Global analysis and correlation of sea-level data. In: Devoy, R. J. N. (ed.): *Sea Surface Studies*: 198–230. London: Croom Helm.
Shennan, I. 1989. Holocene crustal movements and sea-level changes in Great Britain. *Journal of Quaternary Science* 4: 77–89.
Shilts, W. W. 1975. Principles of geochemical exploration for sulphide deposits using shallow samples of glacial drift. *The Canadian Mining and Metallurgical Bulletin* 68: 73–80.
Shilts, W. W. 1976. Glacial till and mineral exploration. In: Legget, R. F. (ed.): *Glacial Till. Royal Society of Canada, Special Publication* 12: 205–223.
Shilts, W. W. 1980. Flow patterns in the central North American ice sheet. *Nature* 286: 213–218.
Shilts, W. W. 1984a. Quaternary events, Hudson Bay Lowland and southern District of Keewatin. In: Fulton, R. J. (ed.): *Quaternary Stratigraphy of Canada – A Canadian Contribution to IGCP Project 24. Geological Survey of Canada Paper* 84–10: 117–126.
Shilts, W. W. 1984b. Till geochemistry in Finland and Canada. *Journal of Geochemical Exploration* 21: 95–117.
Shilts, W. W. & Kettles, I. M. 1990. Geochemical–mineralogical profiles through fresh and weathered till. In: Kujansuu, R. & Saarnisto, M. (eds): *Glacial Indicator Tracing*: 187–216. Rotterdam: Balkema.
Shilts, W. W. & McDonald, B. C. 1975. Dispersal of clasts and trace elements in the Windsor Esker, southern Quebec. *Geological Survey of Canada, Paper* 75–1, Part A: 495–499.
Shilts, W. W., Cunningham, C. M. & Kascycki, C. A. 1979. Keewatin Ice Sheet – re-evaluation of the traditional concept of the Laurentide Ice Sheet. *Geology* 7: 537–541.
Shilts, W. W., Aylsworth, J. M., Kaszycki, C. A. & Klasen, R. A. 1987. Canadian Shield. In: Graf, W. L. (ed.): *Geomorphic Systems of North America. The Geological Society of America, Centennial Special Volume* 2: 119–161.
Shimek, B. 1909. Aftonian sands and gravels in western Iowa. *Geological Society of America Bulletin* 20: 399–408.
Shimek, B. 1910. Geology of Harrison and Mona Counties. *Iowa Geological Survey* 20: 271–485.
Shotton, F. W. 1953. Pleistocene deposits of the area between Coventry, Rugby and Leamington and their bearing on the topographic development of the Midlands. *Philosophical Transactions of the Royal Society of London* B 237: 209–260.
Shotton, F. W. 1983. The Wolstonian Stage of the British Pleistocene in and around its type area of the English Midlands. *Quaternary Science Reviews* 2: 261–280.
Shotton, F. W. 1986. Glaciations in the United Kingdom. *Quaternary Science Reviews* 5: 293–297.
Shreve, R. L. 1984. Glacier sliding at subfreezing temperatures. *Journal of Glaciology* 30: 341–347.
Šibrava, V. 1986. Correlation of European glaciations and their relation to deep-sea record. *Quaternary Science Reviews* 5: 433–441.
Šibrava, V., Bowen, D. Q. & Richmond, G. M. (eds) 1986. Quaternary glaciations in the Northern Hemisphere. *Quaternary Science Reviews* 5: 514 pp.

Siebertz, H. 1990. Die Abgrenzung von äolischen Decksedimenten auf dem Niederrheinischen Höhenzug mit Hilfe von Korngruppenkombinationen. *Decheniana* 143: 476–485.
Siegenthaler, U., Eicher, U., Oeschger, H. & Dansgaard, W. 1984. Lake sediments as continental $\delta^{18}O$ records from the glacial/post-glacial transition. *Annals of Glaciology* 5: 149–152.
Sinn, P. 1972. Zur Stratigraphie und Paläogeographie des Präwürm im mittleren und südlichen Illergletscher-Vorland. *Heidelberger Geographische Arbeiten* 37: 159 pp.
Sissons, J. B. 1978. The parallel roads of Glen Roy and adjacent glens, Scotland. *Boreas* 7: 229–244.
Sissons, J. B. 1980. The Loch Lomond Advance in the Lake District, northern England. *Transactions of the Royal Society of Edinburgh, Earth Sciences* 71: 13–27.
Sissons, J. B. 1981. The last Scottish ice-sheet: facts and speculative discussion. *Boreas* 10: 1–17.
Sjørring, S. 1983. The glacial history of Denmark. In: Ehlers, J. (ed.): *Glacial Deposits in North-West Europe*: 163–179. Rotterdam: Balkema.
Sjørring, S. & Frederiksen, J. 1980. Glacialstratigrafiske observationer i de vestjydske bakkeøer. *Dansk Geologisk Forening, Årsskrift for 1979*: 63–77.
Sjørring, S., Nielsen, P. E., Frederiksen, J., Hegner, J., Hyde, G., Jensen, J. B., Mogensen, A & Vortisch, W. 1982. Observationer fra Ristinge Klint, felt- og laboratorieundersøgelser. *Dansk Geologisk Forening, Årsskrift for 1981*: 135–149.
Skupin, K., Speetzen, E. & Zandstra, J. G. (eds) 1993. *Die Eiszeit in Nordwestdeutschland*. Krefeld: Geologisches Landesamt Nordrhein-Westfalen. 143 pp.
Slater, G. 1926. Glacial tectonics as reflected in disturbed drift deposits. *Proceedings of the Geologists' Association* 37: 392–400.
Slater, G. 1927a. Studies in drift deposits of the southwestern part of Suffolk. *Proceedings of the Geologists' Association* 38: 157–216.
Slater, G. 1927b. Structure of the Mud Buttes and Tit Hills in Alberta. *Geological Society of America Bulletin* 38: 721–730.
Slater, G. 1931. The structure of the Bride Moraine, Isle of Man. *Proceedings of the Liverpool Geological Society* 68: 402–448.
Sloan, D. 1992. The Yerba Buena mud: Record of the last-interglacial predecessor of San Francisco Bay, California. *Geological Society of America Bulletin* 104: 716–727.
Smalley, I. J. 1966. The properties of glacial loess and the formation of loess deposits. *Journal of Sedimentary Petrology* 36: 669–676.
Smalley, I. J. 1972. The interaction of great rivers and large deposits of primary loess. *Transactions of the New York Academy of Sciences* 34: 534–542.
Smalley, I. J. 1990. Possible formation mechanisms for the modal coarse-silt quartz particles in loess deposits. *Quaternary International* 7/8: 23–27.
Smalley, I. J. & Piotrowski, J. A. 1987. Critical strength/stress ratios at the ice–bed interface in the drumlin forming process: From 'dilatancy' to 'cross-over'. In: Menzies, J. & Rose, J. (eds): *Drumlin Symposium*: 81–86. Rotterdam: Balkema.
Smalley, I. J. & Unwin, D. J. 1968. The formation and shape of drumlins and their distribution and orientation in drumlin fields. *Journal of Glaciology* 7: 377–390.
Smart, P. L. & Francis, P. D. 1991. Quaternary Dating Methods – a User's Guide. *Quaternary Research Association, Technical Guide* 4: 233 pp.
Smed, P. 1989. *Sten i det danske landskab*. 2. edition. Brenderup: Geografforlaget. 181 pp.
Smed, P. 1993. Indicator studies: A critical review and a new data-presentation method. *Bulletin of the Geological Society of Denmark* 40: 332–340.
Smed, P. 1994. *Steine aus dem Norden. Geschiebe als Zeugen der Eiszeit in Norddeutschland*. Berlin: Gebrüder Borntraeger. 194 pp.
Smiley, T. L., Bryson, R. A., King, J. E., Kukla, G. J. & Smith, G. I. 1991. Quaternary palaeoclimates. In: Morrison, R. B. (ed.): *Quaternary Nonglacial Geology: Conterminous U.S. Geological Society of America. The Geology of North America*, K-2: 13–44.
Smith, B. J., McGreevy, J. P. & Whalley, W. B. 1987. Silt production by weathering of a sandstone under hot arid conditions: an experimental study. *Journal of Arid Environments* 12: 199–214.
Smith, D. E. & Dawson, A. G. (eds) 1984. *Shorelines & Isostasy*. Orlando (Florida): Academic Press. 387 pp.
Smith, G. A. 1993. Missoula flood dynamics and magnitudes inferred from sedimentology of slack-

water deposits on the Columbia Plateau, Washington. *Geological Society of America Bulletin* 105: 77–100.

Smith, H. T. U. 1965. Dune morphology and chronology in central and western Nebraska. *Journal of Geology* 73: 557–578.

Smith, J. 1838. On the last changes in the relative levels of the land and sea in the British Islands. *Memoirs of the Wernerian Natural History Society* 8: 49–88.

Smol, J. P. 1990. Freshwater algae. In: Warner, B. G. (ed.): *Methods in Quaternary Ecology. Geoscience Canada Reprint Series* 5: 3–14.

Smol, J. P., Battarbee, R. W., Davis, R. B. & Meriläinen, J. (eds) 1986: *Diatoms and Lake Acidity*. Dordrecht: Dr. W. Junk Publishers. 307 pp.

Snowden, J. O., Jr. & Priddy, R. R. 1968. Geology of Mississippi loess. *Mississippi Geological Survey Bulletin* 111: 13–203.

Soergel, W. 1919. *Lösse, Eiszeiten und paläolithische Kulturen. Eine Gliederung und Altersbestimmung der Lösse*. Jena. 177 pp.

Soergel, W. 1924. *Die diluvialen Terrassen der Ilm und ihre Bedeutung für die Gliederung des Eiszeitalters*. Jena: Gustav Fischer. 79 pp.

Soergel, W. 1925. Die Gliederung und absolute Zeitrechnung des Eiszeitalters. *Fortschritte der Geologie und Paläontologie* 13: 125–251.

Soergel, W. 1926. Exkursion ins Travertingebiet von Ehringsdorf. *Paläontologische Zeitschrift* 8: 7–33.

Soergel, W. 1936. Diluviale Eiskeile. *Zeitschrift der Deutschen Geologischen Gesellschaft* 88: 223–247.

Solheim, A., Russwurm, L., Elverhøi, A. & Nyland Berg, M. 1990. Glacial geomorphic features in the northern Barents Sea: direct evidence for grounded ice and implications for the pattern of deglaciation and late glacial sedimentation. In: Dowdeswell, J. A. & Scourse, J. D. (eds): *Glacimarine Environments: Processes and Sediments. Geological Society Special Publication* No. 53: 253–268.

Sollid, J. L. & Reite, A. J. 1983. The last glaciation and deglaciation of Central Norway. In: Ehlers, J. (ed.): *Glacial Deposits in North-West Europe*: 41–59. Rotterdam: Balkema.

Sollid, J. L. & Sørbel, L. 1974. Palsa bogs at Haugtjörnin, Dovrefjell, South Norway. *Norsk Geografisk Tidsskrift* 28: 53–60.

Sollid, J. L. & Sørbel, L. 1979. Deglaciation of western Central Norway. *Boreas* 8: 233–239.

Sollid, J. L. & Sørbel, L. 1981. Kvartærgeologisk verneverdige områder i Midt-Norge. *Miljødepartementet, Avdelingen for naturvern og friluftsliv, Rapport* T-524: 207 pp.

Sollid, J. L. & Sørbel, L. 1992. Rock glaciers in Svalbard and Norway. *Permafrost and Periglacial Processes* 3: 215–220.

Sollid, J. L. & Sørbel, L. 1994. Distribution of glacial landforms in southern Norway in relation to the thermal regime of the last continental ice sheet. *Geografiska Annaler* 76 A: 25–35.

Sommé, J. 1979. Quaternary coastlines in northern France. In: Oele, E., Schüttenhelm, R. T. E. & Wiggers, A. J. (eds): *The Quaternary History of the North Sea. Acta Universitatis Upsaliensis Annum Quingentesimum Celebrantis* 2: 147–158.

Sørensen, R. 1979. Late Weichselian deglaciation in the Oslofjord area, south Norway. *Boreas* 8: 241–246.

Sørensen, R. 1983. Glacial deposits in the Oslofjord area. In: Ehlers, J. (ed.): *Glacial Deposits in North-West Europe*: 19–28. Rotterdam: Balkema.

Sparks, B. W. 1956. The non-marine Mollusca of the Hoxne interglacial. In: West, R. G.: *The Quaternary deposits at Hoxne, Suffolk. Philosophical Transactions of the Royal Society of London* B 239: 351–354.

Sparks, B. W. 1957. The non-marine Mollusca of the interglacial deposits at Bobbitshole, Ipswich. *Philosophical Transactions of the Royal Society of London* B 241: 33–44.

Sparks, B. W. 1980. Land and freshwater Mollusca of the West Runton Freshwater Bed. In: West, R. G.: *The pre-glacial Pleistocene of the Norfolk and Suffolk coasts*: 25–27. Cambridge: Cambridge University Press.

Sparks, B. W., Williams, R. G. B. & Bell, F. G. 1972. Presumed ground-ice depressions in East Anglia. *Proceedings of the Royal Society of London* A 327: 329–343.

Srivastava, S. & Arthur, M. (eds) 1987. *Proceedings of the Ocean Drilling Program, Initial Reports* 105: 917 pp.

Stahr, K. 1979. Bedeutung periglazialer Deckschichten für Bodenbildung und Standorteigenschaften im Südschwarzwald. *Freiburger Bodenkundliche Abhandlungen* 9: 1–273.

Stalder, P. 1985. Glazialmorphologische Untersuchungen zwischen See- und Suhrental. *Physische Geographie* 20: 184 pp.

Stalker, A. MacS. 1973. The large interdrift bedrock blocks of the Canadian Prairies. *Geological Survey of Canada, Paper* 75-1A: 421–422.

Stalker, A. MacS. 1974. Megablocks, or the enormous erratics of the Albertian Prairies. *Geological Survey of Canada, Paper* 76-1C: 185–188.

Stanford, S. D. & Mickelson, D. M. 1985. Till fabric and deformational structures in drumlins near Waukesha, Wisconsin, U.S.A. *Journal of Glaciology* 31: 220–228.

Starkel, L. 1983. Facial differentiation of the Holocene fill in the Wisloka River Valley. *Geologisches Jahrbuch* A 71: 161–169.

Stea, R. R. 1994. Relict and palimpsest glacial landforms in Nova Scotia, Canada. In: Warren, W. P. & Croot, D. G. (eds): *Formation and Deformation of Glacial Deposits*: 141–158. Rotterdam: Balkema.

Stea, R. R. & Mott, R. J. 1989. Deglaciation environments and evidence for glaciers of Younger Dryas age in Nova Scotia, Canada. *Boreas* 18: 169–187.

Stea, R. R., Finck, P. W. & Wightman, D. M. 1986. Quaternary geology and till geochemistry of the western part of Cumberland County, Nova Scotia (sheet 9). *Geological Survey of Canada*, Paper 85-17: 58 pp.

Stea, R. R., Mott, R. J., Belknap, D. F. & Radtke, U. 1992. The pre-late Wisconsinan chronology of Nova Scotia, Canada. In: Clark, P. U. & Lea, P. D. (eds): *The Last Interglacial–Glacial Transition in North America. Geological Society of America Special Paper* 270: 185–206.

Stea, R. R., Boyd, R., Costello, O., Fader, G. B. J. & Scott, D. B. 1995. Deglaciation of the inner Scotian Shelf, Nova Scotia: Correlation of land–sea events. *Geological Society of London, Special Publication* (in press).

Steens, T. N. F., Kroon, D., Ten Kate, W. G. & Sprenger, A. 1990. Late Pleistocene rhythmicities of oxygen isotope ratios, calcium carbonate contents and magnetic susceptibilities of western Arabian sea margin hole 728A (ODP leg 117). *Proceedings of the Ocean Drilling Program, Part B: Scientific Results, Leg 117.*

Steiner, W. 1981. Der Travertin von Ehringsdorf und seine Fossilien. *Neue Brehm-Bücherei* 522: 200 pp. Wittenberg: Ziemsen.

Steinmüller, A. 1967. Eine weichselzeitliche Schichtenfolge in der Goldenen Aue bei Nordhausen. *Jahrbuch für Geologie* 1 (1965): 373–394.

Steno, N. 1669. *De solido intra solidum naturaliter contento dissertationis prodomus*. Florence. 76 pp.

Stephan, H.-J. 1974. Sedimentation auf Toteis in Schleswig-Holstein, diskutiert anhand einiger Beispiele. *Meyniana* 25: 95–100.

Stephan, H.-J. 1981. Eemzeitliche Verwitterungshorizonte im Jungmoränengebiet Schleswig-Holsteins. *Verhandlungen des Naturwissenschaftlichen Vereins in Hamburg (NF)* 24 (2): 161–175.

Stephan, H.-J. 1985. Deformations striking parallel to glacier movement as a problem in reconstructing its direction. *Bulletin of the Geological Society of Denmark* 34: 47–54.

Stephan, H.-J. 1987a. Form, composition, and origin of drumlins in Schleswig-Holstein. In: Menzies, J. & Rose, J. (eds): *Drumlin Symposium*: 335–345. Rotterdam: Balkema.

Stephan, H.-J. 1987b. Moraine stratigraphy in Schleswig-Holstein and adjacent areas. In: Van der Meer, J. J. M. (ed.): *Tills and Glaciotectonics*: 23–30. Rotterdam: Balkema.

Stephan, H.-J. & Ehlers, J. 1983. North German till types. In: Ehlers, J. (ed.): *Glacial Deposits in North-West Europe*: 239–247. Rotterdam: Balkema.

Stephan, H.-J. & Menke, B. 1977. Untersuchungen über den Verlauf der Weichsel-Kaltzeit in Schleswig-Holstein. *Zeitschrift für Geomorphologie N.F., Supplement-Band* 27: 12–28.

Stephan, H.-J., Kabel, Ch. & Schlüter, G. 1983. Stratigraphical problems in the glacial deposits of Schleswig-Holstein. In: Ehlers, J. (ed.): *Glacial Deposits in North-West Europe*: 305–320. Rotterdam: Balkema.

Stepp, R. 1981. Das Böhener Feld – Ein Beitrag zum Altquartär im Südwesten der Iller–Lech–Platte. *Mitteilungen der Geographischen Gesellschaft München* 66: 43–68.

Stewart, M. T. & Mickelson, D. M. 1976. Clay mineralogy and relative age of tills in north-central Wisconsin. *Journal of Sedimentary Petrology* 46: 200–205.

Stingl, H. 1969. Ein periglazialmorphologisches Nord–Süd-Profil durch die Ostalpen. *Göttinger Geographische Abhandlungen* 49: 115 pp.

Stiny, J. 1941. Unsere Täler wachsen zu. *Geologie und Bauwesen* 13: 71–79.
Stoker, M. S. & Bent, A. 1985. Middle Pleistocene glacial and glaciomarine sedimentation in the west central North Sea. *Boreas* 14: 325–332.
Stoker, M. S. & Holmes, R. 1991. Submarine end moraines as indicators of Pleistocene ice-limits off northwest Britain. *Journal of the Geological Society of London* 148: 431–434.
St Onge, D. A. 1987. The Sangamonian Stage and the Laurentide Ice Sheet. *Géographie physique et Quaternaire* 41: 189–198.
Stout, W. & Schaaf, D. 1931. Mintford silts of southern Ohio. *Geological Society of America Bulletin* 42: 663–672.
Strattner, M. & Rolf, C. 1995. Paläo-Magnetostratigraphische Untersuchungen an quartären Deckschichtenprofilen im Bereich der Iller–Lech-Platte und benachbarter Gebiete. *Geologica Bavarica* 99: 55–101.
Strauch, F. 1983. Geological History of the Iceland-Faeroe Ridge and its influence on Pleistocene Glaciations. *NATO Conference Series* IV (8): 601–606. New York: Plenum Press.
Straw, A. 1991. Glacial deposits of Lincolnshire and adjoining areas. In: Ehlers, J., Gibbard, P. & Rose, J. (eds): *Glacial Deposits in Great Britain and Ireland:* 213–221. Rotterdam: Balkema.
Stremme, H. E. 1960. Bodenbildungen auf Geschiebelehmen verschiedenen Alters in Schleswig-Holstein. *Zeitschrift der Deutschen Geologischen Gesellschaft* 112: 299–308.
Stremme, H. E. 1981. Unterscheidung von Moränen durch Bodenbildungen. *Mededelingen Rijks Geologische Dienst* 34: 51–56.
Striegler, R. 1991. Die Europäische Sumpfschildkröte (*Emys orbicularis*) im Eem von Schönfeld. *Natur und Landschaft in der Niederlausitz, Sonderheft*: 130–168.
Strobel, M. L. & Faure, G. 1987. Transport of indicator clasts by ice sheets and the transport half-distance: a contribution to prospecting for ore deposits. *Journal of Geology* 95: 687–697.
Strömberg, B. 1983. The Swedish varve chronology. In: Ehlers, J. (ed.) *Glacial Deposits in North-West Europe*: 97–105. Rotterdam: Balkema.
Strömberg, B. 1989. Late Weichselian deglaciation and clay varve chronology in East-Central Sweden. *Sveriges Geologiska Undersökning* Ca 73: 70 pp.
Strömberg, B. 1990. A connection between the clay varve chronologies in Sweden and Finland. *Annales Academiæ Scientiarum Fennicæ* A III, 154: 32 pp.
Strömberg, B. 1992. The final stage of the Baltic Ice Lake. *Sveriges Geologiska Undersökning* Ca 81: 347–354.
Stuart, A. J. 1974. Pleistocene history of the British vertebrate fauna. *Biological Reviews* 49: 225–266.
Stuart, A. J. 1976. The history of the mammal fauna during the Ipswichian/Last interglacial in England. *Philosophical Transactions of the Royal Society of London* B 276: 221–250.
Stuart, A. J. 1979. Pleistocene occurrences of the European pond tortoise (*Emys orbicularis* L.) in Britain. *Boreas* 8: 359–371.
Stuart, A. J. 1982. *Pleistocene vertebrates in the British Isles*. London: Longman. 212 pp.
Stuart, A. J. 1986. Pleistocene occurrence of *Hippopotamus* in Britain. *Quartärpaläontologie* 6: 209–218.
Stuart, A. J. 1993. The failure of evolution: Late Quaternary mammalian extinctions in the Holarctic. *Quaternary International* 19: 101–107.
Stuiver, M. & Pearson, G. W. 1993. High-precision bi-decadal calibration of the radiocarbon time scale. *Radiocarbon* 35: 1–23.
Suc, J.-P. & Zagwijn, W. H. 1983. Plio-Pleistocene correlations between the northwestern Mediterranean region and northwestern Europe according to recent biostratigraphic and palaeoclimatic data. *Boreas* 12: 153–166.
Sudakova, N. G. 1995 Lithological and palaeogeographical regionalisation of the central Russian Plain. In: Ehlers, J., Kozarski, S. & Gibbard, P. L. (eds): *Glacial Deposits in North-East Europe*: 189–194. Rotterdam: Balkema.
Sudakova, N. G. & Faustova, M. A. 1995. Glacial history of the Russian Plain. In: Ehlers, J., Kozarski, S. & Gibbard, P. L. (eds): *Glacial Deposits in North-East Europe*: 151–156. Rotterdam: Balkema.
Sudakova, N. G., Nemtsova, G. M., Andreicheva, L. N., Bolshakov, V. A. & Glushankova, N. I. 1995. Lithology of Middle Pleistocene tills in the central and southern Russian Plain. In: Ehlers, J., Kozarski, S. & Gibbard, P. L. (eds): *Glacial Deposits in North-East Europe*: 171–178. Rotterdam: Balkema.
Suess, H. E. 1980. Radiocarbon geophysics. *Endeavor, N.S.* 4: 113–117.

Sugden, D. E. 1976. A case against deep erosion of shields by ice sheets. *Geology* 4: 580–582.

Sugden, D. E. & John, B. S. 1976. *Glaciers and Landscape – A Geomorphological Approach.* London: Edward Arnold. 376 pp.

Sutherland, D. G. 1991. The glaciation of the Shetland and Orkney Islands. In: Ehlers, J., Gibbard, P. & Rose, J. (eds): *Glacial Deposits in Great Britain and Ireland*: 121–127. Rotterdam: Balkema.

Sutherland, D. G. & Gordon, J. E. 1993. The Quaternary in Scotland. In: Gordon, J. E. & Sutherland, D. G. (eds): *Quaternary of Scotland. Geological Conservation Review Series* 6: 13–47. London: Chapman & Hall.

Svendsen, J. I. & Mangerud, J. 1987. Late Weichselian and Holocene sea-level history for a cross-section of western Norway. *Journal of Quaternary Science* 2: 113–132.

Svendsen, J. I., Mangerud, J. & Miller, G. H. 1989. Denudation rates in the Arctic estimated from lake sediments on Spitsbergen, Svalbard. *Palaeogeography, Palaeoclimatology, Palaeoecology* 76: 153–168.

Svensson, H. 1976. Relict ice-wedge polygons revealed on aerial photographs from Kaltenkirchen, northern Germany. *Geografisk Tidsskrift* 75: 8–12.

Svensson, H. 1984. The periglacial form group of Southwestern Denmark. *Geografisk Tidsskrift* 84: 25–34.

Svensson, N.-O. 1989. Late Weichselian and early Holocene shore displacement in the Central Baltic, based on stratigraphical and morphological records from eastern Småland and Gotland, Sweden. *Lundqua Thesis* 25: 195 pp.

Synge, F. M. 1980. Quaternary period. In: Naylor, D., Phillips, W. E. A., Sevastopulo, G. D. & Synge, F. M. (eds): *An introduction to the geology of Ireland*: 39–42. Dublin: Royal Irish Academy.

Synge, F. M. 1981. Quaternary glaciation and changes of sea level in the south of Ireland. *Geologie en Mijnbouw* 60: 305–315.

Szabo, J. P. 1992. Reevaluation of early Wisconsinan stratigraphy of northern Ohio. In: Clark, P. U. & Lea, P. D. (eds): *The Last Interglacial–Glacial Transition in North America. Geological Society of America Special Paper* 270: 99–107.

Taramelli, T. 1898. Del deposito lignitico di Leffe in provincia di Bergamo. *Bollettino Società Geologica Italiana* 17: 202–218.

Tavast, E. & Raukas, A. 1982. *The bedrock relief of Estonia* (in Russian). Akademia Nauk Estonskoi SSR, Institut Geologii. 194 pp.

Taylor, F. B. 1895. The Nippissing Beach on the North Superior Shore. *American Geologist* 15: 304–314.

Taylor, F. B. 1897. Notes on the abandoned beaches of the north coast of Lake Superior. *American Geologist* 20: 111–128.

Taylor, K. C., Lamorey, G. W., Doyle, G. A., Alley, R. B., Grootes, P. M., Mayewski, P. A., White, J. W. C. & Barlow, L. K. 1993. The 'flickering switch' of late Pleistocene climate change. *Nature* 361: 432–436.

Teller, J. T. 1985. Glacial Lake Agassiz and its influence on the Great Lakes. In: Karrow, P. F. & Calkin, P. E. (eds): *Quaternary Evolution of the Great Lakes. Geological Association of Canada Special Paper* 30: 1–16.

Teller, J. T. 1987. Proglacial lakes and the southern margin of the Laurentide ice sheet. In: Ruddiman, W. F. & Wright, H. E., Jr. (eds): *North America and adjacent oceans during the last deglaciation. Geological Society of America, Geology of North America* K-3: 39–70.

Teller, J. T. 1990. Volume and routing of Late-Glacial runoff from the Southern Laurentide Ice Sheet. *Quaternary Research* 34: 12–23.

Teller, J. T. & Kehew, A. E. (eds) 1994. Late Glacial history of large proglacial lakes and meltwater runoff along the Laurentide Ice Sheet. *Quaternary Science Reviews* 13 (9/10): 795–981.

Terasmae, J. 1960. Contributions to Canadian palynology no. 2. Part I, A palynological study of post-glacial deposits in the St. Lawrence Lowland; Part II, A palynological study of Pleistocene interglacial beds at Toronto, Ontario. *Geological Survey of Canada, Bulletin* 56: 47 pp.

Ter Wee, M. W. 1962. The Saalian glaciation in the Netherlands. *Mededelingen Geologische Stichting NS* 15: 57–74.

Ter Wee, M. W. 1983a. The Elsterian Glaciation in the Netherlands. In: Ehlers, J. (ed.): *Glacial Deposits in North-West Europe*: 413–415. Rotterdam: Balkema.

Ter Wee, M. W. 1983b. The Saalian Glaciation in the Northern Netherlands. In: Ehlers, J. (ed.): *Glacial Deposits in North-West Europe*: 405–412. Rotterdam: Balkema.

Thenius, E. 1962. Die Großsäugetiere des Pleistozäns von Mitteleuropa. *Zeitschrift für Säugetierkunde* 27: 65–83.

Thenius, E. 1972. *Grundzüge der Verbreitungsgeschichte der Säugetiere.* Stuttgart: Fischer. 345 pp.

Thieme, H. & Veil, S. 1985. Neuere Untersuchungen zum eemzeitlichen Elefantenjagdplatz Lehringen, Landkreis Verden. *Die Kunde NF* 36: 11–58.

Thome, K. N. 1980. Der Vorstoß des nordeuropäischen Inlandeises in das Münsterland in der Elster- und Saale-Kaltzeit – Strukturelle, mechanische und morphologische Zusammenhänge. *Westfälische Geographische Studien* 36: 21–40.

Thompson, R. 1991. Palaeomagnetic dating. In: Smart, P. L. & Frances, P. D. (eds): *Quaternary dating methods – a user's guide. Quaternary Research Association, Technical Guide* 4: 177–198.

Thompson, R. & Turner, G. M. 1979. British geomagnetic master curve 10,000–0 yr B.P. for dating European sediments. *Geophysical Research Letters* 6: 249–252.

Thompson, R. S., Whitlock, C., Bartlein, P. J., Harrison, S. P. & Spaulding, W. G. 1993. Climatic changes in the Western United States since 18,000 yr B.P. In: Wright, H. E., Kutzbach, J. E., Webb III, Th., Ruddiman, W. F., Street-Perrott, F. A. & Bartlein, P. J. (eds): *Global Climates since the Last Glacial Maximum*: 468–513. Minneapolis: University of Minnesota Press.

Thomson, P. W. 1941. Die Klima- und Waldentwicklung des von K. Orviku entdeckten Interglazials von Ringen bei Dorpat/Estland. *Zeitschrift der Deutschen Geologischen Gesellschaft* 93: 274–282.

Thorarinsson, S. 1969. Glacier surges in Iceland, with special reference to the surges of Brúarjökull. *Canadian Journal of Earth Sciences* 6: 875–882.

Thouveny, N., Bonifay, E., Creer, K. M., Guiot, K. M., Icole, M., Johnsen, S., Jouzel, J., Reille, M., Williams, T., Williamson, D. 1994. Climate variations in Europe over past 140 kyr deduced from rock magnetism. *Nature* 371: 503–506.

Thwaites, F. T. & Bertrand, K. 1957. Pleistocene geology of the Door Peninsula, Wisconsin. *Geological Society of America Bulletin* 68: 831–879.

Tillmanns, W. 1977. Zur Geschichte von Urmain und Urdonau zwischen Bamberg und Neuburg/Donau und Regensburg. *Soderveröffentlichungen des Geologischen Instituts der Universität Köln* 30: 198 pp.

Tillmanns, W. 1980. Zur plio-pleistozänen Flußgeschichte von Donau und Main in Nordostbayern. *Jahresberichte und Mitteilungen des Oberrheinischen Geologischen Vereins* 62: 199–205.

Tillmanns, W. 1984. Die Flußgeschichte der oberen Donau. *Jahreshefte des Geologischen Landesamts Baden-Württemberg* 26: 99–202.

Tillmanns, W., Koci, A. & Brunnacker, K. 1986. Die Brunhes/Matuyama-Grenze in Roßhaupten (Bayerisch Schwaben). *Jahresberichte und Mitteilungen des Oberrheinischen Geologischen Vereins N.F.* 68: 241–247.

Tobolski, K. 1986. Paleobotanical studies of the Eemian interglacial and early Vistulian, Władysławów in the vicinity of Turek. *Quaternary Studies in Poland* 7: 91–101.

Todtmann, E. M. 1960. Gletscherforschungen auf Island (Vatnajökull). *Universität Hamburg, Abhandlungen aus dem Gebiet der Auslandskunde* 65 C 19: 95 pp.

Toepfer, V. 1958. Steingeräte und Palökologie der mittelpaläolithischen Fundstelle Rabutz bei Halle (Saale). *Jahresschriften zur Mitteldeutschen Vorgeschichte* 41/42: 140–177.

Toepfer, V. 1970. Stratigraphie und Ökologie des Paläolithikums. In: Richter, H., Haase, G., Lieberoth, I. & Ruske, R. (eds): *Periglazial–Löß–Paläolithikum im Jungpleistozän der Deutschen Demokratischen Republik. Petermanns Geographische Mitteilungen, Ergänzungscheft* 274: 329–422.

Torell, O. 1858. Bref om Island. *Öfversigt af Kongl. Vetenskaps-Akademiens Forhandlingar* XIV: 325–332.

Torell, O. 1865. Foreword. In: Holmström, L. P.: *Iakttagelser öfver märken i Skåne efter istiden*: I-V. Malmö: Chronholmska boktryckeriet.

Torell, O. 1872. Undersökningar öfver istiden del 1. *Öfversigt af Kongl. Vetenskaps-Akademiens Forhandlingar*: 25–66.

Torell, O. 1873. Undersökningar öfver istiden del 2. *Öfversigt af Kongl. Vetenskaps-Akademiens Forhandlingar*: 47–64.

Torell, O. 1875. Schliff-Flächen und Schrammen auf der Oberfläche des Muschelkalkes von Rüdersdorf. *Zeitschrift der Deutschen Geologischen Gesellschaft* 27: 961.

Törnqvist, T. E. 1993a. Fluvial sedimentary geology and chronology of the Holocene Rhine–Meuse delta, The Netherlands. *Nederlandse Geografische Studies* 166: 169 pp.

Törnqvist, T. E. 1993b. Holocene alternation of meandering and anastomosing fluvial systems in the Rhine–Meuse Delta (central Netherlands) controlled by sea-level rise and subsoil erodibility. *Journal of Sedimentary Petrology* 63: 683–693.

Törnqvist, T. E. 1994. Middle and late Holocene avulsion history of the River Rhine (Rhine–Meuse delta, Netherlands). *Geology* 22: 711–714.

Traub, F. & Jerz, H. 1976. Ein Lößprofil von Duttendorf (Oberösterreich) gegenüber Burghausen an der Salzach. *Zeitschrift für Gletscherkunde und Glazialgeologie* 11: 175–193.

Trenhaile, A. S. 1990. *The Geomorphology of Canada. An Introduction:* 240 pp. Toronto: Oxford University Press.

Tricart, J. 1948. Méthode d'etude des terrasses. *Bulletin de la Societé Géologique de France, Ser. 5*, No. 17: 559–575.

Troll, C. 1924. Der diluviale Inn-Chiemseegletscher – Das geographische Bild eines typischen Alpenvorlandgletschers. *Forschungen zur deutschen Landes- und Volkskunde* 23: 1–121.

Troll, C. 1977. Die 'fluvioglaziale Serie' der nördlichen Alpenflüsse und die holozänen Aufschotterungen. In: Frenzel, B. (ed.): *Dendrochronologie und postglaziale Klimaschwankungen. Erdwissenschaftliche Forschung* 13: 181–189.

Turner, Ch. 1970. Middle Pleistocene deposits at Marks Tey, Essex. *Philosophical Transactions of the Royal Society of London* B 257: 373–440.

Turner, Ch. (ed.) 1996. *The Early Middle Pleistocene in Europe*. Rotterdam: Balkema. 329 pp.

Tushingham, A. M. & Peltier, W. R. 1991. Ice 3G: A new global model of Late Pleistocene deglaciation based upon geophysical predictions of postglacial sea-level change. *Journal of Geophysical Research* 96: 4497–4523.

Tyrrell, J. B. 1898. The glaciation of north central Canada. *Journal of Geology* 6: 147–160.

Tyrrell, J. B. 1913. Hudson Bay exploring expedition 1912. *Ontario Bureau of Mines, Annual Report* 22: 161–209.

Tzedakis, P. C., Bennett, K. D. & Magri, D. 1994. Climate and the pollen record. *Nature* 370: 513.

Ullrich, H. 1989. Urmensch, Altmensch und eiszeitlicher Jetztmensch. In: Herrmann, J. (ed.): *Archäologie in der Deutschen Demokratischen Republik – Denkmale und Funde*: 48–54. Stuttgart: Theiss.

Unger, H. J. 1983. *Geologische Karte von Bayern 1:50 000; Erläuterungen zum Blatt Nr. L 7342 Landau an der Isar*: 141 pp. München: Bayerisches Geologisches Landesamt.

Urban, B. 1978. Vegetationsgeschichtliche Untersuchungen zur Gliederung des Altquartärs der Niederrheinischen Bucht. *Geologisches Institut der Universität zu Köln, Sonderveröffentlichungen* 34: 165 pp.

Urban, B. 1983. Biostratigraphic correlation of the Kärlich Interglacial, Northwestern Germany. *Boreas* 12: 83–90.

Urban, B., Thieme, H. & Elsner, H. 1988. Biostratigraphie, quartärgeologische und urgeschichtliche Befunde aus dem Tagebau 'Schöningen', Ldkr. Helmstedt. *Zeitschrift der Deutschen Geologischen Gesellschaft* 139: 123–154.

Urban, B., Lenhard, R., Mania, D. & Albrecht, B. 1991. Mittelpleistozän im Tagebau Schöningen, Ldkr. Helmstedt. *Zeitschrift der Deutschen Geologischen Gesellschaft* 142: 351–372.

Usinger, H. 1985. Pollenstratigraphische, vegetations- und klimageschichtliche Gliederung des 'Bölling–Alleröd-Komplexes' in Schleswig-Holstein und ihre Bedeutung für die Spätglazial-Stratigraphie in benachbarten Gebieten. *Flora* 177: 1–43.

Ussing, N. V. 1903. Om Jyllands hedesletter og teorierne for deres dannelse. *Oversigt over Det Kongelige Danske Videnskabernes Selskabs Forhandlingar* 1903 (2): 1–152.

Vail, P. R., Mitchum, R. M., Jr. & Thompson III, S. 1977. Seismic stratigraphy and global changes of sea level; Part 4. Global cycles of relative changes of sea level. In: Payton, C. E. (ed.): *Seismic stratigraphy – application to hydrocarbon exploration. American Association of Petroleum Geologists Memoir* 26: 83–97.

Van de Meene, E. A. & Zagwijn, W. H. 1978. Die Rheinläufe im deutsch-niederländischen Grenzgebiet seit der Saale-Kaltzeit. Überblick neuer geologischer und pollenanalytischer Untersuchungen. *Fortschritte in der Geologie von Rheinland und Westfalen* 28: 345–359.

Vandenberghe, J. & Pissart, A. 1993. Permafrost changes in Europe during the Last Glacial. *Permafrost and Periglacial Processes* 4: 121–135.

Van den Bogaard, C., Van den Bogaard, P. & Schmincke, H.-U. 1989. Quartärgeologisch-tephrostratigraphische Neuaufnahme und Interpretation des Pleistozänprofils Kärlich. *Eiszeitalter*

und Gegenwart 39: 62–86.
Van der Hammen, Th., Maarleveld, G. C., Vogel, I. C. & Zagwijn, W. H. 1967. Stratigraphy, climatic succession and radiocarbon dating of the Last Glacial in the Netherlands. *Geologie en Mijnbouw N.S.* 46: 79–95.
Van der Kaars, W. A. 1991. Palynology of eastern Indonesian marine piston-cores: A Late Quaternary vegetational and climatic record for Australasia. *Palaeogeography, Palaeoclimatology, Palaeoecology* 85: 239–302.
Van der Meer, J. J. M. 1982. The Fribourg area, Switzerland – a study in Quaternary geology and soil development. Thesis, Universiteit van Amsterdam. 203 pp.
Van der Meer, J. J. M. 1993. Microscopic evidence of subglacial deformation. *Quaternary Science Reviews* 12: 553–588.
Van der Meer, J. J. M. & Laban, C. 1990. Micromorphology of some North Sea till samples, a pilot study. *Journal of Quaternary Science* 5: 95–101.
Van der Meer, J. J. M., Rabassa, J. O. & Evenson, E. B. 1992. Micromorphological aspects of glaciolacustrine sediments in northern Patagonia, Argentina. *Journal of Quaternary Science* 7: 31–44.
Van der Meer, J. J. M., Verbers, A. L. L. M. & Warren, W. P. 1994. The micromorphological character of the Ballycroneen Formation (Irish Sea Till): A first assessment. In: Warren, W. P. & Croot, D. G. (eds): *Formation and Deformation of Glacial Deposits*: 39–49. Rotterdam: Balkema.
Van der Meulen, A. J. & Zagwijn, W. H. 1974. *Microtus (Allophaiomys) pliocaenicus* from the Lower Pleistocene near Brielle, The Netherlands. *Scripta Geologica* 21: 1–12.
Van der Vlerk, I. M. & Florschütz, F. 1950. *Nederland in het Ijstijdvak. De geschiedenis van flora, fauna en klimaat, toen aap en mammoet ons land bewoonden.* Utrecht: de Haan. 287 pp.
Van der Vlerk, I. M. & Florschütz, F. 1953. The paleontological base of the subdivision of the Pleistocene in The Netherlands. *Verhandelingen Koninklijke Nederlandse Akademie van Wetenschappen, afdeling Natuurkunde*, 1e Reeks, deel 20 (2): 1–58.
Van der Wateren, F. M. 1985. A model of glaciotectonics, applied to the ice-pushed ridges in the Central Netherlands. *Bulletin of the Geological Society of Denmark* 34: 55–74.
Van der Wateren, F. M. 1987. Structural geology and sedimentology of the Dammer Berge push moraine, FRG. In: Van der Meer, J. J. M. (ed.): *Tills and Glaciotectonics*: 157–182. Rotterdam: Balkema.
Van der Wateren, F. M. 1992. Structural geology and sedimentology of push moraines. Processes of soft sediment deformation in a glacial environment and the distribution of glaciotectonic styles. Dissertation, Amsterdam. 230 pp.
Van Gijssel, K. 1987. A lithostratigraphic and glaciotectonic reconstruction of the Lamstedt Moraine, Lower Saxony (FRG). In: Van der Meer, J. J. M. (ed.): *Tills and Glaciotectonics*: 145–155. Rotterdam: Balkema.
Van Husen, D. 1968. Ein Beitrag zur Talgeschichte des Ennstales im Quartär. *Mitteilungen der Gesellschaft der Geologie- und Bergbaustudenten* 18 (1967): 249–286.
Van Husen, D. 1971. Zum Quartär des unteren Ennstales von Großraming bis zur Donau. *Verhandlungen der Geologischen Bundesanstalt* 1971 (3): 511–521.
Van Husen, D. 1975. Die quartäre Entwicklung des Steyrtales und seiner Nebentäler. *Jahrbuch des Oberösterreichischen Musealvereines* 120: 271–289.
Van Husen, D. 1976. Zur quartären Entwicklung des Krappfeldes und des Berglandes um St. Veit an der Glan. *Mitteilungen der Gesellschaft der Geologie- und Bergbaustudenten* 23: 55–68.
Van Husen, D. 1977. Zur Fazies und Stratigraphie der jungpleistozänen Ablagerungen im Trauntal. *Jahrbuch, Geologische Bundesanstalt* 120: 1–130.
Van Husen, D. 1979. Verbreitung, Ursachen und Füllung glazial übertiefter Talabschnitte an Beispielen in den Ostalpen. *Eiszeitalter und Gegenwart* 29: 9–22.
Van Husen, D. 1980. *Erläuterungen zu Blatt 160, Neumarkt in Steiermark.* Wien: Geologische Bundesanstalt. 64 pp.
Van Husen, D. 1981. Geologisch-sedimentologische Aspekte im Quartär von Österreich. *Mitteilungen der Österreichischen Geologischen Gesellschaft* 74/75: 197–230.
Van Husen, D. 1983. General sediment development in relation to the climatic changes during Würm in the eastern Alps. In: Evenson, E. B., Schlüchter, Ch. & Rabassa, J. (eds): *Tills and related deposits*: 345–349. Rotterdam: Balkema.

Van Husen, D. 1985a. Preserved strata of synsedimentary rotated loose sediments formed in a dead ice environment. *Bulletin of the Geological Society of Denmark* 34: 27–31.
Van Husen, D. 1985b. Influence of ice discharge of Alpine valley glaciers. *Geological Survey of Finland, Special Paper* 3: 137–142.
Van Husen, D. 1987. *Die Ostalpen in den Eiszeiten*. 24 pp. Wien: Geologische Bundesanstalt.
Van Husen, D. 1989. The last interglacial/glacial cycle in the Eastern Alps. *Quaternary International* 3/4: 115–121.
Van Husen, D. & Draxler, I. 1980. Zur Ausbildung und Stellung der würmzeitlichen Sedimente im unteren Gailtal. *Zeitschrift für Gletscherkunde und Glazialgeologie* 16: 85–97.
Van Kolfschoten, T. 1981. On the Holsteinian? and Saalian mammal fauna from the ice-pushed ridge near Rhenen (The Netherlands). *Mededelingen Rijks Geologische Dienst* 35: 223–251.
Van Kolfschoten, T. 1985. The Middle Pleistocene (Saalian) and Late Pleistocene (Weichselian) mammal faunas from Maastricht-Belvédère (Southern Limburg, The Netherlands). *Mededelingen Rijks Geologische Dienst* 39 (1): 45–74.
Van Kolfschoten, T. 1988. The evolution of the mammal fauna in the Netherlands and the Middle Rhine area (Western Germany) during the Late Middle Pleistocene. Thesis, Rijksuniversiteit Utrecht. 157 pp.
Van Kolfschoten, T. 1990. The evolution of the mammal fauna in the Netherlands and the Middle Rhine area (Western Germany) during the Late Middle Pleistocene. *Mededelingen Rijks Geologische Dienst* 43 (3): 69 pp.
Van Kolfschoten, T. & Van der Meulen, A. J. 1986. Villanyian and Biharian mammal faunas from The Netherlands. *Memorie della Società Geologica Italiana* 31: 191–200.
Van Vliet-Lanoë, B. 1989. Dynamics and extent of the Weichselian permafrost in Western Europe (Substage 5e to Stage 1). *Quaternary International* 3/4: 109–113.
Van Vliet-Lanoë, B. 1991. Differential frost heave, load casting and convection: Converging mechanisms; a discussion of the origin of cryoturbations. *Permafrost and Periglacial Processes* 2: 123–139.
Van Werveke, L. 1927. Norddeutschland war wenigstens viermal vom Inlandeis bedeckt. *Zeitschrift der Deutschen Geologischen Gesellschaft* 79, *Monatsberichte*: 135.
Velichko, A. A. (ed.) 1984. *Late Quaternary Environments of the Soviet Union*. London: Longman. 327 pp.
Velichko, A. A. 1990. Loess-palaeosol formation on the Russian Plain. *Quaternary International* 7/8: 103–114.
Velichko, A. A. & Faustova, M. A. 1986. Glaciations in the East European region of the USSR. *Quaternary Science* Reviews 5: 447–461.
Velichko, A. A., Bugucki, A. B., Morozova, T. D., Udartsev, V. P., Khalcheva, T. A. & Tsatskin, A. I. 1984. Periglacial landscapes of the East European Plain. In: Velichko, A. A. (ed.): *Late Quaternary Environments of the Soviet Union*: 95–118. London: Longman.
Venetz, I. 1830. Sur l'ancienne extension des glaciers, et sur leur retraite dans leurs limites actuelles. *Actes de la Société helvétique des sciences naturelles, 15è réunion à l'Hospice du Grand Saint-Bernard 1829*. Lausanne.
Ventris, P. A. 1985. Pleistocene environmental history of the Nar Valley, Norfolk. Unpublished Ph.D. thesis, University of Cambridge.
Venzo, S. 1950. Rinvenimento di *Anancus arvernensis* nel Villafranchiano dell'Adda di Paderno, di *Archidiskodon meridionalis* e *Cervus* a Leffe. *Societa Italiana di Scienze Naturali, Atti* 89: 43–122.
Venzo, S. 1952. Geomorphologische Aufnahme des Pleistozäns (Villafranchian–Würm) im Bergamasker Gebiet und in der östlichen Brianza: Stratigraphie, Paläontologie und Klima. *Geologische Rundschau* 40: 109–125.
Vierhuff, H. 1967. Untersuchungen zur Stratigraphie und Genese der Sandlößvorkommen in Niedersachsen. *Mitteilungen aus dem Geologischen Institut der Technischen Hochschule Hannover* 5: 99 pp.
Villinger, E. 1985. Geologie und Hydrogeologie der pleistozänen Donaurinnen im Raum Sigmaringen-Riedlingen (Baden-Württemberg). Unter Mitarbeit von J. Werner. *Abhandlungen des Geologischen Landesamts Baden-Württemberg* 11: 141–203.
Villinger, E. 1986. Untersuchungen zur Flußgeschichte von Aare-Donau/Alpenrhein und zur Entwicklung des Malm-Karsts in Südwestdeutschland. *Jahreshefte des Geologischen Landesamts Baden-Württemberg* 28: 297–362.

Villinger, E. 1989. Zur Fluß- und Landschaftsgeschichte im Gebiet von Aare-Donau und Alpenrhein. *Jahreshefte der Gesellschaft für Naturkunde Württembergs* 144: 5–27.

Vincent, J.-S. 1989. Quaternary geology of the southeastern Canadian Shield. In: Fulton, R. J. (ed.): *Quaternary Geology of Canada and Greenland. Geological Society of America. The Geology of North America* K-1: 249–275.

Vincent, J.-S. & Prest, V. K. 1987. The Early Wisconsinan history of the Laurentide Ice Sheet. *Géographie physique et Quaternaire* XLI: 199–213.

Vincent, J.-S., Morris, W.-A. & Occhietti, S. 1984. Glacial and nonglacial sediments of Matuyama paleomagnetic age on Banks Island, Canadian Arctic. *Geology* 12: 139–142.

Viret, J. 1954. Le loess à bancs durcis de Saint-Vallier (Drôme) et sa faune de mammifères villafranchiens avec une analyse granulométrique par Elisabeth Schmid et une analyse pollinique par Charles Krachenbüchl. *Nouvelles Archives du Muséum d'Histoire Naturelle de Lyon* 4: 200 pp. Lyon.

Vogt, P. R., Crane, K. & Sundvor, E. 1994. Deep Pleistocene iceberg ploughmarks on the Yermak Plateau: Sidescan and 3.5 kHz evidence for thick calving ice fronts and a possible marine ice sheet in the Arctic Ocean. *Geology* 22: 403–406.

Volkov, I. A. & Volkova, V. S. 1975. Velikaya prilednikovaya sistema stoka Sibirii (Great glacial system of Siberian run-off). In: Kvasov, D. D. (ed.): *Istoriya ozer v pleistotsene (Lacustrine history of the Pleistocene)*, Vol. 2, 133–140. Leningrad.

Volkov, I. A. & Volkova, V. S. 1979. Sediments from the transgressive phases of the Pleistocene Mansi Lake and great system of glacial runoff in Siberia. *International Geological Correlation Programme, Project 73/1/24, Quaternary Glaciations in the Northern Hemisphere, Report* 5, 236–245.

Volkvov, I. A., Grosswald, M. G. & Troitskiy, S. L. 1978. O stoke prilednikovykh vod vo vremya pocednego oledeneniya Sapadnoi Sibirii (On the periglacial water's run-off during the latest glacial in West Siberia). *Isvestiya Akademii Nauk SSSR, Seriya Geograficheskaya* 1978 (4), 25–35.

Von Buch, L. 1815. Über die Verbreitung großer Alpengeschiebe. *Abhandlungen der Physikalischen Classe der Akademie der Wissenschaften Berlin 1804–1811*: 161–186.

Von Bülow, W. 1967. Zur Quartärbasis in Mecklenburg. *Berichte der Deutschen Gesellschaft für Geologische Wissenschaften* A 12: 375–404.

Von Bülow, W. 1969. Altpleistozäne Schotter (Loosener Kiese) in Südwestmecklenburg mit nordischen und südlichen Geröllen. *Geologie* 18: 563–589.

Von Drygalski, E. 1897. *Grönlandexpedition der Gesellschaft für Erdkunde zu Berlin 1891–1893*, 1. Band. Berlin: Kühl. 555 pp.

Von Gümbel, C. W. 1889. *Kurze Erläuterungen zu dem Blatte Ingolstadt (No. XV) der geognostischen Karte des Königreiches Bayern*: 34 pp. Cassel.

Von Hacht, U. 1979. Neue Beobachtungen an Gesteinen aus Braderup, Sylt. *Natur und Museum* 109: 10–17.

Von Hacht, U. 1987. Spuren früher Kaltzeiten im Kaolinsand von Braderup/Sylt. In: Von Hacht, U. (ed.): *Fossilien von Sylt* II: 269–301. Hamburg: Von Hacht.

Von Humboldt, A. 1845. *Kosmos – Entwurf einer physischen Weltbeschreibung*, Bd. 1. Stuttgart & Augsburg: Cotta. 507 pp.

Von Klebelsberg, R. 1935. *Geologie von Tirol*. Berlin: Gebr. Borntraeger. 872 pp.

Von Klebelsberg, R. 1948/49. *Handbuch der Gletscherkunde und Glazialgeologie* (2 volumes). Wien: Springer. 1028 pp.

Von Klebelsberg, R. 1950. Das Silltal bei Matrei. *Schlern-Schriften* 84: 76–86.

Von Koenigswald, W. 1973. Veränderungen in der Kleinsäugerfauna von Mitteleuropa zwischen Cromer und Eem (Pleistozän). *Eiszeitalter und Gegenwart* 23/24: 159–167.

Von Koenigswald, W. 1988. Paläoklimatische Aussage letztinterglazialer Säugetiere aus der nördlichen Oberrheinebene. In: Von Koenigswald, W. (ed.): *Zur Paläoklimatologie des letzten Interglazials im Nordteil der Oberrheinebene*: 205–314. Stuttgart, New York: Gustav Fischer.

Von Koenigswald, W. 1991. Exoten in der Großsäuger-Fauna des letzten Interglazials von Mitteleuropa. *Eiszeitalter und Gegenwart* 41: 70–84.

Von Koenigswald, W. & Van Kolfschoten, T. 1993. The *Mimomys–Arvicola* boundary and the enemal thickness quotient (SDQ) of *Arvicola* as stratigraphic markers in the Middle Pleistocene. (Cromer Symposium)

Von Leonhard, K. C. 1823/24. *Charakteristik der Felsarten. Für akademische Vorlesungen und zum Selbststudium*. 3 Bände. Heidelberg: J. Engelmann.

Von Morlot, A. 1844. *Ueber die Gletscher der Vorwelt und ihre Bedeutung*. Bern: Rätzer.

Von Morlot, A. 1847. *Erläuterungen zur Geologischen Übersichtskarte der nordöstlichen Alpen*. Wien: Braumüller und Seidel. 208 pp.

Von Morlot, A. 1855. Quartäre Gebilde des Rhonegebiets. *Neues Jahrbuch für Mineralogie und Geologie* (1855): 719–721.

Von Post, L. 1916. Skogsträdpollen i sydsvenska torvmosselagerföldjer. *Geologiska Föreningens i Stockholm Förhandlingar* 38: 384–394.

Von Post, L. 1928. Svea älvs geologiska tidsställning. *Sveriges Geologiska Undersökning* C 347: 132 pp.

Von Richthofen, F. 1877. *China – Ergebnisse eigener Reisen und darauf gegründeter Studien*, Band I. Berlin: Reimer. 758 pp.

Von Senarclens-Grancy, W. 1958. Glazialgeologie des Ötztales und seiner Umgebung. *Mitteilungen der Geographischen Gesellschaft Wien* 49: 257–314.

Von Wettstein, R. 1892. Die fossile Flora der Höttinger Breccie. *Wien, Denkschriften der Mathematisch-naturwissenschaftlichen Klasse der Akademie der Wissenschaften* LIX: 479–524.

Vorren, T. O. 1977. Weichselian ice movements in south Norway and adjacent areas. *Boreas* 6: 247–257.

Vorren, T. O., Hald, M., Edvardsen, M. & Lind-Hansen, O.-W. 1983. Glacigenic sediments and sedimentary environments on continental shelves: General principles with a case study from the Norwegian shelf. In: Ehlers, J. (ed.): *Glacial Deposits in North-West Europe:* 61–73. Rotterdam: Balkema.

Vorren, T. O., Hald, M. & Lebesbye, E. 1988. Late Cenozoic environments in the Barents Sea. *Paleoceanography* 3: 601–612.

Vorren, T. O., Lebesbye, E. & Larsen, K. B. 1990. Geometry and genesis of the glacigenic sediments in the southern Barents Sea. In: Dowdeswell, J. A. & Scourse, J. D. (eds): *Glacimarine Environments: Processes and Sediments. Geological Society Special Publication* No. 53: 269–288.

Vortisch, W. 1982. Clay mineralogical studies of some tills in northern Germany. *Geologica et Palaeontologica* 15: 167–192.

Vuagneux, R. 1983. Glazialmorphologische und gletschergeschichtliche Untersuchungen im Gebiet Flüelapaß (Kt. Graubünden, Schweiz). *Physische Geographie* 10: 249 pp.

Wagenbreth, O. 1978. Die Feuersteinlinie in der DDR, ihre Geschichte und Popularisierung. *Schriftenreihe für Geologische Wissenschaften* 9: 339–368.

Waggoner, P. E. & Bingham, C. 1961. Depth of loess and distance from source. *Soil Science* 92: 396–401.

Wagner, G. A. 1995. *Altersbestimmung von jungen Gesteinen und Artefakten*. Stuttgart: Enke. 277 pp.

Wagner, G. A. & Van den Haute, P. 1992. *Fission-track-dating*. Stuttgart: Enke. 285 pp.

Wahlmüller, N. 1985. Beiträge zur Vegetationsgeschichte Tirols V: Nordtiroler Kalkalpen. *Berichte des Naturwissenschaftlich-Medizinischen Vereins in Innsbruck* 72: 1012–1144.

Wahnschaffe, F. 1882. Über einige glaziale Druckerscheinungen im norddeutschen Diluvium. *Zeitschrift der Deutschen Geologischen Gesellschaft* 34: 562–601.

Wahnschaffe, F. 1891. Die Oberflächengestaltung des Norddeutschen Flachlandes. *Forschungen zur Deutschen Landes- und Volkskunde* VI: 1–166.

Wahnschaffe, F. 1901. *Die Ursachen der Oberflächengestaltung des Norddeutschen Flachlandes*, 2. Auflage. Engelhorn: Stuttgart. 258 pp.

Wahnschaffe, F. & Schucht, F. 1921. *Geologie und Oberflächengestaltung des Norddeutschen Flachlandes*. Stuttgart: J. Engelhorns Nachf. 472 pp.

Walcott, R. I. 1970. Isostatic response to loading of the crust in Canada. *Canadian Journal of Earth Sciences* 7: 716–726.

Walder, J. H. 1986. Hydraulics of subglacial cavities. *Journal of Glaciology* 32: 273–293.

Walder, J. H. & Fowler, A. 1994. Channelized subglacial drainage over a deformable bed. *Journal of Glaciology* 40: 3–15.

Walker, M. J. C., Bohncke, S. J. P., Coope, G. R., O'Connell, M., Usinger, H. & Verbruggen, C. 1994. The Devensian/Weichselian Late-glacial in northwest Europe (Ireland, Britain, north Belgium, The Netherlands, northwest Germany). *Journal of Quaternary Science* 9: 109–118.

Walter, R. C. 1994. Age of Lucy and the First Family: Single-crystal $^{40}Ar/^{39}Ar$ dating of the Denen Dora and lower Kada Hadar members of the Hadar Formation, Ethiopia. *Geology* 22: 6–10.

Walther, M. 1990. Untersuchungsergebnisse zur jungpleistozänen Landschaftsentwicklung Schwansens (Schleswig-Holstein). *Berliner Geographische Abhandlungen* 52: 143 pp.

Wansa, St. 1991. Lithologie und Stratigraphie der Tills bei Gräfenhainichen. *Mauritiana* 13: 189–211.

Wansa, St. & Wimmer, R. 1990. Geologie des Jungpleistozäns der Becken von Gröbern und Grabschütz. *Altenburger Naturwissenschaftliche Forschungen* 5: 49–91.

Warming, E. 1888. Om Grønlands vegetation. *Meddelelser om Grønland* 12: 245 pp.

Warner, B. G. 1990. Plant macrofossils. In: Warner, B. G. (ed.): *Methods in Quaternary Ecology. Geoscience Canada Reprint Series* 5: 53–63.

Warren, W. P. 1979. The stratigraphic position and age of the Gortian interglacial deposits. *Geological Survey of Ireland Bulletin* 2: 315–332.

Warren, W. P. 1985. Stratigraphy. In: Edwards, K. J. & Warren, W. P. (eds): *The Quaternary history of Ireland*: 39–65. London: Academic Press.

Warren, W. P. 1991a. Fenitian (Midlandian) glacial deposits and glaciation in Ireland and the adjacent offshore regions. In: Ehlers, J., Gibbard, P. & Rose, J. (eds): *Glacial Deposits in Great Britain and Ireland*: 79–88. Rotterdam: Balkema.

Warren, W. P. 1991b. Glacial deposits of southwest Ireland. In: Ehlers, J., Gibbard, P. & Rose, J. (eds): *Glacial Deposits in Great Britain and Ireland*: 345–353. Rotterdam: Balkema.

Wascher, H. L., Humbert, R. P. & Cady, J. G. 1948. Loess in the southern Mississippi Valley; Identification and distribution of the loess sheets. *Soil Science Society of America Proceedings* 12: 389–399.

Washburn, A. L. 1979. *Geocryology. A survey of periglacial processes and environments*. London: Edward Arnold. 406 pp.

Washburn, A. L., Burrows, C. & Rein, R., Jr. 1978. Soil deformation resulting from some laboratory freeze–thaw experiments. In: *Third International Conference on Permafrost, Edmonton, Alberta, Proceedings* 1: 756–762.

Wastegård, S. 1995. Late Weichselian – Early Holocene marine stratigraphy in southwestern Värmland and northwestern Dalsland, SW Sweden. *Quaternaria A: Theses and Research Papers* 1.

Wastegård, S., Andrén, T., Sohlenius, G. & Sandgren, P. 1995. Different phases of the Yoldia Sea in the north-western Baltic proper. *Quaternary International* 27: 121–129.

Watts, W. A. 1964. Interglacial deposits at Baggotstown, near Bruff, Co. Limerick. *Proceedings of the Royal Irish Academy* 63 B: 167–189.

Wayne, W. J. 1967. Periglacial features and climatic gradient in Illinois, Indiana, and western Ohio, east-central United States. In: Cushing, E. J. & Wright, H. E., Jr. (eds): *Quaternary Paleoecology*: 393–414. New Haven: Yale University Press.

Wayne, W. J. 1991. Ice-wedge casts of Wisconsinan age in Eastern Nebraska. *Permafrost and Periglacial Processes* 2: 211–223.

Wayne, W. J. & Aber, J. S. 1991. High Plains and Plains Border Sections in Nebraska, Kansas and Oklahoma. In: Morrison, R. B. (ed.): *Quaternary Nonglacial Geology: Conterminous U.S. Geological Society of America. The Geology of North America* K-2: 462–469.

Webb III, Th., Bartlein, P. J., Harrison, S. P. & Anderson, K. H. 1993. Vegetation, lake levels, and climate in eastern North America for the past 18,000 years. In: Wright, H. E., Kutzbach, J. E., Webb II, T., Ruddiman, W. F., Street-Perrott, F. A. & Bartlein, P. J. (eds): *Global Climates since the Last Glacial Maximum*: 415–467. Minneapolis: University of Minnesota Press.

Weber, C. A. 1893. Über die diluviale Vegetation von Klinge in Brandenburg und ihre Herkunft. *Englers Botanisches Jahrbuch* 17, Beiblatt 1.

Weber, C. A. 1896. Über die fossile Flora von Honerdingen und das nordwestdeutsche Diluvium. *Abhandlungen des Naturwissenschaftlichen Vereins zu Bremen* 13: 413–468.

Weber, F., Schmid, Ch. & Figala, G. 1993. Vorläufige Ergebnisse reflexionsseismischer Messungen im Quartär des Inntals/Tirol. *Zeitschrift für Gletscherkunde und Glazialgeologie* 26, 2 (1990): 121–144.

Weber, Th. 1990. Paläolithische Funde aus den Eemvorkommen von Rabutz, Grabschütz und Gröbern. *Altenburger Naturwissenschaftliche Forschungen* 5: 282–299.

Weber, Th. & Litt, Th. 1991. Der Waldelefantenfund von Gröbern, Kr. Gräfenhainichen – Jagdbefund oder Nekrophagie? *Archäologisches Korrespondenzblatt* 21: 17–32.

Weddle, T. K. 1992. Late Wisconsinan stratigraphy in the lower Sandy River valley, New Sharon, Maine. *Geological Society of America Bulletin* 104: 1350–1363.

Weertman, J. 1964. The theory of glacier sliding. *Journal of Glaciology* 5: 287–303.

Weertman, J. 1972. General theory of water flow at the base of a glacier or ice sheet. *Reviews of Geophysics and Space Physics* 10: 287–333.
Wegmüller, S. 1986. Recherches palynologiques sur les charbons feuilletés de la région de Gondiswil/ Ufhausen (Plateau Suisse). *Bulletin de l'Association française pour l'étude de Quaternaire* 1986 (1/2): 29–34.
Weidenbach, F. 1937. Bildungsweise und Stratigraphie der diluvialen Ablagerungen Oberschwabens. *Neues Jahrbuch für Mineralogie und Geologie (Beilagenband B)* 78: 66–108.
Weinberger, L. 1950. Gliederung der Altmoränen des Salzach-Gletschers östlich der Salzach. *Zeitschrift für Gletscherkunde und Glazialgeologie* I: 176–186.
Weinberger, L. 1955. Exkursion durch das österreichische Salzachgletschergebiet und die Moränengürtel des Irrsee- und Attersee-Zweiges des Traungletschers. *Verhandlungen der Geologischen Bundesanstalt D*: 7–34.
Weinhardt, R. 1973. Rekonstruktion des Eisstromnetzes der Ostalpennordseite zur Zeit des Würmmaximums, mit einer Berechnung seiner Flächen und Volumina. *Heidelberger Geographische Arbeiten* 38: 158–178.
Weiss, E. N. 1958. Bau und Entstehung der Sander vor der Grenze der Würmvereisung im Norden Schleswig-Holsteins. *Meyniana* 7: 5–60.
Wellner, R. W., Ashley, G. M. & Sheridan, R. E. 1993. Seismic stratigraphic evidence for a submerged middle Wisconsin barrier: Implications for sea-level history. *Geology* 21: 109–112.
Welten, M. 1944. Pollenanalytische, stratigraphische und geochronologische Untersuchungen aus dem Faulenseemoos bei Spiez. *Veröffentlichungen des Geobotanischen Instituts Rübel, Zürich*, 21: 201 pp.
Welten, M. 1952. Über die spät- und postglaziale Vegetationsgeschichte des Simmentals. *Veröffentlichungen des Geobotanischen Instituts Rübel, Zürich* 26. 135 pp.
Welten, M. 1958. Die spätglaziale und postglaziale Vegetationsentwicklung der Berner Alpen und -Voralpen und des Walliser Haupttales. *Veröffentlichungen des Geobotanischen Instituts Rübel, Zürich* 34.
Welten, M. 1981. Verdrängung und Vernichtung der anspruchsvollen Gehölze am Beginn der letzten Eiszeit und die Korrelation der Frühwürm-Interstadiale in Mittel- und Nordeuropa. *Eiszeitalter und Gegenwart* 31: 187–202.
Welten, M. 1982a. Pollenanalytische Untersuchungen im jüngeren Quartär des nördlichen Alpen-Vorlandes der Schweiz. *Beiträge zur Geologischen Karte der Schweiz N.F.* 156: 174 pp. + diagram volume.
Welten, M. 1982b. Stand der palynologischen Quartärforschung am schweizerischen Nordalpenrand (Überblick, Methodisches, Probleme). *Geographica Helvetica* 37 (2): 75–83.
Welten, M. 1988. Neue pollenanalytische Ergebnisse über das Jüngere Quartär des nördlichen Alpenvorlandes der Schweiz (Mittel- und Jungpleistozän). *Beiträge zur Geologischen Karte der Schweiz N.F.* 162: 40 pp. + diagrams.
Wentworth, C. K. 1919. A laboratory and field study of cobble abrasion. *Journal of Geology* 27: 507–521.
Werth, E. 1912. Die äußersten Jungendmoränen in Norddeutschland und ihre Beziehungen zur Nordgrenze und zum Alter des Löß. *Zeitschrift für Gletscherkunde* 6: 250–277.
West, R. G. 1956. The Quaternary deposits at Hoxne, Suffolk. *Philosophical Transactions of the Royal Society of London* B 239: 265–356.
West, R. G. 1957. Interglacial deposits at Bobbitshole, Ipswich. *Philosophical Transactions of the Royal Society of London* B 246: 1–31.
West, R. G. 1968. *Pleistocene Geology and Biology with especial reference to the British Isles*. Harlow: Longmans. 377 pp.
West R. G. 1972. Relative land–sea level changes in southeastern England during the Pleistocene. *Philosophical Transactions of the Royal Society of London* A 272: 87–98.
West, R. G. 1977. *Pleistocene Geology and Biology with especial reference to the British Isles*, 2nd edition. London: Longman. 440 pp.
West, R. G. 1980a. *The pre-glacial Pleistocene of the Norfolk and Suffolk coasts*. Cambridge: Cambridge University Press. 203 pp. + plates and diagrams.
West, R. G. 1980b. Pleistocene forest history in East Anglia. *New Phytologist* 85: 571–622.
West, R. G. 1991. *Pleistocene palaeoecology of central Norfolk – a study of environments through time*. Cambridge: Cambridge University Press. 110 pp.

West, R. G. 1993. On the history of the Late Devensian Lake Sparks in southern Fenland, Cambridgeshire, England. *Journal of Quaternary Science* 8: 217–234.
West, R. G. & Gibbard, P. L. 1995. Discussion on excavations at the Lower Palaeolithic site at East Farm, Barnham, Suffolk, 1989–1992. *Journal of the Geological Society of London* 152: 570–574
West, R. G. & Wilson, D. G. 1966. Cromer Forest Bed Series. *Nature* 209: 497–498.
West, R. G. & Wilson, D. G. 1968. Plant remains from the Corton Beds at Lowestoft, Suffolk. *Geological Magazine* 105: 116–123.
West, R. G., Dickson, C. A., Catt, J. A., Weir, A. H. & Sparks, B. W. 1974. Late Pleistocene deposits at Wretton, Norfolk. II Devensian deposits. *Philosophical Transactions of the Royal Society of London* B 267: 337–420.
Westgate, J. A. 1968. Linear sole markings in Pleistocene till. *Geological Magazine* 105: 501–505.
Westgate, J. A., Stemper, B. A. & Péwé, T. L. 1990. A 3 m.y. record of Pliocene–Pleistocene loess in interior Alaska. *Geology* 18: 858–861.
Whalley, W. B. & Martin, H. E. 1992. Rock glaciers: a review. Part 2: Mechanisms and models. *Progress in Physical Geography* 11: 127–186.
White, S. E. 1971. Rock glacier studies in the Colorado Front Range, 1961 to 1968. *Arctic and Alpine Research* 3: 43–64.
White, W. A. 1972. Deep erosion by continental ice sheets. *Geological Society of America, Bulletin* 83: 1037–1056.
Whiteman, C. A. 1981. Some aspects of palaeosols in geomorphology and Quaternary studies. *Brighton Polytechnic Geographical Society Magazine* 9: 8–15.
Whiteman, C. A. 1983. Great Waltham. In: Rose, J. (ed.): *The Diversion of the Thames, Field Guide*: 163–169. Cambridge: Quaternary Research Association.
Whiteman, C. A. 1992. The paleogeography and correlation of pre-Anglian glaciation terraces of the River Thames in Essex and the London Basin. *Proceedings of the Geologists' Association* 103: 37–56.
Whiteman, C. A. & Kemp, R. A. 1990. Pleistocene sediments, soils and landscape evolution at Stebbing, Essex. *Journal of Quaternary Science* 5: 145–161.
Whiteman, C. A. & Rose, J. 1992. Thames River Sediments of the British Early and Middle Pleistocene. *Quaternary Science Reviews* 11: 363–375.
Whittecar, G. R. & Mickelson, D. M. 1977. Sequence of till deposition and erosion in drumlins. *Boreas* 6: 213–217.
Whittecar, G. R. & Mickelson, D. M. 1979. Composition, internal structures, and an hypothesis of formation for drumlins, Waukesha County, Wisconsin, U.S.A. *Journal of Glaciology* 22: 357–371.
Whitworth, T., III 1988. The Antarctic Circumpolar Current. *Oceanus* 31: 53–58.
Wickham, S. S., Johnson, W. H. & Glass, H. D. 1988. Regional geology of the Tiskilwa Till Member, Wedron Formation, northeastern Illinois. *Illinois State Geological Survey Circular* 543: 35 pp.
Wiedemann, E. & Schmidt, G. C. 1895. Ueber Lumineszenz. *Ann. Phys. Chem.* 54: 604–625.
Wilcox, R. E. & Naeser, C. W. 1992. The Pearlette Family Ash Beds in the Great Plains: Finding their Identities and their Roots in the Yellowstone Country. *Quaternary International* 13/14: 9–13.
Wiles, G. C. & Calkin, P. E. 1994. Late Holocene, high-resolution glacial chronologies and climate, Kenai Mountains, Alaska. *Geological Society of America Bulletin* 106: 281–303.
Williams, M. A. J., Dunkerley, D. L., De Decker, P., Kershaw, A. P. & Stokes, T. 1993. *Quaternary Environments*. London: Edward Arnold. 329 pp.
Williams, P. J. & Smith, M. W. 1989. *The Frozen Earth. Fundamentals of Geocryology*. Cambridge: Cambridge University Press. 306 pp.
Willman, H. B. & Frye, J. C. 1970. Pleistocene stratigraphy of Illinois. *Illinois State Geological Survey Bulletin* 94: 204 pp.
Willman, H. B., Glass, H. D. & Frye, J. C. 1963. Mineralogy of glacial tills and their weathering profiles in Illinois. Part I. Glacial tills. *Illinois State Geological Survey Circular* 347: 55 pp.
Willman, H. B., Glass, H. D. & Frye, J. C. 1966. Mineralogy of glacial tills and their weathering profiles in Illinois. Part II. Weathering profiles. *Illinois State Geological Survey Circular* 400: 76 pp.
Willman, H. B., Glass, H. D. & Frye, J. C. 1989. Glaciation and origin of the geest in the Driftless Area of Northwestern Illinois. *Illinois State Geological Survey, Circular* 535: 44 pp.
Wilson, P., Bateman, R. M. & Catt, J. A. 1981. Petrography, origin and environment of deposition of the

Shirdley Hill Sand of southwestern Lancashire, England. *Proceedings of the Geologists' Association* 92: 211–229.
Wingfield, R. 1990. The origin of major incisions within the Pleistocene deposits of the North Sea. *Marine Geology* 91: 31–52.
Winker, C. D. 1991. Northwestern Gulf Coastal Plain. In: Morrison, R. B. (ed.): *Quaternary Nonglacial Geology: Conterminous U.S. Geological Society of America. The Geology of North America* K-2: 585–587.
Winters, H. A., Alford, J. A. & Rieck, R. L. 1988. The anomalous Roxana Silt and mid-Wisconsinan events in and near southern Michigan. *Quaternary Research* 29: 25–35.
Wintges, Th. 1984. Untersuchungen an gletschergeformten Felsflächen im Zemmgrund/Zillertal (Tirol) und in Südskandinavien. *Salzburger Geographische Arbeiten* 11: 209 pp.
Wintges, Th. & Heuberger, H. 1982. Untersuchungen an Parabelrissen und Sichelbrüchen im Zemmgrund (Zillertal) und über die damit verbundene Abtragung. *Zeitschrift für Gletscherkunde und Glazialgeologie* 16 (2): 157–170.
Wintle, A. G. 1981. Thermoluminescence dating of Late Devensian loesses in southern England. *Nature* 289: 479–480.
Wintle, A. G. 1986. Thermoluminescence dating of loess at Rocourt, Belgium. *Geologie en Mijnbouw* 66: 35–42.
Wintle, A. G. 1990. A review of current research on TL dating of loess. *Quaternary Science Reviews* 9: 385–397.
Wintle, A. G. 1991. Thermoluminescence dating. In: Smart, P. L. & Frances, P. D. (eds): *Quaternary dating methods – a user's guide. Quaternary Research Association, Technical Guide* 4: 108–127.
Woida, K. & Thompson, M. L. 1993. Polygenesis of a Pleistocene paleosol in southern Iowa. *Geological Society of America Bulletin* 105: 1445–1461.
Woillard, G. 1975. Recherches palynologiques sur le Pleistocène dans l'Est de la Belgique et dans les Vosges Lorraines. *Acta Geographica Lovaniensia* 14: 118 pp.
Woillard, G. 1978. Grande Pile peat bog: A continuous pollen record for the last 140,000 years. *Quaternary Research* 9: 1–21.
Woldstedt, P. 1913. Beiträge zur Morphologie von Nordschleswig. *Mitteilungen der Geographischen Gesellschaft Lübeck* 26: 41–110.
Woldstedt, P. 1925. Die großen Endmoränenzüge Norddeutschlands. *Zeitschrift der Deutschen Geologischen Gesellschaft* 77: 172–184.
Woldstedt, P. 1927a. Über die Ausdehnung der letzten Vereisung in Norddeutschland. *Sitzungsberichte der Preußischen Geologischen Landesanstalt* 2: 115–119.
Woldstedt, P. 1927b. Die Gliederung des Jüngeren Diluviums in Norddeutschland und seine Parallelisierung mit anderen Glazialgebieten. *Zeitschrift der Deutschen Geologischen Gesellschaft, Monatsberichte*, 1927 (3/4): 51–52.
Woldstedt, P. 1929. *Das Eiszeitalter. Grundlinien einer Geologie des Diluviums*. Stuttgart: Enke. 406 pp.
Woldstedt, P. 1938. Über Vorstoss- und Rückzugsfronten des Inlandeises in Norddeutschland. *Geologische Rundschau* 29: 481–490.
Woldstedt, P. 1939. Vergleichende Untersuchungen an isländischen Gletschern. *Jahrbuch der Preußischen Geologischen Landesanstalt für 1938*, 59: 249–271.
Woldstedt, P. 1951. Quartärforschung – einleitende Worte. *Eiszeitalter und Gegenwart* 1: 9–15.
Woldstedt, P. 1952. Die Entstehung der Seen in den ehemals vergletscherten Gebieten. *Eiszeitalter und Gegenwart* 2: 146–153.
Woldstedt, P. 1955. *Norddeutschland und angrenzende Gebiete im Eiszeitalter*. Stuttgart: Koehler. 467 pp.
Woldstedt, P. 1958. *Das Eiszeitalter. Grundlinien einer Geologie des Quartärs*, Band 2, 2. Auflage. Stuttgart: Enke. 438 pp.
Woldstedt, P. 1961. *Das Eiszeitalter. Grundlinien einer Geologie des Quartärs*, Band 1, 2. Auflage. Stuttgart: Enke. 374 pp.
Woldstedt, P. 1965. *Das Eiszeitalter. Grundlagen einer Geologie des Quartärs*, Band 3, 2. Auflage. Stuttgart: Enke. 328 pp.
Woldstedt, P. & Duphorn, K. 1974. *Norddeutschland und angrenzende Gebiete im Eiszeitalter*. 3. Auflage. Stuttgart: Koehler. 500 pp.

Wolff, W. 1907. Der geologische Bau der Bremer Gegend. *Abhandlungen des Naturwissenschaftlichen Vereins zu Bremen* 19: 207–216.

Woodland, A. W. 1970. The buried tunnel valleys of East Anglia. *Proceedings of the Yorkshire Geological Society* 37: 521–578.

Wooldridge, S. W. & Linton, D. L. 1955, *Structure, surface and drainage in south-east England*. London: George Philip. 176 pp.

Worsley, P. 1966. Fossil frost wedge polygons at Congleton, Cheshire, England. *Geografiska Annaler* 48 A: 211–219.

Worsley, P. 1991. Possible early Devensian glacial deposits in the British Isles. In: Ehlers, J., Gibbard, P. & Rose, J. (eds): *Glacial Deposits in Great Britain and Ireland*: 47–51. Rotterdam: Balkema.

Woszidlo, H. 1962. Foraminiferen und Ostracoden aus dem marinen Elster-Saale-Interglazial in Schleswig-Holstein. *Meyniana* 12: 65–96.

Wright, A. E. & Moseley, F. (eds) 1975. Ice Ages – Ancient and Modern. *Geological Journal Special Issue* 6. Liverpool: Seel House Press. 320 pp.

Wright, G. F. 1890. *The ice age in North America*. New York: Appleton. 622 pp.

Wright, G. F. 1911. *The ice age in North America and its bearings upon the antiquity of man*. 5th edition. Oberlin, Ohio: Bibliotheka Sacra Co. 763 pp.

Wright, H. E., Jr. 1973. Tunnel valleys, glacial surges and subglacial hydrology of the Superior Lobe, Minnesota. In: Black, R. F. *et al.* (eds): *The Wisconsinan Stage. Geological Society of America Memoir* 136: 251–276.

Wright, H. E., Jr. 1984. Introduction. In: Wright, H. E. (ed.) *Late Quaternary Environments of the United States, Vol. 2: The Holocene*: XI–XVII.

Wright, H. E., Jr. & Frey, D. G. (eds) 1965a. *The Quaternary of the United States*. Princeton: Princeton University Press. 922 pp.

Wright, H. E., Jr. & Frey, D. G. (eds) 1965b. International Studies on the Quaternary. *Geological Society of America Special Paper* 84: 565 pp.

Wright, H. E., Jr., Cushing, E. J. & Baker, R. G. 1964. Eastern Minnesota. *Midwest Friends of the Pleistocene, 15th Annual Field Conference*: 32 pp.

Wright, H. E., Jr., Kutzbach, J. E., Webb III, Th., Ruddiman, W. F., Street-Perrott, F. A. & Bartlein, P. J. (eds) 1993. *Global Climates since the Last Glacial Maximum*. Minneapolis: University of Minnesota Press. 569 pp.

Wright, W. B. 1914. *The Quaternary ice age*. London: Macmillan. 464 pp.

Wright, W. B. 1937. *The Quaternary ice age*, 2nd edition. London: Macmillan. 478 pp.

Würges, K. 1986. Artefakte aus den ältesten Quartärsedimenten (Schichten A–C) der Tongrube Kärlich, Kreis Mayen-Koblenz/Neuwieder Becken. *Archäologisches Korrespondenzblatt* 16: 1–6.

Wymer, J. J. 1968. *Lower Palaeolithic Archaeology in Britain, as Represented by the Thames Valley*. London: John Baker. 429 pp.

Wymer, J. J. 1981. The Palaeolithic. In: Simmons, I. G. & Tooley, M. J. (eds): *The environment in British prehistory*: 49–81. London: Duckworth.

Wymer, J. J. 1983. The Lower Palaeolithic site at Hoxne. *Suffolk Institute of Archaeology and History* 25: 169–189.

Wymer, J. J. 1988. Palaeolithic archaeology and the British Quaternary sequence. *Quaternary Science Reviews* 7: 79–97.

Wymer, J. J. 1994. The Lower Palaeolithic Period in the London Region. *Proceedings of the London and Middlesex Archaeological Society* 18: 1–15.

Wyssling, L. & Wyssling, G. 1978. Interglaziale Seeablagerungen in einer Bohrung bei Uster, Kt. Zürich. *Eclogae Geologicae Helvetiae* 71: 357–375.

Xiaomin, F., Jijun, L., Derbyshire, E., Fitzpatrick, E. A. & Kemp, R. A. 1994. Micromorphology of the Beiyuan loess–palaeosol sequence in Gansu Province, China: geomorphological and palaeoenvironmental significance. *Palaeogeography, Palaeoclimatology, Palaeoecology* 111: 289–303.

Yaalon, D. H. 1971. Criteria for the recognition and classification of paleosols. A report of the working group on the origin and nature of paleosols, INQUA Commission on paleopedology, 1970. In: Yaalon, D. H. (ed.): *Palaeopedology*: 153–158. Jerusalem: Israel University Press.

York, D., Hall, C. M., Yanase, Y., Hane, J. A. & Kenyon, W. J. 1981. $^{40}Ar/^{39}Ar$ dating of terrestrial minerals with a continuous laser. *Geophysical Research Letters* 8: 1136–1138.

Young, R. R., Burns, J. A., Smith, D. G., Arnold, L. D. & Rains, R. B. 1994. A single, late Wisconsin, Laurentide glaciation, Edmonton area and southwestern Alberta. *Geology* 22: 683–686.
Zagwijn, W. H. 1957. Vegetation, climate and time-correlations in the Early Pleistocene of Europe. *Geologie en Mijnbouw* 19: 233–244.
Zagwijn, W. H. 1961. Vegetation, climate and radiocarbon datings in the late Pleistocene of the Netherlands. I. Eemian and Early Weichselian. *Mededelingen Geologische Stichting N.S.* 14: 15–45.
Zagwijn, W. H. 1963. Pollen-analytical investigations in the Tiglian of the Netherlands. *Mededelingen Geologische Stichting N.S.* 16: 49–71.
Zagwijn, W. H. 1973. Pollenanalytic studies of Holsteinian and Saalian Beds in the Northern Netherlands. *Mededelingen Rijks Geologische Dienst* 24: 139–156.
Zagwijn, W. H. 1974. The palaeogeographic evolution of the Netherlands during the Quaternary. *Geologie en Mijnbouw* 53: 369–385.
Zagwijn, W. H. 1975. Variations in climate as shown by pollen analysis, especially in the Lower Pleistocene of Europe. In: Wright, A. E. & Moseley, F. (eds): *Ice Ages: Ancient and Modern*: 137–152. Liverpool: Seel House Press.
Zagwijn, W. H. 1985. An outline of the Quaternary stratigraphy of the Netherlands. *Geologie en Mijnbouw* 64: 17–24.
Zagwijn, W. H. 1989. Vegetation and climate during warmer intervals in the Late Pleistocene of western and central Europe. *Quaternary International* 3/4: 57–67.
Zagwijn, W. H. & De Jong, J. 1984. Dic Interglaziale von Bavel und Leerdam und ihre stratigraphische Stellung im niederländischen Früh-Pleistozän. *Mededelingen Rijks Geologische Dienst* 37: 155–169.
Zagwijn, W. H. & Zonneveld, J. I. S. 1956. The interglacial of Westerhoven. *Geologie en Mijnbouw* n.s. 18: 37–46.
Zagwijn, W. H., Van Montfrans, H. M. & Zandstra, J. G. 1971. Subdivision of the 'Cromerian' in the Netherlands: pollen-analysis, palaeomagnetism and sedimentary petrology. *Geologie en Mijnbouw* 50: 41–58.
Zalasiewicz, J. A., Mathers, S. J., Hughes, M. J., Gibbard, P. L., Peglar, S. M., Harland, R., Boulton, G. S., Nicholson, R. A., Cambridge, P. & Wealthall, G. P. 1988. Stratigraphy and palaeoenvironments of the Red Crag and Norwich Crag formations between Aldeburgh and Sizewell, Suffolk, England. *Philosophical Transactions of the Royal Society of London* B 322: 221–272.
Zandstra, J. G. 1971. Geologisch onderzoek in de stuwwal van de oostelijke Veluwe bij Hattem en Wapenveld. *Mededelingen Rijks Geologische Dienst N.S.* 22: 215–260.
Zandstra, J. G. 1983. Fine gravel, heavy mineral and grain-size analyses of Pleistocene, mainly glacigenic deposits in the Netherlands. In: Ehlers, J. (ed.): *Glacial Deposits in North-West Europe*: 361–377. Rotterdam Balkema.
Zandstra, J. G. 1988. *Noordelijke kristallijne gidsgesteenten*. Leiden: Brill. 469 pp.
Zandstra, J. G. 1993. Nördliche kristalline Leitgeschiebe und Kiese in der Westfälischen Bucht und angrenzenden Gebieten. In: Skupin, K., Speetzen, E. & Zandstra, J. G. (eds): *Die Eiszeit in Nordwestdeutschland*: 43–106. Krefeld: Geologisches Landesamt Nordrhein-Westfalen.
Zeeberg, J. J. 1995. The nature and distribution of Late Pleistocene dunes in the European lowlands and on the Russian Platform. *Interuniversitair Centrum voor Geo-ecologisch Onderzoek, Rapport* ICG 95/1: 28 pp.
Zelčs, V. S. 1993. Glaciotectonic landforms of divergent type glaciodepressional lowlands (Latvian, English and Russian). Compendium of papers. *University of Latvia, Geological Sciences, Habilitation and Promotion Council, Riga*. 105 pp.
Zeller, E. J., Levy, P. W. & Mattern, P. L. 1967. Geologic dating by electron spin resonance. In: *Radioactive Dating and Methods of Low-Level Counting*: 531–540. Vienna: IAEA.
Zentrales Geologisches Institut (eds) 1972. *Analyse des Geschiebebestandes quartärer Grundmoränen*. Fachbereichsstandard. TGL 25 232.
Zeuner, F. 1945. *The Pleistocene Period, its Climate, Chronology and Faunal Successions*. London: Ray Society. 322 pp.
Zeuner, F. 1949. Frost soils on Mount Kenya. *Journal of Soil Science* 1: 20–30.
Zeuner, F. 1952. Pleistocene shore-lines. *Geologische Rundschau* 40: 39–50.
Zeuner, F. 1959. *The Pleistocene Period, its Climate, Chronology and Faunal Successions*, 2nd edition, London: Hutchinson. 447 pp.

Zheng, B. 1989. Controversy regarding the existence of a large ice sheet on the Qinghai-Xizang (Tibetan) Plateau during the Quaternary period. *Quaternary Research* 32: 121–123.

Ziegler, J. H. 1983. Verbreitung und Stratigraphie des Jungpleistozäns im voralpinen Gebiet des Salzachgletschers in Bayern. *Geologica Bavarica* 84: 153–176.

Ziegler, P. A. 1982. *Geological Atlas of Western and Central Europe*. The Hague: Shell Internationale Petroleum Maatschappij B.V. 130 pp.

Zielinski, T. 1989. Lithofacies and palaeoenvironmental characteristics of the Suwałki outwash (Pleistocene, northeast Poland). *Annales Societatis Geologorum Poloniae* 59: 249–270.

Zolitschka, B., Haverkamp, B. & Negendank, J. W. F. 1992. Younger Dryas oscillation – varve dated palynological, paleomagnetic and microstratigraphic records from Lake Holzmaar, Germany. In: Bard, E. & Broecker, W. S. (eds): *The Last Deglaciation: Absolute and Radiocarbon Chronologies*: 81–102. Springer.

Zöller, L., Stremme, H. & Wagner, G. A. 1988. Thermolumineszenz-Datierungen an Löss-Paläoboden-Sequenzen von Nieder-, Mittel- und Oberrhein/Bundesrepublik Deutschland. *Chemical Geology (Isotope Geoscience Section)* 73: 39–62.

Zollinger, G. 1991. Zur Landschaftsgenese und Quartärstratigraphie im südlichen Oberrheingraben – am Beispiel der Lössdeckschichten der Ziegelei in Allschwil (Kanton Basel-Landschaft). *Eclogae Geologicae Helvetiae* 84: 739–752.

Index